Strength of Materials

Through Questions & Answers

For **ESE, GATE, PSUs**

& Other Competitive Examinations

by

Dr. U.C. Jindal

M.Tech, Ph.D.

Former Professor & Head of the Department

Department of Mechanical Engineering

Delhi College of Engineering, Delhi

MADE EASY Publications

www.madeeasypublications.org

MADE EASY Publications

Corporate Office: 44-A/4, Kalu Sarai (Near Hauz Khas Metro Station), New Delhi-110016
E-mail: infomep@madeeasy.in
Contact: 011-45124660, 8860378007
Visit us at: www.madeeasypublications.org

Strength of Materials through Questions & Answers : Dr. U.C. Jindal

First Edition: 2014
Reprint : 2015
Reprint : 2016
Second Edition (Revised): 2018

PREFACE

I must thank CMD of MADE EASY Group, **Mr. B. Singh** for providing me an opportunity to reach out to the Student Community at large through my present book **"Strength of Materials through Questions & Answers".** Students may be benefitted from my 50 years of teaching /research experience through this book.

Questions in the book are designed on the pattern of questions that are being asked in university examinations and competitive examinations of UPSC/GATE/PSUs.

The book has been thoroughly and questions from competitive examinations for the last 2 years have been added, in this revised and enlarged edition.

Further improvements in the text book will be made after getting the response from the students.

Any error in printing or calculations pointed out by the reader will be acknowledged with thanks by the author.

Dr. U. C. Jindal
Author

CONTENTS

Simple Stresses and Strains

CHAPTER 1

In this chapter we will discuss various types of external loads that can be applied on a body and the deformation produced in body. Due to the applied load, stresses and strains are developed in a body. Relationship between different types of stresses and different types of strains will be developed.

Positive and negative normal stresses and positive and negative shear stresses will also be described.

Question 1.1 **What do you understand by an axial load on a body and axial stress developed in a body.**

Solution: Consider a circular bar *AB*, of diameter d and axial length *L* as shown in figure 1.1. End *A* of the bar is fixed and a load *P* is applied along 00 axis of the bar. **Axis 00 passes through the centroids of all the sections of the bar, as shown.** Load *P* is perpendicular to all sections of the bar. The load applied can be a point load along the axis 00 or a uniformly distributed load over the section of the bar. To maintain equilibrium, a reaction $R = P$ is developed at the fixed end.

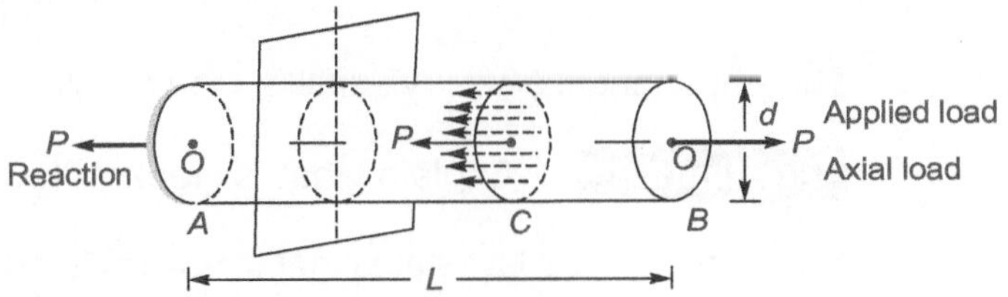

Fig. 1.1

If you consider a portion BC of the bar *AB*, then a force *P* at section *B* is resisted by equal and opposite force at section *C*, this equal and opposite force *on internal section* is known as internal resistance. **This internal resistance *P* per unit area is defined as stress.**

or stress,
$$\sigma = \frac{P}{\text{area of section}} = \frac{4P}{\pi d^2}$$

Note that force *P* is perpendicular to section.

Therefore stress σ is normal to the section.

Stress σ is perpendicular to the section of the bar, and is defined as normal stress. Direct stress is along the axis of bar. If axial load *P* is expressed in N (newton) and area of cross-section in mm^2, then units of stress are N/mm^2 *(newton per mm^2)*. Stress is a second order vector, it has both magnitude and direction.

Moreover load is normal to the section and its direction is away from the plane. Note that load *P* is normal to plane *and is pointing away from the plane.*

Similarly internal resistance *P* is normal to plate at *C* and is pointing away from the plane. **This type of load is known as tensile load and stress produced by tensile load is tensile stress (or positive normal stress).**

Question 1.2 **Explain what is tangential load or shear load and what are shear stress and shear strain?**

Solution: Consider a rectangular block (parallelopiped block) of dimensions l, b, h as shown in figure. 1.2.

Lower face of block, *ABCD* is fixed and a force Q parallel to the upper face *EFGH* is applied.

To maintain equilibrium, reaction Q (equal and opposite to Q on top face) is developed or consider an internal plane IJKL, parallel to top face. Internal resisting force Q, parallel to the internal plane is developed. **This internal resistance per unit area is termed as shear stress, τ**

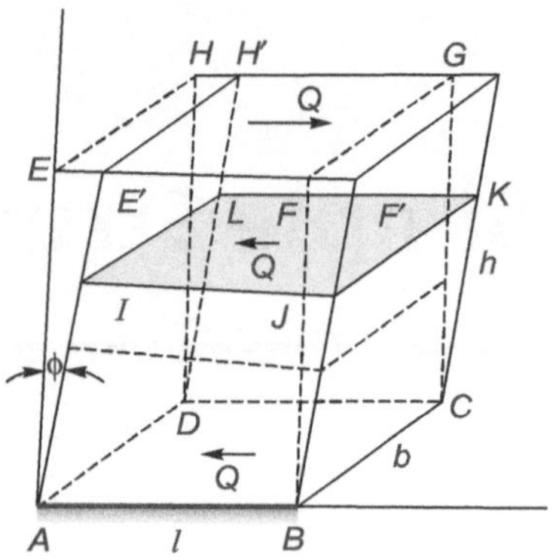

Fig. 1.2

$$\tau\text{, shear stress} = \frac{\text{Tangential force } Q}{\text{Area } l \times b} = \frac{Q}{l.b} \text{ in N/mm}^2$$

If Q is in Newton (N) and area in mm^2.

This ***tangential load on a surface is also known as shear load.*** Due to this shear load, block is deformed i.e.., rectangle *ABFE* is changed to parallelogram ABF′E′.

$$\angle EAE' = \phi\text{, shear angle}$$

$$\phi = \tan^{-1}\frac{EE'}{EA} = \tan^{-1}\frac{EE'}{h}$$

This shear angle ϕ is very small, so $\tan\phi \approx \sin\phi \approx \phi$. That is also known as *shear strain.*

Question 1.3 **Explain what is Hooke's law?**

Solution: Hooke's law states that stress (σ) is proportional to strain (ε)

$$\sigma \propto \varepsilon$$

Take a wire of diameter d, length L, fixed at upper end and a load W is gradually applied at the lower end.

Due to the tensile load the wire gets extended say by an amount δL.

Tensile stress in wire, $\sigma = \dfrac{4W}{\pi d^2}$

Change in length per unit length,

$$\varepsilon = \frac{\delta L}{L} \text{ or } \delta L = \varepsilon L.$$

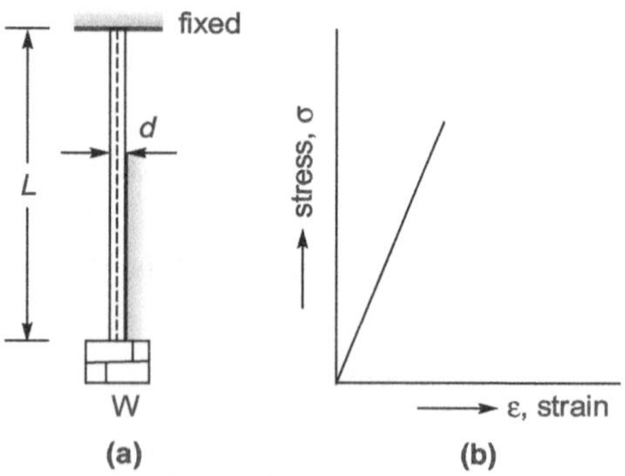

Fig. 1.3

Change in ***length per unit length is termed as strain, ε.***

If a graph is plotted between stress (σ) and strain (ε), for a gradually increasing load W, then within the elastic limit.

$$\sigma \propto \varepsilon$$

Elastic limit, means that within the elastic limit on stress i.e., σ_e, if load is removed from the wire, then **residual strain (residual change in length) will be zero.**

$$\sigma = E\varepsilon$$

where E = Constant of proportionality

= Elastic constant

$$= \text{Young's modulus of elasticity}$$

$$= \sigma/\varepsilon$$

Similarly for the shear load on rectangular block, shear stress α shear strain.

or shear stress, $\tau \propto \phi$, shear strain

$$\tau = G\phi$$

which G is the constant of proportionality

$$G = \frac{\tau}{\phi} = \frac{\text{shear stress}}{\text{shear strain}} = \text{Shear modulus}$$

E and G are elastic constants and are given for the material by the manufacturer.

Question 1.4 **A circular stepped bar made of steel is shown in figure 1.4.**
Lengths AB = 40 mm, BC = 60 mm, CD = 80 mm.
Diameters, d_1 = 20 mm, d_2 = 15 mm, d_3 = 10 mm.
An axial tensile load P = 10 kN is applied on the bar as shown.

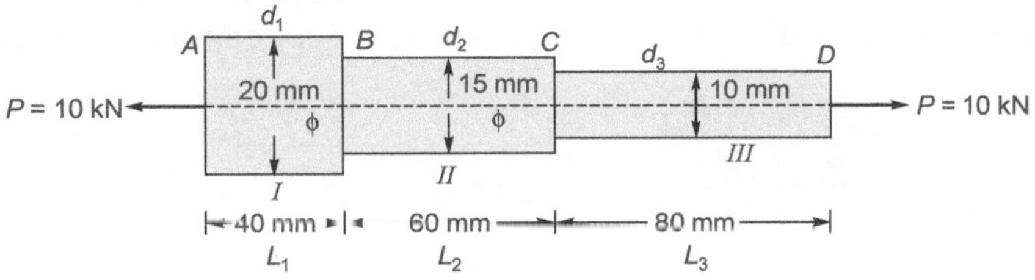

Fig. 1.4

Determine stresses, strains and change in length in each portion. For steel E = 200 GPa = 200 kN/mm^2.

Solution: Axial tensile load is perpendicular to all sections of portions AB, BC and CD. Stress is load per unit area, so maximum stress will occur in portion III **with minimum diameter**. Similarly minimum stress will occur in portion-I of maximum diameter.

Diameters, $d_1 = 20$ mm, $d_2 = 15$ mm and $d_3 = 10$ mm

Areas of cross-section, $A_1 = \frac{\pi}{4} \times 20^2 = 314.16 \text{ mm}^2$

$$A_2 = \frac{\pi}{4} \times 15^2 = 176.7 \text{ mm}^2$$

$$A_3 = \frac{\pi}{4} \times 10^2 = 78.54 \text{ mm}^2$$

Force, $P = 10{,}000$ N

Stresses, $\sigma_1 = \frac{10000}{314.16} = 31.83 \text{ N/mm}^2$

$$\sigma_2 = \frac{10000}{176.7} = 56.59 \text{ N/mm}^2$$

$$\sigma_3 = \frac{10000}{78.54} = 127.32 \text{ N/mm}^2$$

All these stresses are tensile stresses.

Strains $E = 200{,}000 \text{ N/mm}^2$

Using Hooke's law

$$\varepsilon_1 = \frac{\sigma_1}{E} = \frac{31.83}{200000} = 1.59 \times 10^{-4} \text{ tensile strain}$$

$$\varepsilon_2 = \frac{\sigma_2}{E} = \frac{56.59}{200000} = 2.83 \times 10^{-4} \text{ tensile strain}$$

$$\varepsilon_3 = \frac{\sigma_3}{E} = \frac{127.32}{200000} = 6.36 \times 10^{-4} \text{ tensile strain}$$

Chang in length,

$$\delta L_1 = \varepsilon_1 \times L_1 = 1.59 \times 10^{-4} \times 40 = 0.00636 \text{ mm, extension}$$

$$\delta L_2 = \varepsilon_2 \times L_2 = 2.83 \times 10^{-4} \times 60 = 0.01698 \text{ mm, extension}$$

$$\delta L_3 = \varepsilon_3 \times L_3 = 6.36 \times 10^{-4} \times 80 = 0.0508 \text{ mm, extension}$$

Practice Q.1.4 **A circular stepped bar made of copper is shown in figure 1.5. It is subjected to axial load *P* such that maximum stress in bar is not to exceed 100 MPa. What is the magnitude of *P*? What are strains in two portions. What is the total change in length, *E* for copper = 105 GPa = 105 kN/mm².**

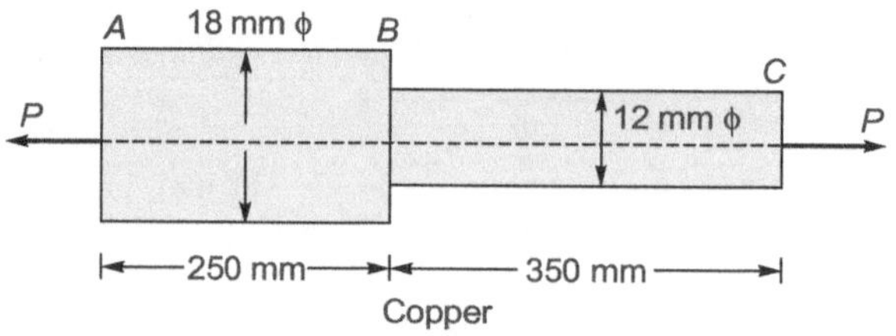

Fig. 1.5

Ans. [11.31 kN, 4.233×10^{-4}, 9.524×10^{-4}, 0.439 mm]

Question 1.5 **Explain (a) Tensile and compressive stresses (b) Positive and Negative shear stresses.**

Solution: Tensile and compressive stresses are normal stresses, i.e., their direction is perpendicular to the plane.

Direction of a tensile stress is away from the plane, while direction of compressive stress is towards the plane.

Shear stress is tangential to the plane, **positive shear stress tends to rotate the body in a clockwise direction, while negative shear stress tends to rotate the body in anticlockwise direction.**

Consider a bar of rectangular section $b \times h$ subjected to axial tensile load P as shown in figure 1.6. Take an inclined plane *abcd*, with side *ad* inclined to axes at an angle α.

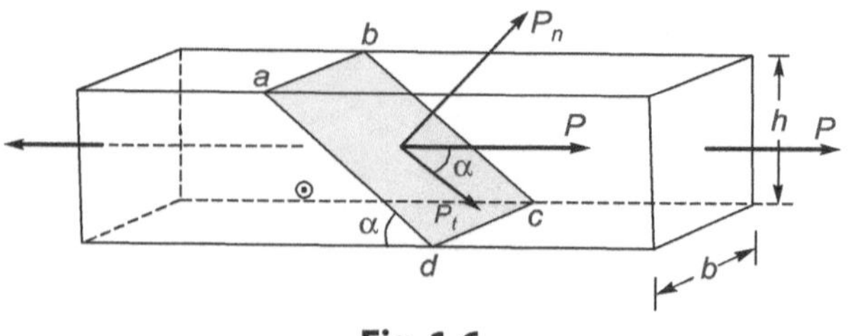

Fig. 1.6

Section *abcd* is an internal section and P is the internal resistance of the body. Force P has two components

$$P \sin \alpha = P_n, \text{ normal component}$$

$$P \cos \alpha = P_t, \text{ tangential component}$$

P_n is pointing away from the plane. It is a **positive normal force** or tensile force.

$$\sigma = \text{normal stress in plane} = \frac{P_n}{\text{area}} = \frac{P_n}{ad \times dc}; \text{ side } ad = \frac{h}{\sin\alpha}$$

$$= \frac{P_n}{\frac{h}{\sin\alpha} \times b} = \frac{P_n \sin\alpha}{hb}, \text{ putting the value of normal force}$$

$$= \frac{P\sin\alpha . \sin\alpha}{hb} = \frac{P\sin^2\alpha}{hb}$$

P_t is a force tangential to the plane *abcd*, if we take moment of the force **about point ⊙ on body**, the force P_t tends to rotate the body in clockwise direction.

This is a positive shear force on the plane

Positive shear stress,
$$\tau = \frac{P\cos\alpha}{ad \times dc} = \frac{P\cos\alpha}{h \times b} \times \sin\alpha$$

$$= \frac{P\sin\alpha\cos\alpha}{bh}; \text{ tends to rotates the body in clockwise direction.}$$

Now consider a rectangular bar of section $b \times h$, subjected to axial compressive load P as shown in figure 1.7.

Take inclined plane *abcd*, side *ad* inclined at an angle α with the axis of the bar

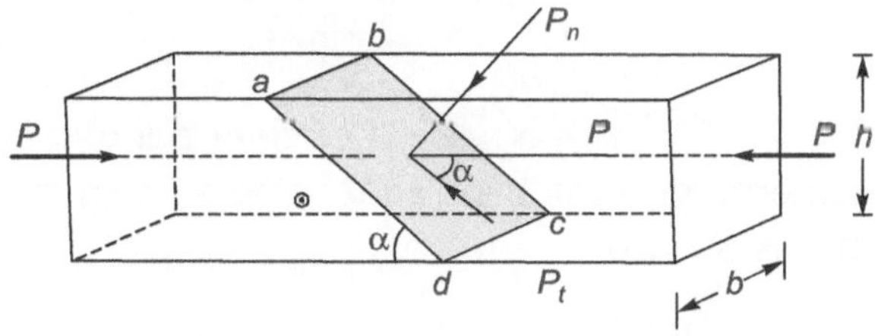

Fig. 1.7

$$\text{side } ad = \frac{h}{\sin\alpha}$$

$$\text{side } ab = b, \text{ breadth}$$

$$\text{area of cross-section} = \frac{h}{\sin\alpha} \times b = \frac{bh}{\sin\alpha}$$

Normal compressive stress on plane

$$\sigma = \frac{P_n}{bh}\sin\alpha$$

where,
$$P_n = P\sin\alpha$$

so
$$\sigma = \frac{P\sin^2\alpha}{bh}, \text{ Note that } P_n \text{ is } \textbf{pointing towards the plane}$$

It is a **compressive force** (a negative normal force), σ is a compressive stress.

Consider the effect of tangential force P_t about a point ⊙ on the body, it tries to rotate the body in anticlockwise direction. It is a negative shear force

Shear stress,
$$\tau = \frac{P_t}{\text{area}} = \frac{P\cos\alpha}{\frac{bh}{\sin\alpha}} = \frac{P\sin\alpha\cos\alpha}{bh}$$

So a negative **shear stress tends to rotate the body in an anticlockwise direction.**

Question 1.6 Consider figure 1.7, with $b = 20$ mm, $h = 30$ mm, and P (compressive force) = 10 kN. Inclined plane at a angle $\alpha = 60°$ with the axis of the bar. Determine normal and shear stresses an inclined plane.

Solution:

$$P = 10 \text{ kNm}, \quad \alpha = 60°$$

$$P_n = P \sin\alpha = 10 \times \sin 60° = 5\sqrt{3} \text{ kN} = 8660.25 \text{ N}$$

Area of cross-section of inclined of plane

$$= \frac{b \times h}{\sin\alpha} = \frac{20 \times 30}{\sin 60°} = 692.82 \text{ mm}^2$$

$$\text{Normal stress on plane} = -\frac{P_n}{692.82} = -\frac{8660.25}{692.82} = -12.5 \text{ N/mm}^2$$

$$\text{Shear stress as plane (negative)}, \tau = \frac{P_t}{bh}\sin\alpha$$

$$P_t = P\cos\alpha = P\cos 60°$$

$$= 10000 \times 0.5 = 5000 \text{ N}$$

$$\text{Shear stress (negative)}, \quad \tau = \frac{P_t}{\text{area}} = \frac{5000}{692.82} = 7.216 \text{ N/mm}^2$$

(tends to rotate the body in anticlockwise direction)

Practice Q.1.6 A circular bar of diameter 20 mm in subjected to an axial tensile force of 15 kN. Determine normal and shear stress on an inclined plane. Angle of inclination of the plane with axes of the bar is 30° (α).

Hints:

$$\text{Area of an ellipse} = \pi \times \text{semi minor axis} \times \text{semi major axis}$$

$$\text{Semi major axis} = \frac{1}{2} \times \frac{d}{\cos\alpha} = \frac{d}{1.732}$$

$$\text{Semi minor axis} = \frac{1}{2}A = 0.5d$$

Ans. [35.81 N/mm², 20.67 N/mm²]

Question 1.7 What do you understand by complementary shear stresses, explain with the help of a neat sketch.

Solution:

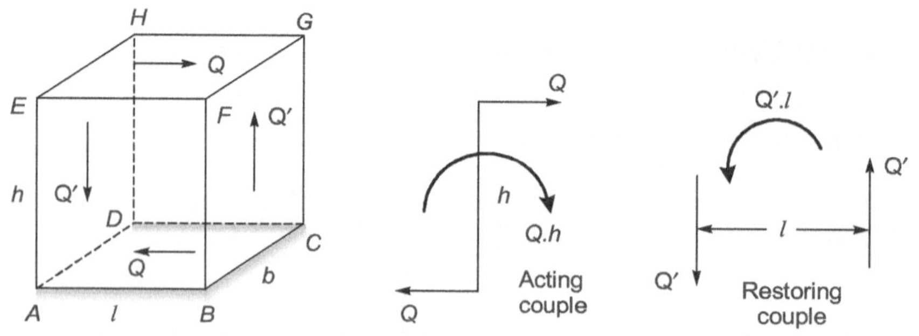

Fig.1.8

Consider a rectangular block of dimensions $l \times b \times h$, fixed at bottom, subjected to a shear force Q on top surface as shown in figure 1.8.

A reaction force Q acts at bottom surface.

There two forces forming a couple of arm h.

Couple moment $\quad M = Q.h$ (clockwise)

Under the action of couple $Q.h$ the block will try to rotate. To maintain equilibrium an equal, and opposite couple will act on the block. Say forces $Q'\uparrow$ and $Q'\downarrow$ act on vertical faces as shown.

Arm of the couple is l.

$Q'.l$ ((anticlockwise couple) reaction couple

$$Q.h = Q'l$$

but $\quad Q = \tau \cdot l \cdot b$, where τ is shear stress on horizontal top and bottom faces

$$\tau.l.b.h = Q'l$$

$$Q' = \tau.b.h = \tau' \times bh$$

where τ' is shear stress on vertical faces, perpendicular to horizontal faces.

$$\tau = \tau'$$

τ' is complementary shear stress to the applied shear stress τ. Note that both these stresses are perpendicular to each other.

Similarly shear stress τ is complementary to shear stress τ'.

Definition: Whenever a shear stress is applied on a certain plane, a complementary shear stress of same magnitude but opposite in nature on a perpendicular plane is developed to maintain equilibrium.

Further note that shear stress τ on horizontal surfaces tends to rotate the body in clockwise direction. Whereas τ', shear stress on vertical faces tends to rotate the body in anticlockwise direction so as to maintain equilibrium.

Question 1.8 **Explain longitudinal strain, lateral strain and Poisson's ratio.**

Solution: When a bar is subjected to an axial tensile force, its length is increased but its diameter is decreased. The change in length per unit length is termed as *longitudinal strain*.

The change in diameter per unit diameter is termed *as lateral strain. The ratio of lateral strain/longitudinal strain is called Poisson's ratio.* Similarly when bar is subjected to axial compressive load, its length is decreased but its diameter is increased. The ratio of change in length per unit length is longitudinal strain (negative) change in diameter (increase) per unit diameter is known as lateral strain (Positive strain). Diameter is in lateral direction to axis of the bar.

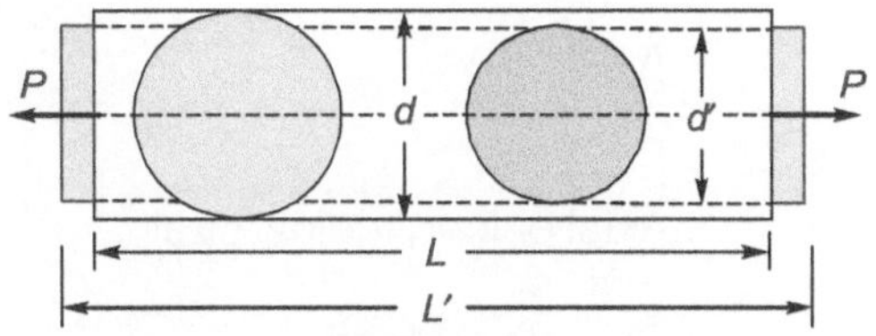

Fig. 1.9

Consider a bar of length L, diameter d subjected to axial tensile load, P. Its length is increased to L', such that $L' > L$, its diameter is decreased, $d' < d$.

Change in length $\quad \delta L = L' - L$

longitudinal strain, $\quad \varepsilon = \dfrac{\delta L}{L} = \dfrac{(L' - L)}{L}$, positive strain

change in diameter, $\quad \delta d = d' - d$

Lateral strain, $\varepsilon' = \frac{\delta d}{d} = \left(\frac{d' - d}{d}\right)$, negative strain

Poisson's ratio, $\nu = \frac{\varepsilon'}{\varepsilon} = -\frac{\delta d}{d} \times \frac{L}{\delta L}$ = **a negative ratio.**

But generally the negative sign is not attached with the Poisson's ratio.

$$\text{Lateral strain} = -\nu \times \text{longitudinal strain}$$

Using Hooke's law

$$\text{Axial tensile load} = P$$

$$\text{Original area of cross-section} = \frac{\pi}{4} d^2$$

Axial stress, $\sigma = \frac{4P}{\pi d^2}$

Longitudinal strain, $\varepsilon = \frac{\sigma}{E} = \frac{\text{Axial stress}}{\text{Young's modulus}}$

Lateral strain, $\varepsilon' = -\nu\varepsilon = -\nu\frac{\sigma}{E}$

Question 1.9 A steel round bar of diameter 16 mm, length 250 mm is subjected to an axial compressive load of 16 kN. (fig. 1.9). If E = 200 GPa, Poisson's ratio = 0.3 for steel, determine (a) axial stress (b) axial or longitudinal strain (c) lateral strain (d) change in length (e) change in diameter. For steel, E = 200 GPa, ν = 0.3 (Poisson's ratio).

Solution:

Original diameter, d = 16 mm

Axial compressive load, P = 16 kN

Length, L = 250 mm

Area of cross-section, $A = \frac{\pi}{4} \times d^2 = \frac{\pi}{4} \times 16^2 = 201.06 \text{ mm}^2$

Axial stress, $\sigma = -\frac{P}{A} = -\frac{16000}{201.06} = -79.57 \text{ N/mm}^2$ (compressive stress)

Axial or longitudinal strain, $\varepsilon = -\frac{\sigma}{E} = -\frac{79.57}{200000} = -0.398 \times 10^{-3}$

Change in length, $\delta L = -\varepsilon L = -0.398 \times 10^{-3} \times 250 = -0.0995$ mm

Lateral strain,, $\varepsilon' = -\nu\varepsilon = +0.3 \times 0.398 \times 10^{-3} = +0.1194 \times 10^{-3}$

Change in diameter, $\delta d = +0.1194 \times 10^{-3} \times 16 = 1.91 \times 10^{-3}$ mm

Practice Q.1.9 A bar of rectangular section 20 × 30 mm is subjected to axial tensile load of 30 kN. Length of bar is 300 mm. Bar is made of copper. For copper E = 105000 N/mm² and Poisson's ratio, ν = 0.35. Determine (a) axial stress (b) axial strain (c) lateral strain (d) change in length (e) changes in breadth and thickness.

Ans. [+50 N/mm², + 0.476 × 10^{-3}, –0.1667 × 10^{-3}, + 0.1428 mm, –3.33 × 10^{-3} mm, –5 × 10^{-3} mm]

Question 1.10 A rectangular block of copper with base 40 × 60 mm and height 80 mm is fixed in the ground. It is subjected to a shear force on top face equal to 24 kN. What are shear stress, shear strain developed in block if G for copper 39 kN/mm². What is displacement AA'?

Solution:

Shear force, $Q = 24$ kN

Area of top face $= 60 \times 40 = 2400 \text{ mm}^2$

Shear stress, $\tau = \dfrac{24000}{2400} = 10 \text{N/mm}^2$

(it tries to rotate the block in clockwise direction about point ⊙)

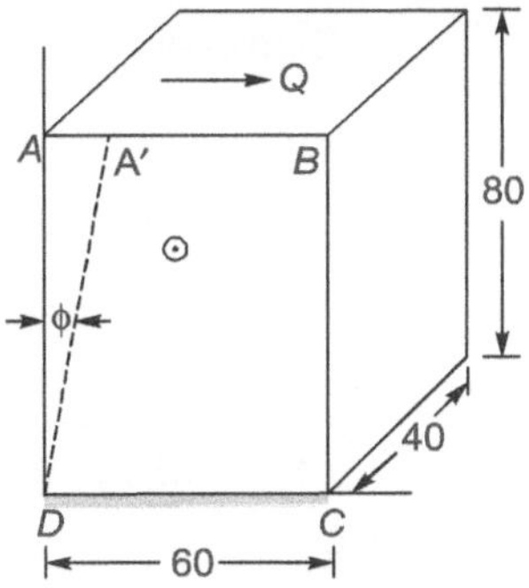

Fig. 1.10

Shear strain, $\phi = \dfrac{\tau}{G} = \dfrac{10}{39000} = 0.2564 \times 10^{-3}$ radian

Displacement, $AA' = \phi h = 0.2564 \times 10^{-3} \times 80 = 0.0205$ mm

Question 1.11 Explain what is stress tensor?

Solution: Stress tensor is a representation of normal and shear stresses on a three dimensional block of size $\Delta x = \Delta y = \Delta z \to 0$. In other words it represents state of stresses at a point along three orthogonal directions.

σ stands for normal stress

τ stands for shear stress

σ_{xx}, I[st] subscript x denotes direction of normal to the plane (yz plane), x second subscript denotes direction of normal stress.

τ_{xy}, I[st] subscript x denote direction of normal to the plane (yz plane) and y second subscript denotes direction of shear stress in y direction. Similarly

τ_{yz}, I[st] subscript y denotes direction of normal to the plane (xz plane) and second subscript z denote direction of shear stress in z-direction.

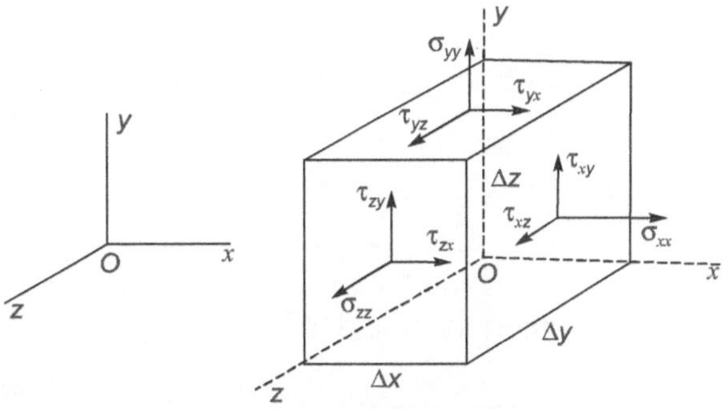

Fig. 1.11 *Normal and shear stresses on 3 orthogonal planes*

Stress tensor is represented in matrix form as follows:

$$\tau_{ij} = \begin{vmatrix} \sigma_{xx} & \tau_{xy} & \tau_{xz} \\ \tau_{yx} & \sigma_{yy} & \tau_{yz} \\ \tau_{zx} & \tau_{zy} & \sigma_{zz} \end{vmatrix}$$

Stresses along diagonal, i.e., σ_{xx}, σ_{yy}, σ_{zz} are normal stresses.

In a two dimensional case, stress tensor,

$$\tau_{ij} = \begin{vmatrix} \sigma_{xx} & \tau_{xy} \\ \tau_{yx} & \sigma_{yy} \end{vmatrix}$$

Shear stress $\tau_{xy} = \tau_{yx}$, $\tau_{zx} = \tau_{xz}$, $\tau_{zy} = \tau_{yz}$ are complementary shear stresses.

Question 1.12 Consider a uniformly tapered circular bar, of axial length L, diameters d and D, subjected to axial load. E is the modulus of elasticity of the material of the bar. Derive expression for change in its axial length.

Solution: Fig. 1.12 shows a circular tapered bar. Take a strip of diameter d_x at a distance of x from end A.

Diameter, $$d_x = d + \frac{D-d}{L}x = d + kx \qquad \text{where } k = \frac{D-d}{L} \text{ a ratio}$$

$$\text{Area of cross-section} = \frac{\pi(d+kx)^2}{4}$$

$$\text{Axial load} = P$$

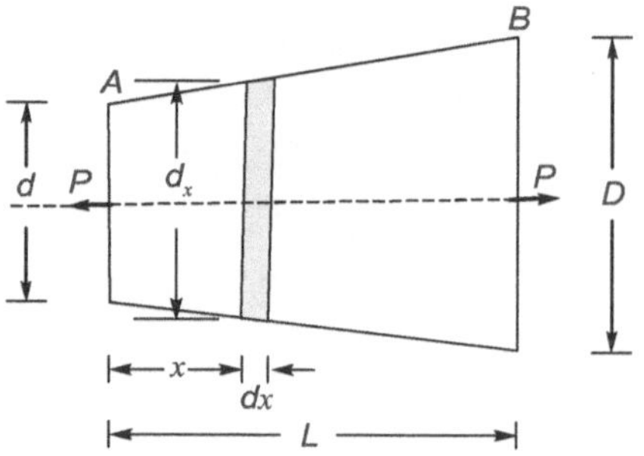

Fig. 1.12

$$\sigma_x,\ \text{stress at elementary strip} = \frac{4P}{\pi(d+kx)^2}$$

Strain at elementary strip, $$\varepsilon_x = \frac{4P}{E\pi(d+kx)^2}$$

Change in length over length dx,

$$\delta dx = \frac{4Pdx}{E\pi(d+kx)^2}$$

Total change in length, $$\delta L = \int_0^L \frac{4Pdx}{E\pi(d+kx)^2}$$

$$= -\left|\frac{4Pdx}{E\pi(d+kx)}\right|_0^L = -\frac{4P}{\pi Ek}\left|\frac{1}{d+kL} - \frac{1}{d}\right|$$

Putting the value of k,

Change in length, $$\delta L = -\frac{4PL}{\pi E(D-d)}\left[\frac{1}{D} - \frac{1}{d}\right]$$

$$= -\frac{4PL}{\pi E(D-d)}\left[\frac{d-D}{dD}\right] = +\frac{4PL}{E\pi Dd}$$

Practice Q.1.12 An aluminium circular bar tapers uniformly from diameter 20 mm to diameter 30 mm over length L = 500 mm. If an axial tensile load of 40 kN in applied on bar. What is the change in length of the bar, if E = 67 GPa.

Ans. [0.633 mm]

Question 1.13 What do you mean by volumetric stress, volumetric strain and bulk modulus of elasticity.

Solution: Volumetric stress is a stress which acts with same magnitude in all directions such as hydrostatic pressure, p. As per Pascal's law, liquid exerts equal pressure in all directions.

As an example, if a ball is immersed in sea water, pressure on ball, $p = wh$, where w is weight density of sea water and h is depth of the ball in sea water.

If the depth of immersion is increased gradually then pressure on ball also increases simultaneously. Diameter of the ball gradually is reduced producing diametral strain in bar and consequently, the volumetric strain, which is three times the diametral strain. It a graph is plotted between pressure, and volumetric strain, ε_v an shown in figure 1.13, then upto proportional limit A, pressure is directly proportional to volumetric strain.

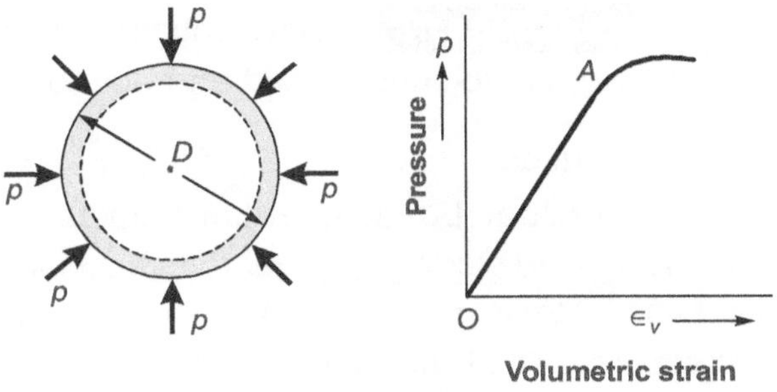

Fig. 1.13

Pressure, $\quad p \propto e_v$, volumetric strain

or $\quad p = Ke_v$

K is proportionality constant, called Bulk modulus

Bulk modulus, $$K = \frac{\text{pressure, } p}{\text{volumetirc strain, } \varepsilon_v}$$

Question 1.14 **A spherical ball of steel of diameter 200 mm goes down to a depth of 1 km in sea water. If specific weight of sea water in 10.2 kN/m³, (and bulk modulus of steel is 170 kN/mm²), determine change in volume of spherical ball.**

Solution:

Depth, $\quad h = 1000 \text{ m}$

Weight density, $\quad w = 10.2 \times 1000 \text{ N/m}^3$

Pressure, $\quad p = wh = 1000 \times 10.2 \times 1000 \text{ N/m}^2$

$\quad = 10.2 \times 10^6 \text{ N/m}^2$

Bulk modulus, $\quad K = 170 \times 10^3 \text{ N/mm}^2 = 170 \times 10^9 \text{ N/m}^2$

Volumetric strain, $$\varepsilon_v = \frac{p}{K} = \frac{10.2 \times 10^6}{170 \times 10^9} = 0.06 \times 10^{-3}$$

Original volume, $$V = \frac{\pi D^3}{6} = \frac{\pi \times 200^3}{6} = 4.188 \times 10^6 \text{mm}^3$$

Change volume, $\quad \delta V = \varepsilon_V \cdot V = 0.06 \times 10^{-3} \times 4.188 \times 10^6$

$\quad = 0.25 \times 10^3 \text{ mm}^3$

$\quad = 250 \text{ mm}^3 = 0.25 \text{ cc}$

Practice Q.1.14 A spherical ball of copper of diameter 300 mm is immersed in sea water specific weight of sea water is 10.2 kN/m³. Bulk modulus of copper is 98 GPa. For how much depth ball must be immersed that change in volume is 10 cc.

Ans. [6.796 km]

Question 1.15 **Differentiate between gradual, sudden and shock loads. Derive expressions of stresses due to sudden load and due to shock load.**

Solution:

Gradual Load: Gradual load is continuously increasing load starting from zero. As the load is gradually increased, stress σ and strain ε in the body also gradually increase. Fig. 1.14 (a) shows a wire subjected to gradually increasing load W on hanger.

$$\sigma_{gradual} = \frac{\text{Load}}{\text{Area}} = \frac{W}{\text{Area}}$$

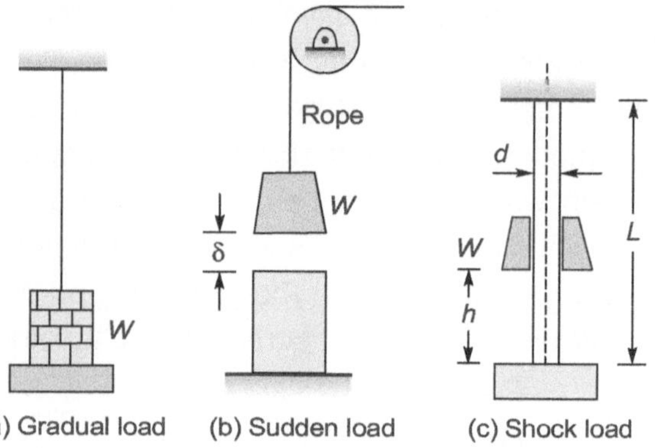

Fig. 1.14

Sudden Load: Load is not gradually increased but **whole of the magnitude of the load acts at once or suddenly** on the body. Fig. 1.14 (b) shows a load W being lowered slowly through a wire rope of a crane. When the gap, δ between body and load tends to become zero, the wire rope breaks and **whole of the load W acts suddenly on the column** (as shown). Say the change in length in body δL.

Work done on body by the sudden load = $W.\delta L$

Internal resistance of the body does not develop suddenly but it takes time and develops gradually **as shown by resistance R in figure 1.15.**

Stress is defined as internal resistance per unit area.

Workdone on the body is absorbed by the body as strain energy.

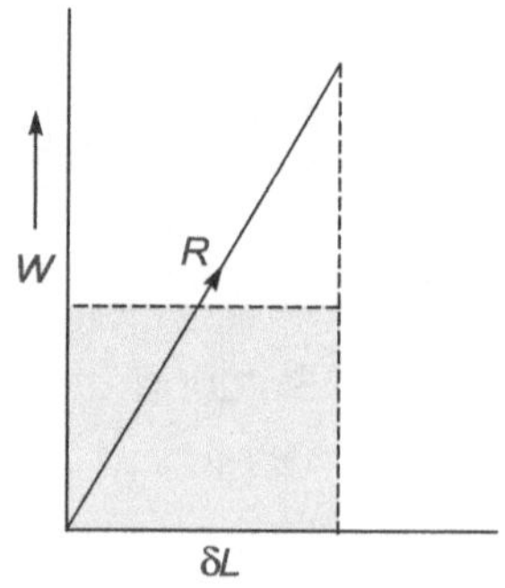

Fig.1.15 *Sudden load*

$$\text{Internal energy} = \frac{1}{2}R.\delta L = W.\delta L$$

(where R denotes internal resistance)

$$\text{Internal resistance,} \quad R = 2W$$

$$\text{Stress due to sudden load, } \sigma_{sudden} = \frac{R}{\text{area}} = \frac{2W}{A} = 2 \times \sigma_{gradual}$$

$\frac{W}{A}$ is the gradual stress as the load W is applied gradually starting from zero.

When the load is applied through some velocity or the load possesses kinetic energy, as the small bullet coming out of the muzzle of a gun with very high velocity produces tremendous stress depending upon the energy.

Figure 1.14 (c) shows a bar of length L, diameter d fixed at upper end and carries a collar at the lower end. A load W is allowed to fall freely on the collar through height 'h'. When the load falls under gravity, velocity $V = \sqrt{2gh}$ is gained by weight, W. Load W is arrested by the collar. Say due to shock load, δL_i is instantaneous extension in bar and say σ_i is the instantaneous stress developed in the bar.

$$\text{Loss of potential energy of weight} = W(h + \delta L_i)$$

$$\text{Gain in strain energy of the bar} = \sigma_i A(\delta L_i)\frac{1}{2}$$

and
$$W(h + \delta L_i) = \frac{\sigma_i A}{2}\delta L_i \quad \text{...(i)}$$

Using Hooke's law

$$\delta L_i = L \times \varepsilon_i' = L \times \text{instantaneous strain}$$

$$= L \times \frac{\sigma_i}{E}$$

Putting these values in equation (i)

$$W\left(h + L\frac{\sigma_i}{E}\right) = \sigma_1 \frac{A}{2} \times L \times \frac{\sigma_i}{E}$$

or

$$Wh + WL\frac{\sigma_i}{E} = \frac{\sigma_i^2}{2E} \times AL$$

or

$$\frac{Wh \times 2E}{AL} + \frac{\sigma_i}{E} \times WL \times \frac{2E}{AL} = \sigma_i^2$$

$$\sigma_i^2 - \frac{2EWh}{AL} - \sigma_i \frac{2W}{A} = 0$$

A quadratic in σ_i, solution of this equation

$$\sigma_i = \frac{1}{2}\left[\frac{2W}{A} \pm \sqrt{\frac{4W^2}{A^2} + \frac{8WEh}{AL}}\right] = \frac{W}{A} \pm \frac{W}{A}\sqrt{1 + \frac{2EAh}{WL}}$$

Negative sign is inadmissible, because the type of the load shown does not produce compressive stress.

$$\sigma_i = \frac{W}{A}\left[1 + \sqrt{1 + \frac{2EAh}{WL}}\right] \quad \text{...(ii)}$$

If you analyse this expression, then σ_i' depends on the magnitude of W and height *h*.

If height $h = 0$, then $\sigma_i = \sigma_{sudden} = \frac{2W}{A}$

Many a times, loss of *PE* or gain in *KE* is so large that stress produced far exceeds σ_{sudden}.

Question 1.16 **Consider a steel bar of diameter 10 mm, length 500 mm, fixed at upper end and carries a collar at the lower end. A load *W* of 1 kN falls through height *h*. Determine *h* if the instantaneous stress developed in bar is 150 N/mm². *E* = 208000 N/mm².**

Solution:

$$\sigma_i = 150 \text{ N/mm}^2$$

$$W = 1000 \text{ N}$$

$$A = \frac{\pi}{4} \times 10^2 = 78.54 \text{ mm}^2$$

$$L = 500 \text{ mm}$$

$$\frac{W}{A} = \frac{1000}{78.54} = 12.73$$

$$E = 208000 \text{ N/mm}^2$$

Putting the values in the expression for instantaneous stress

$$150 = 12.73\left[1 + \sqrt{1 + \frac{2 \times 208000 \times h}{12.73 \times 500}}\right]$$

$$\left(\frac{150}{12.73}-1\right)^2 = 1 + 65.35\,h$$

$$116.277 = 1 + 65.35\,h$$

Height, $$h = \frac{116.277-1}{65.35} = 1.764 \text{ mm}$$

If the load $W = 1$ kN, which can only produce gradual stress of 12.73 N/mm², is allowed to fall through a height of 1.764 mm only then instantaneous stress produced in the bar is 150 N/mm².

Practice Q.1.16 **A copper bar of diameter 15 mm, length 600 mm in fixed at top end and at bottom end a collar is provided. A load of 500 N is allowed to fall through a height of 30 mm on the collar. If *E* for copper is 105000 N/mm², what is the instantaneous stress developed in the bar.**

Ans. [175.2 N/mm²]

Question 1.17 **What do you understand by strain energy absorbed by a body subjected to external loads? Differentiate between resilience, proof resilience, modulus of resilience and toughness.**

Solution: Figure 1.16 (a) shows a bar of length *L*, area of cross-section *A*, subjected to axial load *P*, if *E* is the modulus of elasticity of the material, then,

stress, $$\sigma = \frac{P}{A}$$

Strain, $$\varepsilon = \frac{\sigma}{E} = \frac{P}{AE}$$

Change in length, $$\delta L = \varepsilon L = \frac{PL}{AE}$$

Work done on bar $$= \frac{1}{2}P\delta L$$

= Strain energy absorbed by bar due to strain/deformation developed in bar

Strain energy, $$U = \frac{1}{2}P\delta L = \frac{1}{2}P\times\frac{PL}{AE} = \frac{P^2L}{2AE}$$

$$= \frac{P^2L}{2A^2E}\times A = \frac{1}{2}\frac{P^2}{A^2}\times AL$$

$$= \frac{1}{2}\frac{\sigma^2}{E}AL = \frac{1}{2}\frac{\sigma^2}{E}\times \text{Volume of bar}$$

Strain energy per unit volume

$$= \frac{\sigma^2}{2E}$$

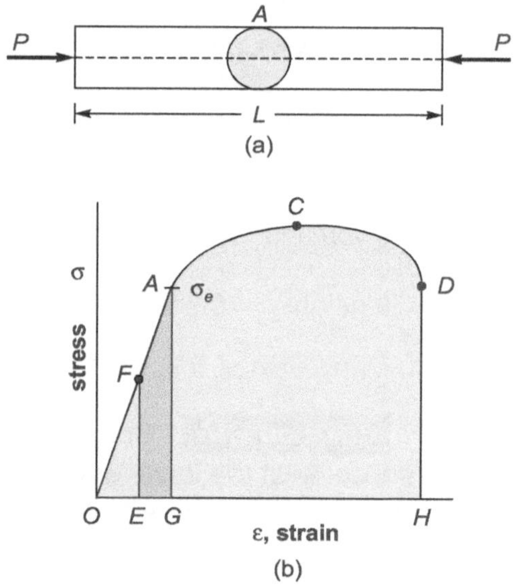

Fig. 1.16

If a graph is plotted between stress σ and strain, ε upto the destruction of the bar, as the shown in figure 1.16 (b) for a ductile material, then

A = limit of proportionality or elastic limit

If load is removed from the bar upto the elastic limit, there is no residual strain remaining in the bar after the removal of the load. Fig. 1.17 shows removal of the load from the bar in the plastic stage.

Point A: limit of proportionality

Point B: elastic limit

Point B is above point A and very close to point A.

If load is removed from point *B* i.e., elastic stage, there will not be any residual strain in bar. If load is removed from the bar in the plastic stage, the unloading curve will be parallel to line *OA*, and there will be residual strain, ε_R remaining in the bar.

Refer to the figure 1.16 (b)

Resilience is the total strain energy absorbed by the body within the elastic limit. Area *OFE* in figure. 1.16(b) shows resilience of the body. If load is removed from point *F*, total energy absorbed by bar is fully recovered.

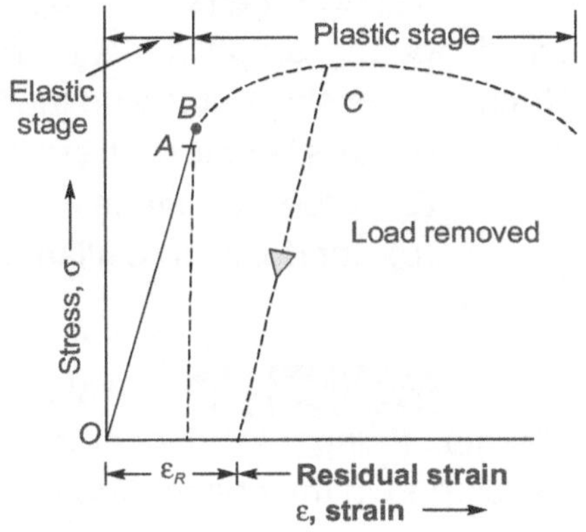

Fig. 1.17

Proof resilience is the total strain energy absorbed by a body upto the elastic limit, i.e., $\frac{\sigma_e^2}{2E} \times$ Volume

Proof resilience is the strain energy absorbed by a body per unit volume upto the elastic limit.

$$\text{Proof resilience} = \frac{\sigma_e^2}{2E},$$ where σ_e is the stress at elastic limit in the body.

Toughness is given by the total energy absorbed by a body upto the stage of breaking i.e., area where the curve *OACDH* fig. 1.16(b). In place of a graph between $\sigma - \varepsilon$, if graph is plotted between load, *P* and extension δL, then total energy given under the curve gives the value of toughness of a material.

Toughness is the most desired mechanical property of a material used in engineering components as crank shaft, connecting rod etc. This property is a combination of strength and ductility of a body.

Question 1.18 **A cube of side 50 mm in subjected to shear stress and a graph between shear stress and shear strain is plotted, it is observed that elastic limit shear stress is τ_e = 230 N/mm². If *G* = shear modulus of material is 83 kN/mm², determine**

(a) proof resilience **(b) modulus of resilience** **(c) resilience at τ = 120 N/mm²**

Solution:

Volume of cube $= 50^3 = 125000 \text{ mm}^3$

(a) Proof resilience = Total strain energy upto elastic limit

$$= \frac{\tau_e^2}{2G} \times \text{Volume} = \frac{230^2}{2 \times 84000} \times 125 \times 10^3$$

$$= 39360 \text{ Nmm} = 39.36 \text{ Nm}$$

(b) $$\text{Modulus of resilience} = \frac{\tau_e^2}{2G} = \frac{230^2}{2 \times 84000} = 0.314 \text{ Nmm/mm}^3$$

(c) $$\text{Resilience at } \tau = 120 \text{ N/mm}^2 = \frac{120^2}{2G} \times \text{Volume} = \frac{120^2}{2 \times 84000} \times 125 \times 10^3$$

$$= 10714 \text{ Nmm} = 10.714 \text{ Nm}$$

Practice Q.1.18 A copper bar of diameter 25 mm and length 500 mm is subjected to an axial tensile load. A graph is plotted between load and extension, it was observed that elastic limit, σ_e = 215 N/mm², in tension determine the following:

(a) resilience at stress of 100 N/mm²
(b) proof resilience
(c) modulus of resilience given that E = 105 kN/mm² for copper

Ans. [11.68 Nm, 54.017 Nm, 0.220 Nmm/mm³]

Question 1.19 Water under a pressure of 4 N/mm² is suddenly admitted on a plunger of diameter 100 mm. Plunger is connected to a steel connecting rod of diameter 20 mm, fig. 1.18. What is the stress developed in the connecting rod.

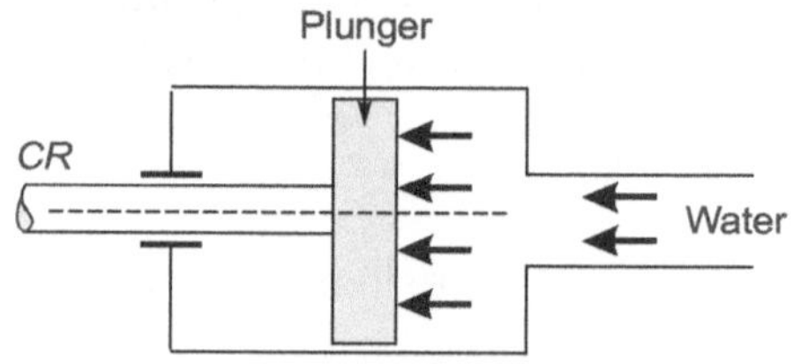

Fig. 1.18

Solution:

Plunger diameter, $D = 100$ mm

Area of cross-section of plunger,

$$A = \frac{\pi}{4}D^2 = \frac{\pi}{4}100^2$$
$$= 7854 \text{ mm}^2$$

Pressure, $p = 4$ N/mm²

Pressure force an plunger, $P = p.A = 4 \times 7854 = 31416$ N
= Sudden load on connecting rod

Diameter of connecting rod, $d = 20$ mm

Area of cross-section of C.R, $A = \frac{\pi}{4} \times 20^2 = 314.16 \text{ mm}^2$

$$\sigma_{\text{sudden,}} \text{ stress in CR} = \frac{2P}{A} = \frac{2 \times 31416}{314.16} = 200 \text{ N/mm}^2$$

Practice Q.1.19 A short column of diameter 0.2 m, length 1 m, made of steel is subjected to a sudden load W such that $\delta \to 0$. If the sudden contraction in column is not to exceed 0.4 mm, what is W? E for column = 210 kN/mm²

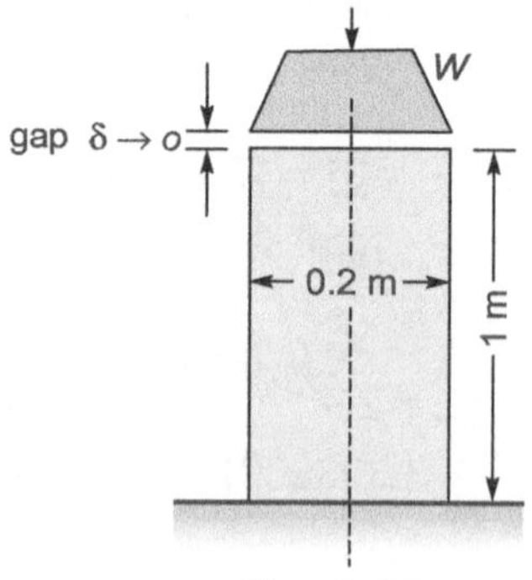

Fig. 1.19

Ans. [1319.47 kN]

Question 1.20 What is a statically indeterminate problem? Explain with the help of an example.

Solution: For a statically determinate problem, equations of equilibrium of forces are sufficient to determine the unknown forces on the body or a structure. But for a statically indeterminate problem, equations of equilibrium are insufficient to determine the unknown forces such as reactions.

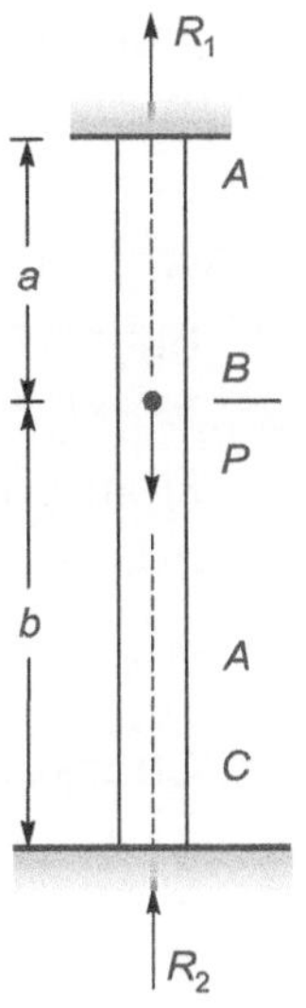

Fig. 1.20

An additional equation involving the geometry of deformed structure is necessary to provide a condition of structural compatibility and then unknown forces are determined.

Example: Figure 1.20 shows a bar *ABC*, held between two rigid supports at ends *A* and *C*, *A* force *P* is applied at section *B*. Such that *AB* = *a*, *BC*= *b*,

[As per equations of equilibrium of forces.]

Force, $P = R_1 + R_2$, sum of reactions at *A* and *C*.

This equation is insufficient to determine reactions under the action of the force *P*, portion *AB* comes under tension and portion BC comes under compression. Then extension in *AB* + contraction in *B* C = 0, and ends are rigidly fixed.

$$\text{Extension in } AB = \frac{R_1 a}{AE}$$

where, *A* = area of cross-section, *E* = modulus of elasticity of bar

$$\text{Contraction in BC} = -\frac{R_2 b}{AE}$$

So $$\frac{R_1 a}{AE} - \frac{R_2 b}{AE} = 0$$

or $$R_1 = R_2 \frac{b}{a} \quad \text{...(ii)}$$

$$P = R_1 + R_2 = R_2 + R_2 \frac{b}{a} = R_2 \frac{(a+b)}{a}$$

$$R_2 = \frac{Pa}{a+b},\ R_1 = \frac{Pb}{a+b}$$

Above Question shows that compatibility condition of involving deformations in a structure is necessary for the solution of a statically indeterminate problem.

Practice Q.1.20 **In the above Question take *a* = 0.4 m, *b* = 0.6 m, diameter of bar 40 mm, force *P* = 50 kN. Determine stresses in portions *AB* and *BC*.**

Ans. [23.87 MPa, –15.91 MPa]

Question 1.21 **A steel stepped bar is shown in figure 1.21. A force *P* = 10 kN in applied at shoulder *B*, gap between lower end of bar and floor is 0.1 mm. What are the stresses developed in bar in portions *AB* and *BC*.**

Solution: Free extension in bar under load *P*

$$\delta L' = \frac{10000}{100} \times \frac{500}{200000}$$

= 0.238 mm (load acting on part AB only)

$\delta L' > 0.1$ mm (gap), so this therefore extension is prevented producing compressive strain in portion *BC*, say stress in *BC* is σ_2 and stress in *AB* is σ_1

Then $$\frac{\sigma_1}{E} \times 500 - \frac{\sigma_2}{E} \times 500 = 0.1 \text{ mm} \quad \text{...(i)}$$

Moreover $$\sigma_1 A_1 = R_1$$

$$\sigma_2 A_2 = R_2$$

$$R_1 + R_2 = P$$

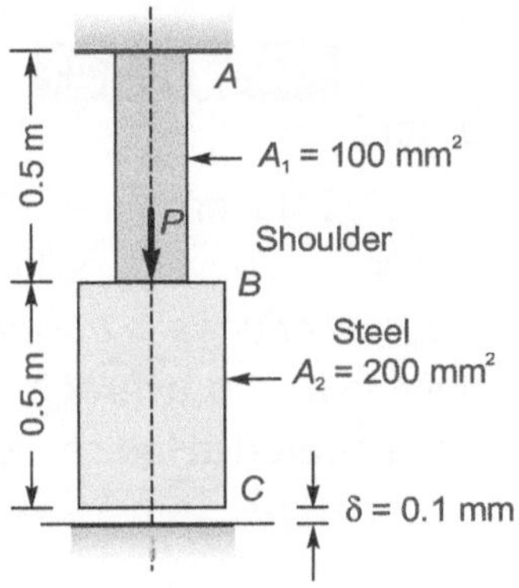

Fig.1.21

From equation (i)

$$\frac{\sigma_1 \times 500}{200000} - \frac{\sigma_2 \times 500}{200000} = 0.1$$

or
$$\sigma_1 - \sigma_2 = 40 \quad \text{...(ii)}$$

But
$$\sigma_1 \times 100 + \sigma_2 \times 200 = 10000$$

$$\sigma_1 + 2\sigma_2 = 100 \quad \text{...(iii)}$$

From, equations (ii) and (iii)

$$3\sigma_1 = 180$$

$$\sigma_1 = 60 \text{ N/mm}^2 \text{ (tensile)}$$

$$\sigma_2 = \sigma_1 - 40 = 60 - 40 = 20 \text{ N/mm}^2 \text{ (compressive)}$$

Checking

Extension in *AB*,
$$\delta L_1 = \frac{60 \times 500}{200000} = 0.15$$

Contraction in *BC*,
$$\delta L_2 = \frac{20 \times 500}{200000} = -0.05$$

Practice Q.1.21 An aluminium bar 10 mm in diameter, 1200 mm long is fixed at top end. At the middle of its length a force $P = 8$ kN is applied. If the gap between lower end and floor is 0.5 mm, what is the reaction at top end and at bottom end. What are the stresses developed in top and bottom portion of the bar. $E = 67000$ N/mm^2.

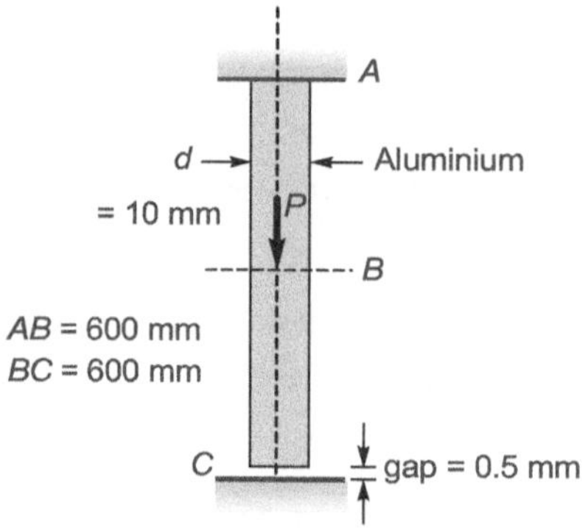

Fig. 1.22

Ans. [$R_1 = 6192.6$ N, $R_2 = 1807.4$ N, $\sigma_1 = +78.846$ MPa, $\sigma_2 = -23.013$ MPa]

Question 1.22 What do you understand by stress concentration and stress concentration factor?

Solution: When a body is subjected to an external load, stresses are developed in the cross-section of a body. When there *is abrupt change in the cross-section*, there is increase in stress, more than the average stress and **this enhanced stress is concentrated at the section of abrupt change.** Figure 1.23 shows a stepped bar, with a sharp corner at the step, subjected to axial load P, Due to the axial load, stress in portion *I* is $\frac{4P}{\pi d_1^2} = \sigma_1$ and

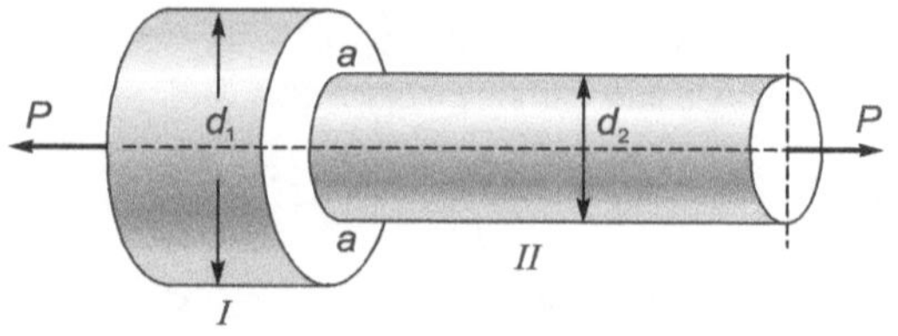

Fig. 1.23

stress in portion II is, $\frac{4P}{\pi d_2^2} = \sigma_2$, so the stress abruptly increases from σ_1 to σ_2 along the inner circular corner of diameter d_2. There is abrupt change in stress from σ_1 to σ_{max} and after some distances from corner, stress becomes σ_2.

Stress at inner corner $\sigma_{max} > \sigma_2$, this effect is called stress concentration.

Stress concentration factor, SCF $= \frac{\sigma_{max}}{\sigma_{av}} = \frac{\text{max. stress}}{\text{average stress}}$.

The effect of stress concentration can be reduced by gradually decreasing the change in dimension i.e., from diameter d_1 to diameter d_2, by providing a **fillet radius at the corner** as shown in figure 1.24.

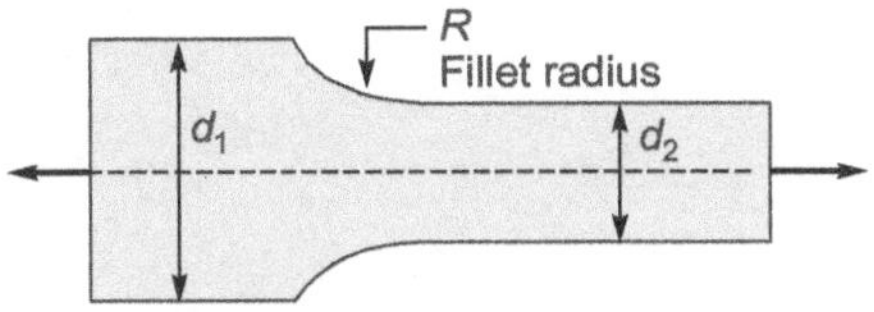

Fig. 1.24

Question 1.23 **A stepped bar *ABCD*, 900 mm long, is subjected to various loads of 12 kN, 3 kN, 9 kN and 6 kN as shown in figure 1.25. Diameters of AB, BC, CD are 20 mm 16 mm and 12 mm respectively. The axial lengths are *AB* = 200 mm, *BC* = 300 mm and *CD* = 400 mm respectively. If the bar is made of steel, what is total change in its length. *E* = 200 GPa.**

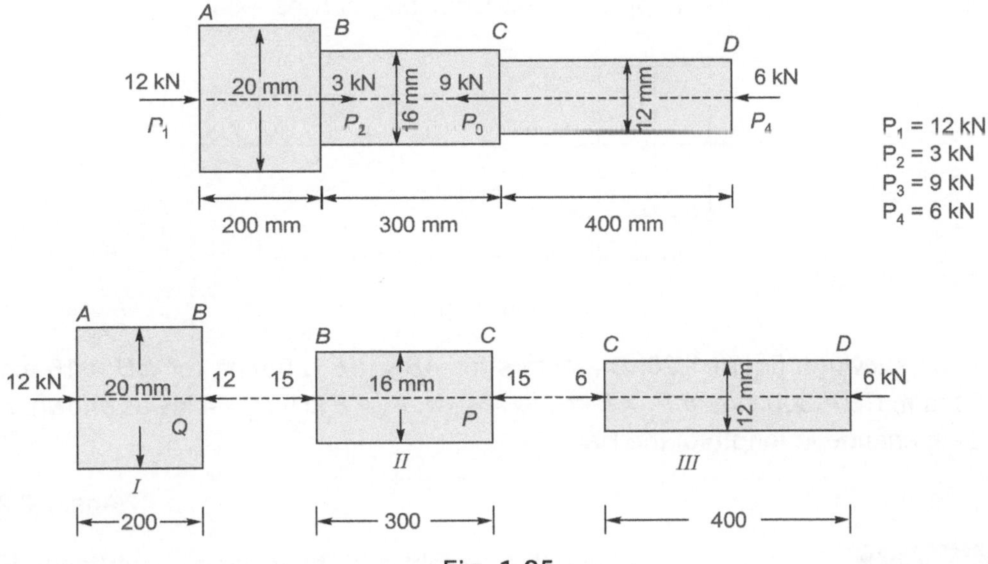

Fig. 1.25

Solution: The bar is in equilibrium because

$$P_1 + P_2 = P_3 + P_4$$

$$12 + 3 = 9 + 6 \text{ kN}$$

Note that axial compression force in *AB* is 12 kN.

Axial compression force in *CD* is 6 kN

Then axial compression force in *BC* = 15 kN

Net force at section *B* = 15 – 12 = 3 kN →

Net force an section C (as shown) = 15 – 6 = 9 kN ←

So all the forces are compressive forces.

Stresses developed in all the three portions are compressive stresses.

Stresses $\sigma_1 = \frac{12000 \times 4}{\pi \times 20^2} = -38.2 \text{ N/mm}^2$

$$\sigma_2 = -\frac{15000 \times 4}{\pi \times 16^2} = -74.60 \text{ N/mm}^2$$

$$\sigma_3 = -\frac{6000 \times 4}{\pi \times 12^2} = -53.05 \text{ N/mm}^2$$

Total change in length $= \delta L_1 + \delta L_2 + \delta L_3$

$$= -\frac{38.2}{E} \times 200 - \frac{74.6 \times 300}{E} - \frac{53.05 \times 400}{E}$$

$$= -\frac{1}{E}[7640 + 22380 + 21220] = -\frac{51240}{E}$$

$$= -\frac{51240}{200000} = -0.2562 \text{ mm (contraction in length)}$$

Practice Q.1.23 A stepped copper bar 600 mm long carries axial.

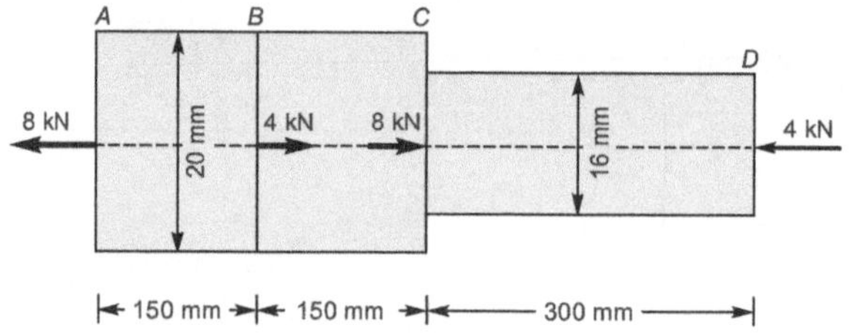

Fig. 1.26

Forces as shown in figure 1.26. Diameters for $AB = BC$, 20 mm, for CD = 16 mm, lengths are $AB = BC$ = 150 mm, CD = 300 mm. If P_1 = 8 kN, P_2 = 4 kN, P_3 = 8 kN, P_4 = 4 kN as shown. E for copper = 105 GPa, what is change in length of the bar.

Ans. [-2.283×10^{-3} mm]

Question 1.24 Consider a bar of length L, weight density w, hanging vertically, with upper end fixed in ceiling. Derive expression for change in its length due to self weight. Take E = Young's modulus of the material of the bar.

Solution: Consider the bar of length L, area of cross-section A, fixed at end B.

weight density of bar $= w$

Take a section cc at a distance y, from lower end.

Weight of portion $CD = wA.y$

Stress at the section 'cc' $= \frac{wAy}{A} = wy$

Strain at section cc $= \frac{wy}{E}$

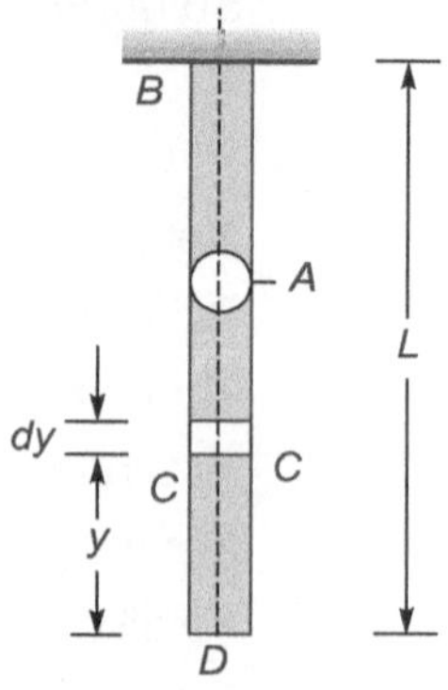

Fig. 1.27

Change in length over length $dy = \frac{wy}{E} \times dy$

Total change in entire length

$$\delta L = \int_0^L \frac{wydy}{E} = \frac{wL^2}{2E}$$

Practice Q.1.24 **A steel wire rope of diameter 8 mm is used to lift load during construction of high rise building. If a 30 m long wire is hanging vertically and a load of 1 kN is being lifted at the lower end at a very slow speed. Determine total extension in wire rope, if**

w = specific weight of wire rope = 0.06×10^{-3} N/mm³, E = 200 GPa.

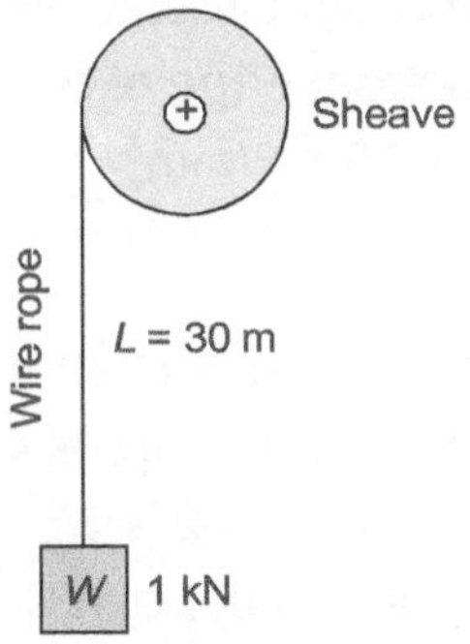

Fig. 1.27

Ans. [0.135, 3.119 mm]

Question 1.25 **What is bar of uniform strength? Derive relation between area of cross-section and uniform strength σ, weight density *w*.**

Solution: A bar of varying cross-section such that under the axial load P applied at the lower end produces the same stress throughout its length L as shown.

The **addition of stress due to self weight is compensated by the increase in area** so that stress remains constant throughout the axial length.

Say A_1 = Area of cross-section at lower end

A_2 = Area of cross-section at upper end

W = total weight of the bar

$$\frac{P+W}{A_2} = \frac{P}{A_1} = \sigma \text{, uniform stress in bar}$$

say w = weight density of bar

Consider a small element *efgh*, as shown in figure 1.28, Area of cross-sections at upper force is $A + dA$

$$\sigma A + wAdy = \sigma(A + dA)$$

$$wAdy = \sigma dA$$

$$\frac{dA}{A} = \frac{w}{\sigma}dy \quad \text{...(i)}$$

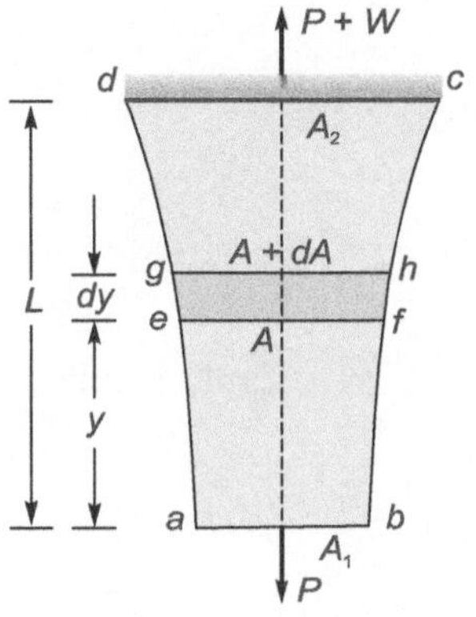

Fig. 1.28

Integrating both the sides of equations (i)

$$\int_{A_1}^{A} \ln A = \int_0^y \frac{w}{\sigma} dy$$

$$\ln \frac{A}{A_1} = \frac{w}{\sigma}, y$$

or
$$A = A_1 e^{wy/\sigma}$$

For section A_2, length of L

$$A_2 = A_1 e^{wL/\sigma}$$

Practice Q.1.25 A 2 m long vertical circular bar is subjected to a uniform stress of 0.2 N/mm² throughout its length. Diameter at the bottom edge of bar is 60 mm. What is diameter at the top section of the bar, if weight density, $w = 0.07644 \times 10^{-3}$ N/mm³.

Ans. [87.93 mm]

Question 1.26 Two wires one of steel and another of copper are attached to a rigid bar. Upper ends of wires are fixed in ceiling. Length of steel wire is 4 m and diameter of steel wire is 2 mm. length of copper wire is 3 m and diameter is 2 mm. $E_s = 2 \times E_{cu} = 210$ kN/mm². Distance between two wires is 100 mm. A load of 600 N is applied on rigid bar at point C. Determine distance 'a' so that rigid bar remains horizontal after the application of the load, (Let to fig. 1.29) by how much distance rigid bar moves downward.

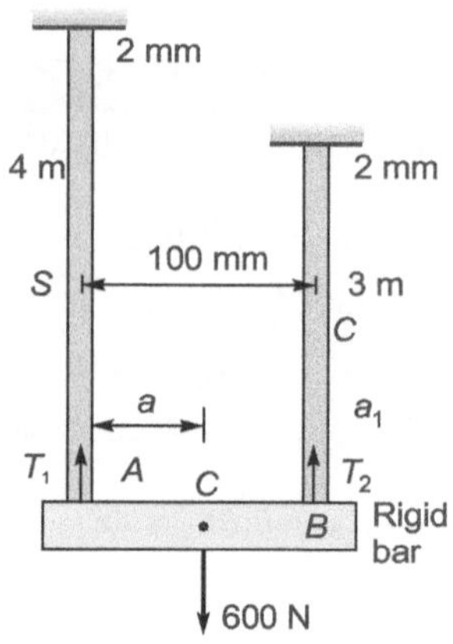

Fig. 1.29

Solution: Say tension in steel wire $= T_1$

Tension in copper wire $= T_2$

(under the action of the load on rigid bar both steel and copper wires come under tension)

$$T_1 + T_2 = 600 \text{ N}$$

Say δL = change in length in steel wire

δL = change in length in copper wire

$$A_s = \frac{\pi}{4} \times 2^2 = \pi \text{ mm}^2$$

$$A_{cu} = \frac{\pi}{4} \times 2^2 = \pi \text{ mm}^2$$

$$\text{Stress in steel wire} = \frac{T_1}{\pi}$$

$$\text{Steel in copper wire} = \frac{T_2}{\pi}$$

Using Hooke's law,
$$\delta L = \frac{T_1}{\pi} \times \frac{L_s}{E_s}$$

$$\delta L = \frac{T_2}{\pi} \times \frac{L_s}{E_{cu}}$$

or
$$\frac{T_1}{\pi} \times \frac{4000}{E_s} = \frac{T_2}{\pi} \times \frac{3000}{E_{cu}}$$

$$\text{or} \quad \frac{T_1}{T_2} = \frac{3}{4}\times\frac{E_S}{E_{cu}} = \frac{3}{4}\times 2 = 1.5$$

$$T_1 = 1.5\,T_2$$

$$\text{or} \quad 1.5T_2 + T_2 = 600 \text{ N}$$

$$T_2 = 240 \text{ N}$$

$$T_1 = 600 - 240 \text{ N} = 360 \text{ N}$$

Taking moment about A

$$600 \times a = T_2 \times 100 = 240 \times 100$$

$$a = 400 \text{ mm}$$

Change in length,

$$\delta L = \frac{T_1}{\pi}\times\frac{400}{E_S} = \frac{360\times 4000}{\pi\times 210000} = 2.18 \text{ mm}$$

= Distance by which rigid bar moves downwards.

Practice Q.1.26 A steel wire of 2 mm diameter, *ABCDE*, is carried over a pulley of radius 0.1 m as shown in figure 1.30. A load of 1.2 kN acts on pulley. After the application of the load, by how much distance centre of the pulley moves down. E = 200 GPa.

Hints: Tension in wire in $T = \frac{W}{2}$, calculate total length of wire

Ans. [2.3 mm]

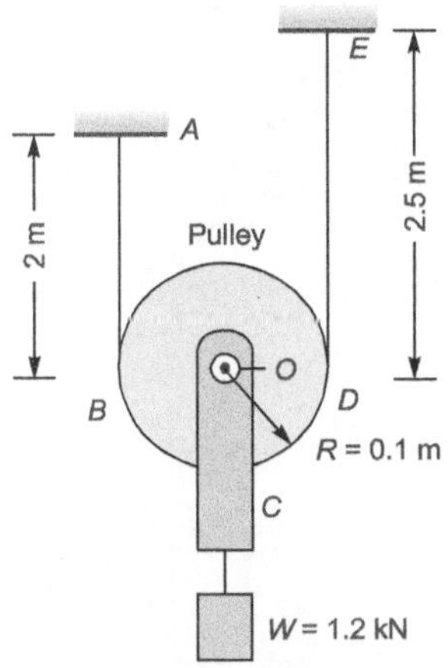

Fig. 1.30

Question 1.27 A rigid plate *ABCD*, 1 m² is supported over four legs of equal lengths of same cross-section and same material. A vertical load 1200 N acts at point *E* such that

$AM = LE = 0.5$ m

$AL = ME = 0.4$ m

Determine reactions R_A, R_B, R_C and R_D.

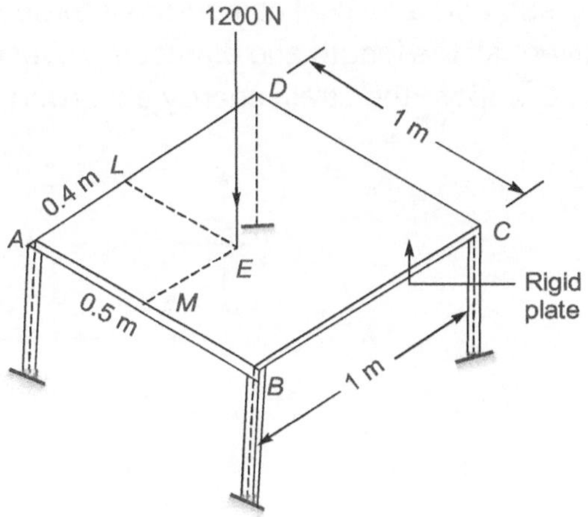

Fig. 1.31

Solution: For equilibrium

$R_A + R_B + R_C + R_D = 1200$ N ...(i)

ABCD is a rigid plate, under load, its plane is changed, it is not deformed. Say under load legs deform by δ_A, δ_B, δ_C, δ_D respectively.

Diagonal *AC* and diagonal *BD* remain straight.

Centre of plate lies on both diagonals *AC* and *BD*.

δ_0 = deflection at centre of plates, O is common on as diagonals AC and BD

$$= \frac{\delta_A + \delta_C}{2} = \frac{\delta_B + \delta_D}{2} \quad \text{...(ii)}$$

Moreover within the elastic limit, deflection is proportional to the load (or reaction) so

$$R_A + R_C = R_B + R_D$$

$$= \frac{1200}{2} = 600 \text{ N} \quad \text{...(iii)}$$

Now taking moments of the forces about the edge AB

$$1200 \times ME = I \times A_C + R_D \times 1$$

$$1200 \times 0.4 = R_C + R_D$$

$$R_C + R_D = 480$$

But $\quad R_B + R_D = 600$

or $\quad R_B - R_C = 120 \quad$ ...(iv)

Taking moments about edge AD

$$1200 \times 0.5 = R_B \times 1 + R_C \times 1$$

$$R_B + R_C = 600$$

$$R_B - R_C = 120$$

$$R_B = 360 \text{ N}$$

$$R_C = 240 \text{ N}$$

$$R_A = 600 - R_C$$

$$R_A = 600 - 240 = 360 \text{ N}$$

$$R_B + R_D = 600$$

$$R_D = 600 - R_B$$

$$= 600 - 360 = 240 \text{ N}$$

Finally the reactions are $\quad R_A = 360 \text{ N}, R_B = 360 \text{ N}, R_C = 240 \text{ N}, R_D = 240 \text{ N}$

Question 1.28 A tapered circular bar, tapers from diameter $2d$ to d over an axial length of L. It is subjected to axial compressive force P. Another bar of same material, but stepped bar of diameters $2d$ over half the length and diameter d over remaining half of the length is also subjected to same axial force P. Compare the strain energy absorbed by the two bars i.e., U_A/U_B. (fig. 1.32)

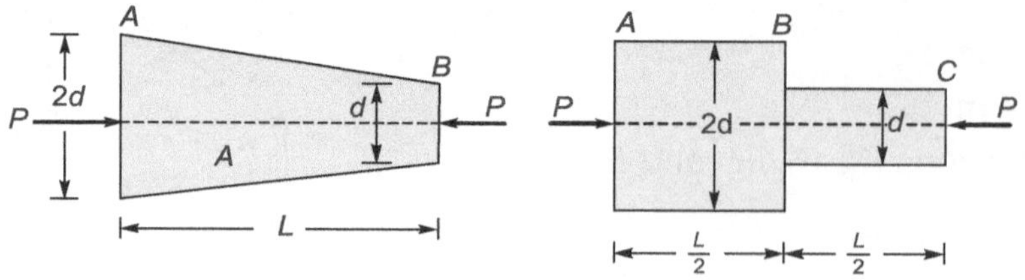

Fig.1.32

Solution:

Say $\quad E$ = Young's modulus

Bar A $\quad \delta L = \dfrac{4PL}{\pi E 2d \times d} = \dfrac{2PL}{\pi E d^2}$

Srain energy, $U_A = \frac{1}{2}P\delta L = \frac{P^2L}{\pi E d^2}$

Bar B $\sigma_{AB} = \frac{4P}{\pi \times (2d)^2} = \frac{P}{\pi d^2}$

$\sigma_{BC} = \frac{4P}{\pi d^2}$

Strain energy, $U_B = \left(\frac{P}{\pi d^2}\right)^2 \times \frac{1}{2E} \times \frac{\pi}{4}(2d)^2 \frac{L}{2} + \left(\frac{4P}{\pi d^2}\right)^2 \times \frac{1}{2E} \times \frac{\pi}{4} \times d^2 \times \frac{L}{2}$

$= \frac{P^2}{2E}\left[\frac{1}{\pi^2 d^4} \times \frac{\pi}{1} d^2 \times \frac{L}{2} + \frac{16}{\pi^2 d^4} \times \frac{\pi}{4} d^2 \frac{L}{2}\right]$

$= \frac{P^2 L}{4E}\left[\frac{1}{\pi d^2} + \frac{4}{\pi d^2}\right] = \frac{P^2 L}{4E} \times \frac{5}{\pi d^2} = \frac{1.25 P^2 L}{\pi E d^2}$

$\frac{U_A}{U_B} = \frac{1}{1.25} = 0.8$

Question 1.29 The cross-section of a bar is given by (169 + 0.01 x^2) mm^2, where x is distance from one end in mm. If the length of the bar is 260 mm and E = 100 GPa, what is change in the length of the bar under an axial load of 4 kN.

Solution: Area of cross-section, A_x = (169 + 0.01 x^2) mm^2

Load P = 4000 N

Stress, $\sigma_x = \frac{P}{A_x} = \frac{4000}{(169 + 0.01x^2)}$

Strain, $\varepsilon_x = \frac{\sigma_x}{E} = \frac{4000}{E(169 + 0.01x^2)}$

Change in length dx of small element considered at x, from one end

$$\delta d_x = \varepsilon_x dx = \frac{4000dx}{E(169 + 0.01x^2)} = \frac{4000dx}{0.01(16900 + x^2)E}$$

$$= \frac{4000dx}{100000 \times 0.01(16900 + x^2)} = \frac{4dx}{(16900 + x^2)}$$

Total change in length, $\delta L = \int_0^{260} \frac{4dx}{(16900 + x^2)} = 4\left|\frac{1}{130}\tan^{-1}\frac{x}{130}\right|_0^{260}$

$$= \frac{4}{130}\left[\tan^{-1}2\right] = \frac{4}{130} \times 1.107 = 0.034 \text{ mm}$$

Question 1.30 Two portions I and II of a rectangular bar 60 × 50 mm (cross-section) are glued along plane abcd making an angle α with cross-section of bar as shown in figure 1.33. An axial load P acts on the bar. The ultimate stresses of glued joint are σ_u = 18 MPa and τ_u = 12 MPa. Determine the range of angle α for which the FOS of the glued is at least 4. P = 45 kN.

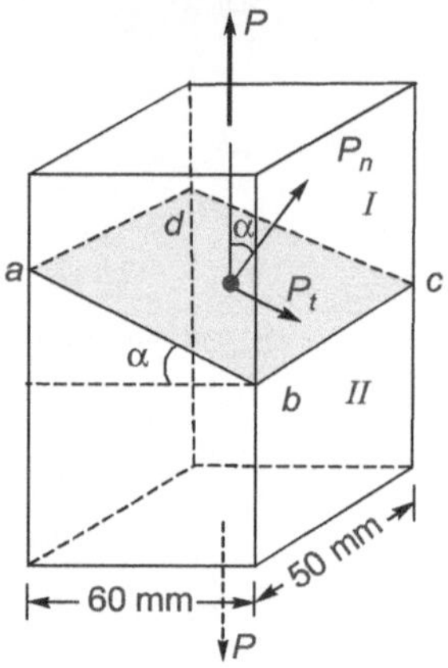

Fig. 1.33

Solution:

$$\text{FOS} = 4$$

Allowable

$$\sigma_a = \frac{18}{4} = 4.5 \text{ N/mm}^2$$

$$\tau_a = \frac{12}{4} = 3.0 \text{ N/mm}^2$$

Cross-sectional area of glued joint,

$$A = \frac{60}{\cos\alpha} \times 50 = \frac{3000}{\cos\alpha}$$

Normal force on plane, $P_n = P\cos\alpha$

Shear force on plane, $P_t = P\sin\alpha$

So

$$\frac{P\cos\alpha}{A} \leq 4.5$$

$$\frac{P\sin\alpha}{A} \leq 3$$

$$P = 45000 \text{ N}$$

or

$$\frac{P\cos\alpha}{3000} \times \cos\alpha = 4.5$$

$$\cos^2\alpha = \frac{4.5 \times 3000}{45000}, \text{ putting the value of } P$$

$$= 0.3$$

$$\cos\alpha = \pm 0.5477$$

$$\alpha = 56.8°, -56.8° \quad \text{...(i)}$$

$$\frac{P\sin\alpha}{A} = 3$$

$$\frac{P\sin\alpha}{3000} \times \cos\alpha = 3$$

$$\sin\alpha\cos\alpha = \frac{9000}{P} = \frac{9000}{45000} = 0.2$$

$$\frac{\sin 2\alpha}{2} = 0.2$$

$$\sin 2\alpha = 0.4$$

$$\alpha = \frac{1}{2}\sin^{-1} 0.4 = 11.8°$$

α varies from 11.8° to 56.8°.

Question 1.31 A stepped steel bar ABCD of length L, has different areas of cross-section. For $\frac{L}{3} = AB = CD$, area A_1. For $BC = \frac{L}{3}$, area is A_2. Area $A_2 = 1.5\,A_1$. Two equal and opposite forces P each are applied at sections B and C as shown in figure 1.34

If $P = 16$ kN, $A_1 = 100$ mm²

What are the stresses developed in sections *AB*, *BC* and *CD* of the bar.

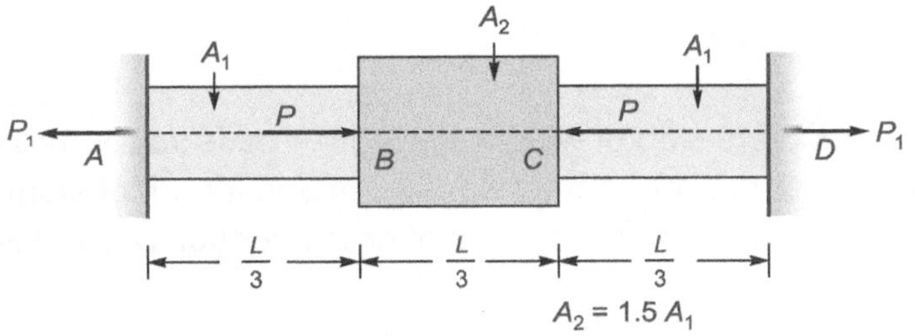

Fig. 1.34

Solution: Taking advantage of symmetry, let us take reactions at *A* and *D* as P_1 and P_1 as shown. Then forces in *AB* and *CD*, P_1 tensile force in *BC* is $(P - P_1)$ compressive.

A and *D* are rigid ends so, net change in length of total length of the bar is zero, so

$$\frac{P_1}{A_1E}\times\frac{L}{3}+\frac{P_1L}{3A_1E}-\frac{(P-P_1)}{A_2E}\frac{L}{3} = 0$$

or
$$\frac{2P_1}{A_1} = \frac{P-P_1}{A_2}$$

or
$$\frac{2P_1}{A_1} = \frac{(P-P_1)}{1.5A_1}$$

or
$$3P_1 = P - P_1$$
$$4P_1 = P$$
$$P_1 = 0.25\,P$$
$$P - P_1 = 0.75\,P$$
$$P = 16\text{ kN}$$
$$P_1 = 4\text{ kN}$$
$$(P - P_1) = 12\text{ kN}$$

Stresses,
$$P_1 = 4\text{ kN}$$

$$\sigma_{AB} = \sigma_{CD} = \frac{P_1}{A_1} = +\frac{4000}{100} = +40\text{ N/mm}^2$$

$$\sigma_{BC} = -\left(\frac{P-P_1}{A_2}\right) = -\frac{12000}{150} = -80\text{ N/mm}^2$$

Practice Q.1.31 A bar *ABCD*, length *L* is made of three portions $AB = BC = CD = \frac{L}{3}$. Portions AB and *CD* are of steel, while portion *BC* is of copper as shown in figure 1.34. Area of cross section throughout is 50 mm². What is force *P* applied at sections *B* and *C*, so that stress in *BC* is equal to 60 MPa. $E_s = 2E_{cu}$.

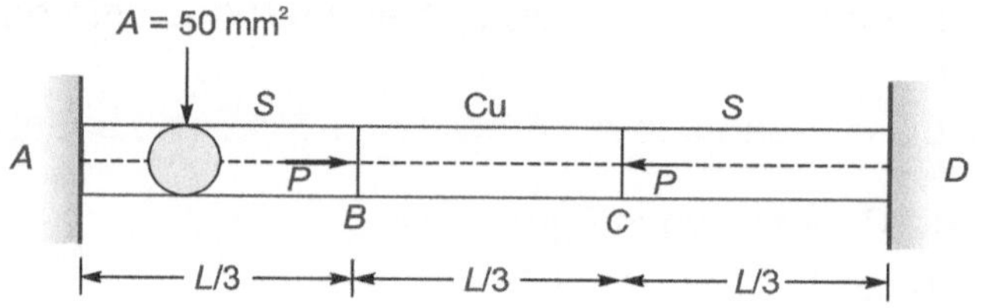

Fig. 1.34

Ans. [P = 4.5 kN]

Question 1.32 A concrete slab of weight W, width 4 m, is supported by wires of aluminium and steel as shown in the figure 1.35. Length of steel wire is 3 m and length of aluminium wire is 2 m as shown in figure 1.35. $E_s = 3E_a$. Determine the ratio of areas of cross-section of steel and aluminium if the slab is to remain horizontal.

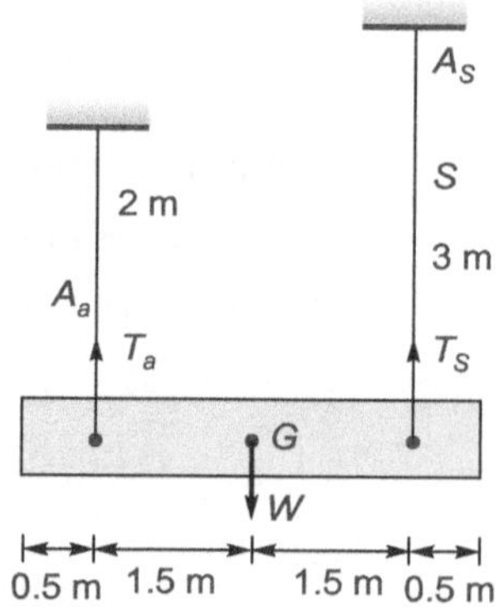

Fig. 1.35

Solution: Say tension in steel wire is T_s and tension in aluminum wire is T_a.
Taking moments about G of slab.

$$1.5T_a = 1.5\,T_s, \quad T_s = T_a$$

Moreover,

$$\delta l_a = \delta l_s, \quad \text{Slab remains horizontal}$$

or

$$\frac{T_a}{E_a} \times \frac{L_a}{A_a} = \frac{T_s L_s}{E_s A_s}$$

or

$$\frac{2}{E_a \times A_a} = \frac{3}{E_s \times A_s} \quad \text{or} \quad 2E_s A_s = 3E_a A_a$$

or

$$\frac{A_a}{A_s} = \frac{E_s}{E_a} \times \frac{2}{3} = 3 \times \frac{2}{3} = 2$$

or

$$E_s = 3E_a$$

$$\frac{A_a}{A_s} = 2 = \frac{\text{Area of cross-section of aluminium wire}}{\text{Area of cross-section of steel wire}}$$

Practice Q.1.32 A rigid beam 2 m long, hinged at end A is supported by two wire hangers at B and D as shown in figure 1.36. A load of 1 kN acts at C. Both the wires are of steel and diameter of wires is 4 mm. If E = 200 GPa for steel, determine stresses in wire at B and D.

Ans. [23.87 MPa, 47.74 MPa]

Hints: [Beam $ABCD$ is rigid, using the principle of similar triangles, $\delta l_D = 2\delta l_B$]

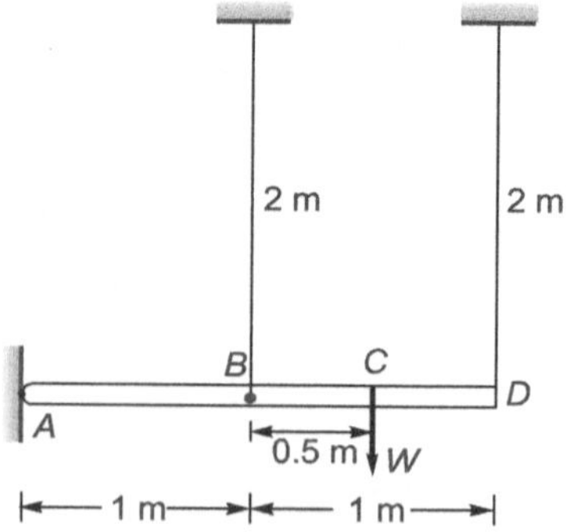

Fig. 1.36

Question 1.33 Two 4 m long copper bars AB and BC are connected at B as shown in figure 1.37 at angle of 90°, diameter of each bar is 4 mm. Load W = 2 kN. E = 105 kN/mm² what is vertical displacement of point B.

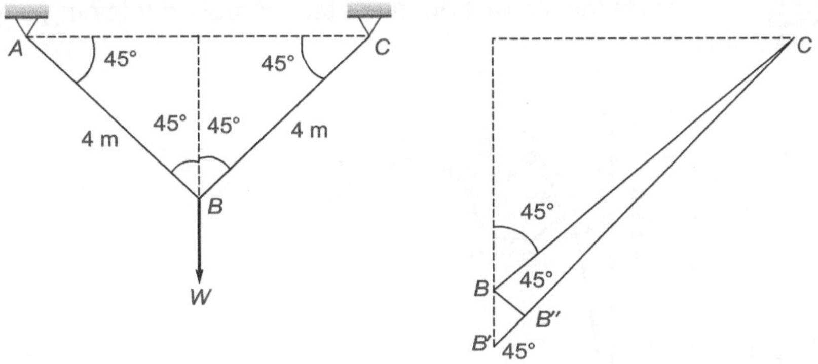

Fig. 1.37

Solution: Vertical load at B = 2000 N

Forces in AB and BC

$$F_{AB} = F_{BC} = \frac{W}{\sin 90^\circ} \times \sin 135^\circ$$

$$= 0.707\text{ W}$$

$$= 0.707 \times 2000 = 1414\text{ N}$$

Area of cross-section of each bar,

$$A = \frac{\pi}{4} \times 4^2 = 4\pi\ \text{mm}^2$$

$$\sigma,\ \text{stress in each bar} = \frac{1414}{4\pi} = 112.54\ \text{N/mm}^2$$

$$\varepsilon,\ \text{strain} = \frac{112.54}{105000} = 1.0718 \times 10^{-3}$$

Change in length, $\delta L = 1.0718 \times 4000 \times 10^{-3} = 4.287$ mm

Refer to enlarged view, displacement

$$B'B'' = 4.287\text{ mm}$$

Vertical displacement, $BB' = \dfrac{B'B''}{\cos 45^\circ} = \dfrac{4.287}{0.707} = 6.0636$ mm

Practice Q.1.33 Two steel bars $AB = BC$ = 3 m connected at B as shown in fig. 1.38 is subjected to load W. The magnitude of load W is such that stress developed in each bar is 100 MPa. If E = 200 GPa, what is vertical displacement of point B. Diameter of each bar is 3 mm.

Ans. [W = 706.8 N, 3 mm]

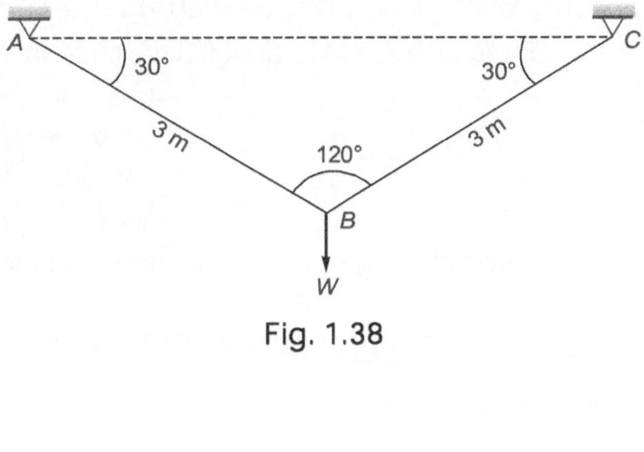

Fig. 1.38

Question 1.34 Describe tensile test on mild steel sample explaining salient points on stress-strain (load extension) curve.

Solution:

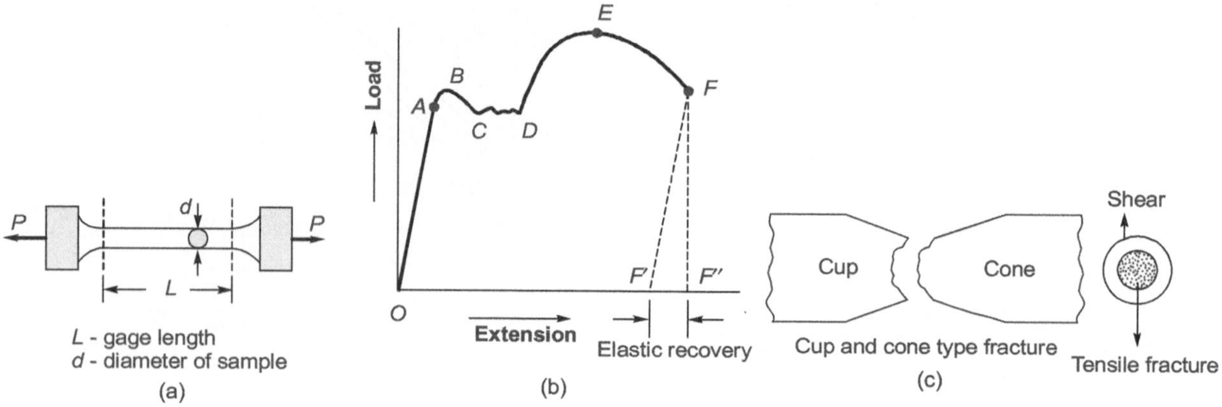

Fig. 1.39

Mild steel is the most commonly used engineering material. In most of engineering colleges, tensile test is performed on mild steel sample using universal testing machine.

A sample of mild steel of standard dimensions is machined on a lathe. Figure 1.39 (a) shows or coller type test sample with gauge length *L* (in which, section remains constant) and diameter *d*. The sample is gripped in the grips of a universal testing machine and machine is turned on keeping the rate of extension very small so that load on sample **is gradually and slowly increased**. The student should continuously record the values of load and extension at regular intervals of change of length.

The student should be careful enough to record the continuous increase in load. At the stage of upper yield point, there is slight decrease in load. Then after some time, load again increases and maximum load is obtained. After reaching the maximum load point; load gradually decreases and the sample fails in two pieces at the breaking load.

At the upper yield point (point *B* in load extension curve) slip in the internal structure has started i.e., the atoms have started slipping on slipping planes. In mild steel, there are carbon atoms (smaller than iron atoms) in addition to iron atoms. Iron atoms slip easily over smaller carbon atoms and internal resistance of the material decreases. During slipping of atoms, dislocations continuously move but get stuck up at harder grain boundaries. The resistance of the material again increases from the point of strain hardening (point *D* in load extension) upto the maximum load point. Change in length is proportional to load. At the point of maximum load, necking takes place in the sample and further extension takes place in the vicinity of the neck. Extension from maximum load point to breaking load is proportional to the lateral dimension of sample i.e., $\sqrt{A}$, where A is area of cross-section.

From *O* to *A*, *OA* is straight line, material obeys Hooke's law.

$$\text{stress} \propto \text{strain}$$

$$\sigma \propto \in$$

$$\frac{P}{\text{area}} \propto \frac{\delta L}{L}$$

Slope of *OA* gives Young's modulus of the material.

$$\text{Upper yield point stress} = \frac{\text{Load at upper yield point B}}{\text{Original area of cross-section}}$$

$$\text{Lower yield point stress} = \frac{\text{Load at lower yield point C}}{\text{Original area of cross-section}}$$

$$\text{Ultimate strength,} \quad \sigma_{ut} = \frac{\text{Maximum load at point E}}{\text{Original area of cross-section}}$$

After the sample breaks into two pieces, the two pieces are joined to measure

(1) diameter at the neck, say d'

(2) final length L' of gauge length

$$\text{original area of cross-section,} \quad A = \frac{\pi}{4}d^2$$

$$\text{area of cross-section at neck,} \quad A' = \frac{\pi}{4}d'^2$$

$$\%\text{ increase in length} = \frac{L'-L}{L}\times 100 = \frac{\delta L}{L}\times 100$$

$$\%\text{ reduction in area} = \frac{A-A'}{A}\times 100$$

Percentage increase in length defines the ductility of the material. Percentage reduction in area defines the degree of cold working of the material.

Note that final extension in sample is OF'.

Extension $F'F''$ is elastic recovery of deformation.

Portion CD of the graph gives yield plateau of the mild steel.

$$\text{Total change in length,} \quad \delta L = bL + c\sqrt{A}$$

Where b and c are known as Barba's constants.

Fractured sample fig. 1.43 (c) shows the cup and cone type fracture in mild steel (a ductile material) outer portion shows shear failure (due to shear stress at ± 45° to the axis, shear stress is maximum) inner portion shows straight fracture due to tensile stress which becomes large in the inner portion.

Question 1.35 **A rectangular bar of 30 mm × 60 mm cross-section and 200 mm length is restrained from expansion along its 30 mm × 200 mm sides by surrounding material. Find the change in dimensions and volume when a compressive force of 180 kN acts in an axial direction. $E = 2 \times 10^5$ MPa; $\nu = 0.3$**

[CE : ESE 2017]

Solution:

Rectangular bar, length = 200 mm

Section 30 × 60 mm

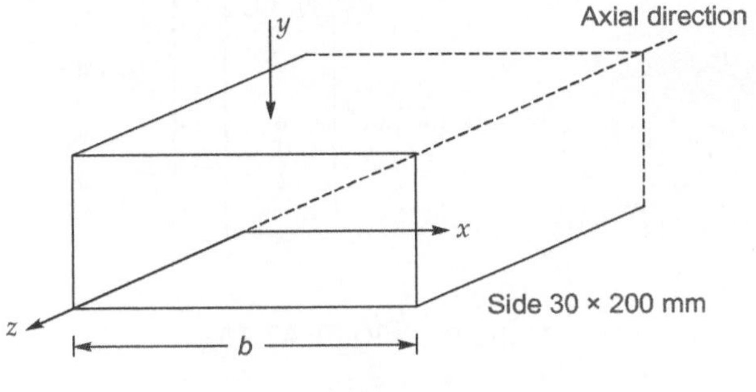

No change in dimension 30 mm

Fig. 1.40

Axial force, $P = 180$ kN (Comp.)

Axial stress, $\sigma_z = -\dfrac{180000}{30 \times 60} = -100$ N/mm^2

Strain in x direction $= 0$

(Stress applied by surrounding material)

say $= \sigma_x$ (Comp.)

$$0 = \frac{\sigma_x}{E} + \frac{\nu\sigma_z}{E} \text{ (lateral strain due to stress, } \sigma_z\text{)}$$

$$0 = -\frac{\sigma_x}{E} + \frac{0.3 \times 100}{E}$$

$$\sigma_x = -30 \text{ N/mm}^2 \text{ (Comp.)}$$

Net strains, $\varepsilon_z = -\dfrac{100}{E} + \dfrac{0.3 \times 30}{E} = -\dfrac{91}{E}$

Strain, $\varepsilon_x = +\dfrac{\nu \times 100}{E} + \dfrac{\nu \times 30}{E} = \dfrac{39}{E}$

$$\varepsilon_x = 0$$

Change in length, $\delta L = -\dfrac{91}{2 \times 10^5} \times 200 = -0.091$ mm

Change in side of 60 mm, $\delta b = +\dfrac{39}{2 \times 10^5} \times 60$

$= +0.0117$ mm

$= +11.7 \times 10^{-3}$ mm

Question 1.36 A non-prismatic rod is attached between two supports as shown in Fig. 1.41. It is subjected to axial loads of 80 kN and 100 kN. Determine the support reactions and elongation. Take $E = 200$ GPa.

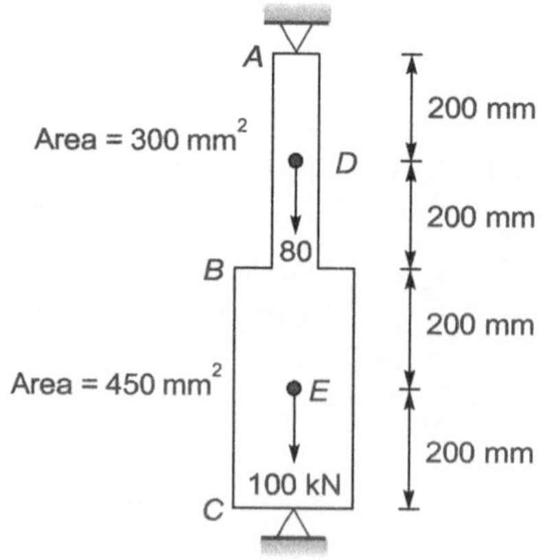

Fig. 1.41 (a)

[CE : ESE 2016]

Solution:

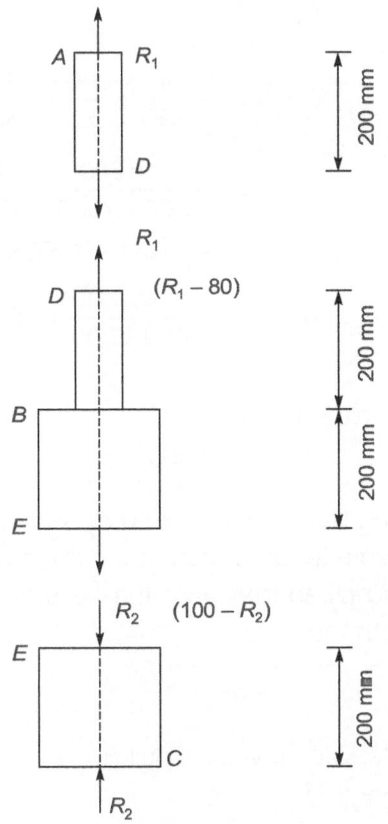

Fig. 1.41 (b)

Say reactions at *A* and *C* are R_1 and R_2 respectively as shown in Fig. 1.41.

Extension in *AD* + Extension in *DBE* + Contraction in *EC* = 0

Moreover,

$$R_1 + R_2 = 80 + 100 = 180 \text{ kN}$$

$$R_1 = 180 - R_2$$

$$R_1 - 80 = 180 - R_2 - 80$$

$$= 100 - R_2$$

$$\frac{R_1}{300} \times \frac{100}{E} + \frac{(100 - R_2)}{300} \times \frac{200}{E} + \frac{(100 - R_2)}{450} \times \frac{200}{E} - \frac{R_2 \times 200}{450E} = 0$$

or $1.5\,R_1 + 1.5\,(100 - R_2) + (100 - R_2) - R_2 = 0$

$$1.5\,R_1 + 150 - 1.5\,R_2 + 100 - R_2 - R_2 = 0$$

or

$$1.5\,R_1 - 3.5\,R_2 = -250$$

and

$$1.5\,R_1 + 1.5\,R_2 = 270$$

$$-5R_2 = -520$$

Reactions

$$R_2 = 104 \text{ kN}$$

$$R_1 = 76 \text{ kN}$$

Changes in length

$$\delta l_{AB} = \frac{76 \times 200}{300 \times 200} = +0.2533 \text{ mm}$$

$$\delta l_{DBE} = -\frac{4}{300} \times \frac{200}{E} - \frac{4}{450} \times \frac{200}{E}$$

$$= -\frac{4 \times 200}{300 \times 2000} - \frac{4}{450} \times \frac{200}{200}$$

$$= -0.0133 - 0.0088 = -0.02213 \text{ mm}$$

$$\delta l_{EC} = -\frac{104 \times 200}{450 \times 200} = -0.2311 \text{ mm}$$

Checking

$$+0.2533 - 0.2213 - 0.2311 = 0$$

$$= 0.00017 \simeq 0$$

Question 1.37 **A tapering bar having d_1 and d_2 as diameters of end sections and another bar of uniform cross-section d are of same length l. Both are subjected to the same axial pull P. What should be the value of d in terms of d_1 and d_2 so that both the bars of same material will have same extension? Valuate maximum stresses in both the bars.**

[ME : ESE 2016]

Solution:

Tapered bar d_1, d_2 diameters, length l axial pull P,

Uniform bar d diameter, length, l

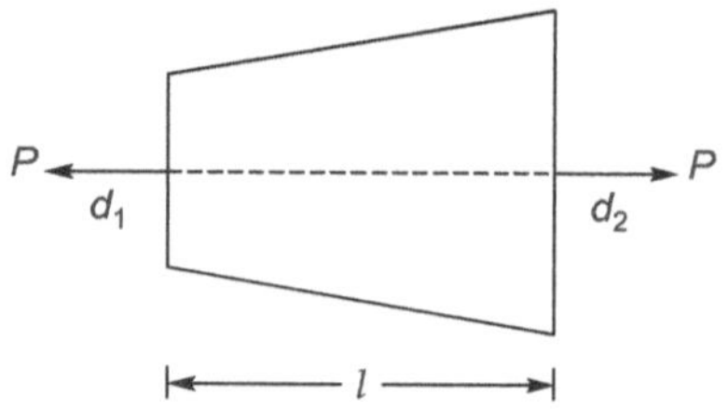

Fig. 1.42

Same extension

$$\frac{4Dl}{\pi E d_1 d_2} = \frac{4Dl}{\pi E d^2}$$

$$d = \sqrt{d_1 d_2} \qquad \ldots\text{(i)}$$

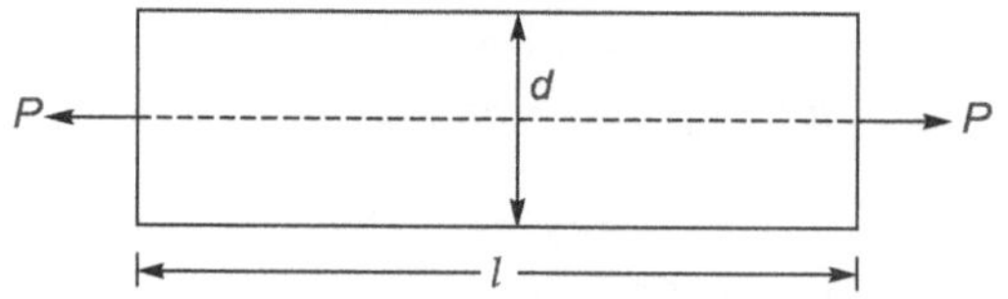

Fig. 1.43

Tapered bar

$$\sigma_{max} = \frac{4P}{\pi d_1^2} \qquad \ldots\text{(ii)}$$

Uniform bar

$$\sigma_{max}' = \frac{4P}{\pi d^2} = \frac{4P}{\pi d_1 d_2} \qquad \text{(in uniform bar)}$$

$$\sigma_{max} > \sigma_{max}' \qquad \text{as } d_2 > d_1$$

Question 1.38 In the figure, the load $P = 1$ N, length $L = 1$ m. Young's modulus $E = 70$ GPa, and the cross-section of the links is a square with dimension 10 mm × 10 mm. All joints are pin joints.

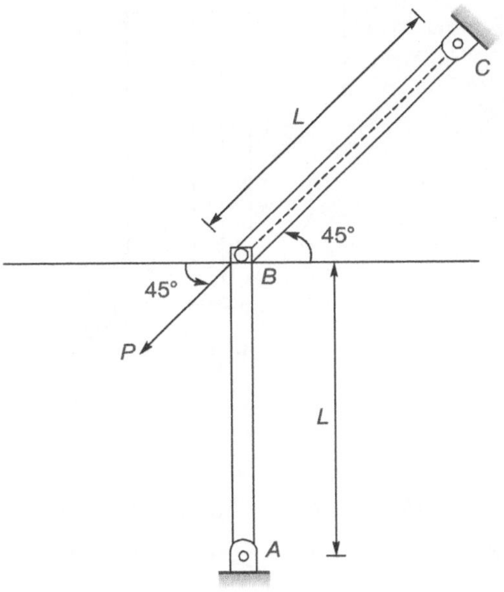

Fig. 1.44

The stress (in Pa) in the link *AB* is ________

(Indicate compressive stress by a negative sign and tensile stress by a positive sign.)

[ME, GATE : 2015 (set-2)]

Solution:

$$\text{Vertical component of } P = \frac{P}{\sqrt{2}} = \frac{1}{\sqrt{2}} \text{ N}$$

$$\text{Area of cross-section} = 100 \text{ mm}^2$$

$$\text{Stress in link } BA = \frac{1}{\sqrt{2}} \times \frac{1}{100} \text{ MPa}$$

$$= \frac{10^6}{\sqrt{2} \times 100} \text{ Pa}$$

$$= 7.07 \times 10^3 \text{ Pa}$$

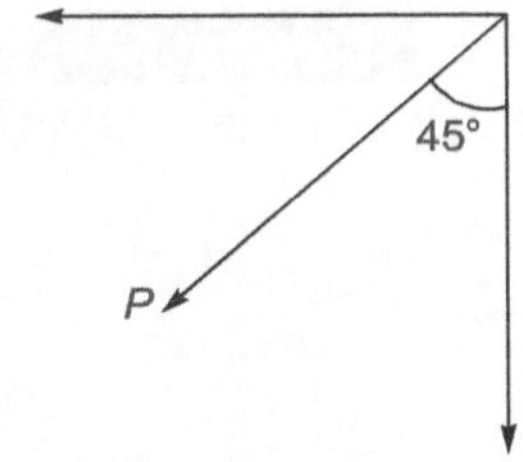

Fig. 1.45

[it is -7.07×10^3 Pa (Compressive)]

Question 1.39 A point mass of 100 kg is dropped onto a massless elastic bar (cross-sectional area = 100 mm^2. Length = 1 m. Young's modulus = 100 GPa) from a height *H* of 10 mm as shown (Figure is not to scale). If $g = 10$ m/s^2, the maximum compression of the elastic bar is ________ mm.

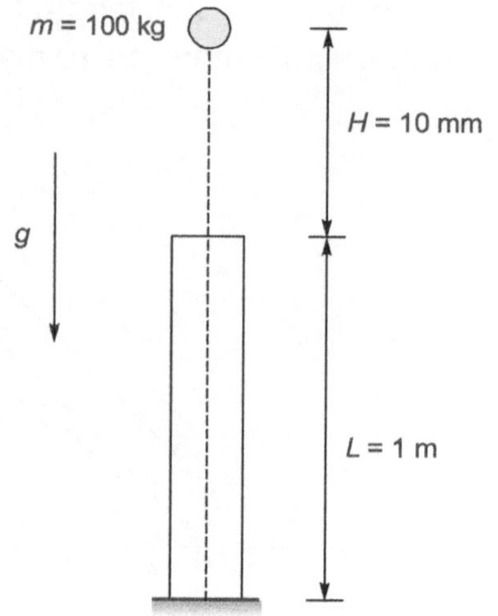

Fig. 1.46

[ME, GATE : 2017 (set-1)]

Solution:

Potential energy, $PE = mgH = 100 \times 10 \times 0.01 = 10 \text{ Nm} = 10^4 \text{ Nmm}$

Volume of the bar $= 100 \times 1000 = 10^5 \text{ mm}^3$

$$\frac{\sigma^2}{2E} \times V = 10^4$$

$$\frac{\sigma^2 \times 10^5}{2 \times 100 \times 10^3} = 10^4$$

$$\sigma = \sqrt{2} \times 100 \text{ MPa}$$

$$\text{Maximum compression} = \frac{\sigma}{E} \times 1000 = \frac{\sqrt{2} \times 100 \times 1000}{100 \times 1000} = \sqrt{2} \text{ mm}$$

Question 1.40 A horizontal bar with a constant cross-section is subjected to loading as shown in the figure 1.47. The Young's modulus for the section *AB* and *BC* are 3*E* and *E*, respectively.

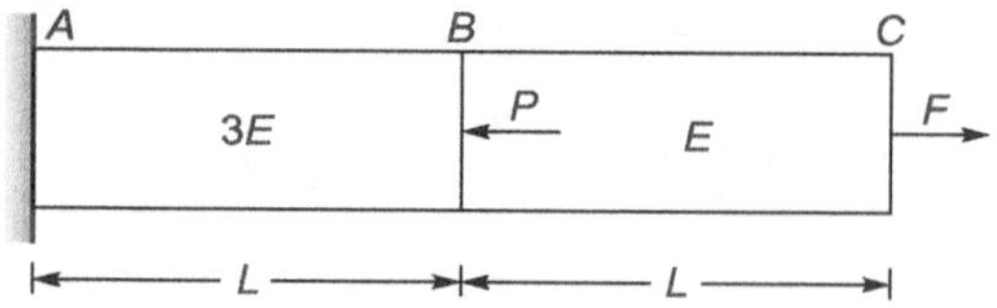

Fig. 1.47

[ME, GATE : 2016 (set-1)]

Solution:

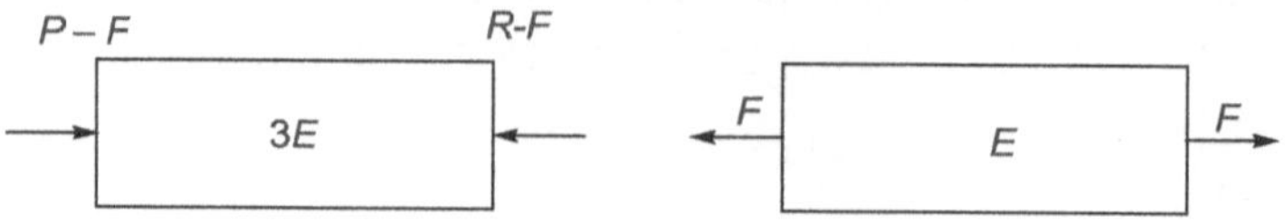

$$\left(\frac{RF}{3E}\right)\times\frac{L}{A} = \frac{F}{E}\times\frac{L}{A}$$

$$P - F = 3F$$

$$P = 4F$$

$$\frac{P}{F} = 4$$

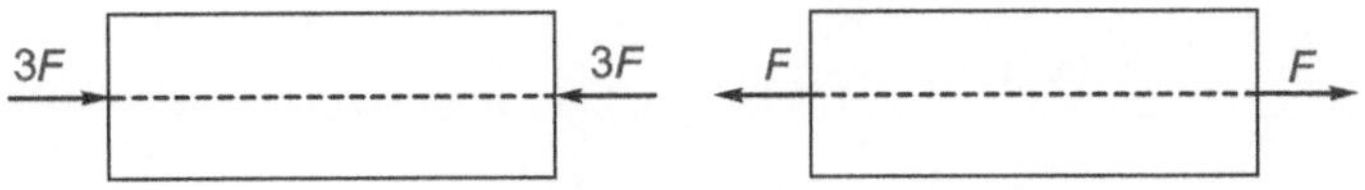

Question 1.41 A tapered circular rod of diameter varying from 20 mm to 10 mm is connected to another uniform circular rod of diameter 10 mm as shown in the following figure. Both bars are made of same material with the modulus of elasticity, $E = 2 \times 10^5$ MPa. When subjected to a load $P = 30$ kN, the deflection at point A is ________ mm.

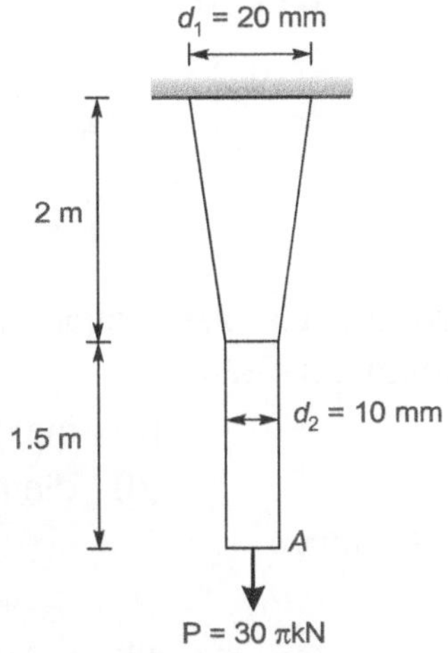

Fig. 1.48

[CE, GATE : 2015 (set-1)]

Solution:

Deflection at A,

$$\delta_A = \frac{30\pi \times 1000 \times 2000}{\pi \times 20 \times 10 \times 2 \times 10^5} + \frac{30\pi \times 1000 \times 1500}{\pi \times 10 \times 10 \times 2 \times 10^5}$$

$$= 1.5 + 2.25 = 3.75 \text{ mm}$$

Question 1.42 Consider the stepped bar made with a linear elastic material and subjected to an axial load of 1 kN shown in the figure 1.49.

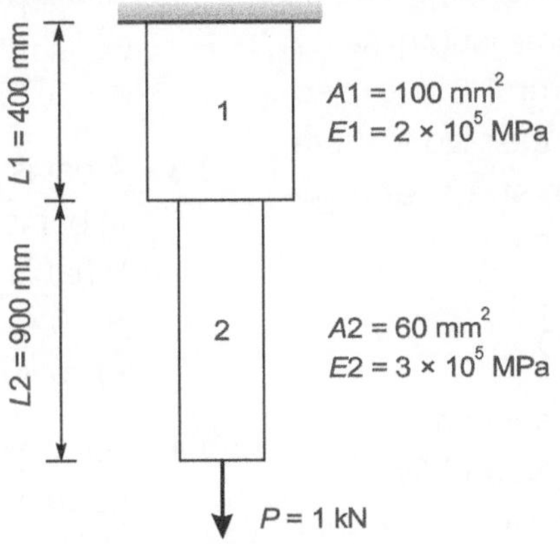

Fig. 1.49

Segments 1 and 2 have cross-sectional area of 100 mm^2 and 60 mm^2. Young's modulus of 2×10^5 MPa and 3×10^5 MPa, and length of 400 mm respectively. The strain energy (in N-mm up to one decimal place) in the bar due to the axial load is ________.

[CE, GATE : 2017 (set-1)]

Solution:

$$\sigma_1 = \frac{1000}{100} = 10 \text{ MPa}$$

$$\sigma_2 = \frac{1000}{60} = 16.667 \text{ MPa}$$

$$V_1 = 100 \times 400 = 4 \times 10^4 \text{ mm}^3$$

$$V_2 = 60 \times 900 = 5.4 \times 10^4 \text{ mm}^3$$

$$\text{Strain energy} = \frac{10^2}{2\times2\times10^5}\times4\times10^4+\frac{(16.667)^2}{2\times3\times10^5}\times5.4\times10^4$$

$$= 10 + 25 = 35 \text{ Nmm}$$

Practice Q.1.42 A wrought iron bar of circular section, diameter 12.5 mm gauge length of 100 mm is tested in tension and following observations are made:

(a) yield load 29.6 kN
(b) maximum load 44.8 kN
(c) load at fracture 37 kN
(d) diameter at neck 9.1 m
(e) total extension in sample 27.6 mm

Determine

(i) yield strength
(ii) ultimate tensile strength
(iii) actual breaking strength
(iv) percentage elongation
(v) percentage reduction in area

Ans. [241.2 N/mm^2, 365.06 N/mm^2, 568.9 N/mm^2, 27.6%, 47%]

Objective Type Questions

Q.1 A circular tapered bar A tapers from diameter $2d$ to d over an axial length L. It is subjected to axial load P. Another bar B of same length, same material but of uniform diameter 1.5 d is also subjected to same load. What is the ratio of strain energy absorbed by A over strain energy absorbed by B, i.e, U_A/U_B

(a) 1.333
(b) 1.125
(c) 0.888
(d) 0.750

Q.2 A steel wire of area of cross-section 3 mm^2, length 3 m, is supported through a pulley. The ends of the wire are fixed in ceiling. A load of 600 N is hung at the centre of the pulley. If E for steel is 200 GPa, by how much distance, centre of the pulley comes down

(a) 3 mm
(b) 1.5 mm
(c) 0.75 mm
(d) 0.3 mm

Q.3 A bar is subjected to uniaxial tension. Axial strain in bar is ε. While volumetric strain in bar is 0.4 ε. What is Poisson's ratio of the material

(a) 0.4
(b) 0.33
(c) 0.30
(d) 0.25

Q.4 Total extension in steel bar shown in figure is equal to, if A = area of cross-section, E = 200 kN/mm^2

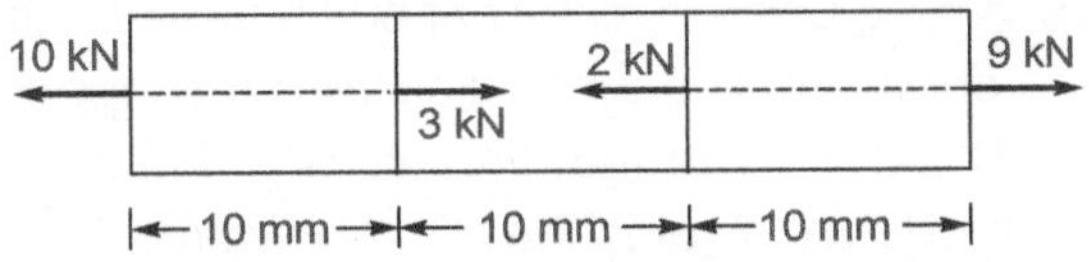

(a) 0.95/A (b) 1.1/A
(c) 1.3/A (d) None of these

Q.5 A bar is hanging vertically with upper end fixed in ceiling. If the deformation in bar due to its own weight is δ, then what is the deformation in the bar if a direct load equal to its own weight is suspended on the bottom of bar, δ′

(a) 2δ (b) δ
(c) δ/2 (d) δ/4

Q.6 In a tensile test, at the yield point

(a) tensile stress increases at a faster rate
(b) tensile stress decreases at a faster rate
(c) tensile stress remains constant for a shorter period
(d) None of these

Q.7 A prismatic bar *ABC*, is rigidly held between two rigid supports at ends *A* and *C*. A force *P* = 9 kN acts at section *B* as shown in figure. What is the reaction at end *C*

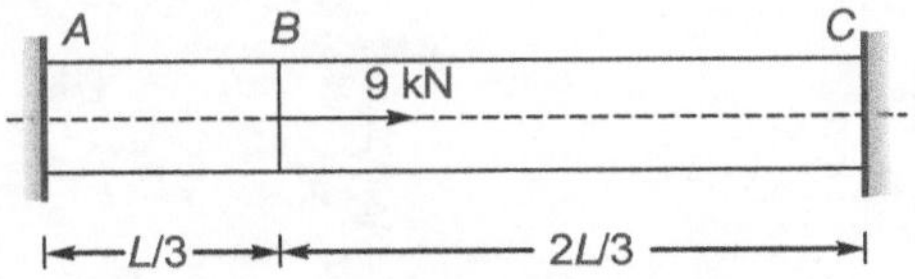

(a) 9 kN (b) 6.0 kN
(c) 4.5 kN (d) 3 kN

Q.8 A prismatic bar *ABC*, 3 m long is shown in figure with a gap of 0.2 mm at end *C* and the wall. Area of cross-section of bar is 1 cm². If E = 200 GPa, what is the value of *P* to fill the gap of 0.2 mm as shown

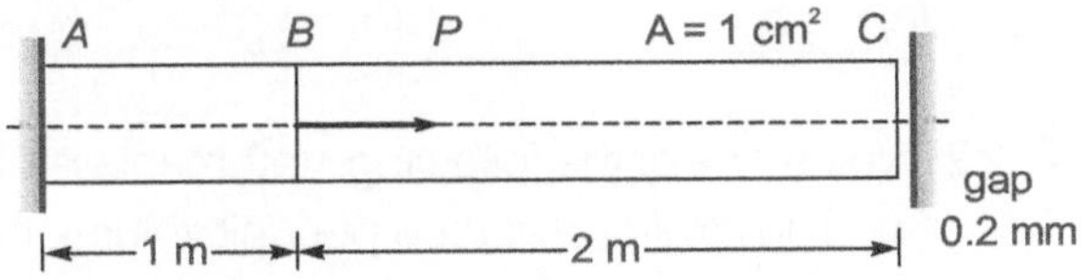

(a) 1.33 kN (b) 2.667 kN
(c) 4.000 kN (d) 6 kN

Q.9 A spherical ball of volume 10000 cc is subjected to a hydrostatic pressure p such that change in volume of the ball is 5 cc and bulk modulus *K* = 160 kN/mm², what is pressure on the ball

(a) 200 N/mm² (b) 160 N/mm²
(c) 80 N/mm² (d) 40 N/mm²

Q.10 In a stress strain curve, area upto elastic limit stress represents which mechanical property of the material

(a) Toughness (b) Resilience
(c) Proof resilience (d) Modulus of resilience

Q.11 What is the stress at the section x-x for the bar ABCD with uniform cross-section 1000 mm²?

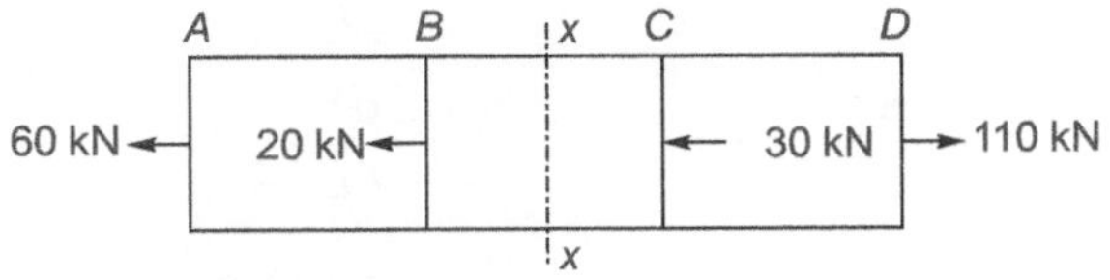

(a) 20 N/mm² (Tensile)
(b) 30 N/mm² (Compressive)
(c) 80 N/mm² (Tensile)
(d) 50 N/mm² (Compressive)

Q.12 The total elongation of the structural element (fixed at one end, free at the other end, and of varying cross-section) as shown in the figure, when subjected to load 2P at the free end is

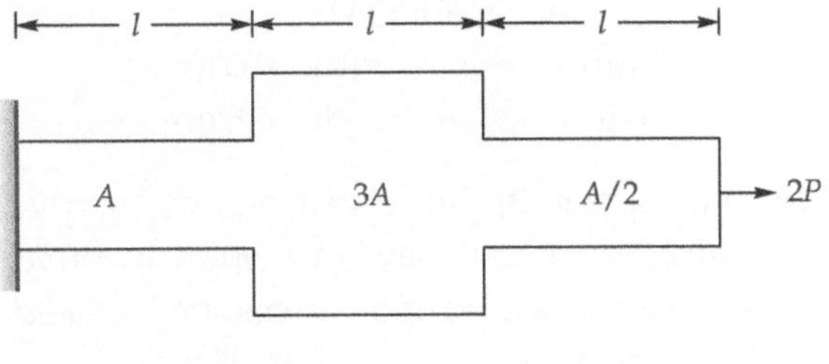

(a) $6.66\frac{Pl}{AE}$ (b) $5.55\frac{Pl}{AE}$

(c) $4.44\frac{Pl}{AE}$ (d) $3.33\frac{Pl}{AE}$

Q.13 A chain, working a crane, has sectional area of 625 mm² and transmits a load of 10 kN. When the load is being lowered at a uniform rate of 40 m/min, the chain gets jammed suddenly at which time the length of the chain unwound is 10 m. Assuming *E* = 200 GPa, the stress induced in the chain due to this sudden jamming is

(a) 100.6 N/mm^2 (b) 120.4 N/mm^2
(c) 140.2 N/mm^2 (d) 160.0 N/mm^2

Q.14 Measured mechanical properties of material are same in a particular direction at each point. This property of the material is known as
(a) isotropy (b) homogeneity
(c) orthotropy (d) anisotropy

Q.15 The strain energy per unit volume of a round bar under uniaxial tension with axial stress σ and modulus of elasticity E is

(a) $\frac{\sigma^2}{E}$ (b) $\frac{\sigma^2}{2E}$

(c) $\frac{\sigma^2}{3E}$ (d) $\frac{\sigma^2}{4E}$

Q.16 A steel rod of cross-sectional area 10 mm^2 is subjected to loads at points *P*, *Q*, *R* and *S* as shown in the figure below:

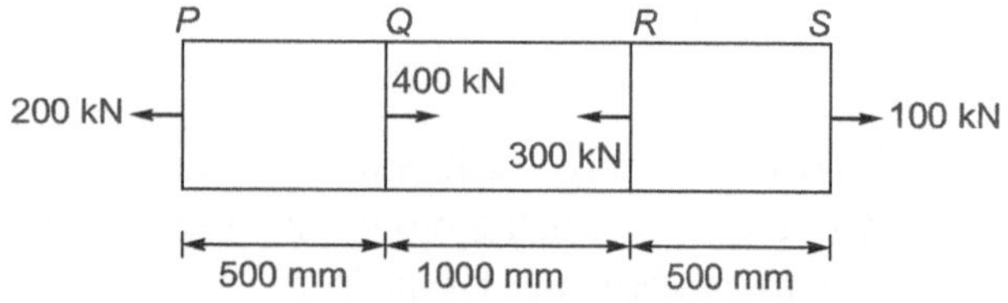

If E_{steel} = 200 GPa, the total change in length of the rod due to loading is
(a) –5 μm (b) –10 μm
(c) –20 μm (d) –25 μm

Q.17 The capacity of a material to absorb energy when deformed elastically and when to have this energy recovered upon unloading is called
(a) endurance (b) resilience
(c) toughness (d) ductility

Q.18 Two steel rods of identical length and material properties are subjected to equal axial loads. The first rod is solid with diameter *d* and the second is a hollow one with external diameter *D* and internal diameter 50% of *D*. If the two rods experience equal extensions, the radio of *d*/*D* is

(a) $\frac{3}{4}$ (b) $\frac{\sqrt{3}}{2}$

(c) $\frac{1}{2}$ (d) $\frac{1}{4}$

Q.19 A solid uniform metal bar of diameter *D* mm and length *l* mm hangs vertically from its upper end. The density of the material is ρ N/mm^3 and its modulus of elasticity is *E* N/mm^2. The total extension of the rod due to its own weight would be

(a) $\frac{\rho l^2}{2E}$ (b) $\frac{\rho l}{2E}$

(c) $\frac{\rho l}{4E}$ (d) $\frac{\rho l^2}{4E}$

Q.20 A mild steel bar, circular in cross-section, tapers from 40 mm diameter to 20 mm diameter over its length of 800 mm. It is subjected to an axial pull of 20 kN. $E = 2 \times 10^5$ N/mm^2. The increase in the length of the rod will be

(a) $\frac{1}{10\pi}$ mm (b) $\frac{2}{5\pi}$ mm

(c) $\frac{4}{5\pi}$ mm (d) $\frac{1}{5\pi}$ mm

Q.21 A uniform T-shaped arm of weight *W*, pinned about a horizontal point *C*, is supported by a vertical spring of stiffness *K*. The extension of the spring is

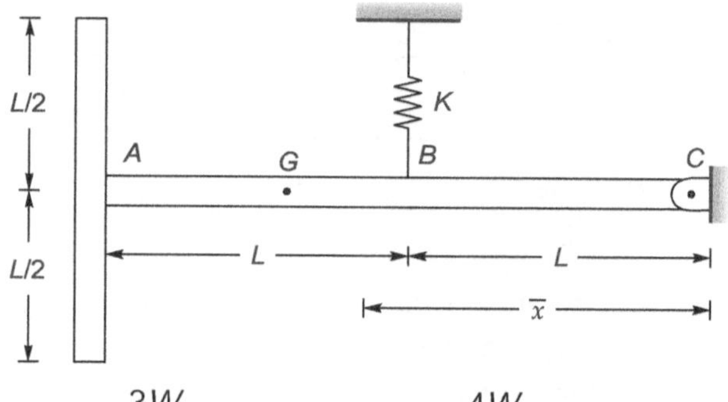

(a) $\frac{3W}{4K}$ (b) $\frac{4W}{3K}$

(c) $\frac{3K}{4W}$ (d) $\frac{4K}{3W}$

Q.22 Which one of the following statements is correct?
(a) The strain produced per unit volume is called resilience.
(b) The maximum strain produced per unit volume is called proof resilience.
(c) The least strain energy stored in a unit volume is called proof resilience.

(d) The greatest strain energy stored in a unit volume of a material without permanent deformation is called proof resilience.

Q.23 Which of the following properties will be the meaningful indicator/indicators of uniform rate of elongation of a test piece of a structural material before necking happens in the test piece?

1. Ductility
2. Toughness
3. Hardness

Select the correct answer using the code given below:

(a) 1 only (b) 2 only
(c) 3 only (d) 1, 2 and 3

Q.24 A 10 mm diameter bar of mild steel of elastic modulus 200×10^9 Pa is subjected to a tensile load of 50000 N, taking it just beyond its yield point. The elastic recovery of strain that would occur upon removal of tensile load will be

(a) 1.38×10^{-3} (b) 3.18×10^{-3}
(c) 2.68×10^{-3} (d) 4.62×10^{-3}

Q.25 A bar produces a lateral strain of magnitude 60×10^{-5} m/m when subjected to a tensile stress of magnitude 300 MPa along the axial direction. What is the elastic modulus of the material if the Poisson's ratio is 0.3?

(a) 200 GPa (b) 150 GPa
(c) 125 GPa (d) 100 GPa

Q.26 A steel wire of 20 mm diameter is bent into a circular shape of 10 m radius. If E, the modulus of elasticity is 2×10^6 kg/cm², then the maximum tensile stress induced in the wire is, nearly

(a) 2×10^3 kg/cm² (b) 4×10^3 kg/cm²
(c) 2×10^4 kg/cm² (d) 4×10^4 kg/cm²

Q.27 The deformation of a vertically held bar of length L and cross-section A is due to its self-weight only. If Young's modulus is E and the unit weight of the bar is γ, the elongation dL is

(a) $\dfrac{\gamma L^3}{2E}$ (b) $\dfrac{EL^2}{2\gamma}$

(c) $\dfrac{\gamma L^2}{2E}$ (d) $\dfrac{\gamma L^2}{2AE}$

Q.28 Which one of the following pairs of tests has been developed to evaluate the fracture resistance of engineering materials, subjected to dynamic loads or impacts?

(a) Tension impacts and Bending impacts
(b) Tensile test and Brinell hardness test
(c) Vickers hardness test and Tensile test
(d) Stereoscope test and File test

Q.29 A mild steel bar of length 450 mm tapers uniformly. The diameters at the ends are 36 mm and 18 mm respectively. An axial load of 12 kN is applied on the bar. $E = 2 \times 10^5$ N/mm². The elongation of the bar will be

(a) $\dfrac{1}{3\pi}$ mm (b) $\dfrac{1}{6\pi}$ mm

(c) $\dfrac{3\pi}{2}$ mm (d) $\dfrac{2}{3\pi}$ mm

Q.30 A rigid beam of negligible weight is supported in a horizontal position by two rods of steel and aluminium, 2 m and 1 m long, having values of cross-sectional areas 100 mm² and 200 mm², and Young's modulus of 200 GPa and 100 GPa, respectively. A load P is applied as shown in the figure below:

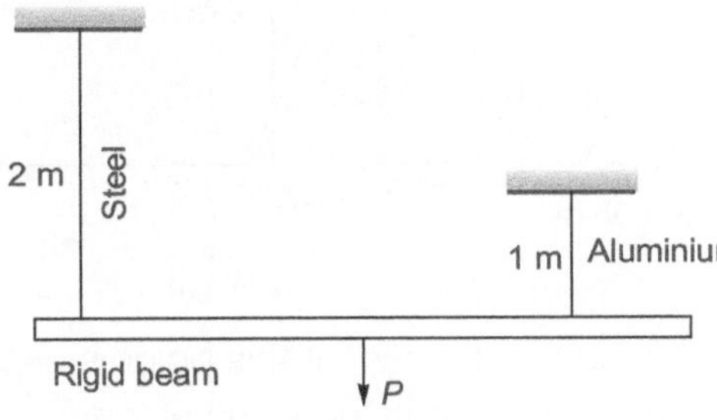

What is the ratio of load shared by steel and aluminium rods?

Q.31 The resilience of steel can be found by integrating stress-strain curve up to the

(a) ultimate fracture point
(b) upper yield point
(c) lower yield point
(d) elastic point

Q.32

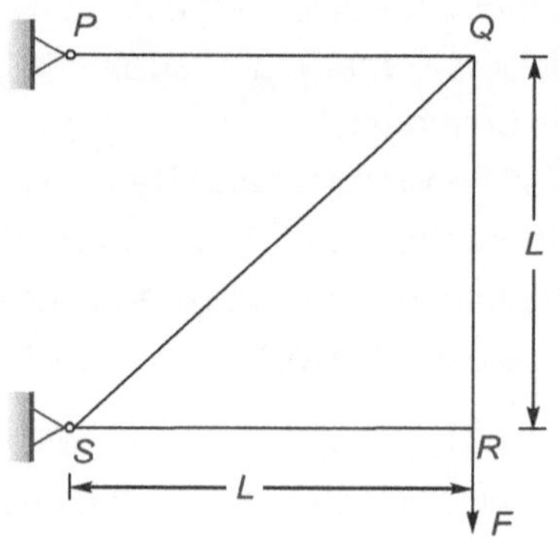

Fort the truss shown in the figure, the force in the member PQ is

(a) F (b) $\frac{F}{\sqrt{2}}$

(c) $\sqrt{2}F$ (d) $2F$

Q.33 In a material under a state of plane strain, a 10 × 10 mm square centered at a point gets deformed as shown in the figure.

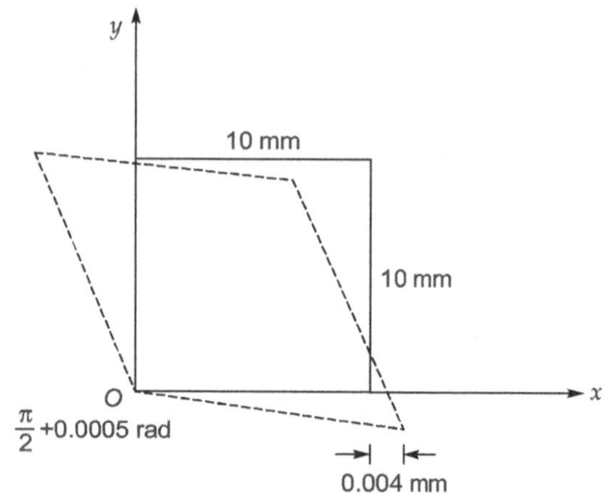

If the shear strain γ_{xy} at this point is expressed as 0.001 k (in rad), the value of k is

(a) 0.50 (b) 0.25
(c) –0.25 (d) –0.50

Q.34 Engineering strain of a mild steel sample is recorded as 0.100%. The true strain is

(a) 0.010% (b) 0.050%
(c) 0.099% (d) 0.101%

Q.35 A bimetallic cylindrical bar of cross-sectional area 1 m^2 is made by binding steel (Young's modulus = 210 GPa) and aluminium (Young's modulus = 70 GPa) as shown in the figure. To maintain tensile axial strain of magnitude 10^{-6} in steel and compressive axial strain of magnitudes 10^{-6} in aluminium bar, the magnitude of the required force P (in kN) along the indicated direction is

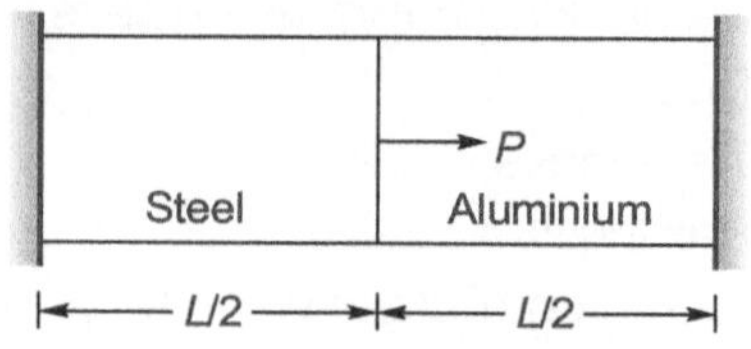

(a) 70 (b) 140
(c) 210 (d) 280

Q.36 A rigid bar ACO as shown is hinged at O and is held in a horizontal position by two identical vertical steel wires AB and CD. A point load of 20 kN is hung at the position shown. The tensions in wires AB and CD are

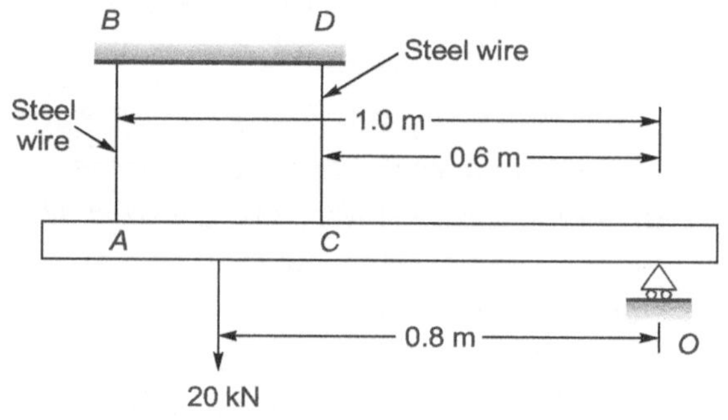

(a) 15.2 kN and 7.1 kN
(b) 11.8 kN and 7.1 kN
(c) 15.2 kN and 5.0 kN
(d) 11.8 kN and 5.0 kN

Q.37 A bar of uniform cross-section and weighing 100 N is held horizontally using two massless and inextensible strings S_1 and S_2 as shown in the figure.

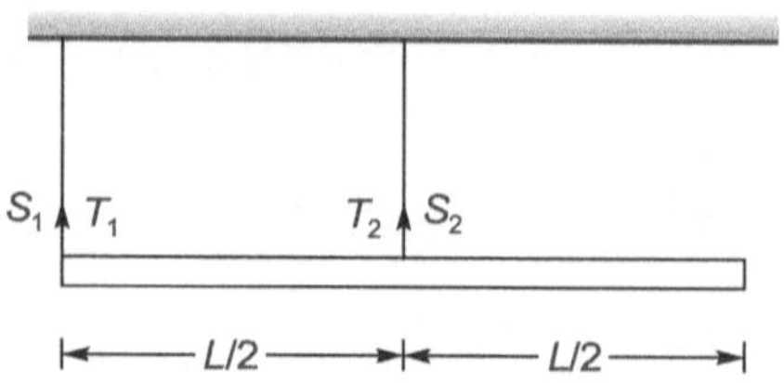

The tension in the string are

(a) $T_1 = 100$ N, $T_2 = 0$ N

(b) $T_1 = 0$, $T_2 = 100$ N

(c) $T_1 = 75$ N, $T_2 = 25$ N

(d) $T_1 = 25$ N, $T_2 = 75$ N

Q.38 A carpenter glues a pair of cylindrical wooden logs by bending the end faces at an angle of $\theta = 30°$ as shown in the figure.

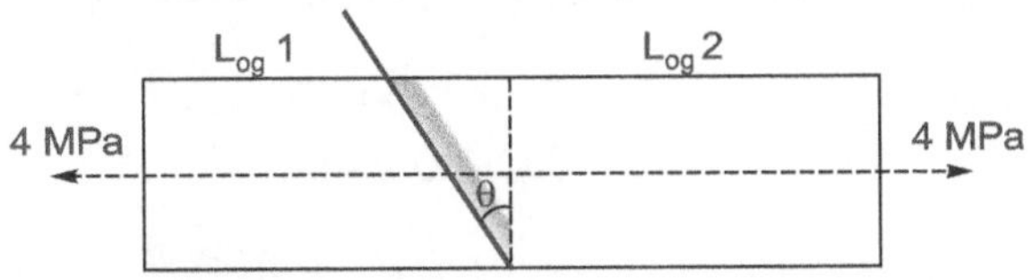

The glue used at the interface fails if

Criterion 1: The maximum normal stress exceeds 2.5 MPa

Criterion 2: The maximum shear stress exceeds 1.5 MPa

Assume at the interface fails before the logs fail, when a uniform tensile stress of 4 MPa is applied the interface

(a) Fails only because of criterion 1

(b) Fails only because of criterion 2

(c) Fails because of both criterion 1 and 2

(d) does not fail

Answers

1. (b)	2. (c)	3. (c)	4. (c)	5. (a)
6. (b)	7. (d)	8. (c)	9. (c)	10. (d)
11. (c)	12. (a)	13. (b)	14. (a)	15. (b)
16. (d)	17. (b)	18. (b)	19. (a)	20. (b)
21. (b)	22. (d)	23. (a)	24. (c)	25. (b)
26. (a)	27. (c)	28. (a)	29. (b)	30. 0.5
31. (d)	32. (a)	33. (a)	34. (c)	35. (d)
36. (b)	37. (b)	38. (c)		

Explanations

1. (b)

Tapered bars, $dL_A = \dfrac{4PL}{\pi E 2d \times d} = \dfrac{4PL}{\pi E 2d^2}$

Uniform bar, $dL_B = \dfrac{4PL}{\pi E 2.25 d^2}$

$$U_A = \frac{1}{2} P dL_A$$

$$U_B = \frac{1}{2} P dL_B$$

$$\frac{U_A}{U_B} = \frac{2.25}{2} = 1.125$$

2. (c)

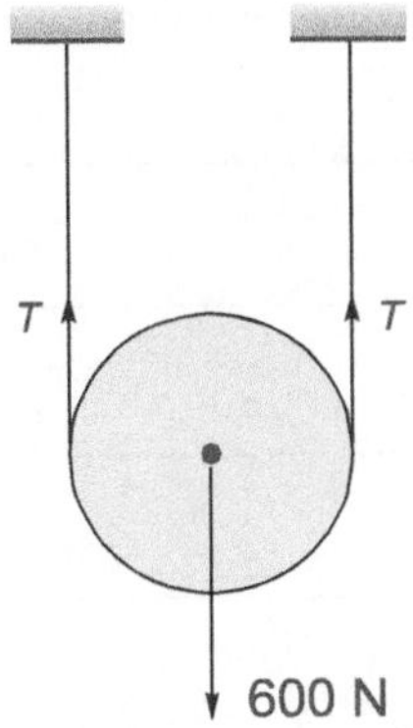

$$T = 300 \text{ N}$$

$$A = 3 \text{ mm}^2$$

$$\sigma = \frac{300}{3} = 100 \text{ N/mm}^2$$

$$\epsilon = \frac{100}{200000} = \frac{1}{2000}$$

$$\delta l = \epsilon L = \frac{3000}{2000} = 1.5 \text{ mm}$$

Vertical displacement of pulley $= \dfrac{dl}{2} = 0.75$ mm

3. (c)

Axial strain $= \varepsilon$

Poisson's ratio $= \nu$

Lateral strain $= -\nu\varepsilon$

Total volumetric strain $= \varepsilon - \nu\varepsilon - \nu\varepsilon$

$= (1 - 2\nu)\varepsilon$

$(1 - 2\nu)\varepsilon = 0.4\varepsilon$

or Poisson's ratio, $\nu = 0.3$

4. (c)

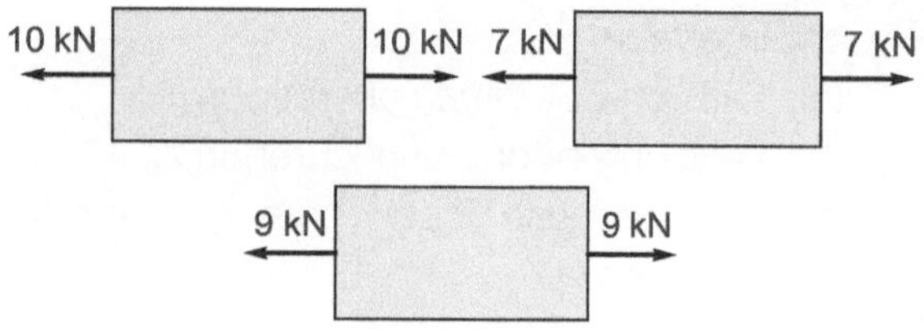

$$\frac{10\times10+7\times10+9\times10}{AE}$$

$$= \frac{100+70+90}{AE}$$

$$= \frac{260}{200\times A}=\frac{1.3}{A}$$

5. (a)

Due to own weight $\frac{wLA}{A}$

Stress, $\sigma = wL$

w = weight density

$$\delta = \frac{wL^2}{2E} = \frac{WL}{2E}$$

L, area A, w, weight density

$$W\text{, weight} = wLA$$

$$\sigma = wL$$

$$\varepsilon = \frac{WL}{A}$$

$$\delta' = \frac{WL}{E}$$

$$\frac{\delta}{\delta'} = \frac{1}{2},\ \delta' = 2\delta$$

6. (b)

Tensile stress decreases at a faster rate with respect to deformation.

7. (d)

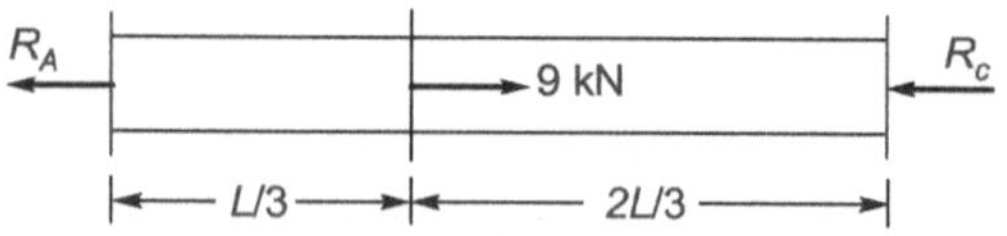

$$R_A + R_C = 9$$

$$\frac{R_A}{EA}\times\frac{L}{3} = R_c\times\frac{2L}{3EA}$$

$$R_A = 2R_C$$

$$2R_C + R_C = 9\text{ kN}$$

$$R_C = 3\text{ kN}$$

8. (c)

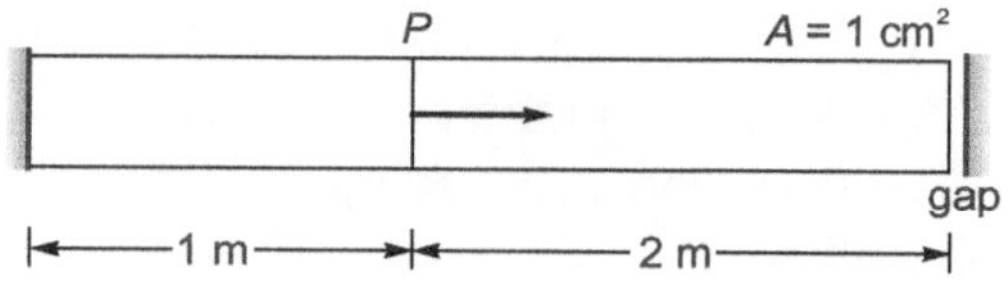

$$E = 200\text{ GPa}$$

$$\delta = 0.2\text{ mm}$$

$$= \frac{P}{100}\times\frac{1000}{200000} = \frac{P}{20000}$$

$$P = 0.2\times 20000 = 4000\text{ N}$$

$$= 4\text{ kN}$$

9. (c)

$$\delta V = e_V \times \frac{p}{k}V$$

$$5 = 10000\times\frac{p}{160000} = \frac{p}{16}$$

$$p = 80\text{ N/mm}^2$$

10. (d)

Area upto elastic limit stress

$$= \frac{\sigma_e^2}{2E}$$

= proof resilience per unit volume

= modulus of resilience

11. (c)

Force on portions DB, BC and CD are as shown

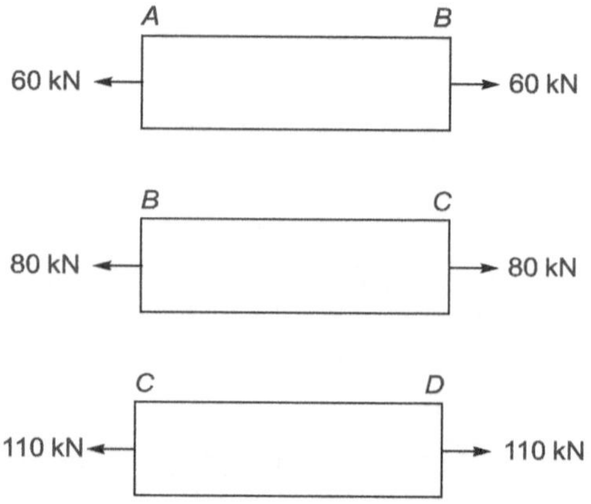

Net force at section B = 20 kN

Net force at section C = 30 kN

Stress in portion BC

$$= \frac{80 \times 1000}{1000} = +80 \text{ N/mm}^2 \text{ (T)}$$

12. (a)

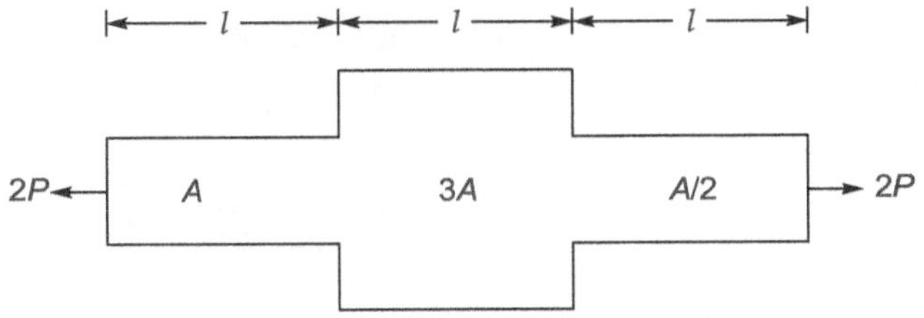

$$\text{Total elongation} = \frac{2PL}{AE} + \frac{2PL}{3AE} + \frac{4PL}{Ae}$$

$$= \frac{PL}{AE}\left[2 + \frac{2}{3} + 4\right] = \frac{20}{3} \times \frac{PL}{AE}$$

$$= 6.66 \frac{PL}{AE}$$

13. (b)

Chain, cross-sectional area,

$$A = 625 \text{ mm}^2$$

$$\text{Load} = 10 \text{ kN} = 10000 \text{ N}$$

$$\text{Speed,} \quad V = \frac{40 \text{ m}}{\text{minutes}} = \frac{2}{3} \text{ m/s}$$

$$KE = \frac{1}{2} mV^2$$

$$= \frac{1}{2} \times \frac{10000}{9.81} \times \left(\frac{2}{3}\right)^2$$

$$= 226.526 \text{ Nm}$$

$$= 226.526 \times 10^3 \text{ Nmm}$$

unwound volume of chain

$$= 625 \times 10000$$

$$= 625 \times 10^4 \text{ mm}^3$$

Say stress induced

$$= \sigma \text{ N/mm}^2$$

$$U, \text{energy} = \frac{\sigma^2}{2 \times 200 \times 1000} \times 625 \times 10^4$$

$$= 15.625 \sigma^2 \text{ Nmm}$$

So, $15.625 \sigma^2 = 3226.526 \times 10^3$

$$\sigma = 120.4 \text{ N/mm}^2$$

14. (a)

Isotrophy

15. (b)

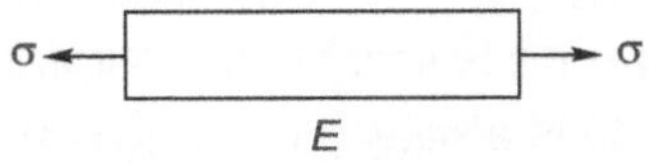

$$\text{Strain energy per unit volume} = \frac{\sigma^2}{2E}$$

16. (d)

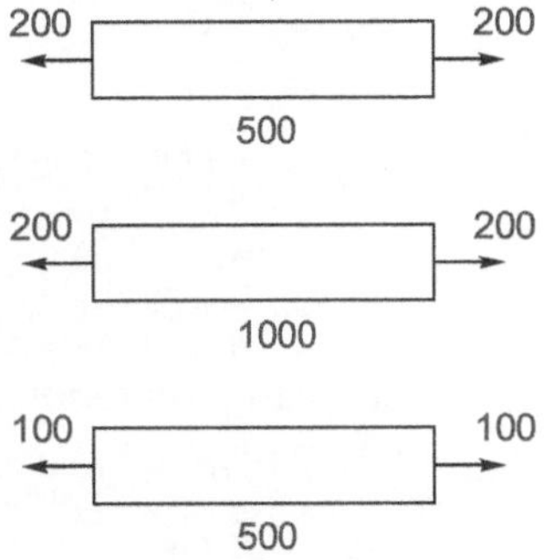

$$A = 10 \text{ mm}^2$$

$$E = 2 \times 10^5 \text{ N/mm}^2$$

$$\delta L = \frac{200 \times 500 - 200 \times 1000 + 100 \times 500}{10 \times 2 \times 10^5}$$

$$= \frac{10^5 - 2 \times 10^5 + 0.5 \times 10^5}{20 \times 10^5}$$

$$= \frac{-0.5}{20} \text{ mm}$$

$$= \frac{-0.5 \times 1000}{20} \mu = -25 \ \mu\text{m}$$

17. (b)

Resilience

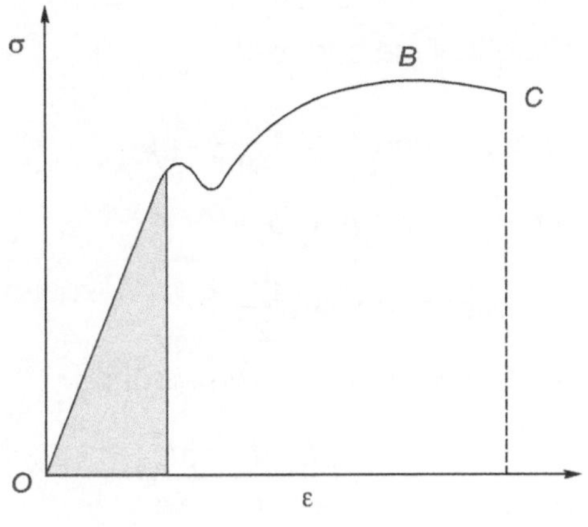

σ_A = Elastic limit strain energy absorbed elastically and then recovered during unloading = Resilience

Total energy indicates toughness $OABC$.

Endurance is limit of stress in fatigue

Ductility is ability to be drawn into wires.

18. (b)

Rod A, diameter = d

Hollow rod, D and 0.5 D

$$A_S = \frac{\pi}{4}d^2$$

$$A_H = \frac{\pi}{4}\left(D^2 - 0.5^2 D^2\right)$$

$$= \frac{3}{4}\pi D^2$$

δl same, length same, material same

$$A_S = A_H$$

$$d^2 = \frac{3}{4}D^2$$

$$d = \frac{\sqrt{3}}{2}D$$

$$\frac{d}{D} = \frac{\sqrt{3}}{2}$$

19. (a)

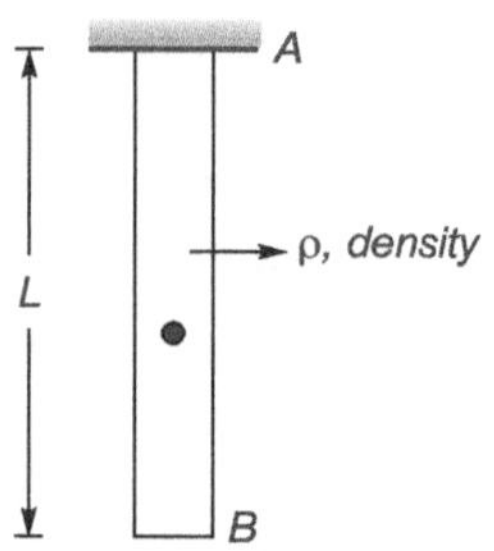

Weight of the bar= ρAL

$$\text{Stress at end } B = \frac{\rho AL}{A} = \rho L$$

$$\text{Stress at end } A = 0$$

$$\text{Mean stress} = \frac{\rho L}{2}$$

δL, extension due to own weight

$$= \frac{\rho L}{2} \times \frac{L}{E} = \frac{\rho L^2}{2E}$$

20. (b)

Tapered bar

$$D = 40 \text{ mm}$$
$$A = 20 \text{ mm}$$
$$L = 800 \text{ mm}$$
$$P = 20 \text{ kN (pull)}$$
$$E = 2 \times 10^5 \text{ N/mm}^2$$

Increase in length of tapered ms bar

$$= \frac{4PL}{\pi EDd}$$

$$= \frac{4 \times 20 \times 10^3 \times 800}{\pi \times 200 \times 10^3 \times 40 \times 20}$$

$$= \frac{16}{\pi \times 40} = \frac{2}{5\pi}$$

21. (b)

Let us calculate distance of CG of T-shaped arm from C

$$\bar{x} = \frac{2L \times L + L \times 2L}{3L} = \frac{4}{3}L$$

Moments about C

$$W \times \bar{x} = R \times L$$

where R is reaction at B

$$W \times \frac{4}{3}L = R \times L$$

$$R = \frac{4}{3}W$$

Extension in spring

$$= \frac{R}{k} = \frac{4W}{3k}$$

22. (d)

Greatest strain energy stored in a unit volume of a material without permanent deformation is called proof resilience $\frac{\sigma_e^2}{2E}$.

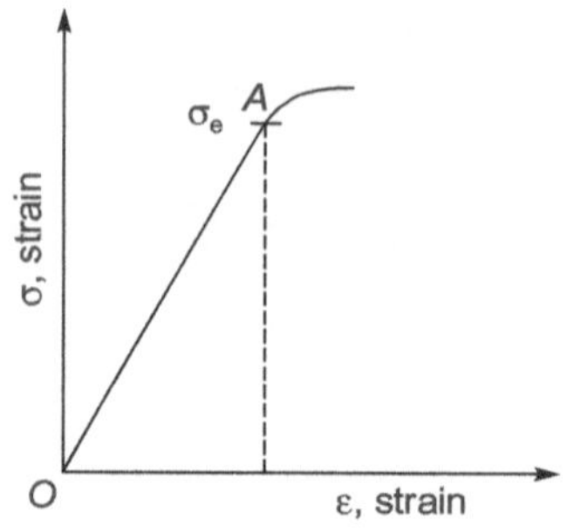

Resilience is the strain energy stored which can be recovered fully during unloading. So (a) and (b) are incorrect (c) is also incorrect.
Only (d) is correct.
Without permanent deformation greatest strain energy stored per unit volume is $\frac{\sigma_e^2}{2E}$, where σ_e is elastic limit stress, E is young's modulus.

23. (a)
Ductility is ability of materials to be drawn into wires, toughness is the indicator of total strain energy observed till breading and hardness is resistance to starting wear and tear, abraism, penetration etc.
Moreover Baba's law states that total elongation

$$\delta L = bL + c\sqrt{A}$$

b and c are constants,
c is the criterion for necking, lateral strain, change In lateral dimension.
δl a l, this necking taken place.
So ductility is indicator for uniform rate of elongation.

24. (c)
Bar diameter,

$$d = 10 \text{ mm}$$

Elastic modulus,

$$E = 200 \text{ GPa}$$

Tensile load, $P = 50$ kN

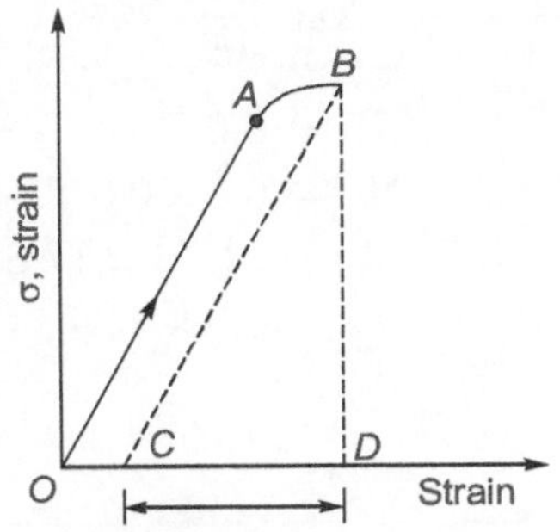

CD = Elastic recovery

Area of cross-section,

$$A = \frac{\pi}{4}\times 10^2 = 25\,\pi \text{ mm}^2$$

Tensile stress, $\sigma = \frac{50\times1000}{25\pi}$

$$= 636.62 \text{ N/mm}^2$$

Elastic modulus,

$$E = 200 \times 10^{-3} \text{ N/mm}^2$$

Elastic strain recovered = CD, as shown in figure.

$$= \frac{636.62}{200\times10^3} = 3.18\times10^{-3}$$

25. (b)
Lateral strain,

$$\varepsilon' = 60 \times 10^{-5} \text{ m/m}$$

Poisson's ratio,

$$\nu = 0.3$$

Axial strain, $\varepsilon = \frac{\varepsilon'}{\nu} = \frac{\sigma_0\times10^{-5}}{0.3}$

$$= 200 \times 10^{-5} \quad \text{...(i)}$$

Axial stress, $\sigma = 300$ MPa
Elastic modulus,

$$E = \frac{\sigma}{\varepsilon} = \frac{300}{200}\times10^5$$

$$= 1.5 \times 10^5 \text{ N/mm}^2$$

$$= 150 \text{ GPa}$$

26. (a)

$$\frac{d}{2} = 10 \text{ mm} = 1 \text{ cm}$$

$$R = 10{,}000 \text{ mm} = 1000 \text{ cm}$$

$$E = 2 \times 10^6 \text{ kg/cm}^2$$

$$\frac{\sigma}{d/2} = \frac{E}{R}$$

$$\sigma = \frac{1\times2\times10^6}{1000}$$

$$= 2 \times 10^3 \text{ kg/cm}^2$$

27. (c)

$$\delta l = \text{due to self weight} = \frac{\gamma l^2}{2E}$$

28. (a)
Resistance to dynamic loads, tension impact test (shock tensile test) and bending impacts.

29. (b)

Uniformly tapered bar

Axial load, $P = 12000$ N

Length, $L = 450$ mm

Diameters, $D = 36$ mm

$d = 18$ mm

Young's modulus,

$E = 2 \times 10^5$ N/mm²

Elongation, $\delta L = \dfrac{4PL}{\pi EDd}$

$$= \frac{4 \times 12000 \times 450}{\pi \times 200 \times 1000 \times 36 \times 18}$$

$$= \frac{1}{6\pi}$$

30. (0.5)

$$\frac{P_S}{100} \times \frac{2}{200} = \frac{P_a}{200} \times \frac{1}{100}$$

$$\frac{P_s}{P_s} = 0.5$$

31. (d)

Resilience = Integrating stress-strain curve upto elastic limit.

(Elastic energy which can be recovered on unloading).

32. (a)

Force in member PQ is F.

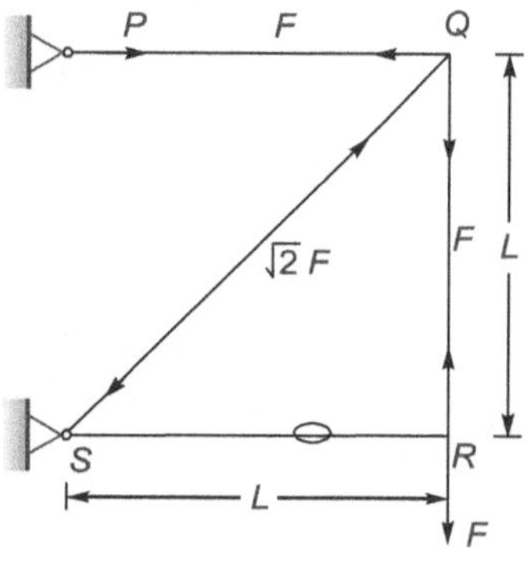

33. (a)

Change in angle of $\dfrac{\pi}{2}$ at corner

$= +\,0.0005$ radians

γ_{xy} = shear strain = + 0.0005 radians

$= +0.0001$ k radians

$k = 0.5$

Note strain $\varepsilon_{xx} = \dfrac{0.004}{10} = 0.0004$

34. (c)

$$\varepsilon_e = 0.100\% = 0.001 = \frac{\delta l}{l}$$

$$\varepsilon_t = \text{true strain} = \frac{\delta l}{l + \delta l} = \frac{0.001}{1.001}$$

$= 9.99 \times 10^{-4} = 9.99 \times 10^{-2}$ %

$= 0.099\%$

35. (d)

$(10^{-6} \times 210 \times 10^6 + 10^{-6} \times 70 \times 10^6) \times 1$ m²

= 280 kN

36. (b)

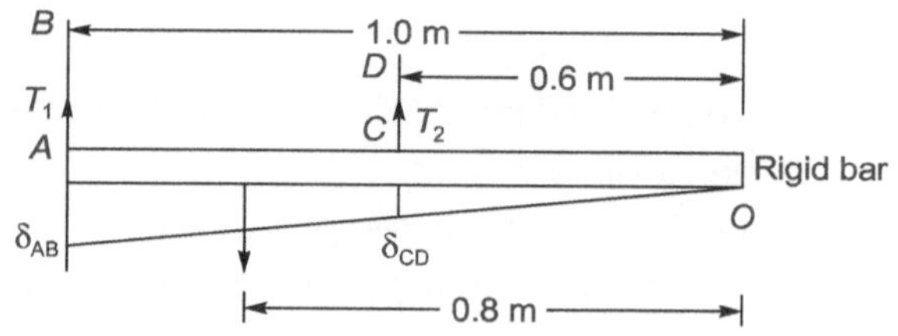

Taking moment about O

$20 \times 0.8 = T_2 \times 0.6 + T_R \times 1.0$

or $16 = 0.6\,T_2 + T_1$ ...(i)

Bar ACO is rigid

Extension in CD = 0.6 × Extension in AB

Say L is length of each wire

$$\frac{T_2}{A} \times \frac{L}{E} = 0.6 \frac{T_1 L}{AE}$$

or $T_2 = 0.6\,T_1$ ...(ii)

$16 = 0.6 \times 0.6\,T_1 + T_2$

$T_1 = 11.76 \simeq 17.8$ kN

$T_2 = 0.6 \times 11.76 = 7.1$ kN

37. (b)

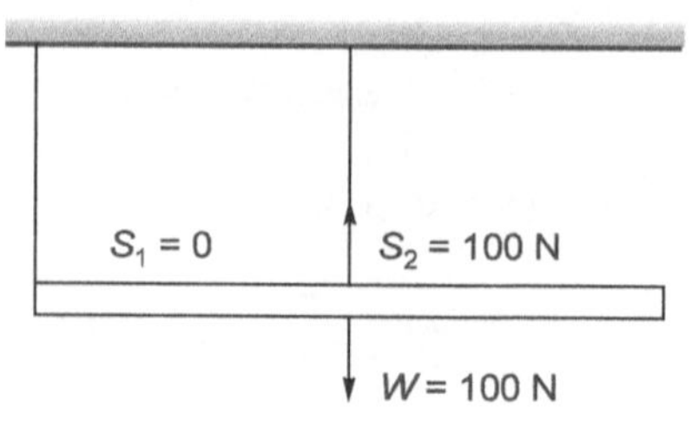

T_2 = 100 N balancing directly the load *W*, weight of bar

$T_1 = 0$

38. (c)

$$P = \frac{\pi}{4} d^2 \times 4 \text{ N}$$

$$P_n = P \cos 30° = \pi \times 0.866 \times d^2$$

$$P_t = P \sin 30° = \pi \times 0.5 \times d^2$$

Area of elliptical surface

$$= \pi ab$$

$$a = \frac{d}{2},\ b = \frac{d}{2} \times \frac{1}{\cos 30°}$$

$$\pi ab = \frac{\pi d^2}{4} \times 1.154$$

$$= \pi \times 0.288\, d^2$$

$$\sigma = \frac{\pi \times 0.866 \times d^2}{\pi \times 0.288 \times d^2} = 3 \text{ MPa}$$

$$\tau = \frac{\pi \times 0.5 \times d^2}{\pi \times 0.288 \times d^2}$$

$$= 1.736 \text{ MPa}$$

Interface fails because of both criteria (1) and (2)

■■■■

Composite Bars and Temperature Stresses

CHAPTER 2

A composite bar can be defined as a bar composed of more than one bars of different materials, perfectly fitted together so that under external load, all the bars of composite bar strain together by same amount.

Composite bars are of engineering importance. Examples are:

(a) Reinforced concrete bar-made of steel bars reinforced in cement concrete, used in building construction.

(b) Bimetallic strip-made of two strips of different materials having different co-efficients of thermal expansion used in thermostat for temperature control.

Question 2.1 **Consider a composite bar made of two bars of different materials subjected to external axial load. Derive expression for stresses in two bars.**

Solution: Take a composite bar consisting of a rod of diameter *d*, material of Young's modulus E_1 and a tube of outer diameter *D* and inner diameter *d*, material of Young's modulus E_2, both rod and tube are perfectly joined together so that under an axial compressive load P(as shown in Fig. 2.1) both strain together and say

$$P_1 = \text{Load shared by rod}$$

$$P_2 = \text{Load shared by tube}$$

Internal load, $P = P_1 + P_2$, for equilibrium

Now both bar and tube strain together

say σ_1 = Stress developed in rod

σ_2 = Stress developed in tube

Compatibility condition is

$$\varepsilon_1 = \varepsilon_2$$

$$\frac{\sigma_1}{E_1} = \frac{\sigma_2}{E_2}, \text{ use Hooke's law}$$

$$\sigma_1 = \sigma_2 \times \frac{E_1}{E_2}$$

$$\sigma_1 = m\sigma_2 \qquad \text{...(ii)}$$

If $\frac{E_1}{E_2} = m$, modular ratio

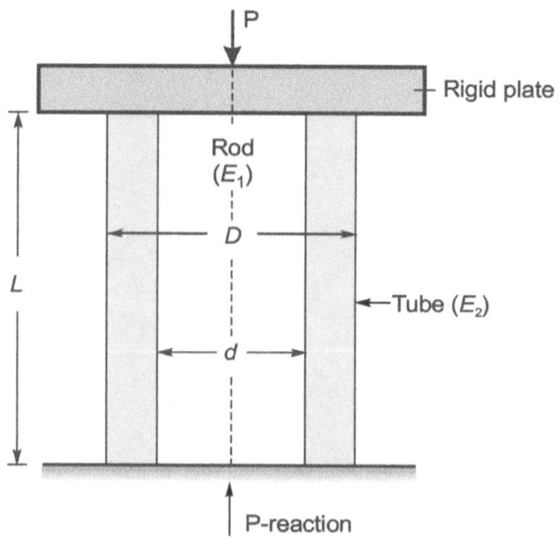

Fig. 2.1

$$\text{Load shared by rod} = \sigma_1 A_1 = \sigma_1 \times \frac{\pi}{4} d^2$$

$$\text{Load shared by tube} = \sigma_2 A_2 = \sigma_2 \times \frac{\pi}{4} (D^2 - d^2)$$

$$P = P_1 + P_2 = \sigma_1 A_1 + \sigma_2 A_2$$

$$= m\sigma_2 A_1 + \sigma_2 A_2 = \sigma_2(A_2 + mA_1) \quad \text{...(iii)}$$

From equation (iii) stress σ_2 can be calculated and

$$\sigma_1 = m\sigma_2$$

Practice Q.2.1.1 A copper bar of diameter 20 mm is completely encased in a steel tube of outer diameter 30 mm and inner diameter 20 mm, so as to make a composite bar. Length of composite bar is 200 mm. If E_c = 105 kN/mm² and E_s = 210 N/mm². What are the stresses developed in steel tube and copper rod under an axial load of 30 kN? What is the change in length of composite bar?

Hints: [$\sigma_s = m\sigma_c$] Ans. [27.28, 54.56 MPa, 0.052 mm]

Practice Q.2.1.2 A brass bar of diameter *d* is encased in a steel tube of outer diameter 35 mm, $E_s = 2E_B$ so as to make a composite bar. External load is applied on composite bar. The external load is equally shared by brass bar and steel tube. What is the diameter of the brass bar?

Hints: [$\sigma_s = 2\sigma_b$, $\sigma_s A_s = \sigma_b A_b$; d = 28.577 mm]

Question 2.2 An aluminium bar of diameter 30 mm and length 200.2 mm is placed coaxially in a steel tube of outer diameter 40 mm and inner diameter 35 mm as shown in Fig. 2.2.

E_s = 204 kN/mm², E_a = 68 kN/mm²

Determine:

(a) The axial force *P* required to just fill the gap of 0.2 mm.

(b) If a force of 100 kN is now applied on the composite system, what will be the stresses developed in steel tube and aluminium bar.

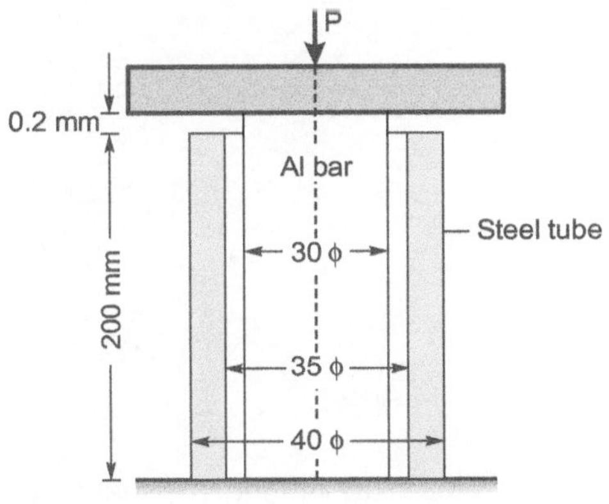

Fig. 2.2

Solution :

(a) When the gap is to be just filled in, *P* load will act only an the aluminium bar of length 200.2 mm.

$$\delta L, \text{ change in length} = 0.2 \text{ mm}$$

$$L, \text{ length of bar} = 200.2 \text{ mm}$$

$$\varepsilon_a, \text{ strain} = \frac{\delta L}{L} = \frac{0.2}{200.2} = \frac{1}{1001}$$

$$\varepsilon_a = 68000 \text{ N/mm}^2$$

Using Hooke's law

$$\text{Stress in aluminium bar} = \varepsilon_a E_a = \frac{1}{1001} \times 68000$$

$$= 67.93 \text{ N/mm}^2$$

Area of cross-section of aluminium bar,

$$A_a = \frac{\pi}{4} \times 30^2 = 706.86 \text{ mm}^2$$

Axial load,
$$P = \sigma A_a = 67.93 \times 706.86$$
$$= 48017 \text{ N}$$
$$= 48.017 \text{ kN (load required to fill the gap)}$$

(b) Load
$$P = 100 \text{ kN}$$
$$= P_a + P_s$$
$$= \sigma_a A_a + \sigma_a A_s \quad \text{...(i)}$$

But
$$\delta l_a = \delta l_s + 0.2$$

or
$$\delta l_a = \frac{\sigma_a}{E_a} \times 200.2$$
$$\delta l_s = \frac{\sigma_s}{E_s} \times 200$$

or
$$\frac{\sigma_a}{E_a} \times 200.2 - \frac{\sigma_s}{E_s} \times 200 = 0.2$$
$$\frac{\sigma_a \times 200.2}{68000} - \frac{\sigma_s \times 200}{204000} = 0.2$$

or
$$\sigma_a \times 3 \times 200.2 - \sigma_s \times 200 = 40800$$
$$3.003\, \sigma_a - \sigma_s = 204$$
$$\sigma_s = 3.003\sigma_a - 204 \quad \text{...(ii)}$$

Area,
$$A_s = \frac{\pi}{4}(40^2 - 35^2) = 294.52 \text{ mm}^2$$

Substituting in equation (i)
$$100{,}000 = \sigma_a \times 706.86 + (3.003\, \sigma_a - 204) \times 294.52$$
$$100{,}000 = \sigma_a(706.88 + 884.44) - 60082.08$$
$$\sigma_a \times 1591.3 = 160082.08$$
$$\sigma_a = 100.6 \text{ N/mm}^2$$
$$\sigma_s = 3.003\, \sigma_a - 204$$
$$= 302.097 - 204$$
$$= 98.097 \text{ N/mm}^2$$

Practice Q.2.2 A steel bar 20 mm diameter is co-axially placed in a copper tube of outside diameter 30 mm and inside diameter 25 mm. Length of tube is 250.3 mm. Length of steel bar is 250 mm. If $E_s = 2E_c = 210$ kN/mm^2. What is the force P required to just fill the gap between steel bar and copper tube as shown in Fig. 2.3? If P is now 45 kN, what will be the stresses in steel bar and copper tube. What is the final change in length of copper tube?

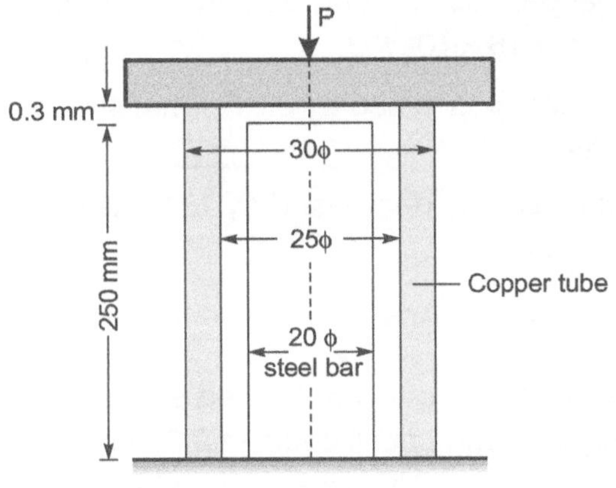

Fig. 2.3

Ans. [27.21 kN, 42.215, 146.93 N/mm^2, 0.35 mm]

Question 2.3 **A metallic bar of length L, area of cross-section A is held between two rigid supports. The temperature of the bar is raised by ΔT. If coefficient of thermal expansion of bar is α. Find the temperature stress developed in the bar. What is the magnitude and the type of the force developed in the bar?**

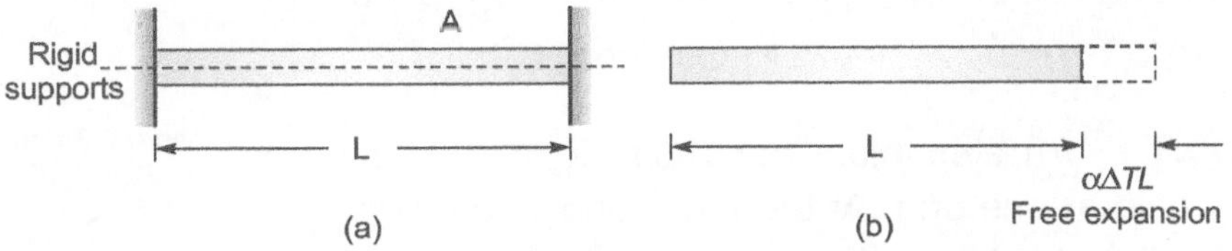

Fig. 2.4

Solution : Fig. 2.4 (a) shows a bar of length L, area of cross-section A fixed at ends in rigid supports. Now the temperature of the bar is raised by ΔT, the bar tends to expand but its free expansion is prevented by rigid supports.

If bar is allowed to expand freely, then

$$\text{Free expansion in bar} = \alpha \cdot \Delta T \cdot L$$

$$\text{Final length of the bar} = L + \alpha \cdot \Delta T \cdot L$$

The end supports provide axial compressive force and the final length $L + \alpha \cdot \Delta T \cdot L$ is compressed to L.

$$\text{contraction in length} = L + \alpha \cdot \Delta T \cdot L - L$$

$$= \alpha \cdot \Delta T \cdot L$$

Strain in bar,

$$\varepsilon = \frac{\alpha \cdot \Delta T \cdot L}{L + \alpha \cdot \Delta T \cdot L} \simeq \frac{\alpha \cdot \Delta T \cdot L}{L}$$

$$\text{Stress in bar, } \sigma = \varepsilon E, \text{ using Hooke's law}$$

$$= \alpha \cdot \Delta T \cdot E$$

$$\text{Axial compressive force in bar, } P_c = \alpha \cdot \Delta T \cdot E \cdot A$$

where,

E = Young's modulus of material

A = Area of cross-section

Note that **if the temperature of the bar is decreased by ΔT, then free contraction of the bar is prevented by end rigid supports** and tensile force ($\alpha \cdot \Delta T \cdot E \cdot A$) is developed in the bar.

Practice Q.2.3 A copper bar 600 mm long, 20 mm in diameter is held between two rigid supports. E for copper is 105 kN/mm^2 and $\alpha = 16 \times 10^{-6}$/°C.

(a) If the temperature of the bar is raised by 25°C, what stress is developed in bar force and what is applied by supports?

(b) If the temperature of two bar is decreased by 15°C, what stress is developed in bar and what force is applied by supports?

Ans. (a) [–42 N/mm^2, –13.195 kN] (b) [+25.2 N/mm^2, +7.917 kN]

Question 2.4 A steel bar of length 1 m is fixed at end B and there is a gap between support and end A equal to 0.2 mm. If the temperature of the bar is raised by 50°C, what is the stress developed in bar, if $\alpha = 11 \times 10^{-6}$/°C. E for steel = 200 GPa.

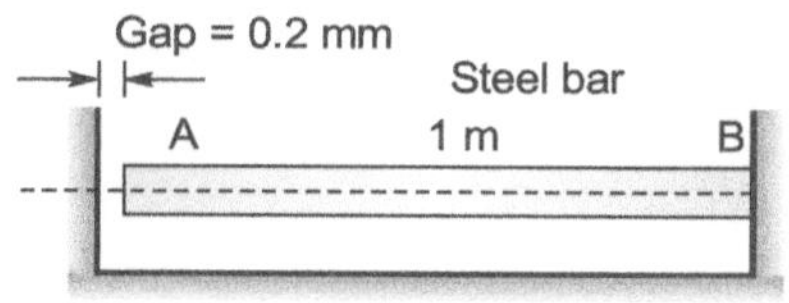

Fig. 2.5

Solution : Prevention of free expansion in bar

$$= \alpha \cdot \Delta T \cdot L - 0.2$$

$$= 11 \times 10^{-6} \times 50 \times 1000 - 0.2 = 0.35 \text{ mm}$$

Strain, $$\varepsilon = \frac{0.35}{L} = \frac{0.35}{1000} = 0.35 \times 10^{-3}$$

Stress, $$\sigma = \varepsilon E = -0.35 \times 10^{-3} \times 200 \times 10^3$$

$$= -70 \text{ N/mm}^2 \text{ (compressive stress)}$$

Practice Q.2.4 An aluminium bar 0.5 m long, 20 mm diameter is rigidly held at one end. At the other end of the bar, there is a gap of 0.3 mm between bar end and rigid support. Temperature of the bar is raised by 60°C, what is the stress developed in bar, if E for aluminium is 68000 N/mm^2? What force is acting on the bar? $\alpha_{al} = 22 \times 10^{-6}$/°C. If the temperature of the bar falls by 60°C, what is the stress developed?

Ans. [–48.96 N/mm^2, –15.38 kN, zero]

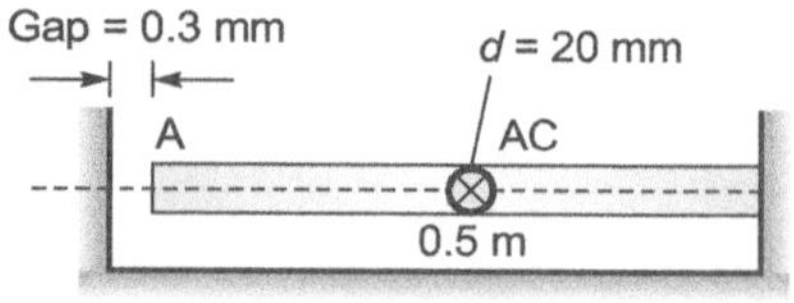

Fig. 2.6

Question 2.5 Explain how stresses are developed in a composite bar, when it is subjected to a temperature change.

Solution : Consider a composite bar of length L, consisting of a bar 1 with area of cross-section A_1, Young's modulus E_1 and coefficient of thermal expansion α_1 and another bar 2 of area of cross-section A_2, Young's modulus E_2 and coefficient of thermal expansion α_2 as shown in fig. 2.7. Say $\alpha_1 > \alpha_2$, composite bar is subjected to change in temperature $+\Delta T$(rise in temperature). As a composite bar, both bars will strain together with change in length equal to δL.

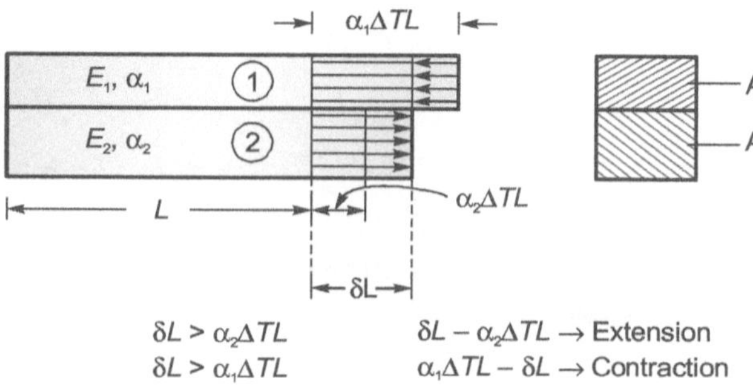

Fig. 2.7

If both bars are free to expand, then

$$\text{Free expansion in bar 1} = \alpha_1 \cdot \Delta T \cdot L$$

Free expansion in bar 2 $= \alpha_2 \cdot \Delta T \cdot L$

$$\alpha_1 \cdot \Delta T \cdot L > \alpha_2 \cdot \Delta T \cdot L$$

But the composite bar changes in length by δL, such that

$$\delta L < \alpha_1 \cdot \Delta T \cdot L$$

$$\delta L > \alpha_2 \cdot \Delta T \cdot L$$

δL_1, contraction in the length of bar 1 $= \alpha_1 \cdot \Delta T \cdot L - \delta L$

δL_2, extension in the length of the bar 2 $= \delta L - \alpha_2 \cdot \Delta T \cdot L$

So contraction in length of bar 1 + extension in the length of bar 2 $= \alpha_1 \cdot \Delta T \cdot L - \alpha_2 \cdot \Delta T \cdot L$

$$= \Delta T \cdot L(\alpha_1 - \alpha_2) \quad \text{...(i)}$$

Say strain in bar 1 is ε_1

Strain in bar 2 in ε_2 then

$$\varepsilon_1 \times (L + \alpha_1 \cdot \Delta T \cdot L) = \delta L_1$$

$$\varepsilon_2 \times (L + \alpha_2 \cdot \Delta T \cdot L) = \delta L_2$$

But $\alpha_1 \cdot \Delta T \cdot L$ and $\alpha_2 \cdot \Delta T \cdot L$ are very small in comparison to length L so

$$\varepsilon_1 = \frac{\delta L_1}{L}; \qquad \varepsilon_2 = \frac{\delta L_2}{L}$$

$$\delta L_1 + \delta L_2 = (\varepsilon_1 + \varepsilon_2) = (\alpha_1 - \alpha_2)\, \Delta T \cdot L$$

From equation (i) $\quad \varepsilon_1 + \varepsilon_2 = (\alpha_1 - \alpha_2)\Delta T$

Say

σ_{1T} — Compressive stress developed in bar 1

σ_{2T} — Tensile stress developed in bar 2

E_1 — Young's modulus of bar 1

E_2 — Young's modulus of bar 2, then

$$\frac{\sigma_{1T}}{E_1} + \frac{\sigma_{2T}}{E_2} = (\alpha_1 - \alpha_2)\, \Delta T \quad \text{...(ii)}$$

These are two unknowns, and only one equation.

Equation of Equilibrium

$$\sigma_{1T} \times A_1 = \sigma_{2T} \times A_2 \quad \text{...(iii)}$$

No external force is applied on composite bar, only the temperature is changed. Therefore, compressive force developed in bar 1 i.e. $\sigma_{1T} \cdot A_1$ is equal to tensile force developed in bar 2 i.e. $\sigma_{2T} \cdot A_2$.

From equations (ii) and (iii) compressive stresses developed in bars due to temperature change can be worked out.

Note that **if the temperature of the composite bar is decreased then tensile stress will be developed in bar 1 with higher coefficient of thermal expansion and compressive stress will be developed in bar 2 with lower coefficient of thermal expansion ($\alpha_2 < \alpha_1$).**

Question 2.6 **A copper rod, 6 cm in diameter is placed within a steel tube, 8 cm external diameter and 6 cm internal diameter of exactly the same length. The two pieces are rigidly fixed together by two transverse pins 20 mm in diameter, one at each end passing through both rod and the tube. Calculate the stresses induced in the copper rod, steel tube and the pins if the temperature of the combination is raised by 50°C.**

Take, $\quad E_S = 210$ GPa, $\alpha_s = 11.5 \times 10^{-6}$/°C

$E_C = 105$ GPa, $\alpha_c = 17.0 \times 10^{-6}$/°C **[IES 2008 : 15 Marks]**

Solution :

Copper rod: Diameter = 6 cm

Steel tube: $D = 8$ cm, $d = 6$ cm

Area of cross-section:

$$A_s = \frac{\pi}{4}(8^2 - 6^2) = 7\pi \text{ cm}^2$$

$$A_c = \frac{\pi}{4}(6^2) = 9\pi \text{ cm}^2$$

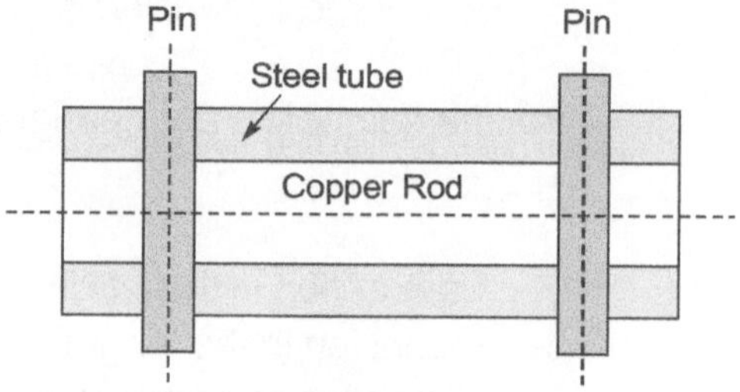

Fig. 2.8

Temperature stresses:

$$\alpha_s = 11.5 \times 10^{-6}/°C$$
$$\alpha_c = 17 \times 10^{-6}/°C$$
$$\Delta T = 50°C$$

$\alpha_c > \alpha_s$, due to rise in temperature, compressive stress will be developed in copper rod (say σ_{cT} and tensile stress will be developed in steel tube say), σ_{sT}.

For equilibrium

$$\sigma_{cT} \times A_c = \sigma_{sT} \times A_s$$
$$\sigma_{cT} \times 9\pi = \sigma_{sT} \times 7\pi$$
$$\sigma_{sT} = 1.2857\,\sigma_{cT} \quad \text{...(i)}$$

$$\frac{\sigma_{sT}}{E_s} + \frac{\sigma_{cT}}{E_c} = (\alpha_s - \alpha_c)\Delta T = 50°C \times (17 - 11.5) \times 10^{-6}/°C$$

$$\frac{1.2857\,\sigma_{cT}}{210 \times 1000} + \frac{\sigma_{cT}}{105000} = 275 \times 10^{-6}$$

$$3.2857\,\sigma_{cT} = 57.75$$
$$\sigma_{cT} = 17.58 \text{ N/mm}^2$$
$$\sigma_{sT} = 1.2857 \times 17.58 = 22.6 \text{ N/mm}^2$$

Tensile force in steel tube = Compressive force in copper rod

$$= 22.6 \times 7\pi \times 100 = 49700 \text{ N} = 49.7 \text{ kN}$$

Tensile force in steel tube and compressive force in copper rod produce shear force for pins.
Pins are in double shear, it may fail along two cross-sections

Pin area, $$A_p = \frac{\pi}{4} \times 20^2 = 100\,\pi \text{ mm}^2$$

$$\tau, \text{ shear stress in pins} = \frac{49700}{2 \times A_p} = \frac{49700}{2 \times 100\pi} = 79.1 \text{ N/mm}^2$$

Practice Q.2.6 Two steel flats of 20 × 10 mm (cross-section) are fixed with one aluminium flat of 20 × 20 mm (cross-section), on top and bottom of aluminium flat. Area of cross-section of composite section is 20 mm wide × 40 mm deep. The two flats are fastened together perfectly, at a temperature of 25°C. Now the temperature of the composite bar is raised to 50°C. Find, stresses in aluminium flat and steel flats.

$\alpha_s = 11.6 \times 10^{-6}/°C$, $\alpha_a = 23.2 \times 10^{-6}/°C$
$E_s = 3E_a = 210$ GPa

Hints: [$A_s = A_a = 200 \text{ mm}^2$, $\sigma_{aT} \times 200 = \sigma_{sT} \times 2 \times 100$]

Ans. [–15.225 N/mm² (in aluminium), +15.225 N/mm² (in steel)].

Question 2.7 A short hollow CI column of external diameter 25 cm and internal diameter 20 cm is filled with concrete. Column carries a total load of 400 kN. If $E_{con} = \frac{1}{6}E_{CI}$. Calculate (a) stresses is concrete and cast iron column. (b) Now the load is increased to 450 kN, what should be the internal diameter of cast iron column if stresses in cast iron column and concrete remain, the same. External diameter of the column also remains unchanged.

Solution:

(a) $\sigma_{CI} = 6\sigma_{con}$ ...(i)

As modular ratio is 6.

External diameter, $D = 25$ cm

Internal diameter, $d = 20$ cm

A_{CI}, area of *CI* column $= \frac{\pi}{4}(D^2 - d^2) = \frac{\pi}{4}(25^2 - 20^2) = 56.25\,\pi\text{ cm}^2$

A_{con}, area of concrete $= \frac{\pi}{4}d^2 = \frac{\pi}{4}\times 20^2 = 100\pi\text{ cm}^2$

External load, $P = 400\text{ kN} = 400{,}000\text{ N}$

$$400{,}000 = \sigma_{CI}\times 56.25\pi\times 100 + \sigma_{con}\times 100\pi\times 100$$

$$4000 = 6\sigma_{con}\times 56.25\pi + \sigma_{con}\times 100\pi$$

$$= \sigma_{con}[337.5 + 100]\pi = 437.5\pi\times\sigma_{con}$$

$$\sigma_{con} = 2.91\text{ N/mm}^2$$

$$\sigma_{CI} = 6\,\sigma_{con} = 17.46\text{ N/mm}^2$$

(b) Say internal diameter $= d$ cm

$$A_{con} = \frac{\pi d^2}{4}$$

$$A_{CI} = \frac{\pi}{4}(25^2 - d^2)$$

$$450{,}000 = \frac{\pi}{4}d^2\times 2.91 + \frac{\pi}{4}(250^2 - d^2)\times 17.46$$

$$572958 = 2.91\,d^2 + 17.46\,(62500 - d^2)$$

$$= -14.55\,d^2 + 1091250$$

$$d^2 = \frac{518292}{14.55} = 35621.44$$

$$d = 188.74\text{ mm} = 18.87\text{ cm}$$

Area of cast iron column is increased.

Practice Q.2.7 A short *CI* column of external diameter 300 mm and wall thickness 25 mm is filled with concrete. If $\sigma_{CI} = 6\,\sigma_{con}$ and maximum stress in concrete is not to exceed 4 N/mm^2, how much compressive load can be safely applied on the column.

Ans. [714.7 kN]

Question 2.8 Consider a tube and bolt assembly, tightened by a nut on bolt and washers on tube. Say p is the pitch of the thread on bolt and nut is tightened by $\frac{p}{n}$. Derive expressive for stresses in bolt and tube and change in the length of bolt and tube.

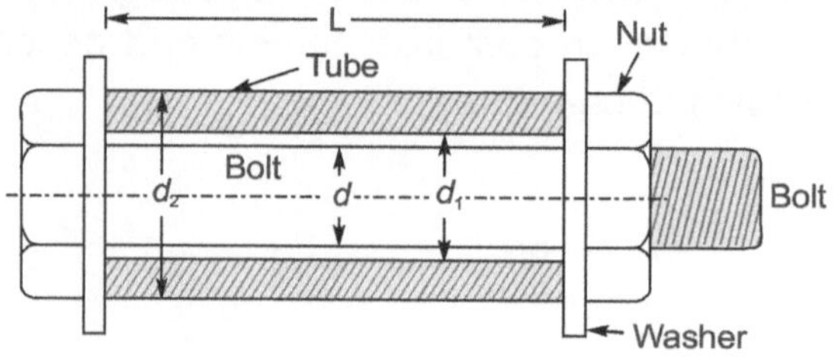

Fig. 2.9 *Bolt and Tube assembly*

Solution: Figure 2.9 shows a bolt of diameter d, co-axially placed in a tube of external diameter d_2 and internal diameter d_1, and nut tightened on tube. By tightening the nut through $\frac{p}{n}$, where p is pitch of thread on bolt.

Due to tightening by nut, bolt gets extended and tube gets contracted.

Net movement of nut on threads = extension in bolt + contraction in tube

$$\frac{p}{n} = \delta l_{bolt} + \delta l_{tube}$$

Equilibrium

Tensile force on bolt = compressive force on tube

$$P_b = P_t$$

$$\sigma_b A_b = \sigma_t A_t$$

where,

$$A_b = \frac{\pi}{4} d^2 \text{, area of cross-section of bolt}$$

$$A_t = \frac{\pi}{4}\left(d_2^2 - d_1^2\right) \text{, area of cross-section of tube}$$

$$\delta l_{bolt} = \frac{\sigma_b}{E_b} \times L$$

$$\delta l_{tube} = \frac{\sigma_t}{E_t} \times L$$

where E_b, E_t are Young's modulus of bolt and tube respectively.

Practice Q.2.8 A steel bolt of 16 mm diameter is passing through a copper tube of inner diameter 20 mm and outer diameter 25 mm, length 200 mm. Pitch of thread on bolt is 2.4 mm. Nut is tightened on bolt by a quarter of a turn. Determine stresses in bolt and tube. If $E_{steel} = 2E_{copper}$. What is the contraction in the length of the tube? E_s = 200 GPa.

Ans. [–217.4 N/mm², +191.1 N/mm², 0.434 mm]

Question 2.9 A steel bolt of 20 mm diameter passes co-axially through a brass tube of inner diameter 25 mm and outer diameter 30 mm. Nut is tightened on tube with the help of washers. Due to tightening, stress developed in steel bolt is 50 MPa. Now the temperature of the assembly is raised by 50°C. What are the resultant stresses developed in bolt and tube. If $E_s = 2 \times E_b$.

$E_s = 2E_b = 200$ kN/mm², $\alpha_s = 11 \times 10^{-6}$/°C, $\alpha_b = 23 \times 10^{-6}$ /°C

Solution: Steel bolt area of cross-section,

$$A_s = \frac{\pi}{4} \times 20^2 = 100\pi \text{ mm}^2$$

Brass tube, area of cross-section,

$$A_b = \frac{\pi}{4}\left(30^2 - 25^2\right) = 68.75\pi \text{ mm}^2$$

$$\sigma_s = \text{stress in bolt} = +50 \text{ MPa}$$

$$\sigma_b = \text{stress in brass tube} = -\sigma_s \times \frac{A_b}{A_s} \quad \text{...(i)}$$

$$= -50 \times \frac{100\pi}{68.75\pi} = -72.73 \text{ N/mm}^2$$

Rise in temperature

$$\Delta T = 50°\text{C}$$

$$\sigma_{sT} A_s = \sigma_{bT} A_b$$

$$\sigma_{sT} = 0.6875\, \sigma_{bT} \quad \text{...(ii)}$$

$$\frac{\sigma_{sT}}{E_s} + \frac{\sigma_{bT}}{E_b} = (\alpha_b - \alpha_s)\Delta T = (23 - 11) \times 10^{-6} \times 50$$

$$\frac{\sigma_{sT}}{200{,}000} + \frac{\sigma_{bT}}{100{,}000} = 600 \times 10^{-6}$$

$$\sigma_{sT} + 2\sigma_{bT} = 120 \quad \text{...(iii)}$$

$$0.6875\, \sigma_{bT} + 2\sigma_{bT} = 120$$

$$\sigma_{bT} = 44.65 \text{ (compressive)}$$

$$\sigma_{sT} = 30.70 \text{ (tensile)}$$

There is increase in temperature, extension in brass tube will be more than extension in steel bolt. Brass tube will try to extend more than the extension in steel bolt. As a result, tube is contracted further and bolt is extended further.

Resultant Stresses

σ_b, tube = −72.73 − 44.65 = −117.38 N/mm² (compressive)

σ_s, bolt = 50 + 30.70 = 80.7 N/mm² (tensile)

Question 2.10 Three vertical wires one of steel, two of copper are suspended in the same vertical plane from a horizontal rigid support as shown in fig. 2.9. All the three bars have same length, same area of cross-section. Load is increased and temperature is changed in such a way that the stress in each wire is increased by 30 N/mm². Find the change in temperature.

$E_s = 2E_{cu} = 200$ GPa, $\alpha_s = 11 \times 10^{-6}$/°C, $\alpha_{cu} = 18 \times 10^{-6}$/°C

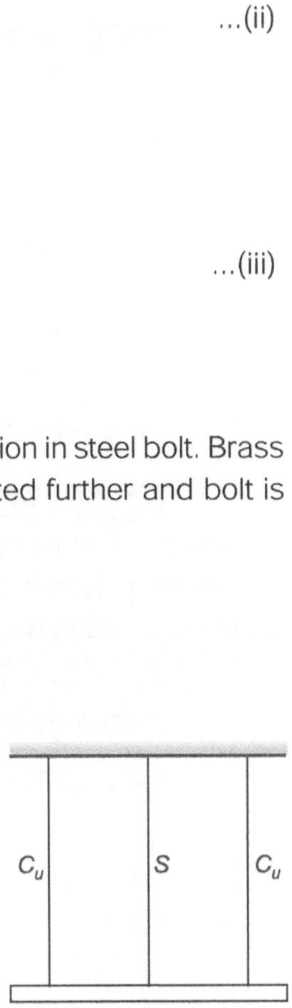

Fig. 2.9

Solution: Say A = Area of cross-section of each wire

In each wire, there is direct stress and a temperature stress.

So, $\sigma_S A + 2\sigma_{cu} A = 3 \times 30 \times A$

$$\sigma_s + 2\sigma_{cu} = 90 \quad \text{...(i)}$$

But, $\dfrac{E_s}{E_{cu}} = 2$

So, $\dfrac{\sigma_s}{\sigma_{cu}} = 2$

$$\sigma_s = 2\sigma_{cu}$$

$$2\sigma_{cu} + 2\sigma_{cu} = 90$$

$$\sigma_{cu} = 22.5 \text{ N/mm}^2$$

$$\sigma_s = 45 \text{ N/mm}^2$$

Say, due to temperature change, stresses in steel and copper wires are σ_{ST}, σ_{CT}

Now, $\sigma_{cu} + \sigma_{CT} = 30$, as given

$$\sigma_{cu} = 22.5 \text{ N/mm}^2$$

$$\sigma_{CT} = 30 - 22.5 = 7.5 \text{ N/mm}^2$$

$$\sigma_s + \sigma_{ST} = 30$$

But $\sigma_s = 45 \text{ N/mm}^2$

$$\sigma_{ST} = 30 - 45 = -15 \text{ N/mm}^2$$

Using the equation of comparability

$$\frac{\sigma_{ST}}{E_S} + \frac{\sigma_{CT}}{E_C} = (\alpha_c - \alpha_s)\Delta T = (18 - 11) \times 10^{-6} \times \Delta T$$

$$\frac{15}{200000} + \frac{7.5}{100000} = 7 \times 10^{-6} \times \Delta T$$

$$15 = 7 \times 100000 \times 10^{-6}\,\Delta T = 0.7\,\Delta T$$

$$\Delta T = \frac{15}{0.7} = 21.43°\text{C}$$

Reduction in temperature, because $\alpha_c > \alpha_s$, and tensile stress is induced in copper wire.

Question 2.11 A circular ring in horizontal plane is suspended by 3 vertical bars A, *B* and *C*. Upper ends of bars are held at different levels. Bar A, 2 m long, 15 mm diameter, made of steel, Bar *B*, 1.8 m long, 12 mm diameter, made of copper, bar C, 1.6 m long, 16 mm diameter, made of aluminium. *A* load of 40 kN is hung on the ring. Calculate how much load is shared by each bar.

$E_s = 2E_c = 3E_a = 210 \text{ kN/mm}^2$

Ring remains horizontal after the application of load.

Solution: Change in length in each bar is same

$$\delta l_s = \delta l_c = \delta l_a \quad \text{...(ii)}$$

Say, W_s, W_c, W_a are the loads shared by steel, copper and aluminium bars respectively.

Total load, $W = W_s + W_c + W_a$ ...(iii)

From eq. (i) $\delta l_s = \dfrac{W_s}{A_s} \times \dfrac{L_s}{E_s}$

$$\delta l_c = \frac{W_c}{A_c} \times \frac{L_c}{E_c}$$

$$\delta l_a = \frac{W_a}{A_a} \times \frac{L_a}{E_a}$$

$$A_s = \frac{\pi}{4} \times 15^2 = 176.7 \text{ mm}^2$$

$$A_c = \frac{\pi}{4} \times 12^2 = 113.1 \text{ mm}^2$$

$$A_a = \frac{\pi}{4} \times 16^2 = 201.06 \text{ mm}^2$$

Putting the values

$$\frac{W_s}{176.7} \times \frac{2000}{210000} = \frac{W_c}{113.1} \times \frac{1800}{105000}$$

$$W_s = 2.812 W_c \qquad \text{...(iii)}$$

$$\frac{W_a}{201.06} \times \frac{1600}{70000} = \frac{W_c \times 1800}{113.1 \times 105000}$$

$$W_a = 1.333\, W_c \qquad \text{...(iv)}$$

$$W = W_s + W_c + W_a$$

$$= 2.812\, W_c + W_c + 1.333\, W_c$$

$$W = 5.145\, W_c$$

$$W_c = \frac{40}{5.145} = 7.774 \text{ kN, load shared by copper bar}$$

$$W_s = 21.862 \text{ kN, load shared by steel bar}$$

$$W_a = 10.362 \text{ kN, load shared by aluminium bar}$$

Practice Q.2.11 A circular ring in horizontal plane is suspended by 3 vertical bars *A*, *B* and *C*. Bars are of same length but their materials and areas of cross-section are different.

Bar *A*, steel, $A_s = 150 \text{ mm}^2$

Bar *B*, copper, $A_c = 120 \text{ mm}^2$

Bar *C*, aluminium $A_a = 200 \text{ mm}^2$

Given: $E_s = 2E_c = 3E_a = 210 \text{ kN/mm}^2$

Determine the forces and stresses developed in each bar if a load of 20 kN is hung on circular ring and the ring remains horizontal.

Ans. [10.844, 4.819, 4.337 kN, 72.3, 36.14, 24.09 N/mm²]

Question 2.12 A rigid bar of negligible weight is supported by steel and copper bars as shown in figure 2.10. Bar carries a load of 50 kN at point C (figure 2.10). Determine the temperature change (ΔT) that will cause the stress in steel to be 50 N/mm². $\alpha_s = 11 \times 10^{-6}/°C$, $\alpha_{cu} = 18 \times 10^{-6}/°C$, $E_s = 2E_{cu} = 210 \text{ kN/mm}^2$.

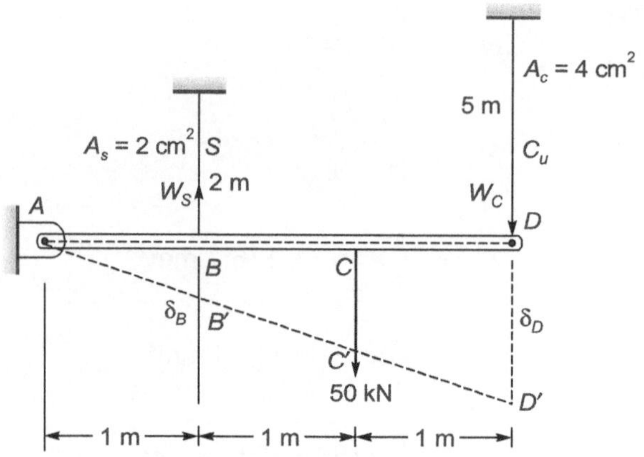

Fig. 2.10

Solution: Due to the effect of direct stress and stress due to temperature change, final stress in steel wire is 50 N/mm^2.

So load, $$W_s = 2 \times 100 \times 50 = 10000 \text{ N} = 10 \text{ kN} \quad \text{...(i)}$$

Say, $$W_c = \text{load in copper wire}$$

Taking moments of the forces about end A,

$$W_c \times 3 + W_s \times 1 = 50 \times 2$$

$$3W_c + 10 \times 1 = 100$$

$$W_c = \frac{100-10}{3} = 30 \text{ kN}$$

Stress in copper bar, $$\sigma_c = \frac{30000}{400} = 75 \text{ N/mm}^2 \quad \text{...(ii)}$$

From the deformation of bars, rigid bar is inclined to $AB'C'D'$ position

$$\delta_B = \frac{1}{3}\delta_D \quad \text{...(iii)}$$

Deflection in steel bar = 1/3 deflection in copper bar

Say $$\Delta T = \text{change in temperature}$$

$$\delta_D = \frac{75}{105000} \times 5000 + 18 \times 10^{-6} \times \Delta T \times 5000$$

$$= 3.571 + 0.09\Delta T$$

$$\delta_B = \frac{50}{210000} \times 2000 + 11 \times 10^{-6} \Delta T \times 2000$$

$$= 0.4762 + 0.022\,\Delta T \quad \text{...(iv)}$$

But $$\delta_B = \frac{1}{3}\delta_D$$

$$0.4762 + 0.022\,\Delta T = \frac{1}{3}[3.571 + 0.09\Delta T] = 1.190 + 0.03\,\Delta T$$

or $$-0.008\,\Delta T = 0.7138$$

$$\Delta T = 89.225°\text{C, reduction in temperature.}$$

Question 2.13 Two steel rods of 20 mm diameter are joined end to end by means of a turn buckle as shown in figure 2.11, other end of each rod is rigidly fixed. Initially, there is no tension in each rod. Effective length of each rod is 5 m. Threads on each rod has a pitch of 2.4 mm. Calculate the increase in tension when the turn buckle is tightened by one third of revolution. E = 208 GPa. If the temperature of the rods is increased, then at what temperature, tension in each rod is reduced to half of its magnitude. $\alpha_s = 11 \times 10^{-6}/°C$.

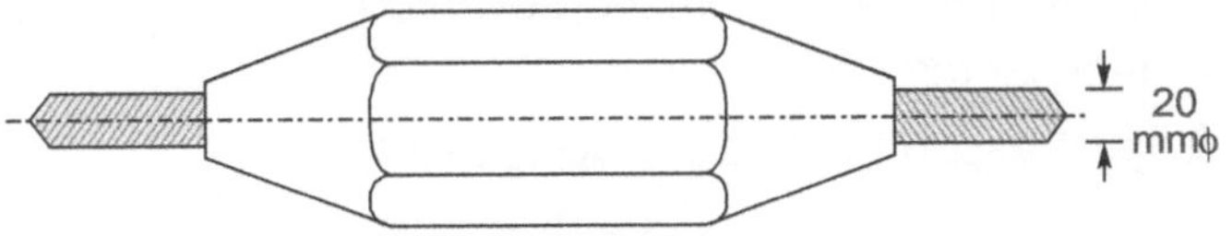

Fig. 2.11

Solution: A turn buckle is a type of a nut having internal threads on two ends but in opposite directions. Ends of tie rods are inserted from both the sides as shown. By rotating the turn buckle, both the tie rods are tightened.

Length of each rod = 5 m

Pitch of thread = 2.4 mm

$$\text{Axial movement of rods on both the sides} = \frac{2.4}{3} = 0.8 \text{ mm}$$

$$\text{Strain in each rod} = \frac{0.8}{5000} = 0.16 \times 10^{-3}$$

$$E = 208000 \text{ N/mm}^2$$

$$\sigma, \text{ stress in each rod} = 0.16 \times 10^{-3} \times 208 \times 1000 = 33.28 \text{ N/mm}^2$$

Area of cross-section of each rod,

$$A = \frac{\pi}{4} \times 20^2 = 100\pi \text{ mm}^2$$

Tension in each rod, $T = 100\pi \times 33.28 = 10455 \text{ N} = 10.455 \text{ kN}$

Tension in rod is to be reduced to half by increasing the temperature of the bar

$$\sigma', \text{ reduction} = \frac{33.28}{2} = 16.64 \text{ N/mm}^2$$

$$\sigma' = E \times \alpha \times \Delta T = 208000 \times 11 \times 10^{-6} \times \Delta T = 16.64$$

$$\Delta T = \frac{16.64}{2.288} = 7.27°C$$

Practice Q.2.13 Two steel rods of 40 mm diameter are joined end to end by means of a turn buckle. The other end of each rod is rigidly fixed at the other end and the rods are just taut. Effective length of each rod is 4 m. Calculate the increase in tension in rods, when the turn buckle is turned by one fourth of a turn. There are threads on each rod with a pitch of 4.2 mm. E = 208 GPa, $\alpha = 11.2 \times 10^{-6}/°C$. What rise in temperature could nullify the increased tension in each rod?

Ans. [68.6 kN, 23.44°C]

Question 2.14 A rigid bar AOB, figure 2.12, pinned at O is attached to two vertical bars of steel and brass as shown. Initially the rigid bar is horizontal and vertical rods are stress free. Determine the stresses developed in brass and steel bars if the temperature of the brass bar only is increased by 50°C. Neglect weight of the bars.

$L_B = 1$, $L_s = 0.7$ m, $A_S = 400$ mm², $A_B = 800$ mm², $\alpha_b = 22.1 \times 10^{-6}$/°C, $E_S = 2E_B = 200$ GPa.

Solution: Free expansion of brass bar is partially prevented, because it is connected to a steel bar through a rigid bar

Say σ_{BT} = compressive stress in brass bar due to length change

$$\delta l_B = 50 \times 22.1 \times 10^{-6} \times 1000 - \frac{\sigma_{BT}}{E_B} \times 1000$$

$$= 1.105 - \frac{\sigma_{BT}}{100,000} \times 1000 = 1.105 - 0.01\sigma_{BT} \quad ...(i)$$

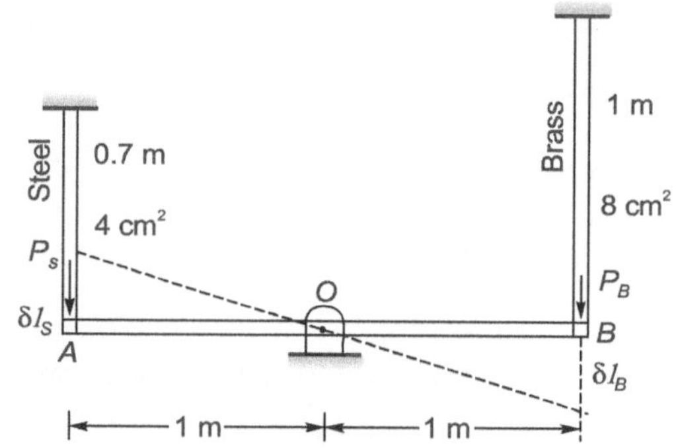

Fig. 2.12

Decrease in the length of steel bar,

$$\delta l_S = \delta l_B$$

as $AO = OB$, (See the deformation triangles)

Say σ_{ST} = compressive stress developed in steel bar

$$\delta l_S = \frac{\sigma_{ST}}{E_S} \times 700$$

$$= \frac{\sigma_{ST} \times 700}{200,000} = 3.5 \times 10^{-3} \sigma_{ST} \quad ...(ii)$$

Compressive force in brass bar = P_B

Compressive force in steel bar = P_S

Moment about O

$$P_B \times 1 = P_S \times 1$$

or $$P_B = P_S \quad ...(iii)$$

$$\sigma_{BT} \times 800 = \sigma_{ST} \times 400$$

or $$\sigma_{BT} = 0.5\sigma_{ST} \quad ...(iv)$$

Now $$\delta l_S = \delta l_B$$

$$1.105 - 0.01\,\sigma_{BT} = 3.5 \times 10^{-3}\,\sigma_{ST}$$

But $$\sigma_{BT} = 0.5\,\sigma_{ST}$$

Therefore $$1.105 - \frac{0.01}{2} \times \sigma_{ST} = 3.5 \times 10^{-3}\,\sigma_{ST}$$

$$1.105 = 8.5 \times 10^{-3}\,\sigma_{ST}$$

Stress $$\sigma_{ST} = 130 \text{ N/mm}^2 \text{ (compressive)}$$

$$\sigma_{BT} = 65 \text{ N/mm}^2 \text{ (compressive)}$$

Question 2.15 A steel tube 2.5 cm external diameter and 1.8 cm internal diameter encloses copper rod 1.6 cm diameter to which it is rigidly joined at each end. If, at a temperature of 20°C there is no longitudinal stress, calculate the stresses in the rod and tube when the temperature is raised to 210°C. Given E_s = 210 GPa, E_c = 100 GPa and $\alpha_s = 12 \times 10^{-6}$/°C, $\alpha_c = 20 \times 10^{-6}$/°C.

[ESE, ME : 2004, 4 Marks]

Solution: Copper rod,

$$A_{cu} = \frac{\pi}{4}\times 16^2 = 201.06 \text{ mm}^2$$

Steel tube,

$$A_s = \frac{\pi}{4}\left(18^2 - 16^2\right) = 53.40 \text{ mm}^2$$

$$\Delta T = \text{Increase in temperature}$$

$$= 210 - 20 = 190°\text{C}$$

$$\alpha_{cu} > \alpha_s$$

Compressive force in copper rod = Tensile force in steel tube, as $\alpha_s < \alpha_c$

$$P_{cu} = P_s$$

$$\sigma_{cu} \times 201.06 = \sigma_s \times 53.40$$

$$\sigma_s = 3.764\, \sigma_{cu}$$

$$\frac{\sigma_s}{E_s} + \frac{\sigma_{cu}}{E_{cu}} = (\alpha_c - \alpha_s) \times \Delta T$$

$$\frac{3.764\sigma_{cu}}{210\times 1000} + \frac{\sigma_{cu}}{100\times 1000} = (20-12)\times 10^{-6} \times 190$$

$$\sigma_{cu}\,[1.793 + 1] \times 10^{-5} = 8 \times 10^{-6} \times 190$$

$$\sigma_{cu} \times 2.793 = 152$$

$$\sigma_{cu} = -54.42 \text{ N/mm}^2 \qquad \text{(Compressive)}$$

$$\sigma_s = 3.764 \times 54.42$$

$$= +204.84 \text{ MPa} \qquad \text{(Tensile)}$$

Question 2.16 A steel bar is held by two fixed supports as shown in the figure 2.13 and is subjected to an increase of temperature ΔT = 100°C. If the coefficient of thermal expansion and Young's modulus of elasticity of steel are 11 × 10⁻⁶/°C and 200 GPa, respectively, the magnitude of thermal stress (in MPa) induced in the bar is ________

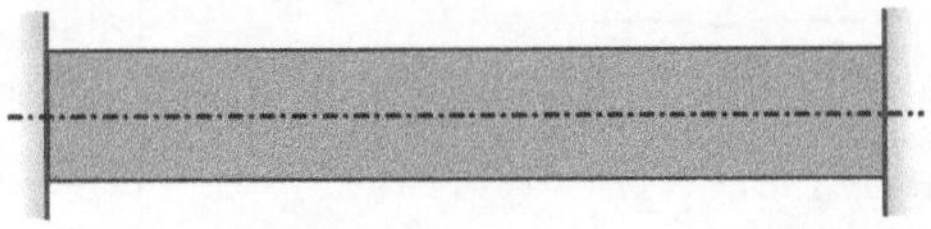

Fig. 2.12

Solution:

$$\Delta T = 100°\text{C}$$

$$\alpha = 11 \times 10^{-6}$$

$$E = 200 \times 10^3 \text{ N/mm}^2$$

$$\sigma, \text{ thermal stress} = E\alpha\Delta T$$

$$= 200 \times 10^{-3} \times 11 \times 10^{-6} \times 100 = 220 \text{ MPa}$$

Question 2.17 A circular metallic rod of length 250 mm is placed between two rigid immovable walls as shown in the figure 2.14. The rod is in perfect contact with the wall on the left side and there is a gap of 0.2 mm between the rod and the wall on the right side. If the temperature of the rod is increased by 200°C, the axial stress developed in the rod is ______ MPa.

Young's modulus of the material of the rod is 200 GPa and the coefficient of thermal expansion is 10^{-5} per °C.

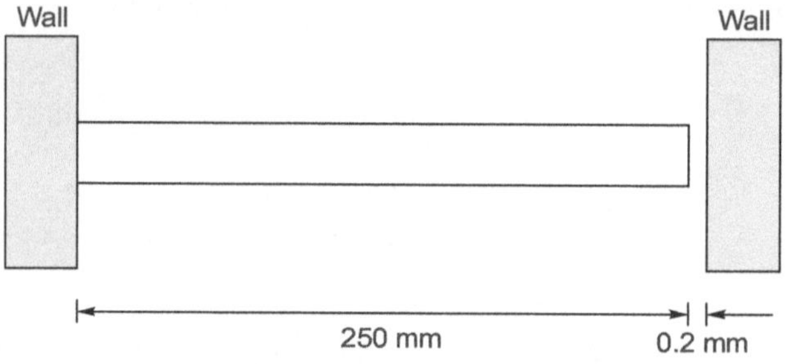

Fig. 2.12

Solution:

$$\text{Free expansion} = \alpha \Delta T L$$
$$= 10^{-5} \times 200 \times 250 = 0.5 \text{ mm}$$
$$\text{Gap} = 0.2 \text{ mm}$$
$$\text{Expansion prevented} = (0.5 - 0.2) = 0.3 \text{ mm}$$
$$\sigma = \frac{0.3}{250} \times 200 \times 1000 = 240 \text{ MPa}$$

Objective Type Questions

Q.1 For a composite bar consisting of a bar enclosed inside a tube of another material, is compressed under a load W as a whole through rigid plates at the end of the bar. The equation of compatability is given by (suffixes1 and 2 refer to bar and tube respectively)

(a) $W_1 + W_2 = W$ (b) $W_1 + W_2 =$ constant

(c) $\frac{W_1}{A_1E_1} = \frac{W_2}{A_2E_2}$ (d) $\frac{W_1}{A_1E_2} = \frac{W_2}{A_2E_1}$

Q.2 A composite bar is made by joining two flats of materials 1 and 2, with Young's modulus E_1 and E_2 respectively. It coefficient of thermal expansion $\alpha_1 > \alpha_2$ and $\alpha_1 = 1.5\ \alpha_2$, ratio

$\frac{E_1}{E_2} = 0.5$ but area of cross-section $A_1 = A_2$ and

stress developed in bar of material 1 is $+50$ N/mm². What is the stress developed in bar of material 2

(a) -100 N/mm² (b) -75 N/mm²

(c) -50 N/mm² (d) -33.33 N/mm²

Q.3 A copper tube of outer diameter 6 cm and inner diameter 4 cm, completely encloses a steel rod of 4 cm diameter. $E_s = 2E_{cu}$, $\alpha_c = 2\alpha_s = 22 \times 10^{-6}$/°C. What is the stress developed in steel rod if temperature of composite bar is raised by 65°C, $E_s = 2E_{cu} = 200$ GPa

(a) -55 N/mm² (b) 55 N/mm²

(c) -44 N/mm² (d) 44 N/mm²

Q.4 A steel rod of area of cross-section 4 cm² is completely encased in a brass tube of 6 cm² cross-section. $E_{brass} = 0.5\ E_{steel}$. A load P acts on the composite bar. If the stress developed in brass bar is 60 MPa, what will be the stress developed in steel bar?

(a) 30 MPa (b) 45 MPa

(c) 90 MPa (d) 120 MPa

Q.5 A circular ring is suspended by 3 bars of steel, copper and aluminium, each of same length and same diameter. Load on ring is 33 kN. If $E_s = 2E_c = 3\ E_a$, how much load is shared by steel bar if the circular ring is to remain horizontal under load

(a) 18 kN (b) 16.5 kN
(c) 12 kN (d) 9 kN

Q.6 A steel rail track is laid by laying rails of 30 m length with a clearance of 4 mm between the two rails. At 20°C, there is no stress in rails. At 50°C, what will be the stress if $\alpha = 11 \times 10^{-6}$/°C for steel and E = 200 GPa

(a) 66.0 N/mm² (b) 39.33 N/mm²
(c) 32.66 N/mm² (d) None of these

Q.7 A bimetallic strip is made of two metals with equal areas of cross-section, but $\alpha_1 = 12 \times 10^{-6}$/°C and $\alpha_2 = 18 \times 10^{-6}$/°C. Due to temperature change, if stress in strip 1 of α_1 is – 40 MPa, what is the stress developed in strip 2 (with α_2) and is there any increase or decrease in temperature is $+\Delta T$ or $-\Delta T$?

(a) 40 MPa, $+\Delta T$ (b) 40 MPa, $-\Delta T$
(c) 60 MPa, $+\Delta T$ (d) –60 MPa, $+\Delta T$

Q.8 A wire of 9 mm² area of cross-section is held between two rigid supports. Temperature of the wire drops by T°C and the wire breaks. If $E = 67000$ N/mm² and $\sigma_{ut} = 210$ N/mm², what is temperature, $\alpha = 23 \times 10^{-6}$/°C

(a) 210°C (b) 175.5°C
(c) 136.3°C (d) 105°C

Q.9 A circular steel rod of 20 cm² cross-sectional area and 10 m length is heated through 50°C with ends clamped before heating. Given, $E = 200$ GPa and coefficient of thermal expansion, $\alpha = 10 \times 10^{-6}$/°C, the thrust thereby generated on the clamp is

(a) 100 kN (b) 150 kN
(c) 200 kN (d) 250 kN

Q.10 A steel rod 10 m long is at a temperature of 20°C. The rod is heated to a temperature of 60°C. What is the stress induced in the rod if it is allowed to expand by 4 mm, when $E = 200$ GPa and $\alpha = 12 \times 10^{-6}$/°C?

(a) 6.4 MPa (b) 48 MPa
(c) 32 MPa (d) 16 MPa

Q.11 An elastic bar of length L, uniform cross-sectional area A, coefficient of thermal expansion α and Young's modulus E is fixed at the two ends. The temperature of the bar is increased by T, resulting in an axial stress σ. Keeping all other parameters unchanged, if the length of the bar is double the axial stress would be

(a) σ (b) 2σ
(c) 0.5σ (d) $0.25\alpha\sigma$

Answers

1. (c) 2. (c) 3. (b) 4. (d) 5. (a) 6. (b) 7. (b) 8. (c) 9. (c) 10. (d) 11. (a)

Explanations

1. (c)

$W_1 + W_2 = W$, equation of equilibrium

Compatibilities equation

Strain in bar = Strain in tube

$$\frac{W_1}{A_1} \times \frac{1}{E_1} = \varepsilon_1 = \frac{W_2}{A_2 E_2} = \varepsilon_2$$

2. (c)

$$\sigma_{1T} A_1 = \sigma_{2T} A_2$$

As, $A_1 = A_2$

$$\sigma_{1T} = +50 \text{ N/mm}^2$$

$$\sigma_{2T} = -50 \text{ N/mm}^2$$

3. (b)

$$A_s = 4\pi$$
$$A_{cuT} = 5\pi$$
$$\sigma_{sT} \times 4\pi = \sigma_{cuT} \times 5\pi$$
$$\sigma_{sT} = 1.25\sigma_{cuT}$$
$$\frac{\sigma_{sT}}{200 \times 1000} + \frac{\sigma_{cuT}}{100 \times 1000} = 11 \times 10^{-6} \times 65$$
$$\sigma_{sT} + 2\sigma_{cuT} = 200 \times 10^3 \times 11 \times 10^{-6} \times 65$$
$$1.25\sigma_{cuT} + 2\sigma_{cuT} = 143$$
$$3.25\sigma_{cuT} = 143$$
$$\sigma_{cuT} = 44 \text{ MPa (compressive)}$$

Now, $\sigma_{sT} = 44 \times 1.25 = 55$ MPa (tensile)

4. (d)

$$\frac{\sigma_s}{\sigma_b} = \frac{E_s}{E_b} = 2$$
$$\sigma_s = 2\sigma_b = 2 \times 60 = 120 \text{ MPa}$$

5. (a)

$$W_s = 3W_a$$
$$W_s = 2W_c$$
$$W_a = \frac{1}{3}W_s$$
$$W_c = \frac{1}{2}W_s$$
$$W = W_s + W_a + W_c$$
$$= N_s + \frac{1}{3}W_s + \frac{1}{2}W_s$$
$$= \frac{11}{6}W_s = 33 \text{ kN}$$
$$W_s = 18 \text{ kN}$$

6. (b)

For $\delta L = 30 \times 11 \times 10^{-6} \times 1000 \times 30$

$= 9.9$ m

Free expension for

$= 9.9 - 4 = 5.9$ mm $= 5.9$

$$\sigma = \frac{59}{30{,}000} \times 200{,}000$$
$$= 39.33 \text{ N/mm}^2$$

7. (b)

$$\alpha_1 < \alpha_2$$

Compressive stress in metal 1

so reduction temperature, $-\Delta T$

$$A_1 = A_2$$
$$\sigma_1 = -\sigma_2$$
$$\sigma_1 = -40$$
$$\sigma_2 = 40 \text{ MPa}$$

8. (c)

$$210 = 23 \times 10^{-6} \times T \times 67000$$
$$= 1.541\,T$$
$$T = 136.27°\text{C}$$

9. (c)

Stress, $\sigma = \alpha \Delta T E$

$= 10 \times 10^{-6} \times 50 \times 200 \times 100$

$= 100$ N/mm^2

Area $= 20$ cm^2 $= 200$ mm^2

Thrust $= 100 \times 2000$

$= 200$ kN

10. (d)

Prevented change,

$$\delta l = 10 \times 1000 \times 40 \times 12 \times 10^{-6} - 4$$
$$= 4.8 - 0.8 \text{ mm}$$
$$= \frac{\sigma}{E} \times L$$
$$0.8 = \frac{\sigma \times 10000}{200 \times 100} = \frac{\sigma}{20}$$
$$\sigma = 16 \text{ MPa}$$

11. (a)

$$\sigma = E\alpha T$$

σ in independent of length L

If L is doubled σ remains the same

■■■■

Principal Stresses and Strains

CHAPTER 3

Determination of principal stresses at the critical section of any structure or any engineering component is the most important step in experimental stress analysis of that structure or engineering component. In this chapter we will learn about (i) analytical determination of principal stresses (ii) graphical method of determining principal stresses (iii) determination of principal strains.

Question 3.1 What are principal stresses? What is the importance of principal stresses?

Solution: At any point in a stressed body, **there exists a set of 3 planes perpendicular to each other, such that on these 3 planes,** there are only normal stresses say p_1, p_2, p_3 and no shear stress exists on these planes. These three normal stresses are called **principal stresses** and three planes on which principal stresses act are called **principal planes**. One of these principal stresses, **is the maximum most normal stress** at the point. Determination of this maximum principal stress is important for designing any structure or any engineering component.

Fig. 3.1 shows a body subjected to external forces, F_1, F_2, F_3, F_4 etc. Say *P* is the critical point, through this point **innumerable number of planes** such as *aa*, *bb* can pass, but there is a set of 3 orthogonal planes on which only the normal stresses act and shear stress is absent on these planes, then these planes are called principal planes and normal stresses on these planes are called principal stresses.

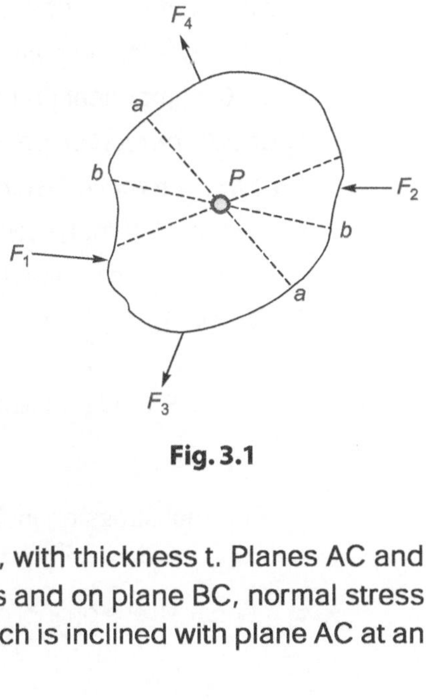

Fig. 3.1

Question 3.2 Figure 3.2 shows a triangular solid block ABC, with thickness t. Planes AC and BC are perpendicular to each other. On plane AC, normal stress σ_1 acts and on plane BC, normal stress σ_2 acts. Determine normal and shear stresses on inclined plane AB, which is inclined with plane AC at an angle θ. Plane AC can be taken as a reference plane

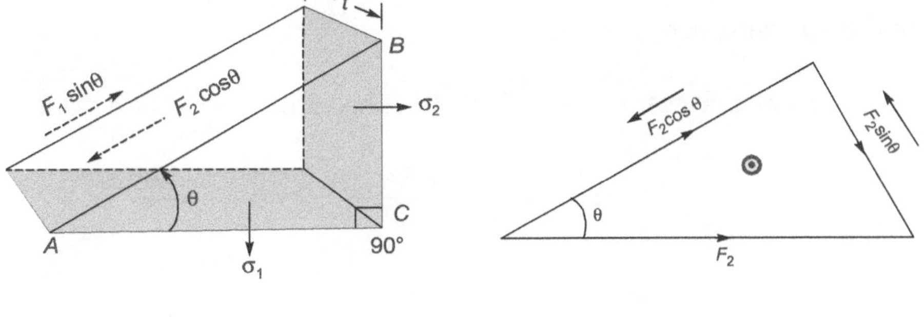

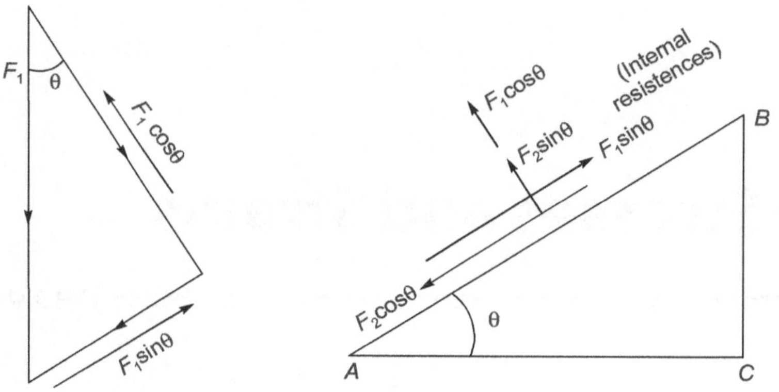

Fig. 3.2

Solution: On planes *AC* and *BC*, there are only normal stresses and there is no shear stress on these planes. These planes are perpendicular to each other. **These two planes are called principal planes.** Third principal plane is the plane of the paper.

F_1, normal force on plane $AC = \sigma_1 \times AC \times t$, where t = thickness of block

$F_1 \cos\theta$, component normal to inclined plane *AB*

$F_1 \sin\theta$, component parallel to the inclined plane *AB*

F_2, normal force on plane $BC = \sigma_2 \times BC \times t$

$F_2 \sin\theta$, component normal to the inclined plane *AB*

$F_2 \cos\theta$, component parallel to the inclined plane *AB*

Note that internal resistance is equal and opposite to the applied force.

F_n = normal internal resistance on plane *AB*

$F_1 \cos\theta + F_2 \sin\theta$, direction away from the plane, i.e., normal tensile force

$F_t = F_1 \sin\theta - F_2 \cos\theta$, shear force

Note that $F_1 \sin\theta$ tends to rotate the body **in the clockwise direction**. It is a positive shear force, Component $F_2 \cos\theta$ tends to rotate the body in the anticlockwise direction, so it is a negative shear force. (about point ⊙)

$$\sigma_\theta = \text{normal stress on inclined plane} = \frac{F_n}{AB \times t} = \frac{F_1 \cos\theta + F_2 \sin\theta}{AB \times t}$$

$$\tau_\theta = \text{shear stress on inclined plane} = \frac{F_t}{AB \times t} = \frac{F_1 \sin\theta - F_2 \cos\theta}{AB \times t}$$

Putting the values of F_1, F_2 and taking $\frac{BC}{AB} = \sin\theta$, $\frac{AC}{AB} = \cos\theta$, area of inclined plane = $AB \times t$

$\sigma_\theta = \sigma_1 \cos^2\theta + \sigma_2 \sin^2\theta$, normal stress.

$\tau_\theta = \sigma_1 \sin\theta \cos\theta - \sigma_2 \sin\theta \cos\theta = \frac{\sigma_1 - \sigma_2}{2} \sin 2\theta$, shear stress on inclined plane.

Now, $\cos^2\theta = \frac{1 + \cos 2\theta}{2}$, $\sin^2\theta = \frac{1 - \cos 2\theta}{2}$, putting these values

$$\sigma_\theta = \sigma_1\left(\frac{1+\cos 2\theta}{2}\right)+\sigma_2\left(\frac{1-\cos 2\theta}{2}\right)$$

$$= \frac{\sigma_1+\sigma_2}{2}+\frac{\sigma_1-\sigma_2}{2}\cos 2\theta \text{, normal stress on inclined plane}$$

Practice Q.3.2 Fig. 3.3 shows normal stresses on two perpendicular planes *AC* and *BC*. What are the normal and shear stresses on inclined plane *AB*, if θ = 15°

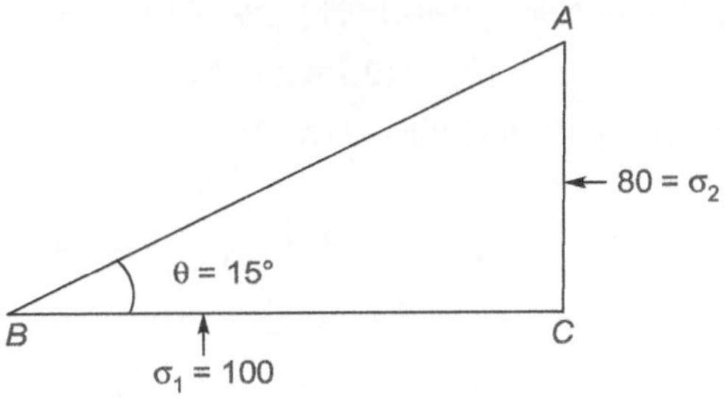

Fig. 3.3

Ans. [σ_θ = –98.66 MPa, τ_θ = –5 MPa]

Question 3.3 Fig. 3.4 shows a triangular block of thickness *t*.

On plane BC (reference plane)

σ_1 = normal tensile stress

τ = shear stress (negative) as it tends to rotate the body in anticlockwise direction

On plane *AC*

σ_2 = normal tensile stress

τ = shear stress (positive) as it tends to rotate the body in clockwise direction

Determine normal and shear stresses on the inclined plane *AB*.

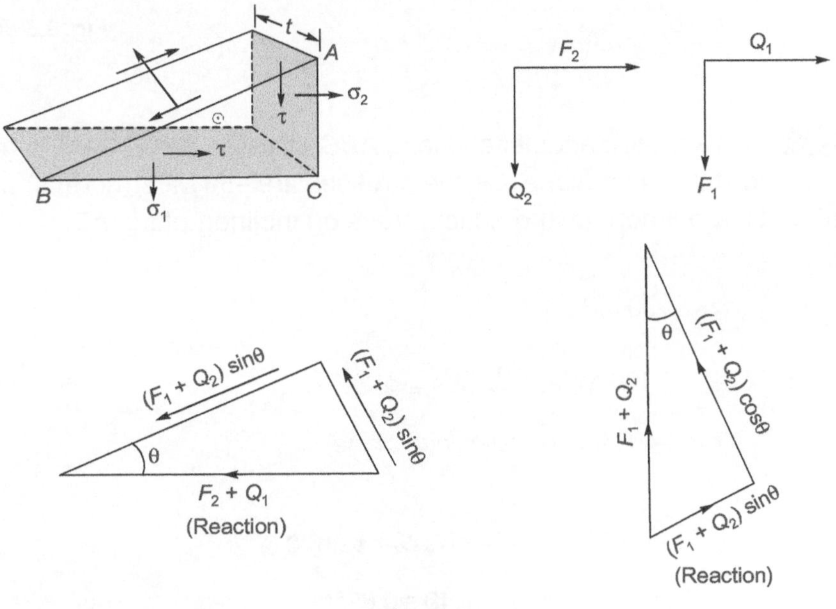

Fig. 3.4

Solution:

Normal force on plane *BC*, $F_1 = BC \times t \times \sigma_1$

Shear force on plane *BC*, $Q_1 = BC \times t \times \tau$

Normal force on plane *AB*, $F_2 = AC \times t \times \sigma_2$

Shear force on plane *AC*, $Q_2 = AC \times t \times \tau$

Component of these forces, perpendicular to plane *AB*

Reaction, $P_n = (F_1 + Q_2)\cos\theta + (F_2 + Q_1)\sin\theta = \sigma_\theta\, AB \times t$ ...(i)

Reaction of components of these force, parallel to the plane *AB*

$$F_t = (F_1 + Q_2)\sin\theta - (F_2 + Q_1)\cos\theta = \tau_\theta\, AB \times t \quad \text{...(ii)}$$

Putting the values of forces in eq. (i) and (ii) and taking

$$\cos\theta = \frac{BC}{AC},\ \sin\theta = \frac{AC}{BC},\ \cos^2\theta = \frac{1+\cos 2\theta}{2},\ \sin^2\theta = 1-\frac{\cos 2\theta}{2}$$

Expression for normal and shear stresses on plane *AB* are

$$\sigma_\theta = \frac{\sigma_1+\sigma_2}{2} + \frac{\sigma_1-\sigma_2}{2}\cos 2\theta + \tau\sin 2\theta \quad \text{...(iii)}$$

$$\tau_\theta = \frac{\sigma_1+\sigma_2}{2}\sin 2\theta - \tau\cos 2\theta \quad \text{...(iv)}$$

Please note that we have taken **negative shear stress on reference plane** *BC*. Angle θ of plane *AB* is taken with reference to this plane.

If **we take shear stress on reference plane as positive**, as shown in figure 3.5. The expressions for normal and shear stresses on inclined plane will be

$$\sigma_\theta = \frac{\sigma_1+\sigma_2}{2} + \frac{\sigma_1-\sigma_2}{2}\cos 2\theta - \tau\sin 2\theta \quad \text{...(v)}$$

$$\tau_\theta = \frac{\sigma_1-\sigma_2}{2}\sin 2\theta + \tau\cos 2\theta \quad \text{...(vi)}$$

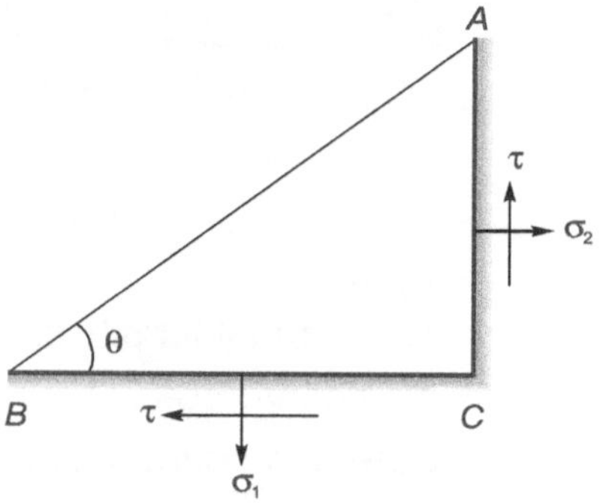

Fig. 3.5 *BC as reference plane*

Question 3.4 **On a small triangular element, *ABC*, stresses are +100 MPa (normal) and –40 MPa (shear) stresses on plane *BC* and on plane *AB*, the stresses are –50 MPa (normal) and +40 MPa (shear) as shown in figure 3.6. Find out normal and shear stress on inclined plane AB, θ = 30°.**

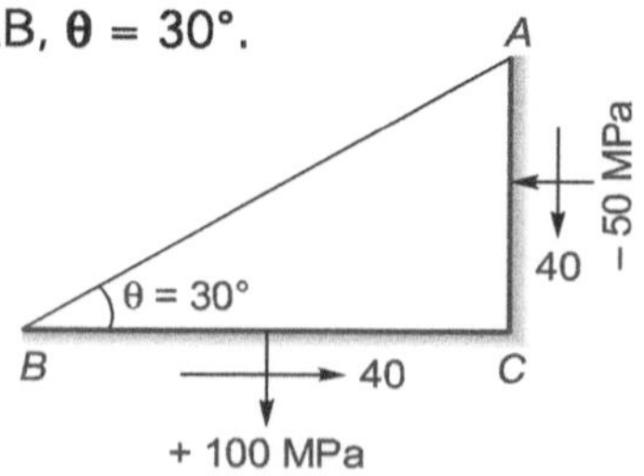

Fig. 3.6

Solution:

$$\sigma_2 = -50 \text{ MPa}$$

$$\sigma_1 = +100 \text{ MPa},\quad \frac{\sigma_1+\sigma_2}{2} = 25,\quad \frac{\sigma_1-\sigma_2}{2} = 75$$

$$\tau = -40 \text{ MPa (on reference plane)}$$

On inclined plane

$$\sigma_\theta = \frac{\sigma_1+\sigma_2}{2} + \frac{\sigma_1-\sigma_2}{2}\cos 2\theta + \tau\sin 2\theta$$

$$= 25 + 75\cos 60^\circ + 40\sin 60^\circ$$

$$= 25 + 75 \times 0.5 + 40 \times 0.866 = 97.14 \text{ N/mm}^2$$

$$\tau_\theta = \frac{\sigma_1 - \sigma_2}{2}\sin 2\theta - \tau\cos 2\theta$$

$$= 75 \sin 60° - 40 \cos 60°$$

$$= 75 \times 0.866 - 40 \times 0.50$$

$$= 64.95 - 20 = 44.95 \text{ MPa}$$

Practice Q.3.4 On a small triangular element ABC, stresses on two perpendicular planes are on plane BC (as reference plane)

$$\sigma_1 = +120 \text{ MPa}, \tau = +30 \text{ MPa}$$

on plane AC

$$\sigma_2 = +40 \text{ MPa}, \tau = -30 \text{ MPa}$$

Determine normal and shear stresses on inclined plane AB, $\theta = 20°$ (fig. 3.7)

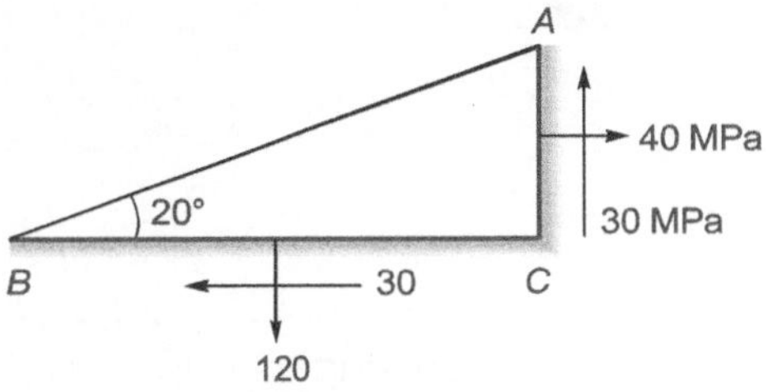

Ans: [+91.356 MPa, +48.69 MPa]

Fig. 3.7

Question 3.5 Derive expression for principal stresses and principal angles with respect to reference plane BC as given in fig. 3.8.

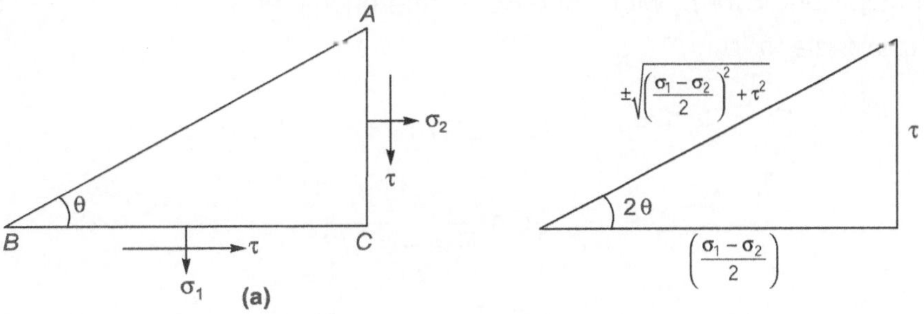

Fig. 3.8

Solution: We know that shear stress on principal plane is zero,
Therefore on principal plane

$$\tau_\theta = \frac{\sigma_1 - \sigma_2}{2}\sin 2\theta - \tau\cos 2\theta = 0$$

or

$$\tan 2\theta = \frac{\tau}{(\sigma_1 - \sigma_2)/2}$$

Triangle shown in fig. 3.8 for values of angle 2θ for which shear stress is zero.

$$\sin 2\theta = \pm\frac{\tau}{\sqrt{\left(\frac{\sigma_1 - \sigma_2}{2}\right)^2 + \tau^2}}$$

$$\cos 2\theta = \pm\frac{(\sigma_1 - \sigma_2)/2}{\sqrt{\left(\frac{\sigma_1 - \sigma_2}{2}\right)^2 + \tau^2}}$$

Putting these values in σ_θ, we will get values of principal stresses

$$p_1 = \frac{\sigma_1+\sigma_2}{2} \pm \frac{[(\sigma_1-\sigma_2)/2]^2}{\sqrt{\left(\frac{\sigma_1-\sigma_2}{2}\right)^2+\tau^2}} \pm \frac{\tau^2}{\sqrt{\left(\frac{\sigma_1-\sigma_2}{2}\right)^2+\tau^2}}$$

Principal stresses

or $$p_1 = \frac{\sigma_1+\sigma_2}{2} + \sqrt{\left(\frac{\sigma_1-\sigma_2}{2}\right)^2+\tau^2}, \quad \text{(taking positive values)}$$

$$p_2 = \frac{\sigma_1+\sigma_2}{2} - \sqrt{\left(\frac{\sigma_1-\sigma_2}{2}\right)^2+\tau^2}, \quad \text{(taking negative values)}$$

$$p_3 = 0$$

Principal angles

For shear stress to be zero on inclined plane

$$\tan 2\theta = \frac{\tau}{(\sigma_1-\sigma_2)/2}$$

say θ_1 = principal angle for p_1 with plane *BC* as reference plane

θ_2 = principal angle for p_2

$$\tan(2\theta_1) = \frac{\tau}{(\sigma_1-\sigma_2)/2}$$

$$\theta_1 = \frac{1}{2}\tan^{-1}\frac{\tau}{(\sigma_1-\sigma_2)/2}$$

$$\tan(2\theta_1 + 180) = \frac{\tau}{(\sigma_1-\sigma_2)/2} = \tan 2\theta_2$$

or $$\theta_2 = \theta_1 + 90$$

(Note that shear stress on reference plane is negative), if shear stress on principal plane is positive.

$$\theta_1 = \frac{1}{2}\tan^{-1}\frac{-\tau}{(\sigma_1-\sigma_2)/2}$$

$$\theta_2 = \theta_1 + 90°$$

i.e., one principal plane is perpendicular to the other principal plane.

Question 3.6 Stresses on two perpendicular planes are as shown in figure 3.9. Determine principal stresses and principal planes with respect to reference plane *BC*.

Solution:

$$\frac{\sigma_1+\sigma_2}{2} = \frac{160+40}{2} = 100 \text{ MPa}$$

$$\frac{\sigma_1-\sigma_2}{2} = \frac{160-40}{2} = 60 \text{ MPa}$$

$$\tau = \pm 30 \text{ MPa}$$

Principal Stresses

$$p_1 = 100 + \sqrt{60^2 + 30^2}$$
$$= 100 + 67.08 = 167.08 \text{ MPa}$$
$$p_2 = 100 - \sqrt{60^2 + 30^2}$$
$$= 100 - 67.08 = 32.92 \text{ MPa}$$

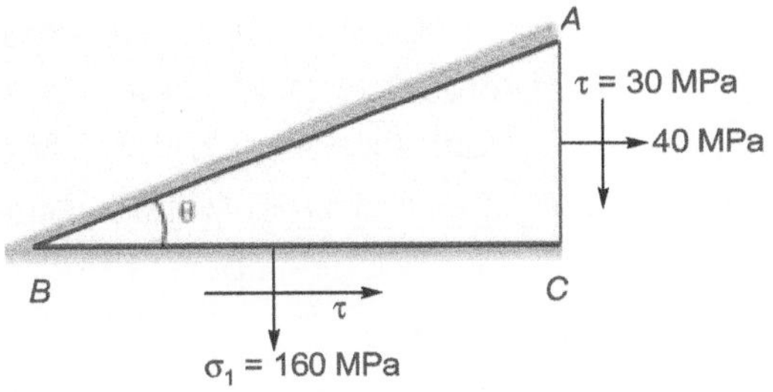

Fig. 3.9

Principal angles

$$\theta_1 = \frac{1}{2}\tan^{-1}\frac{30}{60} = \frac{1}{2} \times 26.56° = 13.28°$$
$$\theta_2 = 90 + 13.28 = 103.28°$$

Practice Q.3.6 Fig. 3.10 shows state of stress on two perpendicular planes *BC* and *AC* of a stressed body. Determine principal stresses and principal angles with respect to the plane BC as reference plane.

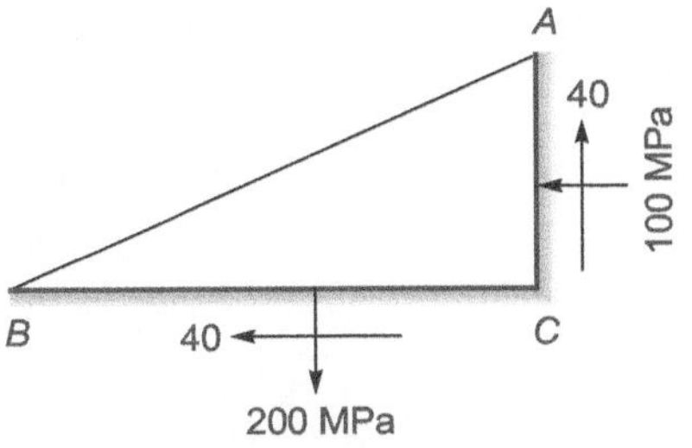

Fig. 3.10

Hints: [Note that shear stress on reference plane is positive]

Ans. [205.24 MPa, – 105.24 MPa, –7.465°, +82.535°]

Question 3.7 Explain the graphical method to determine principal stresses, principal angles and normal and shear stresses on any inclined plane with respect to a given reference plane.

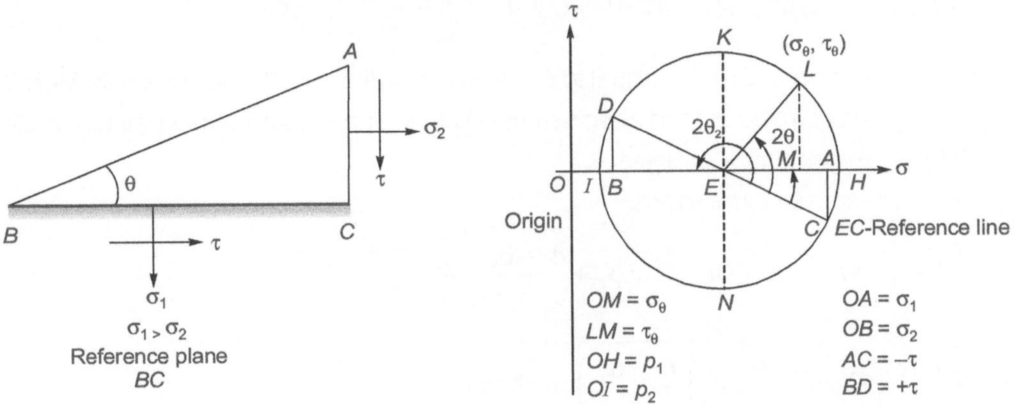

Fig. 3.11

Solution: With the help of a Mohr's Stress Circle, following informations can be easily obtained:

(a) Principal stresses–their magnitude and direction of principal planes.

(b) Normal and shear stresses on any inclined plane

(c) Maximum shear stresses

(d) Direction of planes of maximum shear stresses.

To draw the Mohr's Stress Circle, following guidelines should be taken:

1. Take normal stress along abscissa, tensile normal stress towards right from the origin, compressive normal stress towards left from the origin.
2. Take shear stresses along the ordinate, positive shear stress in the upward direction from the origin and negative shear stress in the downward direction from the origin.

3. Centre of the Mohr's Stress Circle always lies on abscissa.
4. From the centre of the Mohr's Stress Circle upto the point (which gives the state of stress on reference plane) is the reference line for angles for principal stresses, stresses on inclined plane.
5. Angle between principal stresses is 90°, but on Mohr's stress circle, it is represented by 180°, as both the principal stresses are to be represented along abscissa.

Procedure:

1. Draw $\sigma - \tau$, coordinates as shown in fig. 3.11
2. For the given problem, say $\sigma_1 > \sigma_2$

   σ_1 on reference plane, $-\tau$ on reference plane *BC*

   Take $OA = \sigma_1$, $AC = -\tau$

   Point $C(\sigma_1 - \tau)$ gives the state of stress on reference plane *BC*.
3. Take $OB = \sigma_2$, $BD = +\tau$(on plane *AC*)
4. $AC = BD$ (numerically)
5. Join *D* and *C* points, cutting the abscissa at the point *E*.
6. Point *E* is the centre of the Mohr's stress circle.
7. Radius of the circle *ED* or *EC*. Angle between *CED* = 180°, 2 × 90°, as planes *BC* and *AC* are perpendicular to each other.
8. Radius of the circle, $R = \sqrt{(EA)^2 + (AC)^2} = \sqrt{\left(\frac{\sigma_1 - \sigma_2}{2}\right)^2 + \tau^2}$
9. With *E* as centre and radius equal to *EC* or *ED* draw a circle, **this is known as Mohr's stress circle.**
10. Mohr's stress circle cuts the abscissa at points H and I, as on these points, shear stress is zero.
11. $OH = p_1$, major principal stress

    $OI = p_2$ minor principal stress
12. $OH = OE + R = OB + BE + R = \sigma_2 + \frac{\sigma_1 - \sigma_2}{2} + R$

    $$\frac{\sigma_1 + \sigma_2}{2} + R = \frac{\sigma_1 + \sigma_2}{2} + \sqrt{\left(\frac{\sigma_1 - \sigma_2}{2}\right)^2 + \tau^2}$$
13. $OI = OE - R = \frac{\sigma_1 + \sigma_2}{2} - \sqrt{\left(\frac{\sigma_1 - \sigma_2}{2}\right)^2 + \tau^2}$
14. Angle between *HEI* = 180° = 2 × 90°, where as angle between two planes is 90°.
15. $\angle CEH = 2\theta_1$, where θ_1 is the angle of principal plane of principal stress p_1, with respect to reference plane BC
16. $\angle CEI = 2\theta_2 = 2(\theta_1 + 90°)$, where θ_2 is the angle between reference plane and principal plane of stress p_2
17. $\angle CEL = 2\theta$, where θ is the angle of inclined plane AB with respect to reference plane *BC*
18. Coordinates of point $C(\sigma_\theta, \tau_\theta)$ are

    $OM = \sigma_\theta$ (normal stress), $LM = \tau_\theta$ (shear stress)

19. Points K and N give maximum shear stress at the point i.e., τ_{max} = EK = EN = Radius of circle

$$\text{or } \tau_{max} = \pm\sqrt{\left(\frac{\sigma_1 - \sigma_2}{2}\right)^2 + \tau^2}$$

Question 3.8 State of stress at a point in a stressed body is shown in fig. 3.12. Draw Mohr's stress circle, determine principal stresses, principal angles, and stresses on inclined plane. θ = 30°.

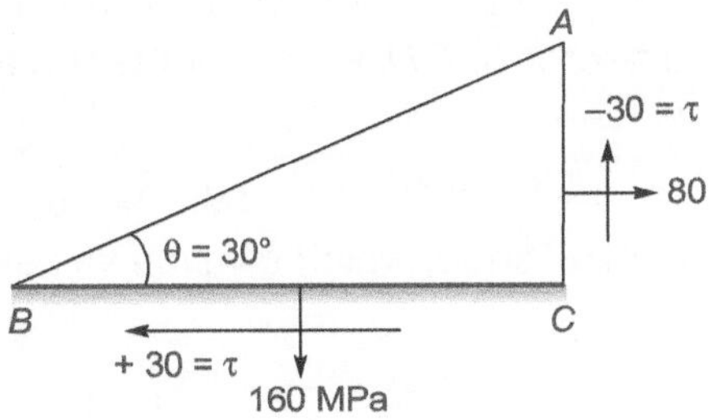

Fig. 3.12

Solution:

1. Draw $\sigma - \tau$ coordinates i.e., normal and shear stress coordinates.

   To some suitable scale, take OA = 160 MPa, AC = +30 MPa

   OB = 80 MPa, DB = −30 MPa

2. Join D and C, line CD cuts the abscissa at E, which is centre of Mohr's circle.

3. With E as centre, radius equal to EC = ED, draw a circle, which cuts the abscissa at H and I

   $OH = p_1$ = 170 MPa (to same scale as taken earlier)

   $OI = p_2$ = 70 MPa

4. Radius EC = ED = 50 MPa

   Note that EC is the reference line.

5. $\angle CEH = 2\theta_1 = -37°$; $\theta_1 = -18.5°$

6. $\angle CEI = +143°$, $\theta_2 = +71.5°$

7. Take $\angle CEL = 2\theta = 2 \times 30° = 60°$

8. Coordinates of L i.e., OM = 114 MPa, ML = +49.5 MPa. are normal and shear stresses on inclined plane

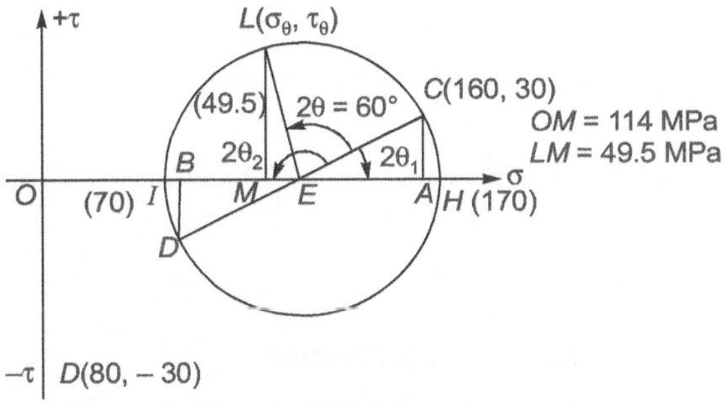

Fig. 3.13

Practice Q.3.8 State of stress at a point of a stressed body is shown in figure 3.14. Draw Mohr's stress circle and determine, principal stresses, principal angles with respect to reference plane BC. What are the normal and shear stresses on inclined plane.

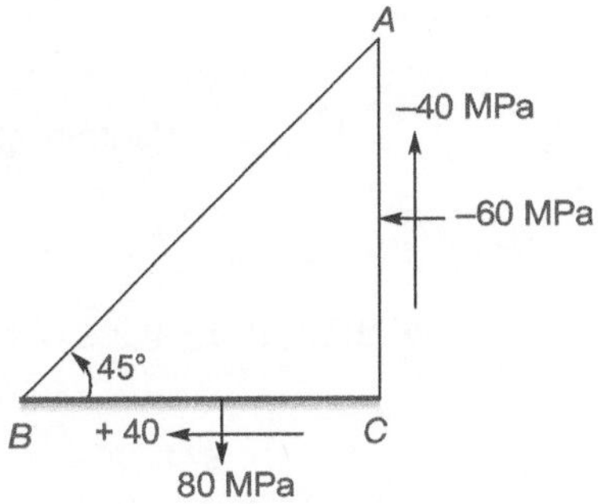

Fig. 3.14

Ans. [91 MPa, −71 MPa, −15°, +75°, −30 MPa, +70 MPa]

Question 3.9 Stresses on two planes OX and OX' are shown in figure 3.15. Determine analytically or graphically the angle between the two planes. What are principal stresses at the point.

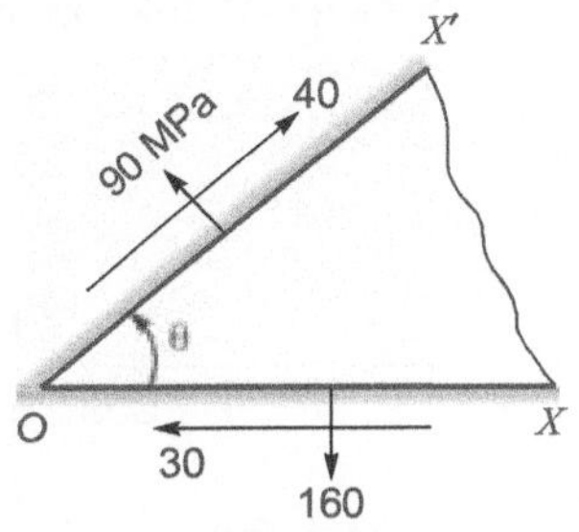

Fig. 3.15

Solution: Draw $\sigma - \tau$ axes, as shown in figure 3.16. To some suitable scale take OA = 160 MPa, AC = +30 MPa. Note that shear stress on both the planes i.e., OX, OX' is positive. Then take OB + 90 MPa.

BD = +40 (MPa). These two points i.e., C and D lie on Mohr's stress circle. **Draw perpendicular bisector of CD, which meets the abscissa at E.** This point E is the centre of Mohr's stress circle. Join CE and ED.

$\angle CED = 2\theta = 90°$, $\theta = 45°$, Angle between the planes is 45°. Mohr's stress circle cuts the abscissa at point H and I, which give the values of principal stresses, p_1 = 170 MPa = OH, p_2 = OI = 70 MPa.

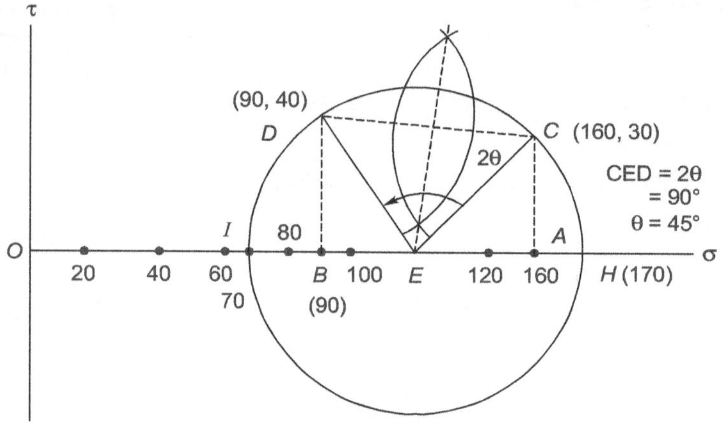

Fig. 3.16

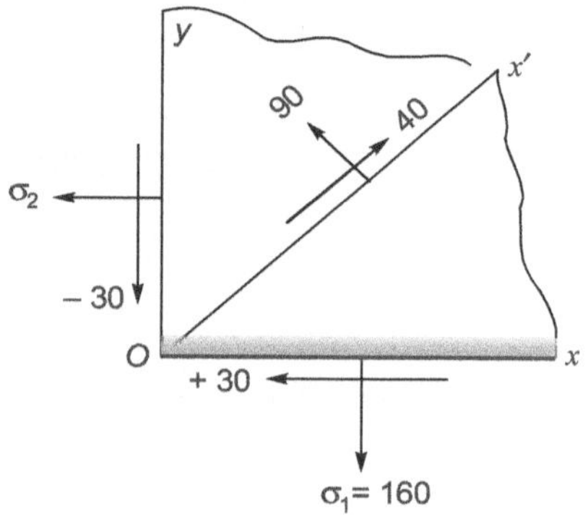

Fig. 3.17

Analysitcal solution

Say OX is the reference plane. On plane OY, perpendicular to OX, shear stress is $-\tau$, on –30 MPa.

On inclined plane

$$\sigma_\theta = \frac{\sigma_1 + \sigma_2}{2} + \frac{\sigma_1 - \sigma_2}{2}\cos 2\theta - \tau \sin 2\theta$$

(Note that shear stress on reference plane is positive)

$$\tau_\theta = \frac{\sigma_1 - \sigma_2}{2}\sin 2\theta + \tau \cos 2\theta$$

Putting the values of σ_1 and τ_1,

$$90 = \frac{160 + \sigma_2}{2} + \frac{160 - \sigma_2}{2}\cos 2\theta - 30 \sin 2\theta \quad \text{...(i)}$$

$$40 = \frac{160 - \sigma_2}{2}\sin 2\theta + 30 \cos 2\theta \quad \text{...(ii)}$$

There are 2 unknowns. i.e., θ and σ_2.

$$90 = 80 + 0.5\,\sigma_2 + 80 \cos 2\theta - 0.5\,\sigma_2 \cos 2\theta - 30 \sin 2\theta \quad \text{...(i)}$$

$$40 = 80 \sin 2\theta - 0.5\,\sigma_2 \sin 2\theta + 30 \cos 2\theta \quad \text{...(ii)}$$

or

$$10 = 0.5\,\sigma_2 (1 - \cos 2\theta) + 80 \cos 2\theta - 30 \sin 2\theta$$

$$40 = 80 \sin 2\theta + 30 \cos 2\theta - 0.5\,\sigma_2 \sin 2\theta$$

$$1 - \cos 2\theta = \sin^2 \theta + \cos^2 \theta - (\cos^2 \theta - \sin^2 \theta) = 2 \sin^2\theta$$

$$10 = 0.5\,\sigma_2 \times 2 \sin^2 \theta + 80 \cos 2\theta - 30 \sin 2\theta$$

$$40 = 80 \sin 2\theta + 30 \cos 2\theta - 0.5\,\sigma_2 \sin 2\theta$$

$$10 = \sigma_2 \sin^2\theta + 80\cos 2\theta - 30\sin 2\theta$$

or $$\sigma_2 \sin^2\theta = 10 - 80\cos 2\theta + 30\sin 2\theta \quad \text{...(i)}$$

$$0.5\,\sigma_2 \sin 2\theta = 80\sin 2\theta + 30\cos 2\theta - 40 \quad \text{...(ii)}$$

or $$\frac{\sigma_2 \sin^2\theta}{\sigma_2 \times \sin\theta\cos\theta} = \frac{10 - 80\cos 2\theta + 30\sin 2\theta}{80\sin 2\theta + 30\cos 2\theta - 40}$$

$$\frac{\sin\theta}{\cos\theta} = \frac{10 - 80\cos 2\theta + 30\sin 2\theta}{80\sin 2\theta + 30\cos 2\theta - 40}$$

or $10\cos\theta - 80\cos 2\theta\cos\theta + 30\sin 2\theta\cos\theta = 80\sin 2\theta\sin\theta + 30\cos 2\theta\sin\theta - 40\sin\theta$

$10\cos\theta + 40\sin\theta - 80\cos 2\theta\cos\theta - 80\sin 2\theta\sin\theta + 30\sin 2\theta\cos\theta - 30\cos 2\theta\sin\theta = 0$

$10\cos\theta + 40\sin\theta - 80(\cos 2\theta\cos\theta + \sin 2\theta\sin\theta) + 30(\sin 2\theta\cos\theta - \cos 2\theta\sin\theta) = 0$

$10\cos\theta + 40\sin\theta - 80\cos\theta + 30\sin\theta = 0$

$70\sin\theta = 70\cos\theta$

$\tan\theta = 1, \quad \theta = 45°$

Putting the value of $\theta = 45°$, $\cos 2\theta = 0$, $\sin 2\theta = 1$ in equation (i)

$90 = 80 + 0.5\,\sigma_2 + 80 \times 0 - 0.5\,\sigma_2 \times 0 - 30 \times 1$

$90 = 80 + 0.5\,\sigma_2 - 30$

$\sigma_2 = 80$ MPa

$$\frac{\sigma_1 + \sigma_2}{2} = \frac{160 + 80}{2} = 120 \text{ MPa}, \quad \frac{\sigma_1 - \sigma_2}{2} = 40 \text{ MPa}$$

$\tau = 30$

Principal stresses are

$$p_1 = 120 + \sqrt{40^2 + 30^2} = 120 + 50 = 170 \text{ MPa}$$

$$p_2 = 120 - 50 = 70 \text{ MPa}$$

Practice Q.3.9 Two planes *OX* and *OX′* are passing through a point *O*. Normal and shear stresses on the planes are

Plane *OX*

$\sigma_1 = 20$ MPa, $\tau_1 = -30$ MPa

Plane *OX′*

$\sigma_2 = 100$ MPa, $\tau_2 = +45$ MPa

Determine angle between the planes and the principal stresses at the point

Ans. [52°, 128 MPa, 32 MPa]

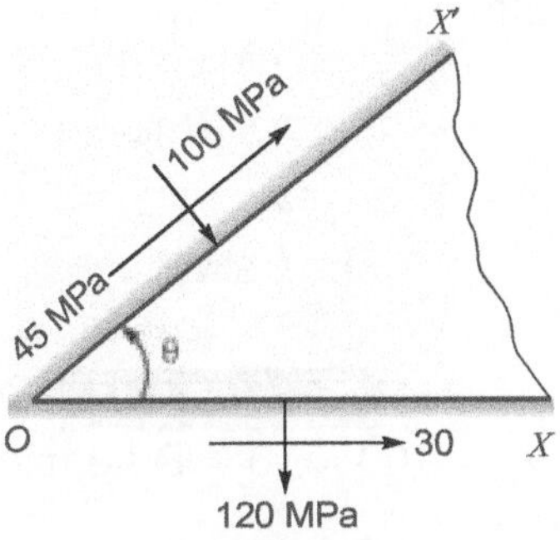

Fig. 3.18

Question 3.10 Explain strain components on a thin plate.

Solution:

Strain Components: In a general case there are normal and shear strains. Normal strains change the volume and shear strains change the shape of the body. Consider a thin elementary plate of dimensions Δx, Δy subjected to strain ε_{xy}, ε_{yy}, and $\gamma_{xy}/2$.

Total shear angle between x and y-axes = $\alpha + \beta = \gamma_{xy}$.

Shear strain, γ_{xy}, (change of angle between x and y-axes) is taken to be equally divided on both the sides.

Normal strain component,

$$\varepsilon_{xx} = \frac{\delta\Delta x}{\Delta x}$$

$$\varepsilon_{yy} = \frac{\delta\Delta y}{\Delta y}$$

Shear strain $\frac{\gamma_{xy}}{2} = \frac{\alpha+\beta}{2}$ as shown in figure 3.19.

Graphical representation of plane strain of state is given in figure 3.20

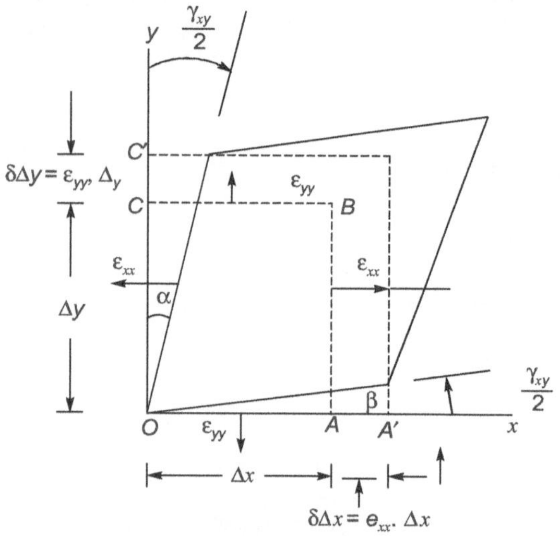

Fig. 3.19

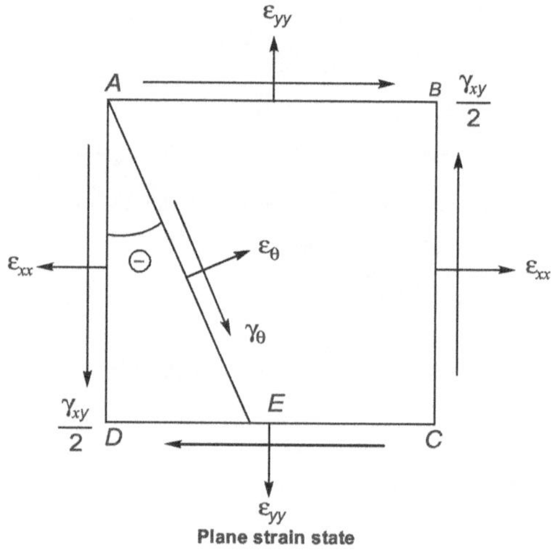

Fig. 3.20

Strain components on an inclined Plane: On the plane AE, inclined at an angle θ with the plane of strain ε_{xx}, normal and shear strains can be represented as follows:

Normal strain,

$$\varepsilon_\theta = \frac{\varepsilon_{xx}+\varepsilon_{yy}}{2} + \frac{\varepsilon_{xx}-\varepsilon_{yy}}{2}\cos 2\theta + \frac{\gamma_{xy}}{2}\sin 2\theta$$

Shear strain,

$$\gamma_\theta = \frac{\varepsilon_{xx}-\varepsilon_{yy}}{2}\sin 2\theta - \frac{\gamma_{xy}}{2}\cos 2\theta$$

Question 3.11 **Plane strains observed on a thin sheet are $\varepsilon_{xx} = 350 \times 10^{-6}$, $\varepsilon_{yy} = -150 \times 10^{-6}$, $\gamma_{xy} = 100 \times 10^{-6}$. Determine ε_θ and γ_θ on a plane inclined at 30° to the plane of ε_{xx}.**

Solution:

$$\varepsilon_{xx} = 350\,\mu \text{ strain}$$

$$\varepsilon_{yy} = -150\,\mu \text{ strain}$$

$$\frac{\gamma_{xy}}{2} = 50\,\mu \text{ strain}, \theta = 30°$$

$$\frac{\varepsilon_{xx}+\varepsilon_{yy}}{2} = 100\,\mu \text{ strain}$$

$$\frac{\varepsilon_{xx}-\varepsilon_{yy}}{2} = 250\,\mu \text{ strain}$$

$$\varepsilon_\theta = \frac{\varepsilon_{xx}+\varepsilon_{yy}}{2}+\frac{\varepsilon_{xx}-\varepsilon_{yy}}{2}\cos 2\theta+\frac{\gamma_{xy}}{2}\sin 2\theta$$

$$= 100 + 250 \cos 60° + 50 \sin 60°$$

$$= 100 + 250 \times 0.5 + 50 \times 0.866 = 100 + 125 + 43.3$$

$$= 268.3\,\mu,\ \text{strain} = 268.3 \times 10^{-6}$$

$$\gamma_\theta = \left(\frac{\varepsilon_{xx}-\varepsilon_{yy}}{2}\sin 2\theta-\frac{\gamma_{xy}}{2}\cos 2\theta\right)$$

$$= 250 \sin 60° - 50 \cos 60° = 250 \times 0.866 - 50 \times 0.5$$

$$= 216.5 - 25 = 191.5\,\mu \text{ strain} = 191.5 \times 10^{-6}$$

Practice Q.3.11 A sheet of metal is deformed uniformly in its own plane such that strain components related to *x*-*y* axes are $\varepsilon_{xx} = -200 \times 10^{-6}$, $\varepsilon_{yy} = 300 \times 10^{-6}$, $\gamma_{xy} = 450 \times 10^{-6}$. Determine the normal and shear strain components on a plane inclined at an angle of 45° to x-axis or plane of ε_{yy} strain.

Ans. [$\varepsilon_\theta = 275\,\mu s$, $\gamma_\theta = -250\,\mu s$]

Question 3.12 Explain Mohr's strain circle.

Solution: Figure 3.21 showsa state of strain on an element, on plane *BC*, with normal strain ε_{xx}, and shear strain $-\frac{\gamma_{xy}}{2}$ (as it tends to rotate the element in the anticlockwise direction so it is negative). On plane *AC*, normal strain ε_{yy} and shear strain $+\frac{\gamma_{xy}}{2}$ are acting.

Take *yox* coordinate axes. Normal strains are represented along abscissa *OX* and shear strains are represented along ordinate *OY*.

Choose some suitable scale and take $OA = \varepsilon_{xx}$, $AB = -\frac{\gamma_{xy}}{2}$; the coordinate of point *B* are ε_{xx}, $-\frac{\gamma_{xy}}{2}$. Then take $OC = \varepsilon_{yy}$, $CD = \frac{\gamma_{xy}}{2}$. Join points *B* and *D*, and the line *BD* intersects the x-axis at the point *E*. With *E* as centre and radius equal to *EB* or *ED*, draw a circle as shown in figure 3.22. This is knows as Mohr's strain circle. This circle cuts the abscissa at points *H* and *I*, where the **shear strain is zero.**

OH = principal strain, ε_{p1}

OI = principal strain, ε_{p2}

$$EB = ED = \text{Radius} = \sqrt{\left(\frac{\sigma_{xx}-\varepsilon_{yy}}{2}\right)^2+\left(\frac{\gamma_{xy}}{2}\right)^2}$$

$$\varepsilon_{p1} = OE + R = OC + CE + R$$

$$= \varepsilon_{yy} + \frac{\varepsilon_{xx}-\varepsilon_{yy}}{2} + \sqrt{\left(\frac{\varepsilon_{xx}-\varepsilon_{yy}}{2}\right)^2+\left(\frac{\gamma_{xy}}{2}\right)^2}$$

$$= \frac{\varepsilon_{xx}+\varepsilon_{yy}}{2} + \sqrt{\left(\frac{\varepsilon_{xx}-\varepsilon_{yy}}{2}\right)^2+\left(\frac{\gamma_{xy}}{2}\right)^2}$$

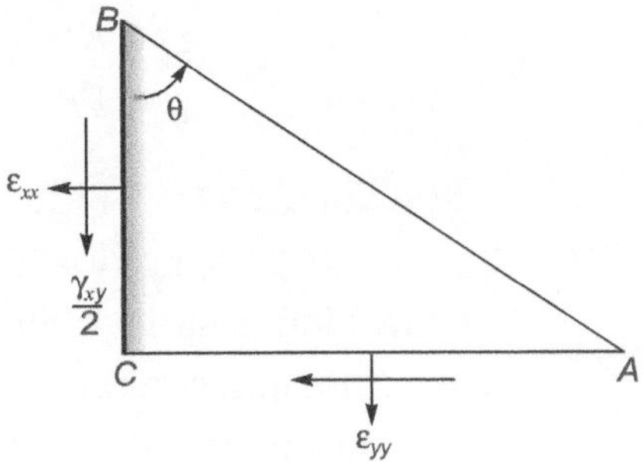

Fig. 3.21

Similarly, $$\varepsilon_{p2} = \frac{\varepsilon_{xx}+\varepsilon_{yy}}{2} - \sqrt{\left(\frac{\varepsilon_{xx}-\varepsilon_{yy}}{2}\right)^2 + \left(\frac{\gamma_{xy}}{2}\right)^2}$$

Principal angles $$\theta_1 = \frac{1}{2}\tan^{-1}\frac{AB}{EA} = \frac{1}{2}\tan^{-1}\frac{\gamma_{xy}}{\varepsilon_{xx}-\varepsilon_{yy}}$$

$$\theta_2 = \theta_1 + 90°$$

For normal and shear strains on inclined AB, take

$\angle BEP = 2\theta$, coordinates of P are OK and KP

$OK = \varepsilon_\theta$, $KP = \gamma_\theta$.

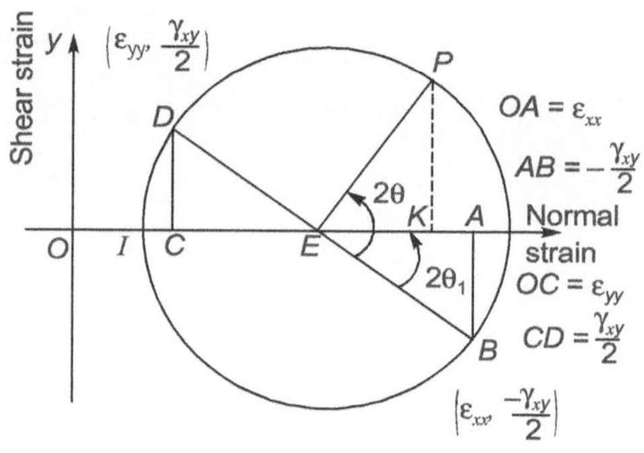

Fig. 3.22

Question 3.13 Figure 3.23 shows state of strain at a point on an element, when

$\varepsilon_{xx} = +120\ \mu s$, $\varepsilon_{yy} = -40\ \mu s$, $\gamma_{xy}/2 = \pm 60\ \mu s$

Draw Mohr's strain circle and find

(i) principal strains

(ii) ε_θ and γ_θ on inclined plane AB as shown $\theta = 30°$

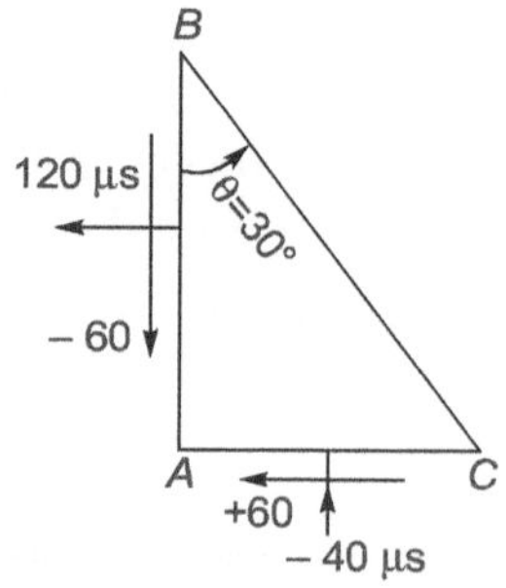

Fig. 3.23

Solution: Choose coordinate axes YOX

From O take

$OA = +120\ \mu s$ (to some suitable scale)

Take $AB = -60\ \mu s$

$OC = -40\ \mu s$

$CD = +60\ \mu s$

Join BD points, intersecting abscissa at E, **which is the centre of the Mohr's strain circle.** With E as center, radius equal to EB or ED draw the circle intersecting abscissa at H and I then

$OH = \varepsilon_{p1} = +140\ \mu s$ (principal strain)

$OS = \varepsilon_{p2} = -60\ \mu s$ (principal strain)

Take EB as reference line, take $\angle BEP = 2\theta = 60°$.

Co-ordinates point P will give normal and shear strains on inclined plane

$\varepsilon_\theta = K = +132\ \mu s$, normal strain

$\gamma_\theta = PK = +42\ \mu s$, shear strain

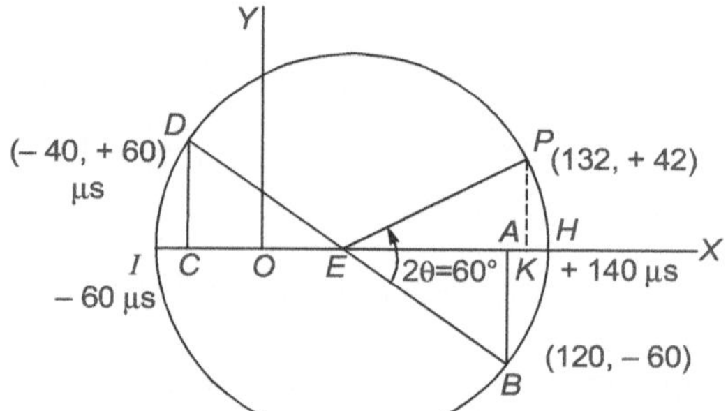

Fig. 3.24

Practice Q.3.13 Figure 3.25 shows state of strain on an element.

$\varepsilon_{xx} = +400\ \mu s$; $\varepsilon_{yy} = +200\ \mu s$; $\gamma_{xy}/2 = \pm 75\ \mu s$.

Draw Mohr's strain circle and determine

(i) principal strains

(ii) normal and shear strains on inclined plane BC

(iii) principal angles with respect to plane BA

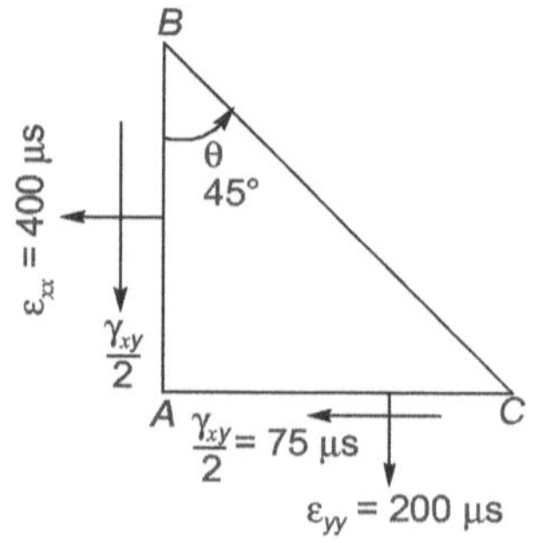

Fig. 3.25

Ans. [425 μs, 175 μs, $\varepsilon_\theta = 375\ \mu s$, $\gamma_\theta = +100\ \mu s$, $\theta_1 = 18.45°$, $\theta_2 = 108.45°$]

Question 3.14 Write down principal strains in terms of principal stresses. What is modified Young's modulus.

Solution: Principal strains in terms of Principal stresses.

If we know principal stresses p_1, p_2, p_3 at a point and E, ν of material, then principal strains are

$$\varepsilon_{p1} = \frac{p_1}{E} - \frac{\nu p_2}{E} - \frac{\nu p_3}{E}$$

$$\varepsilon_{p2} = \frac{p_2}{E} - \frac{\nu p_3}{E} - \frac{\nu p_1}{E}$$

$$\varepsilon_{p3} = \frac{p_3}{E} - \frac{\nu p_1}{E} - \frac{\nu p_2}{E}$$

Where E is Young's modulus and ν is Poisson's ratio of material. Figure 3.26 shows the principal stresses p_1, p_2 and p_3 on three principal planes. On these three planes, there are principal strains ε_{p1}, ε_{p2}, and ε_{p3}, respectively, as shown

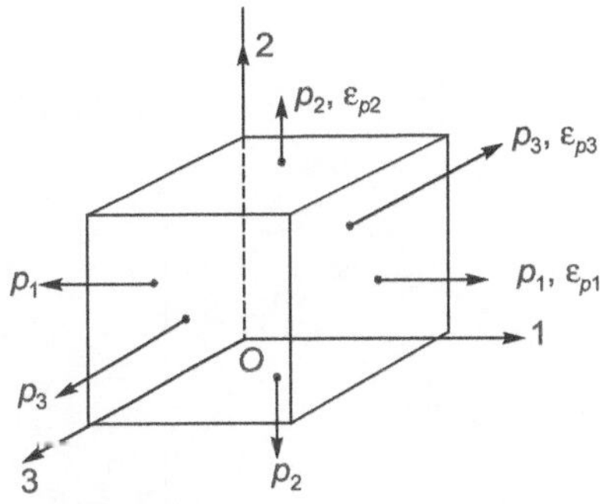

Fig. 3.26 *Principal stresses and principal planes*

Ratio of $\frac{p_1}{\varepsilon_{p1}}$ is termed as **modified Young's modulus** in direction (1). Similarly,

$$E_{m1} = \frac{p_1}{\varepsilon_{p1}},\ E_{m2} = \frac{p_2}{\varepsilon_{p2}},\ E_{m3} = \frac{p_3}{\varepsilon_{p3}},$$

If $\quad p_1 \neq p_2 \neq p_3,$

then $\quad E_{m1} \neq E_{m2} \neq E_{m3}$

Question 3.15 What is a strain gage, describe briefly. What are strain gage rosettes?

Solution: A strain gage is a gage made of strain sensing wire for the measurement of surface strains of a sample. It is made in the form of wire grid, made from very thin wire of diameter 0.02 mm. The alloys used to make wire of strain gage are Isoelastic, Advance, Nichrome and Karma. Figure 3.27 shows a wire grid type, electrical resistance strain gage. This grid is carried between two paper backs or backelits wafers. Size of a strain gage is very small lesser then the size of a postage stamp. It is fixed on the surface of a metallic sample with the help of an adhesive such as cellulose nitrate.

When the sample is subjected to external load, strains developed on sample are transmitted to the strain gage. Resistance of the strain gage changes depending upon the type of strain and change in resistance (ΔR) is measured with the help of an electrical circuit.

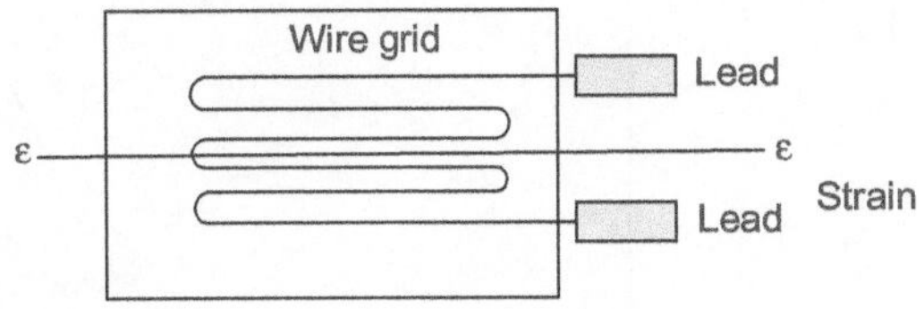

Fig. 3.27 *Electrical resistance strain gage*

Strain, $$\varepsilon = \frac{\Delta R}{R} \times \frac{1}{s_g}$$

where, R is initial resistance of strain gage, s_g is the gage factor provided by the manufacturer of the strain gage.

Strain Gage Rosettes: A combination of strain gages at different angular spacing is known as strain gage rosette.

There are 3 types of rosettes:

(a) Two element rectangular rosette, gages are mounted at 90°, along the directions of principal strains.

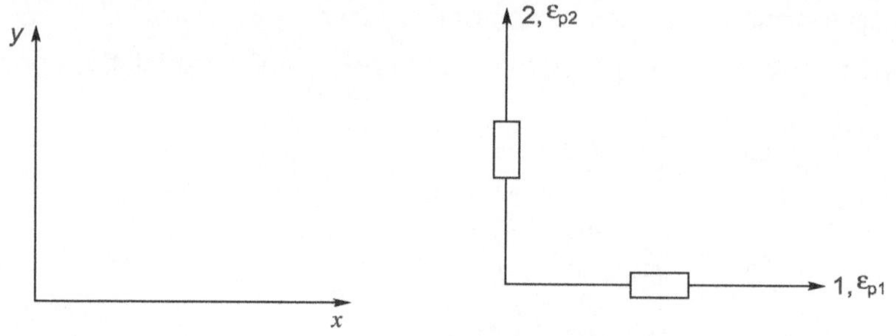

$$\varepsilon_{xx} = \varepsilon_{p1}, \ \varepsilon_{yy} = \varepsilon_{p2}$$

Fig. 3.28

(b) Three-element rectangular rosette

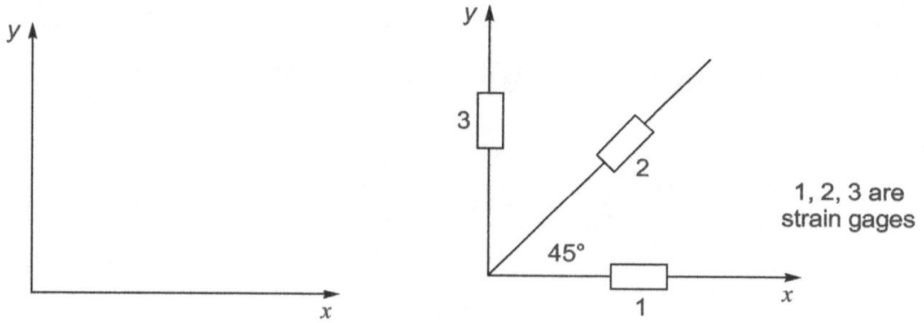

Fig. 3.29

$$\varepsilon_1 = \varepsilon_{xx}; \quad \varepsilon_2 = \frac{\varepsilon_{xx} + \varepsilon_{yy}}{2} + \frac{\gamma_{xy}}{2}; \qquad \varepsilon_3 = \varepsilon_{yy}$$

(c) Three-element delta rosette

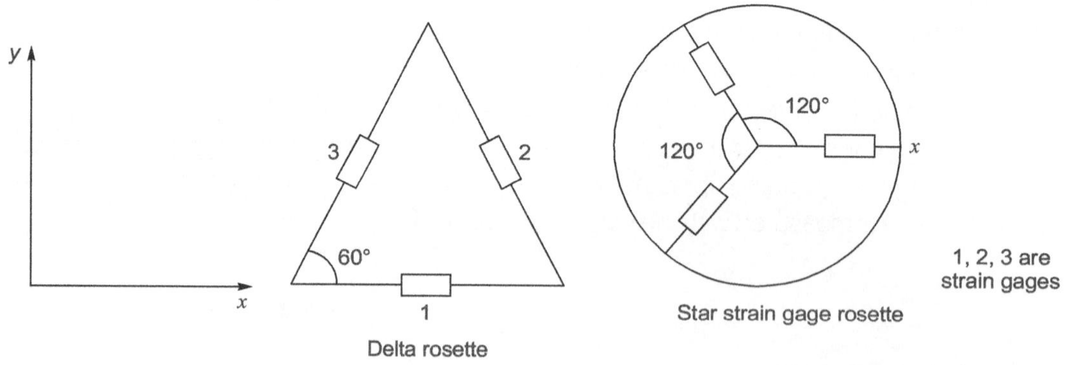

Fig. 3.30

$$\varepsilon_1 = \varepsilon_{xx}; \qquad \varepsilon_2 = \frac{\varepsilon_{xx}}{4} + \frac{3\varepsilon_{yy}}{4} - 0.433\gamma_{xy}; \qquad \varepsilon_3 = \frac{\varepsilon_{xx}}{4} + \frac{3}{4}\varepsilon_{yy} + 0.433\gamma_{xy}$$

Strain gages respond only to normal strains.

$$\varepsilon_\theta = \frac{\varepsilon_{xx}+\varepsilon_{yy}}{2}+\frac{\varepsilon_{xx}-\varepsilon_{yy}}{2}\cos 2\theta+\frac{\gamma_{xy}}{2}\sin 2\theta$$

To determine strain components, ε_{xx}, ε_{yy}, $\frac{\gamma_{xy}}{2}$, it is necessary to determine normal strains.

$\varepsilon_{\theta 1}$, $\varepsilon_{\theta 2}$, $\varepsilon_{\theta 3}$ in 3 directions. Then principal strains can be worked out. From principal strains, using the values of E and ν, principal stresses can be worked out.

Practice Q.3.15 **Two element rectangular rosette is mounted on an aluminium specimen under a load, and following strain readings are recorded.**

$\varepsilon_1 = 200\ \mu s$, $\varepsilon_2 = -150\ \mu s$

If $E = 67$ kN/mm^2, $\nu = 0.33$,

Calculate principal stresses.

Hints: $\left[p_1 = \frac{E}{(1-\nu^2)}+(\varepsilon_1+\nu\varepsilon_2);\ p_2 = \frac{E}{(1-\nu^2)}(\varepsilon_2+\nu\varepsilon_2)\right]$

Ans. [11.316 N/mm^2, –6.316 N/mm^2]

Question 3.16 **Strains at a point on a specimen are recorded with the help of 3 element rectangular rosette. Following observations are made**

$\varepsilon_0 = 400\ \mu s$, $\varepsilon_{45°} = 200\ \mu s$ and $\varepsilon_{90°} = -300\ \mu s$

Determine principal strains.

Solution:

$$\varepsilon_0 = 400\ \mu s = \varepsilon_{xx}$$

$$\varepsilon_{90°} = -300\ \mu s = \varepsilon_{yy}$$

$$\frac{\varepsilon_{xx}+\varepsilon_{yy}}{2} = +50\ \mu s$$

$$\frac{\varepsilon_{xx}-\varepsilon_{yy}}{2} = 350$$

$$\varepsilon_{45°} = \frac{\varepsilon_{xx}+\varepsilon_{yy}}{2}+\frac{\gamma_{xy}}{2} = 200\ \mu s$$

$$50+\frac{\gamma_{xy}}{2} = 200\ \mu s$$

$$\frac{\gamma_{xy}}{2} = 150\ \mu s$$

Principal strains

$$\varepsilon_{p1} = \frac{\varepsilon_{xx}+\varepsilon_{yy}}{2}+\sqrt{\left(\frac{\varepsilon_{xx}-\varepsilon_{yy}}{2}\right)^2+\left(\frac{\gamma_{xy}}{2}\right)^2}$$

$$= 50+\sqrt{350^2+150^2} = 50+380.8 = 430.8\ \mu s$$

$$\varepsilon_{p2} = -330.8\ \mu s$$

Question 3.17 On an aluminium sample, strain readings were recorded on 3-element delta rosette as follows:

$\varepsilon_{0^\circ} = -100\ \mu s$, $\varepsilon_{120^\circ} = -500\ \mu s$ and $\varepsilon_{60^\circ} = +600\ \mu s$

Determine principal strains.

Solution:

$$\varepsilon_0 = \varepsilon_{xx} = -100\ \mu s \quad \text{...(i)}$$

$$\varepsilon_{120^\circ} = -500\mu s = \frac{\varepsilon_{xx}}{4} + \frac{3}{4}\varepsilon_{yy} - 0.433\gamma_{xy}$$

$$-500 = -25 + 0.75\,\varepsilon_{yy} - 0.433\gamma_{xy}$$

or

$$-475 = 0.75\,\varepsilon_{yy} - 0.433\,\gamma_{xy} \quad \text{...(ii)}$$

$$\varepsilon_{60^\circ} = 600 = \frac{\varepsilon_{xx}}{4} + \frac{3}{4}\varepsilon_{yy} + 0.433\gamma_{xy}$$

$$= 25 + 0.75\varepsilon_{yy} + 0.433\gamma_{xy}$$

$$575 = 0.75\,\varepsilon_{yy} + 0.433\,\gamma_{xy} \quad \text{...(iii)}$$

$$-475 = 0.75\,\varepsilon_{yy} - 0.433\,\gamma_{xy} \quad \text{...(iv)}$$

$$100 = 1.5\,\varepsilon_{yy}$$

or

$$\varepsilon_{yy} = +66.67\ \mu s$$

Putting the values in equation (iii)

$$575 = 0.75 \times 66.67 + 0.433\,\gamma_{xy} = 50 + 0.433\,\gamma_{xy}$$

$$\gamma_{xy} = 1212.5\ \mu s$$

$$\frac{\gamma_{xy}}{2} = 606.25\ \mu s$$

Now,

$$\frac{\varepsilon_{xx} + \varepsilon_{yy}}{2} = \frac{-100 + 66.67}{2} = -16.665\ \mu s$$

$$\frac{\varepsilon_{xx} - \varepsilon_{yy}}{2} = \frac{-100 - 66.67}{2} = -83.33\ \mu s$$

$$\frac{\gamma_{xy}}{2} = 606.25\ \mu$$

$$\sqrt{\left(\frac{\varepsilon_{xx} - \varepsilon_{yy}}{2}\right)^2 + (606.25)^2} = \sqrt{(-83.33)^2 + (606.25)^2}$$

$$= \sqrt{6943.9 + 367539} = 611.95\ \mu s$$

Principal strains

$$\varepsilon_{p1} = -16.665 + 611.95 = +595.285\ \mu s$$

$$\varepsilon_{p2} = -16.665 - 611.95 = -628.615\ \mu s$$

Question 3.18 A cantilever is subjected to UDL as shown. Section of the cantilever in circular. Draw Mohr's strain circles for points *A*, *B* and *C* of the section-depth.

Solution:

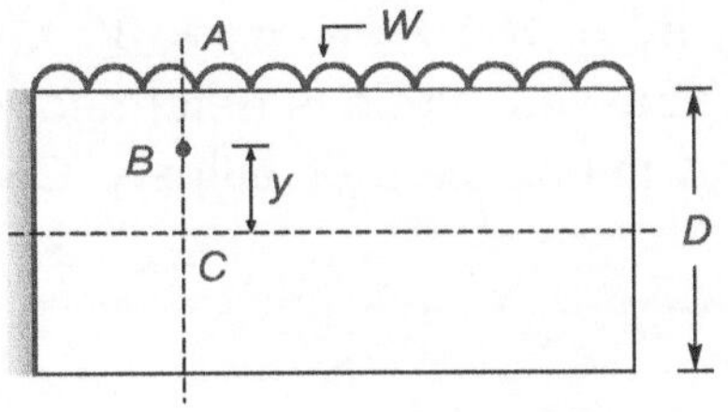

Fig. 3.31

At top point A

$$\tau = 0$$

$$\sigma_b = +\frac{M}{I_{xy}} \times \frac{d}{2}$$

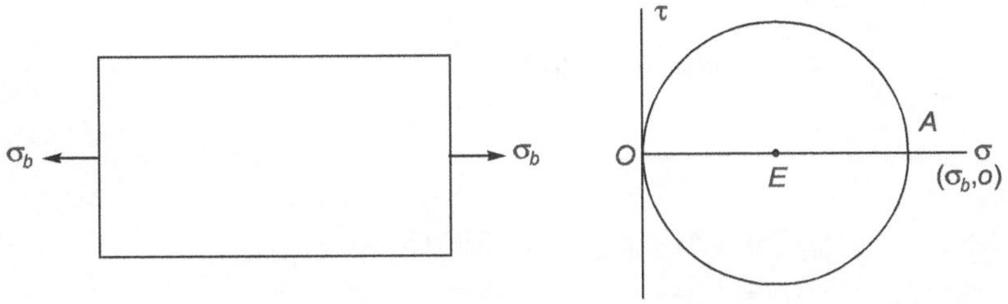

At point B

$$\tau = \frac{F_{a\bar{y}}}{I_b}$$

$$\sigma_b' = \frac{M_x y}{I_{xx}}$$

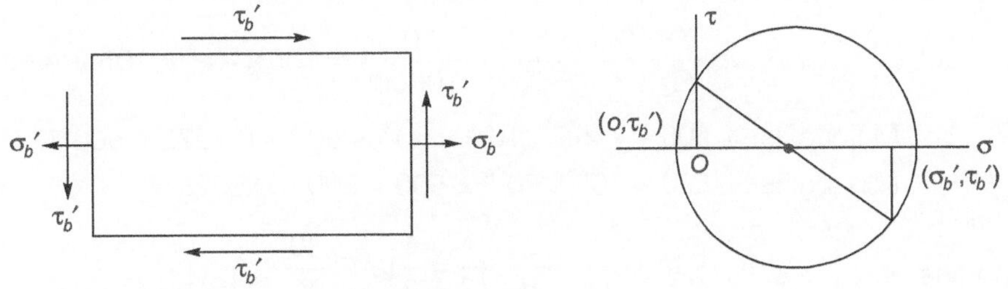

At point C at neutral axis

$$\sigma_c = 0$$

$$\tau = \frac{F_{a\bar{y}}}{Ib}$$

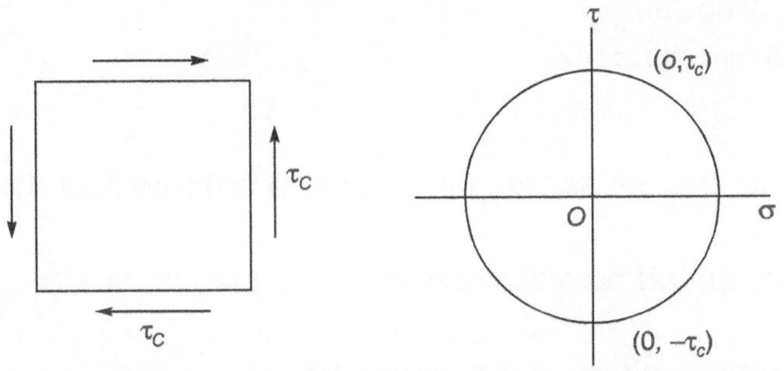

State of stress at points *A*, *B* and *C* are shown alongwith the Mohr's stress circle.

Question 3.19 A circle of diameter 50 mm in inscribed on a square plate of copper, before the application of stresses as shown, σ_1 = 160 MPa, σ_2 = 80 MPa, τ = 30 MPa. After the application of stresses, circle is deformed into an ellipse. What are the major and minor axes of ellipse. Given E = 100 GPa, ν = 0.35.

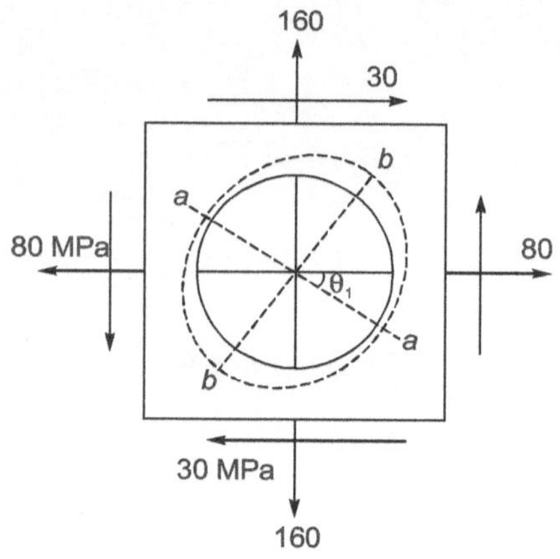

Fig. 3.33

Solution:

$$\frac{\sigma_1+\sigma_2}{2} = \frac{160+80}{2} = 120 \text{ MPa}$$

$$\frac{\sigma_1-\sigma_2}{2} = \frac{160-80}{2} = 40 \text{ MPa}$$

$$\sqrt{\left(\frac{\sigma_1-\sigma_2}{2}\right)^2+\tau^2} \quad \text{where } \tau = 30 \text{ MPa}$$

$$= \sqrt{40^2+30^2} = 50 \text{ MPa}$$

Principal stresses, p_1 = 120 + 50 = 170 MPa

p_2 = 120 – 50 = 70 MPa

Principal strains, $e_1 = \frac{p_1}{E} - \frac{\nu p_2}{E} = \frac{170}{E} - \frac{0.35\times 70}{E}$

$$= \frac{145.5}{100000} = 1455\times 10^{-6} = 1455\ \mu \text{ strain}$$

$$e_2 = \frac{70}{E} - \frac{0.35\times 170}{E} = \frac{10.5}{100000} = 105\times 10^{-6} = 105\ \mu \text{ strain}$$

Major axis = 500 + 1455 × 10^{-6} × 500 = 500 + 0.7275 = 500.7275 mm

Minor axis = 500 + 105 × 10^{-6} × 500 = 500 + 0.0525 = 500.0525 mm

Principal angles, $\theta_1 = \frac{1}{2}\tan^{-1}\frac{\tau}{(\sigma_1-\sigma_2)/2} = \frac{1}{2}\tan^{-1}\frac{30}{40} = 18.44°$

θ_1 = –18.44° as shear stress on reference plane is positive

Note that a – a is principal plane for p_1

p_1, e_1 are along direction *bb*.

p_2, e_2 are along direction *aa*.

Major axis 500.7275 mm along *bb*

Minor axis 500.0525 mm along *aa*.

Question 3.20 State of stress at a point is shown in figure 3.34. Draw Mohr's stress circle and determine.

(a) Plane of abs (τ_{max}) and normal stress on these planes, abs (τ_{max}), direction with respect to plane AB.

(b) Planes of pure shear, direction wrt. plane *AB*

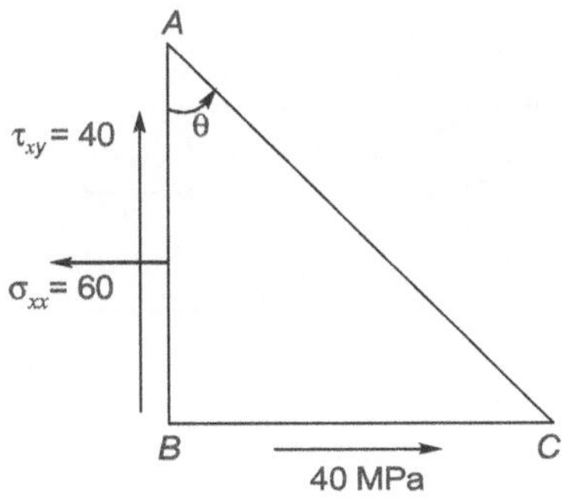

Fig. 3.34

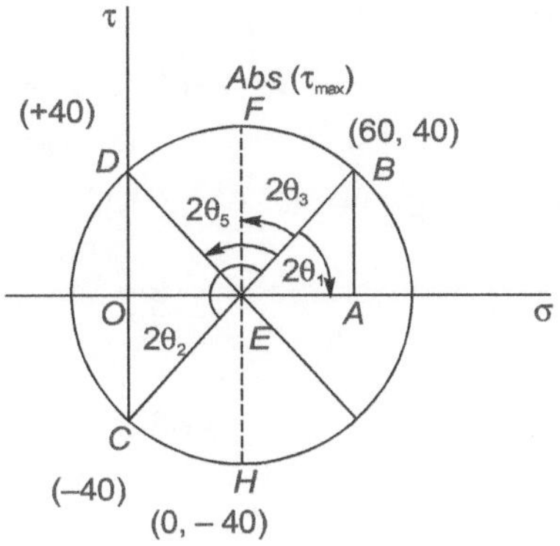

Fig. 3.35

Solution: Choose x-y- coordinates or $\sigma - \tau$ coordinates.

Take a suitable scale for stresses.

Take $OA = +60$ MPa, $AB = +40$ MPa (shear stress).

Take $OC = -40$ MPa (shear stress ⊥ plane BC)

Join B and C, line BC intersecting abscissa at E.

E is the centre of Mohr's stress circle, radius

$$EB = EC = 50 \text{ MPa}.$$

Point F and H show ± abs (τ_{max}) (in plane τ_{max})

$$2\theta_3 = 37° \text{ (from } EB\text{, reference line for reference plane } AB\text{)}$$

$$\theta_3 = +18.5°$$

$$\theta_4 = 90 + 18.5 = 108.5° \text{ [abs } (\tau_{max}) = -50 \text{ MPa]}$$

Point D and C on Mohr's stress circle represent shear stress on planes of pure shear, where $\sigma = O$, normal stress is zero.

$$\tau = \pm 40 \text{ MPa, pure shear}$$

$$2\theta_5 = 73.74°$$

$$\theta_5 = 36.87° \text{ (for point } D\text{)}$$

$$\theta_6 = 126.87° \text{ (for point } C\text{)}$$

Practice Q.3.20 State of stress at a point is shown in figure 3.36. Draw a Mohr's stress circle for the state of stress, determine principal stresses and direction of principal planes with respect to plane *AB*.

Ans: [$p_1 = +50$ MPa, $\theta_1 = -45°$, $p_2 = 50$ MPa, $\theta_2 = +45°$]

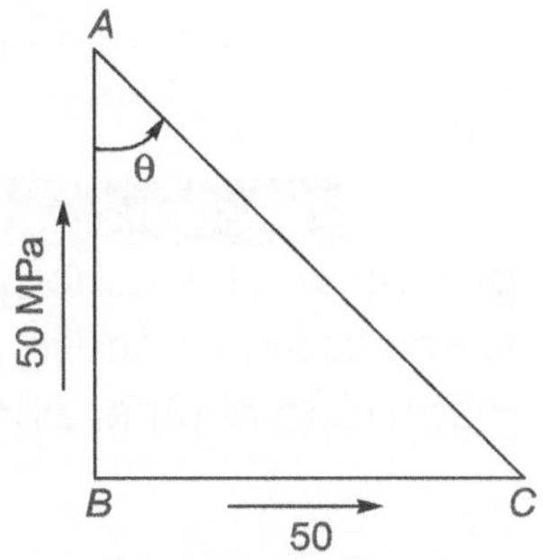

Fig. 3.36

Question 3.21 Resultant stress on reference plane *BC* is inclined at an angle of $\theta = 30°$ as shown. What is the shear stress on plane *AC*. What are principal stresses at the point. What are principal angles w.r.t. plane *BC*.

Solution:

$$\sigma_1 = -100 \cos 30° = -86.6 \text{ MPa (compressive normal stress)}$$

$$\tau = 100 \sin 30° = 50 \text{ MPa (positive shear stress on } BC\text{)}$$

Stresses are

$$\sigma_1 = -86.6,\ \sigma_2 = +50 \text{ MPa}$$
$$\tau = 50 \text{ (+ve on reference plane)}$$

Shear stress on plane $AC = -50$ MPa

$$\frac{\sigma_1+\sigma_2}{2} = -18.3$$

$$\frac{\sigma_1-\sigma_2}{2} = -68.3$$

$$\sqrt{\left(\frac{\sigma_1-\sigma_2}{2}\right)^2+\tau^2} = \sqrt{(-68.3)^2+50^2} = \pm 84.65 \text{ MPa}$$

$$p_1 = -18.3 + 84.65 = +66.35 \text{ MPa}$$
$$p_2 = -18.3 - 84.65 = -102.95 \text{ MPa}$$

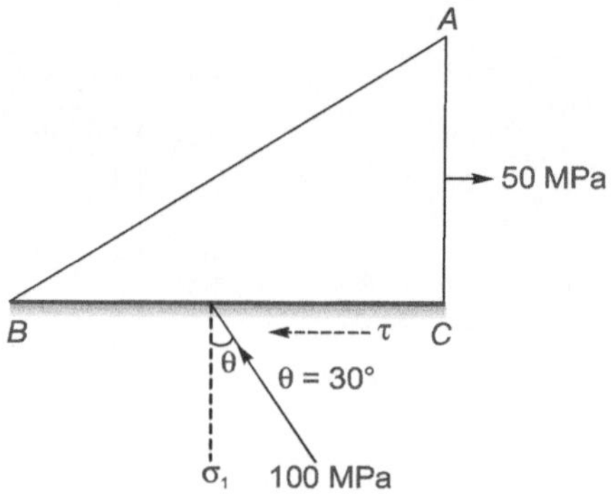

Fig. 3.37

Principal angles

$$\theta_1 = -\frac{1}{2}\tan^{-1}\frac{\tau}{(\sigma_1-\sigma_2)/2}$$

$$= -\frac{1}{2}\tan^{-1}\left(\frac{50}{68.3}\right) = -\frac{1}{2}\times 36.2° = -18.1°$$

$$\theta_2 = -18.1 + 90 = +71.9°$$

Practice Q.3.21 State of stress at a point is given in figure 3.38. Resultant stress on plane *AC* is inclined at an angle 45° as shown. What is the shear stress on plane *BC*. Determine principal stresses and maximum shear stress at the point. What are normal and shear stresses on plane *AB*, θ = 30°.

Ans: [–56.56 MPa, 114.87, 1.69 MPa, ± 56.59 MPa, 108.12, – 26.79 MPa]

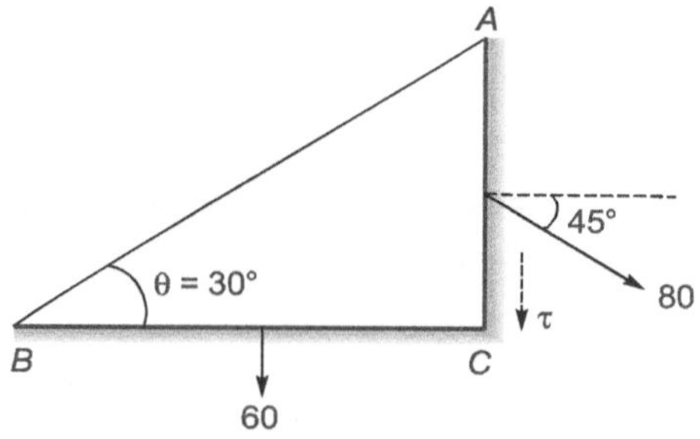

Fig. 3.38

Question 3.22 The magnitude of normal stresses on two mutually perpendicular planes at a point in an elastic body are 60 MPa compressive and 80 MPa tensile respectively. Find the magnitude of shear strain on these planes if magnitude of one of the principal stresses is 100 MPa. Find also the magnitude of other principal stress.

[IES 2002 : 10 Marks]

Solution: State of stress at the point is shown in figure 3.39

$$\sigma_{xx} = +80 \text{ MPa}$$
$$\sigma_{yy} = -60 \text{ MPa}$$

$$\frac{\sigma_{xx}+\sigma_{yy}}{2} = 10 \text{ MPa}$$

$$\frac{\sigma_{xx}-\sigma_{yy}}{2} = 70 \text{ MPa}$$

τ_{xy} = shear stress on plane *AB, AC*

Principal stress, $p_1 = 100 = 10 + \sqrt{70^2 + \tau_{xy}^2}$

$$70^2 + \tau_{xy}^2 = 8100$$
$$\tau_{xy}^2 = 3200$$
$$\tau_{xy} = \pm 56.57 \text{ MPa}$$

Principal stress, $p_2 = 10 - \sqrt{70^2 + \tau_{xy}^2}$

$$= 10 - \sqrt{4900 + 3200}$$
$$= 10 - 90 = -80 \text{ MPa}$$

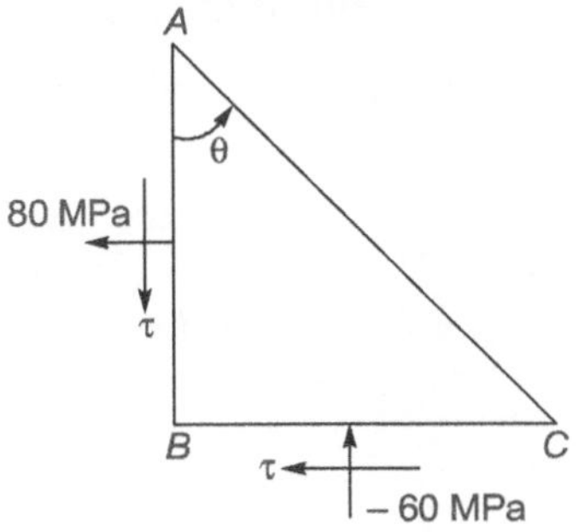

Fig. 3.39

Question 3.23 The data obtained from a rectangular strain gage rosette attached to a stressed steel member are $\varepsilon_0 = -220 \times 10^{-6}$, $\varepsilon_{45°} = 120 \times 10^{-6}$, $\varepsilon_{90°} = +220 \times 10^{-6}$.

Given $E = 2 \times 10^5$ N/mm² and Poisson's ratio = 0.3.

Calculate the values of principal stresses acting at the point and their directions.

[IES 2010 : 10 Marks]

Solution:

$$\varepsilon_{xx} = \varepsilon_0 = -220 \ \mu s$$
$$\varepsilon_{yy} = \varepsilon_{90°} = +220 \ \mu s$$
$$\frac{\varepsilon_{xx} + \varepsilon_{yy}}{2} = 0$$
$$\frac{\varepsilon_{xx} - \varepsilon_{yy}}{2} = -220 \ \mu s$$
$$\frac{\gamma_{xy}}{2} = \varepsilon_{45°} - \frac{\varepsilon_{xx} + \varepsilon_{yy}}{2} = +120 \ \mu s - 0 = 120 \ \mu s$$

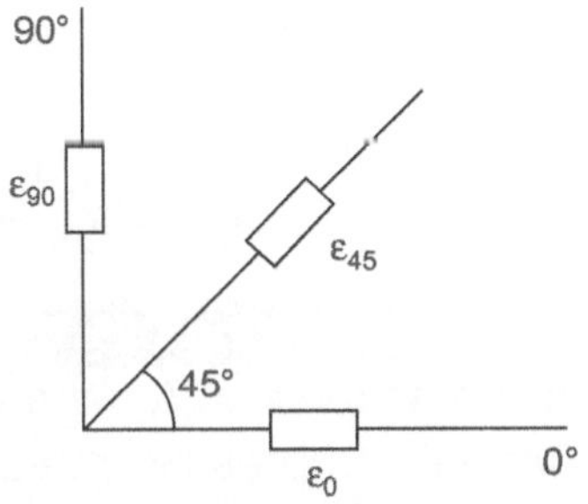

Fig. 3.40

Principal strain, $\varepsilon_1, \varepsilon_2 = \frac{\varepsilon_{xx} + \varepsilon_{yy}}{2} \pm \sqrt{\left(\frac{\varepsilon_{xx} - \varepsilon_{yy}}{2}\right)^2 + \frac{\gamma_{xy}^2}{2}}$

$$\varepsilon_1, \varepsilon_2 = 0 \pm \sqrt{(-220)^2 + (120)^2} = \pm 250.6 \ \mu s$$

$$p_1 = \frac{E}{1-\nu^2}(\varepsilon_1 + \nu\varepsilon_2) = \frac{2 \times 10^5}{1 - 0.3^2}[250.6 - 0.3 \times 250.6]$$
$$= 38.55 \text{ N/mm}^2$$
$$p_2 = -38.55 \text{ N/mm}^2$$

Question 3.24 At a point in a loaded component, the state of stress is given by σ_x = 270 MPa, σ_y = 130 MPa, τ_{xy} = ± 40 MPa. Determine

(i) Maximum and minimum principal stresses and the planes on which these act.

(ii) The maximum shear stress in magnitude and direction.

[IES 2013 : 10 Marks]

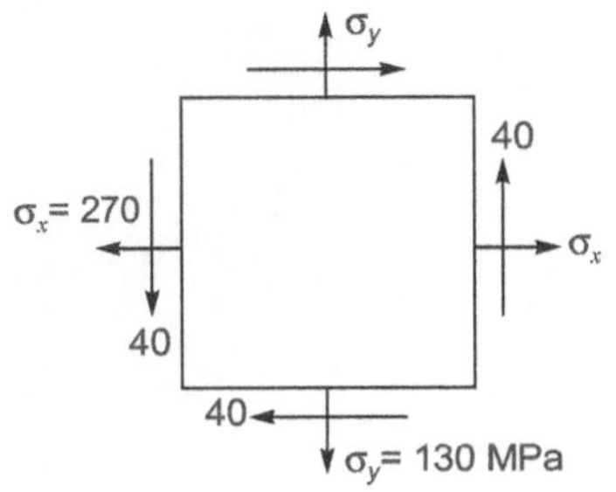

Fig. 3.41

Solution:

$$\sigma_x = 270 \text{ MPa}, \sigma_y = 130 \text{ MPa}, \tau_{xy} = \pm 40 \text{ MPa}$$

$$\frac{\sigma_x + \sigma_y}{2} = \frac{400}{2} = 200 \text{ MPa}$$

$$\frac{\sigma_x - \sigma_y}{2} = 70 \text{ MPa}$$

$$p_1, p_2 = \frac{\sigma_x + \sigma_y}{2} \pm \sqrt{\left(\frac{\sigma_x - \sigma_y}{2}\right)^2 + (\tau_{xy})^2} = 200 \pm \sqrt{70^2 + 40^2}$$

$$= 200 \pm 80.6 = +280.6, +119.4 \text{ MPa}$$

Principal angle

$$\tan 2\theta_1 = \frac{2\tau_{xy}}{\sigma_x - \sigma_y} = \frac{40}{70} = 0.5714$$

$$\theta_1 = 14.87°$$

$$\theta_2 = \theta_1 + 90° = 104.87°$$

$$\text{Maximum shearing stress} = \pm\sqrt{\left(\frac{\sigma_x - \sigma_y}{2}\right)^2 + \tau_{xy}^2} = \pm 80.6 \text{ MPa}$$

$$\theta_3 = \theta_1 + 45° = 14.87 + 45 = 59.87°$$

$$\theta_4 = \theta_3 + 90° = 149.87°$$

Angles of planes which carry maximum shearing stress.

Question 3.25 The state of stress at a point in a structural component of elastic material are given as follows: A normal tensile stress of 160 MPa and a shearing stress of 120 MPa on one plane, a normal compressive stress of 110 MPa and the complementary shearing stress of 120 MPa on a second plane orthogonal to the firs plane, and no stress on the third plane which is orthogonal to the above two planes.

Determine

(i) the principal stresses and the positions of the planes on which they act, and

(ii) the positions of planes on which there is no normal stress.

[ME, ESE 2014 : 10 Marks]

Solution:

$$\sigma_1 = +160 \text{ MPa}$$

$$\sigma_2 = -110 \text{ MPa}$$

$$\tau = 120 \text{ MPa}$$

$$\frac{\sigma_1 + \sigma_2}{2} = +25 \text{ MPa}$$

$$\frac{\sigma_1 - \sigma_2}{2} = 135 \text{ MPa}$$

$$\pm\sqrt{\left(\frac{\sigma_1 - \sigma_2}{2}\right)^2 + \tau^2} = \tau_{max}$$

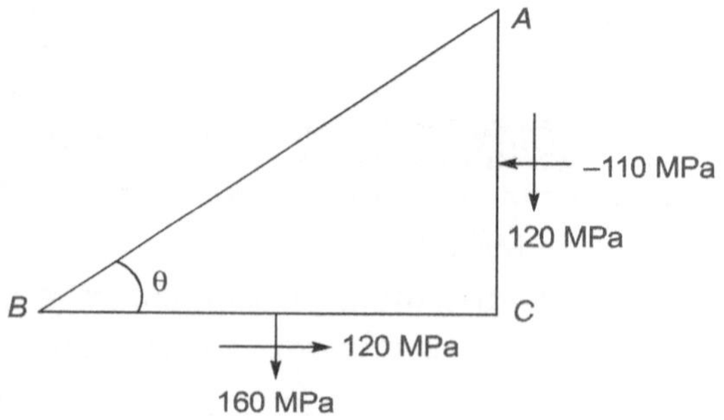

Fig. 3.42

$$\sqrt{135^2 + 120^2} = \pm 180.62 \text{ MPa}$$

Principal stresses

$$p_1 = 25 + 180.62 = +205.62 \text{ MPa}$$

$$p_2 = 25 - 180.62 = -155.62 \text{ MPa}$$

Angles of principal planes w.r.t. plane *BC*

$$\theta_1 = \frac{1}{2}\tan^{-1}\frac{2\tau}{\sigma_1 - \sigma_2}$$

$$= \frac{1}{2}\tan^{-1}\frac{2\times 120}{270} = \frac{1}{2}\times 41.63° = 20.8°$$

$$\theta_2 = \theta_1 + 90° = 110.8°$$

$$\sigma_\theta = \frac{\sigma_1 + \sigma_2}{2} + \frac{\sigma_1 - \sigma_2}{2}\cos 2\theta + \tau \sin 2\theta$$

$$0 = 25 + 135 \cos 2\theta + 120 \sin 2\theta$$

or $5.4 \cos 2\theta + 4.8 \sin 2\theta = -1$

$$\cos 2\theta + 0.888 \sin 2\theta = -0.185$$

$$\cos 2\theta = -0.888 \sin 2\theta - 0.185$$

By trial and error $\theta_1' = 70°$

$$\theta_2' = 160°$$

For planer of no normal stress. Angles with respect to plane *BC*.

Question 3.26 A plane element in a body is subjected to stresses as shown in Fig. 3.43. Determine the principal stresses and maximum shear stresses and the planes on which they act. Indicate them in separate sketches.

[CE, ESE 2017 : 12 Marks]

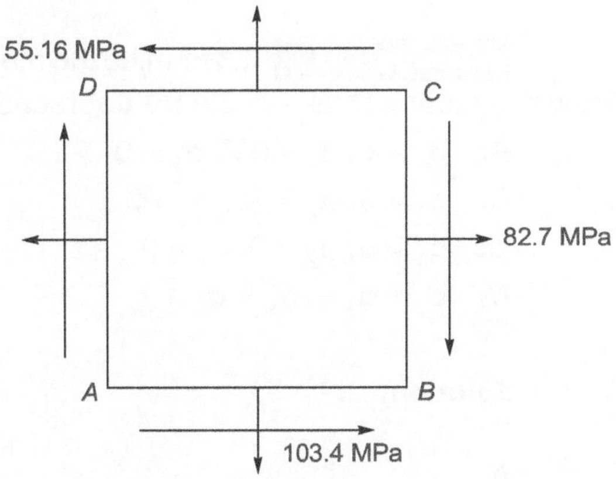

Fig. 3.43

Solution: Let us take AB as reference plane, plane of stress 103.4 MPa

$$\sigma_1 = 103.4 \text{ MPa}$$

$$\sigma_2 = 82.7 \text{ MPa}$$

$$\frac{\sigma_1 + \sigma_2}{2} = 93.05$$

$$\frac{\sigma_1 - \sigma_2}{2} = 10.35$$

$$\tau = \pm 55.16 \text{ MPa}$$

$$\sqrt{\left(\frac{\sigma_1 - \sigma_2}{2}\right)^2 + \tau^2} = \sqrt{10.35^2 + 55.16^2}$$

$$= \sqrt{107.1125 + 3042.6256}$$

$$\tau_{max} = \pm 56.12 \text{ N/mm}^2$$

Principal stresses,

$$p_1 = 93.05 + 56.12 = 149.17 \text{ MPa}$$

$$p_2 = 93.05 - 56.12 = 36.93 \text{ MPa}$$

Angles of principal planes with respect to planes AB

$$\theta_1 = \frac{1}{2}\tan^{-1}\frac{2\tau}{\sigma_1 - \sigma_2}$$

$$= \frac{1}{2}\tan^{-1}\frac{55.16}{16.35} = \frac{1}{2}\times 79.372°$$

$$= 39.686°$$

Plane of p_1 inclined with plane *AB* at angle

$$\theta_1 = 39.686°$$

Plane of p_2 inclined with plane *AB* at

$$\theta_2 = \theta_1 + 90° = 129.686°$$

Angle of planes of τ_{max}

$$\theta_3 = \theta_1 + 45° = 39.686 + 45 = 84.686°$$

$$\theta_4 = \theta_3 + 90° = 174.686° \text{ or } -5.314°$$

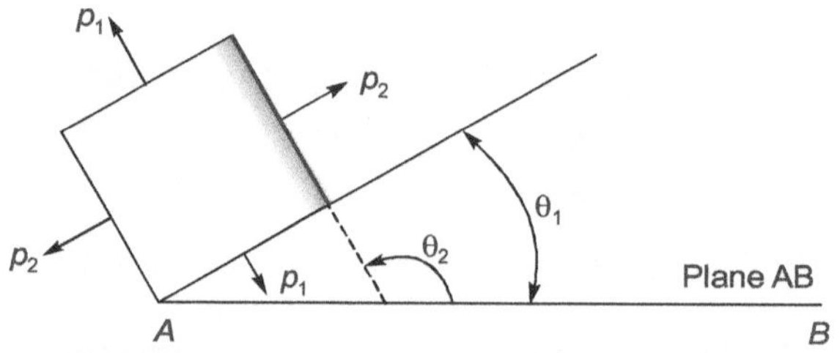

Fig. 3.44

Question 3.27 Determine the maximum shear stress values for the given set of principal stresses. Draw the Mohr's stress circle for each case and show the maximum shear stress on it.

A. $\sigma_1 = \sigma, \sigma_2 = \sigma/2, \sigma_3 = 0$

B. $\sigma_1 = \sigma, \sigma_2 = -\sigma, \sigma_3 = 0$

C. $\sigma_1 = \sigma, \sigma_2 = 0, \sigma_3 = 0$

D. $\sigma_1 = \sigma_2 = \sigma_3 = \sigma$

[ME, ESE 2017 : 10 Marks]

Solution:

A.

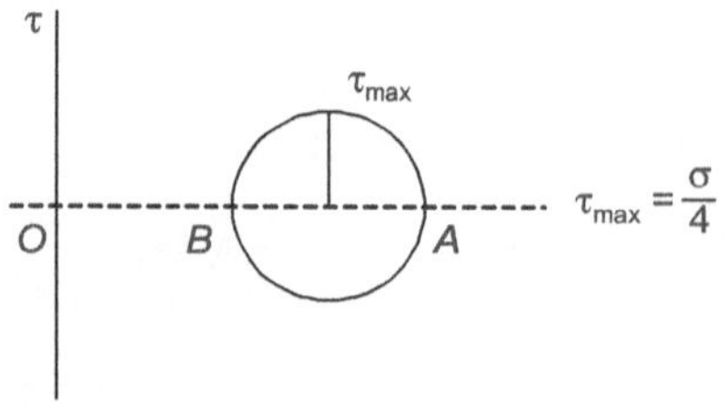

$$OA = \sigma_1 = \sigma \qquad OB = \frac{\sigma}{2} = \sigma_2$$

B.

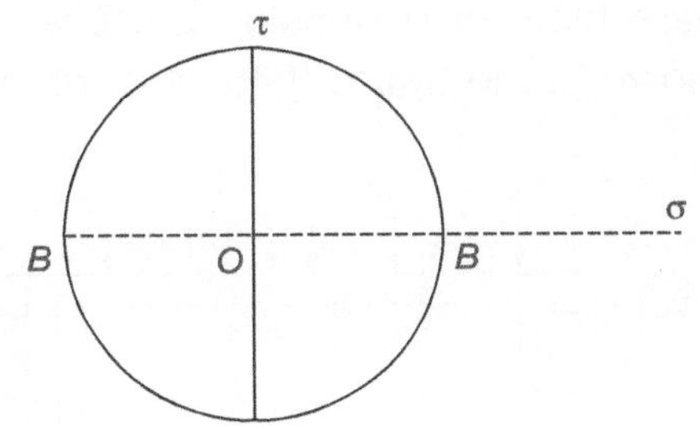

$$\sigma_1 = OA = +\sigma,\ OB = -\sigma = \sigma_2$$
$$\tau_{max} = OC = OC = \sigma$$

C.

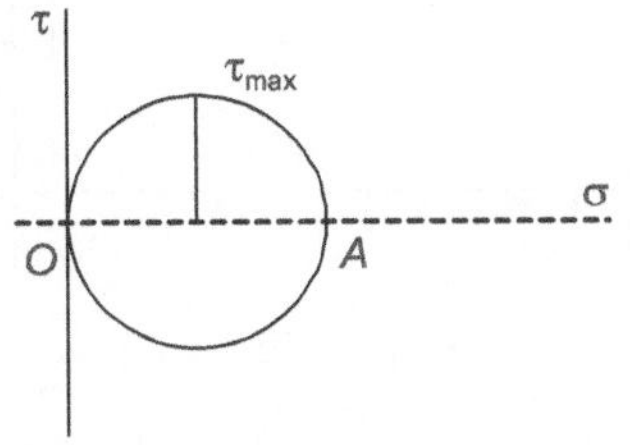

$$\tau_{max} = \frac{\sigma}{2}$$

$$\sigma_1 = OA = \sigma$$
$$\sigma_2 = \sigma_3 = 0$$

D.

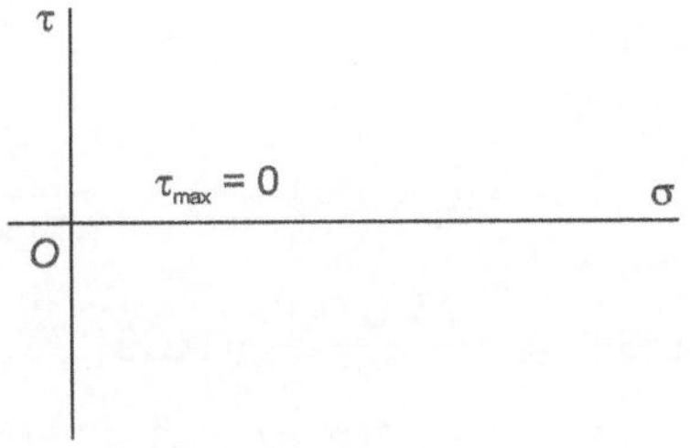

$$OA = OB = OC = 0$$
$$\sigma_1 = \sigma_2 = \sigma_3 = 0$$
$$\tau_{max} = 0$$

Question 3.28 The state of stress at a point is given by σ_x = –6 MPa, σ_y = 4 MPa and τ_{xy} = –8 MPa. The maximum tensile stress (in MPa) at the point is ______.

[ME, GATE : 2014]

Solution:

$$\sigma_x = -6 \text{ MPa}$$
$$\sigma_y = 4 \text{ MPa}$$
$$\tau_{max} = -8 \text{ MPa}$$

$$p_1 = \frac{\sigma_x + \sigma_y}{2} + \sqrt{\left(\frac{\sigma_x - \sigma_y}{2}\right)^2 + \tau_{xy}^2}$$
$$= -1 + \sqrt{(-5)^2 + (-8)^2}$$
$$= -1 + \sqrt{25 + 64}$$
$$= 1 + 9.434 = +8.434 \text{ MPa}$$

Question 3.29 A square plate of dimension $L \times L$ is subjected to a uniform pressure load $p = 250$ MPa on its edges as shown in the figure 3.45. Assume plane stress conditions. The Young's modulus $E = 200$ GPa.

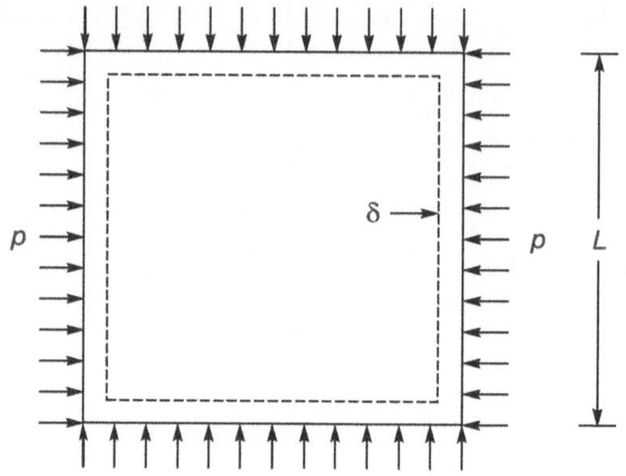

Fig. 3.45

The deformed shape is a square of dimension $L - 2\delta$. If $L = 2$ m and $\delta = 0.001$ m, the Poisson's ratio of the plate material is ________

[ME, GATE : 20016 (set-3)]

Solution:

Principal strains, $\varepsilon_1 = \varepsilon_2 = p(1 - \nu) = \dfrac{250}{200 \times 1000}(1 - \nu)$

$$\text{Compressive strain} = \frac{2 \times 0.001}{2} = 0.001$$

$$250(1 - \nu) = 2 \times 10^5 \times 0.001 = 250$$

$$(1 - \nu) = \frac{200}{250} = 0.8$$

$$\text{Poisson's ratio} = 0.2$$

Question 3.30 For the state of stresses (in MPa) shown in the figure 3.46, the maximum shear stress (in MPa) is _____

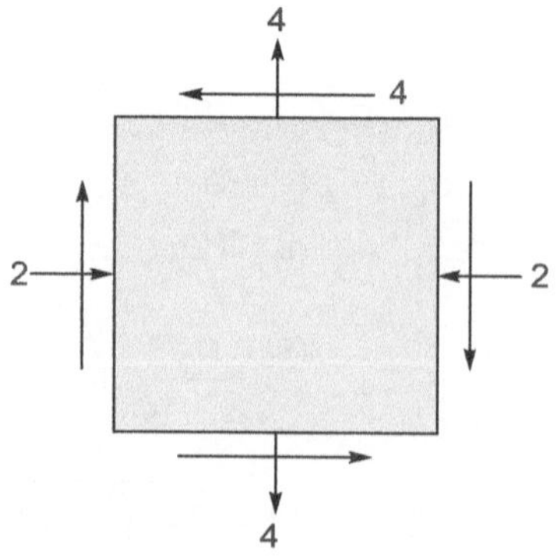

Fig. 3.46

[CE, GATE : 2014]

Solution:

Say, stresses,

$$\sigma_1 = +4$$
$$\sigma_2 = -2$$
$$\tau = 4$$

$$\tau_{max} = \frac{4-2}{2} + \sqrt{\left(\frac{4+2}{2}\right)^2 + 4^2} = 1 + 5 = 6 \text{ MPa}$$

Question 3.31 Two triangular wedges are glued together as shown in the following figure 3.47. The stress acting normal to the interface is ________ MPa.

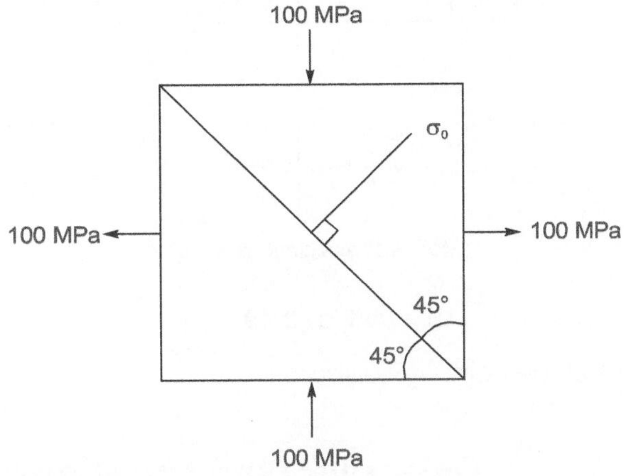

Fig. 3.47

[CE, GATE : 2015]

Solution:

$$p_1 = +100 \text{ MPa}$$
$$p_2 = -100 \text{ MPa}$$
$$\theta = 45°$$

$$\sigma_n = \frac{p_1 + p_2}{2} + \frac{p_1 - p_2}{2}\cos 90°$$
$$= 0 + 100 \times 0 = 0$$

Question 3.32 The uniaxial yield stress of a material is 300 MPa. According to Von Mises criterion, the shear yield stress (in MPa) of the material is ______

[ME, GATE : 2015 (set-2)]

Solution:

Principal stress are $+\tau$, $-\tau$, 0

$$(\tau + \tau)^2 + (-\tau - 0)^2 + (0 - \tau)^2 = 2\sigma_{yp}^2$$
$$6\tau^2 = 2\sigma_{yp}^2$$
$$\tau = \frac{1}{\sqrt{3}}\sigma_{yp}$$
$$= \frac{300}{\sqrt{3}} = 173.2 \text{ MPa}$$

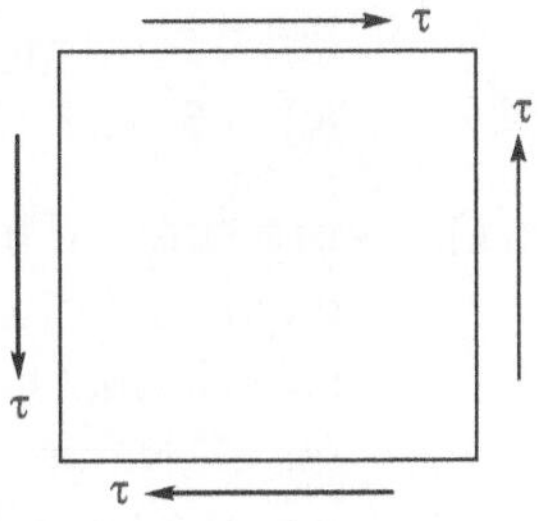

Fig. 3.48

Question 3.33 The state of stress at a point is $\sigma_x = \sigma_y = \sigma_z = \tau_{xz} = \tau_{zx} = \tau_{yz} = \tau_{xy} = 0$ and $\tau_{xy} = \tau_{yx} = 50$ MPa. The maximum normal stress (in MPa) at that point is ______

[ME, GATE : 2017 (set-1)]

Solution:

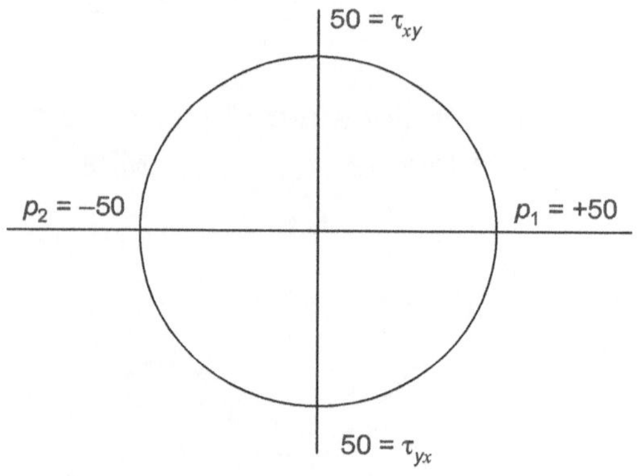

Mohr's stress circle at the point

Fig. 3.49

Maximum normal stress at the point = 50 MPa.

Objective Type Questions

Q.1 State of stress at a point is shown in figure. On inclined plane AC, $\sigma_\theta = +20$ MPa. What is shear stress on the plane. What is maximum principal stress at the point.

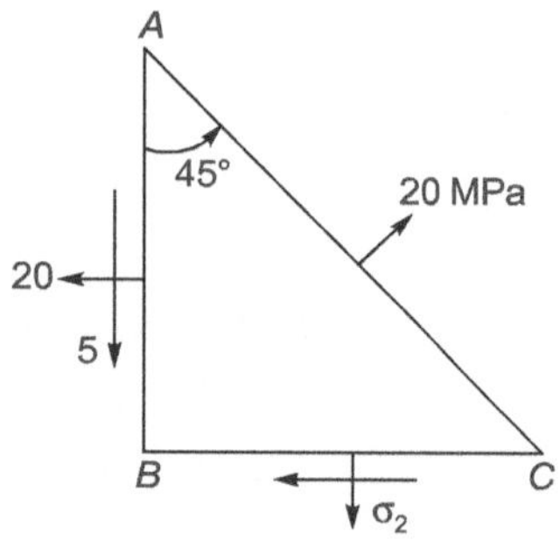

(a) 0.0, 15 MPa (b) +5, +15.07 N/mm^2
(c) +5, +22.07 MPa (d) –5, +22.07 N/mm^2

Q.2 In a strained material at a point, $\tau_{max} = 50$ MPa, major principal stress = 2 × minor principal stress, what is maximum principal stress
(a) 50 N/mm^2 (b) 100 N/mm^2
(c) 150 N/mm^2 (d) 200 N/mm^2

Q.3 Consider the Mohr's stress circle shown in figure. What is the state of stress represented by this circle

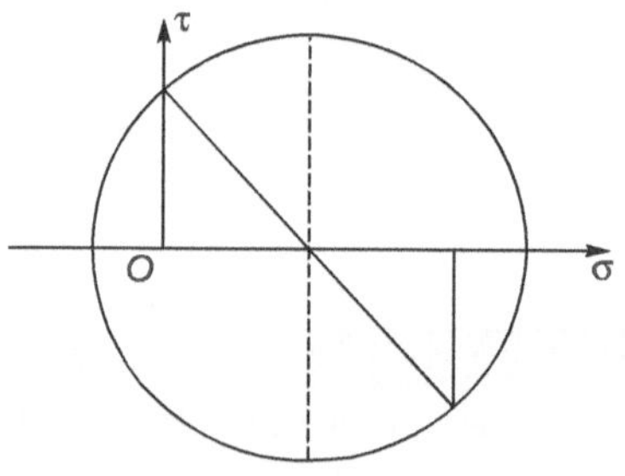

(a) $\tau_{xy} \neq 0$, $\sigma_x \neq 0$, σ_y
(b) $\tau_{xy} \neq 0$, $\sigma_x \neq 0$, $\sigma_y = 0$
(c) $\tau_{xy} = 0$, $\sigma_x \neq 0$, $\sigma_y = 0$
(d) None of these

Q.4 In a strained material, stresses on two mutually perpendicular planes are σ_x, τ_{xy} and σ_y, τ_{xy}. One of the principal stress will be equal to zero, if

(a) $\tau_{xy} = \dfrac{\sigma_x + \sigma_y}{2}$ (b) $\tau_{xy} = \dfrac{\sigma_x - \sigma_y}{2}$

(c) $\tau_{xy} = \sqrt{\sigma_x \times \sigma_y}$ (d) $\tau_{xy} = \sqrt{\sigma_x^2 + \sigma_y^2}$

Q.5 In a strained material, on two mutually perpendicular planes only the normal stress $+\sigma$ and $-\sigma$ act. What is the normal stress on a plane inclined at 45° to the plane of $+\sigma$ and $-\sigma$

(a) 2σ (b) σ

(c) $\dfrac{\sigma}{2}$ (d) zero

Q.6 At a point, if p_1 = maximum principal stress, p_2 = minimum principal stress and τ_{max} = maximum shear stress and ratio p_1/p_2 = 2.5. Then what is the ratio of p_1/τ_{max}

(a) 2.0 (b) $\dfrac{7}{3}$

(c) $\dfrac{10}{3}$ (d) $\dfrac{11}{3}$

Q.7 Principal stresses at a point are +160 MPa, 120 MPa. If Poisson's ratio of material is 0.35. What is the ratio of principal strains $\varepsilon_1/\varepsilon_2$

(a) 1.148 (b) 1.333

(c) 1.84 (d) 1.96

Q.8 At a point in a two dimensional state of strain there is pure shearing strain of magnitude γ_{xy} radians. Which one of the following is the maximum principal strain

(a) γ_{xy} (b) $\dfrac{\gamma_{xy}}{\sqrt{2}}$

(c) $\dfrac{\gamma_{xy}}{2}$ (d) $2\gamma_{xy}$

Q.9 Consider a two dimensional state of stress given for an element as shown in the diagram given below. What are the coordinates of the centre of Mohr's circle?

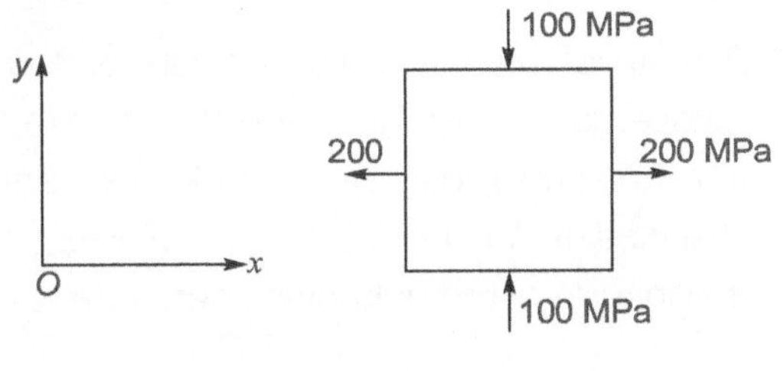

(a) (0.0) (b) (100, 200)

(c) (200, 100) (d) (50, 0)

[IES 2004]

Q.10 A rectangular strain gage rosette, shown in figure, gives following readings in a strain measurement, which are

$\varepsilon_1 = 1000 \times 10^{-6}$, $\varepsilon_2 = 800 \times 10^{-6}$ and

$\varepsilon_3 = 600 \times 10^{-6}$

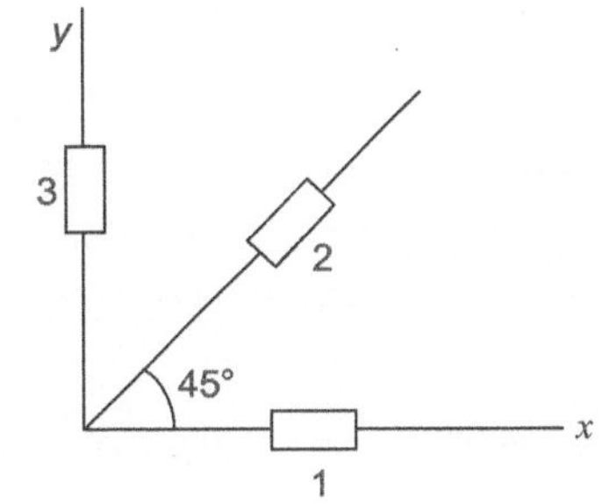

The direction of principal strain with respect to gauge 1 is

(a) 0° (b) 15°

(c) 35° (d) 45°

[IES 20111]

Q.11 If the principal stresses and maximum shearing stresses are of equal numerical value at a point in a stress body. The state of stress can be termed as

(a) isotropic

(b) uniaxial

(c) pure shear

(d) generalised plane stress

[IES 2010]

Q.12 A shaft with a circular cross-section is subjected to pure twisting moment. The ratio of the maximum shear stress to the largest principal stress is

(a) 2.0 (b) 1.0

(c) 0.5 (d) 0

Q.13 The state of stress on an element is as shown in the figure. If E = 2×10^5 N/mm^2 and Poisson's ratio = 0.3, the magnitude of the stress σ for no strain in BC is

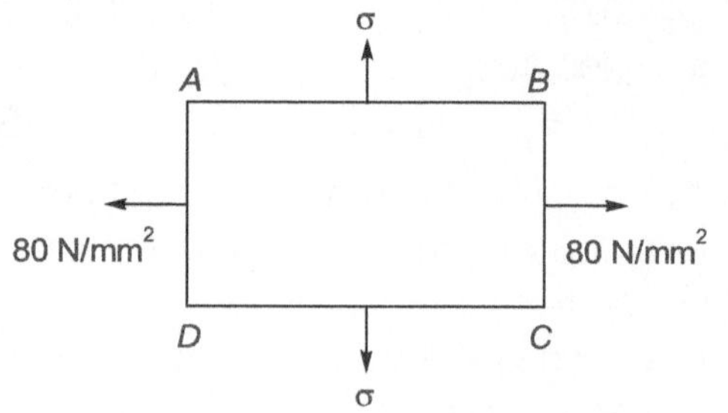

(a) 84 N/mm² (b) 64 N/mm²
(c) 34 N/mm² (d) 24 N/mm²

Q.14 A metal piece under the stress state of three principal stresses 30, 10 and 5 N/mm² is undergoing plastic deformation. The principal strain rates will be in the proportions of
(a) 15, –5 and –10 (b) –15, 5 and –10
(c) 15, 5 and 10 (d) –15, –5 and 10

Q.15 The state of stress at a point in a loaded member is σ_x = 400 MPa, σ_y = –400 MPa and τ_{xy} = ±300 MPa. The principal stresses σ_1 and σ_2 are
(a) 300 MPa and –700 MPa
(b) 400 MPa and –600 MPa
(c) 500 MPa and –500 MPa
(d) 600 MPa and –400 MPa

Q.16 A body is subjected to a direct tensile stress of 300 MPa on one plane accompanied by a simple shear stress of 200 MPa. The maximum normal stress on the plane will be
(a) 100 MPa (b) 200 MPa
(c) 300 MPa (d) 400 MPa

Q.17 The state of stress at a point when completely specified enables one to determine the
1. maximum shearing stress at the point
2. stress components on ay arbitrary plane containing that point

Which of the above is/are correct?
(a) 1 only (b) 2 only
(c) Both 1 and 2 (d) Neither 1 nor 2

Q.18 The principal stresses at a point in a stressed material are σ_1 = 200 N/mm², σ_2 = 150 N/mm², and σ_3 = 200 N/mm². E = 210 kN/mm² and μ = 0.3. The volumetric strain will be
(a) 8.954×10^{-4} (b) 8.954×10^{-2}
(c) 6.54×10^{-3} (d) 6.54×10^{-4}

Q.19 If strains on a piece of metal are ε_x = –120 μm/m, ε_y = –30 μm/m, and γ = 120 μm/m, what is the maximum principal strain?
(a) 0 (b) 50 μm/m
(c) 75 μm/m (d) 150 μm/m

Q.20 A gas is stored in a cylindrical tank of inner radius 7 m and wall thickness 50 mm. The gage pressure of the gas is 2 MPa. The maximum shear stress (in MPa) in the well is
(a) 35 (b) 70
(c) 140 (d) 280

Q.21 The state of stress at a point is given by : σ_x = 80 MPa, σ_y = 100 MPa and τ_{xy} = 60 MPa. If the yield strength for the material is 150 MPa, as determined in tri-axial test, then the maximum shear stress is, nearly
(a) 150.8 MPa (b) 127.4 MPa
(c) 119.3 MPa (d) 104.0 MPa

Q.22 Which of the following statements are correct?
1. Strain in the direction of applied stress is known as longitudinal strain.
2. Tensile stress results in tensile strain in linear and lateral directions.
3. Strains in all directions perpendicular to the applied stress are known as lateral strain.
4. Ratio of change in volume to original volume is known as volumetric strain.

(a) 1, 2 and 3 only (b) 1, 2 and 4 only
(c) 3 and 4 only (d) 1, 2, 3 and 4

Q.23 For a strain gauge (gauge factor = 2.1 and resistance = 50 Ω), subjected to a maximum strain of 0.001, the maximum change is resistance is
(a) 0.084 Ω (b) 0.105 Ω
(c) 0.135 Ω (d) 0.156 Ω

Q.24 A state of plane stress consists of a uniaxial tensile stress of magnitude 8 kPa, exerted on vertical surfaces and of unknown shearing stresses. If the largest stress is 10 kPa, then the magnitude of the unknown shear stress will be

(a) 6.47 kPa (b) 5.47 kPa
(c) 4.47 kPa (d) 3.47 kPa

Q.25 Which of the following statements are correct for stresses acting on mutually perpendicular faces of a plane element?

1. The sum of the normal stresses in mutually perpendicular planes is equal in the sum of the principal stresses.
2. The shearing stresses in two mutually perpendicular planes are equal in magnitude and direction.
3. Maximum shear stress is half of the difference between principal stresses.

(a) 1, 2 and 3 (b) 1 and 2 only
(c) 2 and 3 only (d) 1 and 3 only

Q.26 For the state of stress shown in the figure, the maximum and minimum principal stresses (taking tensile stress as +, and compressive stress as –) will be

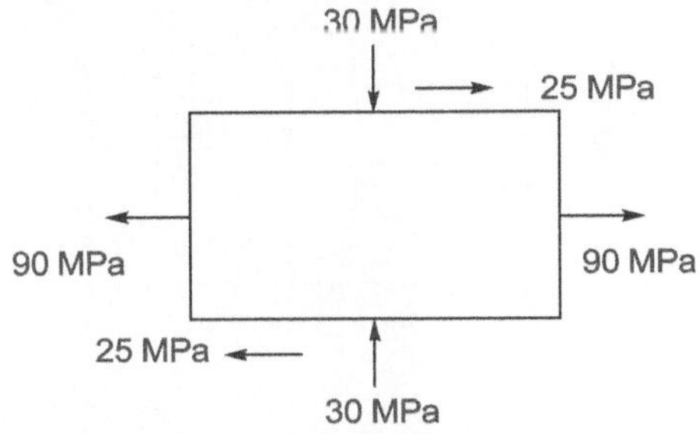

(a) 95 MPa and (–35) MPa
(b) 60 MPa and 30 MPa
(c) 95 MPa and (–30) MPa
(d) 60 MPa and 35 MPa

Q.27 The state of stress at a certain point in a stressed body is as shown in the figure. Normal stress in x-direction is 80 MPa (tensile) and y-direction is 40 MPa (Compressive). The radius of the Mohr's circle for this state of stress will be

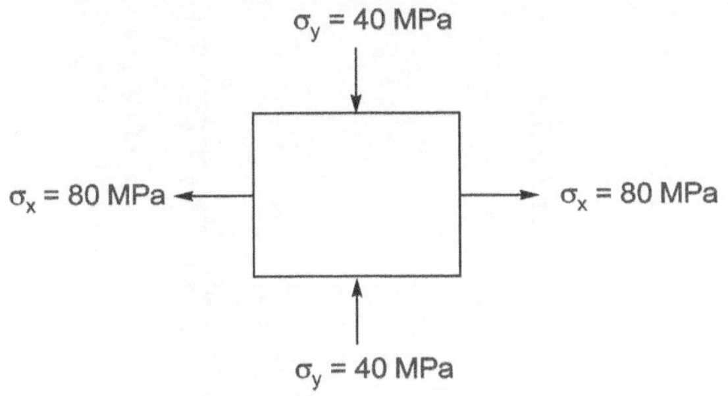

(a) 60 MPa (b) 40 MPa
(c) 20 MPa (d) 10 MPa

Q.28 A square element of a structural part is subjected to biaxial stresses as shown in the figure. On a plane along *BD*, the intensity of the result stress due to these conditions will be

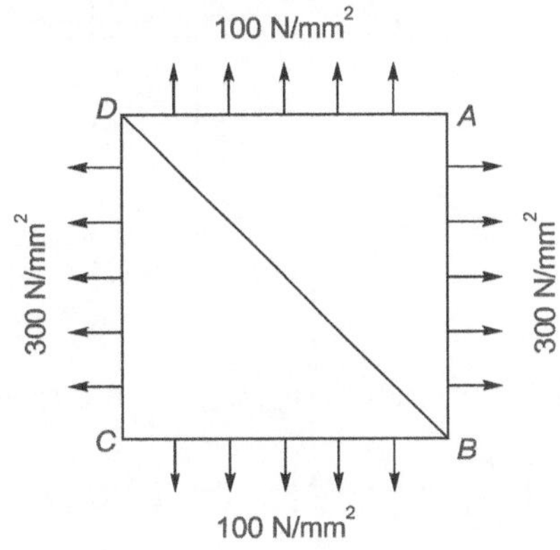

(a) $25\sqrt{5}$ N/mm^2 (b) $50\sqrt{5}$ N/mm^2
(c) $75\sqrt{5}$ N/mm^2 (d) $100\sqrt{5}$ N/mm^2

Q.29 A mild steel bar, 1.5 m long, has a square section 40 mm × 40 mm. The bar is subjected to a two-dimensional stress, σ_x = 310 N/mm^2 (Tensile) and σ_y = 300 N/mm^2 (Compressive). $E = 2 \times 10^5$ N/mm^2, Poisson's ratio μ = 0.3. The elongation of the bar is the direction of σ_x will be

(a) 1.25 mm (b) 1.75 mm
(c) 2.25 mm (d) 3 mm

Q.30 A structural element is subjected to pure shear of 80 N/mm^2, as shown in the figure. The yield stresses both in tension and in compression are 240 N/mm^2. According to the maximum normal stress theory, the factors of safety in tension and compression are, respectively

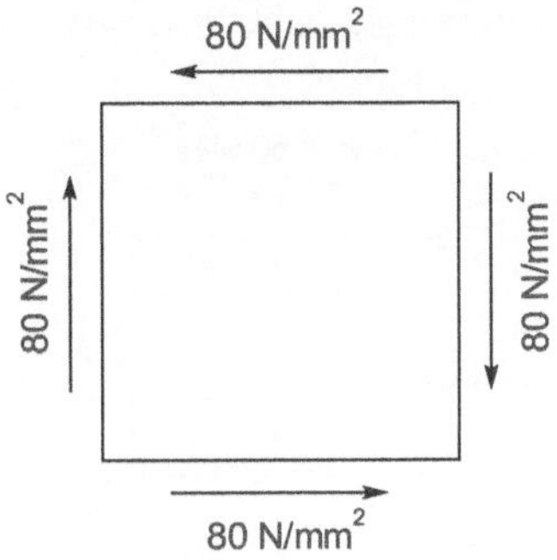

(a) 2 and 2 (b) 2.5 and 2.5
(c) 3 and 3 (d) 4 and 4

Q.31 The state of stress at a point on an element is shown in figure (a). The same state of stress is shown in another coordinate system in figure (b).

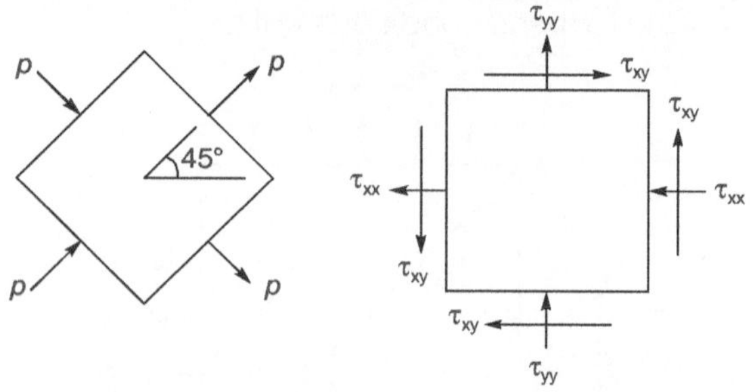

The components $(\tau_{xx}, \tau_{yy}, \tau_{xy})$ are given by

(a) $\left(\frac{P}{\sqrt{2}}, -\frac{P}{\sqrt{2}}, 0\right)$ (b) $(0, 0, p)$

(c) $\left(p, -p, \frac{p}{\sqrt{2}}\right)$ (d) $\left(0, 0, \frac{p}{\sqrt{2}}\right)$

Q.32 For the stress state (in MPa) shown in the figure, the major principal stress is 10 MPa.

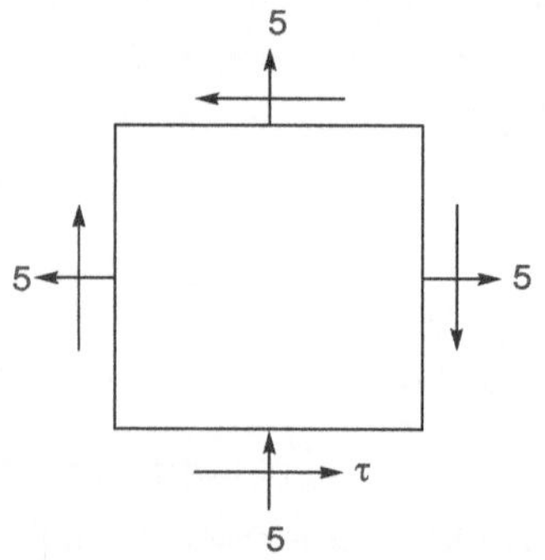

The shear stress τ is

(a) 10.0 MPa (b) 5.0 MPa
(c) 2.5 MPa (d) 0.0 MPa

Q.33 For the plane stress situation shown in the figure, the maximum shear stress and the plane on which it acts are

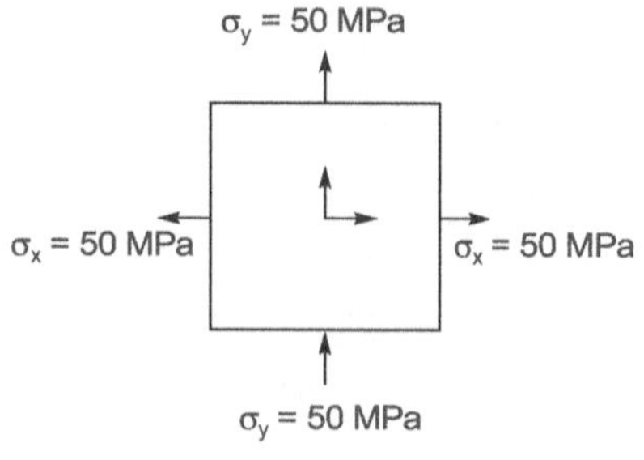

(a) –50 MPa, on a plane 45° clockwise w.r.t. x-axis
(b) –50 MPa, on a plane 45° anti-clockwise w.r.t. x-axis
(c) 50 MPa, at all orientations
(d) Zero, at all orientations

Q.34 In a plane stress condition, the components of stress at a point are σ_x = 20 MPa, σ_y = 80 MPa and τ_{xy} = 40 MPa. The maximum shear stress (in MPa) at the point is

(a) 20 (b) 25
(c) 50 (d) 100

Q.35 The state of stress at a point for a body in plane stress in shown in the figure below. If the minimum principal stress in 10 kPa, then the normal stress σ_y (in kPa) is

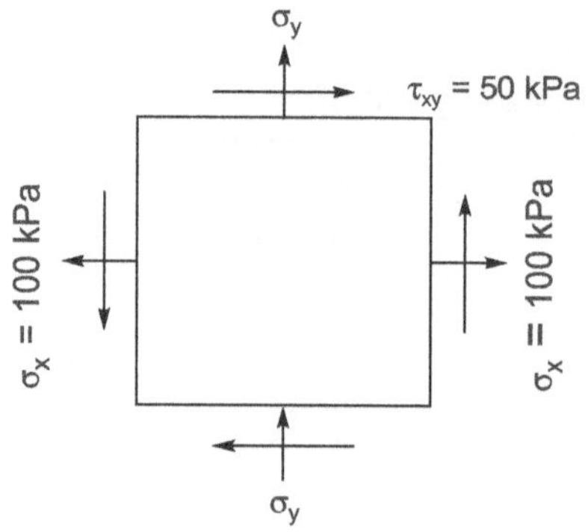

(a) 9.45 (b) 18.88
(c) 37.78 (d) 78.50

Q.36 A plate in equilibrium is subjected to uniform stresses along its edges with magnitude σ_{xx} = 30 MPa and σ_{yy} = 50 MPa as shown in figure.

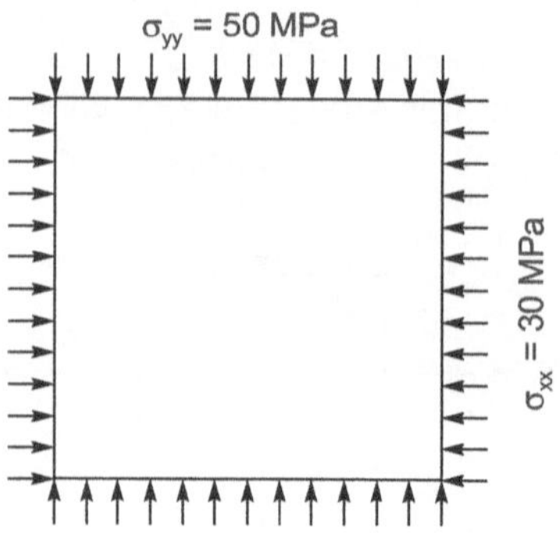

The Young's modulus of the material is 2×10^{11} N/m^2, and Poisson's ratio 0.3. If σ_{zz} is negligible and assumed to be zero, what is strain ε_{zZ}

(a) -120×10^{-6} (b) -60×10^{-6}

(c) 0.0 (d) $+120 \times 10^{-6}$

Answers

1. (c)	2. (d)	3. (b)	4. (c)	5. (d)
6. (c)	7. (c)	8. (c)	9. (d)	10. (a)
11. (c)	12. (b)	13. (d)	14. (a)	15. (c)
16. (d)	17. (c)	18. (1.0476×10^{-3})	19. (d)	
20. (c)	21. (a)	22. (b)	23. (b)	24. (c)
25. (d)	26. (a)	27. (a)	28. (d)	29. (d)
30. (c)	31. (b)	32. (b)	33. (d)	34. (c)
35. (c)	36. (a)			

Explanations

1. (c)

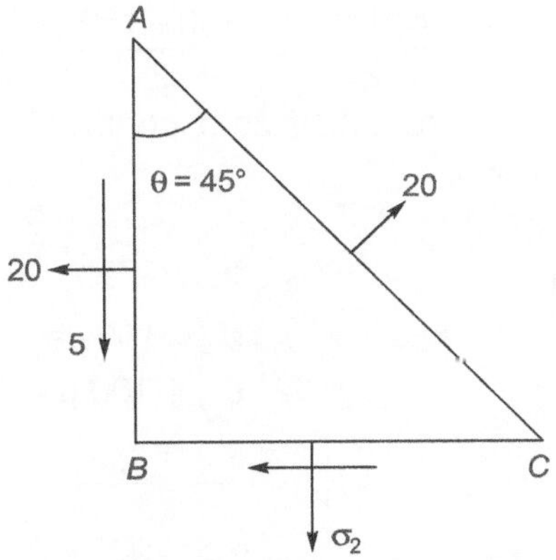

Say σ_2 is normal stress as plane *BC*, then

$$20 = \frac{20+\sigma_2}{2} + \frac{20-\sigma_2}{2}\cos 90° + 5\sin 90° = 10 + \frac{\sigma_2}{2} + 5$$

$$\sigma_2 = 10 \text{ N/mm}^2$$

$$\tau_\theta = \frac{\sigma_1 - \sigma_2}{2}\sin 2\theta - \tau\cos 2\theta = \frac{20-10}{2}\sin 90° - 5\cos 90°$$

$$= +5 \text{ N/mm}^2$$

$$p_{max} = \frac{\sigma_1 + \sigma_2}{2} + \sqrt{\left(\frac{\sigma_1 - \sigma_2}{2}\right)^2 + \tau^2}$$

$$= 15 + \sqrt{5^2 + 5^2}$$

$$= 22.07 \text{ N/mm}^2$$

2. (d)

$$p_{max} = \frac{\sigma_1 + \sigma_2}{2} + \tau_{max}$$

$$p_{min} = \frac{\sigma_1 + \sigma_2}{2} - \tau_{max}$$

$$p_{max} - p_{min} = 2\,\tau_{max} \qquad \text{but } p_{max} = 2p_{min}$$

$$p_{min} = \frac{1}{2}p_{max}$$

So $\frac{1}{2}p_{max} = 2\tau_{max}$

$$p_{max} = 4\,\tau_{max}$$

$$= 4 \times 50 = 200 \text{ N/mm}^2$$

3. (b)

$\tau_{xy} \neq 0,\ \sigma_x \neq 0\ ,\ \sigma_y =$

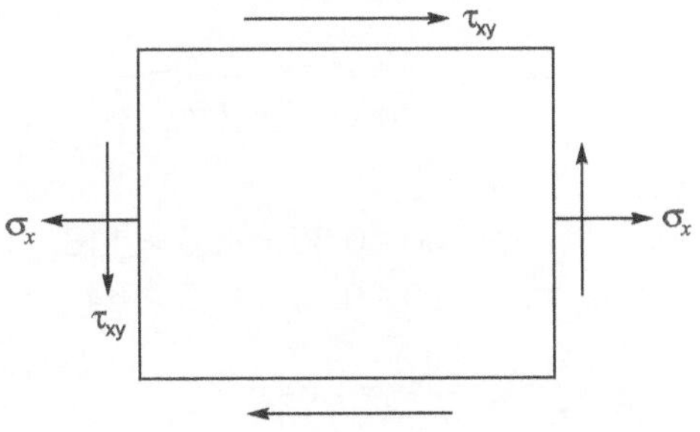

4. (c)

$$p_{min} = 0 = \frac{\sigma_x + \sigma_y}{2} - \sqrt{\left(\frac{\sigma_x - \sigma_y}{2}\right)^2 + \tau_{xy}^2}$$

or $$\frac{\sigma_x + \sigma_y}{2} = \sqrt{\left(\frac{\sigma_x - \sigma_y}{2}\right)^2 + \tau_{xy}^2}$$

or $$\left(\frac{\sigma_x + \sigma_y}{2}\right)^2 = \left(\frac{\sigma_x - \sigma_y}{2}\right)^2 + +\tau_{xy}^2$$

or $$\sigma_x\,\sigma_y = \tau_{xy}^2$$

or $$\tau_{xy} = \sqrt{\sigma_x \times \sigma_y}$$

5. (d)

$$\sigma_\theta = \frac{\sigma - \sigma}{2} + \frac{\sigma - (-\sigma)}{2}\cos 90^\circ$$
$$= 0 + 0 = 0$$

6. (c)

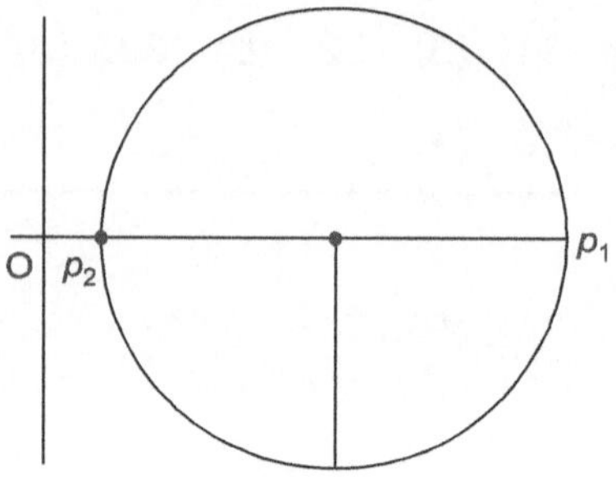

$$\frac{p_1}{p_2} = 2.5$$
$$p_2 = 0.4\,p_1$$
$$\frac{p_1 - p_2}{2} = \tau_{max}$$
$$p_1 - 0.4p_1 = 2\tau_{max}$$
$$p_1 \times 0.6 = 2\,\tau_{max}$$
$$\frac{p_1}{\tau_{max}} = \frac{2}{0.6} = \frac{10}{3}$$

7. (c)

$$\varepsilon_1 = 160 - 0.35 \times 120 \propto 118$$
$$\varepsilon_2 = 120 - 0.35 \times 160 \propto 64$$
$$\frac{\varepsilon_1}{\varepsilon_2} = \frac{118}{64} = 1.84$$

8. (c)

$$\varepsilon_1 = \frac{\gamma_{xy}}{2}$$

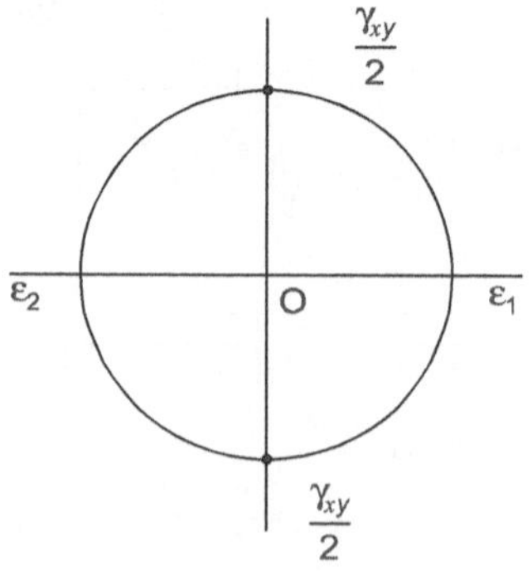

9. (d)

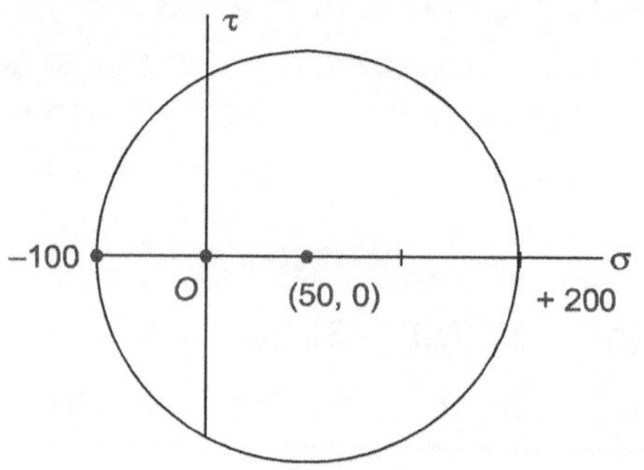

$$\sigma_1 = +200\text{ MPa}$$
$$\sigma_2 = -100\text{ MPa}$$

x-coordinates of centre of Mohr's stress circle

$$\frac{\sigma_1 + \sigma_2}{2} = \frac{200 - 150}{2} = +50$$

So, y-coordinate is zero, centre of circle lies on abscissa.

10. (a)

$$\varepsilon_1 = \varepsilon_{xx} = 100\ \mu s$$
$$\varepsilon_3 = \varepsilon_{yy} = 100\ \mu s$$
$$\varepsilon_{45^\circ} = \frac{\varepsilon_{xx} + \varepsilon_{yy}}{2} + \frac{\gamma_{xy}}{2}$$

[as cos 90° = 0, sin 90° = 1]

$$800 = \frac{1000 + 600}{2} + \frac{\gamma_{xy}}{2}$$
$$\frac{\gamma_{xy}}{2} = 0$$

No shear stress, x and y directions give principal directions 1 and 2 respectively.

11. (c)

Pure shear

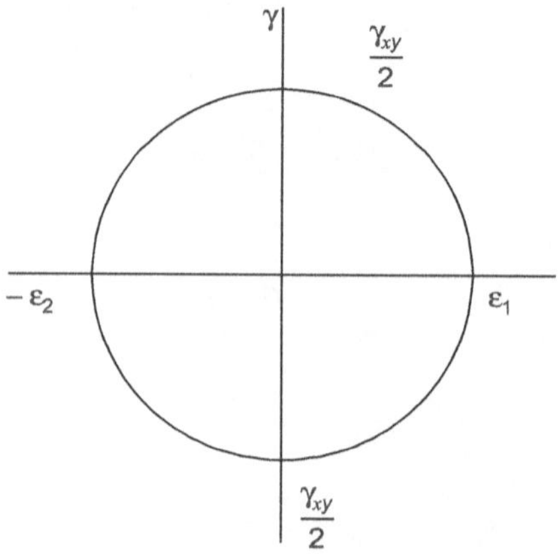

12. (b)

Shaft with pure twisting moment

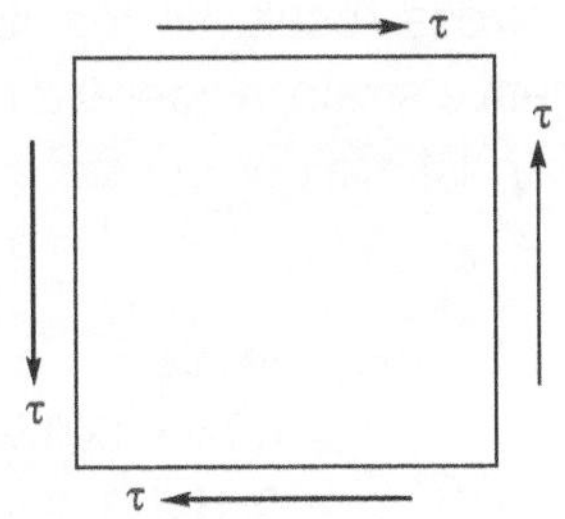

τ is maximum shear stress on principal shaft stress,

$$p_1 = +\tau$$

$$\frac{p_1}{\tau} = 1$$

13. (d)

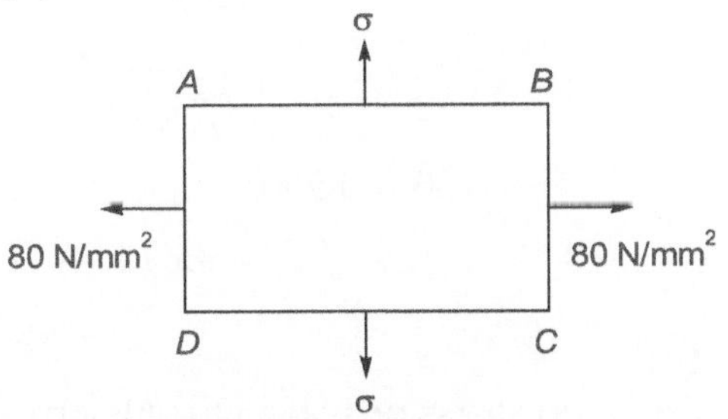

$$E = 200 \text{ GPa}$$

Possion's ratio, $\nu = 0.3$

For no strain is direction BC

$$\frac{\sigma}{e} - \frac{\nu \times (80)}{E} = 0$$

$$\sigma = \nu \times 80$$
$$= 0.3 \times 80$$
$$= 24 \text{ N/mm}^2$$

14. (a)

$\varepsilon_1 \propto 30 - 7.5 = 22.5$

$\varepsilon_2 \propto 10 - 17.5 = -7.5$

$\varepsilon_3 \propto 5 - 20 = -15$

$\varepsilon_1 : \varepsilon_2 : \varepsilon_3 \quad 3 - 1 - 1$

or $\quad 15, -5, -10$

15. (c)

Principal stresses

$$\sigma_x = 400$$
$$\sigma_y = -400$$
$$\tau_{xy} = +300$$
$$p_1 = 500 \text{ MPa}$$
$$p_2 = -500 \text{ MPa}$$
$$\frac{\sigma_x + \sigma_y}{2} = 0$$
$$\frac{\sigma_x - \sigma_y}{2} = 400$$
$$\sqrt{\left(\frac{\sigma_x - \sigma_y}{2}\right)^2 + \tau_{xy}^2} = \pm 500$$

So, $p_1 = +500$, $p_2 = -500$ MPa

16. (d)

Simple stresses

$$p = \frac{300}{2} + \sqrt{\left(\frac{300}{2}\right)^2 + 200^2}$$
$$= 150 + 250$$
$$= 400 \text{ MPa}$$
$$= \text{Max. principal stress}$$
$$= \frac{\sigma}{2} + \sqrt{\left(\frac{\sigma}{2}\right)^2 + \tau^2}$$

17. (c)

Principal stresses

State of stress when completely specified at a point, enables one to determine maximum shearing stress and stress components an any arbitrary plane containing that point.

18. (*)

Principal stresses

$$\sigma_1 = 200 \text{ MPa}$$
$$\sigma_2 = 150 \text{ MPa}$$
$$\sigma_3 = 200 \text{ MPa}$$
$$\sigma_1 + \sigma_2 + \sigma_3 = 550 \text{ MPa}$$

Poisson's ratio, $\nu = 0.3$

Volumetric strain,

$$\varepsilon_v = \frac{\sigma_1 + \sigma_2 + \sigma_3}{E}(1 - 2\nu)$$
$$= \frac{550(1 - 2 \times 0.3)}{E}$$
$$= \frac{550 \times 0.4}{210000} = \frac{220}{210000}$$
$$= 1.0476 \times 10^{-3}$$

Wrong question.

19. (d)

$$\varepsilon_x = -120\,\mu,\ \text{microstrain}$$
$$\varepsilon_y = -30\,\mu$$
$$\frac{\gamma}{2} = 60\,\mu$$
$$\frac{\varepsilon_x + \varepsilon_y}{2} = 75$$
$$\frac{\varepsilon_x - \varepsilon_y}{2} = -45$$

ε_1 = maximum principal strain

$$= \frac{\varepsilon_x + \varepsilon_y}{2} + \sqrt{\left(\frac{\varepsilon_x - \varepsilon_y}{2}\right)^2 + \left(\frac{\gamma}{2}\right)^2}$$
$$= -75 + \sqrt{(-45)^2 + (60)^2}$$
$$= -75 + 75 = 0$$
$$\varepsilon_2 = -75 - 75 = -150\ \mu\text{m/m}$$

20. (c)

$$D = 14000\ \text{mm}$$
$$t = 50\ \text{mm}$$
$$p = 2\ \text{MPa}$$
$$\sigma_c = \frac{pD}{2t} = \frac{2 \times 14000}{2 \times 50}$$
$$= 280\ \text{MPa}$$
$$\sigma_a = \frac{1}{2}\sigma_c = 140\ \text{MPa}$$

Principal stress at inner radius
280, 140, –2 MPa

$$\tau_{max} = \frac{280 + 2}{2} = 141\ \text{MPa}$$
$$\simeq 140\ \text{MPa}$$

21. (a)

$$\sigma_x = +80\ \text{MPa}$$
$$\sigma_y = 100\ \text{MPa}$$
$$\tau_{xy} = \pm\,60\ \text{MPa}$$
$$\tau_{max} = 90 + \sqrt{10^2 + 60^2}$$
$$= 90 + 60.8$$
$$= 150.8\ \text{MPa}$$

22. (b)
Statement (3) incorrect
Tensile stress results in tensile stain in linear but negative strain in lateral direction.
(b) 1, 2 and 4 are correct statements.

23. (b)

$$\Delta R = R \varepsilon S_g$$
$$= 50 \times 0.001 \times 2.1$$
$$= 0.105\ \Omega$$

24. (c)

$$\sigma = 8\ \text{kPa}$$
$$\frac{\sigma}{2} = 4$$
$$p_1 = \frac{\sigma}{2} + \sqrt{\left(\frac{\sigma}{2}\right)^2 + \tau^2}$$
$$10 = 4 + \sqrt{(4)^2 + \tau^2}$$
$$36 = 16 + \tau^2$$
$$\tau = \sqrt{20} = 4.47\ \text{kPa}$$

25. (d)
Shearing stress on two mutually perpendicular planes are equal in magnitudes but opposite in direction

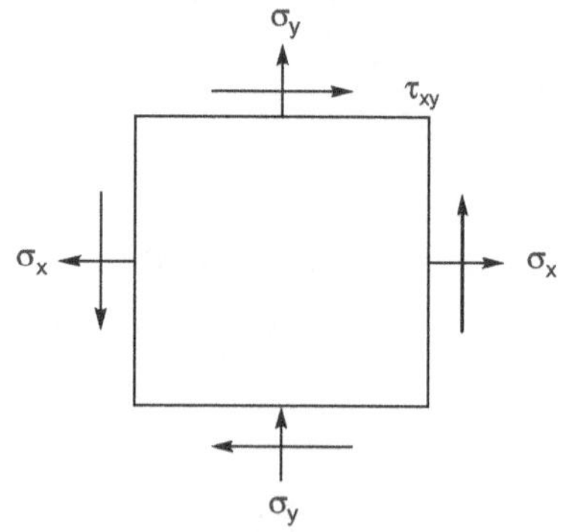

Principal stress, p_1 and p_2
1. $\sigma_x + \sigma_y = p_1 + p_2$ correct
2. $\tau_{max} = \frac{1}{2}(p_1 - p_2)$

26. (a)
Please note there will be complementary shear stresses as vertical planes also as shown.

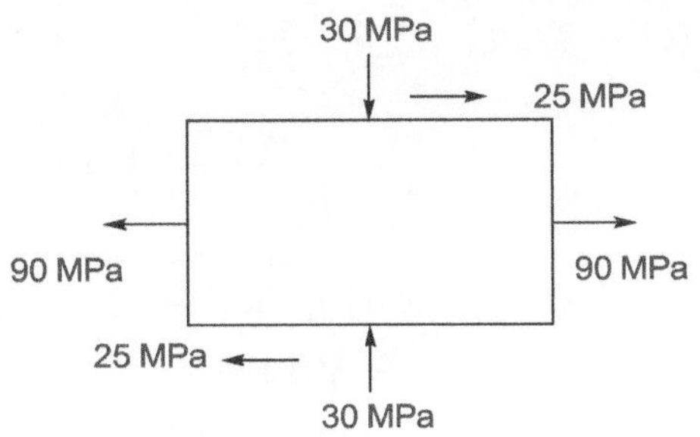

$$\sigma_1 = +90$$
$$\sigma_2 = -30$$
$$\tau = \pm 25$$
$$\frac{\sigma_1 + \sigma_2}{2} = 30$$
$$\frac{\sigma_1 - \sigma_2}{2} = 60$$

Principal stress are

$$p_1, p_2 = \frac{\sigma_1 + \sigma_2}{2} \pm \sqrt{\left(\frac{\sigma_1 - \sigma_2}{2}\right)^2 + \tau^2}$$
$$= 30 \pm \sqrt{60^2 + 25^2} = \pm 65 \text{ MPa}$$
$$p_1 = 95$$
$$p_2 = -35$$

27. (a)

Radius of Mohr's stress circle,

$$R = \frac{p_1 - p_2}{2}$$

p_1 = major principal stress

p_2 = minor principal stress

In the sure σ_x and σ_y are not accompanied by shear strain

So, $p_1 = \sigma_x = +80$ MPa (tensile)

$p_2 = \sigma_y = -40$ MPa (comp.)

R, radius of Mohr's stress circle

$$= \frac{80 - (-40)}{2} = 60 \text{ MPa}$$

28. (d)

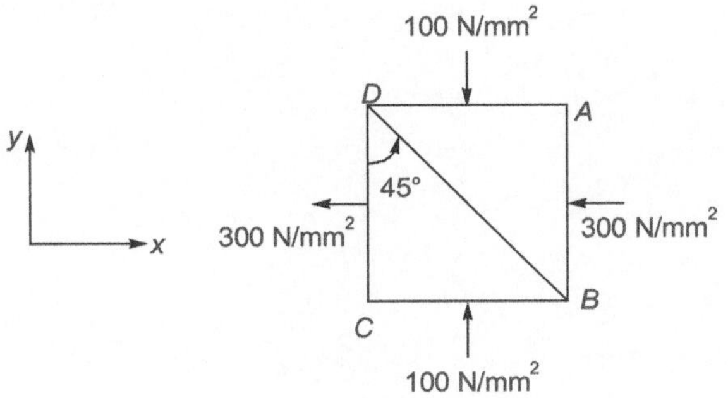

$$\sigma_x = 30 \text{ N/mm}^2$$
$$\sigma_y = 100 \text{ N/mm}^2$$

Plane BD

σ, normal stress

$$= \frac{300 + 100}{2} + \frac{300 - 100}{2} \times \cos 90°$$
$$= 200 \text{ N/mm}^2$$

τ_s shear stress

$$= \frac{300 - 100}{2} \times \sin 90° = 100 \text{ N/mm}^2$$

τ_r = resultant stress

$$= \sqrt{\sigma^2 + \tau^2} = \sqrt{200^2 + 100^2}$$
$$= 100\sqrt{5}$$

29. (d)

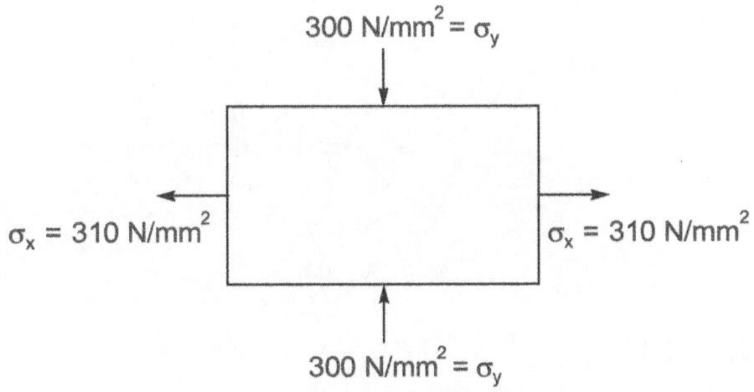

μ, Poisson ratio = 0.3

Strain in x-direction

$$\varepsilon_x = \frac{310}{E} - \frac{0.3(-300)}{E}$$
$$= \frac{400}{E} = \frac{400}{2 \times 10^5} = 2 \times 10^{-3}$$

Length, $L = 1.5$ m

$\delta L = 2 \times 10^{-3} \times 1500$ mm

$= 3$ mm

30. (c)

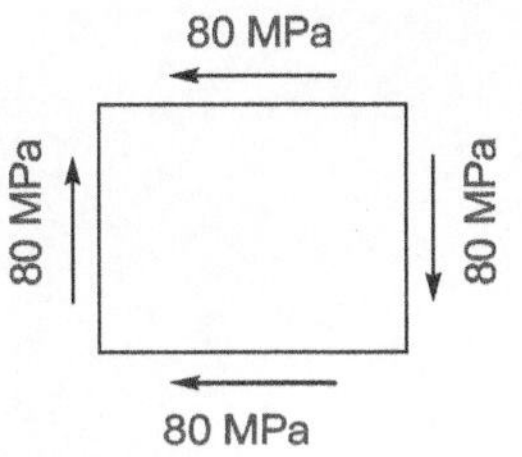

Principal stresses,

$p_1 = +80$ MPa

$$p_2 = 80 \text{ MPa}$$
$$\sigma_y = +240 \text{ MPa (tensile)}$$

Maximum normal stress theory

$$\text{(In tensile) FOS} = +\frac{240}{80} = 3$$

$$\text{(In comp.) FOS} = \frac{-240}{-80} = 3$$

(c) 3 and 3, correct answer.

31. (b)

When, $\tau_{xx} = \tau_{yy} = 0$

$$\tau_{xy} = p$$

Principal sress are +p, –p

at ± 45° as shown

$$\tau_{xx} = 0,\ \tau_{yy} = 0,\ \tau_{xy} = p$$

32. (b)

Major principal,

$$p_1 = 10 \text{ MPa}$$
$$= \frac{5+5}{2} + \sqrt{\left(\frac{5-5}{2}\right)^2 + \tau^2}$$
$$= 5 + \tau$$
$$\tau = 5 \text{ MPa}$$

33. (d)

Under plane stress coordinates

$$p_1 = p_2 = +50 \text{ MPa}$$

$$\tau_{max} = \frac{p_1 - p_2}{2} = 0 \text{ at all directions}$$

Mohr's stress circle is a point.

34. (c)

$$\tau_{max} = \sqrt{\left(\frac{\sigma_x - \sigma_y}{2}\right)^2 + \tau_{xy}^2}$$
$$= \sqrt{(-30)^2 + 40^2}$$
$$= 50 \text{ MPa}$$

35. (c)

$$p_{min} = \frac{\sigma_x + \sigma_y}{2} - \sqrt{\left(\frac{\sigma_x - \sigma_y}{2}\right)^2 + \tau^2}$$

$$10 = \frac{100 + \sigma_y}{2} - \sqrt{\left(\frac{100 - \sigma_y}{2}\right)^2 + 50^2}$$

$$\left(40 + \frac{\sigma_y}{2}\right)^2 = \left(50 - \frac{\sigma_y}{2}\right)^2 + 50^2$$

$$1600 + 40\sigma_y + \frac{\sigma_y^2}{4} = 2500 - 50\sigma_y + \frac{\sigma_y^2}{4} + 2500$$

$$90\,\sigma_y = 3400$$
$$\sigma_y = 37.78 \text{ MPa}$$

36. (a)

$$\varepsilon_{zz} = -\frac{\nu\sigma_{xx}}{E} - \frac{\nu\sigma_{yy}}{E} = -\frac{0.3 \times 80}{E}$$
$$= -\frac{24}{2 \times 10^5} = -12 \times 10^{-5} = -120 \times 10^{-6}$$

■■■■

Elastic Constants

CHAPTER 4

It is necessary to determine the values of elastic constants accurately because **all deformations** as deflection and slope in a beam subjected to bending moment, angular twist in a shaft, subjected to twisting moment and extension/contraction in a bar subjected to axial load, all **depend on elastic constants**. Two basic elastic constants E (Young's modulus), ν, (Poisson's ratio) can be determined by using a sample with axial load and using electrical resistant strain gauges, for the measurement of axial and lateral strains, other elastic constants as G (Shear modulus) and K (Bulk modulus) can be expressed in terms of E and ν.

Question 4.1 **Describe an experiment for the determination of Young's modulus and Poisson's ratio of a metallic sample.**

Solution: A sample of dog-bone shape is made from a flat of the metal as shown in fig. 4.1. strain gauges SG1 and SG2, SG3 and SG4 are mounted on sample in axial and transverse directions. Gauges SG1 and SG2 are axial direction gages and SG3 and SG4 in transverse directions are mounted on two faces of the sample. Strain gages are mounted away from the fillet portion (where stress distribution is non uniform).

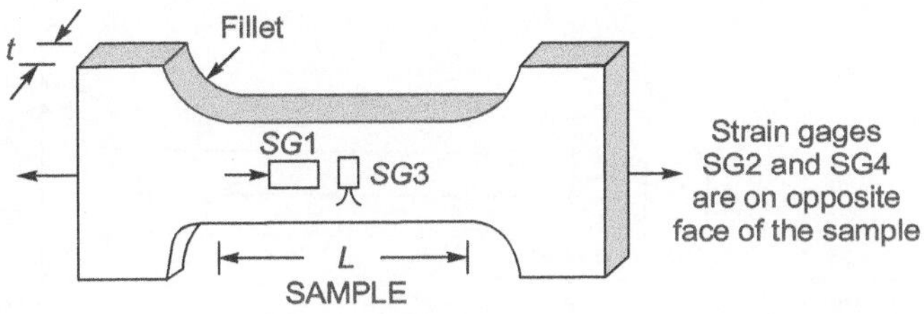

Fig. 4.1

The specimen is fitted in the grips of a universal testing machine. The strain gages SG1 and SG2 are connected in one Wheatstone bridge circuit and strain gages SG3 and SG4 are connected in another Wheatstone bridge circuit in strain indicators for the recording of axial strain ε_a and lateral strain ε_l respectively. Universal testing machine is put on and a gradually and slowly increasing tensile load is applied on the sample. **Reading of load W, axial strain, ε_a and lateral strain ε_l are continuously recorded.** Graphs between W and ε_a, and between ε_l and ε_a are plotted.

A = Area of cross-section of the sample in uniform portion

Slope of W vs. ε_a curve, $E = \dfrac{W}{A\varepsilon_a}$ = Young's modulus of elasticity

From slope of ε_l vs. ε_a curve, Poisson's ratio, $\nu = \dfrac{\varepsilon_l}{\varepsilon_a}$.

Following precautions must be taken while performing the experiment:

(i) The strain gages must be cured properly before the readings are started.

(ii) Apply the load slowly and gradually

(iii) Pass the straight line between the mean positions of points in graphs (W vs. ε_a), (ε_l vs. ε_a).

(vi) Sample must be of uniform thickness t.

(v) Take gauges length L of sample and width of the sample, fillet radius as per standard specifications for material testing.

(vi) Strain gauges must be connected in such a way so to have temperature compensation of the signal from strain gauges.

Practice Q.4.1 **A cooper bar of 10 mm in diameter and 100 mm in gauge length is tested under tension. A tensile force of 10.6 kN produces an axial change in length of 0.128 mm, while diameter of the bar is reduced by 0.0043 mm.**

Determine E and ν for copper

Ans: [105.4 kN/mm², 0.336]

Question 4.2 **Describe an experiment to determine modulus of rigidity of a metallic sample.**

Solution: A sample for torsion test of standard size is made in workshop as per figure 4.2. Diameter of sample is d and gauge length is L.

There are tapered circular ends with key ways as shown in the sample. The ends are inserted in the grips of the machine and in the grips, keys are provided so that the sample is firmly gripped in the machine chucks. One end of the specimen is fitted in a stationary grip.

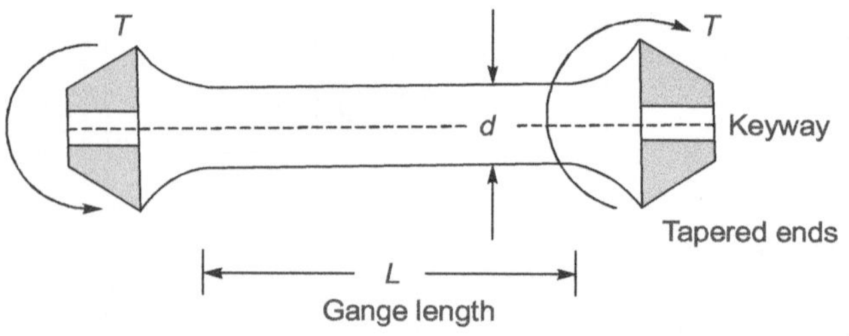

Fig.4.2

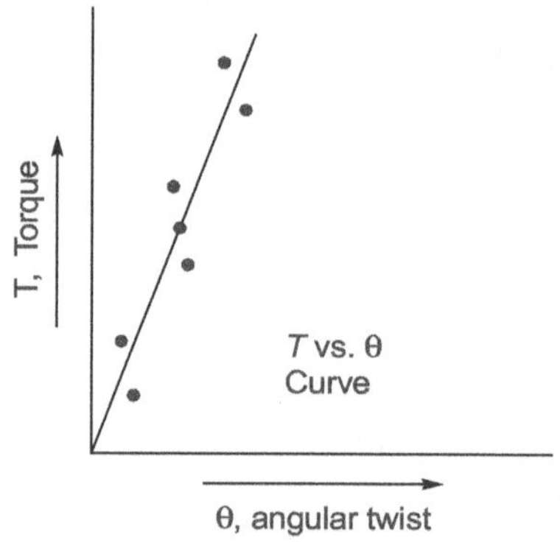

Fig.4.3

A torque is applied on the grip at the other end of the sample. Gradually torque is increased and reading of angular twist developed in the shaft is measured on a scale. Readings of torque and angular twist are simultaneously recorded and plotted as shown in figure 4.3.

From the graphs a reading of $\frac{T}{\theta}$ is obtained.

$$T = \frac{\pi}{16} \times d^3 \times \tau$$

where d is shaft diameter and τ is maximum shear stress developed on shaft.

Shear stress,
$$\tau = \frac{16T}{\pi d^3} \quad \ldots(i)$$

Moreover,
$$\frac{d}{2} \times \theta = L.\phi$$

or shear angle/shear strain $= \dfrac{\theta d}{2L}$

Ratio of $\dfrac{\tau}{\phi} = \dfrac{\text{shear stress}}{\text{shear strain}} = G = \text{shear modulus}$

$$\frac{16T}{\pi d^3} \times \frac{2L}{\theta d} = \frac{32TL}{\theta \pi d^4} = \text{modulus of rigidity} = G$$

Practice Q.4.2 **An aluminium specimen with gauge lengths of 250 mm and diameter 25 mm is tested in torsion. A torque of 17.8 Nm produces an angular twist of 0.24° in the specimen. Determine the modulus of rigidity of aluminum.**

Ans. [27.7 kN/mm²]

Question 4.3 **Derive relationship between E, G and ν i.e., Young's modulus, modulus of rigidity and Poisson's ratio of a material.**

Solution: Consider a cubical block of side 'a' fixed at bottom and subjected to a shear stress τ on the top surface. Complementary shear stresses τ are developed on vertical surfaces of the block to maintain equilibrium as shown in fig. 4.4.

Block is deformed into shape *ABC′D′* as a parallelogram. Diagonal AC is increased to *AC′* and diagonal *BD* is reduced to *BD′*.

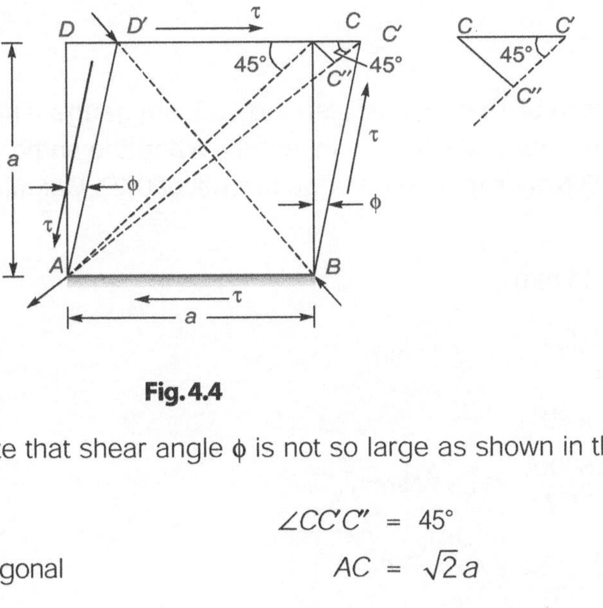

Fig. 4.4

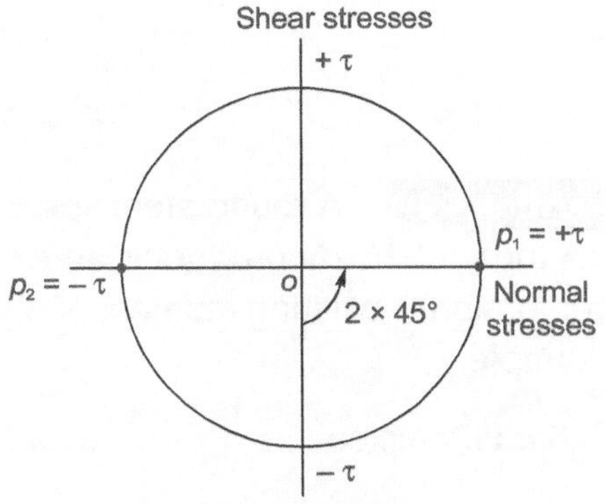

Mohr's stress Circle

Fig. 4.5

Note that shear angle φ is not so large as shown in the figure. For most of the metals φ is much less than 1°.

So $\angle CC'C'' = 45°$

Diagonal $AC = \sqrt{2}a$

Increase is diagonal $= C'C'' = \dfrac{CC'}{\sqrt{2}}$

ε, strain along diagonal $AC = \dfrac{\text{change in length}}{\text{original length}} = \dfrac{c'c''}{\sqrt{2}a}$

$$\frac{cc'}{\sqrt{2} \times \sqrt{2}a} = \frac{cc'}{2a} = \varepsilon$$

But shear strain, $\phi = \frac{cc'}{a}$

So, $\varepsilon = \frac{\phi}{2}$ ...(i)

If a Mohr's stress circle is drawn at the point with shear stress τ only on two perpendicular planes, it will be as shown in figure. 4.5.

Principal stress, $p_1 = \tau$

Similarly principal stress, $p_2 = -\tau$

If E = Young's modulus of the material

ε, strain along diagonal $AC = \frac{p_1}{E} - \frac{\nu p_2}{E}$, where ν is Poisson's ratio

$$\varepsilon = \frac{\tau}{\varepsilon} + \frac{\nu\tau}{\varepsilon} = \frac{\tau}{\varepsilon}(1+\nu) \quad \text{...(ii)}$$

From equation (i)

$$\varepsilon = \frac{\phi}{2} = \frac{\tau}{E}(1+\nu)$$

$$E = \frac{\tau}{\phi} \times 2(1+\nu)$$

But $\frac{\tau}{\phi}$ = G, Modulus of rigidity

$$E = 2G(1+\nu)$$

Question 4.4 **A round steel specimen of 15 mm diameter and 150 mm gauge length is tested in tension. A force of 16 kN produced an extension of 0.068 mm in length. When the same specimen is tested under torsion, a twisting moment of 8.93 Nm produced an angular twist of 0.2°. Calculate E, G and ν for steel sample.**

Solution: Diameter, d = 15 mm

Area of cross-section, $A = \frac{\pi}{4} \times 15^2 = 176.71 \text{ mm}^2$

Axial load, P = 16000 N

Axial stress, $\sigma = \frac{16000}{176.71} = 90.54 \text{ N/mm}^2$

Axial strain, $\varepsilon_a = \frac{0.068}{150} = 4.53 \times 10^{-4}$

Young's modulus, $E = \frac{\sigma}{\varepsilon_a} = \frac{90.54}{4.53 \times 10^{-4}} = 199.85 \text{ kN/mm}^2$

Torque, T = 8.93 Nm = 8930 Nmm

$$\theta = 0.2° = \frac{\pi}{900} \text{rad}$$

Modulus of rigidity, $G = \frac{32TL}{\pi d^4\theta} = \frac{32TL}{\pi d^4} \times \frac{900}{\pi}$

$$= \frac{32 \times 8930 \times 150 \times 900}{\pi \times 15^4 \times \pi}$$

$$= 77209 \text{ N/mm}^2 = 77.209 \text{ kN/mm}^2$$

$$\nu = \frac{E}{2G} - 1 = \frac{199.85}{2 \times 77.2} - 1$$

$$= 0.294, \text{ Poisson's ratio.}$$

Practice Q. 4.4 **A copper bar specimen of 15 mm diameter and 150 mm gauge length when tested in tension, a force of 7.81 kN produced an extension of 0.066 mm. When the same specimen is tested under torsion, a twisting moment of 4.35 Nm produced an angular twist of 0.2°. Determine *E*, *G* and ν of the material.**

Ans.: [100.45 kN/mm^2, 37.6 kN/mm^2, 0.335]

Question 4.5 **Derive relationship between *E*, ν and *K* i.e., Young's modulus, Poisson's ratio and Bulk modulus of a material.**

Solution: Consider a cubical block of side 'a' subjected to hydrostatic pressure p on all sides as shown in figure 4.6.

If E is the Young's modulus of the material, ν is the Poisson's ratio, then

Strain in direction, 1 $\varepsilon_1 = \frac{p}{E} - \frac{\nu p}{E} - \frac{\nu p}{E}$

$$= \frac{p}{E}(1 - 2\nu) \text{ (compressive strain)}$$

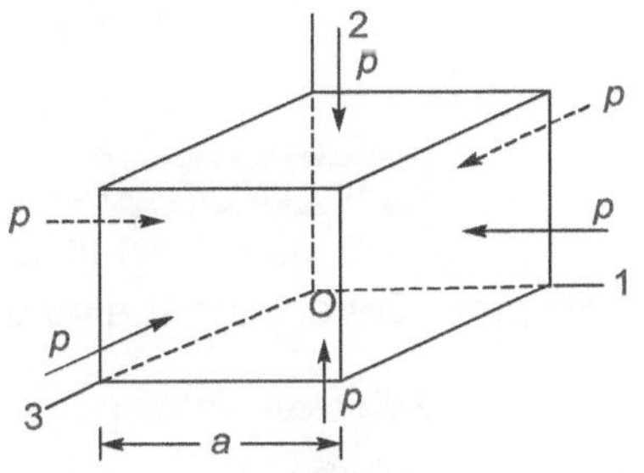

Fig. 4.6

Similarly strain in other two directions.

$$\varepsilon_2 = \frac{p}{E}(1 - 2\nu), \ \varepsilon_3 = \frac{p}{E}(1 - 2\nu)$$

Volumetric strain, $\varepsilon_v = \varepsilon_1 + \varepsilon_2 + \varepsilon_3 = \frac{3p}{E}(1 - 2\nu)$

But, $\frac{p}{\varepsilon_v} = \frac{\text{Volumetric stress}}{\text{Volumetric strain}}$ = K, Bulk modulus

So, $K = \frac{E}{3(1 - 2\nu)}$

or $E = 3K(1 - 2\nu)$

Young's modulus, E = 3 × Bulk modulus (1 – 2ν), where ν is Poisson's ratio

Question 4.6 **What change in volume a copper sphere of 120 mm diameter would suffer at a depth of 2 km in sea water. *E* for brass = 102 kN/mm^2. ν for brass = 0.34, weight density of sea water = 10.06 kN/m^3.**

Solution:

Depth, $h = 2000$ m

Weight density, $w = 10.06$ kN/m^3

Pressure, $p = wh = 2000 \times 10.06 = 20120$ kN/m^2

$$K = \frac{E}{3(1-2\nu)} = \frac{102 \text{ GPa}}{3(1-0.68)} = 106.25 \text{ GPa}$$

$$= 106.25 \times 10^9 \text{ N/m}^2$$

$$p = 21200 \times 10^3 \text{ N/m}^2$$

Volumetric strain, $$\varepsilon_v = \frac{p}{K} = \frac{20120\times10^3}{106.25\times10^9} = 189.36\times10^{-6}$$

δV, change in volume $= \varepsilon_v \times V$

$$= 189.36\times10^{-6} \times \frac{\pi}{6} \times 120^3 = 171.33 \text{ mm}^3$$

$$= 0.171 \text{ cm}^3 = 0.171 \text{ cc}$$

Practice Q.4.6 What change in volume would a 300 mm cube of steel would suffer at a depth of 3 km in sea water? For steel, E = 208 GPa, ν = 0.3, weight density of sea water, $w = 1 \times 10^{-5}$ N/mm^3.

Ans. [4671 mm^3]

Question 4.7 A steel bar 250 mm long of square section 50 × 50 mm is subjected to stresses +100 N/mm^2 along the axis and –30 N/mm^2 along the sides. The increase in volume was observed to be 50 mm^3. Determine the values of Poisson's ratio, modulus of rigidity, bulk modulus if E = 200 GPa.

Solution: Principal stress are +100 N/mm^2, –30 N/mm^2, –30 N/mm^2

Volume, $V = 250 \times 50 \times 50 = 625 \times 10^3$ mm^3

Change in volume, $\delta V = 50$ mm^3

Volumetric strain, $$\varepsilon_v = \frac{\delta V}{V} = \frac{50}{625\times10^3} = 8\times10^{-5}$$

$$= \frac{1}{E}\left[(p_1+p_2+p_3) - 2\nu(p_1+p_2+p_3)\right] = \varepsilon_v$$

$$p_1 + p_2 + p_3 = 100 - 30 - 30 = 40 \text{ N/mm}^2$$

So $$8 \times 10^{-5} = \frac{1}{200000}\left[40 - 2\nu(40)\right] = \frac{40(1-2\nu)}{200000}$$

$$(1 - 2\nu) = 0.4$$

$$\nu = 0.3, \text{ Poisson's ratio}$$

Shear modulus, $$G = \frac{E}{2(1+\nu)} = \frac{200}{2(1+0.3)} = 76.92 \text{ GPa}$$

Bulk modulus, $$K = \frac{E}{3(1-2\nu)} = \frac{200}{3(1-0.6)} = 1\,66.66 \text{ GPa}$$

Practice Q.4.7 A copper bar 200 mm long, diameter 50 mm, is subjected to +120 N/mm² along the axis and –30 N/mm² along lateral directions. The increase in volume was observed to be 72 mm³. Determine the values of Poisson's ratio, modulus of rigidity, bulk modulus, if E = 105 kN/nm².

Ans. [0.34, 39.18 kN/mm², 109.37 kN/mm²]

Question 4.8 The modulus of rigidity of a material is 24.8 kN/mm². A 10 mm diameter rod of the material is subjected to an axial tensile force of 5 kN and change in its diameter is observed to be 0.0032 mm. Calculate Poisson's ratio and modulus of elasticity of the material.

Solution: Modulus of rigidity, $G = 24.8 \text{ kN/mm}^2$

Diameter, $d = 10 \text{ mm}$

Tensile force, $P = 5 \text{ kN}$

Tensile stress, $\sigma = \frac{P}{A} = \frac{5000}{\pi} \times \frac{4}{10^2} = 63.66 \text{ N/mm}^2$

Change in diameters, $\delta d = 0.0032 \text{ mm}$

Lateral strain, $\varepsilon_l = \frac{\delta d}{d} = \frac{0.0032}{10} = 0.00032$

Axial strain, $\varepsilon_a = \frac{\sigma}{E} = \frac{63.66}{E}$

Shear modulus, $G = \frac{E}{2(1+\nu)} = 24800 \text{ N/mm}^2$ ...(i)

$$\frac{\varepsilon_l}{\varepsilon_a} = \nu \quad \text{or} \quad E = \frac{\nu \times 63.66}{0.0032} \quad \text{...(ii)}$$

From equation (i) and (ii)

$$2(1+\nu)\,24800 = \frac{63.66 \times \nu}{0.00032}$$

$$(1+\nu) = 4.010\,\nu$$

or $\nu = \frac{1}{3.010} = 0.332 = \text{Poisson's ratio}$

Modulus of rigidity, $G = 24800 \text{ N/mm}^2$

$$E = 2(1+\nu) \times 24800$$

$$= 2(1.332) \times 24800 = 66067.2 \text{ N/mm}^2$$

$$= 66.067 \text{ kN/mm}^2 = 66.067 \text{ GPa}$$

Practice Q.4.8 The modulus of rigidity of a material is 39.1 kN/mm². A 10 mm diameter rod of the material is tested in tension and a load of 4 kN is applied, change in its diameter is observed to be 0.0016 mm. Calculate the Poisson's ratio and modulus of elasticity fo the material.

Ans. [0.325, 103.615 kN/mm²]

Question 4.9 A round bar 15 mm in diameter and 150 mm in length is tested in tension and it was observed that change in the length was 30.2 times the change in its diameter. Calculate modulus of rigidity of the bar, if E = 70 GPa. What twisting moment is required to produce an angular twist of 0.2° in the same bar?

Solution: Diameter, $d = 15$ mm

Length, $L = 150$ mm

$$\frac{\delta L}{\delta d} = 30.2$$

Poisson's ratio $$\nu = \frac{\delta d}{d} \times \frac{L}{\delta L} = \left(\frac{\delta d}{\delta L}\right) \times \frac{L}{d} = \frac{1}{30.2} \times \frac{150}{15} = 0.331$$

$$E = 70.0 \text{ GPa}$$

Modulus of rigidity, $$G = \frac{E}{2(1+\nu)} = \frac{70}{2(1.331)} = 26.29 \text{ GPa}$$

$$\theta = 0.2° = \frac{\pi}{900} \text{ rad}$$

Torque, $$T = \frac{G\pi d^4\theta}{32L} = \frac{26290 \times \pi \times 15^4 \times \pi}{32 \times 150 \times 900}$$

$$= 3041.4 \text{ Nmm} = 3.04 \text{ Nm}$$

Practice Q.4.9 A round bar of 15 mm in diameter and 150 mm in length is tested in tension. It was observed that change in length was 33.2 times the change in its diameter. If $E = 210$ GPa, what is the modulus of rigidity of the bar? What twisting moment is required to produce an angular twist of 0.25° in the bar.

Ans. [80.77 GPa, 11.667 Nm]

Question 4.10 A small piston of area 200 mm^2 compresses oil in a rigid container of 25 litre capacity. A weight of 120 N is gradually applied to the piston. If the bulk modulus of oil is 2970 N/mm^2, find by how much distance the weight will move down the piston.

Solution: Weight, $W = 120$ N

Piston area, $A = 200$ mm^2

Pressure develop, $$p = \frac{W}{A} = \frac{120}{200} = 0.6 \text{ N/mm}^2$$

Bulk modulus, $K = 2970$ N/mm^2

Volumetric strain, $$\varepsilon_v = \frac{p}{K} = \frac{0.6}{2970} = 0.202 \times 10^{-3}$$

$$V = 25 \text{ litre} = 25000 \text{ cc} = 25 \times 10^6 \text{ mm}^3$$

$$\delta V = \varepsilon_v \times V = 0.202 \times 10^{-3} \times 25 \times 10^6 = 5050 \text{ mm}^3$$

Piston area = 200 mm^2

Movement of the piston, $$\delta = \frac{\delta V}{A} = \frac{5050}{200} = 25.25 \text{ mm}$$

Practice Q.4.10 A small piston of area 150 mm^2 compresses oil in a rigid container of 20 liters capacity. When a weight of 90 N is gradually applied to the piston its movement is observed to be 28 mm. Determine bulk modulus of oil.

Ans. [2857 N/mm^2]

Question 4.11 In a torsion test, the specimen is a hollow shaft with 50 mm external and 30 mm internal diameter. An applied torque of 1.6 kNm is found to produce an angular twist of 0.4° measured on a length of 0.2 m of the shaft. The Young's modulus of elasticity obtained from a tensile test has been found to be 200 GPa, find the value of

(i) Modulus of rigidity, and (ii) Poisson's ratio

[IES 2009 : 10 Marks]

Solution:

Torque, $T = 1.6 \times 10^6$ Nmm

Length, $L = 200$ mm

Outer dia. $D = 50$ mm

Inner dia. $d = 30$ mm

Polar moment of inertia, $J = \frac{\pi}{32}(D^4 - d^4) = \frac{\pi}{32}(50^4 - 30^4)$

$= 53.407 \times 10^4$ mm^4

Angular twist, $\theta = 0.4°$

$$= \frac{0.4 \times \pi}{180} = 6.98 \times 10^{-3} \text{ rad}$$

$$\frac{T}{J} = \frac{G\theta}{L}$$

$$G = \frac{TL}{J\theta}$$

$$= \frac{1.6 \times 10^6 \times 200}{53.407 \times 10^4 \times 6.98 \times 10^{-3}}$$

$= 85.8 \times 10^3$ N/mm^2 = 85.8 GPa

Modulus of elasticity, $E = 200$ GPa

Poisson's ratio, $\nu = \frac{E}{2G} - 1 = \frac{200}{2 \times 85.8} - 1 = 0.165$

Question 4.12 A rod is subjected to a uni-axial load within linear elastic limit. When the change in the stress is 200 MPa, the change in the strain is 0.001. If the Poisson's ratio of the rod is 0.3, the modulus of rigidly (in GPa) is ______.

Solution:

$$E = \frac{200}{0.001} = 200 \times 10^3 \text{ N/mm}^2$$

$\nu = 0.3$

G = Modulus of rigidity

$$= \frac{200 \times 10^3}{2(1+0.3)} = 76.92 \times 10^3 \text{ MPa}$$

= 76.92 GPa

Objective Type Questions

Q.1 The number of independent elastic constants required to express stress-strain relationship for a linearly elastic isotropic materials is
(a) one (b) two
(c) three (d) four
[IES 1998]

Q.2 The number of elastic constants for a completely anisotropic elastic material is
(a) 2 (b) 3
(c) 9 (d) 21
[IES 1999]

Q.3 Which of the following is correct is respect of Poisson's ratio (ν) limits for an isotropic elastic solid?
(a) $-\infty \le \mu \le \infty$ (b) $\frac{1}{4} \le \mu \le \frac{1}{3}$
(c) $-1 \le \mu \le \frac{1}{2}$ (d) $-\frac{1}{2} \le \mu \le \frac{1}{2}$
[IES 2004]

Q.4 If E, G and K denote Young's modulus, modulus of rigidity and bulk modulus, respectively for an elastic materials, then which one of the following can be possibly true for certain value of Poisson's ratio
(a) $G = 2K$ (b) $G = E$
(c) $K = E$ (d) $G = K = E$
[IES 2005]

Q.5 A bar produces a lateral strain of magnitude -60×10^{-5}, when subjected to tensile stress of magnitude 300 MPa along the axial direction. What is the elastic modulus of the material, if the Poisson's ratio is 0.3?
(a) 100 GPa (b) 150 GPa
(c) 200 GPa (d) 400 GPa
[IES 2009]

Q.6 If a piece of material neither expands nor contracts in volume when subjected to stress, the Poisson's ratio of the material must be
(a) zero (b) 0.25
(c) 0.33 (d) 0.50
[IES 2011]

Q.7 A 16 mm diameter bar elongates by 0.04% under a tensile force of 16 kN. The average decrease in diameter is found to be 0.01%. Then Which statements are correct?
1. E = 210 GPa, G = 77 GPa
2. E = 199 GPa, ν = 0.25
3. E = 199 GPa, ν = 0.30
4. E = 199 GPa, G = 80 GPa

(a) 3 and 4 (b) 2 and 4
(c) 1 and 3 (d) 1 and 4
[IES 2013]

Q.8 Arrange the following materials with decreasing order of their Poisson ratio.
I. Glass II. Phosphor bronze
III. Mild steel IV. Cast iron

(a)	I	II	IV	III
(b)	II	III	IV	I
(c)	IV	III	II	I
(d)	II	I	III	IV

Q.9 For an elastic material, Poisson's ratio is μ, Modulus of Elasticity is E, Modulus of Rigidity is C and Bulk Modulus is K. μ is expressible in terms of K and C as
(a) $\frac{6K-2C}{3K-2C}$ (b) $\frac{6K+2C}{3K-2C}$
(c) $\frac{3K-2C}{6K+2C}$ (d) $\frac{3K+2C}{6K+2C}$

Q.10 An isotropic material is characterized by
(a) two independent moduli of elasticity along two mutually perpendicular directions
(b) two independent moduli of elasticity along two mutually perpendicular directions and Poisson's ratio
(c) a modulus of elasticity, a modulus of rigidity and Poisson's ratio
(d) any two out of a modulus of elasticity, and modulus of rigidity and Poisson's ratio

Q.11 The modulus of rigidity of an classic material is found to be 38.5% of the value of its Young's modulus. The Poisson's ratio μ of the material is nearly
(a) 0.28 (b) 0.30
(c) 0.33 (d) 0.35

Q.12 For a material, the modulus of rigidity is 100 GPa and the modulus of elasticity is 250 GPa. The value of the Poisson's ratio is

(a) 0.20 (b) 0.25
(c) 0.30 (d) 0.35

Q.13 The longitudinal strain of a cylindrical bar of 25 mm diameter and 1.5 m length is found to be 3 times its lateral strain in a tensile test. What is the value of Bulk Modulus by assuming $E = 1 \times 10^5$ N/mm²?

(a) 2×10^5 N/mm² (b) 1.1×10^5 N/mm²
(c) 1×10^5 N/mm² (d) 2.1×10^5 N/mm²

Q.14 The Poisson's ratio for a perfectly incompressible linear elastic material is

(a) 1 (b) 0.5
(c) 0 (d) Infinity

Q.15 An elastic isotropic body is in a hydrostatic state of stress as shown in the figure. For no change in the volume to occur, what should be its Poisson's ratio?

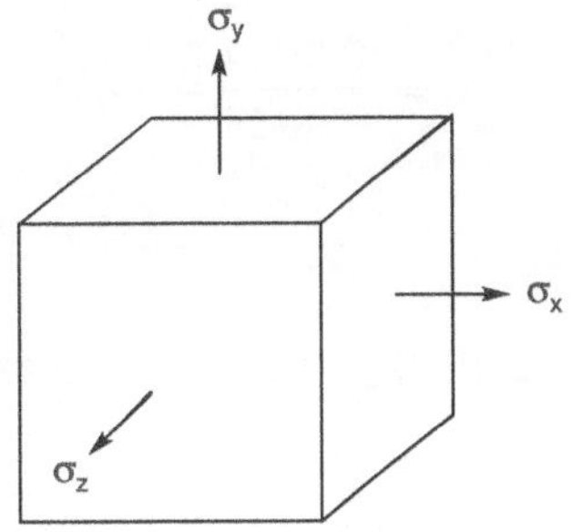

(a) 0.00 (b) 0.25
(c) 0.50 (d) 1.00

Answers

1. (b)	2. (d)	3. (b)	4. (c)	5. (b)
6. (d)	7. (b)	8. (b)	9. (c)	10. (d)
11. (b)	12. (b)	13. (c)	14. (d)	15. (c)

Explanations

1. (b)

E, ν are two independent elastic constants.

2. (d)

Type of the material	Number of independent elastic constants
1. Isotropic and homogeneous	2 (*E*, ν)
2. Orthotropic	9
3. Anisotropic	21

3. (b)

For isotropic material, Poisson's ratio, ν = 1/4
on the basis of an atomic model, ν = 1/3
For most metals ν = 1/4 to 1/3
But for other material

Materials	ν
Cork	0
Concrete	0.1-0.2
Rubber	0.5

4. (c)

$$E = K \text{ for } \nu = 1/3$$

5. (b)

$$\text{Axial strain} = +\frac{60\times10^{-5}}{0.3} = +200\times10^{-5}$$

$$\text{Axial stress} = 300 \text{ N/mm}^2$$

$$E = \frac{300}{200\times10^{-3}} = 1.5\times10^5$$

$$= 150 \text{ kN/mm}^2 = 150 \text{ GPa}$$

6. (d)

$$e_v = \frac{\delta V}{v} = 0$$

when ν = 0.5

$$\delta V = Ve_v = V\frac{p}{K} = \frac{Vp}{3E(1-2\nu)}$$

$\delta V \to 0$, when ν = 0.5

7. (b)

$$\sigma = \frac{10000\times4}{\pi\times256} = 79.57 \text{ N/mm}^2$$

$$\varepsilon = 0.04\% = 0.0004$$

$$E = \frac{79.57}{0.0004} = 199000 \text{ N/mm}^2$$
$$= 199 \text{ GPa}$$
$$\nu = \frac{0.0001}{0.0004} = 0.25$$
$$G = \frac{199}{2(1+0.25)} = 79.6 \simeq 80 \text{ GPa}$$

Both options (2) and (4) are correct.

8. (b)

II.	Phosphor bronze	0.35
III.	Mild steel	0.30
IV.	Cast iron	0.25
I.	Glass	0.20

9. (c)

$$K = \frac{E}{3(1-2\nu)}$$
$$C = \frac{E}{2(1+\nu)}$$
$$3K = \frac{E}{1-2\nu}$$
$$2C = \frac{E}{1+\nu}$$
$$3K - 2C = \frac{E}{1-2\nu} - \frac{E}{(1-\nu)} = \frac{3\nu E}{(1-2\nu)(1+\nu)}$$
$$6K + 2C = \frac{2E}{1-2\nu} + \frac{E}{(1+\nu)} = \frac{(2+2\nu+1-2\nu)E}{(1-2\nu)(1+\nu)}$$
$$= \frac{3E}{(1-2\nu)+(1+\nu)}$$
$$\frac{3K-2C}{6K+2C} = \frac{3\nu}{3} = \nu$$

10. (d)

Elastic constants,

E, G, K, ν are

Elastic constant; E, G, K related to each other internal of γ any two as

$$E = 2G(1+\nu)$$
$$E = 3K(1-2\nu)$$

E and ν can surface.

E or G and ν

11. (b)

Elastic constants

Shear modulus,

$$G = 0.385 \times E$$

where E is Young modulus

Poisson's ratio,

$$\mu = \frac{E}{2G} - 1 = \frac{1}{20.385} - 1$$
$$= 1.3 - 1 = 0.3$$

12. (b)

$$E = 2G(1+\nu)$$
$$250 = 100 \times 2(1+\nu)$$
$$\nu = 0.25 \text{ Poisson's ratio.}$$

13. (c)

Longitudinal strain,

$$\varepsilon = 3 \times \varepsilon', \text{ lateral strain}$$

ν, Poisson's ratio $= \frac{1}{3}$

Bulk modulus, $K = \frac{E}{3(1-2\gamma)}$

where E is Young's modulus

$$= \frac{E}{3\left(1-\frac{2}{3}\right)} = E$$
$$= 1 \times 10^5 \text{ N/mm}^2$$

14. (d)

Perfectly incompressible,

$\delta V = 0$, change volume is zero

$$\delta V = \frac{p}{k} \times V = 0$$
$$K = E(1-2\nu)$$

If V-infinity

$$K \simeq (-2\nu)$$
$$K = \text{Infinity}$$
$$\delta V = 0$$

15. (c)

Bulk Modulus,

$$K = E(1-2\nu)$$

Where ν is Poisson's ratio

For no change in volume, K = 0

$$E(1-2\nu) = 0$$

Poisson's ratio

$$\nu = 0.5$$

Thin Cylindrical and Spherical Shells

CHAPTER 5

Any shell cylindrical or spherical when subjected to external or internal fluid pressure, stresses are developed in the wall of the shell. If the variation in radial and hoop stresses along the radial thickness of the shell is negligible, then it is termed as a thin shell. **Generally if the ratio of *D/t* i.e., diameter/thickness is greater than 20, then shell is classified as a thin shell.** In such a shell, radial stress is much less than the hoop stress. Therefore effect of radial stress in the calculation of strains is generally neglected. In a cylindrical shell, if ratio of *D/t* is equal to 20, then hoop stress is 10 times the radial stress, i.e., pressure *p*.

Question 5.1 **Consider a thin cylindrical shell, of diameter, *D* thickness, *t* subjected to internal pressure *p* and derive expression for axial and circumferential stresses developed in shell.**

Solution:

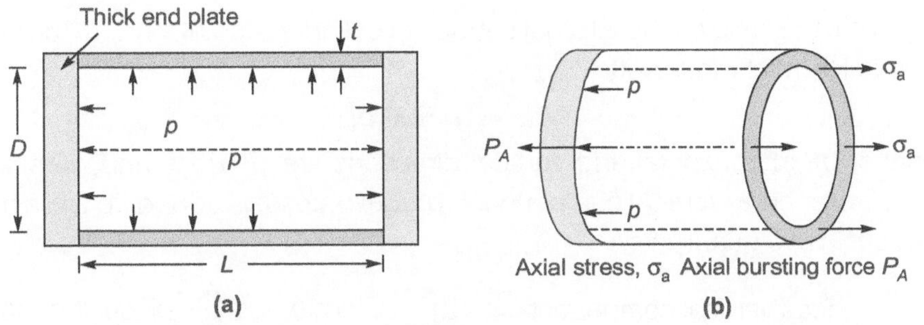

Fig. 5.1

Fig. 5.1 (a) shows a thin cylindrical shell of internal diameter *D*, wall thickness *t*, length *L*, with thick end plates. Cylindrical shell is subjected to internal pressure *p*. The internal pressure may be developed in cylinder by pumping extra volume δV of liquid in addition to original volume $V\left(=\frac{\pi}{4}D^2L\right)$. Due to internal fluid pressure:

(a) Length *L*, of the cylinder will increase, developing axial tensile stress in cylinder

(b) Diameter *D*, of the cylinder will increase or consequently the circumference of the cylinder will increase, developing hoop tensile stress.

Axial Stress: Due to internal pressure, axial bursting force P_A developed in cylinder is

Axial bursting force, $$P_A = p\times\frac{\pi}{4}D^2 \qquad \text{...(i)}$$

This axial bursting force tries to break the cylinder in two halves as shown in figure 5.1 (b). Internal resistance offered by cylinder wall = P_A

$$= \text{axial stress} \times \text{resisting area of cross-section}$$

$$P_A = \sigma_a \times \pi D t \quad \text{...(ii)}$$

Axial stress
$$\sigma_a = \frac{pD}{4t} \quad \text{...(iii)}$$

Circumferential Stress (Hoop Stress)

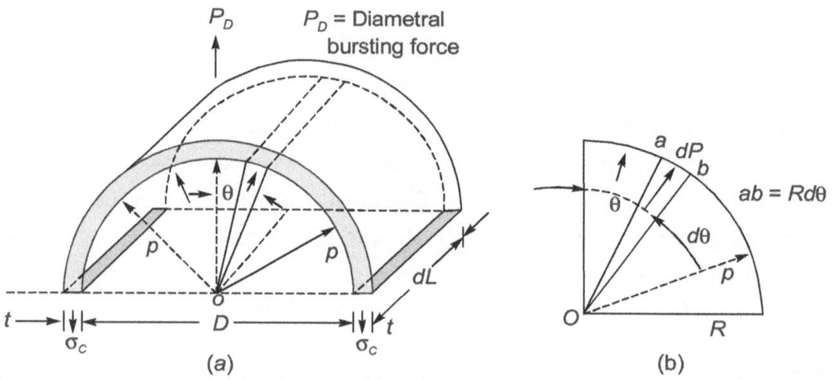

Fig. 5.2

Consider an elementary length δL of cylindrical shell subjected to internal pressure p as shown in figure 5.2 (a).

Radial pressure p is acting on the internal surface of cylinder. Radial force dP on small elementary area of length dL and circumferential width $Rd\theta$

$$dP = p.Rd\theta.dL$$

Components of dP in horizontal and vertical directions are $dP\cos\theta$ and $dP\sin\theta$. When the force is integrated over semicircular area, vertical components ($p.Rd\theta dL\cos\theta$) is added up while horizontal component $dP\sin\theta$ is cancelled out in two quadrants.

$$\text{Total vertical component} = 2\int_0^{90°} dP\cos\theta = 2\int_0^{\pi/2} pRd\theta dL\cos\theta$$

$$= 2pRdL\int_0^{\pi/2}\cos\theta d\theta = 2pRdL|\sin\theta|_0^{\pi/2}$$

$$= 2pRdL = p\cdot D\cdot dL, \text{ as } 2R = D \quad \text{...(i)}$$

$$= \text{Diametral bursting}$$

$$= \text{Resisting area} \times \text{Stress}$$

$$= 2 \times t \times dL \times \sigma_c$$

$$\sigma_c, \text{ circumferential stress} = \frac{pDdL}{2tdL} = \frac{pD}{2t} \quad \text{...(ii)}$$

Note that circumferential stress, $\sigma_c = 2 \times \sigma_a$, axial stress.

Question 5.2 **A thin cylindrical shell of diameter 200 mm is subjected to an internal pressure of 2.5 N/mm². Efficiency of longitudinal joint is 75%. What is the minimum thickness of shell required so that maximum hoop stress developed in cylinder is not to exceed 80 MPa.**

Solution: To develop a thin cylindrical shell, a rectangular sheet of thickness t of the shell, length L and width πD is taken and bent by cold rolling as shown in figure 5.3

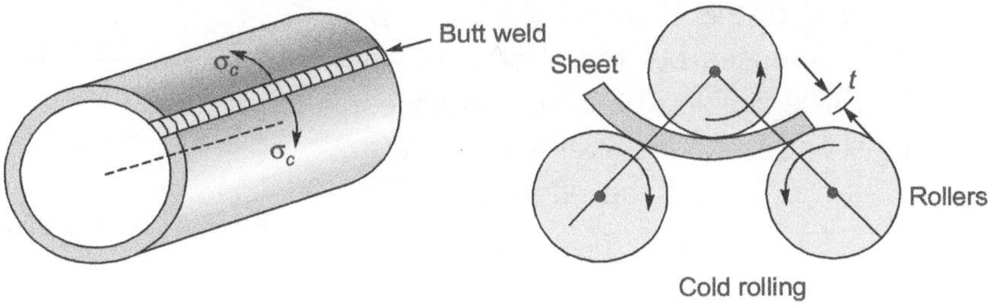

Fig. 5.3

When the sheet is bent into required circular shape, **edges are butt welded. This is known as longitudinal joint as it is along the length of the cylinder.**

say η = efficiency of the longitudinal joint

Hoop stress, $$\sigma_c = \frac{pD}{2t}$$

Allowable hoop stress, $$\sigma_c \times \eta = \frac{pD}{2t}$$

or thickness, $$t = \frac{pD}{2\sigma_c \eta}$$

p = 2.5 N/mm^2

D = 200 mm

η = 75% = 0.75

σ_c = 80 MPa, hoop stress

Thickness, $$t = \frac{2.5 \times 200}{2 \times 80 \times 0.75} = 4.166 \text{ mm}$$

Practice Q.5.2 **A thin cylindrical shell of 500 mm diameter is subjected to an internal pressure p. Efficiency of circumferential joint is 60%. Wall thickness of cylinder is 12 mm. What is the maximum pressure to which the shell can be subjected, if allowable axial stress in cylinder is 60 MPa?**

Ans. [3.5 N/mm^2]

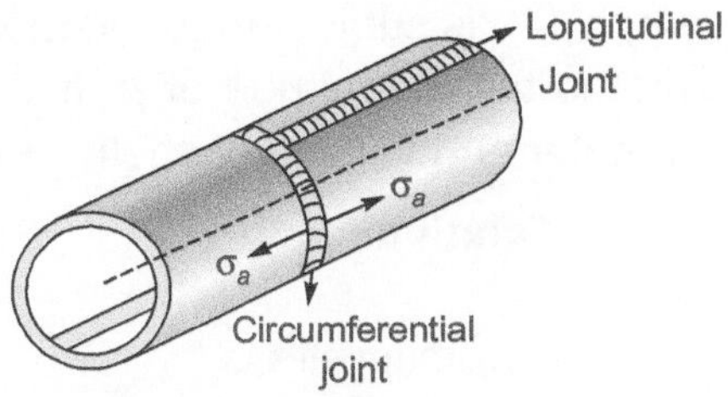

Fig. 5.4

Note: If the length of thin cylinder is large, then many cylinders as obtained by making longitudinal joints and welded together at circumferential joint as shown in figure 5.4.

Two cylinders are joined by a circumferential joint. To make the circumferential joint fail, axial stress σ_a acts on the joint.

Question 5.3 **Consider a thin cylindrical shell of diameter D, wall thickness t, length L subjected to internal pressure p. Knowing the expressions for axial and hoop stress, derive expression for (i) axial strain (ii) circumferential strain (iii) volumetric strain, if E = Young's modulus, ν = Poisson's ratio of the material.**

Solution: Consider an element of cylindrical shell subjected to internal pressure p as shown in figure 5.5

Three stresses are acting on this element i.e., σ_c, σ_a, $-p$, where σ_c and σ_a are tensile stresses but radial stress p is compressive. Radial stress p is negligible. These three stresses are perpendicular to each other and are normal stresses, therefore these are principal stresses. Neglecting the effect of radial stress $-p$.

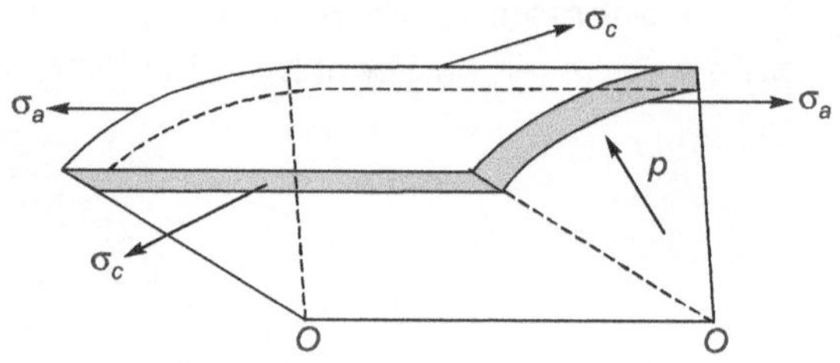

Fig. 5.5 *Element of a thin cylindrical shell*

Axial strain,

$$\varepsilon_a = \frac{\sigma_a}{E} - \frac{\nu\sigma_c}{E} = \frac{pD}{4tE} - \frac{\nu pD}{2tE} = \frac{pD}{4tE}(1-2\nu) \quad \text{...(i)}$$

Circumferential strain,

$$\varepsilon_c = \frac{\sigma_c}{E} - \frac{\nu\sigma_a}{E} = \frac{pD}{2tE} - \frac{\nu pD}{4tE} = \frac{pD}{4tE}(2-\nu) \quad \text{...(ii)}$$

Volume of shell, $$V = \frac{\pi}{4}D^2L$$

Taking partial derivative of eq. (iii)

$$\partial V = \frac{\pi}{4}\times 2DL\times\partial D + \frac{\pi}{4}D^2\partial L$$

or

$$\frac{\partial V}{V} = \frac{2\partial D}{D} + \frac{\partial L}{L} = 2\times\text{diametral strain + axial strain}$$

But diametral strain $\frac{\partial D}{D} = \frac{\pi\partial D}{\pi D}$ = circumferential stain

So volumetric strain, $\varepsilon_v = 2\varepsilon_c + \varepsilon_a$

Putting the values from eq. (i) and (ii)

Volumetric strain,

$$\varepsilon_v = 2\times\frac{pD}{4tE}(2-\nu) + \frac{pD}{4tE}(1-2\nu) = \frac{pD}{4tE}(5-4\nu) \quad \text{...(iv)}$$

Question 5.4 **A thin cylindrical shell made of steel, of diameter 250 mm, wall thickness 6 mm, length 1 m is subjected to internal pressure p such that maximum stress developed in shell is 120 MPa. What is the magnitude of p. If E = 200 GPa and Poisson's ratio is 0.3. Determine changes in length, diameter and volume of shell.**

Solution:

Maximum stress,

$$\sigma_c = \frac{pD}{2t} = 120 \text{ N/mm}^2$$

$$p = \frac{120\times 2t}{D} = \frac{120\times 2\times 6}{250} = 5.76 \text{ N/mm}^2$$

$$E = 200000 \text{ N/mm}^2,\ \nu = 0.3$$

Axial strain,

$$\varepsilon_a = \frac{pD}{4tE}(1-2\nu) = \frac{5.76\times 250}{4\times 6\times 200000}(1-2\times 0.3) = 1.2\times 10^{-4}$$

Circumferential strain,

$$\varepsilon_c = \frac{pD}{4tE}(2-\nu) = \frac{5.76\times 250}{4\times 6\times 200000}(2-0.3) = 5.1\times 10^{-4}$$

Volumetric strain, $\varepsilon_v = 2\varepsilon_c + \varepsilon_a = 2 \times 5.1 \times 10^{-4} + 1.2 \times 10^{-4} = 11.4 \times 10^{-4}$

Change in length, $\delta L = \varepsilon_a L = 1.2 \times 10^{-4} \times 1000 = 0.12$ mm

Change in diameter, $\delta D = \varepsilon_c D = 5.1 \times 10^{-4} \times 250 = 0.1275$ mm

Original volume, $V = \frac{\pi}{4} \times D^2 \times L = \frac{\pi}{4} \times 250^2 \times 1000 = 49.0874 \times 10^6 \text{ mm}^3$

Change in volume, $\delta V = \varepsilon_v \times V = 11.4 \times 10^{-4} \times 49.0874 \times 10^6$

$= 55960 \text{ mm}^3 = 55.96 \text{ cm}^3 = 55.96$ cc

Practice Q.5.4 **A thin copper cylinder of diameter 150 mm, length 500 mm, wall thickness 5 mm is subjected to an internal pressure of 5 N/mm². If E and ν for copper are 102 GPa and 0.35 respectively, determine change in length, diameter and volume of the shell.**

Ans. [0.055 mm, 0.091 mm, 11.69 cc]

Question 5.5 **Consider a thin spherical shell of diameter *D*, wall thickness t subjected to internal pressure, p. Derive expressions for (a) hoop stress, (b) hoop strain, (c) volumetric strain in shell, if *E* = Young's modulus of the material and ν = Poisson's ratio of the material.**

Solution: A thin spherical shell of diameter, *D*, wall thickness *t*, subjected to internal fluid pressure *p* is shown in figure 5.6 (a). Internal pressure p can be developed in shell by pumping extra fluid under pressure inside the shell and the shell tries to expand. Due to expansion, diameter of the shell or consequently the circumference of the shell tends to expand developing circumferential stress of tensile nature in the shell. Due to pressure p, a diametral bursting force P_D is developed in shell which tries to break the shell in two hemispherical halves. Stress developed in shell is hoop stress, σ_c as shown.

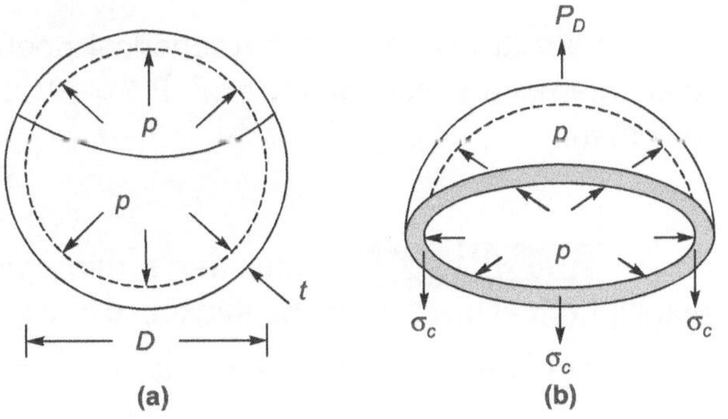

Fig. 5.6

P_D = Diametral bursting force

= Pressure × Projected area of hemispherical surface

$$= p \times \frac{\pi}{4} D^2$$

= Resisting force = Stress × Resisting area

$$= \sigma_c \times \pi D t \quad \text{...(i)}$$

or, circumferential stress,

$$\sigma_c = \frac{pD}{4t} \quad \text{...(ii)}$$

Take an element of thin spherical shell, subjected to stresses as shown in figure 5.7. Stresses acting on these element are σ_c, σ_c, $-p$. These stresses are normal stresses and perpendicular to each other. There are three principal stresses acting on the element. Hoop stress σ_c is much larger than radial stress $-p$.

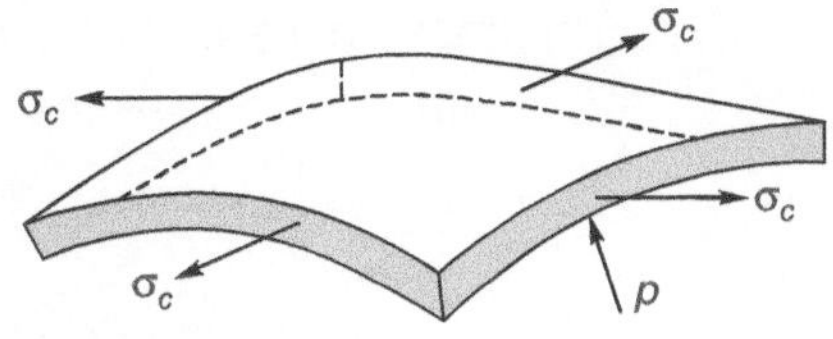

Fig. 5.7 *Element of thin spherical shell*

Circumferential strain,

$$\varepsilon_c = \frac{\sigma_c}{E} - \frac{\nu\sigma_c}{E} = \frac{pD}{4tE}(1-\nu),\ (\text{putting the value of } \sigma_c) \qquad \text{...(iii)}$$

Volumetric Strain

Volume of thin spherical shell,
$$V = \frac{\pi D^3}{6} \qquad \text{...(iv)}$$

Taking partial derivative of eq. (iv)

$$\partial V = \frac{3\pi D^2}{6}\partial D$$

$$\frac{\partial V}{V} = \frac{3\pi D^2 \partial D}{6} \times \frac{6}{\pi D^3} = \frac{3\partial D}{D}$$

$$= 3\times\frac{\partial D\times\pi}{\pi D} = 3\times \text{ circumferential strain}$$

Volumetric strain,
$$\varepsilon_V = 3\varepsilon_c = 3\frac{pD}{4tE}(1-\nu) \qquad \text{...(v)}$$

Practice Q.5.5 A thin spherical steel shell of diameter 400 mm, well thickness 12 mm is subjected to an internal pressure of 3 N/mm^2. If E = 210 GPa, ν = 0.295, determine change in diameter and change in volume.

Ans. [0.0335 mm; 8.435 cc]

Question 5.6 **Consider a thin cylindrical shell, of diameter *D*, length *L*, thickness t with hemispherical ends. How the thickness of hemispherical ends is decided, discuss.**

Solution: In some storage vessels, hemispherical ends are provided at the ends of a thin cylindrical shell.

Thin cylindrical shell with hemispherical ends.

When this vessel is subjected to internal pressure,

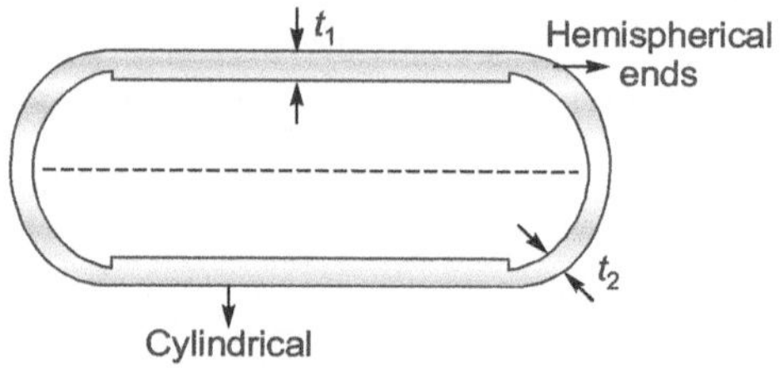

Fig. 5.8

σ_c', hoop stress developed in cylindrical portion $= \dfrac{pD}{2t_1}$

σ_c'', hoop stress developed in hemispherical portion $= \dfrac{pD}{4t_2}$

ε_c', hoop strain in cylindrical portion $= \dfrac{pD}{4t_1E}(2-\nu)$

ε_c'', hoop strain developed in hemispherical portion $= \dfrac{pD}{4t_2E}(1-\nu)$

Thickness of hemispherical ends i.e., t_2 is decided on two considerations which are

(i) Equal hoop stresses i.e., $\sigma_c' = \sigma_c''$.

$$\frac{pD}{2t_1} = \frac{pD}{2t_2}$$

or
$$t_2 = \frac{1}{2}t_1 \qquad \text{...(i)}$$

(ii) equal hoop strains i.e., $\varepsilon_c' = \varepsilon_c''$

$$\frac{pD}{4t_1E}(2-\nu) = \frac{pD}{4t_2E}(1-\nu)$$

or

$$t_2 = t_1 \times \frac{(1-\nu)}{(2-\nu)}$$

Practice Q.5.6 A thin cylindrical shell of diameter 200 mm wall thickness 6 mm is provided with hemispherical ends. Determine the thickness of hemispherical ends if (i) hoop stresses developed in cylindrical portion and hemispherical portions are the same (ii) hoop strains developed in cylindrical portion and hemispherical portion are the same. Poisson's ratio of material is 0.298.

Ans. [(i) 3 mm (i) 2.48 mm]

Question 5.7 What is the purpose of wire winding a thin cylindrical shell? Consider a thin cylindrical shell of diameter *D*, wall thickness, *t* wound with a layer of wire of diameter d_w under tension, σ_w. Determine the initial circumferential stress developed in cylinder. Now the cylinder is subjected to internal pressure *p*. Derive expressions for resultant hoop stress developed in cylinder and resultant stress in wire after cylinder is subjected to internal pressure. Take E_c, E_w, Young's modulus of cylinder and wire respectively: ν_c and ν_w Poisson's ratio of cylinder and wire respectively.

Solution:

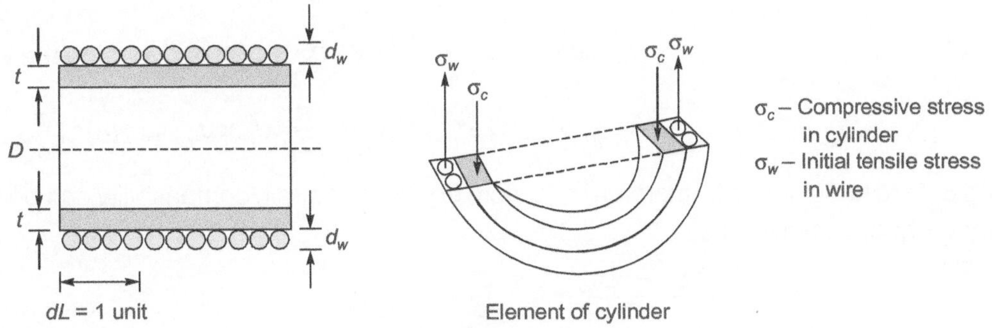

Fig. 5.9

In a thin cylindrical shell, subjected to internal pressure, hoop stress = 2 axial stress. So cylinder is not equally strong in both axial and hoop directions. By wire winding, initial compressive hoop stress is developed in cylinder.

Advantage is that it increases pressure bearing capacity of cylinder.

A cylinder of diameter *D*, wall thickness t wound with wire of diameter d_w, under tensile stress σ_w.

When steel wire is tightly wound over the outer surface of cylinder, cylinder comes under diametral compression.

Take a unit length of cylinder $dL = 1$

Number of wires per unit length, $n = \dfrac{1}{d_w}$

Tensile force in wires, $P_w = \dfrac{\pi}{4} d_w^2 \times \sigma_w \times 2 \times n$

Per unit length of cylinder $$P_w = \frac{\pi}{2} \times d_w^2 \sigma_w \times \frac{1}{d_w}, \text{ (putting the value of } n)$$

$$= \frac{\pi}{2} \times d_w \sigma_w$$

Compressive force developed in cylinder,

$$P_c = P_w \quad \text{(for equilibrium)}$$

$$P_c = 2 \times t \times 1 \times \sigma_c$$

where σ_c is initial compressive stress developed in cylinder.

$$2t\sigma_c = \frac{\pi}{2} \times d_w \sigma_w$$

$$\sigma_c = \frac{\pi}{4t} \times d_w \sigma_w, \text{ (compressive)} \quad \text{...(i)}$$

where σ_w initial tension in wire.

Internal pressure p acting on cylinder with wire winding.

Consider unit length of cylinder.

Diametral bursting force $= p \times D \times l$

= resisting force due to (shell thickness + wires wound)

Say tensile stresses developed in cylinder is σ_c' and in wire it is σ_w'.

$$pD = 2 \times t \times \sigma_c' + 2 \times n \times \frac{\pi}{4} d_w^2 \times \sigma_w', \text{ putting the value of } n$$

$$pD = 2t\sigma_c' + 2 \times \frac{1}{d_w} \times \frac{\pi}{4} \times d_w^2 \times \sigma_w' = 2t\sigma_c' + \frac{\pi}{2} \times d_w \sigma_w' \quad \text{...(ii)}$$

σ_c' and σ_w' are two unknowns, therefore one more equation of strain compatibility can be used i.e., hoop strain in cylinder = hoop strain in wire

$$\frac{\sigma_c' - \nu_c \sigma_a'}{E_c} = \frac{\sigma_w'}{E_w} \quad \text{...(iii)}$$

where $\sigma_a' = \text{axial stress} = \dfrac{pD}{4t}$ in cylinder

From eq. (ii) and (iii) stress σ_c' and σ_w' can be worked out.

Final stresses or resultant stresses in cylinder and wire are

$$\sigma_{CR} = \sigma_c' - \sigma_c \quad \text{...(iv)}$$

$$\sigma_{WR} = \sigma_w' + \sigma_w \quad \text{...(v)}$$

Question 5.8 **A bronze cylinder of diameter 200 mm, wall thickness 8 mm is wound with steel wire of diameter 4 mm under initial tension of 20 N/mm^2. The wire wound cylinder is subjected to internal pressure p such that resultant hoop and axial stresses developed in cylinder are the same. What is magnitude of p. Given E_w = 210 GPa, E_c = 105 GPa, ν_c = 0.35. What is the final stress developed in wire.**

Solution:

$$\sigma_w = 20 \text{ N/mm}^2$$

$$d_w = 4 \text{ mm}$$

$$t = 8 \text{ mm}$$

$$\sigma_c = \text{Initial compressive stress in cylinder.}$$

$$= \frac{\pi}{4} \times \frac{d_w}{t} \sigma_w = \frac{\pi}{4} \times \frac{4}{8} \times 20 = 7.854 \text{ N/mm}^2 \text{ (comp.)}$$

Say internal pressure = p

Axial stress, $\sigma_a' = \frac{pD}{4t} = \frac{p \times 200}{4 \times 8} = 6.25p$

Hoop stress $= \sigma_c'$.

$$\sigma_{CR} = (\sigma_c' - 7.854) = 6.25p, \text{ (as given)} \quad \text{...(ii)}$$

or $\sigma_c' = 6.25p + 7.854$ ...(iii)

Using equilibrium and comparability conditions.

$$pD = 2t\sigma_c' + \frac{\pi}{2} d_w \sigma_w' \quad \text{...(iv)}$$

$$\frac{\sigma_c' - \nu_c \sigma_a'}{E_c} = \frac{\sigma_w'}{E_w} \quad \text{...(v)}$$

From eq. (v) $\sigma_c' - \nu_c \sigma_c' = \sigma_w' \times \frac{E_c}{E_w} = \sigma_w' \times \frac{105}{210} = 0.5\sigma_w'$

or $6.25p + 7.854 - 0.35\,(6.25p) = 0.5\,\sigma_w'$

or $4.0625p + 7.854 = 0.5\,\sigma_w'$

$$\sigma_w' = 8.125p + 15.708$$

Putting the values of σ_w' and σ_c' in eq. (iv)

$$p \times 200 = 2 \times 8 \times (6.25p + 7.854) + \frac{\pi}{2} \times 4 \times (8.125p + 15.708)$$

$$200p = 100p + 125.664 + 6.283\,(8.125p + 15.708)$$

$$100p = 125.664 + 51.05p + 98.69$$

$$48.95p = 224.354$$

$$p = 4.58p \text{ N/mm}^2$$

$$\sigma_{aR} = 6.25p = 28.65 \text{ N/mm}^2$$

$$\sigma_{cR} = 6.25p + 15.71 - 15.71 = 6.25p$$

$$= 28.65 \text{ N/mm}^2$$

So internal pressure, $p = 4.58 \text{ N/mm}^2$

$$\sigma_w' = 8.125p + 15.708 = 8.125 \times 4.58 + 15.708$$

$$= 52.92 \text{ N/mm}^2$$

Final stress in wire, $\sigma_{WR} = 52.92 + 20 = 72.92 \text{ N/mm}^2$

Practice Q.5.8 A thin cylindrical shell of an internal diameter 250 mm and wall thickness of 10 mm is closely wound around its circumference by a 3 mm diameter steel wire under an initial tension of 10 N/mm^2. Cylinder is further subjected to an internal pressure of 3 N/mm^2. Determine the resultant hoop stress developed in the cylinder and wire. Cylinder is also made of steel. E = 208 GPa, ν = 0.3.

Ans. [29.07 N/mm^2, 35.8 N/mm^2]

Question 5.9 A thin bronze cylinder of 300 mm internal diameter and 8 mm wall thickness is wound with a single layer of 2 mm thick steel tape under a tensile stress of 60 N/mm^2. Find the maximum

pressure in cylinder if the hoop stress in cylinder is not to exceed 50 N/mm². Determine also the final stress in steel tape E_{bronze} = 105 kN/mm², E_{steel} = 210 kN/m², ν_{bronze} = 0.33.

Solution: t = thickness of cylinder = 8 mm

t' = thickness of tape = 2 mm

σ_w = initial tension in tape = 60 N/mm²

σ_c = initial compressive stress in cylinder

$$= \frac{t'}{t} \times \sigma_w = \frac{2}{8} \times 60 = 15 \text{ N/mm}^2 \text{ (comp.)} \quad \ldots(i)$$

Now consider unit length of cylinder.

Note that width of the tape is not given, so

$$pD = 2t\sigma_c' + 2t' \times \sigma_w'$$

$$p \times 300 = 2 \times 8 \times \sigma_c' + 2 \times 2 \times \sigma_w'$$

$$300p = 16\sigma_c' + 4\sigma_w'$$

Moreover

$$\frac{\sigma_c' - \nu_c \sigma_a'}{E_c} = \frac{\sigma_w'}{E_w} \quad \ldots(ii)$$

$$\sigma_c' - 0.33\sigma_a' = \sigma_w'\left(\frac{E_c}{E_w}\right) = 0.5\sigma_w'$$

$$\sigma_a' = \frac{pD}{4t} = \frac{p \times 300}{4 \times 8} = 9.375p \quad \ldots(iii)$$

or

$$\sigma_c' - 0.33 \times 9.375p = 0.5\,\sigma_w'$$

$$\sigma_c' = 0.5\,\sigma_w' + 3.09375p \quad \ldots(iv)$$

Putting the value of σ_c' in eq. (ii)

$$300p = 16[0.5\,\sigma_w' + 3.09375p] + 4\,\sigma_w'$$

$$= 8\sigma_w' + 49.5p + 4\sigma_w' = 12\sigma_w' + 49.5p \quad \ldots(v)$$

or

$$\sigma_w' = 20.875p \quad \ldots(vi)$$

$$\sigma_c' = 0.5\sigma_w' + 3.09375p = 0.5 \times 20.875p + 3.09375p$$

$$= 13.53p$$

$$\sigma_c = \sigma_c' - 15 = 50 \text{ N/mm}^2 \text{ (as given)}$$

$$\sigma_c' = 65 \text{ N/mm}^2$$

$$p = \frac{65}{13.53} = 4.80 \text{ N/mm}^2$$

$$\sigma_w' = 20.875p = 20.875 \times 4.80 = 100.286 \text{N/mm}^2$$

Find stress in tape,

$$\sigma_{WR} = \sigma_w + \sigma_w' = 60 + 100.286 = 160.286 \text{ N/mm}^2$$

Practice Q.5.9 **A thin copper cylinder of 200 mm bore and 5 mm thickness is wound with a single layer of 1.6 mm thick 5 mm wide steel tape, wound under tension of 50 N/mm² throughout the length of the cylidner. Now the cylinder is subjected to internal pressure of 3.6 N/mm². Determine the final stresses developed in steel tape and copper cylinder. $E_s = 2E_{cu}$ = 208 kN/mm², ν_{cu} = 0.34.**

Ans. [32.68 N/mm², 122.88 N/mm²]

Question 5.10 A thin walled pressure vessel made of aluminium alloy, 250 mm diameter and 4 mm thick is subjected to an internal pressure of 2.5 N/mm². Strain gauges mounted on the surface of vessel in circumferential and axial directions give readings of 932 μs and 190 μs respectively, under full pressure. Determine, E, ν of the aluminium alloy.

Solution:

$$p = 2.5 \text{ N/mm}^2$$
$$D = 250 \text{ mm}$$
$$t = 4 \text{ mm}$$

$$\varepsilon_c, \text{ circumferential stress} = \frac{pD}{4tE}(2-\nu) = 932\times10^{-6} \quad \text{...(i)}$$

$$\varepsilon_a, \text{ axial strain} = \frac{pD}{4tE}(1-2\nu) = 190\times10^{-6} \quad \text{...(ii)}$$

Dividing eq. (i) by eq. (ii)

$$\frac{2-\nu}{1-2\nu} = \frac{932}{190} = 4.905$$

or

$$2 - \nu = 4.905 - 9.81\nu$$
$$8.81\nu = 2.905$$

Poisson's ratio,

$$\nu = \frac{2.905}{8.81} = 0.33 \quad \text{...(iii)}$$

From eq. (i)

$$E = \frac{pD(2-\nu)}{4t\times932\times10^{-6}} = \frac{2.5\times250(2-0.33)}{4\times4\times932\times10^{-6}}$$
$$= 69994 \text{ N/mm}^2 = 69.99 \text{ kN/mm}^2$$

Practice Q.5.10 A thin walled copper alloy pressure vessel of diameter 250 mm and a wall thickness of 5 mm is subjected to an internal pressure of 4 MPa. The strain gauges that are bonded on the surface in the hoop and axial directions give readings of 790 μs and 153 μs at full pressure, respectively. Determine E and ν for the material.

Ans. [105 kN/mm², 0.34]

Question 5.11 A thin cylindrical shell of diameter 200 mm, thickness 8 mm is held between two rigid supports as shown in figure 5.10. An internal pressure of 5 N/mm² is applied on inner surface of cylinder. Determine resultant hoop and axial stresses developed in cylinder, ν for material = 0.35.

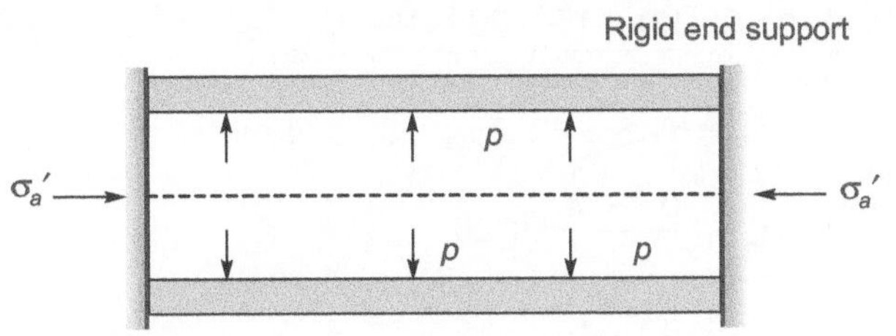

Fig. 5.10

Solution:

Stresses developed $\sigma_c = \frac{pD}{2t} = \frac{5\times200}{2\times8} = 62.5\ \text{N/mm}^2$

$\sigma_a = \frac{pD}{4t} = 31.25\ \text{N/mm}^2$

Due to rigid end supports, axial strain, $\varepsilon_a = 0$

Say σ_a' is axial compressive stress provided by rigid end supports.

Resultant axial stress, $\sigma_{aR} = 31.25 - \sigma_a'$

Axial strain, $\varepsilon_a = \frac{31.25-\sigma_a'}{E} - \frac{\nu\times62.5}{E} = 0$

or $31.25 - \sigma_a' - 0.35(62.5) = 0$

$\sigma_a' = 31.25 - 21.875$

$= 9.375\ \text{N/mm}^2$

Resultant axial stress, $\sigma_{aR} = 31.25 - 9.375 = 21.875\ \text{N/mm}^2$

Resultant hoop stress, $\sigma_{cR} = 62.5\ \text{N/mm}^2$

Practice Q.5.11 **A 200 mm long copper tube with closed ends is 100 mm in diameter and has a wall thickness of 4 mm with no internal pressure. The tube just fits between the two rigid end walls, Calculate the axial and hoop stresses in the cylinder for an internal pressure of 3 N/mm². E = 105 GPa, ν = 0.33. What is volumetric strain?**

Ans. [12.375 N/mm², 37.5 N/mm², 636.5 μs]

Question 5.12 **A closed pressure vessel of diameter 120 mm, wall thickness 4 mm is subjected to an internal pressure of 6 N/mm². Determine normal and shear stresses on an element of cylindrical wall at 30° to the longitudinal axis of cylinder.**

Solution: Element *ab* is inclined at an angle of 30° to the axis 00 of cylinder.

Hoop stress, $\sigma_c = \frac{pD}{2t} = \frac{6\times120}{2\times4} = 90\ \text{N/mm}^2$

Axial stress, $\sigma_a = \frac{pD}{4t} = 45\ \text{N/mm}^2$

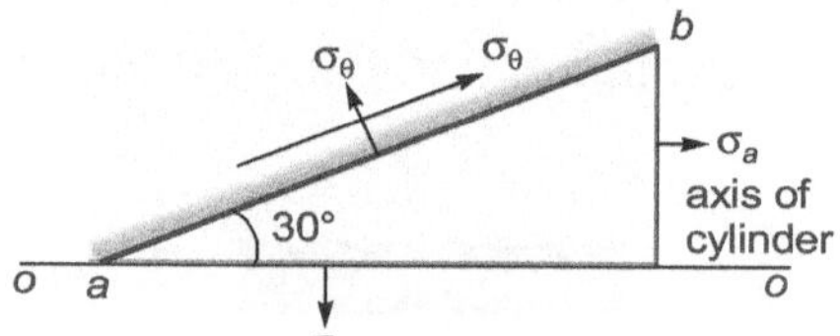

Fig. 5.11

Stresses on element

Normal stress, $\sigma_\theta = \frac{\sigma_c+\sigma_a}{2} + \frac{\sigma_c-\sigma_a}{2}\cos2\theta$

$= \frac{90+45}{2} + \frac{90-45}{2}\times\cos60°$

$= 67.5 + 11.25 = 78.75\ \text{N/mm}^2$

Shear stress, $\tau_\theta = \frac{\sigma_c-\sigma_a}{2}\sin2\theta$

$= \frac{90-45}{2}\sin60° = 19.485\ \text{N/mm}^2$

Question 5.13 For a hydraulic test, a steel tube of internal diameter 100 mm and a wall thickness of 2.5 mm and a length of 1.0 m is fitted with end plugs and filled with oil at a pressure of 2 MPa. Determine the volume of oil leakage which would cause the pressure to fall to 1.5 MPa.

For oil $K = 2.8$ GPa, $K = 2800$ N/mm^2

For steel, $E = 208$ GPa, ν for steel $= 0.29$

Solution: $D = 100$ mm, $t = 2.5$ mm, $L = 1000$ mm

$$\text{Original volume} = \frac{\pi}{4}\times 100^2 \times 1000 = 7.854\times 10^6 \text{ mm}^3$$

Change in volume due to pressure

$$\delta V = \delta V_1 + \delta V_2$$

$$= \text{Increase in volume of cylinder + Decrease in volume of oil}$$

$$= e_v V + \frac{\Delta p}{K}.V$$

$$\Delta p = 2 - 1.5 = 0.5 \text{ MPa} = 0.5 \text{ N/mm}^2$$

$$\varepsilon_v = \frac{\Delta p D}{4tE}(5-4\nu) = \frac{0.5\times 100}{4\times 2.5\times 208000}(5-4\times 0.29) = 92.3\times 10^{-6}$$

$$\delta V = 92.3\times 10^{-6}\times 7.854\times 10^6 + \frac{0.5}{2800}\times 7.854\times 10^6$$

$$= 724.92 \text{ mm}^3 + 1402.5 \text{ mm}^3$$

$$= 2127.42 \text{ mm}^3 = 2.12 \text{ cc}$$

Practice Q.5.13 For a hydraulic test, a steel tube of an internal diameter of 80 mm, a wall thickness of 2 mm and a length of 1.2 m is fitted with end plugs and filled with oil at a pressure of 2 MPa. Determine the value of oil leakage which would cause the pressure to fall to 1.5 MPa.

For oil K for oil $= 2.8$ GN/m^2

For steel $E = 208$ GPa, ν for steel $= 0.29$

Ans. [1.633 cc]

Question 5.14 A thin tyre is to be shrunk on a rigid wheel of diameter 1 m. Determine the necessary internal diameter of the tyre if the maximum hoop stress in tyre is 100 N/mm^2. Also determine the least temperature to which the tyre must be heated above the temperature of the wheel before the tyre can be slipped on wheel.

α for tyre $= 11\times 10^{-6}$/°C, $E = 204$ GPa.

Solution:

$$\sigma_c, \text{ Hoop stress} = 100 \text{ N/mm}^2$$

$$E = 204000 \text{ N/mm}^2$$

$$\sigma_a = 0, \text{ Axial stress is zero}$$

$$\text{Hoop strain} \quad \frac{\delta D}{D} = \frac{100}{204000} = 490\times 10^{-6} = \varepsilon_c$$

$$\delta D = \frac{100}{204000}\times 1000 = 0.49 \text{ mm}$$

$$\text{Internal diameter of tyre} = D - \delta D$$

$$= 1000 - 0.49 = 999.51 \text{ mm}$$

$$\varepsilon_c = \alpha \Delta T = 11 \times 10^{-6} \times \Delta T = 490 \times 10^{-6}$$

Rise in temperature $\Delta T = 44.54°C$

Question 5.15 A steel tyre of thickness 15 mm, a width of 75 mm, and an inside diameter of 1.5 m is heated and shrunk onto a steel wheel of diameter 1501 mm. If the coefficient of static friction between the tyre and steel is 0.25, what twisting moment is required to twist the tyre relative to the wheel? Neglect deformation of wheel, E = 200 GPa.

Solution:

Width of tyre, $b = 75$ mm

thickness of tyre, $t = 15$ mm

Tyre diameter, $D = 1500$ mm

Wheel diameter, $D' = 1501$ mm

Circumferential strain/diametral strain, $\varepsilon_c = \dfrac{1501-1500}{1500} = \dfrac{1}{1500}$

$E = 2 \times 10^5 \text{ N/mm}^2$

Hoop stress in tyre, $\sigma_c = \varepsilon_c E = \dfrac{1}{1500} \times 2 \times 10^5$

$$= 133.33 \text{ N/mm}^2 = \frac{pD}{2t}$$

p = junction pressure between tyre and wheel.

So junction pressure, $p = \dfrac{133.33 \times 2 \times t}{D} = \dfrac{133.33 \times 2 \times 15}{1500}$

$$= 2.666 \text{ N/mm}^2$$

Twisting Moment

Total radial force on tyre $= p \times \pi D b$ where b is breadth

$N = 2.666 \times \pi \times 1500 \times 75 = 942242$ N

Coefficient of friction, $\mu = 0.25$

Force of friction, $F = \mu N = 0.25 \times 942242 = 235560$ Newton

Twisting moment, $T = F \times R$ (radius of steel)

$= 235560 \times 0.75$ Nm

$= 176670$ Nm $= 176.67$ kNm

Practice Q.5.15 A steel tyre of thickness 12 mm, a width of 80 mm and an inside diameter of 1000 mm is heated and shrunk onto a steel wheel of diameter 1001 mm. If the coefficient of static friction between the tyre and the wheel is 0.30, what twisting moment is required to twist the tyre relative to the wheel? Neglect deformation in wheel. E = 200 GPa.

Ans. [181 kNm]

Question 5.16 A cylindrical shell has the following dimensions. Length = 3 m. Inside diameter = 1 m. Thickness of metal = 10 mm. Internal pressure = 1.5 MPa. Calculate the change in dimensions of the shell and maximum intensity of shear stress induced. Take E= 200 GPa and Poisson's ratio, ν = 0.3.

Solution:

Pressure, $p = 1.5\ \text{N/mm}^2$,

$D = 1000\ \text{mm}$

$t = 10\ \text{mm}$

Axial stress, $\sigma_a = \dfrac{pD}{4t} = \dfrac{1.5\times1000}{4\times10} = 37.5\ \text{N/mm}^2$

Hoop stress, $\sigma_c = \dfrac{pD}{2t} = 75\ \text{N/mm}^2$

Principal stresses at inner radius of shell +75, +37.5, −1.5 N/mm²

Maximum shear stress, $\tau_{max} = \dfrac{\sigma_c-(-p)}{2} = \dfrac{75+1.5}{2} = 38.25\ \text{N/mm}^2$

$E = 200\times10^3\ \text{N/mm}^2,\ \nu = 0.3$

$$\frac{pD}{4tE} = \frac{37.5}{200\times1000} = 1.875\times10^{-4}$$

Axial strain, $\varepsilon_a = \dfrac{pD}{4tE}(1-2\nu) = 0.75\times10^{-4}$

Hoop stress, $\varepsilon_c = \dfrac{pD}{4tE}(2-\nu) = 3.1875\times10^{-4}$

Volume strain, $\varepsilon_v = \dfrac{pD}{4tE}(5-4\nu) = 7.125\times10^{-4}$

Change in length, $\delta L = \varepsilon_a\times L = 0.75\times10^{-4}\times3000 = 0.225\ \text{mm}$

Change in diameter, $\delta D = \varepsilon_c\times D = 3.1875\times10^{-4}\times1000 = 0.31875\ \text{mm}$

Volume, $V = \dfrac{\pi}{4}\times1000^2\times3000 = 2.356\times10^9\ \text{mm}^3$

$\delta V = \varepsilon_v\times V = 2.356\times10^9\times7.125\times10^{-4}$

$= 16.788\times10^5\ \text{mm}^3 = 16.788\times10^2\ \text{cm}^3$

Question 5.17 A thin cylindrical shell with hemispherical ends is subjected to internal fluid pressure. For equal maximum stress to occur in both the cylindrical and spherical portions, what would be the ratio of the thicknesses of spherical portion to that of the cylindrical portion?

[IES : 2013, 4 Marks]

Solution:

Hoop stress in cylindrical stress $= \dfrac{pD}{2t_1}$

Hoop stress in hemispherical shell $= \dfrac{pD}{4t_2}$

$$\frac{pD}{2t_1} = \frac{pD}{4t_2}$$

$$\frac{t_2}{t_1} = \frac{1}{2}$$

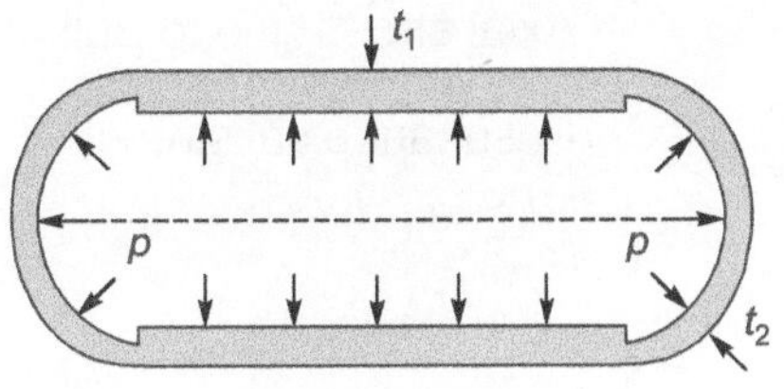

Fig. 5.12

Question 5.18 A cylindrical pressure vessel 200 cm in diameter and 350 cm in length is made of 1.30 cm thick plates. It is subjected to an internal pressure of 10 kg/cm². Calculate the longitudinal and circumferential stresses developed in the vessel.

[ME, ESE 2015 : 10 Marks]

Solution:

$$p = 10 \text{ kg/cm}^2$$
$$D = 200 \text{ cm}$$
$$t = 1.3 \text{ cm}$$
$$\frac{D}{t} = \frac{200}{1.3} = 153.85 > 20, \text{ so, thin pressure vessel}$$

Circumferential stress,
$$\sigma_c = \text{Hoop stress} = \frac{pD}{2t}$$
$$= \frac{10 \times 200}{2 \times 1.3} = 769.2 \text{ kg/cm}^2$$
$$\sigma_a = \text{Longitudinal stress} = \frac{\sigma_c}{2}$$
$$= 384.6 \text{ kg/cm}^2$$

Question 5.19 A thin cylinder with closed ends has an internal pressure of 6 MN/m². The cylinder is of 50 mm internal diameter and 2.5 mm thick. It is also subjected to an axial pull of 10 kN and a torque of 500 Nm.

Determine the principal stresses in the cylinder and maximum shear stress.

[ME, ESE 2016 : 10 Marks]

Solution:

Thin cylinder

Internal pressure, $p = 6 \text{ N/mm}^2$

Diameter, $D = 50 \text{ mm}$

Thickness, $t = 2.5 \text{ mm}$

Hoop stress,
$$\sigma_c = \frac{pD}{2t} = \frac{6 \times 60}{2 \times 2.5} = +72 \text{ N/mm}^2$$

Radial stress, $\sigma_r = -6 \text{ N/mm}^2$

Axial stress, $\sigma_a = +36 \text{ N/mm}^2$

Axial pull, $P = 10 \text{ kN}$

Axial stress due to pull,
$$\sigma_a' = \frac{10000}{\pi \times D \times t} = \frac{10000}{\pi \times 50 \times 2.5} = +25.46 \text{ N/mm}^2$$

Resultant axial stress, $\sigma_{ar} = 36 + 25.46 = 61.46 \text{ N/mm}^2$

Torque, $T = 500 \text{ Nm} = 500 \times 1000 \text{ Nmm}$

$$T = \pi Dt\frac{D^2}{4} = \frac{\pi D^3 t}{4} = \frac{\pi \times 50^3 \times 2.5}{4} = 24.54 \times 10^4 \text{mm}^4$$

Shear,
$$\tau = \frac{T}{J} \times R = \frac{5 \times 10^5 \times 25}{24.54 \times 10^4} = 50.93 \text{ N/mm}^2$$

Total tangential stress, $\sigma_{cR} = 72 + 50.93 = 122.93\ \text{N/mm}^2$

Now the, principal stresses are +122.93 N/mm² and 61.46 N/mm² and p (= –6 N/mm²)

Maximum shear stress, $$\tau_{max} = \frac{122.53+6}{2} = 64.465\ \text{N/mm}^2$$

Question 5.20 The internal diameter of the cylinder of a hydraulic ram is 10 cm. Find the thickness required to withstand an internal pressure of 500 atm (1 atm = 98.07 kPa), if the yield point for the material (in tension as well as compression) is σ_y = 500 MPa. Use a Factor of Safety of 2.

[ME, ESE 2014 : 10 Marks]

Solution:

$$\sigma_y = 500\ \text{MPa}$$

$$\text{FOS} = 2$$

$$\text{Allowable stress} = \frac{500}{2} = 250\ \text{MPa}$$

Hydraulic ram, $$D = 100\ \text{mm}$$

$$p = 98.07 \times 500\ \text{kPa}$$

$$= 98.07 \times 500 \times 10^{-3}\ \text{MPa}$$

$$= 49.035\ \text{MPa}$$

$$\sigma_c = \text{Hoop stress} = \frac{pD}{2t}$$

$$250 = \frac{49.035 \times 100}{2 \times t}$$

Thickness $$t = \frac{49.035 \times 100}{2 \times 250} = 9.807\ \text{mm}$$

Question 5.21 A cylindrical tank with closed ends is filled with compressed air at a pressure of 500 kPa. The inner radius of the tank is 2 m, and it has wall thickness of 10 mm. The magnitude of maximum in-plane shear stress (in MPa) is ________

[ME, GATE : 2015 (set-3)]

Solution:

Pressure, $p = 500\ \text{kPa} = 0.5\ \text{MPa}$

Inner diameter, $D = 4\ \text{m} = 4000\ \text{mm}$

Wall thickness, $t = 10\ \text{mm}$

Hoop stress, $$\sigma_c = +\frac{pD}{2t} = \frac{0.5 \times 4000}{2 \times 10} = 100\ \text{MPa}$$

Axial stress, $$\sigma_a = \frac{1}{2}\sigma_c = 50\ \text{MPa}$$

Principal stress are 100, 50, –0.5 MPa

$$\text{Maximum in plane shear stress} = \frac{100+0.5}{2} = 50.25\ \text{MPa}$$

Question 5.22 A thin-walled cylindrical can with rigid end caps has a mean radius, R = 100 mm and a wall thickness of t = 5 mm. The can is pressurised and an additional tensile stress of 50 MPa is imposed along the axial direction as shown in the figure. Assume that the state of stress in the wall is uniform along its length. If the magnitude of axial and circumferential components of stress in the can are equal, the pressure (in MPa) inside the can is ________ (correct to two decimal place).

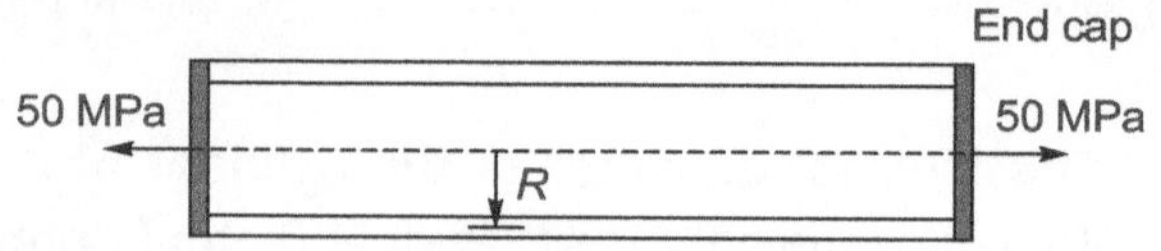

Fig. 5.13

Solution:

$$\sigma_a + 50 = \sigma_c$$

$$\sigma_c = 2\sigma_a$$

$$\sigma_a + 50 = 2\,\sigma_a$$

$$\sigma_a = 50 \text{ MPa} = \frac{pR}{2t}$$

$$p = \frac{50 \times 2 \times 5}{100} = 5 \text{ MPa}$$

Objective Type Questions

Q.1 Hoop stress and longitudinal stress in a boiler shell under internal pressure are 100 MN/m^2 and 50 MN/m^2 respectively. Young's modulus of elasticity and Poisson's ratio of the shell material are 200 GN/m^2 and 0.3 respectively. The hoop strain in boiler shell is

(a) 0.425×10^{-3} (b) 0.5×10^{-3}
(c) 0.585×10^{-3} (d) 0.75×10^{-3}

[IES 1996]

Q.2 From the design point of view, spherical pressure vessels are preferred over cylindrical pressure vessels, because they

(a) are cost effective in fabrication
(b) have uniform higher circumferential stress
(c) uniform lower circumferential stress
(d) have a larger volume for the same quantity of material used

Q.3 When a thin cylinder of diameter d and thickness t is pressurized with an internal pressure of p (1/m is the Poisson's ratio) and E is the modulus of elasticity, then out of the following, which statement is correct

(a) circumferential strain is equal to $\frac{pd}{2tE}\left(\frac{1}{2}-\frac{1}{m}\right)$

(b) longitudinal strain will be equal to $\frac{pd}{2tE}\left(1-\frac{1}{2m}\right)$

(c) the longitudinal stress is equal to $\frac{pd}{2t}$

(d) ratio of longitudinal strain to circumferential strains is equal to $\frac{m-2}{2m-1}$

[IES 1998]

Q.4 A thin cylinder with closed ends is subjected to internal pressure and supported at ends as shown in figure. What is the state of stress at point x?

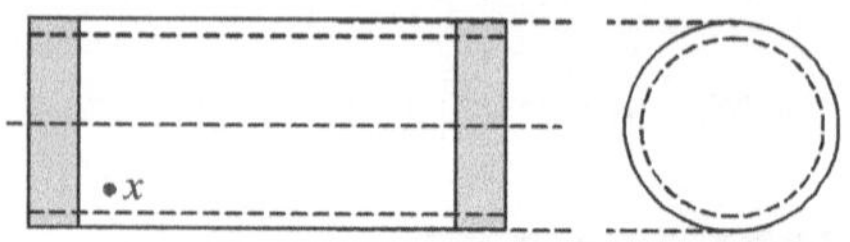

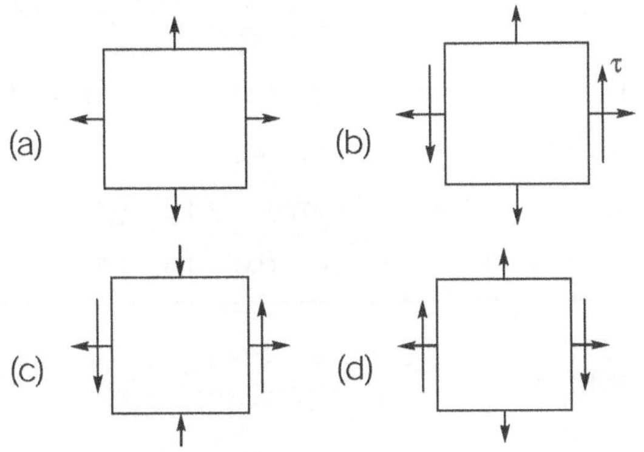

[IES 1999]

Q.5 A thin cylinder with both ends closed is subjected to internal pressure p. The longitudinal stress at the surface has been calculated as σ_0. Maximum shear stress at the surface will be

(a) $2\sigma_0$ (b) $1.5\sigma_0$
(c) σ_0 (d) None of these

[IES 1998]

Q.6 A thin cylinder contains fluid at a pressure of 500 MN/m^2, internal diameter of the shell is 0.6 m and tensile stress in the materials is to be limited to 9000 N/m^2. The shell must have a minimum thickness of nearly

(a) 9 mm (b) 11 mm
(c) 17 mm (d) 21 mm

[IES 2000]

Q.7 A thin cylinder of radius '*r*' and thickness *t*, when subjected to an internal pressure of *p* causes a radial displacement '*u*', then the tangential strain caused is

(a) $\frac{du}{dr}$ (b) $\frac{1}{r}\frac{du}{dr}$

(c) $\frac{u}{r}$ (d) $\frac{2u}{r}$

[IES 2002]

Q.8 What is the safe working pressure for a spherical pressure vessel 1.5 m internal diameter and 1.5 cm wall thickness if the maximum allowable tensile stress in vessel is 45 MPa?

(a) 0.9 MPa (b) 3.6 MPa
(c) 2.7 MPa (d) 1.8 MPa

[IES 2013]

Q.9 A thin cylinder of 100 mm diameter and 5 mm thickness is subjected to an internal pressure of 10 MPa, and a torque of 2000 Nm. Calculate the magnitude of the principal stresses.

[GATE 1996, 2 Marks]

Q.10 A thin walled spherical vessel is subjected to an internal pressure. If the radius of the shell is increased by 1% and thickness is reduced by 1% with the internal pressure remaining the same, the percentage change in circumferential (hoop stress) is

(a) 0 (b) 1
(c) 1.08 (d) 2.02

[GATE 2012, 1 Marks]

Q.11 In a thin cylindrical shell maximum stress developed is 120 MPa. What is the normal stress on an element of the cylinder on a plane inclined at 30° to the longitudinal axis of the cylinder, in N/mm^2

(a) 120 MPa (b) 105 MPa
(c) 75 MPa (d) None of these

Q.12 A thin cylindrical shell is subjected to internal pressure. Ratio of volumetric strain/hoop strain is 9/4, what is Poisson's ratio of the material?

(a) $\frac{2}{10}$ (b) $\frac{2}{9}$

(c) $\frac{2}{7}$ (d) $\frac{2}{11}$

Q.13 A thin cylindrical shell, diameter 200 mm, thickness *t* is subjected to internal pressure of 2 N/mm^2. It efficiency of longitudinal joint is 70% and the maximum allowable stress is 100 MPa, what is thickness in mm

(a) 2 mm (b) 2.54 mm
(c) 2.86 mm (d) 3.0 mm

Q.14 A water main of 1 m diameter contains water at a pressure head of 100 m. The permissible tensile stress in the material of the water main is 25 MPa. What is the minimum thickness of the water main? [Take g = 10 m/s^2]

(a) 10 mm (b) 20 mm
(c) 50 mm (d) 60 mm

[IES 2009]

Q.15 A thin cylindrical pressure vessel with closed ends is subjected to internal pressure. The ratio of circumferential (hoop) stress to the longitudinal stress is

(a) 0.25 (b) 0.50
(c) 1.0 (d) 2.0

[GATE 2015]

Answers

1. (a) 2. (c) 3. (d) 4. (a) 5. (c)
6. (c) 7. (c) 8. (d)
9. (109.80, 40.2 N/mm²) 10. (d) 11. (b)
12. (c) 13. (c) 14. (b) 15. (d)

Explanations

1. (a)

$$\varepsilon_c = \frac{\sigma_c}{E} - \frac{\nu\sigma_a}{E} = \frac{100 - 0.3 \times 50}{200 \times 1000}$$

$$= \frac{85}{200} = 0.425 \times 10^{-3}$$

2. (c)

Have uniform lower circumferential stress

$$(\sigma_c)_{spherical} = \frac{1}{2}\sigma_{c(cylindrical)}$$

Spherical vessels are costly to fabricate.

3. (d)

$$\varepsilon_a = \frac{pD}{4tE}\left(1 - \frac{2}{m}\right), \quad \varepsilon_c = \frac{pD}{4tE}\left(2 - \frac{1}{m}\right)$$

$$\frac{\varepsilon_a}{\varepsilon_c} = \frac{m-2}{2m-1}$$

4. (a)

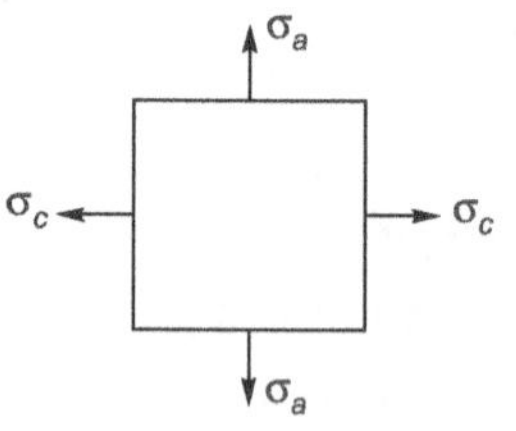

No. shear stress, τ

Hoop and axial stresses act, both are tensile.

5. (c)

At the surface

$$\sigma_a = \sigma_0$$
$$\sigma_c = 2\sigma_0$$
$$p = 0 \text{ (radial stress)}$$

$$\tau_{max} = \frac{2\sigma_0 - 0}{2} = \sigma_0$$

6. (c)

$$\sigma_c = 9000 = \frac{500 \times 600}{2 \times t}$$

$$t = 16.66 \simeq 17 \text{ mm}$$

7. (c)

$$\varepsilon_c = \frac{2\pi(r+u) - 2\pi r}{2\pi r} = \frac{u}{r}$$

8. (d)

$$\sigma_{all} = \frac{pD}{4t}$$

$$p = \frac{4t\sigma_{att}}{D} = \frac{4 \times 1.5 \times 45}{1500} = 1.8 \text{ MPa}$$

9. (109.80, +40.2 N/mm²)

$$\sigma_c = \frac{10 \times 100}{2 \times 5} = 100 \text{ MPa} \quad \text{...(i)}$$

$$\sigma_a = 50 \text{ MPa} \quad \text{...(ii)}$$

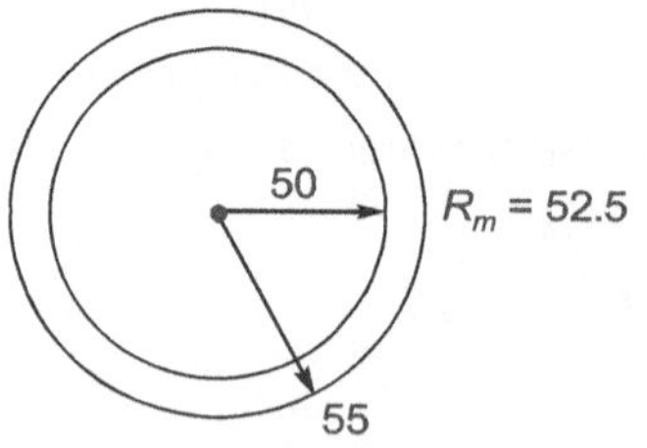

$$J = 2\pi \times 5 \times 52.5^3 = 4.546 \times 10^6 \text{ mm}^4$$

$$\tau_{max} = \frac{2 \times 10^6}{4.546 \times 10^6} \times 55 = 24.2 \text{ N/mm}^2$$

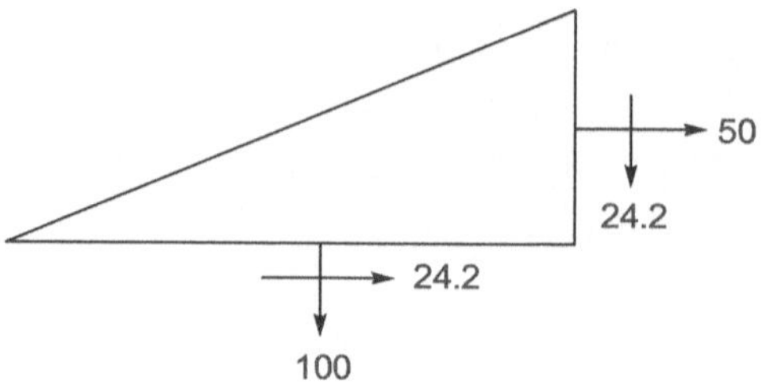

$$p_1, p_2 = 75 \pm \sqrt{25^2 + 24.2^2}$$
$$= 75 \pm \sqrt{1210.64}$$
$$= 75 + 34.8$$
$$= 109.80, + 40.2 \text{ N/mm}^2$$

10. (d)

$$\sigma_c = \frac{p \times r}{2t}$$

$$\sigma_c' = \frac{p \times 1.01r}{2 \times 0.99t}$$

$$\frac{\sigma_c}{\sigma_c'} = 0.9802$$

$$\frac{\sigma_c'}{\sigma_c} = 1.0202$$

% change in σ_c

$$= \frac{1.0202 - 1}{1} \times 100 = 2.02\%$$

11. (b)

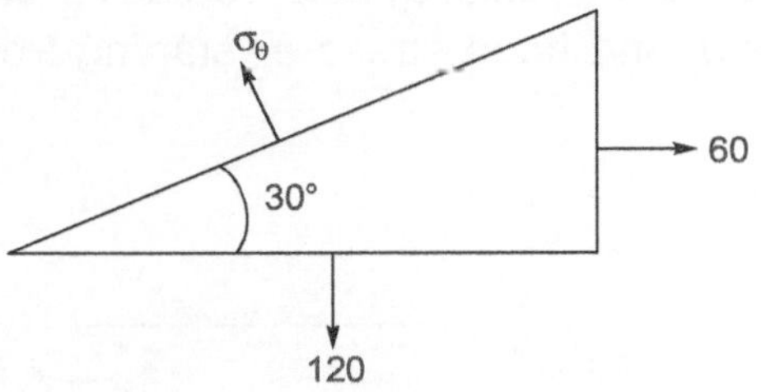

$$\sigma_\theta = \frac{120 + 60}{2} + \frac{120 - 60}{2}\cos 60°$$
$$= 90 + 15 = 105 \text{ MPa}$$

12. (c)

$$\frac{\varepsilon_v}{\varepsilon_c} = \frac{5 - 4\nu}{2 - \nu} = \frac{9}{4}$$

$$20 - 16\nu = 18 - 9\nu$$
$$7\nu = 2$$
$$\nu = \frac{2}{7}$$

13. (c)

$$100 \times 0.70 = \frac{pD}{2t} = \frac{2 \times 200}{2 \times t} = 2.86 \text{ mm}$$

14. (b)

$$p = \rho g h = 1000 \times 10 \times 100$$
$$= 10^6 \text{ N/m}^2 = 1 \text{ N/mm}^2$$

$$t = \frac{pD}{2\sigma} = \frac{1 \times 1000}{2 \times 25} = 20 \text{ mm}$$

15. (d)

$$\frac{\sigma_c}{\sigma_a} = 2$$

or,
$$\sigma_c = \frac{pD}{2t}$$

$$\sigma_a = \frac{pD}{4t}$$

■■■■

Thick Shells

CHAPTER 6

In thick shells, the wall thickness is considerable in comparison to inner radius. There is large variation of radial stress and hoop stress along the thickness, in such shells. In thick cylindrical shells, axial stress is independent of radius and remains constant from inner radius to outer radius of the shell.

Thick tubes are used for the transmission of very high pressure as high as 1000 atmospheres as in this case of pressure gauges for the measurement of pressures in transmission lines which are under high pressure liquids.

Question 6.1 **Consider a thick cylindrical shell with inner radius R_1, outer radius R_2, subjected to internal pressure p. Derive expressions for radial stress σ_r and hoop stress σ_c starting from basic principles. Clearly state the assumptions used.**

Solution:

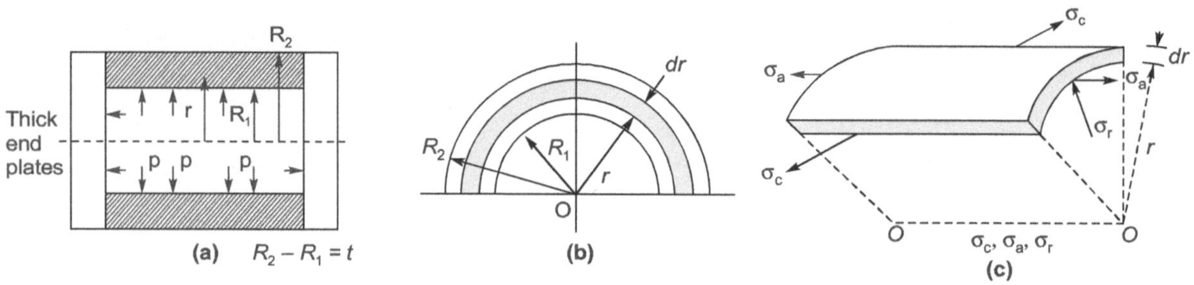

Fig. 6.1

A thick cylindrical shell, with inner radius R_1, outer radius R_2 subjected to internal pressure p is shown in figure 6.1 (a).

There are thick plates at the ends. Under pressure p, cylinder tends to expand introducing σ_a, tensile axial stress and σ_c tensile circumferential stress in the shell. To develop expressions for σ_r and σ_c, following assumptions are taken

(i) Axial strain ε_a, remains constant, along the length of the shell

(ii) No distortion of end plates

Say at a particular radius r, the stresses are σ_c, σ_r, σ_a

These are three principal stresses, perpendicular to each other. If E is Young's modulus and ν is Poisson's ratio, then axial strain,

$$\varepsilon_a = \frac{\sigma_a}{E} + \frac{\nu\sigma_r}{E} - \frac{\nu\sigma_c}{E}, \text{ Note } \sigma_r \text{ is compressive}$$

$$= \frac{\sigma_a}{E} + \frac{\nu}{E}(\sigma_r - \sigma_c) \qquad \text{...(i)}$$

As per assumptions, ε_a is constant, E and ν are elastic constants.

Moreover σ_a, axial stress $= \dfrac{p(\pi R_1^2)}{\pi(R_2^2 - R_1^2)} = \dfrac{pR_1^2}{R_2^2 - R_1^2}$, is also constant.

From eq. (i) it is obvious that

$\sigma_c - \sigma_r$ = difference of radial and hoop stresses is constant at any radius

$\sigma_c - \sigma_r = 2A$ (say)

Consider unit length of cylinder say at a particular radius r radial stress is σ_r and at radius is $r + dr$, radial stress is $\sigma_r + d\sigma_r$. Consider equilibrium of elementary ring as shown in fig. 6.2

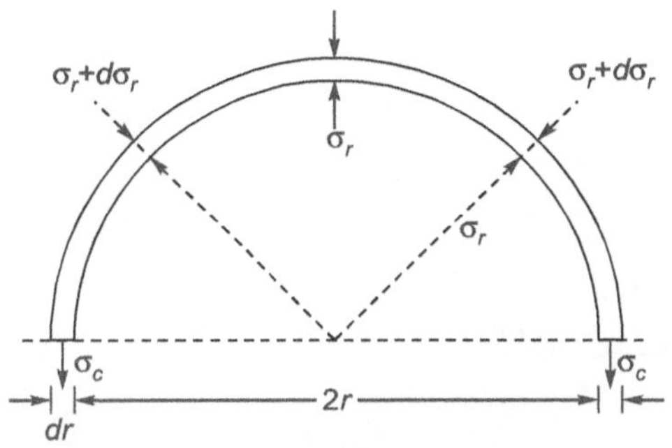

Fig. 6.2

$$2(\sigma_r + d\sigma_r)(r + dr) + 2\sigma_c\, dr = 2\sigma_r r$$

$$\sigma_r r + r d\sigma_r + \sigma_r dr + d\sigma_r dr + \sigma_c\, dr = \sigma_r r$$

Neglecting very small quantities $d\sigma_r\, dr$

We get $\quad r d\sigma_r + \sigma_r dr + \sigma_c dr = 0$

or $\quad r d\sigma_r + (\sigma_c + \sigma_r) dr = 0$

or
$$\sigma_c + \sigma_r = -r\frac{d\sigma_r}{dr} \qquad \text{...(ii)}$$

From eq. (i) and (ii)

$$2\sigma_r = -r\frac{d\sigma_r}{dr} - 2A$$

$$2(A + \sigma_r) = -r\frac{d\sigma_r}{dr}$$

or
$$\frac{d\sigma_r}{A + \sigma_r} = -2\frac{dr}{r} \qquad \text{...(iii)}$$

Integrating eq. (iii)

$$\ln(A + \sigma_r) = -2 \ln r + \ln B$$

Where B is another constant

or
$$A + \sigma_r = \frac{B}{r^2}$$

Radial stress,
$$\sigma_r = \frac{B}{r^2} - A \qquad \text{...(iv)}$$

Hoop stress,
$$\sigma_c = \sigma_r + 2A = \frac{B}{r^2} + A$$

These two equations are known as Lame's equations and constants A and B are Lame's constants.

At any radius $\quad \sigma_c + \sigma_r = 2A$

Both σ_c and σ_r depend on the radius r. Using boundary conditions, constants A and B are determined. For both σ_r and σ_c, hyperbolic curves give the stress distribution along radial thickness of shell.

Question 6.2 Take a thick cylinder of inner radius R_1 and outer radius R_2 subjected to internal pressure p. Determine Lame's constants A and B. Find values of hoop stress at inner and outer radii.

Solution:

Radial stress, $$\sigma_r = \frac{B}{r^2} - A$$

Boundary condition

Radial stress, $$\sigma_r = p \quad \text{at inner radius } R_1$$
$$= 0 \quad \text{at centre radius } R_2$$

So $$p = \frac{B}{R_1^2} - A$$

$$o = \frac{B}{R_2^2} - A$$

or $$A = \frac{B}{R_2^2}$$

or $$p = \frac{B}{R_1^2} - \frac{B}{R_2^2}$$

or constant, $$B = \frac{pR_1^2 R_2^2}{R_2^2 - R_1^2}$$

$$A = \frac{B}{R_2^2} = \frac{pR_1^2}{R_2^2 - R_1^2}$$

Hoop stress $$\sigma_c = \frac{B}{r^2} + A = \frac{p(R_1^2 R_2^2)}{R_2^2 - R_1^2} \times \frac{1}{r^2} + \frac{pR_1^2}{R_2^2 - R_1^2}$$

$$= \frac{pR_1^2}{R_2^2 - R_1^2} \times \left[\frac{R_2^2}{r^2} + 1\right]$$

at $r = R_1$, $$\sigma_c = p\frac{R_1^2 + R_2^2}{R_2^2 - R_1^2}, \text{ maximum hoop stress}$$

at $r = R_2$, $$\sigma_c = \frac{2pR_1^2}{R_2^2 - R_1^2}, \text{ minimum hoop stress}$$

Question 6.3 A thick cylindrical shell of inner radius 120 mm and outer radius 200 mm is subjected to internal pressure of 50 N/mm². Determine Lame's constants A, B and plot stress distribution for both σ_c not σ_r, by calculating values at radii 120, 150, 180, 200 mm.

Solution: $R_1 = 120$ mm, $R_2 = 200$ mm, $p = 50$ MPa

Constant, $$A = p\frac{R_1^2}{R_2^2 - R_1^2} = 50 \times \frac{120^2}{200^2 - 120^2}$$

$$= 28.125 \text{ N/mm}^2, \quad 2A = 56.25$$

$$B = p\frac{R_1^2 R_2^2}{R_2^2 - R_1^2} = 50\frac{120^2 \times 200^2}{200^2 - 120^2} = 112.5\times10^4 \text{ N}$$

Stresses

$r = 120$ mm, $\sigma_r = \dfrac{112.5\times10^4}{120^2} - 28.125 = 78.125 - 28.125 = 50 \text{ N/mm}^2$

$\sigma_c = \sigma_r + 2A = 50 + 56.25$

$= 106.25 \text{ N/mm}^2$

$r = 150$ mm, $\sigma_r = \dfrac{112.5\times10^4}{150^2} - 28.125$

$= 21.875 \text{ N/mm}^2$

$\sigma_c = 78.125 \text{ N/mm}^2$

$r = 180$ mm, $\sigma_r = \dfrac{112.5\times10^4}{180^2} - 28.125$

$= 6.6 \text{ N/mm}^2$

$\sigma_c = 6.6 + 56.25$

$= 62.85 \text{ N/mm}^2$

$r = 200$ mm, $\sigma_r = \dfrac{12.5\times10^4}{200^2} - 28.125 = 0$

$\sigma_c = 0 + 56.25 = 56.25 \text{ N/mm}^2$

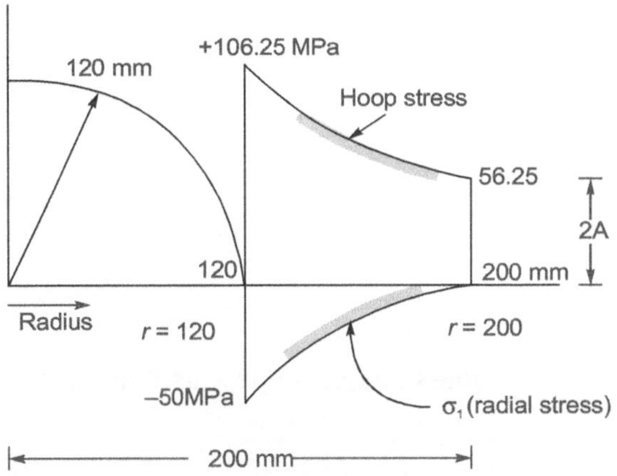

Fig. 6.3

Fig 6.3 shows variation of hoop stress σ_c and radial stress σ_r along the thickness ($R_2 - R_1$)

Practice Q.6.3 A thick cylindrical shell of inner radius 150 mm and outer radius 200 mm is subjected to internal pressure of 40 N/mm². Determine maximum and minimum values of hoop stress. Draw distribution of stresses along the thickness of 50 mm, by taking values of radii of 150, 175, 200 mm.

Ans. [142.86, 118.6, 102.86 N/mm², 40, 15.742, 0 N/mm²]

Question 6.4 Consider a thick cylindrical shell of inner radius, R_1 outer radius R_2 subjected to external pressure p. What are 3 principal stresses at the inner radius of the shell. Draw variation of hoop stress along the thickness of the shell. What is the axial stress developed in the shell.

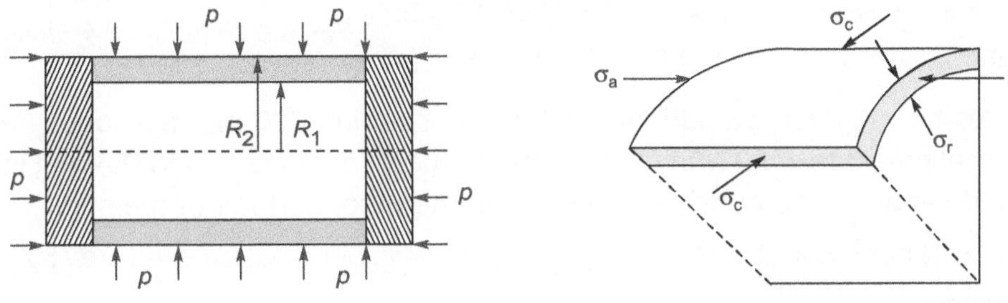

(a) *Thick cylindrical shell subjected to external pressure* **(b)** *Stresses on an element*

Fig. 6.4

Solution: Fig. 6.4 (a) shows the thick cylindrical shell subjected to external pressure p.

Uniform axial stress, $\sigma_a = \dfrac{pR_2^2}{R_2^2 - R_1^2}$, a compressive stress developed along the length

Lame's equations, $\sigma_r = \dfrac{B}{r^2} - A$, radial stress, $\sigma_c = \dfrac{B}{r^2} + A$, hoop stress

Boundary conditions

$$\sigma_r = p, \quad \text{at } r = R_2$$
$$\sigma_r = 0, \quad \text{at } r = R_1$$

$$0 = \frac{B}{R_1^2} - A \quad \text{or} \quad A = \frac{B}{R_1^2}$$

$$p = \frac{B}{R_2^2} - A = \frac{B}{R_2^2} - \frac{B}{R_1^2} = \frac{B(R_1^2 - R_2^2)}{R_1^2 R_2^2}$$

$$B = -p\frac{R_1^2 R_2^2}{R_2^2 - R_1^2}, \qquad A = -p\frac{R_2^2}{R_2^2 - R_1^2}$$

Note that both constants A and B are negative.

$$\sigma_c = -\frac{p}{r^2}\frac{R_1^2 R_2^2}{(R_2^2 - R_1^2)} - p\frac{R_2^2}{R_2^2 - R_1^2}$$

at $r = R_1$, inner radius $\sigma_{c\,max} = -\dfrac{2pR_2^2}{R_2^2 - R_1^2}$

at $r = R_2$, outer radius $\sigma_{c\,min} = -p\dfrac{(R_1^2 + R_2^2)}{R_2^2 - R_1^2}$

At the inner radius of the shell, there are two principal stresses, both are compressive i.e.,

$$-p\frac{R_2^2}{R_2^2 - R_1^2}, \; -\frac{2pR_2^2}{R_2^2 - R_1^2}; \; 0$$

Figure 6.5 shows variations of hoop stress σ_c along the thickness $R_2 - R_1$ of the shell.

Note that in this case all the stresses i.e., radial, axial and hoop are negative stresses i.e., compressive stresses.

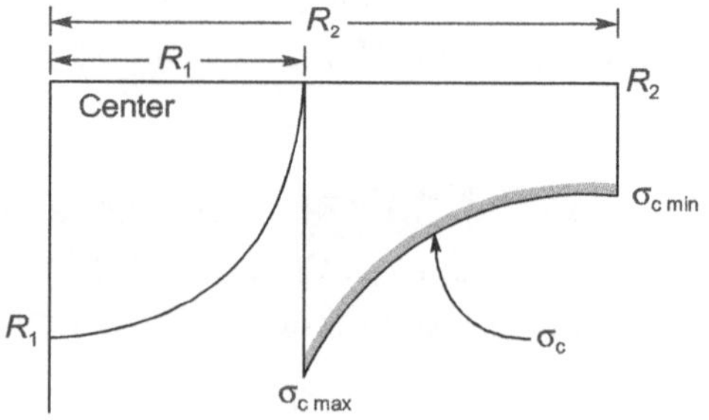

Fig. 6.5 *Variation of σ_c along thickness ($R_2 - R_1$)*

Practice Q.6.4 **A thick cylindrical shell of inner radius 200 mm and outer radius 300 mm is subjected to external pressure of 50 N/mm². What are Lame's constants A and B. What is maximum value of hoop stresses. What are three principal stresses on the outer surface of the shell.**

Ans. [$A = -90$ N/mm², $B = -3.6 \times 10^6$ N; $\sigma_{c\,max} = -180$ MPa, principal stresses are –50, –90, –130 N/mm²]

Question 6.5 **What is compounding of cylinders? What is the basic purpose of compounding? What type of stresses are developed at junction radius in inner and outer cylinders.**

Solution: In a thick cylindrical shell, subjected to internal pressure, variation of hoop stress from inner surface to outer surface of cylinder is large. The maximum hoop stress at the inner surface cannot be greater than the allowable stress. **So the hoop stress at the outer surface is much less than the allowable stress.** Consequently while the inner region of the shell is put to allowable stress and so material is fully utilized but the outer region of shell is not economically utilized. The first purpose of compounding the cylinders is to reduce the order of variation in hoop stresses and second purpose is to increase the pressure bearing capacity of the inner cylinder.

For compounding, two cylinders are taken. The outer diameter of the inner cylinder is slightly more than inner, diameter of the outer cylinder, **this difference in diameters is known as shrinkage allowance.** The outer cylinder is heated to the stage, when its inner diameter expands due to thermal expansion and becomes equal to outer diameter of the inner cylinder. At this stage the outer cylinder is slipped on the inner cylinder. Outer cylinder tries to cool down. The inner diameter of outer cylinder starts decreasing exerting radial pressure on the inner cylinder. At the same time there is equal and opposite reaction offered by inner cylinder and inner cylinder exerts equal radial pressure on the inner surface of outer cylinder. Net result of this shrinkage fitting is

(i) inner cylinder is subjected to external junction pressure producing compressive hoop stress in inner cylinder.

(ii) outer cylinder is subjected to internal junction pressure, producing tensile hoop stress in outer cylinder

(iii) since due to shrinkage fitting compressive hoop stresses are developed in inner cylinder, its pressure bearing capacity is increased

(iv) when compound cylinder is subjected to internal pressure, the initial compressive hoop stress in inner cylinder superimposed on the tensile hoop stress produced by inner fluid pressure.

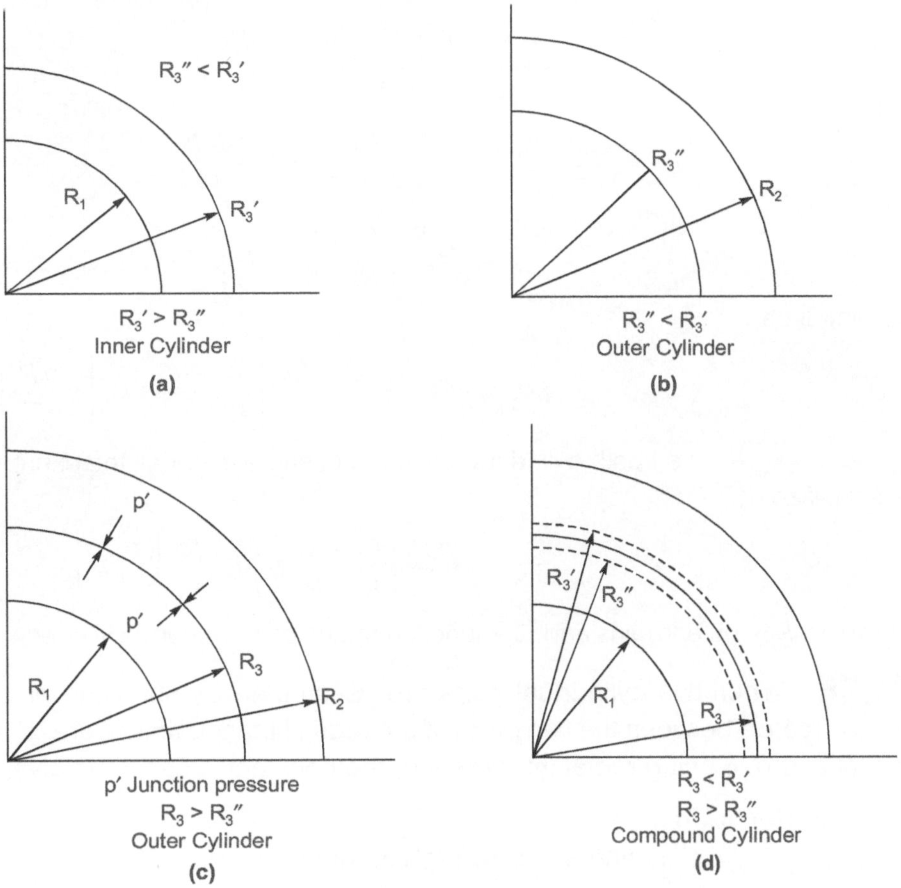

Fig. 6.6

Final junction radius R_3 in compound cylinder is less than R_3' but greater than R_3''.
Reduction in radius of inner cylinder = $R_3' - R_3$.
Increase in radius of outer cylinder = $R_3 - R_3''$.
Shrinkage allowance on radius, $\delta R_3 = R_3' - R_3''$.
= reduction in junction radius of inner cylinder + increase in junction radius of outer cylinder.

Hoop stress in inner cylinder at R_3, $\sigma_{ciR3} = -p \times \dfrac{R_3^2 + R_1^2}{R_3^2 - R_1^2}$

Hoop stress in outer cylinder at R_3, $\sigma_{coR3} = p \times \dfrac{R_2^2 + R_3^2}{R_2^2 - R_3^2}$, (refer to question 6.2, 6.4)

Say E_o, E_i – Young's modulus of outer, inner cylinders
ν_o, ν_i – Poisson's ratio of outer, inner cylinders.
Circumferential strain in outer cylinder at R_3

$$\varepsilon_{coR3} = \frac{p'}{E_o}\left(\frac{R_2^2 + R_3^2}{R_2^2 - R_3^2}\right) + \frac{\nu_o p'}{E_o}$$

$$R_3 - R_3'' = \frac{p'R_3}{E_o}\left(\frac{R_2^2 + R_3^2}{R_2^2 - R_3^2} + \nu_o\right) \qquad \text{...(i)}$$

Circumferential strain in inner cylinder at R_3

$$\varepsilon_{ci'R3} = \frac{p'}{E_i}\left(\frac{R_3^2 + R_1^2}{R_3^2 - R_1^2}\right) - \frac{\nu_i p'}{E_i}$$

$$R_3' - R_3 = \frac{p'R_3}{E_i}\left(\frac{R_3^2 + R_1^2}{R_3^2 - R_1^2}\right) - \frac{\nu_i p' R_3}{E_i}$$

$$= \frac{p'R_3}{E_i}\left(\frac{R_3^2 + R_1^2}{R_3^2 - R_1^2} - \nu_i\right) \qquad \text{...(ii)}$$

Shrinkage allowance

$$\delta R_3 = R_3' - R_3'' = \frac{p'R_3}{E_o}\left[\frac{R_2^2 + R_3^2}{R_2^2 - R_3^2} + \nu_o\right] + \frac{p'R_3}{E_i}\left[\frac{R_3^2 + R_1^2}{R_3^2 - R_1^2} - \nu_i\right]$$

If $E_1 = E_o = E$, $\nu_i = \nu_o = \nu$, i.e., both cylinders are made of same material, the expression for shrinkage allowance can be simplified as

$$\delta R_3 = \frac{p'R_3}{E}\left[\frac{R_2^2 + R_3^2}{R_2^2 - R_3^2} + \frac{R_3^2 + R_1^2}{R_3^2 - R_1^2}\right], \text{ where}$$

R_1 – inner radius, R_2 – outer radius and R_3 – junction radius of compound cylinder.

Question 6.6 When two cylindrical parts are assembled by shrinking or by press fitting, a constant pressure is created between the two parts. If the radii of inner cylinder are '*a*' and '*c*' and that of outer cylinder ($c - \delta$) and *b*, δ being radial interference, the constant pressure is given by

$$p = \frac{E\delta}{c}\left[\frac{(b^2 - c^2)(c^2 - a^2)}{2c^2(b^2 - a^2)}\right], \text{ where E is Young's modulus .}$$

[IES : 2001, 10 Marks]

Solution:

Material is the same for outer and inner cylinders

$$E = \text{Young's modulus for both cylinder}$$

$$\frac{\delta}{c} = \frac{\text{radial interference}}{\text{junction radius}}$$

$$= \frac{p}{E}\left[\frac{b^2+c^2}{b^2-c^2}+\frac{c^2+a^2}{c^2-a^2}\right]$$

where p = junction pressure

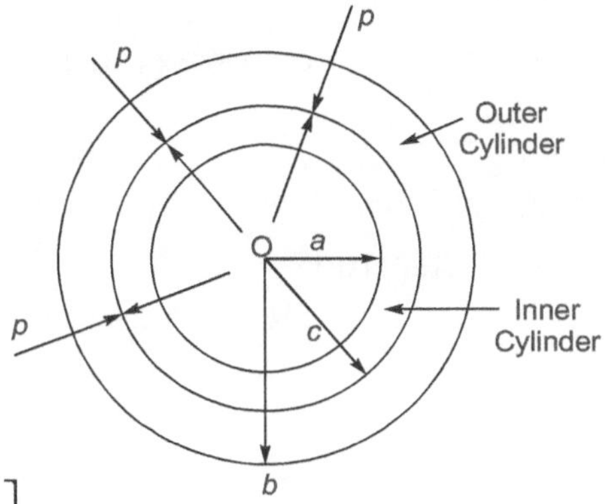

Fig. 6.7

$$p = \frac{E\delta}{c}\left[\frac{(b^2-c^2)(c^2-a^2)}{(b^2+c^2)(c^2-a^2)+(b^2-c^2)(c^2+a^2)}\right]$$

$$p = \frac{E\delta}{c}\left[\frac{(b^2-c^2)(c^2-a^2)}{b^2c^2+c^4-b^2a^2-c^2a^2+b^2c^2-c^4+b^2a^2-c^2a^2}\right]$$

$$= \frac{E\delta}{c}\left[\frac{(b^2-c^2)(c^2-a^2)}{2b^2c^2-2c^2a^2}\right]$$

$$= \frac{E\delta}{c}\left[\frac{(b^2-c^2)(c^2-a^2)}{2c^2(b^2-a^2)}\right]$$

Question 6.7 A steel rod of diameter 50 mm is forced into a bronze casing of outside diameter 90 mm, producing a tensile hoop stress of 30 MPa at the outside diameter of the casing. Find

(i) The radial pressure between rod and the casing

(ii) Interference allowance

(iii) The rise in temperature which would just eliminate the force fit.

Assume the following properties

$E_s = 2 \times 10^5$ MPa, $\nu_s = 0.25$

$\alpha_s = 12 \times 10^{-6}/°C$, $E_b = 1 \times 10^5$ MPa, $\nu_b = 0.30$.

$\alpha_b = 19 \times 10^{-6}/°C$

[IES : 2003, 20 Marks]

Solution:

Bronze casing

$$\sigma_c \text{ at outer radius} = +p\frac{2R_1^2}{R_2^2-R_1^2}$$

$$30 = p\times\frac{2\times 25^2}{45^2-25^2} = p\times\frac{1250}{1400}$$

Junction pressure, $$p = \frac{30\times 1400}{1250} = 33.6 \text{ N/mm}^2$$

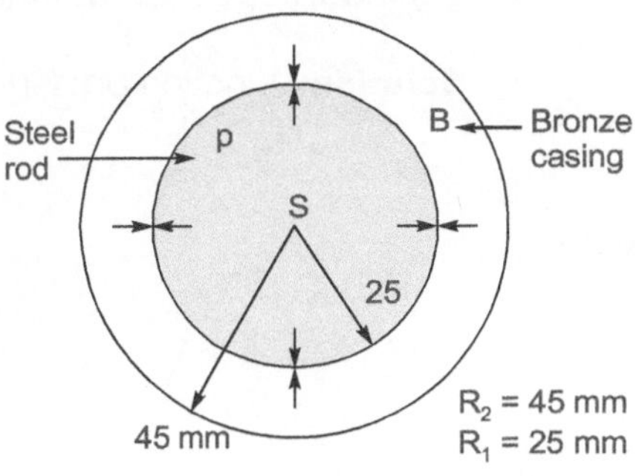

Fig 6.8

$$\sigma_c \text{ at inner radius} = +p\frac{R_2^2+R_1^2}{R_2^2-R_1^2} = 33.6\times\frac{45^2+25^2}{45^2-25^2} = 33.6\times\frac{2650}{1400}$$

$$\sigma_{ci} = 63.6 \text{ N/mm}^2$$

p = junction pressure (compressive)

Steel rod

Radial pressure = p(comp.)

$\sigma_{c'}$ hoop stress = –p(comp.)

$$\frac{\delta R_1}{R_1} = \left[\frac{\sigma_{ci}}{E_b}+\frac{\nu_B p}{E_b}\right]+\left[\frac{p}{E_s}-\frac{p\nu_s}{E_s}\right]$$

$$= \frac{1}{1\times10^5}[63.6+0.30\times33.6]+\frac{1}{2\times10^5}[33.6-0.25\times33.6]$$

$$= \frac{73.68}{1\times10^5}+\frac{12.6}{1\times10^5}=\frac{86.28}{1\times10^5}$$

$$\delta R_1 = \frac{86.28}{1\times10^5}\times25 = 0.02157 \text{ mm}$$

= Radial interference allowance

Rise in temperature

Due to rise in temperature, bronze will expand more and steel will expand less

$$\frac{\delta R_1}{R_1} = (\alpha_b-\alpha_s)\Delta T$$

$$\frac{0.02157}{25} = (1.9-1.2)\times10^{-5}\times\Delta T = 0.7\times10^{-5}\Delta T$$

$$\Delta T = \frac{0.02157}{25}\times\frac{10^5}{0.7} = 123.26°C$$

Question 6.8 A compound cylinder is obtained by shrinking one steel cylinder over another steel cylinder such that junction pressure developed is 14 N/mm². It inner, outer and junction radii are 120, 200 and 160 mm respectively. Determine

(a) shrinkage allowance on radius

(b) hoop stresses at inner, outer radii of inner cylinder

(c) hoop stresses at inner and outer radii of outer cylinder

Solution: Junction pressure, $p' = 14 \text{ N/mm}^2$

$E = 200000 \text{ N/mm}^2$

$R_1 = 120$ mm, $R_3 = 160$ mm, $R_2 = 200$ mm

$$\delta R_3 = \frac{p'R_3}{E}\left[\frac{R_3^2+R_1^2}{R_3^2-R_1^2}+\frac{R_2^2+R_3^2}{R_2^2-R_3^2}\right], \text{ putting the values}$$

$$= \frac{14\times160}{200000}\left[\frac{160^2+120^2}{160^2-120^2}+\frac{200^2+160^2}{200^2-160^2}\right]$$

$$= 11.2 \times 10^{-3}\,[3.571 + 4.555]$$
$$= 11.2 \times 10^{-3} \times 8.126 = 0.091 \text{ mm}$$
= Shrinkage allowance on radius

Inner cylinder

Hoop stresses,
$$\sigma_{ciR_1} = \frac{2p'R_3^2}{R_2^2 - R_1^2} = -2\times 14\times\frac{160^2}{160^2-120^2} = -64 \text{ MPa}$$

$$\sigma_{ciR_3} = -p'\times\frac{R_3^2+R_1^2}{R_3^2-R_1^2} = -14\times 3.571 = -50 \text{ MPa}$$

Outer cylinder

Hoop stresses,
$$\sigma_{coR_3} = p'\times\frac{R_2^2+R_3^2}{R_2^2-R_3^2} = 14\times 4.555 = 63.77 \text{ N/mm}^2$$

$$\sigma_{coR_2} = p'\times\frac{2R_3^2}{R_2^2-R_3^2} = 14\times\frac{2\times 160^2}{200^2-160^2} = 49.78 \text{ N/mm}^2$$

Question 6.9 A compound cylinder is obtained by compounding one cylinder over another, such that junction pressure is 3 N/mm^2. Inner, junction and outer radii are 100 mm, 150 mm and 200 mm. The compound cylinder is subjected to an internal pressure of 30 N/mm^2. Draw approximate sketch of resultant hoop stresses in inner and outer cylinders.

Solution: Junction pressure, $p' = 3$ N/mm^2, $R_1 = 100$ mm, $R_3 = 150$ mm, $R_2 = 200$ mm

Inner cylinder

$$\sigma_{ciR_1} = \frac{2p'\times R_3^2}{R_3^2-R_1^2} = -\frac{2\times 3\times 150^2}{150^2-100^2} = -10.8 \text{ MPa}$$

$$\sigma_{ciR_3} = p'\times\frac{R_3+R_1^3}{R_3-R_1^2} = -3\times\left(\frac{1.5^2+1^2}{1.5^2-1}\right) = -7.8 \text{ MPa}$$

Outer Cylinder

$$\sigma_{coR_3} = p'\times\frac{R_2^2+R_3^2}{R_2^2-R_3^2} = 3\times\frac{2^2+1.5^2}{2^2-1.5^2} = 10.71 \text{ N/mm}^2$$

$$\sigma_{coR_2} = p'\times\frac{2R_3^2}{R_2^2-R_3^2} = 3\times 2\times\frac{1.5^2}{2^2-1.5^2} = 7.71 \text{ N/mm}^2$$

Stresses due to internal pressure p

$p = 30$ N/mm^2,

$R_1 = 100$ mm

$R_2 = 200$ mm

$$B = 30\times\frac{R_1^2R_2^2}{R_2^2-R_1^3} = 30\times\frac{100^2\times 200^2}{200^2-100^2} = 4\times 10^5 \text{N}$$

$$A = \frac{p\times R_1^2}{R_2^2-R_1^2} = 30\times\frac{100^2}{200^2-100^2} = 10 \text{ N/mm}^2$$

Hoop stress

at $r = R_1 = 100$ mm, $\sigma_{cR1} = \dfrac{400000}{100^2} + 10 = 50 \text{ N/mm}^2$

$r = R_3 = 150$ mm, $\sigma_{cR2} = \dfrac{400000}{150^2} + 10 = 27.77 \text{ N/mm}^2$

$r = R_2 = 200$ m, $\sigma_{cR2} = \dfrac{400000}{200^2} + 10 = 20 \text{ N/mm}^2$

Resultant Stresses

Inner cylinder, $\sigma_{cRR_1} = 50 - 10.8 = 39.2$ MPa

$\sigma_{cRR_2} = 27.77 - 7.8 = 19.97$

Outer cylinder

$\sigma_{cRR_3}{}' = 27.77 + 10.71 = 38.48 \text{ N/mm}^2$

$\sigma_{cRR_2} = 20 + 7.71 = 27.71 \text{ N/mm}^2$

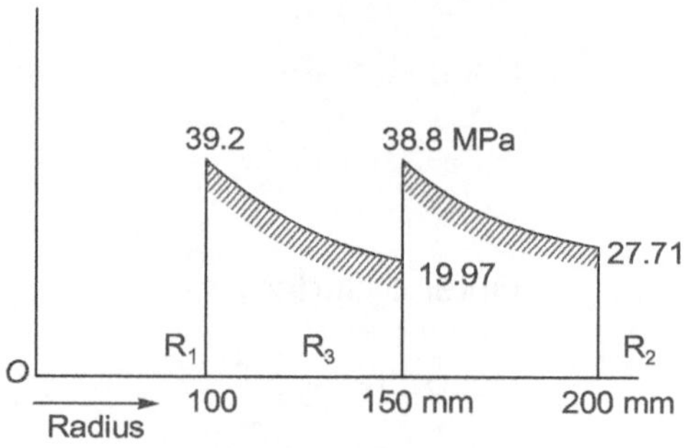

Fig. 6.9 *Resultant hoop stresses in compound cylinder*

Fig. 6.9 shows the variation of hoop stress in compound cylinder. If there was no compounding then hoop stress would vary from 50 MPa at inner radius to 20 MPa at outer radius.

Question 6.10 **A steel cylinder of outer diameter 180 mm is shrunk on another steel cylinder of inner diameter 120 mm. The common diameter is 150 mm. If after shrinkage, the radial pressure at the common surface is p'. The compound cylinder is subjected to an internal pressure of 100 N/mm². Determine the magnitude of p'. So that maximum hoop tension in inner and outer cylinder is equal.**

Solution: Stresses due to internal pressure

$R_1 = 120$ mm, $R_2 = 180$ mm, $p = 100 \text{ N/mm}^2$

Lame's constants

$$B = p\frac{R_1^2 R_2^2}{R_2^2 - R_1^2} = 100 \times \frac{180^2 \times 120^2}{180^2 - 120^2}$$

$$= 2592000 \text{ N} = 2.592 \times 10^6 \text{ N}$$

$$A = p\frac{R_1^2}{R_2^2 - R_1^2} = 100 \times \frac{120^2}{180^2 - 120^2} = 80 \text{ N/mm}^2$$

at $r = 120$ mm, $\sigma_c = \dfrac{2592000}{120^2} + 80 = 260 \text{ N/mm}^2$

$r = 150$ mm, $\sigma_c = \dfrac{2592000}{150^2} + 80 = 195.2 \text{ N/mm}^2$

Stresses due to junction pressure

Inner cylinder at R_1, $\sigma_{ciR1} = -\dfrac{2 \times R_3^2 p'}{R_3^2 - R_1^2} = -2p' \times \dfrac{150^2}{150^2 - 120^2} = -5.555p'$

Outer cylinder at R_3, $\sigma_{coR_3} = p' - \dfrac{R_2^2 + R_3^2}{R_2^2 - R_3^2} = p' \times \dfrac{180^2 + 150^2}{180^2 - 150^2} = p' \times \dfrac{549}{99} = 5.545p'$

Maximum hoop tension in outer cylinder $= 195.2 + 5.545p'$

Maximum hoop tension in inner cylinder $= 260 - 5.555p'$

As per this question

$$260 - 5.555p' = 195.2 + 5.545p'$$

or

$$p' = \frac{64.8}{11.1} = 5.84 \text{ N/mm}^2$$

Junction pressure p' should be of the order of 5.84 N/mm^2.

Practice Q.6.10 A steel cylinder of an outer diameter 200 mm is shrunk on another steel cylinder of an inner diameter 120 mm. The common diameter is 160 mm. If after shrinkage, the radial pressure at the common surface is 5 N/mm^2, determine the magnitude of internal pressure to which the compound cylinder can be subjected so that the maximum hoop tension in the inner and order cylinder is the same.

Ans. [64.2 N/mm^2]

Question 6.11 A sleeve of outer radius R is force fitted over a shaft of radius r, such that junction pressure p' is developed at the junction radius of the two. Derive expression for force fit allowance in terms of r and R and E, if material of both shaft and sleeve is the same.

Solution:

Junction pressure $= p$

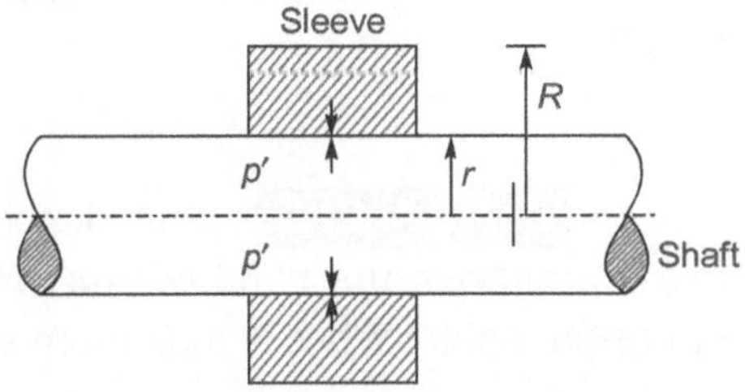

Fig. 6.10

Sleeve

$$\sigma_r = \frac{B_1}{r^2} - A_1$$

$$\sigma_c = \frac{B_1}{r^2} + A_1$$

where B_1 and A_1 are Lame constants.

$$\sigma_r = 0, \quad \text{at } r = R$$

$$\sigma_r = p' \text{ at radius } r$$

$$B_1 = p' \times \frac{R^2 r^2}{R^2 - r^2}$$

$$A_1 = p' \times \frac{r^2}{R^2 - r^2}$$

σ_c in sleeve at radius r,

$$\sigma_{cs} = p' \times \frac{R^2}{R^2 - r^2} + \frac{p' r^2}{R^2 - r^2} = p' \times \frac{p' + r^2}{R^2 - r^2} \text{(tensile)}$$

Shaft

$$\sigma_r = \frac{B_2}{r^2} - A_2$$

shaft is solid, it exists at $r = 0$, so stresses cannot be infinite, therefore constant $B_2 = 0$

$$\sigma_r = A_2 = p' \text{ at any radius}$$

$$\sigma_c = A_2 \quad \text{but } \sigma_r \text{ is compressive}$$

σ_c is taken as tensile,

$$\sigma_c = p'$$

$$\sigma_r = p'$$

both the stresses are compressive

At junction

$$\varepsilon_{c\ sleeve}, \text{ hoop strain in sleeve} = \frac{p'}{E} \times \frac{R^2 + r^2}{R^2 - r^2} + \frac{\nu p'}{E}$$

$$\varepsilon_{c\ shaft} = \frac{p'}{E} - \frac{\nu p'}{E}$$

Where E = Young's modulus, ν = Poisson's ratio.

Force fit allowance = reduction in radius of shaft + increase in radius of sleeve at junction

$$\delta r = r\left[\frac{p'}{E} - \frac{\nu p'}{E}\right] + r\left[\frac{p'}{E} \times \frac{R^2 + r^2}{R^2 - r^2} + \frac{\nu p'}{E}\right]$$

$$= \frac{rp'}{E} + \frac{rp'}{E} \times \frac{R^2 + r^2}{R^2 - r^2} = \frac{rp'}{E}\left[1 + \frac{R^2 + r^2}{R^2 - r^2}\right]$$

$$= \frac{rp'}{E} \times \frac{2R^2}{R^2 - r^2}$$

Practice Q.6.11 A steel sleeve of 200 mm outer diameter is force fitted over a steel shaft of 100 mm diameter. If the force fit allowance on radius is 0.001 × 50 = 0.050 mm provided on sleeve. How much junction pressure is developed in sleeve and shaft. What is the hoop stress developed in sleeve. E= 200 GPa.

Ans. [75 N/mm², 125 N/mm²]

Question 6.12 A thick spherical shell of inner radius 80 mm and outer radius 120 mm is subjected to internal pressure of 60 N/mm². What is the maximum and minimum hoop stresses developed in thick spherical shell? What is maximum stress at inner surface?

Solution: Lame's equation for thick spherical shell are

$$\sigma_c = \frac{B}{r^3} + A(\text{tensile})$$

$$\sigma_r = \frac{2B}{r^3} - A(\text{comp.})$$

Boundary conditions are

$\sigma_r = p$ at $r = R_1$

$\sigma_r = 0$ at $r = R_2$

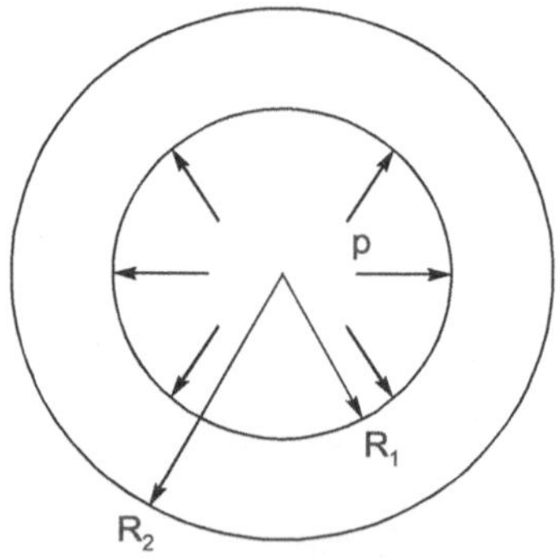

Fig. 6.11 *Thick spherical shell under internal pressure*

So, $$60 = \frac{2B}{80^3} - A \qquad o = \frac{2B}{120^3} - A$$

$$A = \frac{2B}{120^3} \qquad 60 = \frac{2B}{80^3} - \frac{2B}{120^3}$$

$$60 = 2B\left[\frac{120^3 - 80^3}{120^3 \times 80^3}\right]$$

$$2B = \frac{60 \times 120^3 \times 80^3}{120^3 \times 80^3} = \frac{60 \times 1728 \times 512 \times 10^6}{(1728 - 512) \times 10^3} = 43654.73 \times 10^3$$

$$A = \frac{2B}{120^3} = 25.263 \text{ N/mm}^2$$

$$B = 21827.36 \times 10^3$$

Hoop stresses

At inner radius of 80 mm, $\sigma_{c\,max} = \frac{B}{R_1^3} + A = \frac{21827.36 \times 10^3}{80^3} + 25.263$

$$= 42.63 + 25.263 = 67.893 \text{ N/mm}^2$$

At outer radius of 120 mm, $\sigma_{c\,min} = \frac{B}{R_2^3} + A = \frac{21827.36 \times 10^3}{120^3} + 25.263$

$$= 12.63 + 25.263 = 37.895 \text{ N/mm}^2$$

At the inner surface of thick spherical shell, three principal stresses are

$$+ 67.893, 67.893, -60 \text{ N/mm}^2$$

τ_{max}, maximum shear stress $= \frac{67.893 - (-60)}{2} = 63.95 \text{ N/mm}^2$

Practice Q.6.12 A thick spherical cylinder of inner radius 50 mm is subjected to internal pressure p. What should be the outer diameter of the shell it maximum hoop stress developed in shell is $2p$.

Ans. [$R_2 = 63$ mm]

Question 6.13 A thick cylindrical shell made of steel has inner radius 60 mm and outer radius 90 mm. It is subjected to internal pressure of 50 N/mm². Two strain gauges are mounted on the outer surface of the cylinder one in axial direction and other in hoop direction. What are hoop and axial strains recorded by strain gauges if E = 210 GPa and ν = 0.3.

Solution:

$$p = 50 \text{ N/mm}^2$$

$$R_1 = 60 \text{ mm}$$

$$R_2 = 90 \text{ mm}$$

Axial stress, $\sigma_a = p\frac{R_1^2}{R_2^2 - R_1^2} = 50 \times \frac{60^2}{90^2 - 60^2} = 40 \text{ N/mm}^2$

(which is independent of radius)

Hoop stress at outer surface, $\sigma_c = \frac{2pR_1^2}{R_2^2 - R_1^2} = 80 \text{ N/mm}^2$

$$E = 210000 \text{ N/mm}^2, \nu = 0.3$$

Hoop strain, $\varepsilon_c = \frac{\sigma_c}{E} - \nu\frac{\sigma_a}{E} = \frac{80}{210000} - \frac{0.3 \times 40}{210000}$

$$= 323.8 \times 10^{-6} = 323.8\ \mu s$$

Axial strain, $\varepsilon_a = \frac{\sigma_a}{E} - \frac{\nu\sigma_c}{E} = \frac{40}{E} - \frac{0.3 \times 80}{210000}$

$$= \frac{16}{E} = \frac{16}{210000} = 76.2 \times 10^{-6} = 76.2\ \mu s$$

Practice Q.6.13 A thick cylindrical shell of outer diameter two times the inner diameter is subjected to internal pressure *p*. What is the magnitude of p if hoop strain recorded at outer surface of shell is 680 μs. Given E = 105 GPa, ν = 0.35.

Ans. [129.82 N/mm²]

Question 6.14 A steel plug of diameter 60 mm is forced into a copper ring of an external diameter 100 mm and width 40 mm. From a strain gauge fixed on the outer surface of the ring in the circumferential direction, strain is found to be 650 μs. Considering that coefficient of friction between the mating surface is 0.25, determine the axial force required to push the plug out of the ring. E_{cu} = 100 GPa.

Solution: Put radius, $R_1 = 30$ mm

Copper ring, outer radius, $R_2 = 50$ mm

ε_c on surface of ring = 650 μs

$E_{copper} = 100000$ N/mm²

There is no axial stress in this case, so

σ_c, hoop stress on outer surface of ring $= 650 \times 10^{-6} \times 100000 = 65$ N/mm²

Say, p' = junction pressure

$$\sigma_c, \text{ at outer surface } = 2p' \times \frac{R_1^2}{R_2^2 - R_1^2} = 2p' \times \frac{30^2}{50^2 - 30^2}$$

$$= p' \times 1.125 = 65 \text{ N/mm}^2$$

Junction pressure, $$p' = \frac{65}{1.125} = 57.8 \text{ N/mm}^2$$

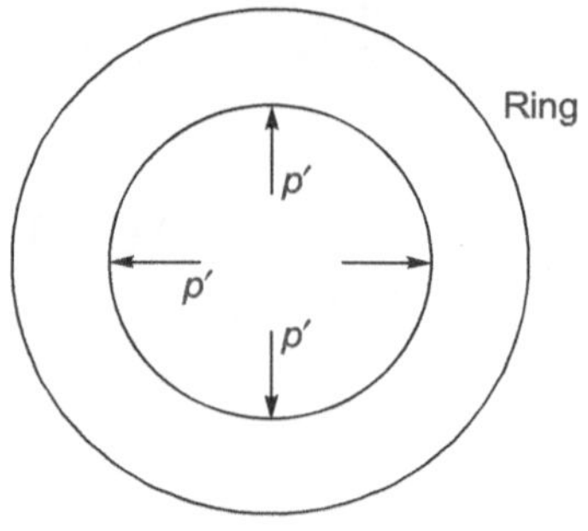

Fig. 6.12

Breadth of the ring, $b = 40$ mm

Inner circumferential stress area $= 2\pi \times 30 \times 40 = 7540$ mm²

Total normal radial force, $P_R = p' \times \text{area}$

$= 57.8 \times 7540 = 435800$ N

Coefficient of friction, $\mu = 0.25$

Force of friction, $F = \mu P_R = 0.25 \times 435800$

= 108950 N

= Axial force required to push the plug out.

Practice Q.6.14 A steel plug of 50 mm diameter is forced into a bronze ring of external diameter 80 mm and a width of 30 mm. From a strain gauge fixed on the outer surface of ring in the circumferential direction, the strain is found to be 85 μs. Taking the coefficient of friction between the mating surfaces as 0.2, determine the axial force required to push the plug out of the ring.

Given E for steel = 210 GPa, = 2 E_{bronze}

Ans. [6.56 kN]

Question 6.15 A thick cylinder subjected to both internal and external pressures. Inner radius of shell is 75 mm and outer radius of shell is 125 mm. If maximum permissible stress in cylinder is 50 N/mm² and external radial pressure is 10 N/mm², determine the intensity of internal fluid pressure.

Solution:

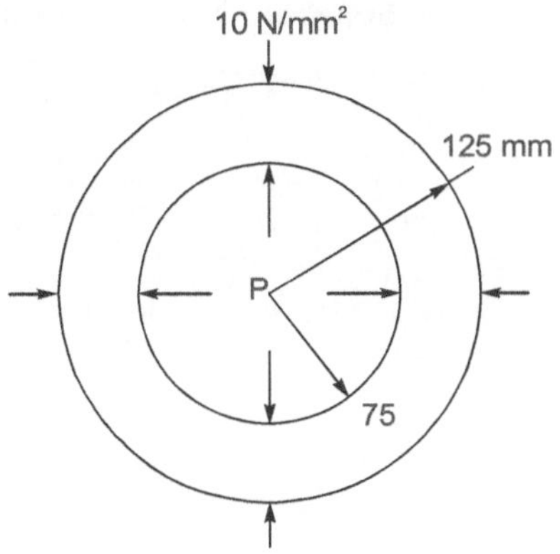

Fig. 6.13

$$\sigma_r = \frac{B}{r^2} - A; \quad \sigma_c = \frac{B}{r^2} + A$$

$$\sigma_r = p, \quad \text{at } r = 75 \text{ mm}$$

$$\sigma_r = 10, \quad \text{at } r = 125 \text{ mm}$$

$$10 = \frac{B}{125^2} - A$$

$$p = \frac{B}{75^2} - A$$

$$p - 10 = \frac{B}{75^2} - \frac{B}{125^2} = \frac{B(125^2 - 75^2)}{75^2 \times 125^2}$$

$$B = (p-10) \times \frac{75^2 \times 125^2}{125^2 - 75^2}$$

$$B = (p-10) \times \frac{87890625}{10000} = (p-10) \times 8789.0625$$

$$A = \frac{B}{125^2} - 10 = 0.5625(p-10) - 10$$

$$= 0.5625p - 5.625 - 10 = 0.5625p - 15.625$$

Maximum σ_c occurs at inner radius

$$\sigma_{c\,max} = \frac{B}{R_1^2} + A = \frac{(p-10) \times 8789.025}{75^2} + (0.5625p - 15.625)$$

$$= (p - 10)\,(1.5625) + (0.5625p - 15.625)$$

$$\sigma_{c\,max} = 1.5625p - 15.625 + 0.5625p - 15.625$$

$$50 = 2.125p - 31.25$$

Internal pressure, $p = \dfrac{81.25}{2.125} = 38.23 \text{ N/mm}^2$

Practice Q.6.15 A thick cylinder is subjected to both internal and external pressure. Inner radius of shell is 60 mm and outer radius of shell is 120 mm. If the maximum permissible stress in cylinder is 100 N/mm^2 and internal pressure is 80 N/mm^2, determine the intensity of external pressure.

Ans. [12.50 N/mm^2]

Question 6.16 A spherical shell of 150 mm internal diameter has to withstand an internal pressure of 30 MN/m^2. Calculate the thickness of the shell if the allowable stress is 80 MN/m^2. Assume the stress distribution in the shell to follow the law, $\sigma_r = a - \dfrac{2b}{r^3}$, $\sigma_\theta = a + \dfrac{b}{r^3}$.

[IES : 2010, 10 Marks]

Solution: Note that both σ_r and σ_θ are maximum at the inner radius i.e., 7.5 cm

$$\sigma_r = -3000 \text{ N/cm}^2 = a - \frac{2b}{R_1^3}$$

$$\sigma_\theta = 8000 \text{ N/cm}^2 = a + \frac{b}{R_1^3}$$

where R_1 = inner radius = 7.5 cm

$$8000 = a + \frac{b}{R_1^3}$$

$$-3000 = a - \frac{2b}{R_1^3}$$

$$+ \quad - \quad +$$

$$11000 = \frac{3b}{R_1^3}$$

$$b = \frac{11000 \times R_1^3}{3} = \frac{11000 \times 7.5^3}{3} = 1546875$$

$$\sigma_\theta = 8000 = a + \frac{b}{R_1^3} = a + \frac{1546875}{7.5^3}$$

$$8000 = a + 3666.667$$

$$a = 4333.33$$

But $\sigma_r = 0$, at outer radius.

So

$$O = 4333.33 - \frac{2b}{R_2^3}$$

$$R_2^3 = \frac{2b}{4333.33} = \frac{2 \times 1546875}{4333.33} = 713.943, \; R_2 = 8.9374 \text{ cm}$$

thickness, $t = 8.9374 - 7.5 = 1.438 \text{ cm} = 14.38 \text{ mm}$

Objective Type Questions

Q.1 In a thick cylinder subjected to internal and external pressure, let r_1 and r_2 be the internal and external radii respectively. Let element at radius r_1, $r_2 \geq r \geq r_1$, identifying the cylinder axis as z-axis, the radial strain component e_{rr} is

(a) $\frac{u}{r}$ (b) $\frac{u}{\theta}$

(c) $\frac{du}{dr}$ (d) $\frac{du}{d\theta}$

[IES 1991]

Q.2 Auto frettage is the method of

(a) joining thick cylinder

(b) calculating stress in thick cylinder

(c) prestressing thick cylinder

(d) increasing the life of thick cylinder

[IES 1996]

Q.3 A thick walled hollow cylinder having outside and inside radii of 90 mm and 40 mm respectively is subjected to an external pressure of 800 MN/m^2. The maximum circumferential stress in the cylinder will occur at a radius of

(a) 40 mm (b) 60 mm

(c) 65 mm (d) 90 mm

[IES 1998]

Q.4 In a thick cylinder, pressurized from inside, the hoop stress is maximum at

(a) the centre of wall thickness
(b) the outer radius
(c) the inner radius
(d) both the inner and outer radius

[IES 1998]

Q.5 If a thick cylindrical shell is subjected to internal pressure, the hoop stress, radial stress and longitudinal stress at point in the thickness will be
(a) tensile, compressive, compressive respectively
(b) all compressive
(c) all tensile
(d) tensile, compressive and tensile respectively

[IES 2000]

Q.6 A thick cylinder is subjected to internal pressure of 100 N/mm^2. If hoop stress developed at the outer radius of the cylidner is 100 N/mm^2, the hoop stress developed at the inner radius is
(a) 100 N/mm^2 (b) 200 N/mm^2
(c) 300 N/mm^2 (d) 400 N/mm^2

[IES 2001]

Q.7 A thick cylinder of internal radius '*a*' and external radius '*b*' is subjected to internal pressure *p* as well as external pressure '*p*'. The magnitude of circumferential stress developed is
(a) maximum at radius *a*
(b) maximum at radius *b*
(c) maximum at radius $r = \sqrt{ab}$
(d) constant

Q.8 Consider the following statements for a pressure vessel which is subjected to internal pressure
1. radial stress at inner radius in always zero
2. radial stress at outer radius is always zero
3. tangential stress is always higher than other stresses
4. tangential stress is always lower than other stresses

Which of these statement are correct?
(a) 1 and 2 (b) 1 and 4
(c) 2 and 3 (d) 2 and 4

[IES 2005]

Q.9 Consider the following statements at given point in the case of thick cylinder subjected to fluid pressure
1. Radial stress is compressive
2. Hoop stress is tensile
3. Hoop stress is compressive
4. Longitudinal stress is tensile and it varies, along the length
5. Longitudinal stress is tensile and remains constant along the length of the cylinder

Which of these statements are correct?
(a) 1, 2 and 4 (b) 3 and 4
(c) 1, 2 and 5 (d) 1, 3 and 5

[IES 2006]

Q.10 Where does the maximum hoop stress in a thick cylinder under external pressure occur
(a) at the outer surface
(b) at the inner surface
(c) at the mid thickness
(d) at the 2/3rd outer radius

[IES 2008]

Q.11 A seamless pipe of diameter *d* m (metre) is to carry fluid under a pressure of *p* kN/cm^2. The necessary thickness *t* of metal in cm, if the maximum stress is not the exceed σ kN/cm^2 is

(a) $t \geq \dfrac{pd}{2\sigma}$cm (b) $t \geq \dfrac{100pd}{2\sigma}$cm

(c) $t \leq \dfrac{pd}{2\sigma}$cm (d) $t \leq \dfrac{100pd}{2\sigma}$cm

[IES 2012]

Q.12 In a thick cylindrical shell, $R_2 = 1.5\ R_1$, maximum allowable stress is 260 MPa, what is the value of maximum internal pressure
(a) 150 MPa (b) 130 MPa
(c) 100 MPa (d) 90 MPa

Q.13 A bronze sleeve of outer diameter 100 mm is forced over a solid shaft of diameter 60 mm. If junction pressure developed is 32 N/mm^2, E = 100 GPa, what is hoop strain at outer radius of sleeve
(a) 320 μs (b) 360 μs
(c) 720 μs (d) None of these

Q.14 Consider the following statements
1. the amount of interference needed to create a tight joint varies with the diameter of the shaft.
2. an interference fit creates no stress state in the shaft
3. the stress state in the hub is similar to of a thick walled cylinder with internal pressure

Which of these statements are correct?
(a) 1, 2 and 3 (b) 1 and 2 only
(c) 2 and 3 only (d) 1 and 3 only

Q.15 Consider the following statements:
1. In case of a thin spherical shell of diameter *d* and thickness *t*, subjected to internal pressure *p*, the principal stress at any point equal $\frac{pd}{4t}$.
2. In case of thin cylinders, the hoop stress is determined assuming it to be uniform across the thickness of the cylinder.
3. The thick cylinders, the hoop stress is not uniform across the thickness but it varies from a maximum value at the inner circumference to a minimum value at the outer circumference.

Which of the above statements are correct?
(a) 1 and 2 only (b) 1 and 3 only
(c) 2 and 3 only (d) 1, 2 and 3

[ESE 2018]

Q.16 Consider the following statements for a thick walled cylinder, subjected to an internal pressure:
1. Hoop stress is maximum at the inside radius.
2. Hoop stress is zero at the outside radius.
3. Shear stress is maximum at the inside radius.
4. Radial stress in uniform throughout the thickness of the wall.

Which of the above statements are correct?
(a) 1 and 4 (b) 1 and 3
(c) 2 and 3 (d) 2 and 4

[ESE 2016]

Answers

1. (c)	2. (c)	3. (a)	4. (c)	5. (d)
6. (b)	7. (d)	8. (c)	9. (c)	10. (b)
11. (b)	12. (c)	13. (b)	14. (d)	15. (d)
16. (b)				

Explanations

1. (c)

$$r = \text{changer to } r + u$$
$$r + dr = \text{changes to } r + u + dr + du$$

Radial strain $\frac{du}{dr}$.

2. (c)

Prestressing thick cylinders: Auto frettage in one of the oldest methods of **prestressing the cylinder** i.e, prestressing the inner surface of the cylinder upto a stress ***beyond elastic limit stress***, so that the inner surface of the cylinder is deformed into plastic state.

The internal pressure *p* is applied in such a way that σ_c at inner radius > σ_e, elastic limit stress of the material. The outer surface of the cylinder remains in elastic range of stress. When the pressure is released, residual compressive stresses are induced at the inner surface of the cylinder. This compressive residual stress at inner surface increases the pressure bearing capacity of the cylinder.

This method is commonly used in gun-barrels, it increases the pressure bearing capacity of the gun-barrel.

3. (a)

$$\sigma_{c\,max} = -p \times \frac{2R_2^2}{R_2^2 - R_1^2}$$

where, R_1 and R_2 are internal and external radii, occurs at inner radius R_1.

4. (c)

$$\text{At inner radius } R_1,\ \sigma_{c\,max} = p\frac{R_2^2+R_1^2}{R_2^2-R_1^2}$$

R_2 outer radius, p = pressure at R_1

5. (d)

σ_c (tensile), σ_r (compressive), σ_a (tensile)

6. (b)

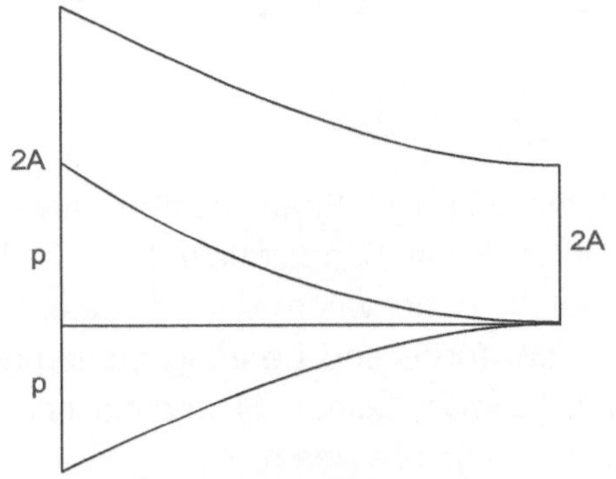

$$p = \frac{B}{R_1^2} - A$$

$$\frac{B}{R_1^2} + A = p + 2A$$

$$2A = 100$$

$$p = 100$$

$$\sigma_{cR_1} = p + 2A$$

$$= 100 + 100 = 200 \text{ N/mm}^2$$

7. (d)

$$\sigma_r = p, \quad \text{at } r = a$$

$$\sigma_r = p, \quad \text{at } r = b$$

σ_r remains constant

$$\sigma_r = \frac{B}{r^2} - A, \quad B = 0$$

$$\sigma_r = -A$$

$$\sigma_c = +A \quad \text{constant.}$$

8. (c)

2. Radial stress in always zero at outer radius
3. The tangential stress is always higher than other stress.

9. (c)

1. Radial stress is compressive
2. Hoop stress in tensile
5. Longitudinal stress is tensile and remains constant along the length of the cylinder.

10. (b)

At the inner surface.

11. (b)

d in m $=100\,d$ in cm

$$t \geq \frac{p\times 100d}{2\sigma}\text{cm}, \text{Hoop stress is maximum}$$

12. (c)

$$260 = \sigma_{c\,max} = p\frac{R_2^2+R_1^2}{R_2^2-R_1^2}$$

$$= p\times\frac{2.25+1}{2.25-1} = \frac{3.25}{1.25}\times p$$

$$p = \frac{260}{3.25}\times 1.25 = 100 \text{ MPa}$$

13. (b)

σ_c, at outer radius

$$= p\times\frac{2R_1^2}{R_2^2-R_1^2} = 32\times\frac{2\times 30^2}{50^2-30^2} = 36 \text{ MPa}$$

$$\sigma_r = 0$$

$$\varepsilon_c = \frac{36}{100000} = 360\times 10^{-6} = 360\ \mu\text{s}$$

14. (d)

1 and 3 statements are correct.

15. (d)

All statements 1, 2 and 3 are correct.

16. (b)

1. Hoop stress maximum at inside radius – correct.
2. Hoop stress is not zero at outside radius – incorrect statement.
3. Shear stress is maximum at the insie radius–correct.
4. Radial stress is not unifrom throughout the thickness of the wall–incorrect.

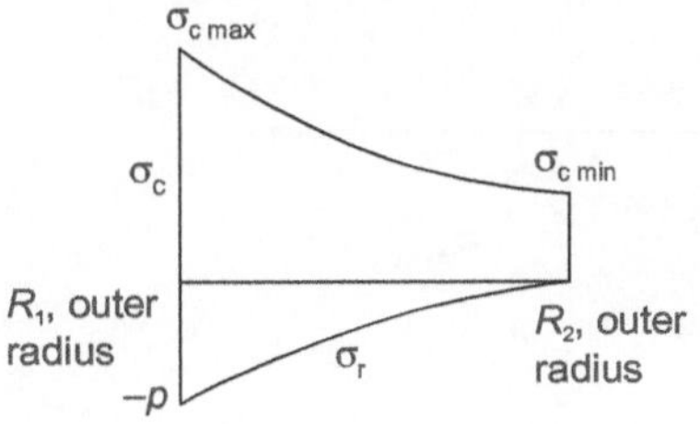

$$\text{Shear stress at inside radius} = \frac{\sigma_{c\,max}+p}{2}$$

Shear Force and Bending Moment Diagrams

CHAPTER 7

In any civil engineering structure, there are beam and column members. Beam members are subjected to transverse loads and moments applied along the axis of the beam. Beam bends due to bending moments produced by the transverse loads and moments. Cross-sections of the beam where shear force and bending moment are maximum are determined. **Continuous graphs of shear force and bending moments** varying along the length of the beam are plotted. From these SF and BM diagrams maximum SF and maximum BM are determined and beam section as a whole is determined for efficient design of a beam.

Question 7.1 Describe the characteristics of following supports used to support beams and cantilevers.

(a) Roller support (b) Hinged support and (c) Fixed support

Solution:

(a) **Roller Support** (a simple support): beam is supported over a roller or a number of rollers.
- roller support can move along the length of the beam
- it can rotate about the axes of the beam
- it can move in direction parallel to beam axis
- its motion in vertical direction is restricted (fig. 7.1)

(b) **Hinged Support** (a simple support) beam is hinged (through a pin joint at hinged support) support can rotate about axis of the beam. Support cannot have any vertical motion. There can be two components of reaction at hinged support i.e., R_{AH}, R_{AV}, resultant $R_A = \sqrt{R_{AH}^2 + R_{AV}^2}$

$$\tan\theta = \frac{R_{AH}}{R_{AV}}$$

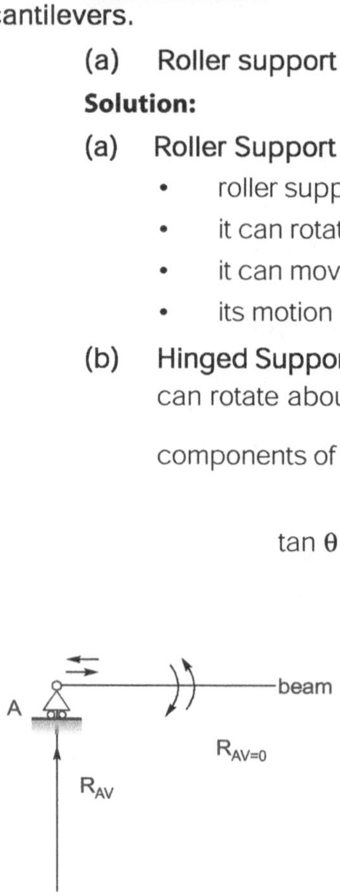

Fig. 7.1 *Roller support*

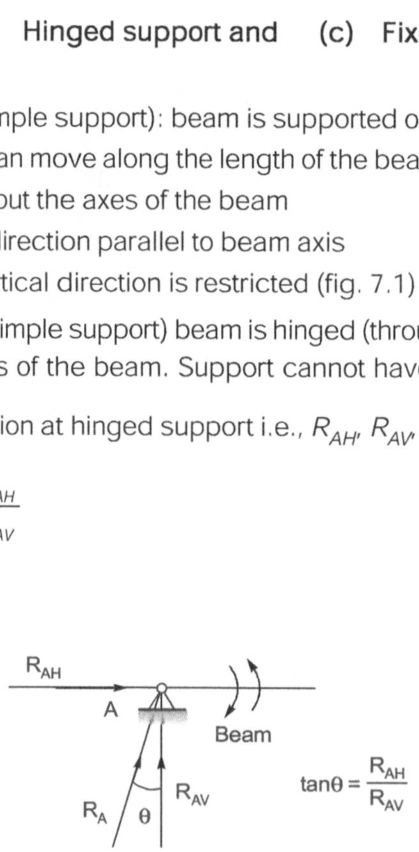

Fi. 7.2 *Hinged support*

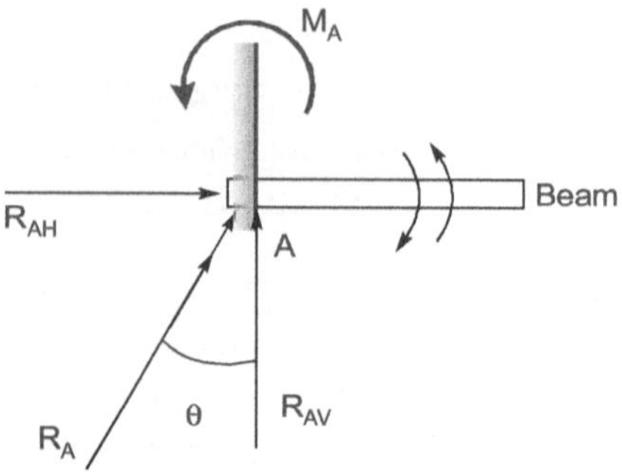

Fig. 7.3 *Fixed support*

(c) **Fixed support** (Built in support) End of the beam is built in the wall (fixed) support. End of the beam cannot move axially, cannot move vertically and cannot rotate. Reaction components are R_{AH}, R_{AV}, fixing couple M_A. Slope and deflection in cantilever/beam will be zero at fixed support.

There can be two components of reaction R_{AH}, R_{AV} resultant $R_A = \sqrt{R_{AV}^2 + R_{AH}^2}$

Question 7.2 **With the help of simple example, explain what are (i) positive and negative shear forces (ii) positive and negative bending moments or sagging or hogging bending moments.**

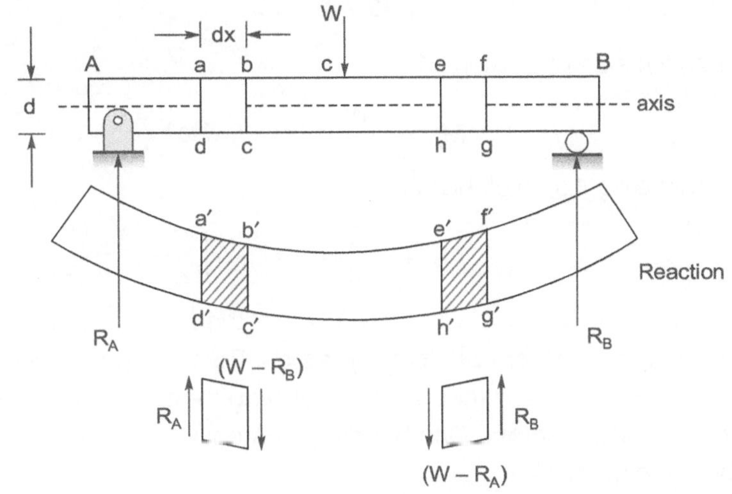

Fig. 7.4

Solution: Consider a beam *AB* of length *L*, depth *d*, hinged at end *A* and roller supported at *B*, subjected to a vertical central load *W* as shown in figure 7.4. After the application of load *W*, transverse to the axis of the beam, the beam bends and elements *abcd* deformed to *a′ b′ c′ d′* and element *efgh* derormed to *e′ f′ g′ h′* respectively as shown in a rectangular section *abcd* is deformed in to a parallelogram *a′ b′ c′ d′*.

Load *W* is at the centre, so reaction at ends

$$R_A = R_B = \frac{W}{2}$$

On element *abcd*, net force on left side of the element is R_A, upwards and on the right side of the element, net force is $W - R_B = R_A = \frac{W}{2}$. These two forces constitute **a couple with clockwise moment** i.e., couple tends to deform the element as shown. This force, R_A on left side is known as **positive shear force as tends to rotate the element in the clockwise direction.** For simplicity let us say if the net force on the left side of the element is upwards, then it is a positive shear force.

Similarly the element *efgh* (in portion *CB* of beam) deforms to *e′ f′ g′ h′*. Net force on left side of the section is $W - R_A = R_B$, downwards and net force on the right side of the element is R_B upwards. This couple tends to rotate the element in the anticlockwise direction. In other words, ***if the net force on the left side of the section is downwards then it is a negative shear force.***

Bending Moment: Consider a beam *AB*, of length *L*, hinged at end *A*, roller supported at *B* with a central load *W*. At a section XX, at a distance of x from *A*.

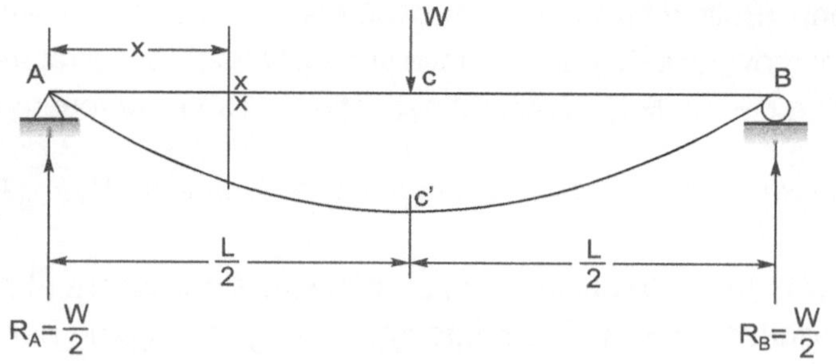

Fig. 7.5

Bending moment on left side of section is

$$M_x = \frac{W}{2}x, \text{ (clockwise)}$$

Bending moment on the right side of the section is

$$M_x = \frac{W}{2}(L-x) - W\left(\frac{L}{2}-x\right) = \frac{Wx}{2}, \text{ (anticlockwise)}$$

This type of bending moment i.e., clockwise on left side of the section and anticlockwise on the right side of the section produces ***sagging in beam or concavity upwards***. The bending moment which produces sagging in beam is a positive bending moment. Or in other words **a bending moment which is clockwise** on the left side of the section is **a positive bending moment**. Or a bending moment which is anticlockwise on the right side of the section is a positive bending moment.

Negative Bending Moment:

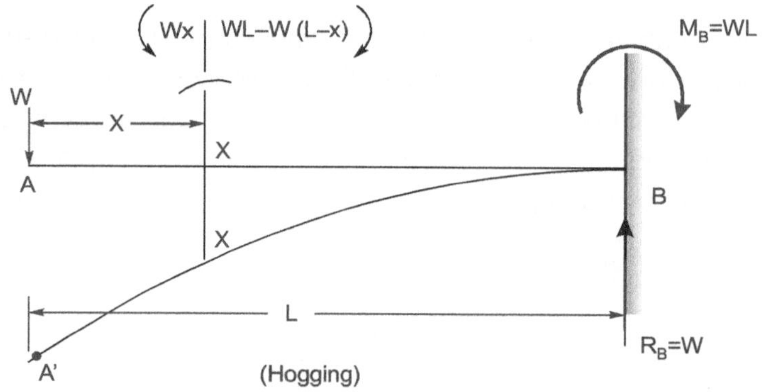

Fig. 7.6

Consider a beam *AB*, free at end *A*, fixed at end *B* carrying a load *W* at the free end. This type of beam which is fixed at one end and free at the other end is **known as cantilever**. Under the action of *W*, cantilever *AB* bends to a shape as *A'B*. (hogging shape or showing convexity upwards).

At a section XX at a distance of x from end *A*.

Bending moment on left side of the section

$$M_x = W.x \text{ (anticlockwise)}$$

Bending moment on the right side of the section

$$M_x = WL - W(L-x)$$

$$= Wx. \text{ (clockwise)}$$

Note that at end *B*, reaction $R_B = W$, upwards and a fixed couple $M_B = WL$ (clockwise) to maintain equilibrium.

This type of bending moment i.e., anticlockwise on the left side and clockwise on the right side of the section is known as Negative bending moment. This type of bending moment produces **hogging in beam or convexity upwards**.

In simple words clockwise moment on the left side of the section is a positive bending moment and anticlockwise moment on the left side of the section is a negative bending moment.

Question 7.3 **A beam *ABCD*, 6 m long hinged at end *A*, roller supported at *C* carries point loads of 6 kN at *B*, 2 m from *A* and 2 kN at *D*. Draw SF and BM diagrams for the beam.**

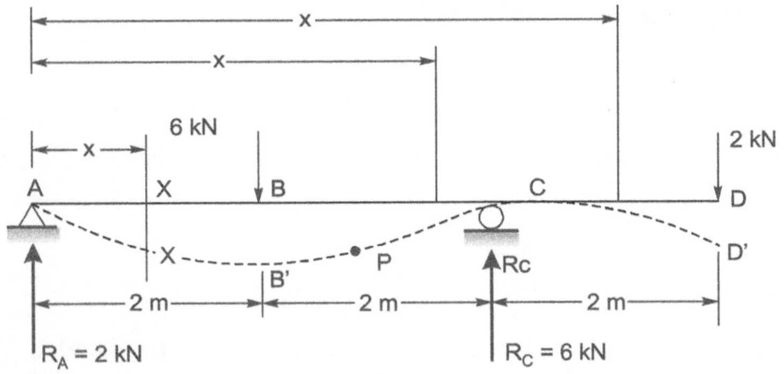

Fig. 7.7

Solution: There is overhang portion *CD* in the beam. The beam bends as shown by *AB' CD'* i.e.,. Some portion shows sagging (or concavity upwards) while other portion shows hogging (convexity upwards). ***AB'P* sagging, *PCD'*- *hogging, P is the point of contraflexure***, where bending moment changes sign from +ve to –ve. Let us first calculate support reactions.

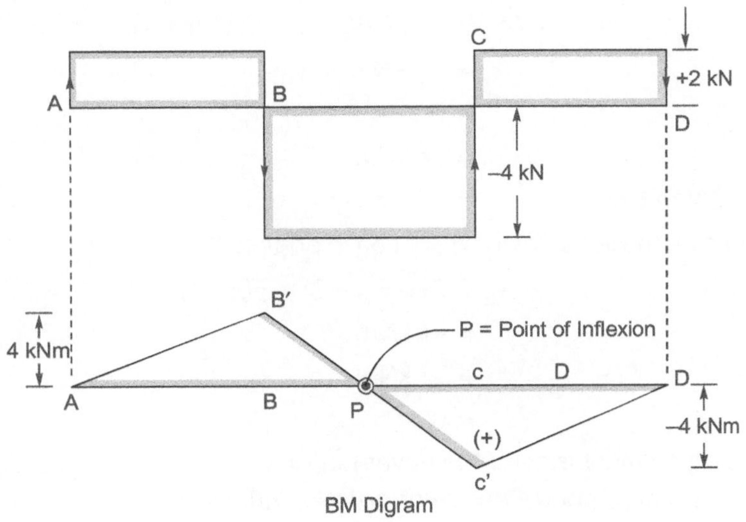

Fig. 7.8

Support reactions: Taking moment of forces about end *A*

$$6 \times 2 + 2 \times 6 = 4R$$

Reaction, $R_C = 6$ kN

Total vertical load $= 6 + 2 = 8$ kN

Reaction, $R_A = 8 - 6 = 2$ kN

SF Diagram: There are 3 portions in beam, i.e., *AB, BC, CD,* shear force remains constant in each portion, in this case taking upward forces on the left side of a section to be positive.

Portion *AB* (x = 0 to 2 m)

$F_x = R_A = +$ 2kN, (remains constant from *A* to *B*)

Portion *BC*(x = 2 to 4 m)

$F_x = +2 - 6 = -4$ kN, (remains constant from *B* to *C*)

Portion *CD*(x = 4 m to 6 m)

$F_x = +2 - 6 + 6 = +2$ kN

(**Note:** that there are 3 forces on left sides of section in portion *CD*) (F_x remains constant from *C* to *D*)

Fig. 7.4 (b) shows the SF diagram. As a standard convention, the boundary of the SF diagram is hatched.

BM Diagram: Portion *AB* (x = 0 to 2 m)

Taking a section X, a distance of x from *A*

Bending moment, $M_x = +R_A x = 2x$ kNm

(Taking clockwise moments on the left side of the section to be positive)

$M_x = 0$ at x = 0 m

$= 2$ kNm at x = 1 m

$= 4$ kNm at x = 2 m

Portion *BC*(x = 2m to 4 m)

Taking a section XX at a distance of x from *A*, in portion *BC*. There are two forces on the left side of the section.

$M_{2x} = +2x - 6(x - 2)$, Note that distance of 6 kN from x is (x – 2m)

$= +4$ kNm at x = 2 m

$= 0$ at x = 3 m

$= -4$ kNm at x= 4 m

Portion *CD* (x = 4 m to 6 m)

There are 3 forces on the left side of the section in portion *CD*

$M_x = 2x - 6(x - 2) + 6(x - 4)$

$= -4$ kNm at x = 4 m

$= -2$ kNm at x = 5 m

$= 0$ at x = 6 m

At point *P*, bending moment is zero. Moreover bending moment changes sign from +ve to –ve at this section. ***This is known as point of contraflexure or point of inflexion.***

Practice Q.7.3 A beam *ABCD*, 7 m long, hinged at *D* and roller supported at *B* carries point loads of 5 kN at *A*, 10 kN at *C*, 3 m from *D*, as shown in figure 7.9. Determine (i) support reactions (ii) Draw SF and BM diagram (iii) Locate the position of the point of contraflexure.

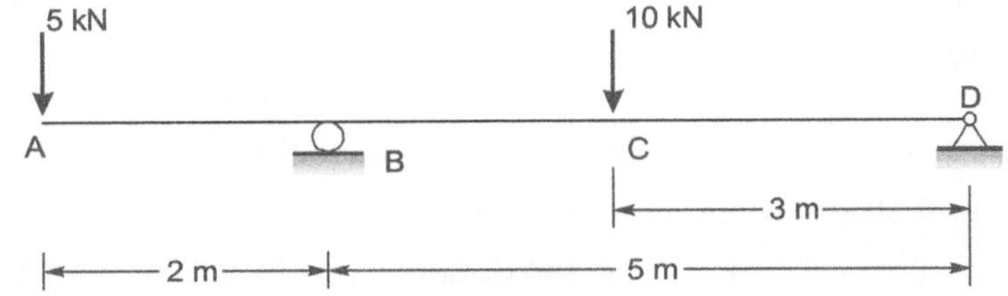

Fig. 7.9

Ans . [R_B = 13 kN, R_D = 2 kN, F_{AB} = –5 kN, F_{BC} = 8 kN, F_{CD} = –2 kN; M_A = 0, M_B = –10 kNm, M_C = 6 kNm, M_D = 0, point of contraflexure lies at 3.25 m from *A*]

Note: A student is advised to use cm, graph paper for drawing SF and BM diagrams to scale.

Question 7.4 A beam *ABCD*, 1.5 *L* long carries a uniformly distributed load *w* throughout its length. The beam is supported over span. *L*, with equal overhang on both the sides as shown in figure 7.10. Draw SF and BM diagrams of the beam and determine position of points of contraflxrure, if any.

Solution:

Let us first calculate the support reactions

$$\text{Total load on beam} = w \times 1.5\,L \qquad \ldots\text{(i)}$$

The beam is symmetrically loaded about its centre at *E*. Therefore the reactions are equal

$$R_B = R_C = \frac{3wL}{4} \qquad \ldots\text{(ii)}$$

There are 3 portions in the beam i.e., *AB*, *BC* and *CD* as shown.

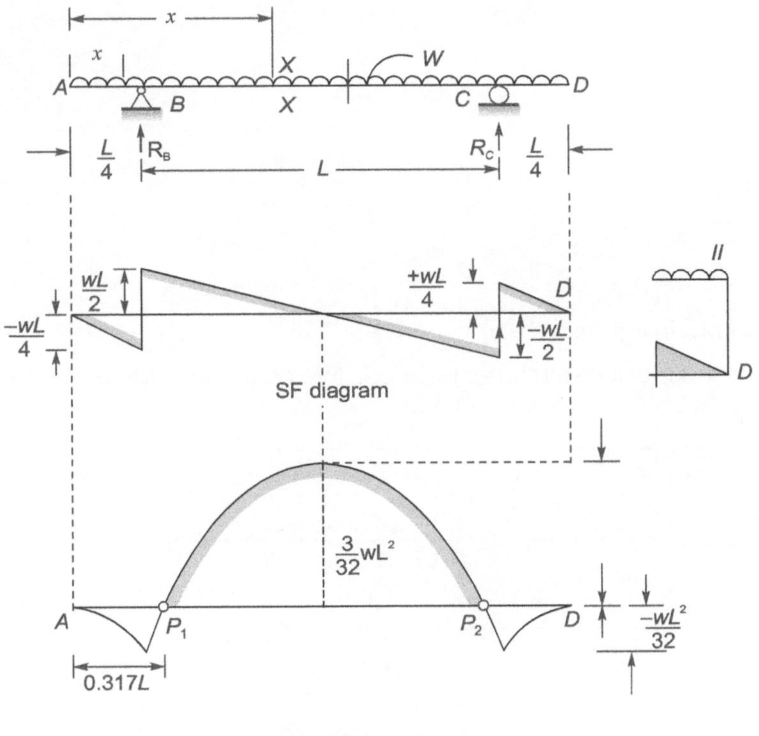

Fig. 7.10 *SF Diagram*

Portion $AB\left(x=0 \text{ to } \frac{L}{4}\right)$

Taking upward forces on the left side of the section as positive.

Taking section XX at a distance x from A, shear force.

$$F_x = -wx \text{ (an equation of straight line)}$$

$$= 0 \qquad \text{at } x = 0$$

$$= -\frac{wL}{4} \qquad \text{at } x = \frac{L}{4}$$

Portion $BC\left(x=\frac{L}{4} \text{ to } \frac{5L}{4}\right)$

$$F_x = wx + R_B = -wx + \frac{3wL}{4} \text{ (at equation of straight line)}$$

$$= \frac{wL}{4} + \frac{3wL}{4} = \frac{wL}{2} \qquad \text{at } x = \frac{L}{4}$$

$$= 0 \text{ (at center)} \qquad \text{at } x = \frac{3L}{4}$$

$$= -w\times\frac{5L}{4} + \frac{3wL}{4} = -\frac{wL}{2} \qquad \text{at } x = \frac{5L}{4}$$

Portion $CD\left(x=\frac{5L}{4} \text{ to } \frac{3L}{2}\right)$

Shear force,

$$F_x = -wx + \frac{3wL}{4} + \frac{3wL}{4}$$

$$= -wx + \frac{3wL}{4}$$

$$= -\frac{5wL}{4} + \frac{3wL}{2} = \frac{wL}{4} \qquad \text{at } x = \frac{5L}{4}$$

$$= 0 \qquad \text{at } x = \frac{3L}{2}$$

SF diagram is shown in figure 7.10 (b)

BM diagram: Taking clockwise moments as positive on the left side of a section.

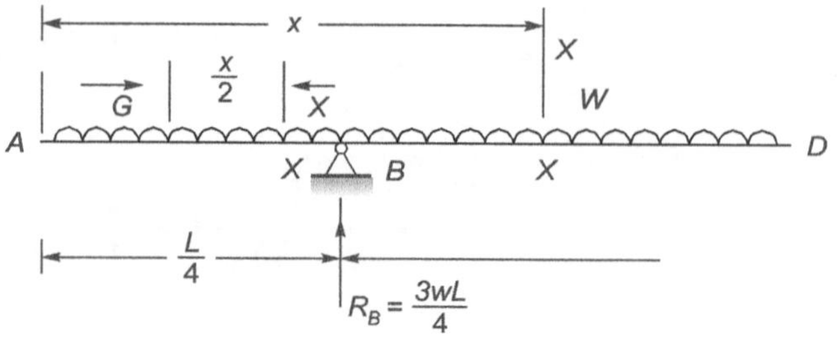

Fig. 7.11

Portion *AB* $\left(x = 0 \text{ to } \frac{L}{4}\right)$

$$M_x = -wx\frac{x}{2} = -\frac{wx^2}{2}$$

Note that total load upto x = $-wx$ (downward load)

CG of this load acts at a distance of $\frac{x}{2}$ from section XX

So bending moment, $M_x = -\frac{wx^2}{2}$

(it is negative as it is anticlockwise moment on the left side of section) (eqaution of parabola)

$$M_x = 0 \quad \text{at } x = 0$$

$$= -\frac{wL^2}{128} \quad \text{at } x = \frac{L}{8}$$

$$= -\frac{wL^2}{32} \quad \text{at } x = \frac{L}{4}$$

Portion *BC* $\left(x = \frac{L}{4} \text{ to } \frac{5L}{4}\right)$

$$M_x = -\frac{wx^2}{2} + R_B\left(x - \frac{L}{4}\right) = -\frac{wx^2}{2} + \frac{3wL}{4}\left(x - \frac{L}{4}\right)$$

(equation of parabola)

$$= -\frac{wL^2}{32} \quad \text{at } x = \frac{L}{4}$$

$$= -\frac{wL^2}{8} + \frac{3wL^2}{16} \quad \text{at } x = \frac{L}{2}$$

$$= +\frac{wL^2}{16} \quad \text{at } x = \frac{L}{2}$$

$$= -\frac{9wL^2}{32} + \frac{3wL}{4}\left(\frac{L}{2}\right)$$

$$= -\frac{9wL^2}{32} + \frac{3wL^2}{8} = +\frac{3wL^2}{32} \quad \text{(at the centre of the beam)}$$

As the beam is symmetrically loaded about its centre. BM diagram will be symmetrical as shown in figure 7.10 (c)

There are two points of contraflexure in beam i.e., P_1 and P_2, where BM = 0 and changes sign.

To determine position of P_1 and P_2 let us put, equation of bending moment

$$-\frac{wx^2}{2} + \frac{3wL}{4}\left(x - \frac{L}{4}\right) = 0$$

$$-\frac{x^2}{2} + \frac{3L}{4}x - \frac{3L^2}{16} = 0$$

or $$x^2 - 1.5Lx + \frac{3}{8}wL^2 = 0$$

$$x = \frac{1.5L \pm \sqrt{(1.5L)^2 - 4 \times \frac{3}{8} \times L^2}}{2} = \frac{1.5L \pm \sqrt{2.25L^2 - 1.5L^2}}{2}$$

$$= \frac{1.5L \pm 0.866L}{2} = 1.183\ L,\ 0.317\ L$$

Points of contraflexure P_1 and P_2 lie at distance of 0.317 L and 1.183 L from A.

Practice Q.7.4 A beam *AB*, length *L*, simply supported at ends *A* and *B* carries a uniformly distributed load of intensity *w* throughout its length. Draw SF and BM diagrams of the beam.

Ans. [$R_A = R_B = \frac{wL}{2}$, $F_x = \frac{wL}{2} - wx$, $M_x = \frac{wL}{2}x - \frac{wx^2}{2}$, $M_{max} = \frac{wL^2}{8}$ at the centre, no point of contraflexure]

Question 7.5 A beam *ABCD*, 35 m long, roller supported at *B*, 10 m from *A*, and hinged at *D* carries a load of 300 N at *A* and a *CW* moment of 2000 Nm at point *C* as shown in figure 7.12 (a). Determine support reaction and draw BMD only.

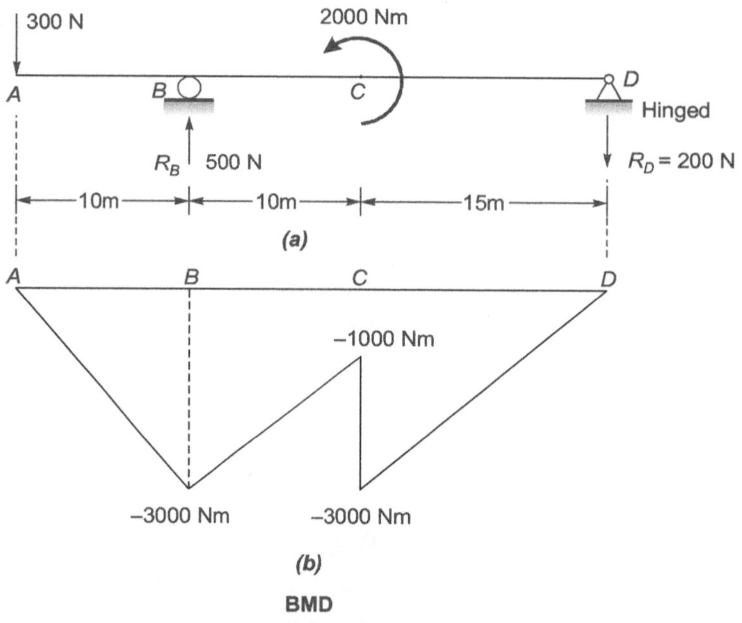

Fg. 7.12 (a) (b)

Solution: Taking moments about *B*

$$300 \times 10 + 2000 = R_D \times 25$$

$$R_D = 200\ \text{N}\downarrow$$

Reaction, $$R_B = 300 + 200 = 500\ \text{N}\uparrow$$

BM diagram

$$M_A = 0$$
$$M_B = -300 \times 10 = -3000 \text{ Nm (CCW as left side)}$$
$$M_C = -300 \times 20 + 500 \times 10 = -6000 + 5000 = -1000 \text{ Nm}$$
$$M_C' = -1000 \text{ Nm} - 2000 \text{ Nm} = -3000 \text{ Nm}$$
$$M_D = 0$$

BMD is shown in figure 7.12 (b)

There is no point of contraflexure in beam.

Question 7.6 **A beam *ABCDE*, 5 m long, roller supported at *B*, 1 m from *A* and hinged at *E* carries loads and a moment as shown in figure 7.13 (a) Determine support reactions draw SF and BM diagrams of the beam. Determine the location of the point of contraflexure.**

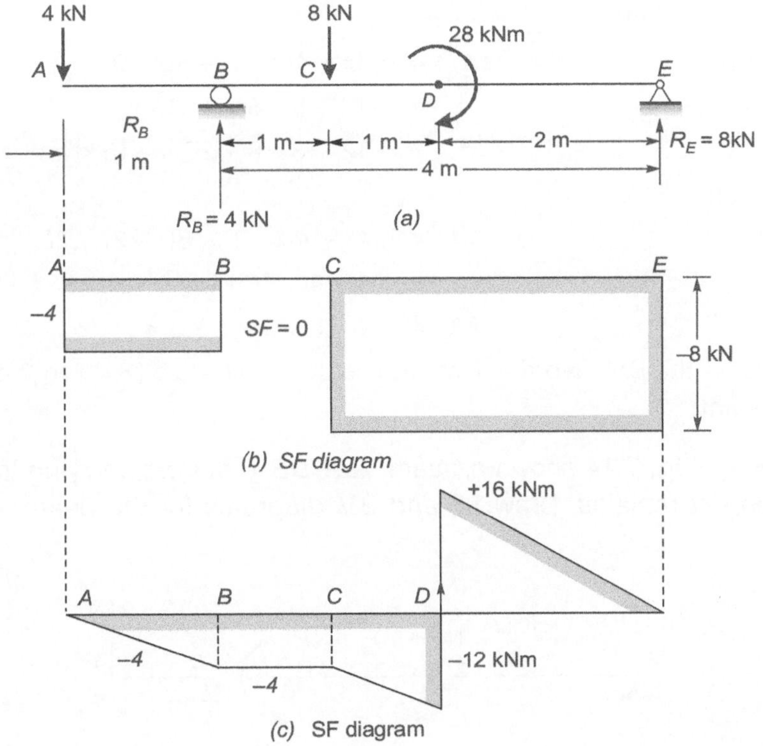

Fig. 7.13

Solution: Taking moments about point *B*

$$8 \times 1 + 28 = 4R_E + 4$$

Reaction, $R_E = 8$ kN

Total vertical load $= 8 + 4 = 12$ kN

Reaction, $R_B = 4 + 8 - 8 = 4$ kN

SF diagram: Taking upward forces on the left side of the section to be positive

Portion *AB* ($x = 0 - 1$ m)

$$F_{AB} = -4 \text{ kN (remaining constant)}$$

Portion *BC* (x = 1 to 2 m)

$$F_{BC} = -4 + 4 = 0 \text{ (remains constant)}$$

Portion *CDE* (x = 2 – 5 m)

$$F_{CD} = -4 + 4 - 8 = -8 \text{ kN (remains constant upto } E)$$

BM diagram: Taking clockwise moment on the left side of a section to be positive

Portion *AB* (x = 0 – 1 m)

$$M_x = -4x$$
$$= 0 \quad \text{at } x = 0 \text{ m}$$
$$= -4 \text{ kNm} \quad \text{at } x = 1 \text{ m}$$

Portion *BC* (x = 1 – 2 m)

$$M_x = -4x + 4(x - 1)$$
$$= -8 + 4 = -4 \text{ kNm} \quad \text{at } x = 2 \text{ m}$$

Portion *CD* (x = 2 m – 3 m)

$$M_x = -4x + 4(x - 1) - 8(x - 2)$$
$$= -12 + 8 - 8 = -12 \text{ kNm}$$
$$M_x' = -12 + 8 - 8 + 28 = +16 \text{ kNm} \quad \text{at } x = 3 \text{ m}$$

Portion *DE* (x = 3 m – 5 m)

$$M_x = -4x + 4(x - 1) - 8(x - 2) + 28$$
$$= -4 \times 4 + 4(3) - 8(2) + 28 = +8 \text{ kNm} \quad \text{at } x = 4 \text{ m}$$
$$Mx = -4 \times 5 + 4 \times 4 - 8 \times 3 + 28 = 0 \quad \text{at } x = 5 \text{ m}$$

Fig. 7.13 (c) shows the BM diagram. Note that in the portion *BC* (x = 1 to 2 m) *SF* = **0, the bending moment remains constant.**

Practice Q.7.6 Fig. 7.14 shows a beam, *ABCDE*, 6 m long carrying loads and a moment as shown. Determine support remains. Draw SF and BM diagrams for the beam, where lies the point of contraflexure.

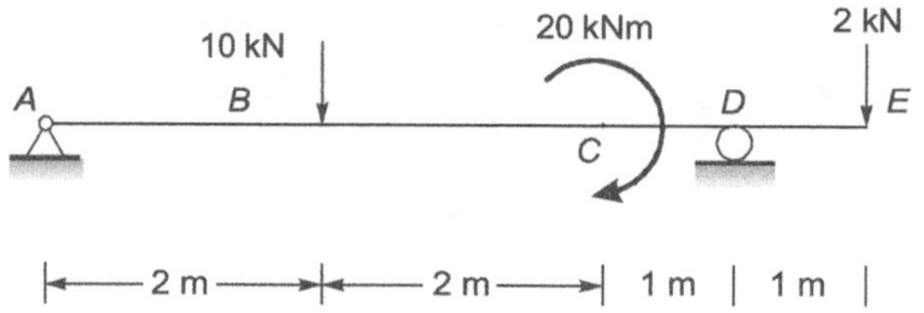

Fig. 7.14

Ans. [R_A = 1.6 kN, R_D = 10.4 kN, F_{AB} = +1.6 kN, F_{BD} = –8.4 kN, F_{DE} = +2 kN, M_A = 0, M_B = +3.2 kNm, M_C = –13.6 kNm, M_C' = +6.4 kNm, M_D = –4 kNm, points of contraflexure lie at distances of 2.38 m from *A* and 4.76 m from *A*]

Question 7.7 A beam *ABCD*, 7 m long carries a uniformly distributed load of intensity 6 kN/m from *A* to *C*, 4 m. Determine support reactions, draw SF and BM diagrams. Locate the position of point of contraflexure. Fig. 7.15 (a)

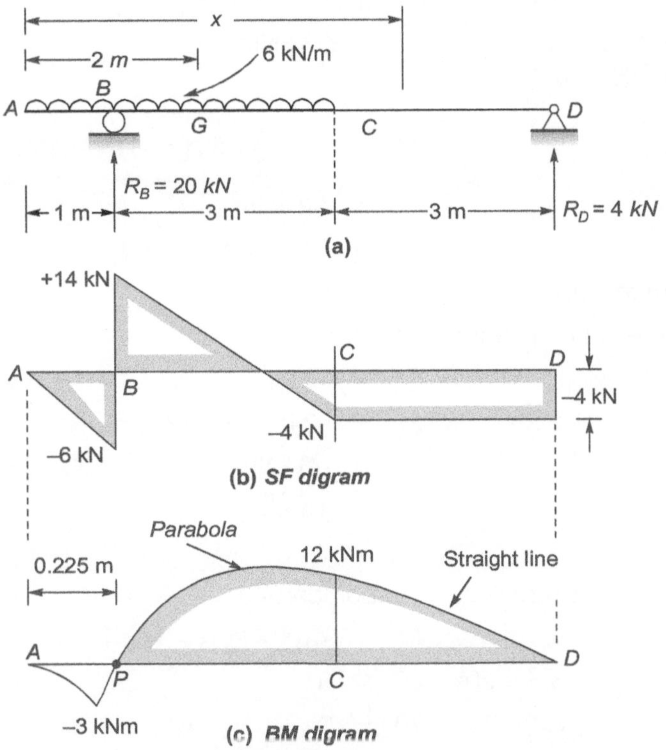

Fig. 7.15

Solution: Total load on beam $= 6 \times 4 = 24$ kN

CG of this load acts at a distance of 2 m from A or 1 m from B. Taking moment of the forces about A

$$24 \times 1 = 6\,R_D$$

Reaction, $R_D = 4$ kN

Reaction, $R_B = 24 - 4 = 20$ kN

SF diagram: (Taking upward forces on the left side of a section to be positive)

$$F_x = -wx = -6x$$

$= 0$ at $x = 0$

$= -6$ kN at $x = 1$ m

Portion BC ($x = 1$ to 4 m)

$$F_x = -wx + R_A = -6x + 20$$

$= +14$ kN at $x = 1$ m

$= +8$ kN $x = 2$ m

$= -4$ kN $x = 4$ m

Portion CD

$$F_x = -24 + 20 = -4 \text{ kN (constant)}$$

Fig. 7.10 (b) shows SF diagram.

BM diagram: Portion AB ($x = 0$ to 1 m)

$$M_x = -wx\frac{x}{2} = -\frac{w}{2}x^2 = -\frac{6}{2}x^2 = -3x^2 \text{ (eq. of a parabola)}$$

$$= 0 \quad \text{at } x = 0 \text{ m}$$

$$= -3 \text{ kNm} \quad \text{at } x = 1 \text{ m}$$

Portion *BC* (x = 1 m to 4 m)

$$M_x = -wx\frac{x}{2} + R_A(x-1) = -3x^2 + 20(x-1) \text{ (eq. of parabola)}$$

$$= +8 \text{ kNm} \quad \text{at } x = 2 \text{ m}$$

$$= -48 + 60 = 12 \text{ kNm} \quad \text{at } x = 4 \text{ m}$$

Portion *CD* (x = 4 m to 7 m)

G of UDL of 24 kN lies at a distance of 2 m from *A*

$$M_x = -24(x-2) + R_A(x-1)$$

$$M_x = -24(x-2) + 20(x-1), \quad \text{equation for straight line}$$

$$= +8 \text{ kNm} \quad \text{at } x = 5 \text{ m}$$

$$= +4 \text{ kNm} \quad \text{at } x = 6 \text{ m}$$

$$= 0 \quad \text{at } x = 7 \text{ m}$$

Note that point of contraflexure lies at *P* in portion *BC*

$$M_x = -3x^2 + 20(x-1) = 0$$

$$-3x^2 + 20x - 20 = 0$$

$$3x^2 - 20x + 20 = 0$$

$$x = \frac{+20 \pm \sqrt{20^2 - 4 \times 3 \times 20}}{6}$$

$$= \frac{20 - \sqrt{400 - 240}}{6} = 1.225 \text{ m}$$

Point of contraflexure lies at a distance of 1.225 m from *A* at *P* as shown in BM diagram.

Practice Q.7.7 **A beam *ABCD*, 6 m long hinged at end *A*, roller supported at *C* carries at UDL of 6 kNm/m over *AB* = 3 m and a point load 2 kN at *D*, as shown in figure 7.16. Determine support reactions and draw SF and BM diagram. Where lies the point of contraflexure.**

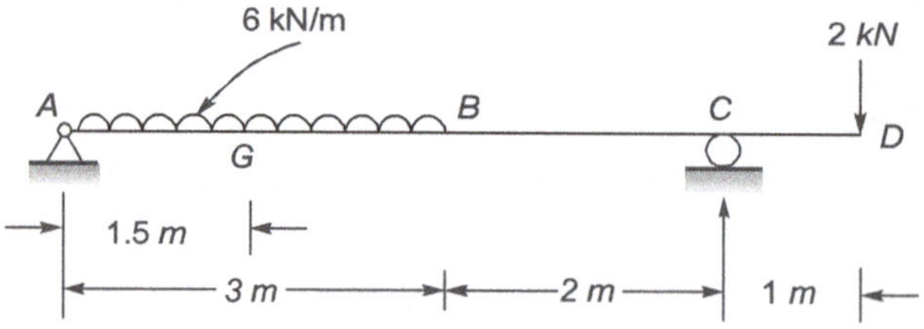

Fig. 7.16

Ans. [R_A = 12.2 kN, R_C = 7.8 N, F_{AB} = 12.2 – 6x, F_{BC} = –5.8 kN, F_{CD} = +2 kN, M_A = 0, M_1 = 9.2 kNm, M_2 = 12.4 kNm, M_3 = +9.6 kNm, M_C = –2 kNm, Point of contraflexure lies at a distance of 4.65 m from *A*]

Question 7.8 **A cantilever *ABC*, 5 m long carries a clockwise moment 10 kNm at *A*, a UDL of intensity 4 kN/m over *BC* = 3 m. Draw SF and BM diagram of the beam. What is support reaction and support moment at *C*?**

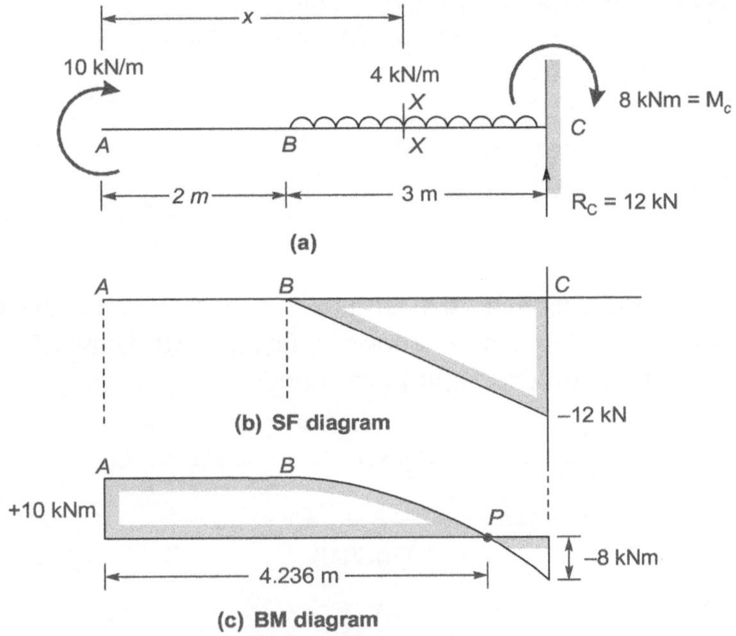

Fig.7.17

Solution: Support reaction, R_C = $4 \times 3 = 12$ kN↑ (upwards)

Support moment, M_C = $10 - 4 \times 3 \times 1.5 = -8$ kNm

Fixing couple at C = +8 kNm

SF diagram: Taking upward forces on the left side of the section as positive.

F_{AB} = 0

(no force on left side of the section, there is only one moment)

Portion BC (x = 2 to 5 m)

F_{BC} = $-w(x-2)$ kN

= $-4(x-2)$ kN

= 0 at x = 2 m

= –4 kN at x = 3 m

= –8 kN at x = 4 m

= –12 kN at x = 5 m

Fig. 7.17 (b) shows the SF diagram for the cantilever.

BM diagram (x = 0 to 2 m)

M_{AB} = +10 kNm remains constant from A to B

$$M_x = +10 - \frac{w}{2}(x-2)^2$$

Note that CG of the load $w(x-2)$ lies at a distance of $\frac{(x-2)}{2}$ from section XX

M_x = $10 - 2(x-2)^2$ putting the value of w

= 8 kNm at x = 3 m

= 2 kNm at x = 4 m

= –8 kNm at x = 5 m

Point of contraflexure P, lies in portion BC (fig. 7.17 (c)

$$M_x = 10 - 2(x-2)^2 = 0$$
$$10 - 2(x^2 - 4x + 4) = 0$$
$$10 - 2x^2 + 8x - 8 = 0$$
$$x^2 - 4x - 1 = 0$$
$$x = \frac{4 \pm \sqrt{16+4}}{2} = \frac{4+4.472}{2} = 4.236 \text{ m}$$

Practice Q.7.8 A cantilever ABC, 6 m long fixed at end C is subjected to a moment of 12 kNm at A and a UDL of 5 kN/m over BC = 4 m as shown in figure 7.18. Draw SF and BM diagrams of the cantilever. What are support reaction and support moment?

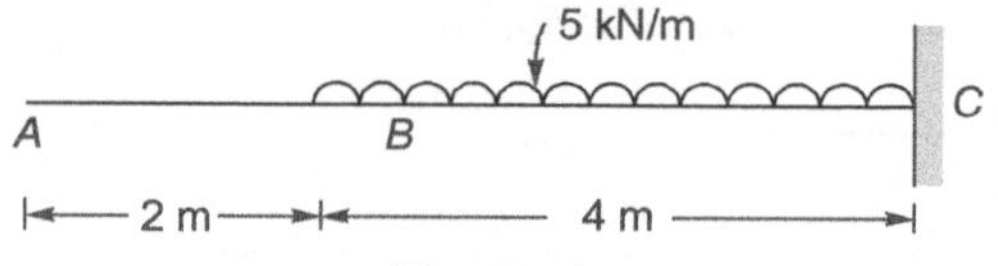

Fig. 7.18

Ans. [R_C = 20 kN, M_C = 52 kNm, F_{AB} = 0, $F_{BC} = -w(x-2)$, $F_{BC} = -5(x-2)$, M_{AB} = −12 kNm, $M_{BC} = -12 - \frac{w}{2}(x-2)^2 = -12 - 2.5\,(x-2)^2$]

Question 7.9 A beam $ABCD$, 6 m long carries a load of 12 kN at lever BEF and a UDL of 3 kN/m over CD = 4 m as shown in figure 7.19. Draw SF and BM diagram of the beam.

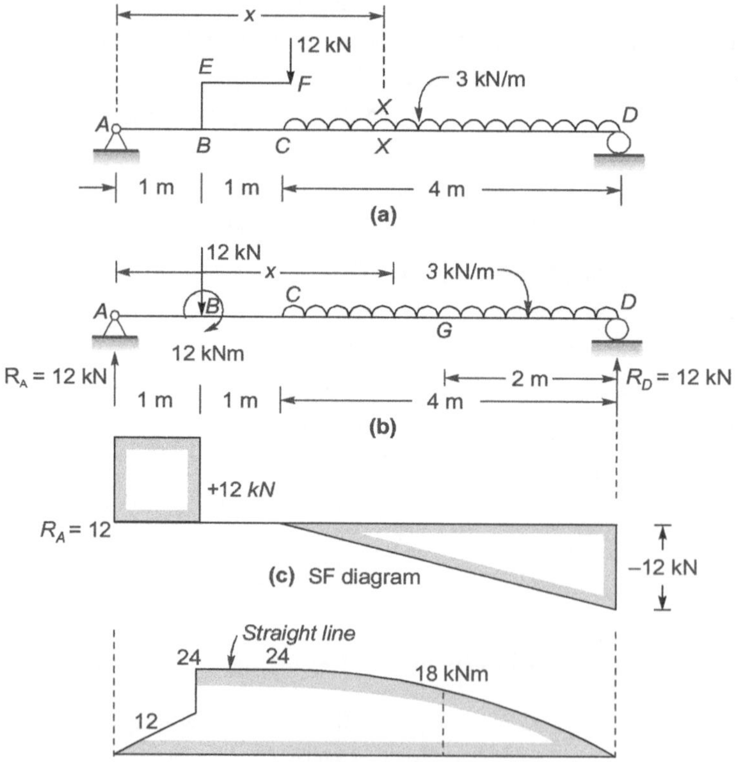

Fig. 7.19

Solution: A force of 12 kN at *F* can be replaced by a force of 12 kN and a clockwise moment of 12 kNm at *B* as shown in figure 7.19 (b)

Support reactions: Taking moments about *A*,

$$12 + 12 \times 1 + 3 \times 4(2 + 2) = 6R_D$$

Note that CG of UDL lies at *G*, 2 m from *D* or 4 m from *A*

Reaction, $R_D = \frac{72}{6} = 12$ kN

Total load on beam $= 12 + 3 \times 4 = 24$ kN

Reaction, $R_A = 24 - 12 = 12$ kN

SF diagram:

Portion *AB* (x = 0 to 1 m)

$F_{AB} = +R_A = +12$ kN (remains constant)

Portion *BC* (x = 1 to 2 m)

$F_{BC} = 12 - 12 = 0$ kN (constant for portion *BC*)

Portion *CD* (x = 0 to 6 m)

$F_x = 12 - 12 - w(x - 2) = -3(x - 2)$

$= 0$ at $x = 2$ m

$= -6$ kN at $x = 4$ m

$= -12$ kN at $x = 6$ m

BM diagram: Portion *AB* (x = 0 to 1 m) Taking clockwise moment on left side of a section to be positive.

$M_x = 12x$

$= 0$ at $x = 0$

$= 12$ kNm at $x = 1$ m

Portion *BC* (x = 1 to 2 m)

$M_x = 12x + 12 - 12(x - 1)$

$M_2 = 24 + 12 - 12 = 24$ kNm at $x = 2$ m

$M_{1.5} = 12 \times 1.5 + 12 - 12\ (0.5)$

$= 18 + 12 - 6 = 24$ kNm

(remains constant over *BC* as shear force is zero)

Portion *CD* (x = 2 to 6 m)

$$M_x = 12x + 12 - 12(x-1) - \frac{w}{2}(x-2)^2$$

Note that UDL of $w(x - 2)$, its CG lies at $\frac{x-2}{2}$ from section XX

$M_x = 24 - 1.5(x - 2)^2$, (equation of parabola)

$= +18$ kNm at $x = 4$ m

$= 0$ at $x = 6$ m

Fig. 7.19 (d) shows the BM diagram. There is no point of contraflexure in this beam.

Practice Q.7.9 **A beam *ABCDE*, 6 m long carries at point load of 3 kN at *E*, UDL of 6 kN/m over CD = 3 m , 6 kN/m over *FH*, Lever *BFH* as shown in figure 7.20. Draw SF and BM diagrams of the beam, where lies the point of contraflexure. What is maximum BM in beam?**

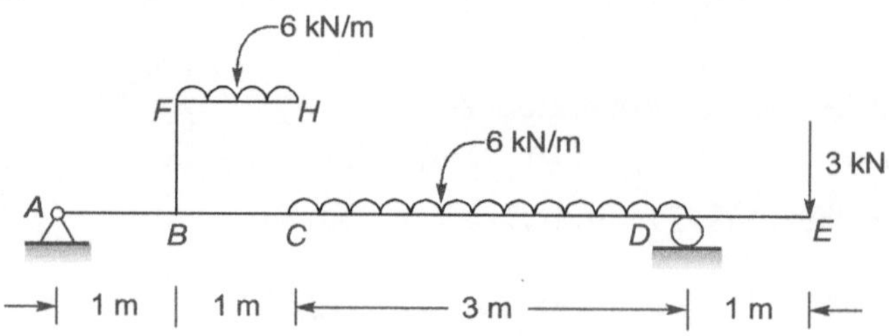

Fig. 7.20

Hints: [Load of 6 kN over *FH* is equivalent to a load of 6 kN and a clockwise moment of 3 kNm at *B*. CG of UDL 6 kN/m lies at 0.5 m from *F*]

Ans. [R_A = 9 kN, R_D = 18 kN, F_{AB} = +9.0 kN, F_{BC} = +3 kN, $F_{CD} = 3 - 6(x - 2)$, F_{DE} = +3 kN, M_A = 0, M_B = 9 kNm, M_B' = 12 kNm, M_C = +15 kNm, M_D = −3 kNm, M_E = 0, M_{max} = 15.75 kNm, where SF = 0 at x = 2.5 m]

Question 7.10 A cantilever 6 m long, fixed at end *C*, carries a load of 3 kN at end of a lever *ADE*, as shown in figure 7.21. A load of 4 kN acts at *B* and a UDL of 4 kN/m acts over *BC* = 4 m. Draw SF and BM diagrams of the cantilever. What are support reaction and support moment?

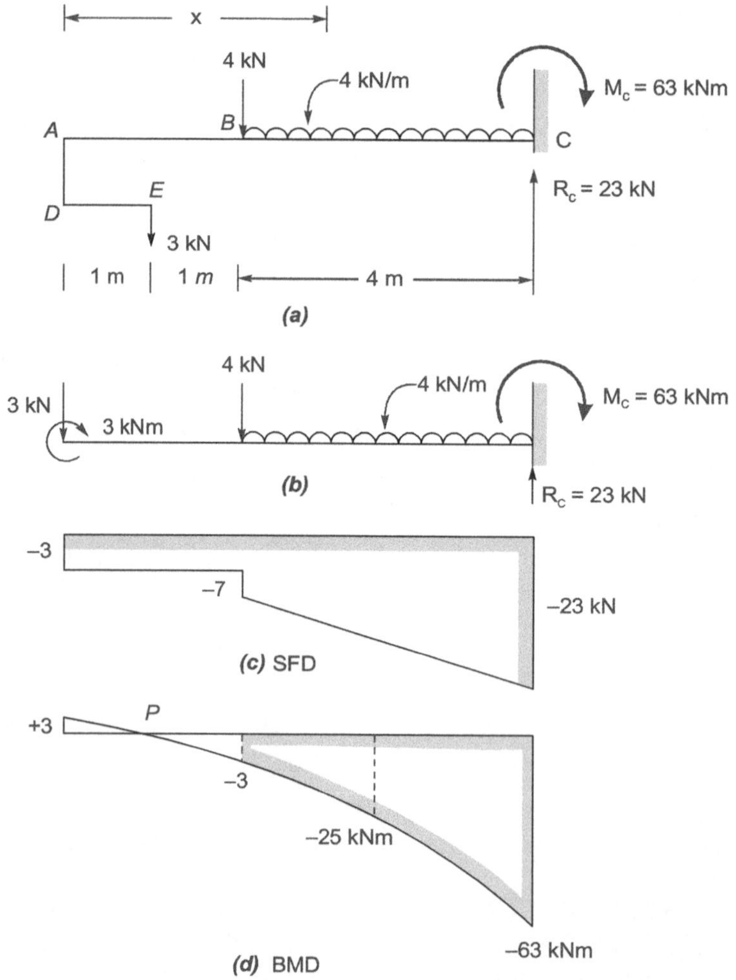

Fig. 7.21

Solution: A load of 3 kN at *E* can be replaced by a couple of 3 kNm (clockwise) and a load 3 kN at *A* as shown in figure 7.21 (b) equivalent loading diagram.

Support reaction

$$R_C = 3 + 4 + 4 \times 4 = 23 \text{ kN}$$

$$M_C = +3 - 3 \times 6 - 4 \times 4 - 4 \times 4 \times 2 = 3 - 18 - 16 - 32 = -63 \text{ kN}$$

Fixing couple is equal and opposite at *C*, provided by the wall, in which the cantilever is fixed, shown by clockwise couple 63 kNm.

SF diagram: (Taking upward forces on the left side of the section as positive)

$$F_{AB} = -3 \text{ kN (remains constant)}$$

$$F_{BC} = -3 - 4 - w(x - 2) = -7 - 4(x - 2)$$

$$= -7 \text{ kN} \qquad \text{at } x = 2 \text{ m}$$

$$= -15 \text{ kN} \qquad \text{at } x = 4 \text{ m}$$

$$= -23 \text{ kN} \qquad \text{at } x = 6 \text{ m (at } C)$$

Fig. 7.16 (c) shows SF diagram.

BM diagram

Portion *AB* (x = 0 to 2 m)

$$M_x = +3 - 3x$$

$$= +3 \text{ kMm} \qquad \text{at } x = 0$$

$$= 0 \qquad \text{at } x = 1 \text{ m (point of contraflexure)}$$

$$= -3 \text{ kNm} \qquad \text{at } x = 2 \text{ m}$$

Portion *BC* (x = 2 m – 6 m)

$$M_x = 3 - 3x - 4(x-2) - \frac{w}{2}(x-2)^2$$

$$= 3 - 3x - 4(x - 2) - 2(x - 2)^2$$

$$= 3 - 12 - 8 - 8 = -25 \text{ kNm} \qquad \text{at } x = 4 \text{ m}$$

$$= 3 - 18 - 16 - 32 = -63 \text{ kNm} \qquad \text{at } x = 6 \text{ m at } C$$

Figure 7.21 (a) shows BM diagram with point of contraflexure *P* at 1 m from *A*.

Practice Q.7.10 **A cantilever *ABC*, 8 m long fixed at end *C*, carries a UDL of intensity 2 kN/m over *BC* = 4 m and a load of 4 kN at *E*, end of lever *ADE*, above *DE* = 1 m as shown in figure 7.22. What are support reaction and support moment. Draw SF and BM diagram of the beam, where lies the point of contraflexure.**

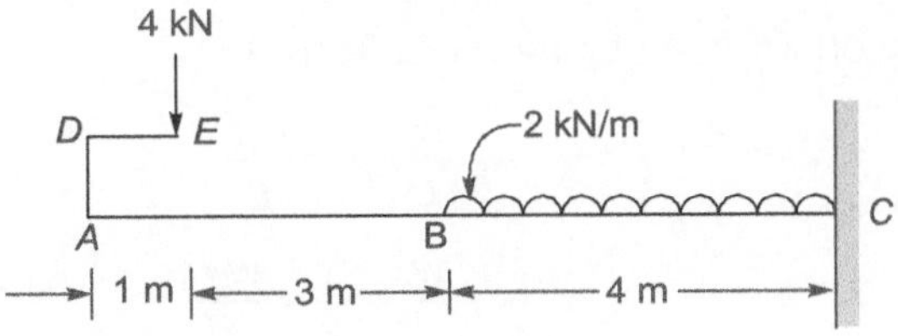

Fig. 7.22

Ans. [R_C = 12 kN, M_C = 44 kNm, F_{AB} = –4 kN, F_{BC} = –4 – 2(x – 4), Point of contraflexure lies at 1 m from *A*]

Question 7.11 A beam *AB* of length *L*, carries a linearly varying load of maximum intensity *w* kN/m length. Determine support reactions. Draw SF and BM diagrams. What is M_{max} and where it occurs?

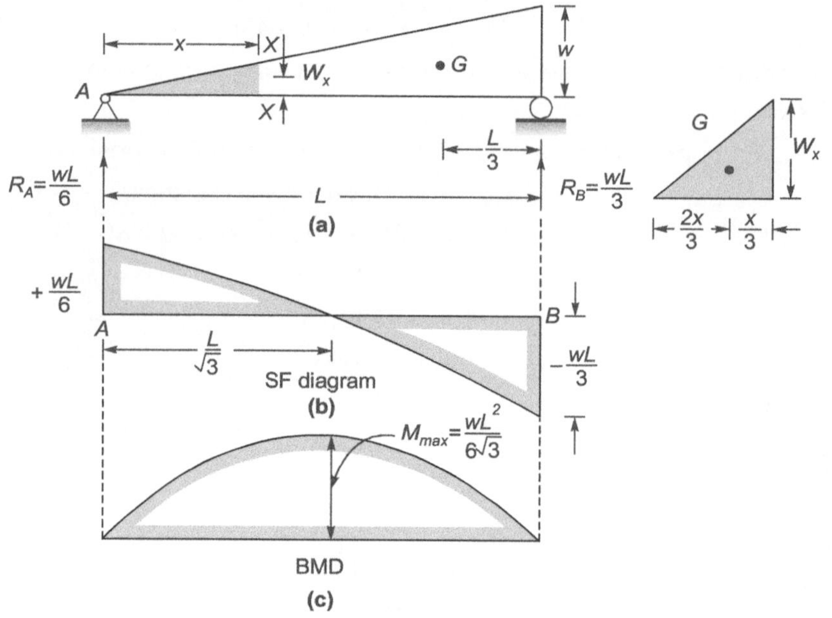

Fig. 7.23

Solution:

$$\text{Total load on beam} = \frac{wL}{2}$$

CG of the load is at a distance of $\frac{2L}{3}$ from A as shown in figure 7.23 (a)

Taking moments about *A*

$$R_B \cdot L = \frac{wL}{2} \times \frac{2L}{3} = \frac{wL^2}{3}$$

Reaction, $$R_B = \frac{wL}{3}$$

Reaction, $$R_A = \frac{wL}{2} - \frac{wL}{3} = \frac{wL}{6}$$

SF diagram: Take a section XX at a distance of x from *A*

w_x, rate of loading at $$x = \frac{w}{L} \times x = \frac{wx}{L}$$

Total load upto, $$x = \frac{wx}{L} \times \frac{x}{2} = \frac{wx^2}{2L}$$

Shear force at section XX

$$F_x = \frac{wL}{6} - \frac{wx^2}{2L}$$

$$= \frac{wL}{6} \qquad \text{at } x = 0$$

$$= \frac{wL}{9} \qquad \text{at } x = \frac{L}{3}$$

$$= \frac{wL}{24} \qquad \text{at } x = \frac{L}{2}$$

$$= -\frac{wL}{18} \qquad \text{at } x = \frac{2L}{3}$$

$$= -\frac{wL}{3} \qquad \text{at } x = L$$

SF diagram is shown in figure 7.18(b)

For maximum BM, $F_x = 0 = \frac{wL}{6} - \frac{wx^2}{2L}$

$$x^2 = \frac{L^2}{3}, \quad x = \frac{L}{\sqrt{3}} = 0.577L$$

BM diagram: BM at the section

$$M_x = R_A x - \frac{wx^2}{2L}\left(\frac{x}{3}\right)$$

(Note that CG of the load upto X lies at $\frac{x}{3}$ from section X)

$$= \frac{wL}{6}x - \frac{wx^3}{6L}$$

$$= 0 \qquad \text{at } x = 0$$

$$= \frac{4}{81}wL^2 = 0.0494wL^2 \qquad \text{at } x = \frac{L}{3}$$

$$= \frac{5}{81}wL^2 = 0.0617wL^2 \qquad \text{at } x = \frac{2L}{3}$$

$$= 0 \qquad \text{at } x = L$$

$$M_{max} = \frac{wL^2}{6\sqrt{3}} = 0.0962wL^2 \qquad \text{at } x = \frac{L}{\sqrt{3}}$$

Fig. 7.23 (c) shows the BM diagram of the beam.

Practice Q.7.11 A beam *AB* of length 6 m carries a linearly varying load of maximum intensity of loading equal to 6 kN/m at other end starting from zero from one end. Determine support reactions. Draw SF an BM diagrams. What is M_{max} and where it occurs.

Ans. [R_A = 6 kN, R_B = 12 kN, M_{max} = 20.785 kNm at x = 3.464 m]

Question 7.12 A beam *ABCDE*, 6 m long, hinged at *A*, roller supported at *D*, 1 m from *E* as shown in figure 7.24 (a). The beam carries an inclined load of 10 kN at $\theta = \tan^{-1}\frac{4}{3}$, and a UDL of 6 kN/m over *CDE* = 3 m. Determine support reactions, Draw SF and BM diagram and find the location of point of contraflexure.

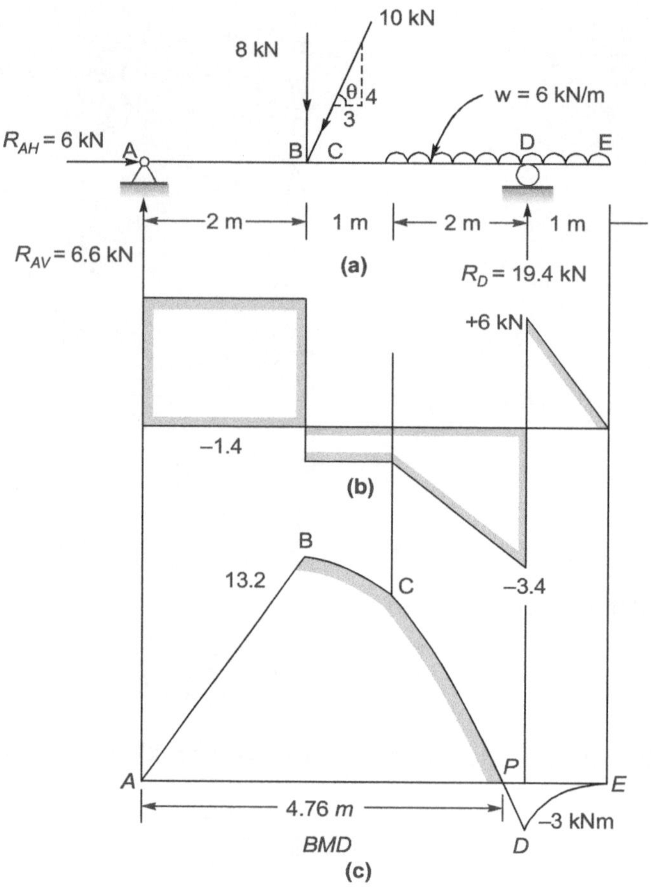

Fig. 7.24

Solution: Component of 10 kN load at B, are 8 kN↓ (vertical) and 6 kN ← (horizontal).

Due to 6 kN component, there will be horizontal reaction R_{AH} = 6 kN→ as shown.

Taking moments about A

$$8 \times 2 + 6 \times 3 \times 4.5 = 5R_D$$

Note that CG of UDL are CDE lies at a distance of 4.5 m from A.

$$16 + 81 = 97 = 5\,R_D$$

Reaction, $R_D = 19.4$ N

Reaction, $R_{AV} = 8 + 18 - 19.4 = 6.6$ kN

SF diagram

$$F_{AB} = +6.6 \text{ kN}$$

$$F_{BC} = 6.6 - 8 = -1.4 \text{ kN}$$

Portion CD (x = 3 to 5)

$$F_x = 6.6 - 8 - 6(x - 3)$$

$$F_4 = -7.4 \text{ kN} \qquad \text{at } x = 4 \text{ m}$$

$$F_5 = -13.4 \text{ kN} \qquad \text{at } x = 5 \text{ m}$$

$$F_5' = -13.4 + 19.4 = +6 \text{ kN}$$

Portion DE (x = 5 to 6 m)

$$F_x = 6.6 - 8 - 6(x - 3) + 19.4$$

$$= 18 - 6(x - 3)$$

$$F_5 = +6 \qquad \text{at } x = 5 \text{ m}$$
$$= +3 \qquad \text{at } x = 5.5 \text{ m}$$
$$= 0 \qquad \text{at } x = 6 \text{ m (fig. 7.4 (b)}$$

BM diagram

$$M_A = 0$$
$$M_B = +6.6 \times 2 = +13.2 \text{ kNm}$$
$$M_C = 6.6 \times 3 - 8 \times 1 = +11.8 \text{ kNm}$$
$$M_4 = 6.6 \times 4 - 8 \times 2 - 6 \times 1 \times 0.5$$
$$= 26.4 - 16 - 3 = 7.4 \text{ kNm}$$
$$M_5 = 6.6 \times 5 - 8 \times 3 - 6 \times 2 \times 1$$
$$= 33 - 24 - 12 = -3 \text{ kNm}$$
$$M_6 = 0 \qquad \text{[BMD shown in figure 7.24 (c))]}$$

Point of contraflexure lies in portion *CD*

$$M_x = 6.6x - 8(x-2) - \frac{w}{2}(x-3)^2$$
$$= 6.6x - 8(x-2) - \frac{6}{2}(x-3)^2$$
$$= 6.6x - 8(x - 2) - 3(x^2 - 6x + 9)$$
$$= 6.6x - 8x + 16 - 3x^2 + 18x - 27$$
$$= -3x^2 + 16.6x - 11 = 0$$
$$3x^2 - 16.6x + 11 = 0$$
$$x = \frac{16.6 + \sqrt{16.6^2 - 4 \times 3 \times 11}}{6}$$
$$= \frac{16.6 + \sqrt{275.56 - 132}}{6}$$
$$= \frac{16.6 + 11.98}{6} = 4.76 \text{ m}$$

Point of contraflexure at 4.76 m from *A*.

Practice Q.7.12 A beam *ABCD*, 6 m long carries an inclined load $\theta = \tan^{-1}\frac{4}{3}$ of magnitude 15 kN and a load 3 kN at *D*. Determine support reactions. Draw SF and BM diagrams. Where lies the point of contraflxrure.

Ans. [R_{AH} = 9 kN, R_{AV} = 4.5 kN, R_C = 10.5 kN, F_{AB} = +4.5 kN, F_{BC} = −7.5 kN, F_{CD} = +3 kN, M_A = 0, M_B = 9 kNm, M_C = −6 kNm, M_D = 0]

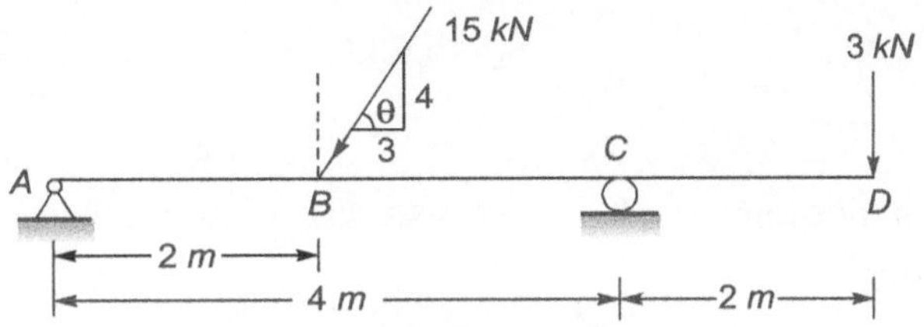

Fig. 7.25

Point of contraflexure lies at 3.2 m from *A*.

Question 7.13 A beam *ABCD*, 6 m long, hinged at end *A* roller supported at *C*, carries a UDL of intensity 2 kN/m over *BC* = 2 m as shown in fig 7.26 (a). Determine (i) support reactions (ii) draw SF and BM diagrams (iii) where lies the point of contraflexure?

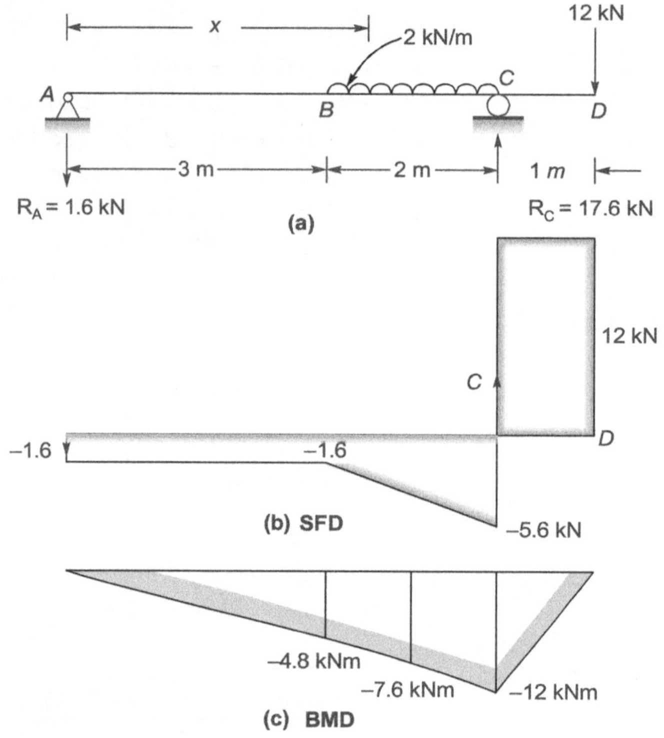

Fig. 7.26

Solution: Taking moments about *A*

(CG of UDL lies at 4 m from *A*)

$$2 \times 2 \times 4 + 12 \times 6 = 5 R_C$$

$$16 + 72 = 5 R_C$$

Reaction, $R_C = 17.6$ kN

Total load on beam $= 2 \times 2 + 12 = 16$ kN

Reaction at *A*, $R_A = 16 - 17.6 = -1.6$ kN (downwards)

SF diagram: Portion *AB* ($x = 0 - 3$ m) (taking upward forces on left side of the section as positive)

$$F_{AB} = -1.6 \text{ kN (remains constant)}$$

Portion *BC* ($x = 3 - 5$ m)

$$F_x = -1.6 - w(x - 3) = -1.6 - 2(x - 3)$$

$= -3.6$ kN at $x = 4$ m

$= -5.6$ kN at $x = 5$ m

Portion *CD*

$$F_{CD} = -1.6 - 4 + 17.6 = +12 \text{ kN (remains constant)}$$

BMD (taking clockwise moment on left side of a section to be positive)

Portion *AB* ($x = 0 - 3$ m)

$$M_x = -1.6x$$

$= 0$ at $x = 0$ m

$= -4.8$ kNm at $x = 3$ m

Portion *BC* (x = 3 – 5 m)

$$M_x = -1.6x - \frac{w}{2}(x-3)^2 = -1.6x - (x-3)^2, \text{ as } w = 2 \text{ kN/m}$$

Putting the value of x, (equation of parabola)

$$= -6.4 - 1 = -7.6 \text{ kNm} \qquad \text{at } x = 4 \text{ m}$$

$$= -8 - 4 = -12 \text{ kNm} \qquad \text{at } x = 5 \text{ m}$$

Portion *CD*

$$M_C = -12 \text{ kNm}$$

$$M_D = 0$$

There is no point of contraflxrure in this beam. Throughout there is negative bending moment. Beam bends showing convexity upwards.

Practice Q.7.13 **A beam *ABCD*, 6 m long, hinged at *A*, roller supported at *C*, carries point loads as shown in the figure 7.27. Determine support reactions. Draw SF and BM diagrams, where lies the point of contraflexure?**

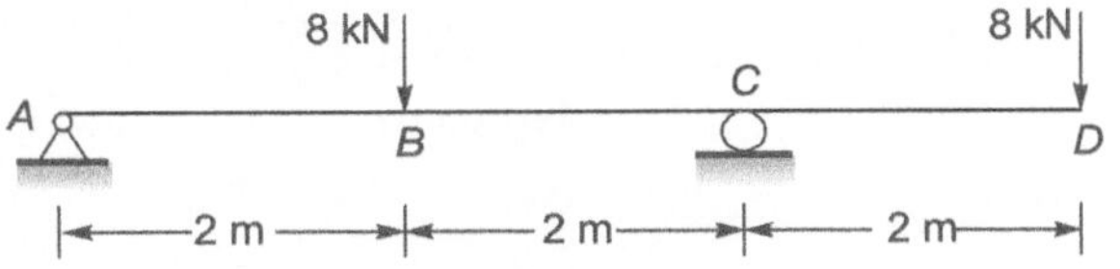

Fig. 7.27

Ans. [R_A = 0, R_C = 16 kN; F_{AB} = 0, F_{BC} = –8 kN, F_{CD} = +8 kN, M_A = 0, M_B = 0, M_C = –16 kNm, M_D = 0, No point of contraflexure]

Question 7.14 **A beam *AB*, 3 m long is hinged at end *B* to another beam *BCDE*, 5 m long as shown in figure 7.28 (a). Beam *AB* carries a UDL of 2 kN/m over its full length, while beam *BCDE*, carries a point load of 6 kNm, 1m from *C* at *D*. Determine support reactions and draw BM diagram of each beam.**

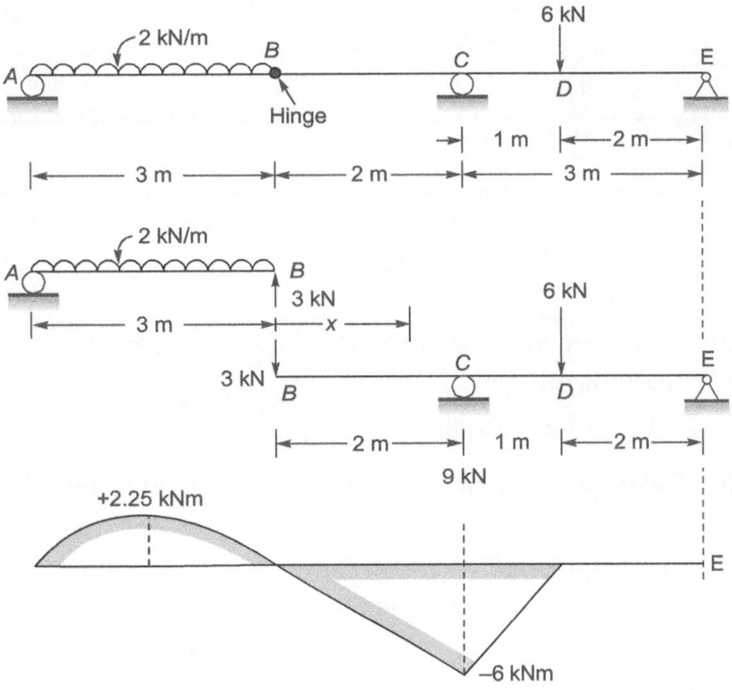

Fig. 7.28

Solution: Say reaction at B is R_B, for beam AB this reaction will act load for beam $BCDE$.

$$R_B = \frac{2\times 3}{2} = 3 \text{ kN}, \quad R_A = 3 \text{ kN}$$

$$M_C = \frac{wL^2}{8} = \frac{2\times 3^2}{8} = 2.25 \text{ kNm at centre}$$

Beam *BCDE:* Taking moments about point *E*.

$$6 \times 2 + 3 \times 5 = 3R_C$$

Reaction, $R_C = 9$ kN

Reaction, $R_E = 3 + 6 = 9 \text{ kN} = 0 \text{ kN}$

Portion *BC* (x = 0 to 2 m)

$$M_x = -3x$$

$= 0$ at x = 0 m

$= -3$ kNm at x = 1 m

$= -6$ kNm at x = 2 m

Portion *CDE* (x = 2 to 3 m)

$$M_x = -3x + 9(x - 2)$$

$M_D = 0$ at x = 1 m

There is no bending moment, no shear force in portion *DE* as reaction, $R_E = 0$.

Practice Q.7.14 **A beam *ABC*, 4 m long, simply supported at *A* carries a point load of 4 kN at its centre. It is hinged at end *C* to another beam *CDE* which is hinged at *E* and roller supported at *D* as shown in figure 7.29. The beam carries a UDL of 4 kN/m over *DE* = 4 m. Determine support reactions. Draw BM diagram of each beam.**

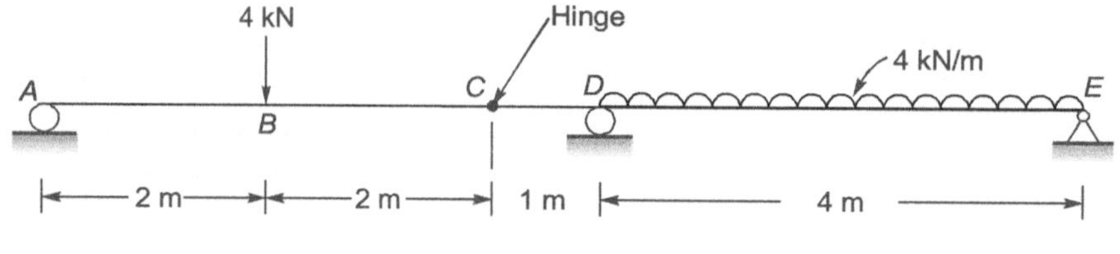

Fig. 7.29

Ans. [R_A = 2 kN, R_C = 2 kN, R_D = 10.5 kN, R_E = 7.5 kN, Beam *ABC*, M_B = +4 kNm, Beam *CDE*, M_D = −2 kNm, M_3 = +7 kNm (parabola)]

Point of contraflexure lies at 1.25 m from *C*.

Question 7.15 **Develop relationship between rate of loading, shear force and bending moment in a beam.**

Solution: Consider a beam *ABCDEF*, hinged at *F* and roller supported at *B* carrying point loads and UDL as shown in figure 7.30. Taking a small element of the beam of length dx as shown. On the left side of element shear force is F.

Bending moment is M

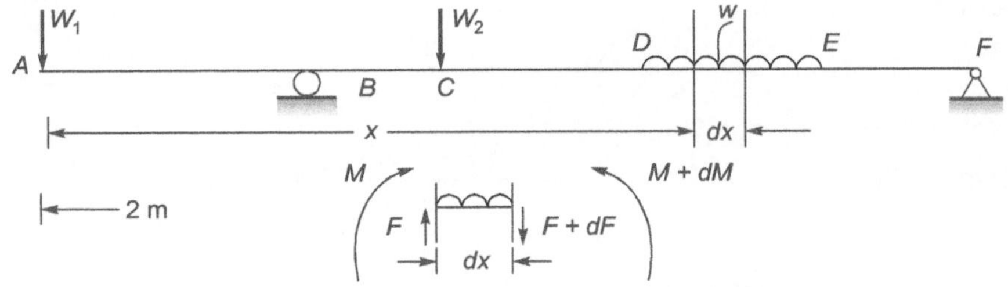

Fig. 7.30

On the right side of the element, shear force = $F + dF$

Bending moment = $M + dM$

For equilibrium

$$F = F + dF + wdx \quad \text{or} \quad dF = -wdx$$

$$w, \text{ rate of loading} = -\frac{dF}{dx}$$

i.e., rate of change of shear force at a section is numerally equal to rate of loading.

Now taking moments about the right hand end of the element.

$$M + dM = M + Fdx - wdx\frac{dx}{2}$$

$$M + dM = M + Fdx - \text{neglecting higher order terms}$$

or

$$\frac{dM}{dx} = F$$

i.e., rate of change of bending moment at a section is equal to the shear force at the section.

Question 7.16 **SF diagram of a beam $ABCD$, 6 m long, hinged at A and roller supported at C, is shown in figure 7.31 (a). Draw its loading diagram and BM diagram. What is M_{max} and where lies the point of contraflexure?**

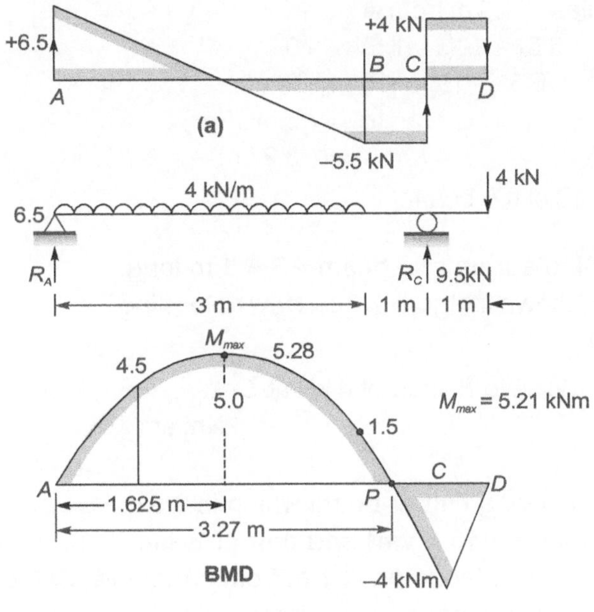

Fig. 7.31

Solution:

From SF diagram

Reaction at A, $R_A = 6.5$ kN↑

Reaction at C, $R_C = 5.5 + 4 = 9.5$ kN↑

Rate of loading, $w = \dfrac{6.5-(-5.5)}{3} = 4$ kN/m

Over $AB = 3$ m

Overhang, $CD = 1$ m

Load at point, $D = 4.0$ kN↓

Fig. 7.31 (b) shows the loading diagram of the beam.

BM diagram

Portion AB ($x = 0 - 3$ m)

$$Mx = 6.5x - \frac{w}{2}(x)^2 = 6.5x - 2x^2 \text{, as } w = 4 \text{ kN/m}$$

$= 0$ at $x = 0$

$= 4.5$ kNm at $x = 1$ m

$= +5$ kNm at $x = 2$

$= +1.5$ kNm at $x = 3$ m

$M_D = 0$

$M_C = -4$ kNm

M max.

SF in portion AB, $6.5 - 4(x) = 0 =$ SF

$x = 1.625$ m

$M_{max} = 6.5 \times 1.625 - 2 \times 1.625^2 = 10.5625 - 5.28125$

$= 5.28$ kNm

Point of contraflexure

Lies in portion BC $M_x = 6.5x - 4 \times 3\,(x - 1.5)$

Note that CG of UDL lies at 1.5 m from A

$6.5x - 12(x - 1.5) = 0$

$6.5x - 12x + 18 = 0$

$18 = 5.5x$

$x = 3.27$ m

Fig. 7.31 (c) shown BMD of the beam.

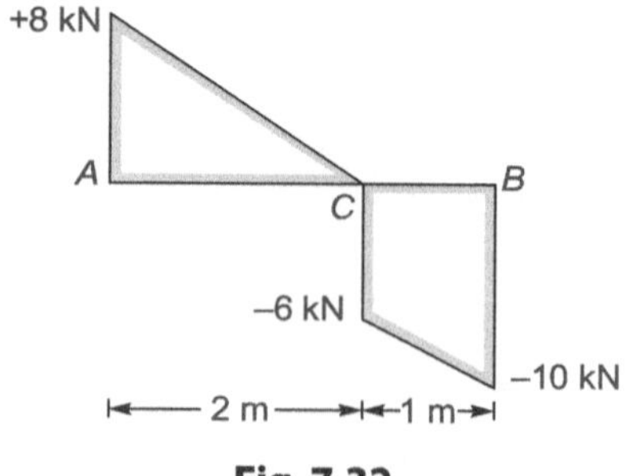

Fig. 7.32

Practice Q.7.16 SF diagram of a beam AB = 3 m long, supported at ends A and B is shown in figure 7.32. Draw to loading diagram and BMD. What is M_{max}?

Ans. [w = 4 kN/m from A to B, load of 6 kN at C, M_{max} = 8 kNm at C]

Question 7.17 A 6 m long timber beam which is 0.5 m square in section floats horizontally in sea water. The weight of timber is 640 kg/m³ and that of water is 1025 kg/m³. Two equal weights just sufficient to immerse the beam are placed on it, 1 m from each end. Calculate the value of each weight. What is maximum bending moment in beam. Draw its BM diagram.

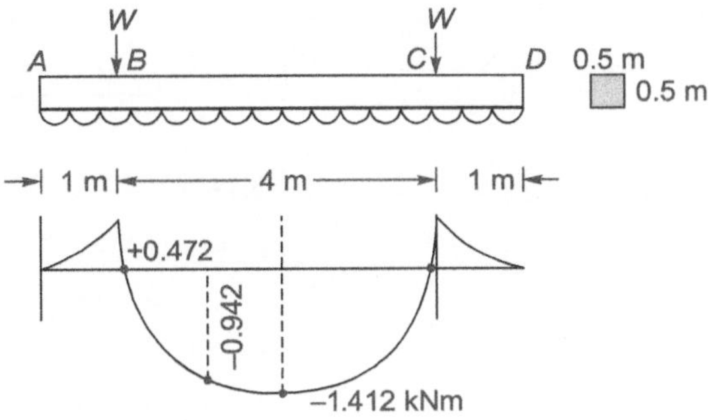

Fig. 7.33

Solution:

Area of cross-section of beam = 0.5 × 0.5 = 0.25 m^2

Length of beam = 6 m

Volume of beam = 0.25 × 6 = 1.5 m^2

Weight of beam = 640 × 1.5 = 960 kg

Weight of water displaced = 960 kg

Now the beam is completely immersed in sea water

Weight of water displacement = 1.5 × 1025 = 1537.5 kg

Extra weight to be placed = 1537.5 – 960 = 577.5 kg

Each weight, $W = \frac{577.5}{2} \times 9.81 = 2832.6 \text{ N} = 2.8326 \text{ kN}$

Net upward buoyancy force = 577.5 kg = 5665.275 N = 5.665 kN

UDL, in upward direction $= \frac{5.665}{6} = 0.944 \text{ kN/m}$

BM Diagram

Portion *AB*, (x = 0 – 1 m)

$M_x = +\frac{wx^2}{2}$ (taking upward force on left side of section to be positive)

$= \frac{0.944 \times x^2}{2}$

= 0.472 x^2 kNm (parabolic equation)

= 0 at x = 0 m

= 0.472 kNm at x = 1 m

Portion *BC* (x = 1 to 5 m)

$M_x = \frac{wx^2}{2} - W(x-1) = 0.472x^2 - 2.83(x-1)$ parabolic eq.

= 0.472 × 2^2 – 2.83 × 1 at x = 2 m

= –0.942 kNm at x = 2 m

= –1.412 kNm at x = 3 m

Beam is symmetrically loaded about its centre.

Maximum bending moment occurs at the centre. i.e., 1.412 kNm.

Practice Q.7.17 An 8 m long timber beam which is 0.4 m × 0.4 m in section floats horizontally in sea water. The weight of the timber is 640 kg/m^3 and that of water is 1020 kg/m^3. Two equal weights just sufficient to immerse the beam in sea are placed on it, 2 m from each side. Calculate the magnitude of each weight. What is M_{max}?

Ans. [2.385 kN each, 1.2 kNm under the load]

Question 7.18 A cantilever *ABC*, 5 m long is propped by a force of 8 kN at free end as shown in figure 7.34 (a). It carries a UDL of 4 kN/m over *BC* = 3 m. Draw SF and BM diagram of the beam. Where lies the point of contraflexure.

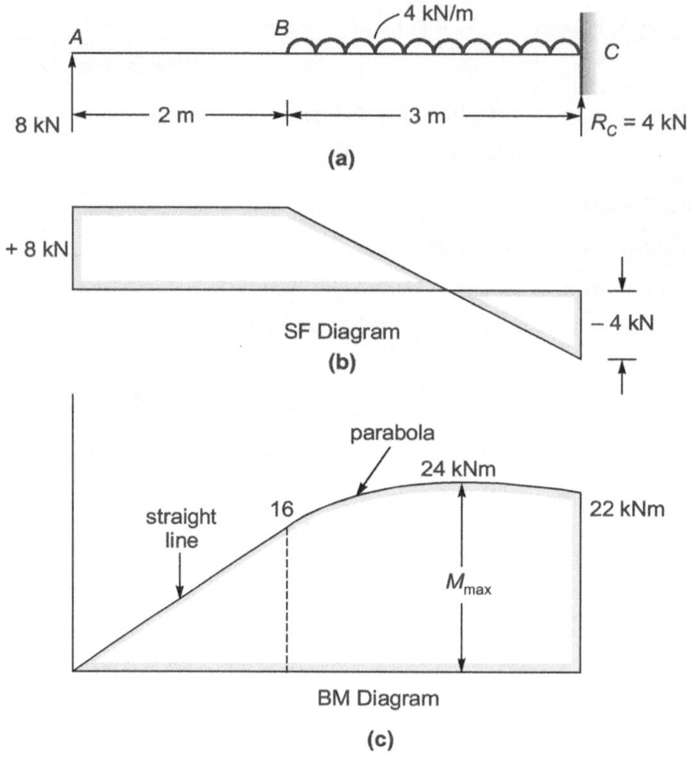

Fig. 7.34

Solution:

Reaction at *C*, $R_C = -8 + 12 = +4$ kN (upwards)↑

SF diagram: Portion *AB* (x = 0 to 2m)

$$F_{AB} = +8 \text{ kN (constant)}$$

(Taking upward forces on the left side of the section to be positive)

Portion *BC* (x = 2 to 5m)

$$F_x = 8 - 4(x - 2)$$

$= 0$ at $x = 4$ m

$= -4$ kN at $x = 5$

Fig. 7.34 (b) shows SF diagram.

BM diagram: Portion *AB* (x = 0 to 2m)

$$M_x = +8x$$

$= 0$ at $x = 0$ m

$= 16$ kNm at $x = 2$ m

Portion *BC* (x = 2 to 5 m)

$$M_x = 8x - \frac{w(x-2)^2}{2} = 8x - 2(x-2)^2$$

$= 22$ kNm at $x = 3$ m

$= 24$ kNm at $x = 4$ m

$= 40 - 18 = 22$ kNm at $x = 5$ m

There is no point of contraflexure in the cantilever.

Practice Q.7.18 A cantilever *ABCD*, 6 m long is propped at free end *A*. It is fixed at end *D*. Propping force at *A* is 4 kN↑. Two forces of 6 kN each are applied at *B* and *C* as shown in figure 7.35. Determine support moment, support reaction. Draw SF and BM diagram. Where lies the point of contraflexure.

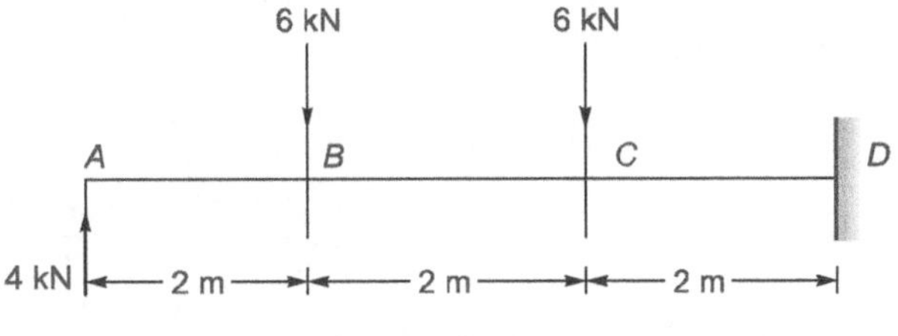

Fig. 7.35

Ans. [R_D = 8 kN (upward), M_D = 12 kNm (clockwise) fixing couple. F_{AB} = +4 kN, F_{BC} = −2 kN, F_{CD} = −8 kN, M_A = 0, M_B = +8 kNm, M_C = +4 kNm, M_D = −12 kNm, Point of contraflxrure lies at a distance of 4.5 m from *A*]

Question 7.19 A simply supported beam of length 10 m carries a uniformly varying load, whose intensity varies from a maximum value of 5 kN/m at both the ends to zero at the centre of the beam. Draw the variation of SF and BM of beam.

[IES 2005]

Solution:

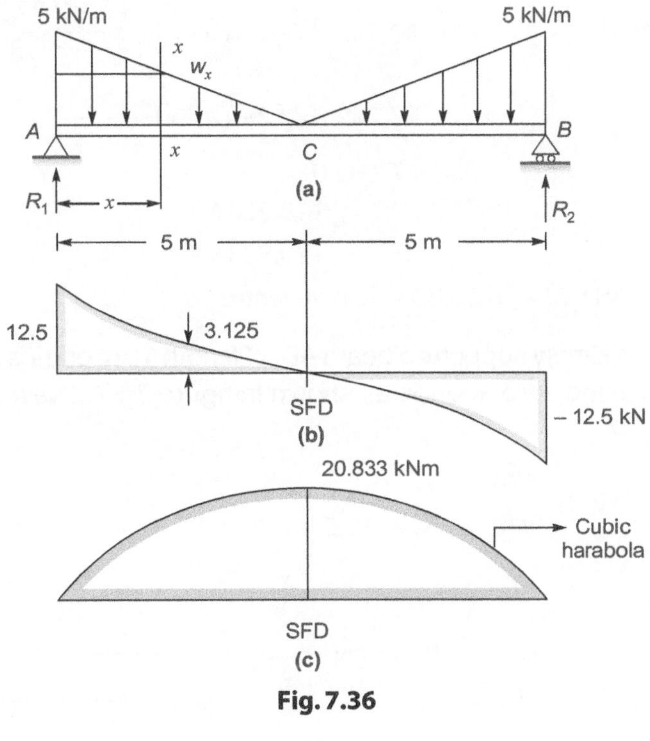

Fig. 7.36

Total load of beam $= \dfrac{5\times 5}{2}\times 2 = 25$ kN

Symmetrical loading $R_1 = R_2 = 12.5$ kN, reactions

Consider a section XX at a distance x from end A.

$$w_x = \text{ratio of loading} = \frac{5(5-x)}{5} \text{ kN/m} = (5-x) \text{ kN/m}$$

Load upto XX from A

$$= w_x x + (5 - w_x)\frac{x}{2}$$

$$= (5-x)x + (5-5+x)\frac{x}{2}$$

$$= (5-x)x + \frac{x^2}{2}$$

Shear force at section XX,

$$F_x = 12.5 - (5-x)x - \frac{x^2}{2} \quad \text{...(i)}$$

$$= 12.5 \quad \text{at } x = 0$$

$$= 12.5 - (5-2.5)2.5 - \frac{2.5^2}{2}$$

$$= 12.5 - 6.25 - 3.125 = 3.125 \text{ kNm} \quad \text{at } x = 2.5 \text{ m}$$

$$= 0 \quad \text{at } x = 5 \text{ m}$$

SF diagram will repeat from C to B in a negative direction a shown in the figure 7.36 (b) Bending moment at section XX

$$M_x = 12.5x - (5-x)x\frac{x}{2} - \frac{x^2}{2} \times \frac{2x}{3}$$

$$= 12.5x - (5-x)\frac{x^2}{2} - \frac{x^3}{3} \quad \text{...(ii)}$$

$$M_x = 0 \quad \text{at } x = 0, x = 10 \text{ m}$$

$$= +18.23 \text{ kN/m} \quad \text{at } x = 2.5 \text{ mm}, 7.5 \text{ m}$$

$$= +20.833 \text{ kNm} \quad \text{at } x = 5 \text{ m}$$

Fig. 7.36 (c) shows BMD, M_{max} = 20.833 kNm at centre.

Practice Q.7.19 **A simply supported beam AB, of length 10 m carries uniformly varying distributed load of 5 kN/m at the centre and zero at ends as shown in figure 7.37. Draw SFD and BMD of the beam.**

Hints:

$$w_x = \frac{5}{5}x = x$$

Reaction

$$R_A = R_B = 12.5 \text{ kN}$$

$$F_x = w_x = 12.5 - \frac{x^2}{2}$$

$$F_{2.5} = 9.375 \text{ kN}$$

$$F_0 = 12.5 \text{ kN}$$

$$M_x = 12.5x - \frac{x^3}{6}$$

[M_{max} = 41.667 kNm at centre].

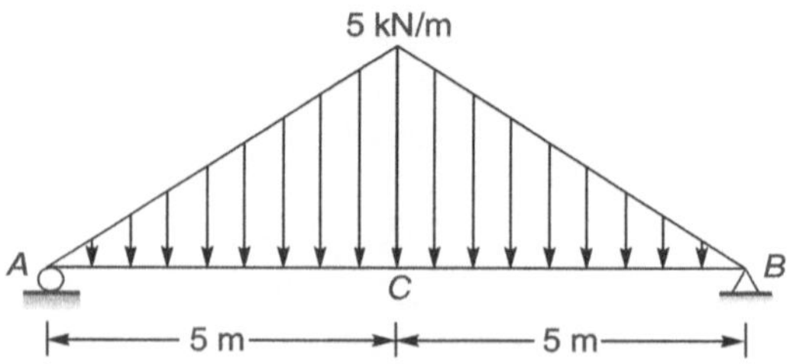

Fig. 7.37

Question 7.20 A beam *AB*, 8 m long hinged at end A and roller supported at end *B* carries a point load of 8 kN of crank *ED* = 1 m and uniformly distributed load of 5 kN/m over *CB* = 4 m. Determine (i) reaction at ends, (ii) bending moment where SF is zero, (iii) draw BM diagram only

[IES 2012, 20 Marks]

Solution: Load of 8 kN at *E* is equivalent to a load 8 kN at *C* and a CCW moment of 8 kN at *C* as shown in equivalent loading diagram as shown in figure 7.38(b)

Reaction

Moments about *A*

$$\begin{aligned} 8R_B + 8 &= 8 \times 4 + 5 \times 4 \times (4 + 2) \\ &= 32 + 20 \times 6 = 152 \\ R_B &= 18 \text{ kN} \\ R_A &= 20 + 8 - 18 = 10 \text{ kN} \end{aligned}$$

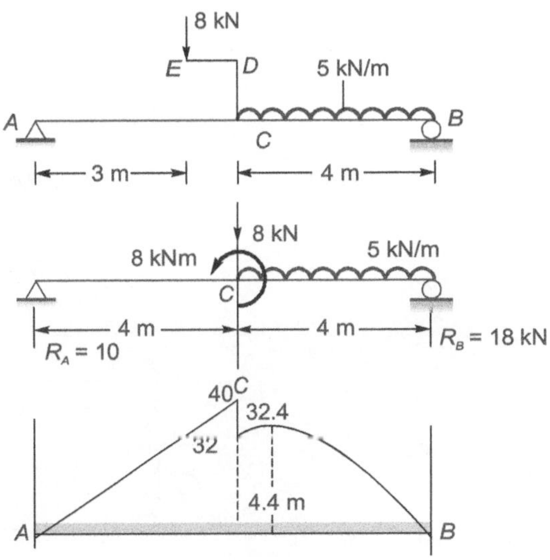

Fig. 7.38

BM diagram

$$\begin{aligned} M_A &= 0 \\ M_C &= 10 \times 4 = +40 \text{ kN} \\ M_C' &= 40 - 8 = 32 \text{ kN} \\ M_6 &= 10 \times 6 - 8 - 8 \times 2 - 5 \times 2 \times 1 \\ &= 60 - 8 - 16 - 10 = 26 \text{ kNm} \\ M_8 &= 0 \end{aligned}$$

In portion *CB*

$$\begin{aligned} F_x &= 10 - 8 - (x - 4) = 0 \\ 2 - 5(x - 4) &= 0 \\ 2 &= 5(x - 4) \\ x &= 4.4 \text{ m} \\ M &= \text{bending moment where SF} = 0 \\ &= 10 \times 4.4 - 8(4.4 - 4) - 8 - 5(4.4 - 4)(0.2) \\ &= 44 - 3.2 - 8 - 5(0.4)(0.2) \\ &= 32.8 - 0.4 = 32.4 \text{ kNm} \end{aligned}$$

Question 7.21 Draw the shear force and bending moment diagrams for the given simply supported beam with moment loads as shown in the figure. Show the magnitude of shear force and bending moments at the respective points of the beam i.e., at A, B, C, D, and E.

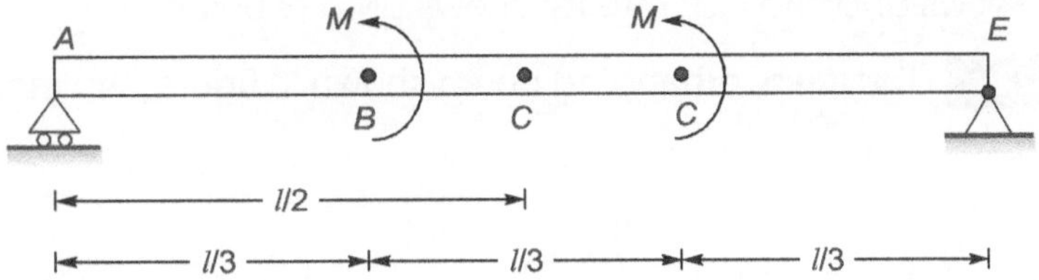

Fig. 7.39

[ME, ESE 2017 : 10 Marks]

Solution:

Taking moments about A

$$2M = R_E \times l$$

Reaction, $$R_E = \frac{2M}{l} \downarrow$$

Reaction, $$R_A = \frac{2M}{l} \uparrow \quad \text{(balancing } R_E\text{)}$$

BMD $$M_A = 0$$

$$M_B = \frac{2M}{l} \times \frac{l}{3} = \frac{2M}{3}$$

$$M_B' = \frac{2M}{3} - M = -\frac{M}{3}$$

$$M_C = \frac{2M}{l} \times \frac{l}{2} - M = 0$$

$$M_D = \frac{2M}{l} \times \frac{2l}{3} - M = +\frac{M}{3}$$

$$M_D' = \frac{4M}{3} - 2M = -\frac{2M}{3}$$

$$M_E = 0$$

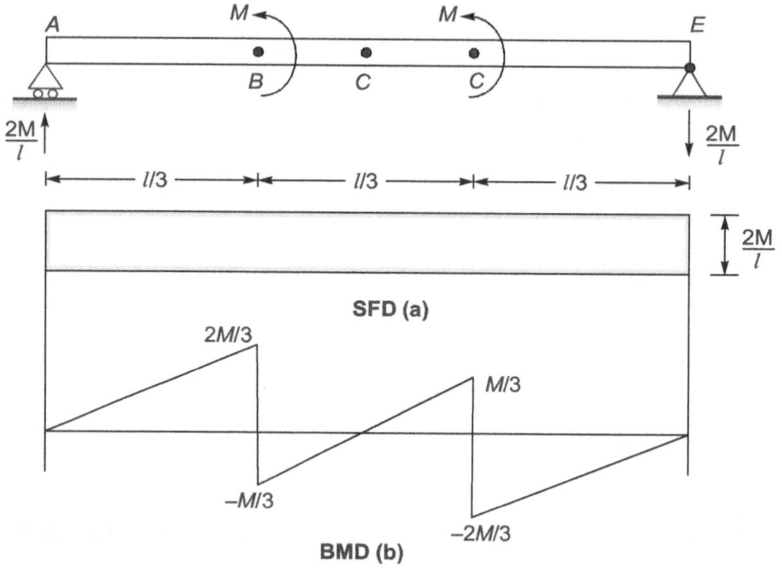

Fig. 7.40

Fig. 7.40 (a) shows SFD and Fig. 7.40 (b) shows BMD of beam.

Question 7.22 For the overhanging beam shown in figure, the magnitude of maximum bending moment (in kN-m) is _______

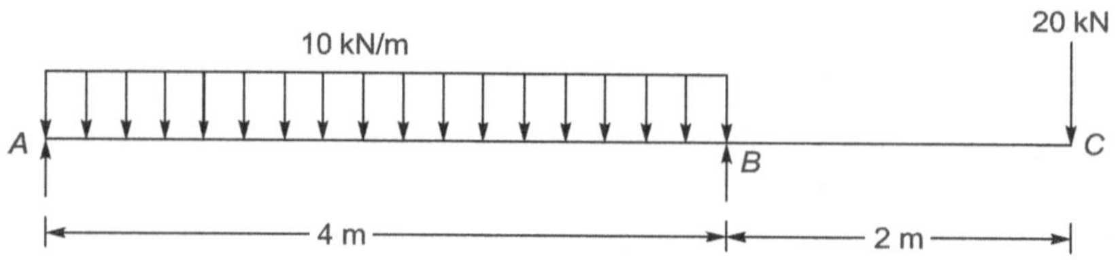

Fig. 7.41

[ME, GATE : 2015 (set-3)]

Solution:

Moment about A

$$10 \times 4 \times 2.0 + 20 \times 6 = 4R_B$$

Reaction, $R_B = 50$ kN

Reaction, $R_A = 10$ kN

Bending moment at C_1 $M_C = 0$

Bending moment at B, $M_B = -20 \times 2 = -40$ kNm

Bending moment at x from A

$$M_x = 10x - \frac{10}{2}(x^2) = 10x - 5x^2$$

$$M_z = 20 - 20 = 0 \text{ kNm}$$

$$M_{max} = -40 \text{ kNm at point } B$$

In portion AB, bending moment will be can theory 40 kNm.

Question 7.23 For a loaded cantilever beam of uniform cross-section, the bending moment (in Nmm) along the length $M(x) = 5x^2 + 10x$, where x is distance (in mm) measured from the free end of the beam. The magnitude of shear force (in N) in the cross-section at $x = 10$ mm, is ________

[ME, GATE : 2017 (set-1)]

Solution:

$$M = 5x^2 + 10x \text{ Nmm}$$

$$\frac{dM}{dx} = 10x + 10 = F_x, \text{ shear force}$$

$$x = 10 \text{ mm}$$

$$F = 10 \times 10 + 10 = 110 \text{ N}$$

Objective Type Questions

Q.1 A beam 6 m long, simply supported at ends carries uniformly varying distributed load, which is 4 kN/m at ends, decreasing linearly to zero at the centre. What is bending moment at the centre of the beam?

(a) 12 kNm (b) 9 kNm
(c) 6 kNm (d) 4.5 kNm

Q.2 For a simply supported beam on two end supports (carrying loads), the bending moment is maximum

(a) usually at the supports
(b) always at mid span
(c) where there is no shear force
(d) where the deflection is maximum

[GATE 1989, 1 Marks]

Q.3 A concentrated load P acts on a simply supported beam of span L, at a distance $L/3$ from the left support. The bending moment at the point of application of load is given by

(a) $\frac{PL}{3}$ (b) $\frac{2PL}{3}$
(c) $\frac{PL}{9}$ (d) $\frac{2PL}{9}$

Q.4 A mass of 35 kg is suspended from a weightless bar AB, which is supported by a cable CB and a pin at A as shown in figure. The pin reactions at A on the bar AB are

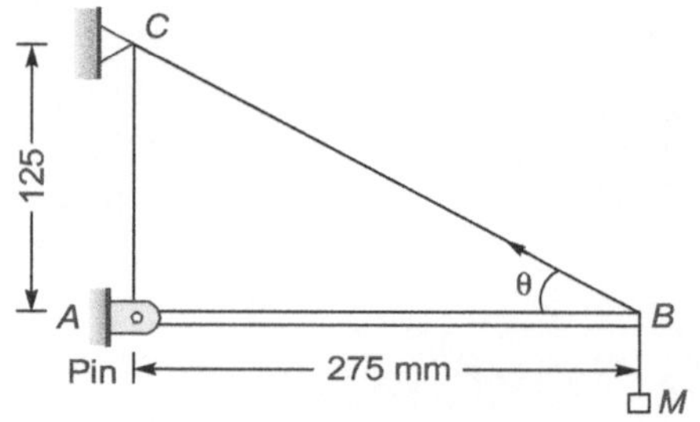

(a) $R_y = 343.4$ N, $R_y = 755.4$ N
(b) $R_x = 343.4$ N, $R_y = 0$
(c) $R_x = 755.4$ N, $R_y = 343.4$ N
(d) $R_x = 755.4$ N, $R_y = 0$

Q.5 A cantilever beam carries the antisymmetric load shown in figure 7.40. Bending moment at *A* is M_A and at *C* it is M_C. What is the ratio of M_A/M_C?

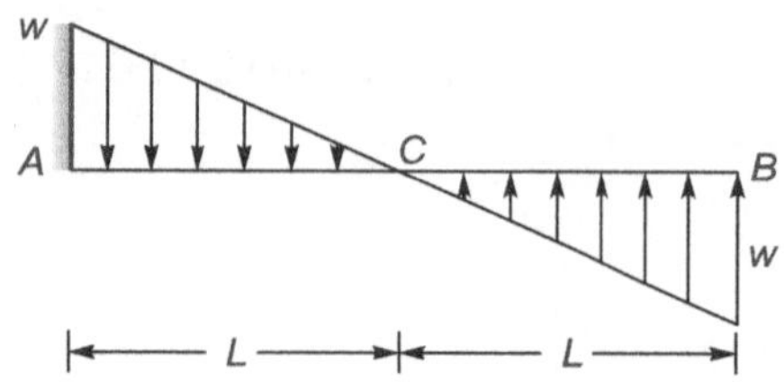

(a) +2 (b) +1.5
(c) –2 (d) –1.5

Q.6 A simply supported beam of length *L* is subjected to a varying distributed load $\sin\frac{(3\pi x)}{L}$ N/m, where the distance x is measured from left support. The magnitude of vertical reaction force in N at the left support is

(a) zero (b) $\frac{L}{3\pi}$
(c) $\frac{L}{\pi}$ (d) $\frac{2L}{\pi}$

[GATE 2013, 2 Marks]

Q.7 The given figure shows a beam *BC*, simply supported at *C* and hinged at *B* (free end) a cantilever *AB*, the beam and the cantilever carry force of 100 kg) and 200 kg respectively. The bending moment at *B* is

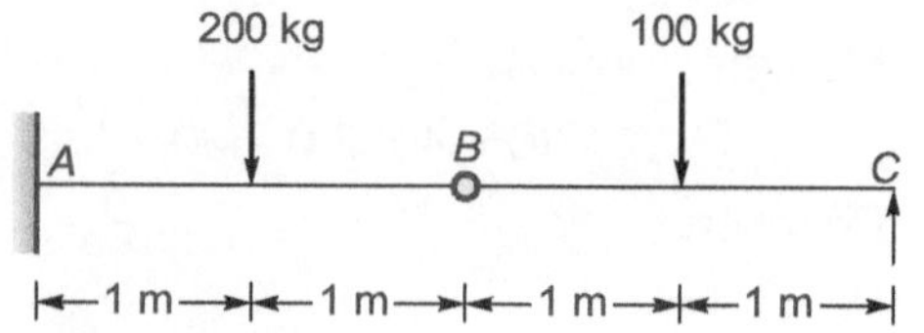

(a) zero (b) 100 kg-m
(c) 150 kg-m (d) 200 kg-m

[IES 1995]

Q.8 Consider the following statements:
If at a section at a distance x from one of the ends of the beam, *M* represents the bending moment, *V* the shear force and w the intensity of loading, then

1. $\frac{dM}{dx} = V$
2. $\frac{dV}{dx} = w$
3. $\frac{dw}{dx} = y$ (deflection of the beam at the section)

Which of these statements are correct?
(a) 1 and 3 (b) 1 and 2
(c) 2 and 3 (d) 1, 2 and 3

[IES 1995]

Q.9 A cantilever beam having 5 m length is so loaded that it develops a shearing force of 20 *T* and a bending moment of 20 T-m at a section 2 m from the free end. Maximum SF and maximum BM developed in the beam under this load, are respectively 50 T and 125 T-m. The load on the beam is
(a) 25 T concentrated load at free end
(b) 20 T concentrated load at free end
(c) 5 T concentrated load at free end, 2 T/m load over another length
(d) 10 T/m UDL over entire length

[IES 1995]

Q.10 The bending moment (*M*) is constant over a length segment (*L*) of a beam, the shearing force will also be constant over this length and is given by

(a) $\dfrac{M}{L}$ (b) $\dfrac{M}{2L}$

(c) $\dfrac{M}{4L}$ (d) None of these

[IES 1996]

Q.11 If the shear force acting at every section of a beam is of the same magnitude and of the same direction, then it represents a

(a) a simply supported beam with a concentrated load at the centre

(b) overhang beam having equal overhang at both the supports and carrying equal concentrated loads acting in the same direction

(c) cantilever subjected to a concentrated load at free end

(d) simply supported beam having concentrated loads of equal magnitudes and of the same direction acting at equal distances from the ends

[IES 1996]

Q.12 The figure shows the shear force diagram for the beams *ABCD*, bending moment in the portion *BC* of the beam

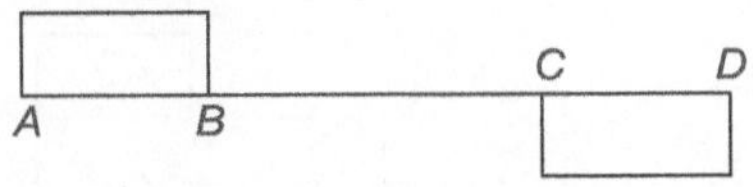

(a) is a nonzero constant

(b) is zero

(c) varies linearly from *B* to *C*

(d) varies parabolically from *B* to *C*

[IES 1996]

Q.13 A beam subjected to load *P* is shown in figure. The bending moment at the support *A* of the beam will be

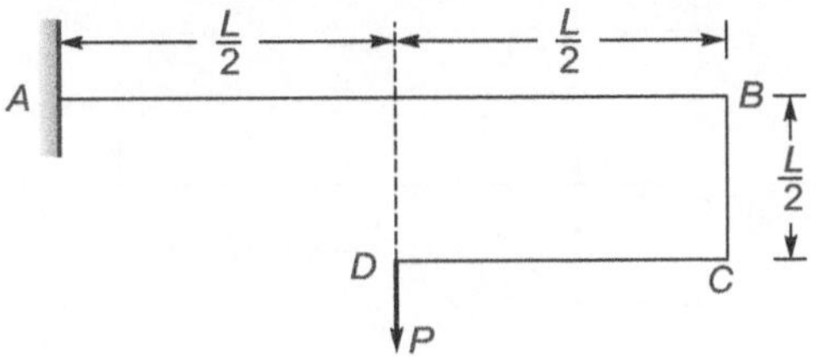

(a) *PL* (b) $\dfrac{PL}{2}$

(c) 2*PL* (d) zero

[IES 1997]

Q.14 A 2 m long *BC* beam, carries a single concentrated load at the mid span and is simply supported at ends by two cantilevers *AB*, 1 m long and *CD*, 2 m long as shown in the figure. Shear force at end *A* of cantilever *AB* will be

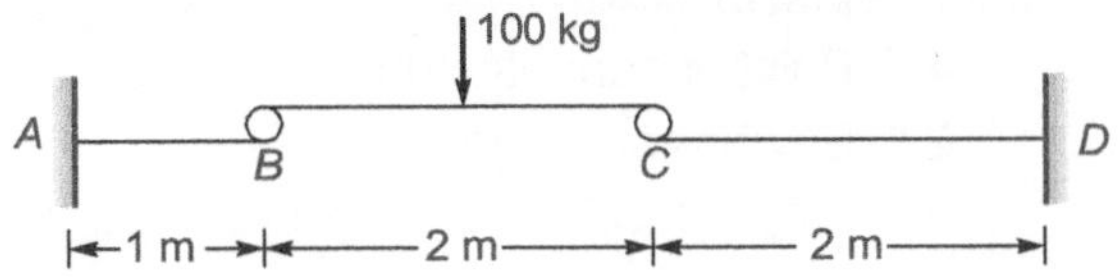

(a) zero (b) 40 kg

(c) 50 kgf (d) 60 kgf

[IES 1997]

Q.15 For the beam shown in the figure. The elastic curve between the supports *B* and *C* will be

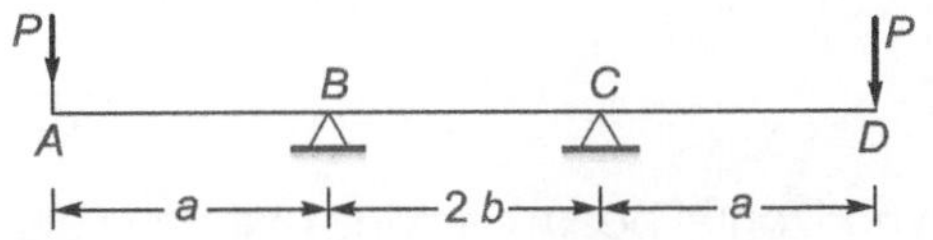

(a) circular (b) parabolic

(c) elliptic (d) a straight line

Q.16 A simply supported beam is loaded as shown in the figure. The maximum shear force in the beam will be

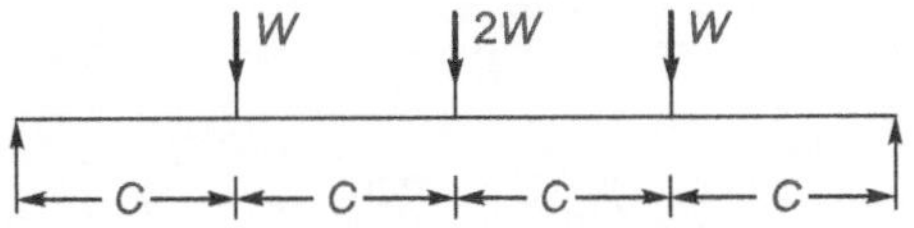

(a) zero (b) *W*

(c) 2*W* (d) 4*W*

Q.17 A lever is supported on two hinges at *A* and *C*. It carries a force of 3 kN as shown in figure. The bending moment at B will be

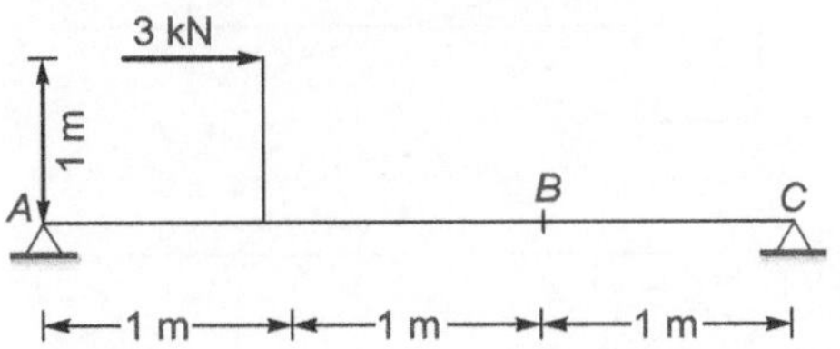

(a) 3 kNm (b) 2 kNm

(c) 1 kNm (d) zero

[IES 1998]

Q.18 Match **List-I** with **List-II** and select the correct answer using the codes below the lists:

List-I

A. BM is constant
B. BM is maximum or minimum
C. BM is zero
D. Loading is constant

List-II

1. Point of contraflexure
2. SF changes sign
3. Slope of SFD is zero over the portion of the beam
4. SF is zero over the portion of the beam

Codes:

	A	B	C	D
(a)	4	1	2	3
(b)	3	2	1	4
(c)	4	2	1	3
(d)	3	1	2	4

[IES 2000]

Q.19 A horizontal beam carrying uniformly distributed load is supported with equal overhang as shown in figure. The resultant BM at the mid span shall be zero, if a/b is

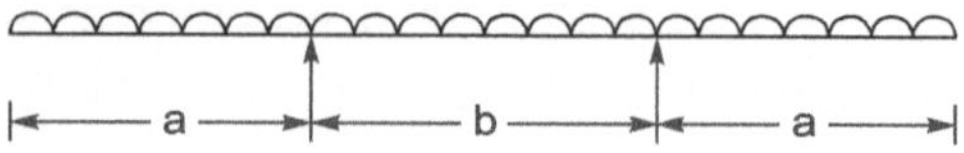

(a) $\frac{3}{4}$ (b) $\frac{2}{3}$

(c) $\frac{1}{2}$ (d) $\frac{1}{3}$

[IES 2001]

Q.20 The shear force diagram is shown below for a loaded beam. The corresponding bending moment diagram is represented by

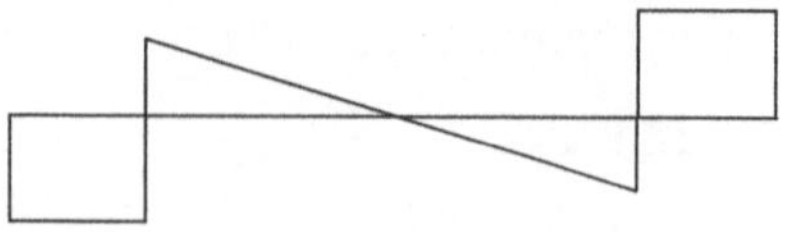

(a) 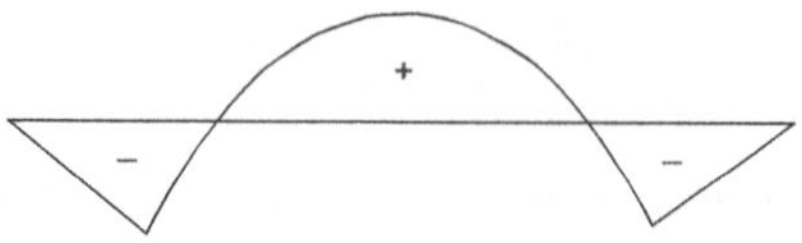

(b) 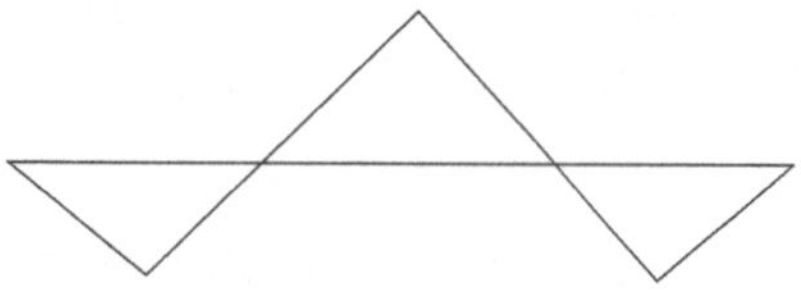

(c) 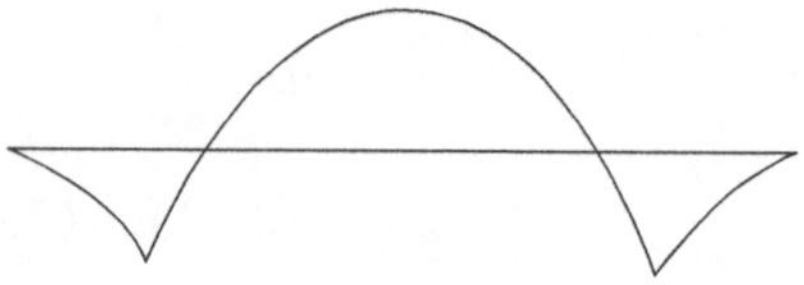

(d) 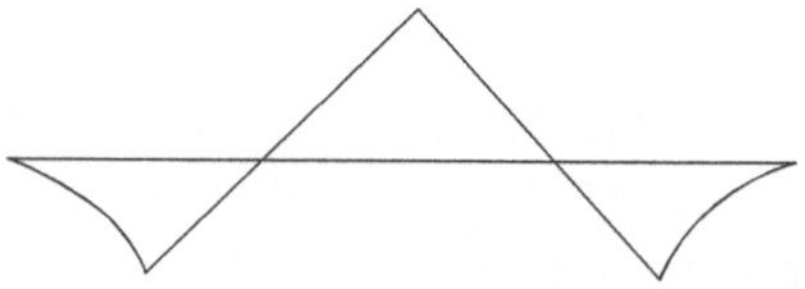

[IS 2003]

Q.21 A simply supported beam of span L is subjected to a uniformly varying load having zero intensity at the left support and w N/m at the right hand support. The reaction at the right support is

(a) $\frac{wL}{2}$ (b) $\frac{wL}{5}$

(c) $\frac{wL}{4}$ (d) $\frac{wL}{3}$

Q.22 A beam of length $4L$, is simply supported on two supports with equal overhangs of L on either side and carries 3 equal loads, one each at free ends and the third at the mid span. Which one of the following diagrams represents correct distribution of shearing force as the beam?

(a) 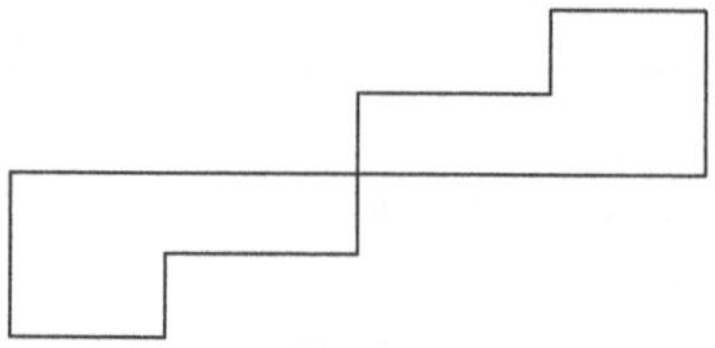

(b) 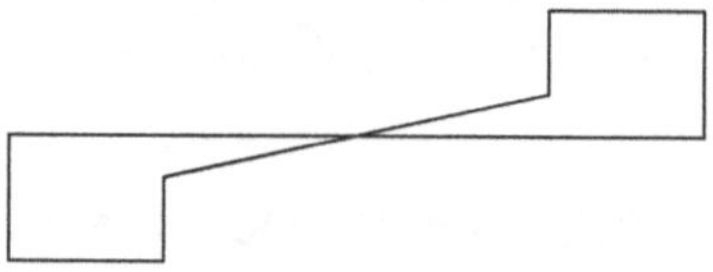

(c) 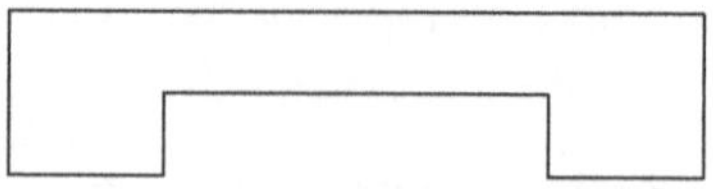

(d) 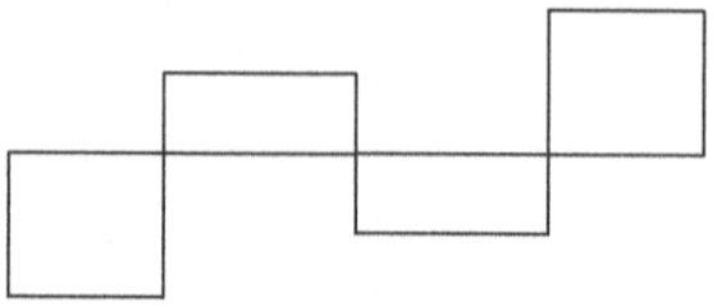

[IES 2004]

Q.23 The point of contraflexure is a point where
(a) Shear force diagram
(b) BM changes sign

(c) SF is maximum
(d) BM is maximum

Q.24 Figure shows the BM diagram for a beam simply supported at ends. The beam is subjected to which one of the following

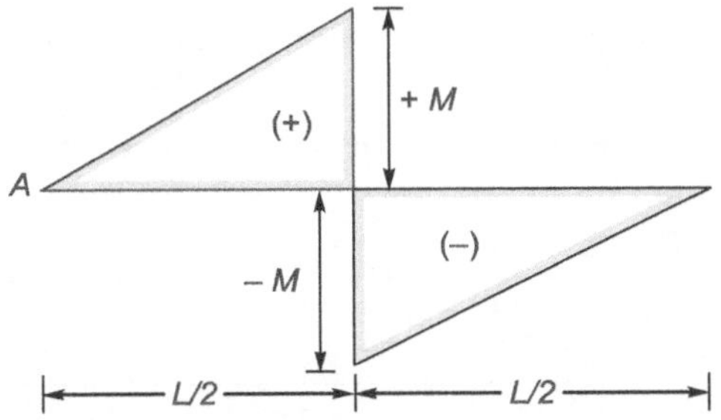

(a) a concentrated load at mid span
(b) a UDL over its length
(c) a couple at its mid length
(d) couple at one fourth of the span from each end

[IES 2006]

Q.25 A cantilever beam of 2 m length supports a triangularly distributed load over its entire length, the maximum of which is at the free end. The total load is 37.5 kN. What is the bending moment at the fixed end

(a) 50×10^6 Nmm (b) 12.5×10^6 Nmm
(c) 100×10^6 Nmm (d) 25×10^6 Nmm

[IES 2007]

Q.26 A UDL *w*(in kN/m) is acting over the entire length of a 3 m long cantilever beam. The shear force at mid point of cantilever is 6 kN, what is the value of *w*?

(a) 2 kN/m (b) 3 kN/m
(c) 4 kN/m (d) 5 kN/m

[IES 2009]

Q.27 An overhang beam *ABC*, is supported at points *A* and *B* as shown in figure. Find the maximum bending moment and the point where it occurs

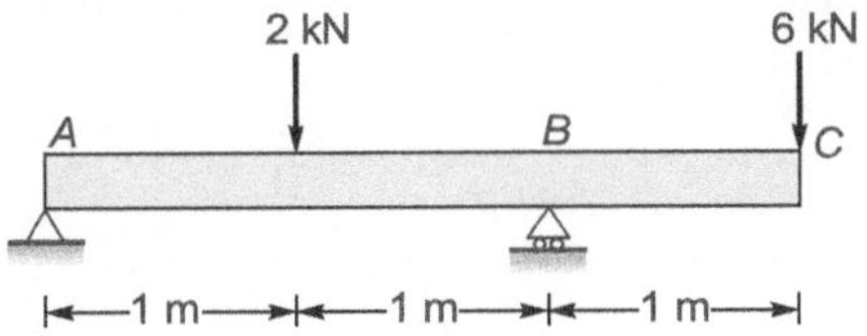

(a) 6 kNm at right support
(b) 6 kNm at left support
(c) 4.5 kNm at right support
(d) 4.5 kNm at mid point between the support

[IES 2009]

Q.28 A freely supported beam at its ends carries a central concentrated load, and maximum bending moment is M, if the same load be uniformly distributed over the beam length, then what is the maximum bending moment

(a) M (b) $\frac{M}{2}$

(c) $\frac{M}{3}$ (d) $2M$

[IES 2009]

Q.29 SFD and BMD for a beam *ABCD* are shown in the figure. It can be concluded that

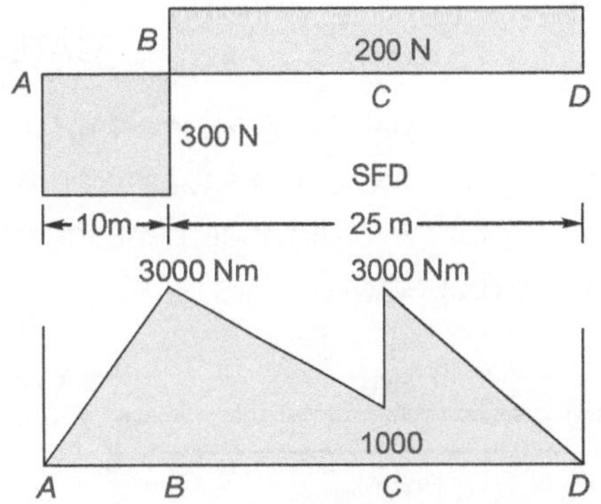

(a) the beam has three supports
(b) end *A* is fixed
(c) A couple 2000 Nm acts at *C*
(d) UDL in confined to portion *BC* ends

Q.30 **Statement (I):** If the bending moment along the length of a beam is constant, then the beam cross-section will not experience any shear stress.

Statement (II): The shear force along the beam will be zero everywhere along the length.

(a) Both statement (I) and statement (II) are individually true and statement (II) is the correct explanation of statement (I)

(b) Both statement (I) and statement (II) are individually true but statement (II) is not the correct explanation of statement (I)
(c) Statement (I) is true but statement (II) is false
(d) Statement (I) is false but statement (II) is true

[IES 2012]

Q.31 A cantilever beam OP is connected to another beam PQ with a pin joint as shown in the figure. A load of 10 kN is applied at the mid-point of PQ. The magnitude of bending moment (in kN-m) at fixed end O is

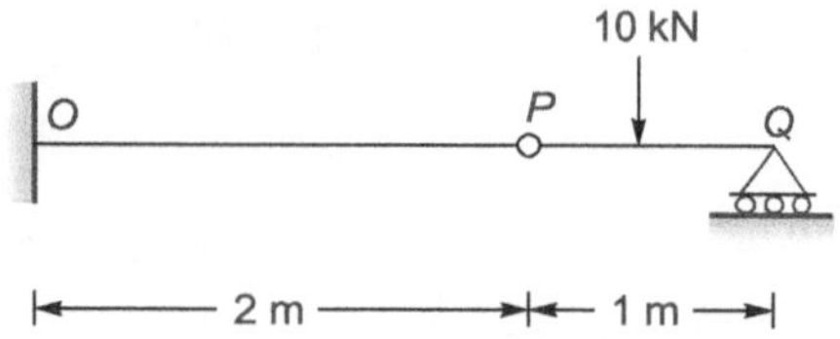

(a) 2.5 (b) 5
(c) 10 (d) 25

[GATE 2015]

Q.32 The values of axial stress (σ) in kN/m^2, bending moment (M) in kNm, and shear force (V) in kN acting at point P for arrangement shown in the figure are respectively.

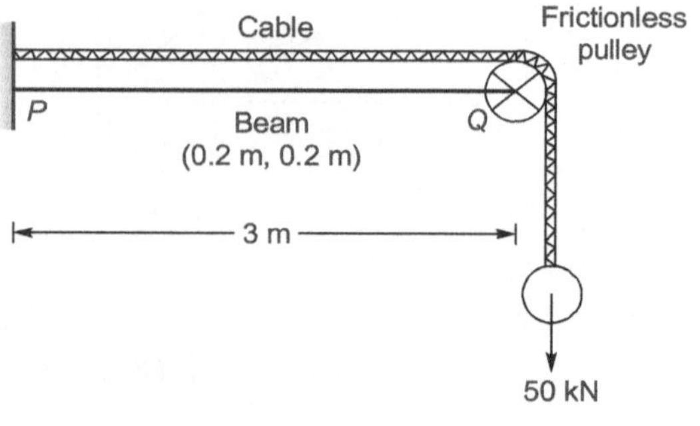

(a) 1000, 75 and 25
(b) 1250, 150 and 50
(c) 1500, 225 and 75
(d) 1750, 300 and 100

[GATE 2016]

Q.33 The bending moment diagram for the beam shown below is

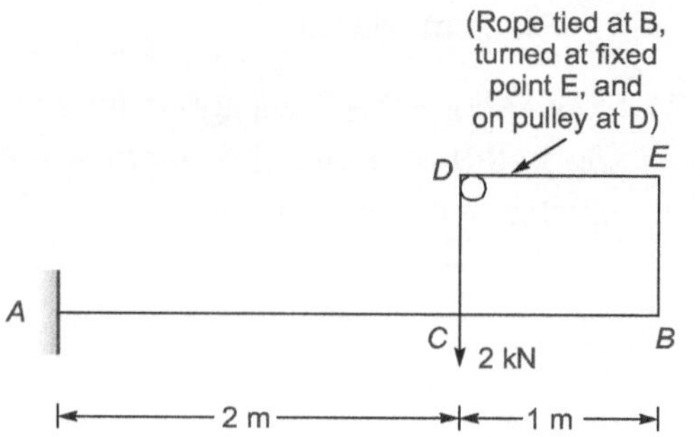

(a)
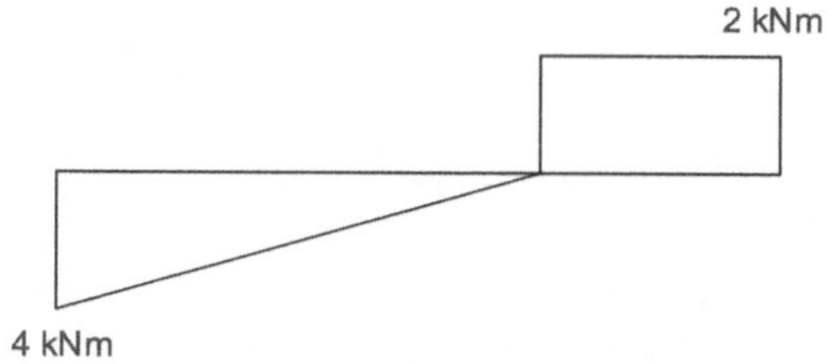

(b)
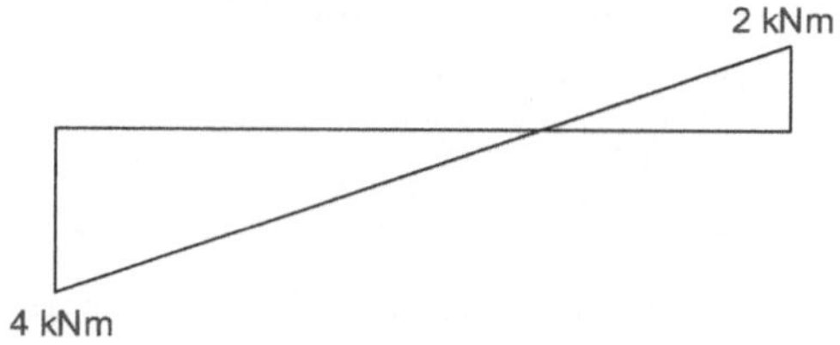

(c)
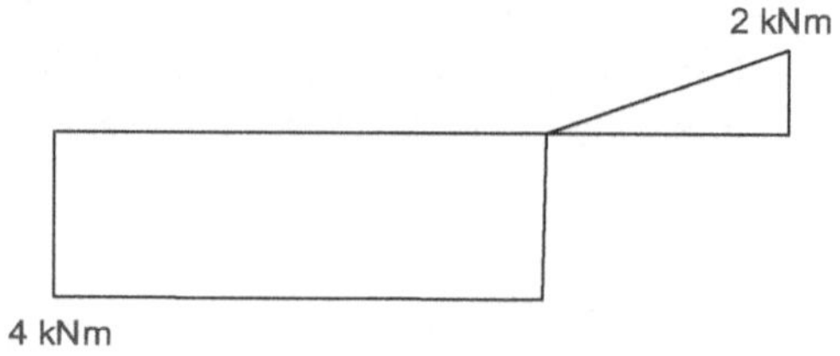

(d)
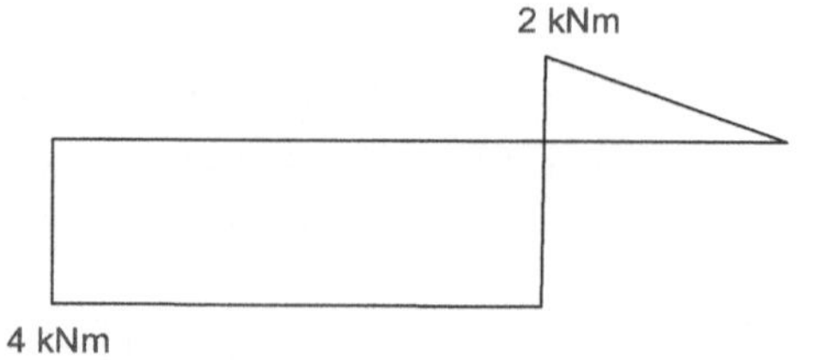

[ESE 2016]

Q.34 Which one of the following is the correct bending moment diagram for a beam which is hinged at the ends and is subjected to a clockwise couple acting at the mid-span?

(a)
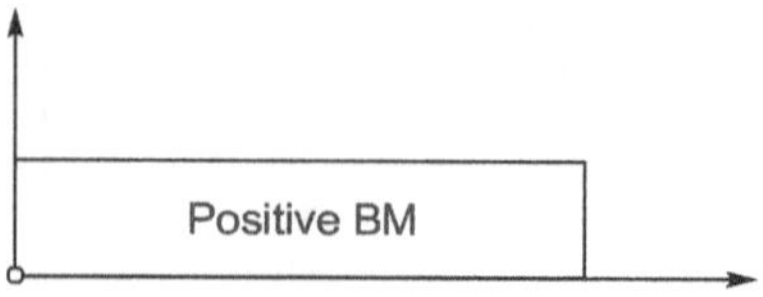

(b) 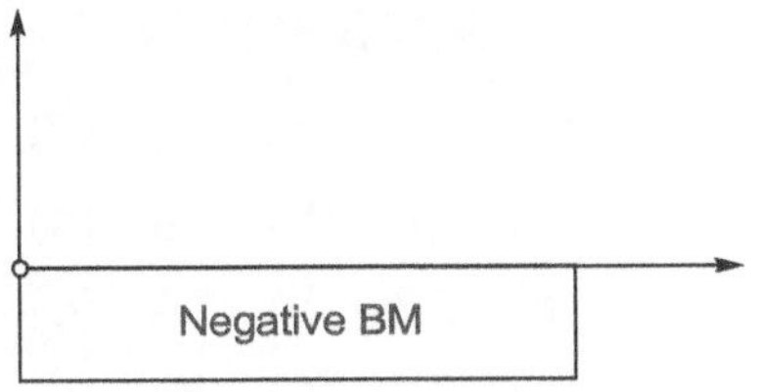

(c) 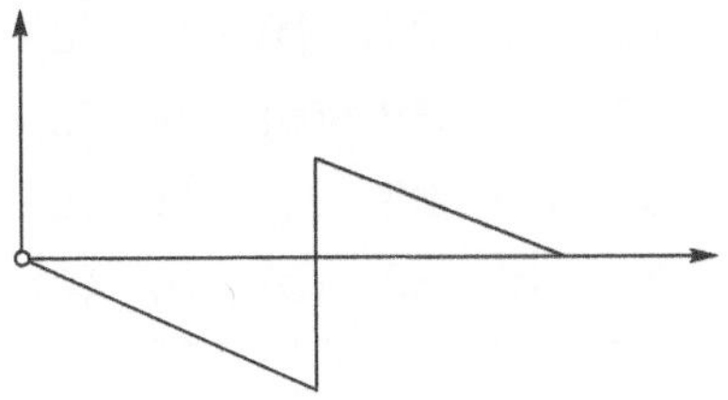

(d) 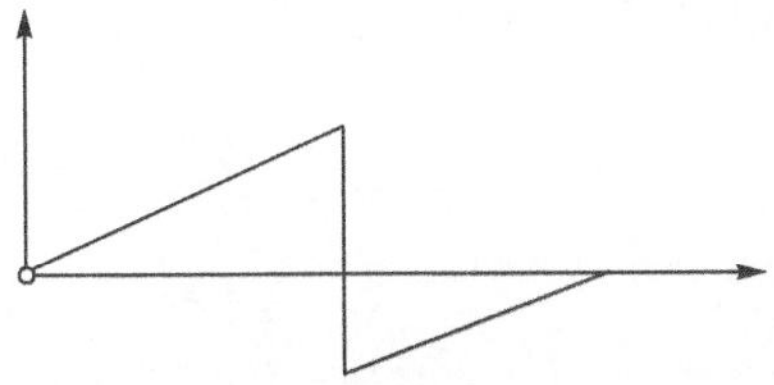

[ESE 2018]

Q.35 The shear forced diagram of a single overhanging beam is shown in figure. One simple support is at end A. The 'total' downward load acting on the beam is

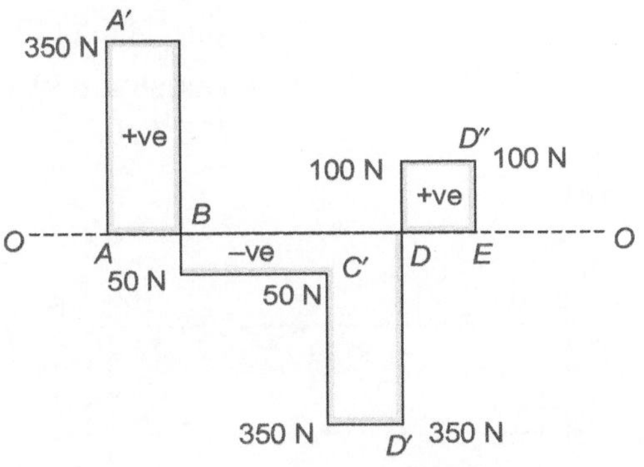

(a) 800 N (b) 600 N
(c) 400 N (d) 200 N

[ESE 2018]

Q.36 Two persons weighting W each are sitting in a plank of length L floating on water, at L/4 from either end. Neglecting the weight of the plank, the bending moment at the middle point of the plank is

(a) $\frac{WL}{16}$ (b) $\frac{WL}{64}$
(c) $\frac{WL}{8}$ (d) Zero

[ESE 2018]

Q.37 Each span of a two-span continuous beam of uniform flexural rigidity is 6 m. All three supports are simple supports. It carries a uniformly distributed load of 20 kN/m over the left span only. The moment at the middle support is
(a) 90 kNm Sagging
(b) 45 kNm Hogging
(c) 90 kNm Hogging
(d) 45 kNm Sagging

[ESE 2018]

Q.38 The bending moment at A for the beam shown below (with BD being a rigid bar) is

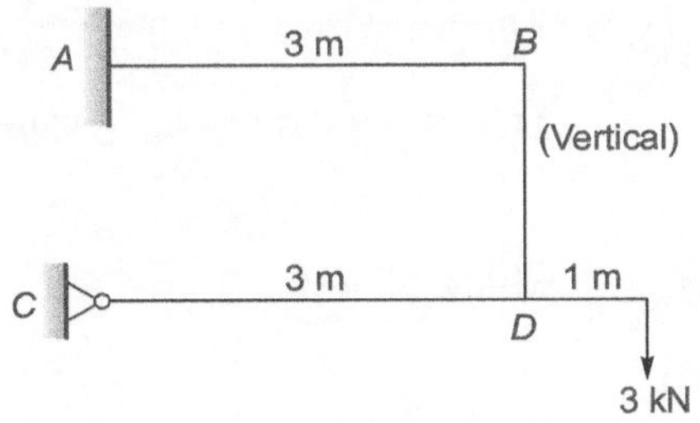

(a) Zero (b) 12 kNm
(c) 8 kNm (d) 6 kNm

[ESE 2016]

Q.39 The bending moment at A for the beam shown below (not to scale) is

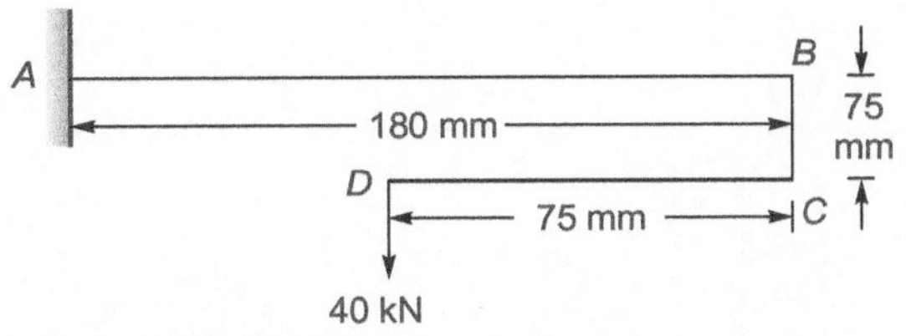

(a) 3200 kNmm (b) 3600 kNmm
(c) 4200 kNmm (d) 4800 kNmm

[ESE 2016]

Q.40 A vertical load of 10 kN acts on a hinge located at a distance of L/4 from the roller support Q of a beam of length L, (see figure)

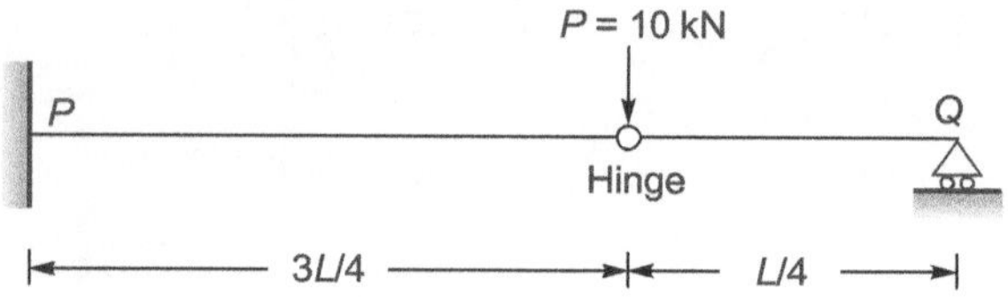

The vertical reaction at support Q is

(a) 0.0 kN (b) 2.5 kN
(c) 7.5 kN (d) 10 kN

Answers

1. (c)	2. (c)	3. (d)	4. (d)	5. (a)
6. (b)	7. (a)	8. (b)	9. (d)	10. (d)
11. (c)	12. (a)	13. (b)	14. (c)	15. (a)
16. (c)	17. (c)	18. (c)	19. (c)	20. (a)
21. (d)	22. (d)	23. (b)	24. (c)	25. (a)
26. (c)	27. (a)	28. (b)	29. (c)	30. (a)
31. (c)	32. (b)	33. (b)	34. (c)	35. (a)
36. (c)	37. (b)	38. (b)	39. (c)	40. (a)

Explanations

1. (c)

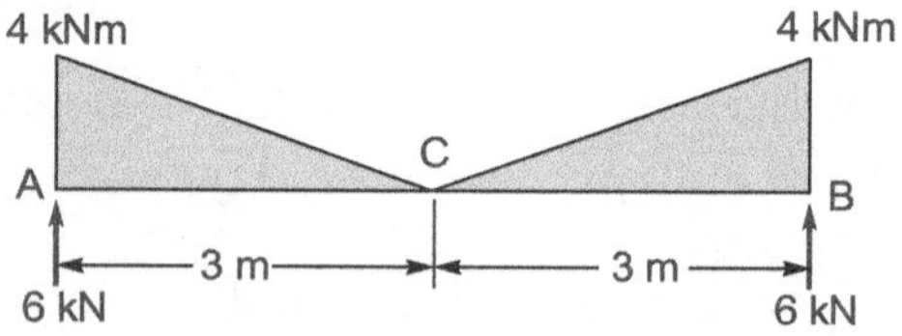

$$M_C = 6 \times 3 - 6 \times 2 = +6 \text{ kNm}$$

2. (c)

M_{max}, where SF = 0

3. (d)

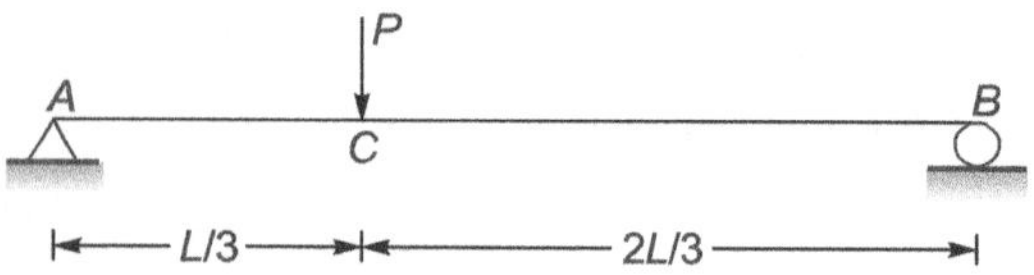

$$M_C = P \times \frac{L}{3} \times \frac{2L}{3} \times \frac{1}{L} = \frac{2PL}{9}$$

4. (d)

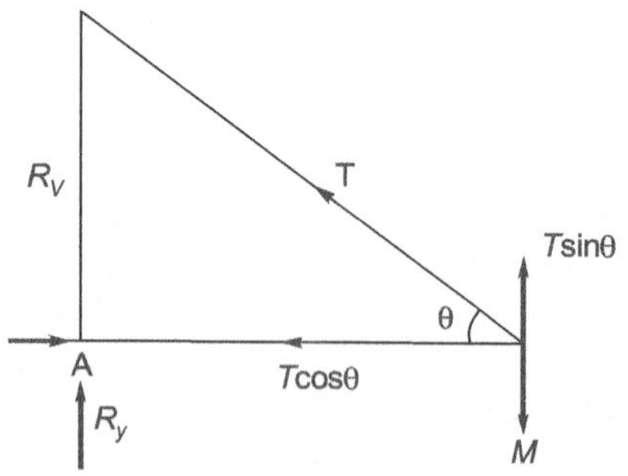

$$\tan\theta = \frac{125}{275}$$

$$\theta = 24.44°$$

$$\sin\theta = 0.4133$$

$$R_y + T\sin\theta = 35 \times 9.81$$

$$R_x = T\cos\theta$$

$$\cos\theta = 0.91$$

Moment about A

$$T\sin\theta \times L = 35 \times 9.81 \times L$$

$$T\sin\theta = 343.35 \text{ N}$$

$$T = 830 \text{ N}$$

$$R_y = 343.35 - 343.35 = 0$$

$$R_x = T\cos\theta = 830 \times 0.91$$

$$= 755.4 \text{ N}$$

5. (a)

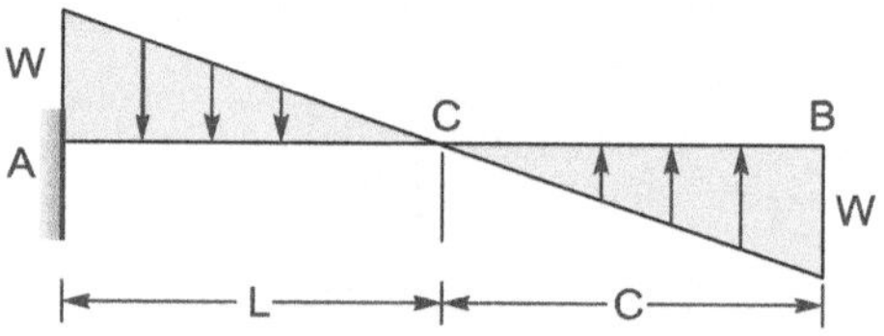

$$M_C = +\frac{wl}{2}\left(\frac{2l}{3}\right) = \frac{wl^2}{3}$$

$$M_A = \frac{wl}{2}\left(l + \frac{2l}{3}\right) - \frac{wl}{2}\left(\frac{l}{3}\right)$$

$$= \frac{5wl^2}{6} - \frac{wl^2}{6} = \frac{2}{3}wl^2$$

$$\frac{M_A}{M_C} = +2$$

6. (b)

$$w = \sin\left(\frac{3\pi x}{L}\right) \text{ N/m}$$

$$\text{Total load} = \int_0^L \sin\left(\frac{3\pi x}{L}\right)dx$$

$$= \left| -\cos\left(\frac{3\pi x}{L}\right) \times \frac{L}{3\pi}\right|_0^L$$

$$= -\frac{L}{3\pi}\left[\cos\frac{3\pi x}{L}\right]_0^L = -\frac{L}{3\pi}[-1-1] = \frac{2L}{3\pi}$$

$$\text{Reactions} = \frac{2L}{3\pi} \times \frac{1}{2} = \frac{L}{3\pi}$$

7. (a)
BM is zero at hinged end *B*.

8. (b)
1, 2 are correct statement.

$$\frac{dw}{dx} \neq y$$

9. (d)

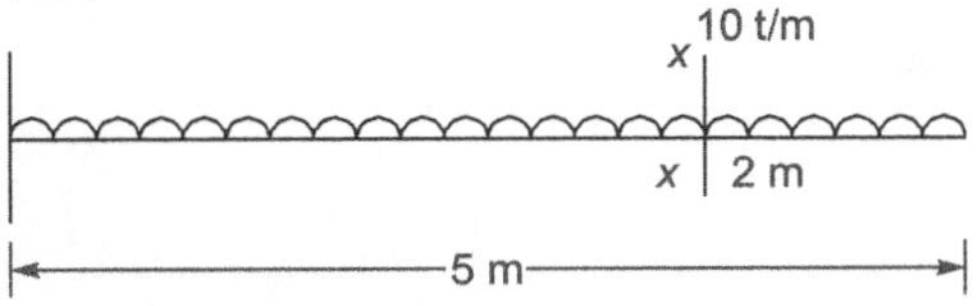

Section XX

$V = 20$ T

$M = 20 \times 1 = 20$ T-m

$M_{max} = 50 \times 2.5 = 125$ T-m

$V_{max} = 50$ T

10. (d)

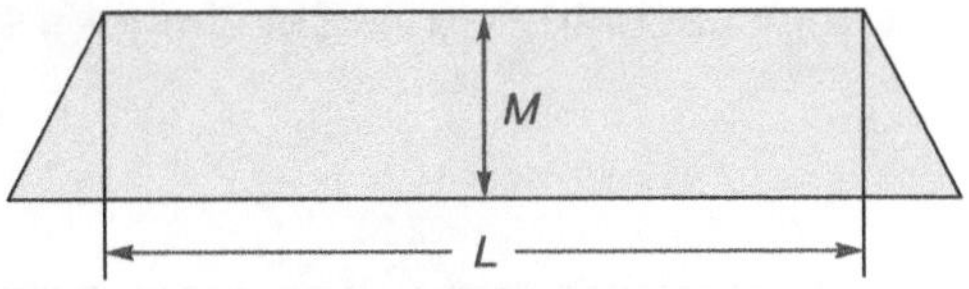

M constant shearing force is zero.

11. (c)

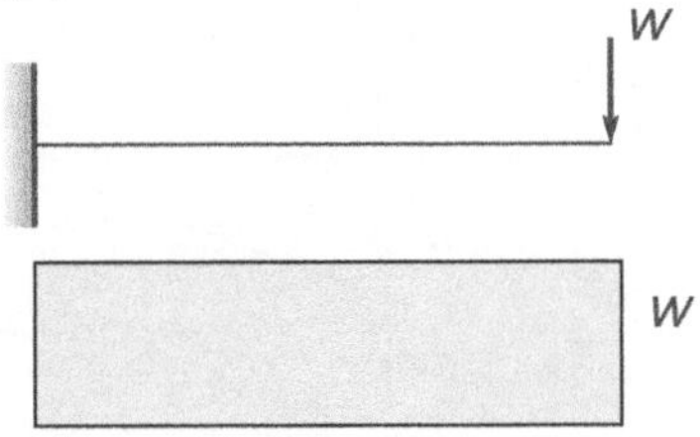

SFD

12. (a)
BM is a nonzero constant.

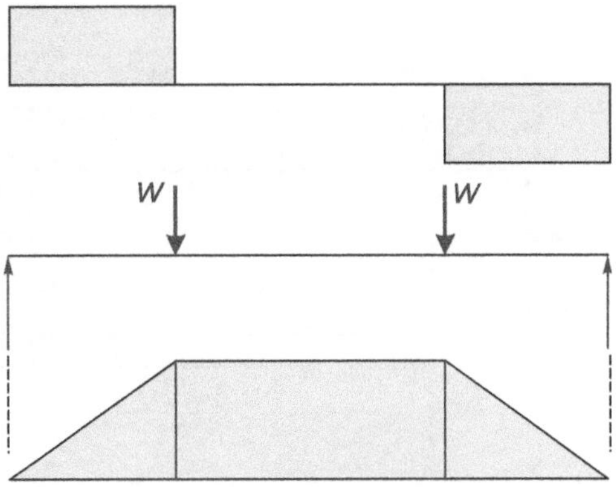

BM

13. (b)

$$\text{BM at support}, A = \frac{PL}{2}$$

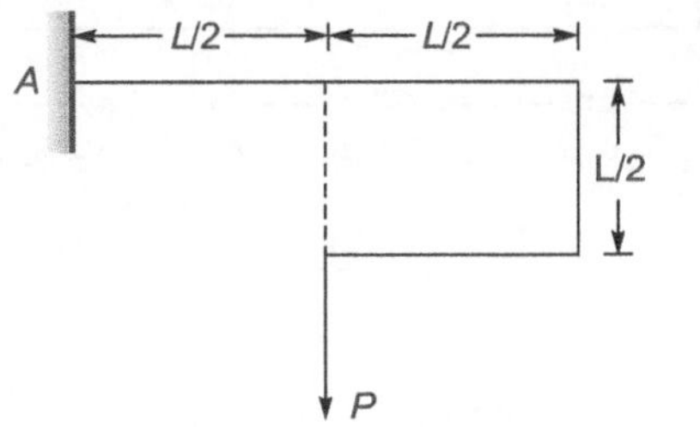

14. (c)

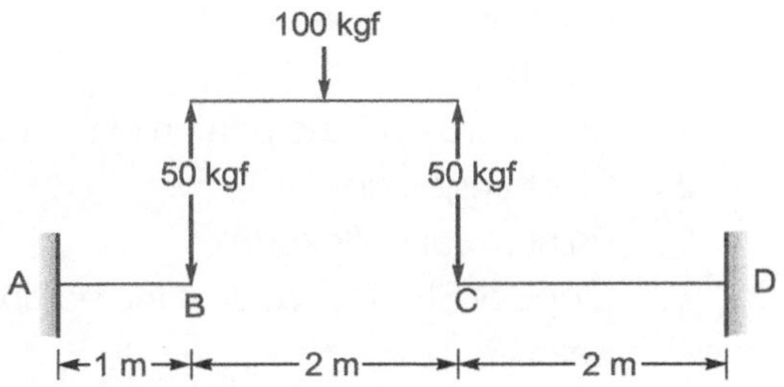

15. (a)

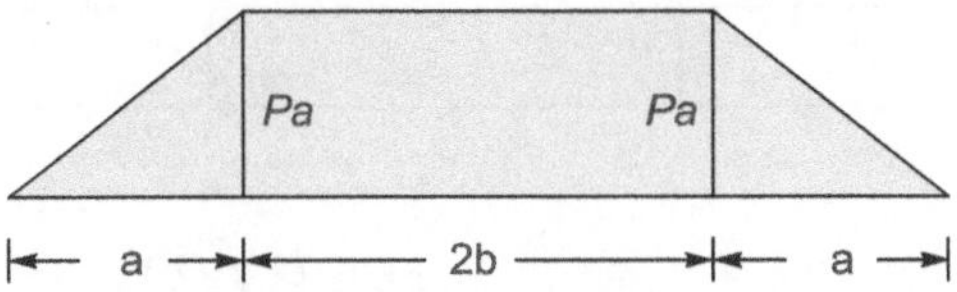

$$\text{Constant BM}, \frac{M}{I} = \frac{E}{R}$$

Circular elastic curve.

16. (c)

Maximum SF = $2W$

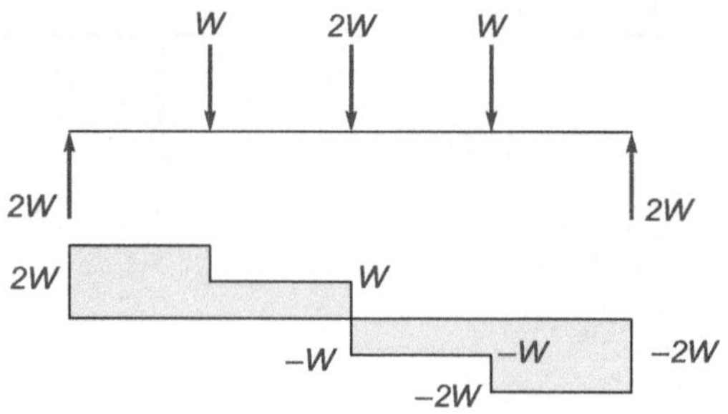

17. (c)

$M_B = +1$ kNm

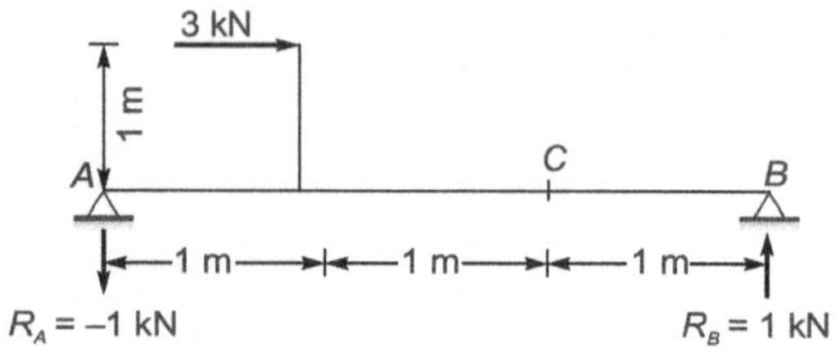

18. (c)

List-I

A. BM is constant

B. BM is maximum or minimum

C. BM is zero

D. Loading is constant

List-II

4. SF is zero over the portion of the beam

2. SF changes sign

1. Point of contraflexure

3. Slope of SFD is zero over the portion of the beam

19. (c)

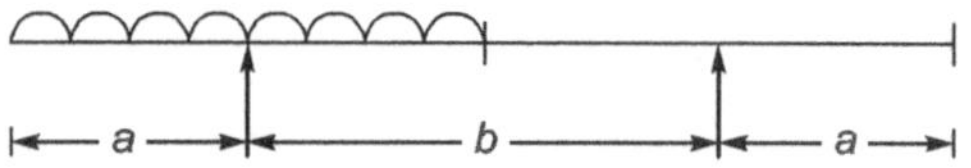

$$\frac{w}{2}\left(a+\frac{b}{2}\right)^2 = w\left(a+\frac{b}{2}\right)\left(\frac{b}{2}\right)$$

$$\left(a+\frac{b}{2}\right) = 2\times\frac{b}{2} = b$$

$$a = b - \frac{b}{2} = \frac{b}{2}$$

$$\frac{a}{b} = \frac{1}{2}$$

20. (a)

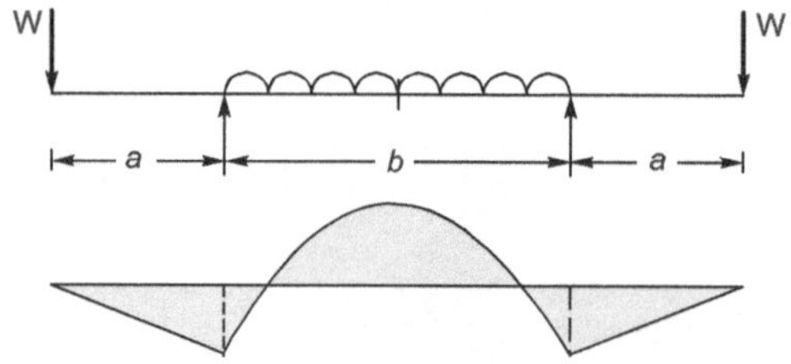

BM diagram

21. (d)

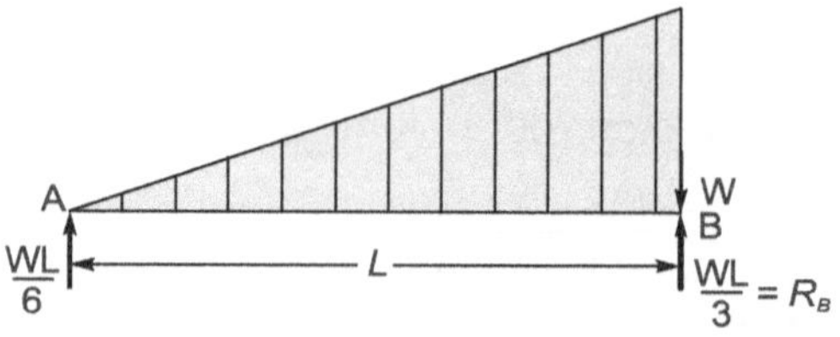

22. (d)

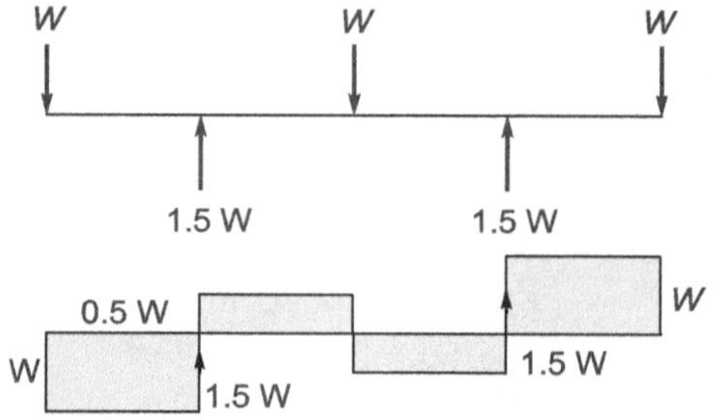

23. (b)

At point of contraflexure BM changes sign.

24. (c)

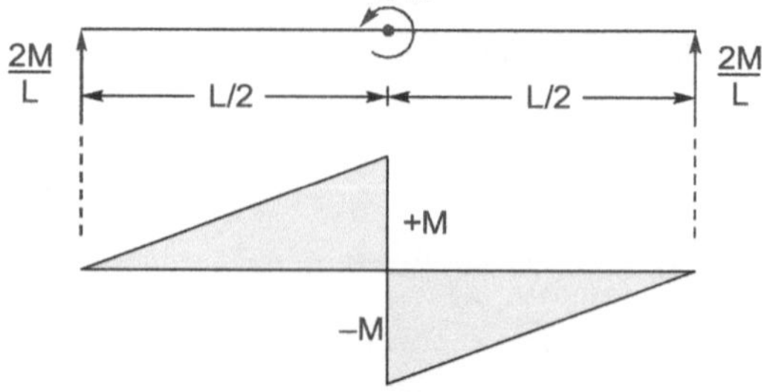

BMD

25. (a)

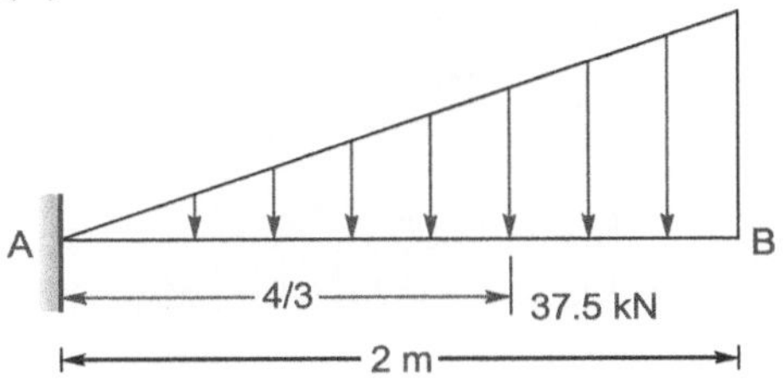

BM at $A = 37.5 \times \frac{4}{3} = 50$ kNm

$= 50 \times 10^6$ Nmm

26. (c)

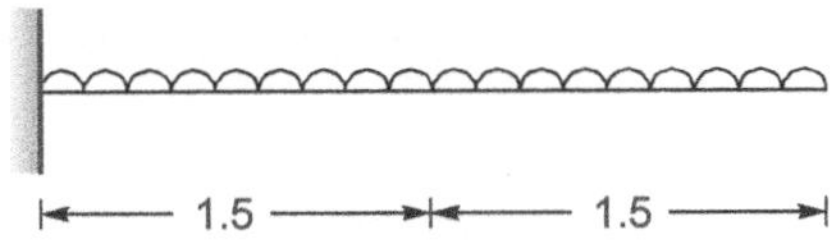

$SF_{1.5} = 6 \text{ kN} = 4 \times 1.5$

$= 4 \times 1.5 - 6$

$w = 4$ kN/m

27. (a)

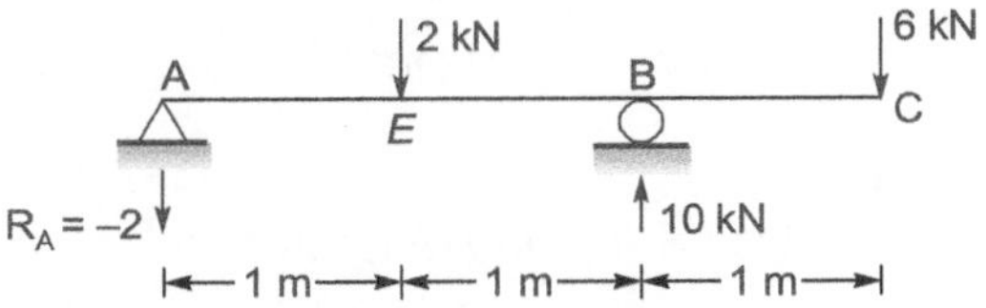

$18 + 2 = \frac{20}{2} = 10 \text{ kN} = R_B$

$M_B = 6$ kNm (max.)

$M_E = -2$ kNm

28. (b)

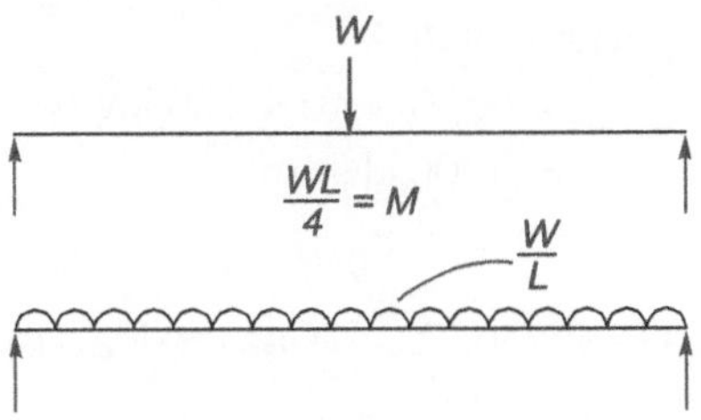

$$M_{max} = \left(\frac{W}{L}\right)\left(\frac{L^2}{8}\right) = \frac{WL}{8} = \frac{1}{2}M$$

29. (c)

In the BM diagram BM changes in step from 3000 Nm to 1000 Nm, so a couple of 2000 Nm acts at point C.

30. (a)

Both statement (I) and statement (II) are individually true and statement (II) is the correct explanation of statement (I).

31. (c)

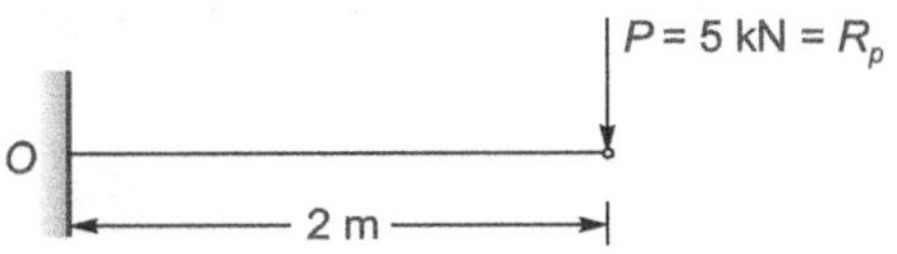

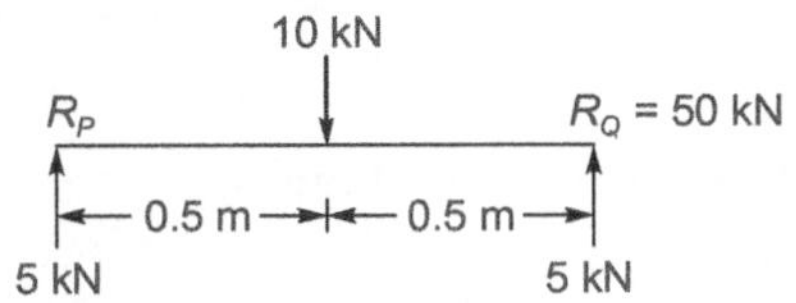

Equivalent diagram.

Bending moment at 0,

$M_0 = 5 \times 2 = 10$ kNm

32. (b)

At point P

Tension in cable = 50 kN

Axes of cross-section of beam

$= 0.2 \times 0.2 = 0.4 \text{ m}^2$

σ, axial stress $= \frac{50}{0.04} = 1250 \text{ kN/m}^2$ (due to cable tension)

Bending moment,

$M = 50 \times 3 = 150$ kNm

Shear force, V = 50 kN

33. (b)

Equivalent loading diagram

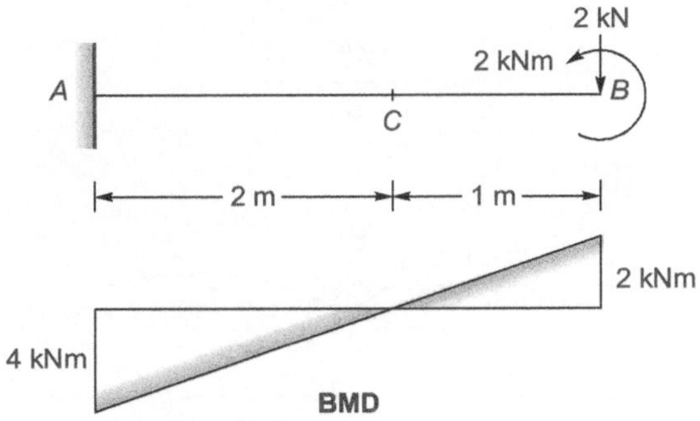

34. (c)

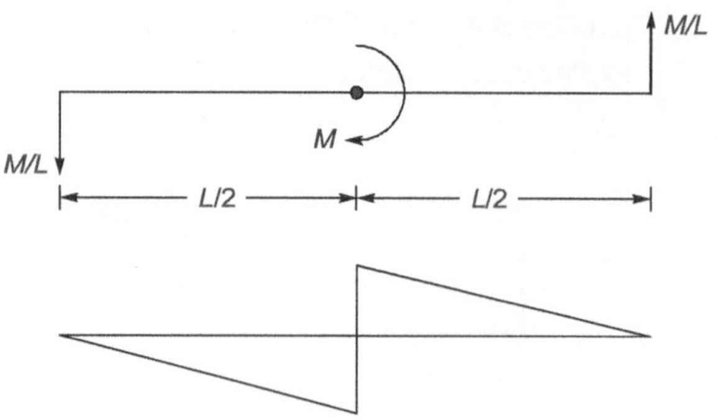

35. (a)

Simple support at end A,

Reaction R_A

Overhang DE

Reaction at D, R_D

$$R_A = 350 \text{ kN } (= AA')$$

$$R_D = 450 \text{ kN } (= D'\,DD'')$$

Total downward load

$$= 350 + 450 = 800 \text{ N}$$

36. (c)

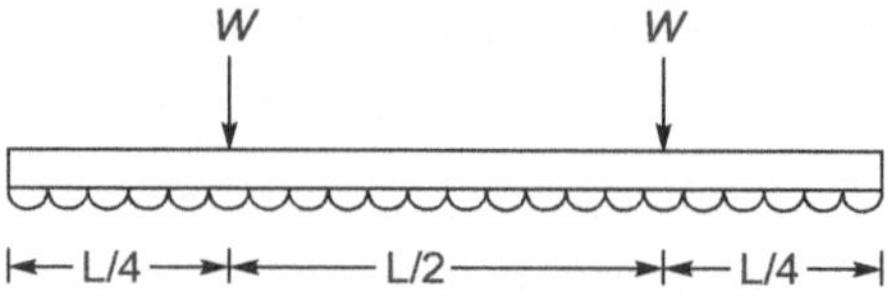

$$w = \frac{W}{L}, \text{ upward threat}$$

$$M = -\frac{wL^2}{8} + W\left(\frac{L}{2} - \frac{L}{4}\right)$$

$$= -\frac{W}{L} \times \frac{L^2}{8} + \frac{WL}{4}$$

$$= -\frac{WL}{8} + \frac{WL}{4}$$

$$= +\frac{WL}{8}$$

37. (b)

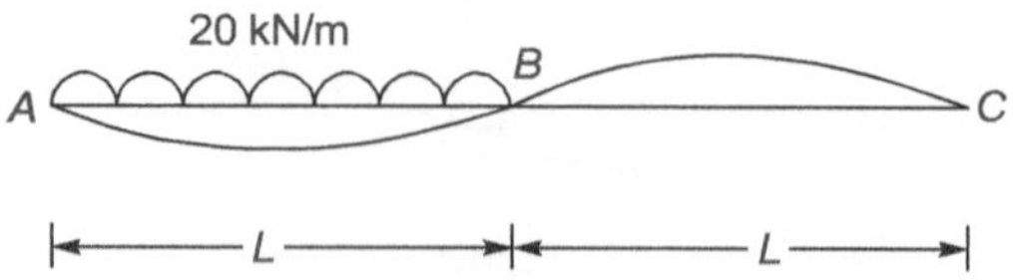

$$M_A = M_C = 0$$

$$4M_B L = \frac{wL^3}{4} = \frac{20 \times 6^3}{4} + 0$$

$$24M_B = 1080$$

$$M_B = 45 \text{ kNm (Hogging)}$$

38. (b)

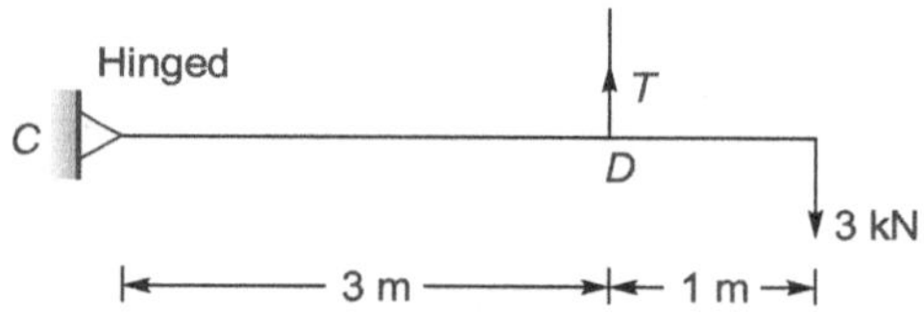

Moment about $C = 0$ or

$$3T = 4 \times 3 = 12 \text{ kNm}$$

$$T = 4 \text{ kN}$$

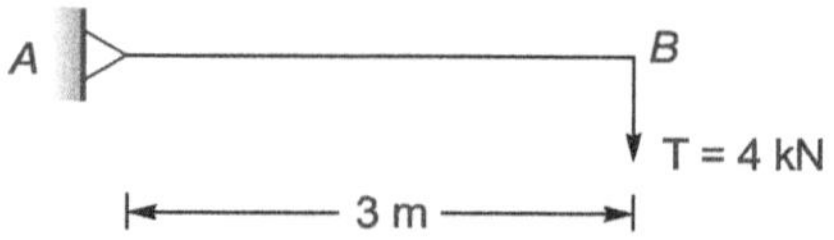

$$M_A = 12 \text{ kNm}$$

39. (c)

Equivalent loading diagram

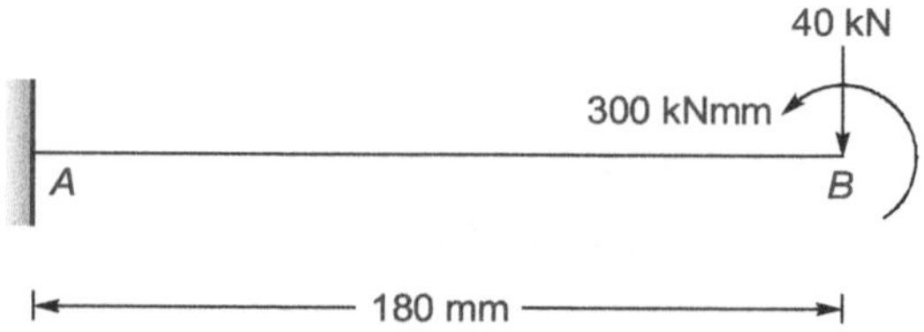

Bending moment at A

$$= -300 + 40 \times 180 \text{ kNm}$$

$$= 4200 \text{ kNmm}$$

40. (a)

Support reaction, $R_Q = 0$, as the load is applied on hinge, roller support becomes unstable and can not carry any load. Load is carried fully by portion of $\frac{3}{4}L$.

Reaction, $R_Q = 0$

■■■■

Theory of Simple Bending

CHAPTER 8

In this chapter, a relationship between bending moment at a particular section, second moment of area of the section and the direct stress due to bending will be developed. The stress is not constant over the section but varies along the depth of the section. Moreover in a section, there is a neutral axis, on which there is no stress, no strain. For the efficient design of a beam section, it is necessary to determine the maximum stress developed in the section due to a particular moment.

Question 8.1 **Consider an element of a beam subjected to a bending moment *M*, and derive expressions (a) between bending moment and stress (b) between bending moment and radius of curvature. Explain the assumptions taken to develop the relationship.**

Solution: Following assumptions are taken to derive relationship between bending moment, stress and radius of curvature.

1. Beam is initially straight, before the application of bending moment.
2. Material of the beam is homogeneous i.e., it possesses same physical properties throughout its volume.
3. Material of the beam is isotropic, i.e., it possesses same elastic constants throughout its volume in all directions.
4. Young's modulus of material in tension, E_t is same as Young's modulus in compression, E_c.
5. Stress developed in any layer does not exceed the elastic limit of the material.
6. Each layer of the beam is independent to extend or to contract irrespective of the layers above or below this layer.
7. Beam section is symmetrical about the plane of bending as shown in figure 8.1.
8. Transverse sections (i.e., perpendicular to axis of the beam) which are plane before bending remain plane after bending. The direction of the plane is changed. In other words there is no distortion of transverse section, as shown in figure 8.2

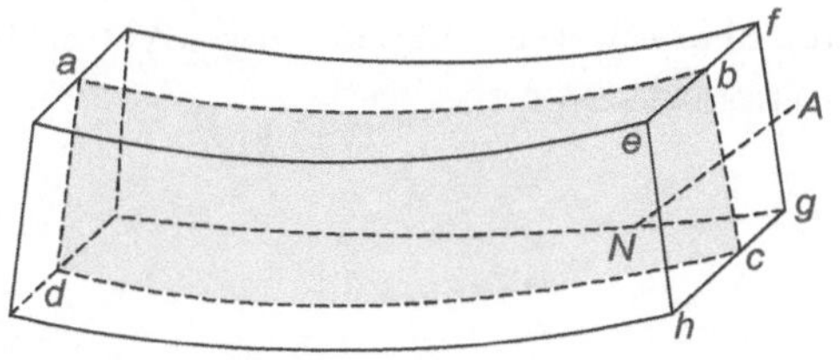

Fig. 8.1 *abcd-plane of bending, section efgh section symmetrical about plane of bending*

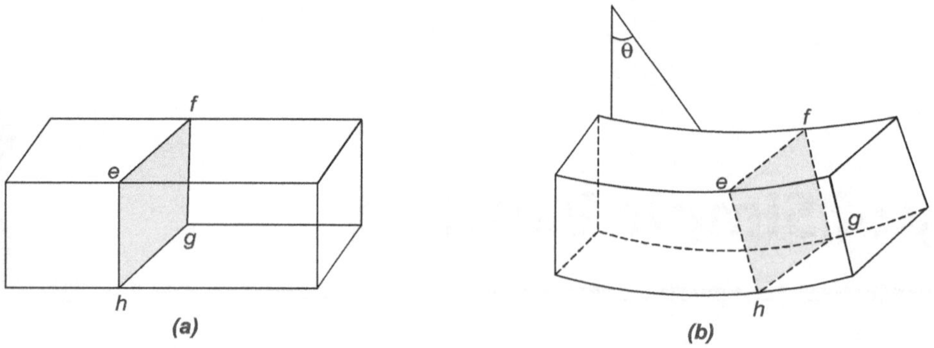

Fig. 8.2 *(a) Plane section efgh before bending, (b) Plane section efgh changes direction after bending.*

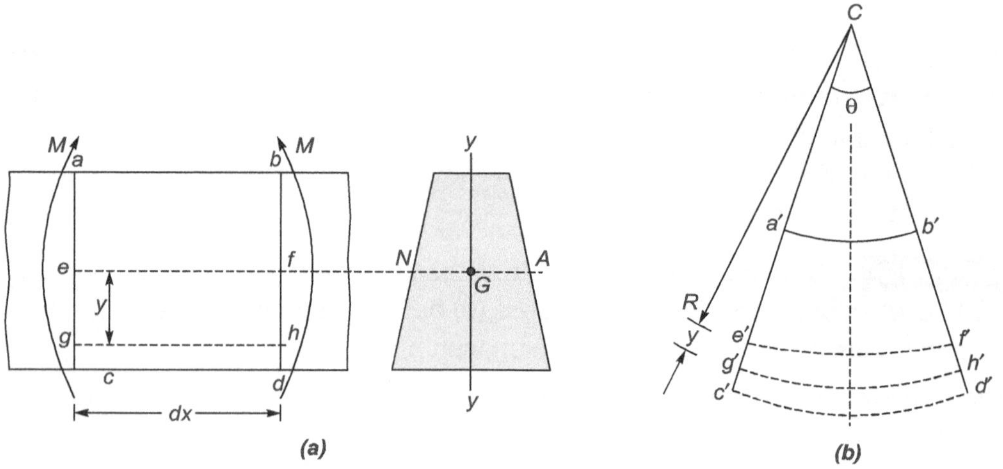

Fig. 8.3 *(a) ef - centroidal layer gh-layer at a radial distance y from centroidal layer, NA-Neutral Axis (b) Element abcd changes of a′b′c′d′*

Consider a small length *dx* of the beam along the axis of the beam or element *abcd*. After the application of the bending moment, element *abcd* changes to *a′b′c′d′* i.e.,

$$a'b' < ab \text{ (negative strain, compressive stress)}$$

$$c'd' > cd \text{ (positive strain, tensile stress)}$$

This shows that there is gradual change in strain (at *ab*) from maximum negative strain to maximum positive strain (at *cd*). There has to be a layer *ef* where there is no strain, i.e., $e'f' = ef$. **This layer is known as neutral layer**. In the section, this layer is NA, (neutral axis). Say R - radius of curvature from centre of curvature *C* to neutral layer *ef*.

$$e'f' = ef = R\theta \text{ (at neutral layer)}$$

$$g'h' = (R + y)\theta, \text{ but } gh = ef = R\theta$$

$$\text{Strain in layer } gh,\ \varepsilon = \frac{g'h' - gh}{gh} = \frac{(R+y)\theta - R\theta}{R\theta}$$

$$\text{Strain, } \varepsilon = \frac{y}{R} = \frac{\text{distance from neutral layer}}{\text{radius of curvature}}$$

Using Hooke's law, stress $\propto$ strain(stress is proportional to strain)

$$\text{Stress, } \sigma = E.\varepsilon$$

$$\sigma = \frac{Ey}{R}$$

or $$\frac{\sigma}{y} = \frac{E}{R}$$

Stress in layer *GH*, $\sigma = \frac{yE}{R}$, at distance *y* from neutral axis

where *E* and *R* are constants.

$\sigma \propto y$ (stress is proportional to y-distance from neutral layer)

stress in a layer is proportional to its distance from neutral layer.

where *y* is positive, (away from centre of curvature) stress σ is positive (tensile stress)

if *y* is negative (towards the centre of curvature from neutral layer) then σ is negative (compressive stress)

Section is divided into two zones i.e.,

1. Compressive zone - on one side of neutral axis.

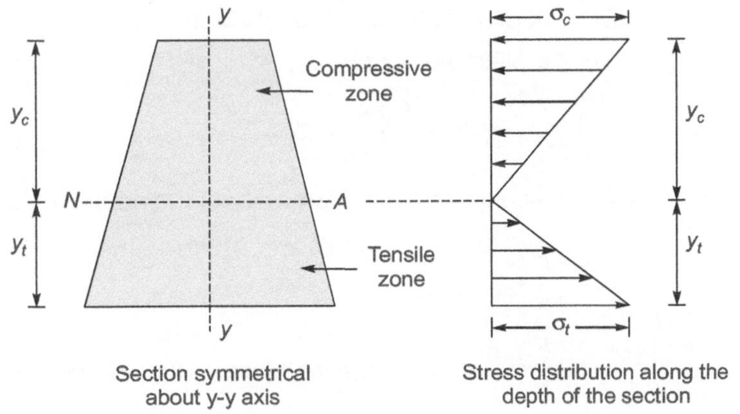

Fig. 8.4

2. Tension zone - on other side of neutral axis

   σ_c = maximum compressive stress

   σ_t = maximum tensile stress

   y_c = distance of extreme layer in compression from neutral layer

   y_t = distance of extreme layer in tension from neutral layer

Neutral Axis:

Consider a layer at a distance *y* from neutral layer, area of layer = *dA*

Stress, $$\sigma = \frac{yE}{R}$$

Force in layer, $$dP = \sigma dA = \frac{yE}{R} dA$$

P, total force on the section $$= \frac{E}{R}\int_{-y_c}^{+y_t} y\,dA$$

Note that we have not applied any force, we have applied only the moment M. **Total force in compression zone will be equal to the total force in tension zone,** to achieve equilibrium.

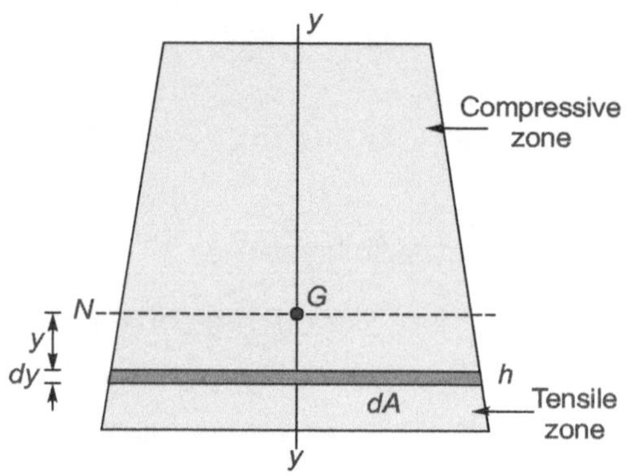

Fig. 8.5

So, $P = 0 = \frac{y}{E}\int_{-yc}^{yt} ydA = 0$

or $\int_{-yc}^{yt} ydA = 0$, or first moment of area about neutral axis is zero.

But the first moment of area is zero about centroidal axis. Therefore it proves **that neutral axis passes through the centroid** G **of the section as shown.**

Moment of Resistance

$$dP = \sigma dA = \frac{yE}{R}dA = \text{force on small element of area dA}$$

dM = moment of force on small element about neutral or centroidal axis

Total moment, M = applied moment

$$= \int y^2 \frac{E}{R} dA = \frac{E}{R}\int_{-yc}^{+y_t} y^2 dA$$

But $\int_{-yc}^{+y_t} y^2 dA$ = second moment of area about the neutral axis, NA = I_{NA}

or $$M = \frac{E}{R} I_{NA}$$

or $$\frac{M}{I_{NA}} = \frac{E}{R} = \frac{\sigma}{y},$$ **This is the well known flexure formula of the beam.**

Note that $\frac{M}{I_{NA}} = \frac{E}{R}$ is used only when the radius of curvature remains constant, or the beam element forms the part of a circle of radius R.

Moreover, $$\frac{M}{I_{NA}} = \frac{\sigma_c}{y_c} = \frac{\sigma_t}{\sigma_t}$$

or $$M = \frac{I_{NA}}{y_c} \times \sigma_c = Z_c \times \sigma_c$$

Z_c = section modulus in compression = $\frac{I_{NA}}{y_c}$

Z_t = section modulus in tension = $\frac{I_{NA}}{y_t}$

or $$M = \sigma_c Z_c = \sigma_t Z_t$$

Practice Q.8.1 A copper strip of width 20 mm, thickness 10 mm is bent around a drum of radius 2500 mm. What are the maximum stresses in tension and compression developed in copper strip. E for copper = 105 GPa.

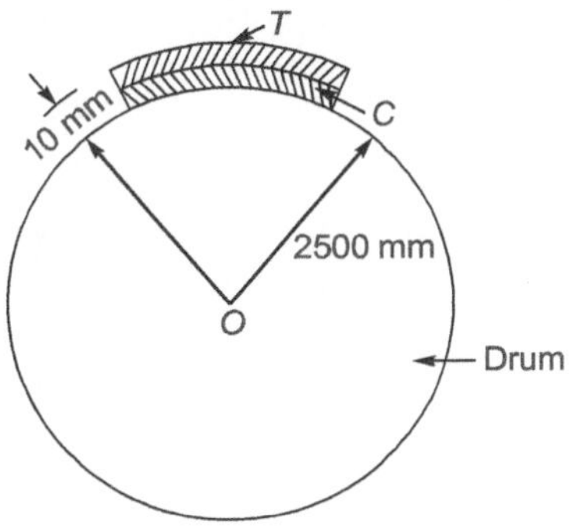

Fig. 8.6

Ans. [±209.6 N/mm^2]

Question 8.2 Tabulate the section modulii of important structural sections.

Solution:

Solid circular section

$$Z_1 = Z_2 = \frac{\pi d^3}{32}$$

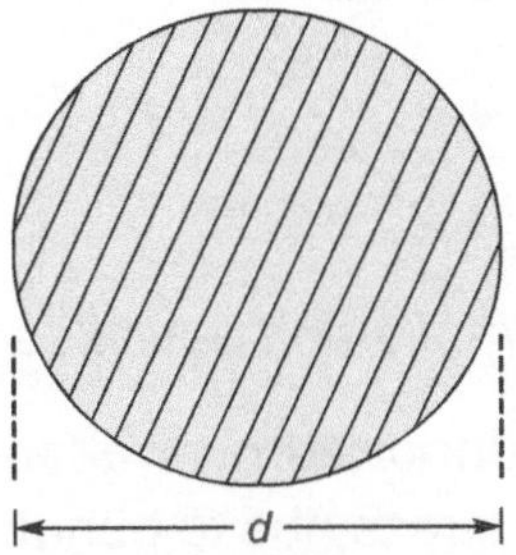

Fig. *Solid circular section*

Rectangular section

$$Z_1 = Z_2 = \frac{bd^2}{6}$$

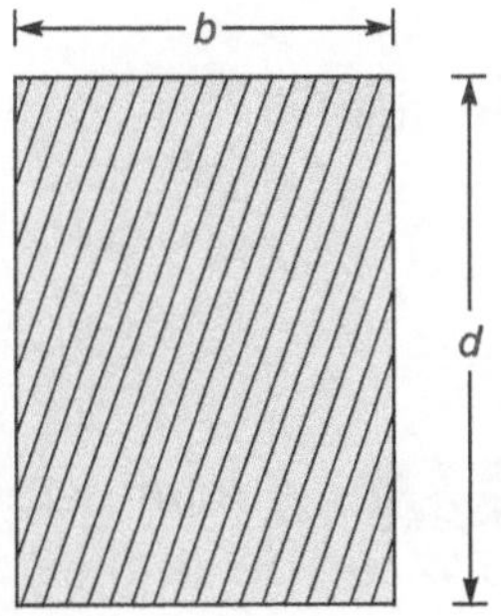

Fig. *Rectangular section*

Hollow circular section

$$Z_1 = Z_2 = \frac{\pi}{32D}\left(D^4 - d^4\right)$$

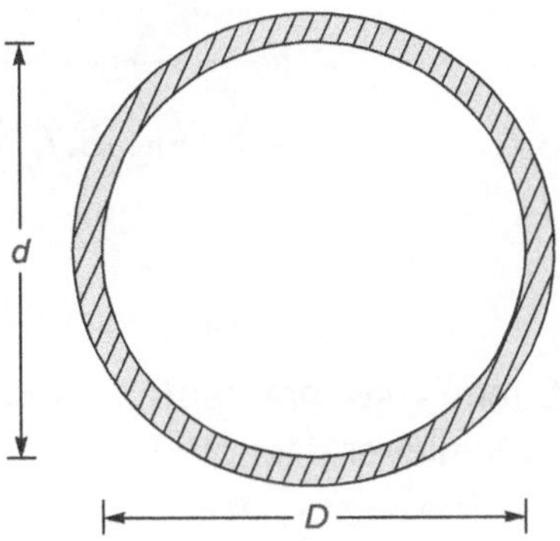

Fig. *Hollow Circular Section*

Hollow square section

$$Z_1 = Z_2 = \frac{\left(A^4 - a^4\right)}{6A}$$

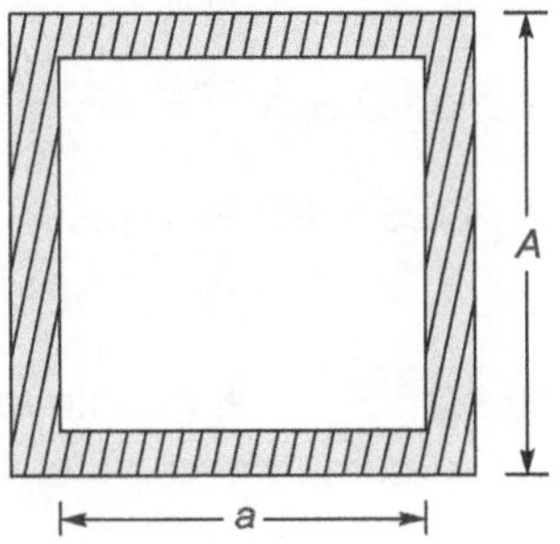

Fig. *Hollow square section*

Symmetrical I-section

$$Z_1 = Z_2 = \frac{1}{6D}\left[BD^3 - (B-b)d^3\right]$$

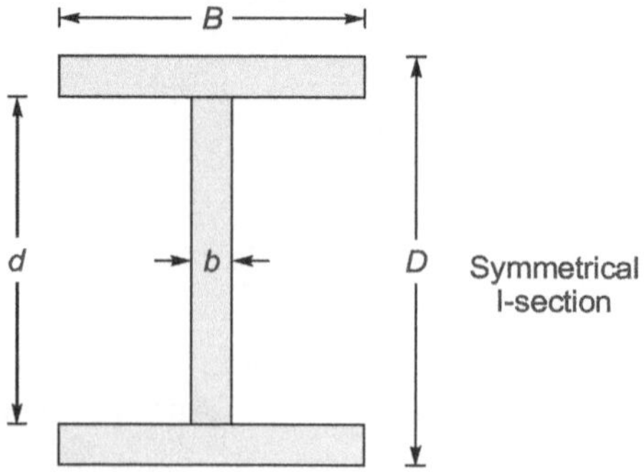

Fig. *Symmetrical I-section*

Triangular section

$$Z_1 = \frac{bh^2}{12},\ Z_2 = \frac{bh^2}{24}$$

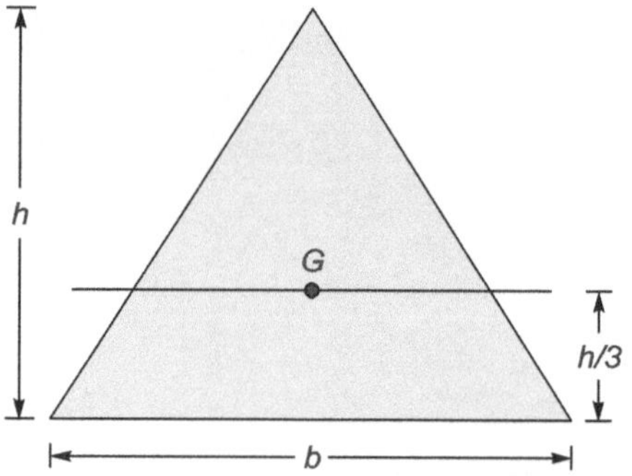

Fig. *Triangular section*

Diagonals horizontal and vertical

(Square section) $$Z_1 = Z_2 = \frac{a^3}{6\sqrt{2}}$$

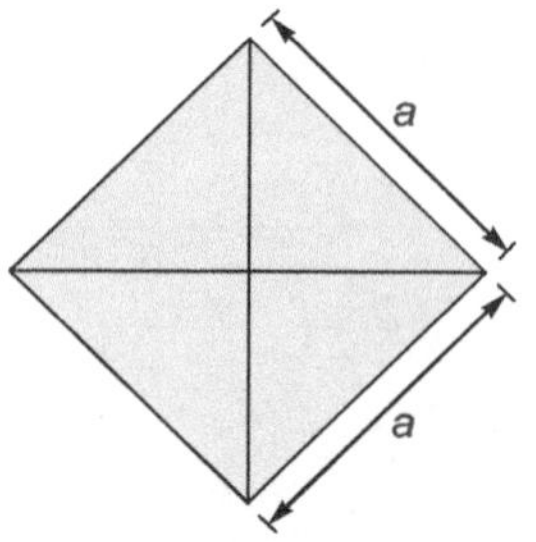

Fig. *Diagonals horizontal and vertical section*

Question 8.3 A round steel beam 50 mm diameter is supported over a length of 1 m. How much central load W can be applied on the beam, so that maximum stress due to bending in beam is not exceed 120 MPa.

Solution: Maximum bending moment occurs at the centre,

$$M_{max} = \frac{WL}{4} = \frac{W \times 1}{4} = 0.25\text{ W kNm}$$

(Note that W is taken in kN)

$$= 0.25 \times 10^6\ W\text{ Nmm}$$

$$= \sigma Z$$

Section modulus, $$Z = \frac{\pi \times 50^3}{32} = 12.27 \times 10^3\ \text{mm}^3$$

$$0.25 \times 10^6\ W = 120 \times 12.27 \times 10^3$$

$$W = 5.89\text{ kN}$$

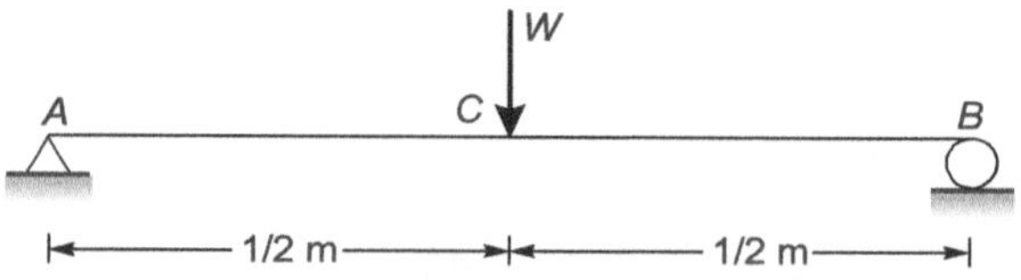

Fig. 8.7

Practice Q.8.3 A cantilever of square section 40 mm × 40 mm carries a load of 1 kN at free end. How much maximum stress is developed in section of cantilever if length is 1 m.

Hint: (maximum bending moment occurs at fixed end]

Ans. [± 93.75 N/mm^2]

Question 8.4 A beam is of I-section with flanges 200 × 20 mm and web 160 × 10 mm, so that overall depth of section is 200 mm, what is section modulus of the I-section. How much bending moment can be safely applied if maximum stress is not to exceed 160 MPa.

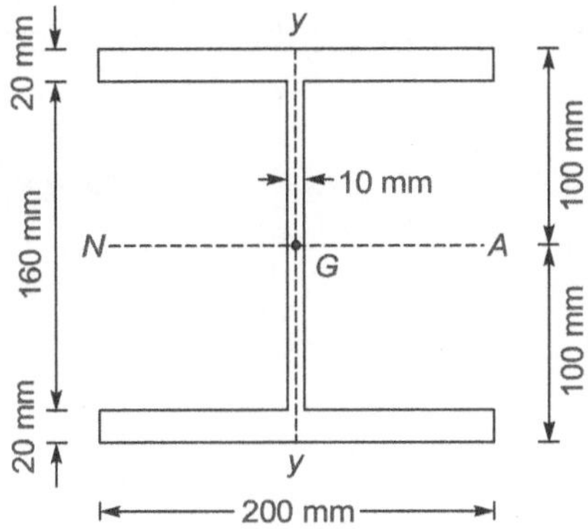

Fig. 8.8

Solution: Fig. 8.8 shows I-section, symmetrical about yy-axis and Neutral axis passes through centroid G such that $y_c = y_t = 100$ mm.

Moment of inertial, $$I_{NA} = \frac{200 \times 200^3}{12} - \frac{190 \times 160^3}{12}$$

$$= 1.333 \times 10^8 - 0.6485 \times 10^8 \text{ mm}^4$$

$$= 0.6844 \times 10^8 \text{ mm}^4$$

$$Z_c = Z_t = \frac{0.6844 \times 10^8}{100}$$

$$= 0.6844 \times 10^6 \text{ mm}^3$$

Safe bending moment, $M = Z \times \sigma$, where σ is safe maximum stress

$$= 0.6844 \times 10^6 \times 160$$

$$= 109.5 \times 10^6 \text{ Nmm} = 109.5 \text{ kNm}$$

Practice Q.8.4 A beam is of I-section given in question 8.4. It is supported over a span of 4 m. How much central load W can be applied so that maximum stress does not exceed 160 MPa.

Hints: $\left[M_{max} = \frac{WL}{4}\right]$ Ans. [$W = 109.5$ kN]

Question 8.5 Explain what are parallel axes and perpendicular axes theorems?

Solution: In engineering mechanics, a student learns about these theorems. However for revision, the final expressions are reproduced here. Fig. 8.10 shows a plane lamina of area A, G is the centroid of the area and Gx and Gy are centroidal axes passing through G of the area

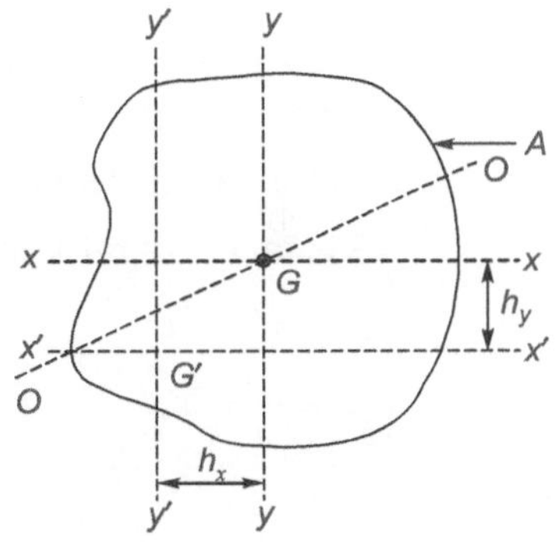

Fig. 8.10 *Gxy-centroidal axis*

I_{xx} = moment of inertia of section about x-x axis

I_{yy} = moment of inertial of section about y-y axis

Polar moment of inertia

$$I_{00} = I_{xx} + I_{yy} \quad \text{...(i)}$$

This is known as perpendicular axes theorem.

Axis $y'y'$ is parallel to yy axis, at a distance of h_x as shown.

Moment of inertia, $$I_{y'y'} = I_{yy} + A.h_x^2 \quad \text{...(ii)}$$

Similarly $x'x'$ is parallel to xx -axis at a distance of h_y as shown.

Moment of inertia, $$I_{x'x'} = I_{xx} + Ah_y^2 \quad \text{...(iii)}$$

This is known as parallel axes theorem i.e., if moment of inertial about a centroidal axis is known, then moment of inertia about a parallel axis can be calculated by using the parallel axes theorem.

Question 8.6 A T-section with flange 150 × 20 mm and web 180 mm × 10 mm is shown in figure 8.11. Locate the centroid of the section. Calculate I_{NA}, moment of inertia about x-x centroidal axis. Determine also the section modulii of this T-section.

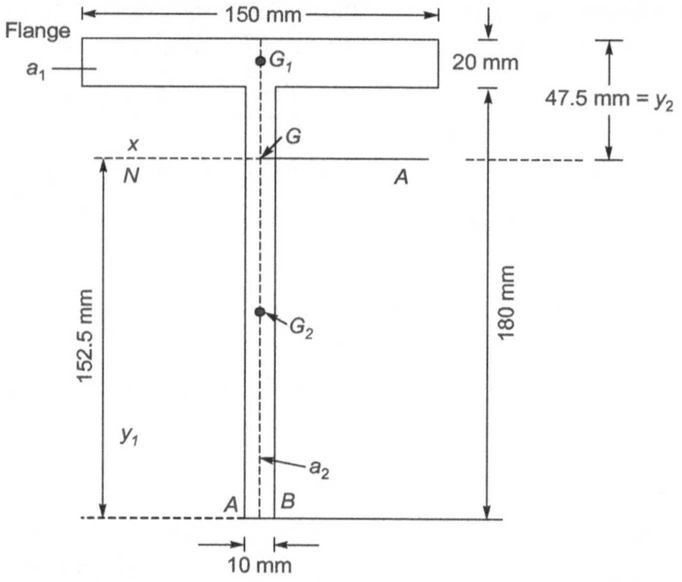

Fig. 8.11

Solution:

Flange area, $a_1 = 150 \times 20 = 3000 \text{ mm}^2$

Web area, $a_2 = 180 \times 10 = 1800 \text{ mm}^2$

G_1 of a_1 lies at 180 + 10 = 190 mm from AB, and G_2 of a_2 lies at 90 mm from edge AB

AB is taken as reference axis.

y_1 = distance of centroid of T-section from AB

$$= \frac{3000\times190+1800\times90}{3000+1800} = \frac{57\times10^4+16.2\times10^4}{4800}$$

$$= \frac{73.2\times10^4}{4800} = 152.5 \text{ mm}$$

$$y_2 = 200 - 152.5 = 47.5 \text{ mm}$$

$$I_{NA} = \frac{150\times20^3}{12} + 3000(47.5-10)^2 + \frac{10\times180^3}{12} + 1800(152.5-90)^2$$

Note that $\frac{150\times20^3}{12}$ – moment of inertia of flange about its own centroidal axis

h_1 = 37.5 mm, distance of centroidal axis of flange from NA

Similarly $\frac{10\times180^3}{12}$ = moment of inertia of web about its own centroidal axis.

$h_2 = 152.5 - 90 = 62.5$ mm, distance of centroidal axis of web from NA

$I_{NA} = 10 \times 10^4 + 421.875 \times 10^4 + 486 \times 10^4 + 703.125 \times 10^4 \text{ mm}^4$

$= 1621 \times 10^4 \text{ mm}^4$

Section modulus, $Z_1 = \frac{I_{NA}}{y_1} = \frac{1621 \times 10^4}{152.5} = 106.295 \times 10^3 \text{ mm}^3$

$Z_2 = \frac{I_{NA}}{y_2} = \frac{1621 \times 10^4}{47.5} = 341.263 \times 10^3 \text{ mm}^3$

Practice Q.8.6 **A beam is of T-section as shown in figure 8.11. T-section beam is used as a cantilever with flange at top position. Length of cantilever is 2 m and load at free end is 12 kN. Determine maximum tensile and compressive stresses developed in T-section.**

Hints: Maximum bending moment occurs at fixed section = *WL*, flange will be in tension and lower portion of web will be in compression.

Ans. [$\sigma_t = 70.327$ N/mm², $\sigma_c = 225.85$ N/mm²]

Question 8.7 **Section of a beam is equilateral triangle with side '*a*'. It is supported over a length of 2 m, carries UDL of 2 kN/m. What should be the side '*a*' of section, if maximum stress is not to exceed 80 MPa.**

Solution:

Rate of loading, $w = 2$ kN/m

Length, $L = 2$ m

$M_{max} = \frac{wL^2}{8}$, occurs at centre of the beam

$= \frac{2 \times 2^2}{8} = 1 \text{ kNm} = 1 \times 10^6 \text{ Nmm}$

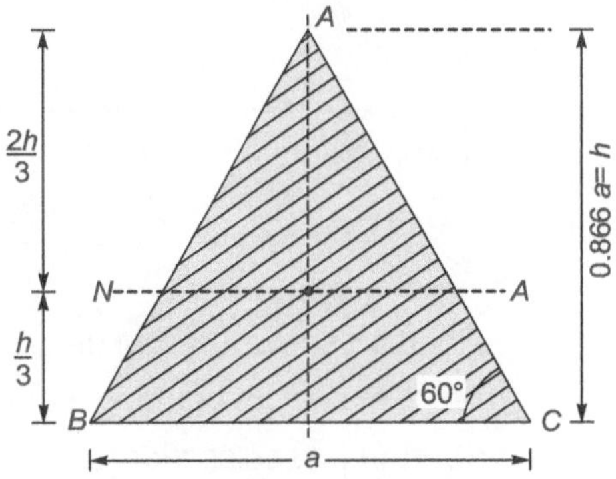

Fig. 8.12

Section: Equilateral triangle, of side *a*

base = *a*

Altitude, $h = 0.866$ a

Moment of inertial, $I_{NA} = \frac{bh^3}{36} = \frac{a \times (0.866a)^3}{36} = 0.0180\, a^4$

Maximum stress will occur at edge *A* as **stress in a layer is linearly proportional to its distance from neutral axis**

$y_1 = \frac{h}{3}, \quad y_2 = \frac{2h}{3}$

$y_2 = \frac{0.866 \times a \times 2}{3} = 0.5773\, a$

Section modulus, $Z_2 = \frac{I_{NA}}{y_2} = \frac{0.0180a^4}{0.5773a} = 0.03118a^3$

σ = maximum stress at *A* = 80 MPa

= 80 N/mm²

$$M_{max} = \sigma \times z$$

$$1 \times 10^6 = 80 \times 0.03118\, a^3$$

$$a^3 = 0.4009 \times 10^6 \text{ mm}^3$$

$$\text{Side } a = 0.7374 \times 10^2 \text{ mm}$$

$$= 73.74 \text{ mm side of equilateral triangle is } 73.74 \text{ mm}$$

Question 8.8 A beam section is isosceles triangle with base 60 mm and altitude 60 mm. It is supported over a span of 6 m. It carries two loads of 2 kN each at a distance of 1 m from each end. What are skin stresses produced in the section at maximum bending moment.

Solution:

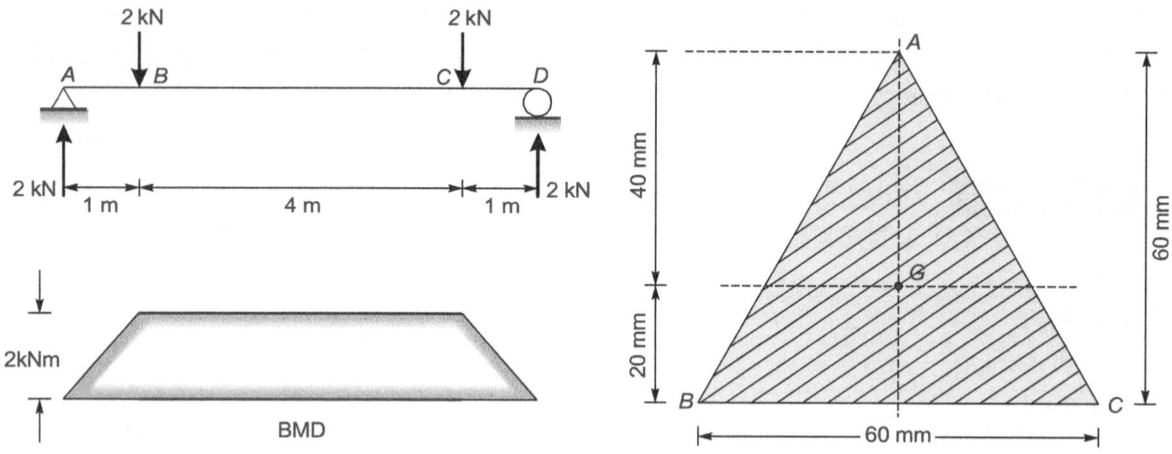

Fig. 8.13

Fig. 8.13 shows the BM diagram of beam.

M_{max}, maximum bending moment = 2 × 1 = 2 kNm = 2 × 10^6 Nmm

Section Isosceles triangle, 60 mm base × 60 mm altitude

σ_1 at $A > \sigma_2$ at BC

$$I_{NA} = \frac{bh^3}{36} = \frac{60 \times 60^3}{36} = 36 \times 10^4 \text{ mm}^4$$

$$Z_1 = \frac{36 \times 10^4}{40} = 9 \times 10^3 \text{ mm}^3$$

$$Z_2 = \frac{36 \times 10^4}{20} = 18 \times 10^3 \text{ mm}^3$$

Skin stress (stresses at extreme fibres of the beam)

$$\sigma_1 = \sigma_A = \frac{M}{Z_1} = \frac{2 \times 10^6}{9 \times 10^3} = 222.22 \text{ MPa}$$

$$\sigma_2 = \sigma_{BC} = \frac{M}{Z_2} = \frac{2 \times 10^6}{18 \times 10^3} = 111.11 \text{ MPa}$$

Practice Q.8.8 A beam section is isosceles triangle with base '*a*' and altitude '*a*'. Bending moment acting on the section is 1.2 kNm. Determine base 'a' of triangle if maximum stress in beam section is not to exceed 100 MPa.

Ans. [66 mm]

Question 8.9 A rolled steel section ISLB 400 has following specifications.

D, depth = 400 mm, *B*, breadth = 165 mm, I_{xx} = 1930.3 × 10^4 mm^4. This is subjected to a sagging moment of 2 kNm. (ISLB: I-section as per Indian Standards with depth 400 mm)

Draw stress distribution along the depth of the section.

Solution: *Sagging Moment*

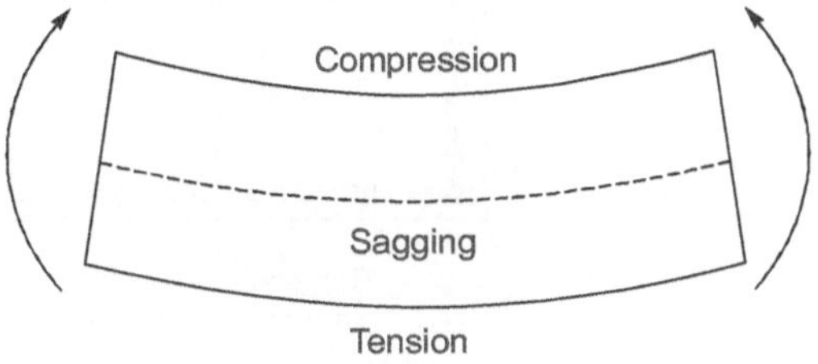

G of the section lies at CG of this web.

$$I_{xx} = 1930.3 \times 10^4 \text{ mm}^4$$

$$\frac{D}{2} = 200 \text{ mm}$$

Section modulus,
$$Z_x = \frac{1930.3\times10^4}{200} = 965.15\times100 \text{ mm}^3$$

$$M = 2 \text{ kNm} = 2 \times 10^6 \text{ Nmm}$$

Skin stresses
$$\pm\, \sigma_c, \sigma_t = \frac{2\times10^6}{965.15\times100} = +20.72 \text{ N/mm}^2$$

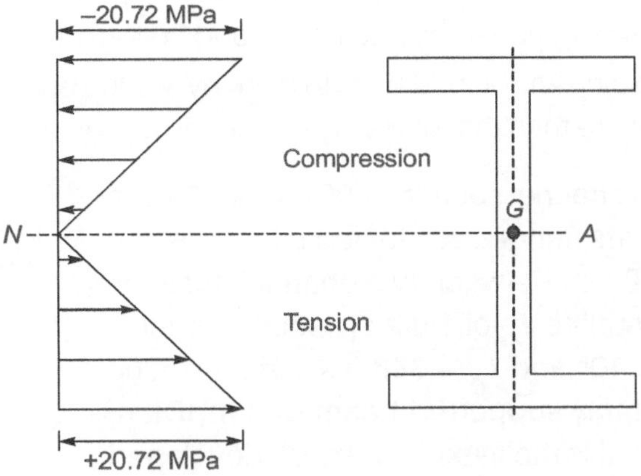

Fig. 8.14 *Stress distribution in I-section under sagging moment*

Practice Q.8.9 An I-section girder is simply supported over a span of 8 m. It is subjected to a UDL of 2.4 kN/m. Depth of section is 400 mm, width of the section is 165 mm. If Z_x section modulus of section is 960 × 10^3 mm^3, draw stress distribution along depth of the section for the section which carries the maximum bending moment.

Hints: $\left[M_{max} = \frac{wL^2}{8}\right]$

Ans. [σ_c = –20 MPa, σ_t = +20 MPa]

Question 8.10 What is a built up section. Give three examples. What is the importance of a built up section?

Solution: Several sections are joined together by welding or by riveting to increase moment of inertia about neutral axis and consequently the section modulus is increased.

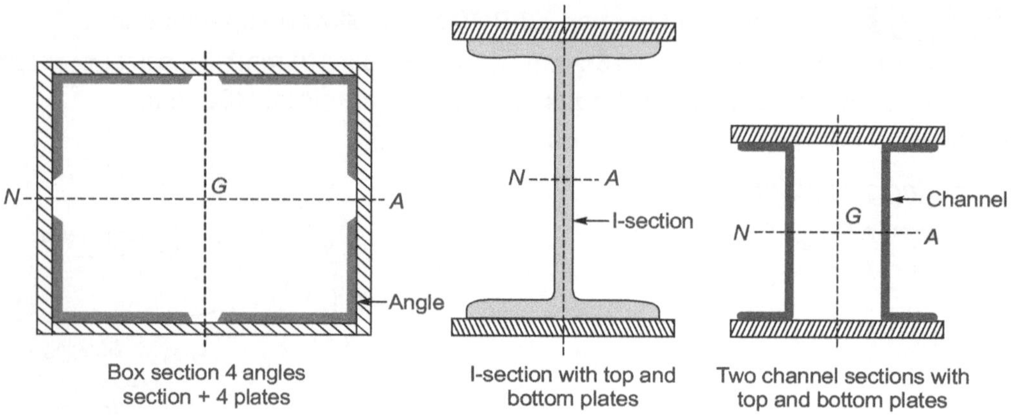

Fig. 8.15

For the same maximum allowable stress, the moment bearing capacity of the beam is increased. Section modulus of a built up section is much more than the section modulus of individual component section. Figure 8.15 shows three built up sections as follows:

8.15 (a) Four angle sections are joined by four plates to obtain a box section. Moment of inertial I_{NA} of built up box section is much more than that of an angle section or a plate section.

8.15 (b) An I-section is strengthened by adding top and bottom plates to enhance moment of inertial I_{NA}

8.15 (c) Two channel sections are joined back to back by two top and bottom plates to enhance the value of I_{NA} of built up section much more than that of a single channel section.

Question 8.11 Two channel sections 300 mm × 100 mm are placed back to back and the flanges are joined by plates 250 mm × 10 mm, with a gap of 50 mm between two channel sections as shown in figure 8.16. Determine I_{xx} of built up section if for a channel sections I_{xx} = 6066 × 10^4 and I_{yy} = 346.9 × 10^4 mm^4. The compound section forms a simply supported beam with a UDL of 100 kN/m, span length 2 m. What is the maximum stress developed in compound section?

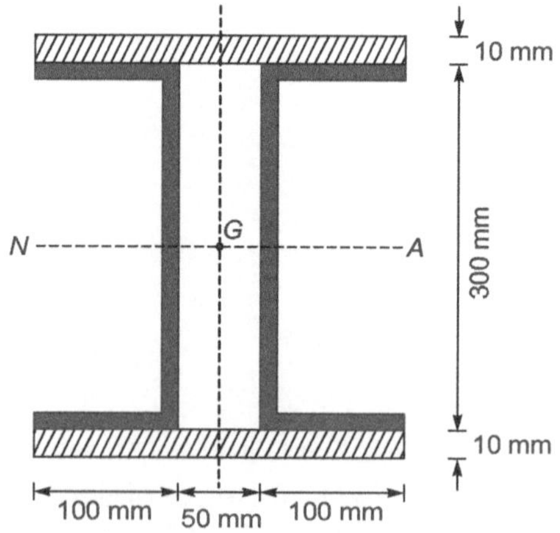

Fig. 8.16

Solution: Fig. 8.16 shows the compound section.

NA passes through G as shown

$$I_{NA} = 2 \times I_{x'x'} + \frac{2 \times 250 \times 10^3}{12} + 2 \times 250 \times 10 \times (155)^2$$

Note that second term gives moment of inertia of plates about its own horizontal centroidal axis, area of the plate 250 × 10 mm^2 and distance of G of plate from neutral axis is 150 + 5 = 155 mm. We have used parallel axis theorem

$$I_{NA} = 2 \times 6066 \times 10^4 + 4.1666 \times 10^4 + 120.125 \times 10^6$$

$$= [12132 + 4.1666 + 12012.5] \times 10^4$$
$$= 24148.666 \times 10^4 \text{ mm}^4$$
$$= 2.4148 \times 10^8 \text{ mm}^4$$

Overall depth of the compound section = 320

$$y_1 = y_2 = 160 \text{ mm} = (320/2)$$

$$Z_x, \text{ section modulus} = \frac{2.4148 \times 10^8}{160} = 1.5093 \times 10^6 \text{ mm}^3$$

Bending Moment

Rate of loading, $w = 100$ kN/m

Span length, $L = 2$m

Maximum bending moment, $$\frac{wL^2}{8} = \frac{100 \times 2^2}{8} = 50 \text{ kNm} = 50 \times 10^6 \text{ Nmm}$$

Maximums stress, $$\sigma = \pm\frac{M}{Z_x} = \frac{50 \times 10^6}{1.5093 \times 10^6} = \pm 33.128 \text{ N/mm}^2$$

Practice Q.8.11.1 An I-section girder is simply supported over a span of 8 m, It is subjected to a central load of 14 kN. Properties of I-section are $Z_x = 960 \times 10^3$ mm^3, depth of section is 400 mm, width of the I-section is 165 mm. It is strengthened by flange plates on both the sides 20 mm thick, 200 mm width. What is Z_x of composite section? What is the maximum stress developed in composite section due to BM?

Hints: [$Z_x \times 200 = I_{xx}$ of I-section]

Ans. [±11.30 N/mm^2]

Practice Q.8.11.2 A T-section girder with flange $B \times 20$ mm and web 130×10 mm is used as a cantilever with flange at top. Length of cantilever is 2 m and a point load W is applied at the free and maximum stress developed in tension = $\frac{1}{2}$ maximum stress developed in compression. If the allowable stress in tension is 80 MPa, determine the magnitude of load *W*, width of the flange, *B*.

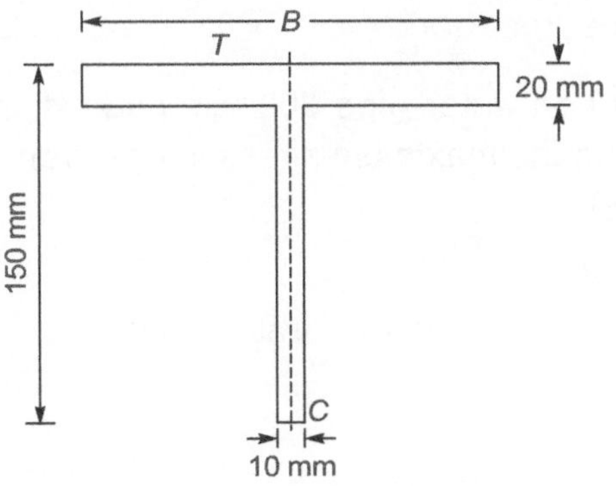

Fig. 8.18

Hints: $\left[\sigma_t = \frac{1}{2}\sigma_c, y_t = \frac{1}{2}y_c, y_t = 50 \text{ mm}, \; y_c = 100 \text{ mm}\right]$

Ans. [$B = 56.875$ mm, $W = 4.22$ kN]

Question 8.12 A beam is of rectangular section 200 mm wide × 300 mm deep subjected to a positive bending moment of 180 kNm. What is the force acting on the shaded area *abcd* of beam section.

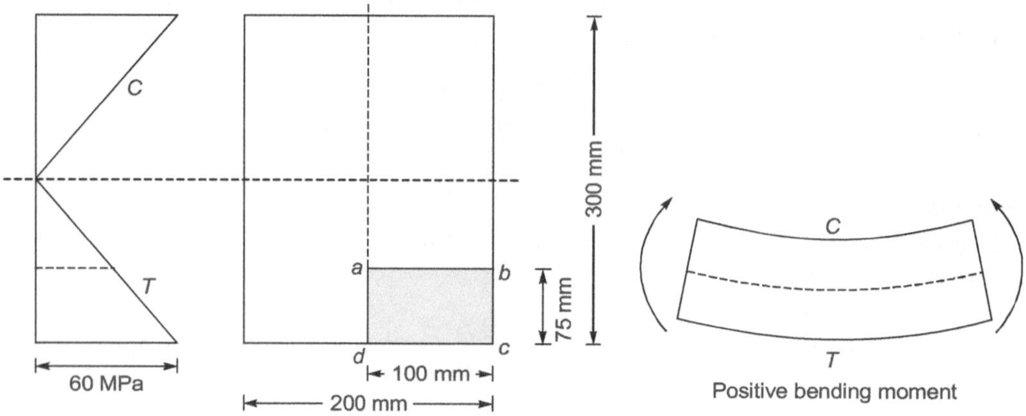

Fig. 8.19

Solution: Positive bending moment is a sagging bending moment, producing concavity upwards. Compression on upper surface, tension on lower surface. So the area *abcd* (75 × 100 mm) is in tension zone.

$$M = 180 \text{ kNm} = 180 \times 10^6 \text{ Nmm}$$

$$\text{Section modulus, } Z = \frac{bd^2}{6} = \frac{200 \times 300^2}{6} = 3 \times 10^6 \text{ mm}^3$$

$$\sigma_t = \text{maximum tensile stress} = \frac{M}{Z} = \frac{180 \times 10^6}{3 \times 10^6} = 60 \text{ MPa}$$

$$\text{Stress along edge, } cd = 60 \text{ MPa}$$

$$\text{Stress along edge, } ab = 30 \text{ MPa}$$

$$\text{Average stress, } \sigma_{av} = \frac{60 + 30}{2} = 45 \text{ MPa} \quad \text{(over area } abcd\text{)}$$

$$\text{Force in the area, } 'abcd' = 45 \times 75 \times 100 = 337500 \text{ N} = 337.5 \text{ kN (tension)}$$

Question 8.13 A cast iron water pipe 300 mm outer diameter with 15 mm wall thickness is supported over a span of 8 m. Find the maximum stress in cast iron when pipe is running full of water.

Density of *CI* = 71.6 kN/m^3

Density of water = 9.8 kN/m^3

Solution: Let us consider one meter length of water pipe

CI Pipe

Outer diameter, $D = 0.3$ m

$t = 0.015$ m

Inside diameter, $d = 0.27$ m

Area of cross-section (*CI*) $= \frac{\pi}{4}(0.3^2 - 0.27^2) = 0.01343 \text{ m}^2$

Area of cross-section of water $= \frac{\pi}{4} \times 0.27^2 = 0.05725 \text{ m}^2$

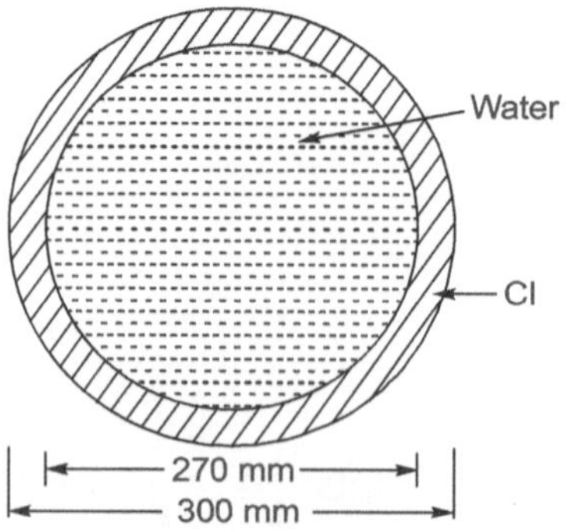

CI Pipe with water

Fig. 8.21

Weight of *CI* per meter length $= 0.01343 \times 71.6 = 0.9616$ kN

Weight of water per meter length $= 0.05725 \times 9.8 = 0.5610$ kN

Total weight of beam per metre length,

$$w = 0.9616 + 0.5610 = 1.5226 \text{ kN/m}$$

Length of the beam, $L = 8$ m

Maximum bending moment, $M_{max} = \frac{wL^2}{8} = \frac{1.5226 \times 8^2}{8} = 12.1808 \text{ kNm} = 12.1808 \times 10^6 \text{ Nmm}$

Section modulus of *CI* pipe, $Z = \frac{\pi}{32D}(D^4 - d^4) = \frac{\pi(300^4 - 270^4)}{32 \times 300} = 91.158 \times 10^4 \text{ mm}^4$

Maximum stress in *CI* pipe, $\sigma = \frac{M}{Z} = \frac{12.1808 \times 10^6}{91.158 \times 10^4} = \pm 13.36 \text{ N/mm}^2$

Practice Q.8.13 **A steel tube of outer diameter 20 mm, wall thickness 1 mm charged with mercury, is supported over a span of 1.6 m. Tube is plugged at the ends. What is the maximum stress developed in steel tube.**

Given, density of mercury = 138 kN/m³ = 1.38×10^{-4} N/mm³

Density of steel = 74 kN/m³ = 0.74×10^{-4} N/mm³

Ans. [± 46.83 N/mm²]

Question 8.14 **What is a beam of uniform strength? Take a beam simply supported at ends, rectangular section, how the depth can be varied so as to obtain a beam of uniform strength. Load on the beam is a UDL (uniformly distributed load)**

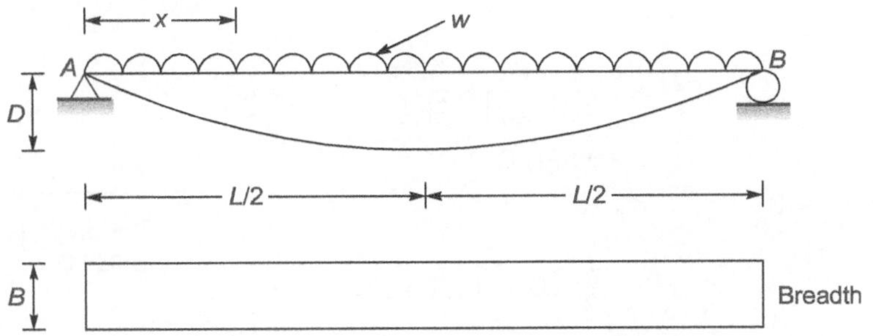

Fig. 8.22

Solution: When a beam is subjected to a bending moment, skin stresses are developed in the section. If the section is rectangular, square or circular, $\sigma_t = \sigma_c$, maximum stress in tension = maximum stress in compression

$$M = \sigma Z$$

If $\frac{M}{Z}$ is maintained constant, then σ will remain constant, where σ is the skin stress. **This type of beam is known as a beam of uniform strength.**

Beam *AB* of length *L*, simply supported at ends carries a UDL, equation of bending moment.

$$M_x = \frac{wLx}{2} - \frac{wx^2}{2}, \text{ is parabolic}$$

$$Z_x = \frac{BD_x^2}{6}, \text{ where } D_x \text{ is depth at a section } x.$$

If $\frac{M_x}{Z_x}$ remains constant, then σ will be same throughout the length.

at $$x = \frac{L}{2},\ M_x = \frac{wL^2}{8}.$$

$$\frac{wL^2}{8} = \sigma\frac{BD^2}{6}$$

or $$D^2 = \frac{6}{8}\times\frac{wL^2}{\sigma B} = \frac{3}{4}\times\frac{wL^2}{B\sigma}$$

At $$x = \frac{L}{4},\ M_x = \frac{wL}{2}\times\frac{L}{4} - \frac{wL^2}{32} = \frac{3}{32}wL^2$$

$$Z_x = \frac{BD_x^2}{6}$$

$$\frac{3}{32}wL^2 = \sigma\frac{BD_x^2}{6}$$

$$D_x^2 = \frac{9}{16}\times\frac{wL^2}{\sigma B}$$

D_x at $\frac{L}{4}$ $$D_x^2 = \frac{9}{16}\times\frac{wL^2}{\sigma B}\times\frac{4B\sigma}{3wL^2}D^2 = \frac{3}{4}\times D^2$$

$$D_x = 0.866\ D$$

So at $x = 0$ m, $D_x = 0$

$$x = \frac{L}{4},\ D_x = 0.866\ D$$

$$x = \frac{L}{2},\ D_x = D$$

$$x = \frac{3L}{4},\ D_x = 0.866\ D$$

$$x = L,\ D_x = 0$$

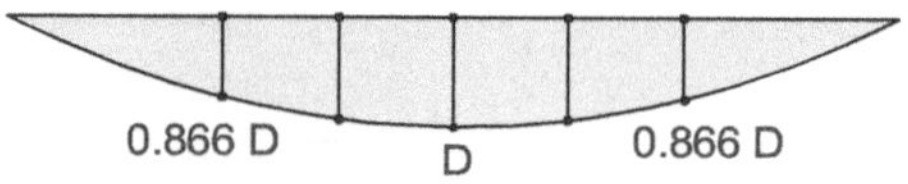

Fig. 8.23 *Variation of depth along length*

Practice Q.8.14 A beam supported over a length of 2 m carries a central load, Beam section is rectangular with uniform depth of 20 mm throughout. For a central load of 2 kN, what is breadth at the centre for a uniform strength of 100 N/mm^2. Draw variation of breadth with length.

Ans. [B = 150 mm at centre and 0 at ends]

Question 8.15 A cantilever of length L, carries UDL of intensity w throughout its length. If breadth of the cantilever remains constant, show the variation of depth so as to achieve a cantilever of uniform strength σ. Section of beam is rectangular

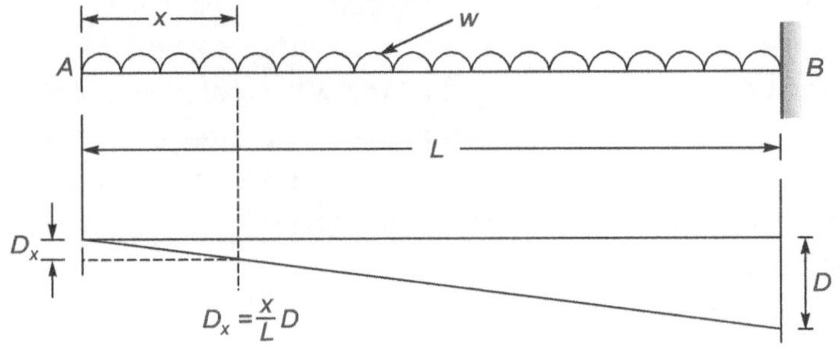

Fig. 8.24

Solution: A cantilever of length L carrying UDL of intensity w is shown in figure 8.24

Bending moment at any section XX $= \dfrac{wx^2}{2}$

Uniform breadth $= B$

Say depth $= D_x$

Section modulus $= \dfrac{BD_x^2}{6}$

Stress, $\sigma = \dfrac{M_x}{Z_x} = \dfrac{wx^2}{2} \times \dfrac{6}{BD_x^2} = \dfrac{3w}{B}\dfrac{x^2}{D_x^2}$

But σ is constant so $\dfrac{x^2}{D_x^2}$ = C, a constant

$$x^2 = CD_x^2$$

At $x = L,\ D_x = D$

$$L^2 = CD^2$$

$$C = \frac{L^2}{D^2}$$

$$D_x^2 = \frac{x^2}{C} = \frac{x^2}{L^2} \times D^2$$

$$\frac{x^2}{D_x^2} = \frac{L^2}{D^2}$$

$\dfrac{x}{D_x} = \dfrac{L}{D}$, $D_x = \dfrac{x}{L}.D$, linear variation of depth as shown.

Practice Q.8.15 A cantilever of length 1.2 m, carries a point load at free end W = 1 kN. Depth of beam remains constant as 60 mm. Strength of the cantilever remain constant as 100 N/mm^2. Show the variation of breadth along the length of the cantilever.

Ans. [x = 0, B = 0, x = 0.6 m, B = 10 mm, x = 1.2 m, B = 20 mm]

B_x is linearly proportional to x, $B_x = \dfrac{x}{60}$mm

Question 8.16 **What is a composite beam? What is the purpose of providing steel flitches in wooden beams? Give some examples of composite beam.**

Solution: A composite beam is made of more than one beam of different materials. Most common example of a composite beam is a wooden beam in between a steel flitch is used as shown in figure 8.24 (a) or there are two steel flitches on either side of the beam as shown in fig. 8.24 (b). Moreover steel plates can be used at top and bottom of wooden beam as shown in fig. 8.24 (c), with steel plates of size $B \times t$ on top and bottom. In 8.24 (d) a composite beam is shown with unequal size of plates at top and bottom i.e., $B_1 \times t_1$ and $B_2 \times t_2$ over and below a wooden beam of side $B \times D$. Basic purpose of a composite beam is to make the stiffer beam to share more bending moment. Since the modular ratio of E_{steel}/E_{wood} is very high as high as 15-20, the major portion of the applied moment is shared by steel plates.

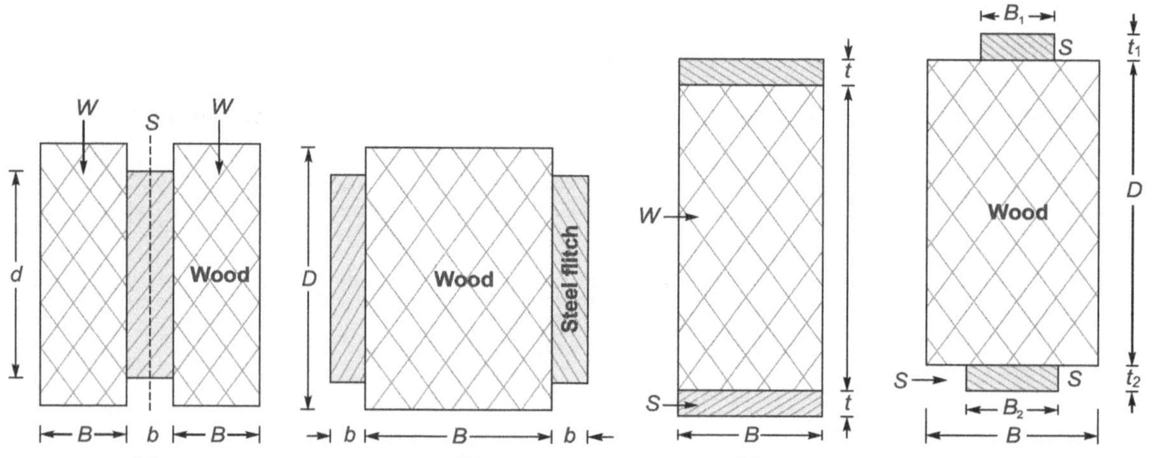

Fig. 8.24

For the determination of stresses in components of composite beam as shown in fig. 8.24 (a) and (b), simple strain compatibility equations can be used. But for the determination of stresses in wooden and steel beams as shown in fig. 8.24 (c) and (b) an equivalent steel beam section is made and stresses are determined in the equivalent steel section.

Total bending moment $M = M_s + M_w$

= bending moment shared by steel beam + bending moment shared by wooden beam.

Question 8.17 **A steel flitch of depth 160 mm and width 20 mm is sandwiched between two wooden beams of depth 200 mm and width 50 mm each as shown in figure 8.25.**

If the maximum stress in steel section is not to exceed 160 MPa and maximum stress in wooden beam section is not to exceed 10 MPa, calculate the safe bending moment for this composite section.

Solution:

$$\text{Modular ratio} = \frac{E_s}{E_w} = 16$$

Note that maximum stress in wooden beam will occur on edges *cc* and *dd*, but maximum stress in steel would occur at edge *aa* or *bb* as shown in figure 8.25.

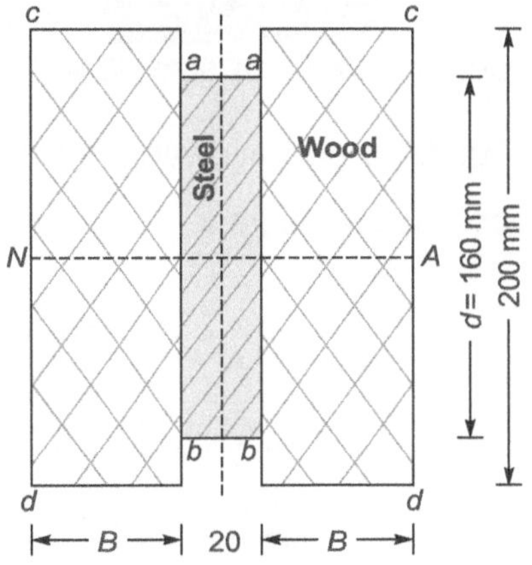

Fig. 8.25

But strains at *aa* or at *bb* will be the same in steel beam and in wooden beam.

Strain in a beam at a particular layer is proportional to its distance from the neutral layer.

At edge *aa*

$$e_s = e_w$$

$$\frac{\sigma_s}{E_s} = \frac{\sigma_w'}{E_w}$$

$$\sigma_w' = \frac{E_w}{E_s} \times \sigma_s \text{, at edge } aa. = \frac{\sigma_s}{m},$$

$$\text{where } m = 16 = \frac{E_s}{E_w}$$

$$\sigma_w' = \frac{\sigma_s}{m} = \frac{160}{16} = 10 \text{ MPa}$$

But $$\sigma_w \text{ at } cc = \sigma_w \times \frac{100}{80} = 10 \times 1.25 = 12.5 \text{ N/mm}^2$$

But allowable stress in wood, σ_w = 10 N/mm². Therefore stress in steel will be decided as per this condition

$$\sigma_s' \text{ at } cc = \sigma_w \times m = 10 \times 16 = 160 \text{ MPa}$$

$$\sigma_s \text{ at } aa = 160 \times \frac{80}{100} = 128 \text{ N/mm}^2$$

So the maximum stress in steel σ_s = 128 N/mm²

in wood, σ_w = 10 N/mm²

$$\text{Bending moment shared by wooden beams} = 2 \times \frac{BD^2}{6} \times \sigma_w = 2 \times \frac{50 \times 200^2}{6} \times 10$$

$$M_w = 6.667 \times 10^6 \text{ Nmm}$$

$$\text{Bending moment shared by steel flitch} = \frac{bd^2}{6} \times \sigma_s = \frac{20 \times 160^2}{6} \times 128 = 10.9226 \times 10^6 \text{ Nmm}$$

$$\text{Total bending moment, } M = M_s + M_s = (6.667 + 10.9226) \times 10^6 \text{ Nmm} = 17.5896 \text{ kNm}$$

Practice Q.8.17 **A wooden beam of breadth 150 mm, depth 250 mm is reinforced symmetrically on either side by two steel plates of depth 200 mm and thickness 20 mm each. Allowable stresses in steel and wood are 120 MPa and 8 MPa respectively. Modular ratio is $\frac{E_s}{E_w} = 20$. What safe bending moment can be applied on composite beam?**

Hints: $\left[\sigma_s = m \times \frac{d}{D} \times \sigma_w = 20 \times \frac{200}{250} \times \sigma_w = 16\sigma_w\right]$ [σ_s = 120, σ_w = 7.5]

Ans. [32 + 11.71875 = 43.71875 kNm]

Question 8.18 A wooden beam 200 mm deep × 160 mm wide is strengthened by top and bottom plates of size 60 × 12 mm and 120 × 20 mm as shown in figure 8.26. Allowable stress in steel 100 MPa and allowable stress in wood is 8 MPa. If $\frac{E_s}{E_w} = 16$. Determine the safe bending moment on composite beam.

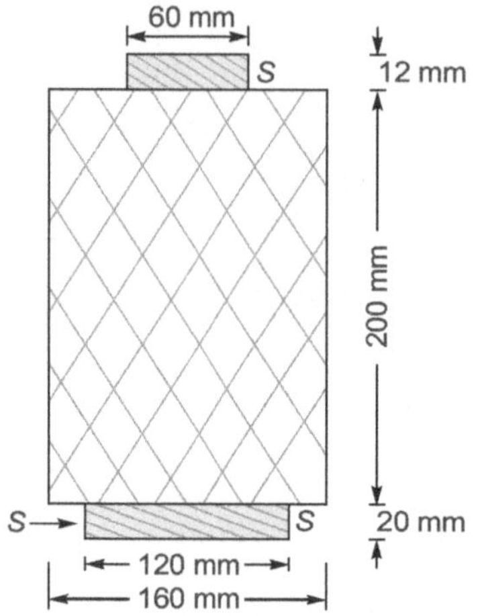

Fig. 8.26

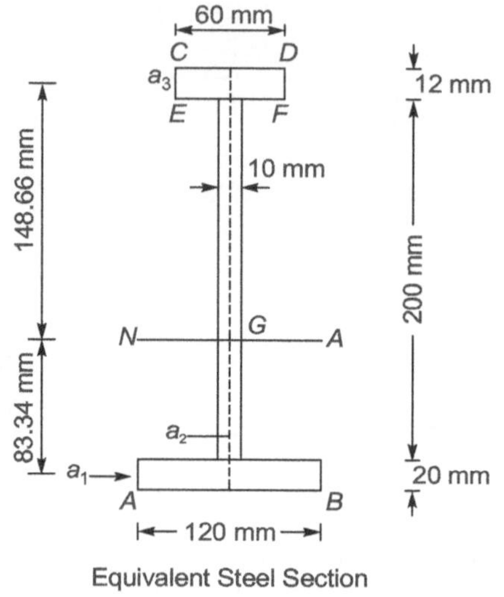

Fig. 8.27

Solution:

Modular ratio = 16 = m

Width of the wooden beam will be reduced by m, i.e., $b = \frac{B}{16} = \frac{160}{16} = 10$ mm

$$y_1 = \frac{120\times20\times10+200\times10\times(20+100)+60\times12(220+6)}{120\times20+200\times10+60\times12}$$

$$= \frac{24000+240000+162720}{5120} = \frac{426720}{5120} = 83.34 \text{ mm}$$

$$y_2 = 200 + 20 + 12 - 83.34 = 148.66 \text{ mm}$$

Moment of inertia,

$$I_{NA} = \frac{120\times20^3}{12} + 120\times20\times(83.34-10)^2 + \frac{10\times200^3}{12}$$

$$+2000(120-83.34)^2 + \frac{60\times12^3}{12} + 720(148.66-6)^2$$

$$= 80000 + 12.909\times10^6 + 6.6667\times10^6 + 2.688\times10^6 + 8640 + 14.653\times10^6$$

$$= 36.9147\times10^6 + 88640 = 37.0\times10^6 \text{ mm}^4$$

Note that maximum stress in steel will occur at edge cd, $\sigma_s = 100$ MPa.

But maximum stress in wood will occur at edge EF

$$\sigma_w = 8 \text{ MPa}$$

$$m = 16$$

$$\sigma_s' \text{ at } EF = 16 \times 8 = 128 \text{ MPa}$$

$$\sigma_s' \text{ at } CD = 128\times\frac{148.66}{136.66} = 139.24 \text{ N/mm}^2$$

This is greater than 100 MPa.

So, σ_s = 100 MPa at edge *CD*

$$\sigma_w \text{ at } EF = 100\times\frac{136.66}{148.66}\times\frac{1}{16} = 5.75 \text{ N/mm}^2$$

$$5.75 < 8 \text{ MPa}$$

So let us take σ_s = 100 MPa

$$M = \sigma_s\times\frac{I_{NA}}{148.66} = \frac{100\times37\times10^6}{148.66} = 24.89\times10^6 \text{ Nmm}$$

$$= 24.89 \text{ kNm, safe bending moment.}$$

Practice Q.8.18 **A wooden beam 200 mm wide × 240 mm deep is strengthened by top and bottom steel plates 200 × 20 mm as shown. Allowable stresses in steel and wood are 80 MPa and 5 MPa.**

If modular ratio $\frac{E_s}{E_w} = 20$, what is the safe bending moment for this beam section.

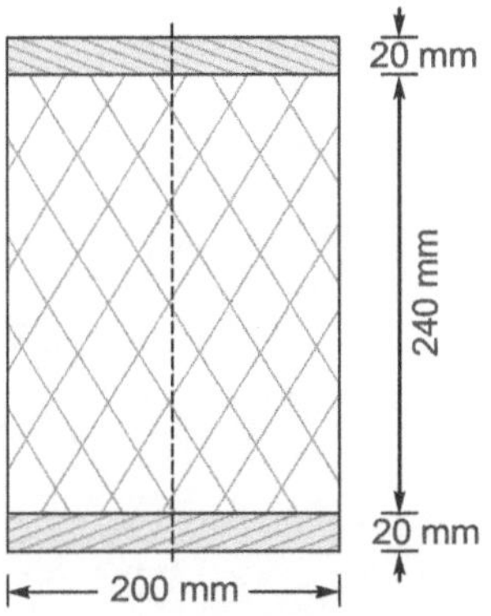

Fig. 8.28

Hints: $\sigma_s = 5\times\frac{140}{120}\times20 = 116.66 \text{ MPa} > 80 \text{ MPa}$

$\sigma_w = 80\times\frac{120}{140}\times\frac{1}{20} = 3.428 \text{ MPa} < 5 \text{ MPa}$, σ_s at extreme edge

Taking σ_w as 3.428 MPa at lower edge

σ_s allowable $= 3.428\times\frac{140}{120}\times20 = 80$ MPa

[M_{safe} = 83.99 kNm]

Question 8.19 **Three steel tubes of diameter 20 mm and inner diameter 16 mm are welded together so that their centres make an equilateral triangle of 20 mm side. This combination is used as a beam over a length of 2 m. How much central load can be applied on the beam if σ_{max} is not to exceed 60 MPa.**

[IES, 2011, 10 Marks]

Solution: Fig. 8.29 shows the section of the beam.

$$A, \text{ area } = \frac{\pi}{4}(20^2 - 16^2) = 36\pi \text{ mm}^2$$

I of each tube about its own *x′x′* axes

$$= \frac{\pi}{64}(20^4 - 16^4)$$

$$= 0.4637\times10^4 \text{ mm}^4$$

Centres of tubes make an equilateral triangle of side 20 mm.

G lies at $\frac{h}{3}$ from *ab*,

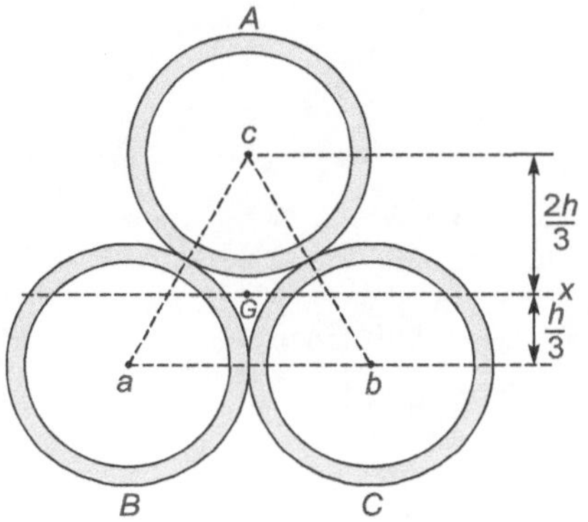

Fig. 8.29

where $h = \frac{20}{2} \times \tan 60° = 17.32$ mm

$$\frac{h}{3} = 5.77 \text{ mm}$$

$$\frac{2h}{3} = 11.55 \text{ mm}$$

I_{xx} of composite section $= 0.4637 \times 10^4 + 36\pi \times (11.55)^2 + 2 \times 0.4637 \times 10^4 + 36\,\pi \times 2(5.77)^2$

$$= 1.3911 \times 10^4 + 1.5087 \times 10^4 + 0.753 \times 10^4 \text{ mm4}$$

$$= 3.6528 \times 10^4 \text{ mm}^2$$

σ_{max} will occur at top edge A

$$y = \frac{2h}{3} + 10 = 11.55 + 10 = 21.55 \text{ mm}$$

$$Z = \frac{3.6528 \times 10^4}{21.55} = 1.695 \times 10^3 \text{ mm}^3$$

$$\sigma = 60 \text{ MPa}$$

$$M = 60 \times 1.695 \times 10^3 \text{ Nmm} = 101.7 \times 10^3 \text{ Nmm} = \frac{WL}{4}$$

$$L = 2000 \text{ mm}$$

$$101.7 \times 10^3 = \frac{W \times 2000}{4}$$

$$W = \frac{101.7 \times 10^3}{500} = 203.4 \text{ N}$$

Central load.

Question 8.20 Find the slope and deflection at free end of a cantilever beam of length 6 m as loaded shown in figure 8.30 using the method of superposition, evaluate their numerical values, if E = 200 GPa, $I = 1 \times 10^{-4}$ m^4, W = 1 kN.

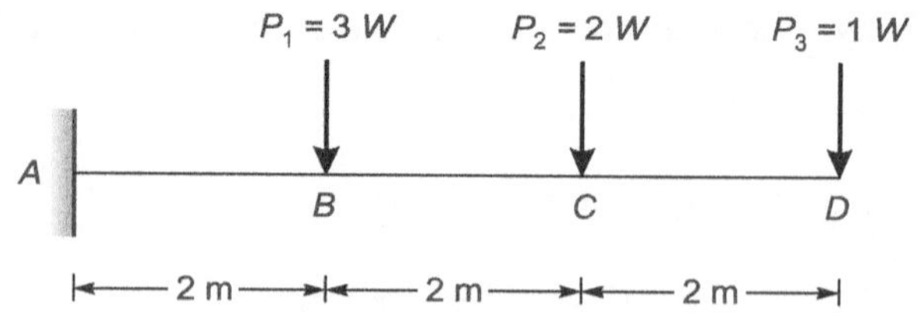

Fig. 8.30

[IES 2003, 10 Marks]

Solution:

Load, $P_3 = 1$ kN

$$\delta_D = \frac{1 \times 6^3}{3EI}, \; i_D = \frac{1 \times 6^2}{2EI}$$

Load, $P_2 = 2W = 2\text{ kN}$

$$\delta_D = \frac{2\times 4^3}{3EI}+\frac{2\times 4^2}{2EI}\times 2=\frac{128}{3EI}+\frac{64}{2EI}=\frac{256+192}{6EI}=\frac{448}{6EI}$$

$$i_C = \frac{2\times 4^2}{2EI}=\frac{16}{EI}$$

Load, $P_3 = 3\text{ kN}$

$$\delta_D = \frac{3\times 2^3}{3EI}+\frac{3\times 2^2}{2EI}\times 4=\frac{24}{3EI}+\frac{48}{2EI}=\frac{48+144}{6EI}=\frac{192}{6EI}$$

$$i_B = \frac{3\times 2^2}{2EI}=\frac{12}{2EI}=\frac{6}{EI}$$

$$\delta_{D'\ total} = \frac{216}{3EI}+\frac{448}{6EI}+\frac{192}{6EI}=\frac{1072}{6EI}=\frac{176.8667}{EI}$$

Slope, $i_{D\ total} = \dfrac{36}{2EI}+\dfrac{32}{2EI}+\dfrac{12}{2EI}=\dfrac{80}{2EI}=\dfrac{40}{EI}$

$$E = 20\text{ kN/mm}^2 = 200\times 10^6\text{ kN/m}^2$$

$$EI = 200\times 10^6\times 1\times 10^{-4} = 20000\text{ kNm}^2$$

Deflection, $\delta_{D\ total} = \dfrac{176.8667}{20000}=8.93\times 10^{-3}\text{m}=8.93\text{ mm}$

Total slope, $i_D = \dfrac{40}{EI}=\dfrac{40}{20000}=2\times 10^{-3}\text{ rad}$

Question 8.21 A beam of rectangular cross-section 50 mm wide and 100 mm deep is simply supported over a span of 1500 mm. It carries a concentrated load of 50 kN, 500 mm from the left support. Calculate (i) The maximum tensile stress in the beam and indicate where it occurs, (ii) The vertical deflection of the beam at a point 500 mm from the left support. *E* for the material of the beam = 2 × 10^5 MPa.

[IES : 2002, 20 Marks]

Solution:

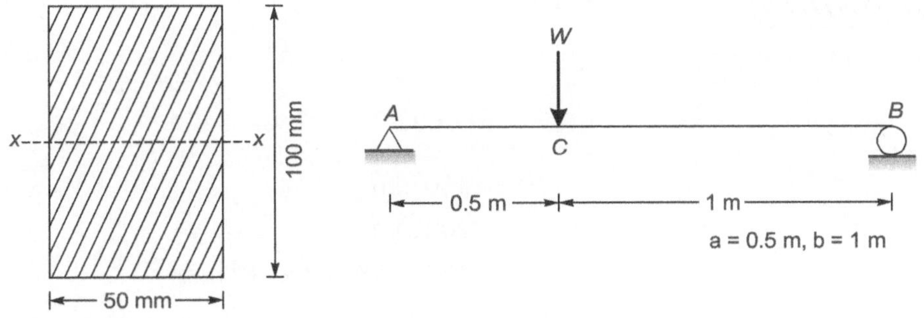

Fig. 8.31

Section $b = 50$ mm

$d = 100$ mm

$$Z = \frac{bd^2}{6} = \frac{50\times100^2}{6} = 8.33 \times 10^4 \text{ mm}^3$$

$$I_{xx} = \frac{bd^3}{12} = \frac{50\times100^3}{12} = 4.166\times10^6 \text{ mm}^4 = 4.166\times10^{-6} \text{ m}^4$$

$$E = 2 \times 10^5 \times 10^6 \text{ N/m}^2 = 2 \times 10^8 \text{ kN/m}^2$$

$$EI = 4.166 \times 10^{-6} \times 2 \times 10^8 = 8.33 \times 10^2 \text{ kNm}^2 = 833 \text{ kNm}^2$$

$$W = 50 \text{ kN}$$

$$M_{max} = \frac{Wab}{L} = \frac{50\times0.5\times1}{1.5} = 16.667 \text{ kNm} = 16.667\times10^6 \text{ Nmm}$$

$$\sigma_{max} = \frac{M_{max}}{Z} = \frac{16.66\times10^6}{8.33\times10^4} = 200 \text{ N/mm}^2$$

(Maximum tensile stress at the bottom of the section) at point *C*, i.e., under load.

δ_C, deflection under load $= \dfrac{Wa^2b^2}{3EIL} = \dfrac{50\times0.5^2\times1^2}{3\times833\times1.5}$

$= 3.334 \times 10^{-3}$ m $= 3.334$ mm (under load)

Question 8.22 A timber beam 15 cm wide and 20 cm deep carries a uniformly distributed load over a span of 4 m and is simply supported. If the permissible stresses are 30 N/mm^2 longitudinally and 3 N/mm^2, transverse shear, calculate the maximum load which can be carried by the timber beam.

[IES 2006, 20 Marks]

Solution:

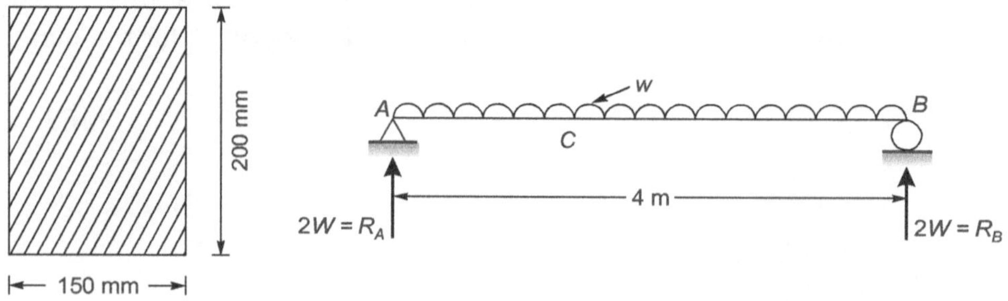

Fig. 8.32

Section $b = 150$ mm

$d = 200$ mm

Area, $A = 150 \times 200 = 3 \times 10^4$ mm^2

$$Z = \text{section modulus} = \frac{bd^2}{6} = \frac{150\times200^2}{6} = 1\times10^6 \text{ mm}^3$$

$$L\text{, span length} = 4 \text{ m}$$

$$M_{max} = \text{maximum bending moment} = \frac{wL^2}{8} = \frac{w \times 4^2}{8} = 2w \text{ kNm}$$

$$\sigma_{max} = \frac{2w \times 10^6}{1 \times 10^6} = 30 \text{ N/mm}^2$$

$$w = 15 \text{ kN/m} \qquad \text{...(i)}$$

Maximum shear force, $F = R_A = R_B = 2w \text{ kN} = 2w \times 10^3 \text{ N}$

$$A = 3 \times 10^4 \text{ mm}^2$$

$$\tau_{max} = 1.5\, \tau_{av} \text{ (rectangular section)}$$

$$3 = 1.5 \times \frac{2w \times 10^3}{3 \times 10^4}$$

Rate of loading, $w = 30$ kN/m

Permissible rate of loading $= 15$ kN/m

Question 8.23 A simply supported beam *AB* of span length 4 m supports a uniformly distributed load of intensity $w = 4$ kN/m spread over the entire span and a concentrated load $P = 2$ kN placed at a distance of 1.5 m from left end *A*. The beam is constructed of a rectangular cross-section with width = 10 cm and depth, $d = 20$ cm. Determine the maximum tensile and compressive stresses developed in the beam due to bending.

[IES 2008, 15 Marks]

Solution:

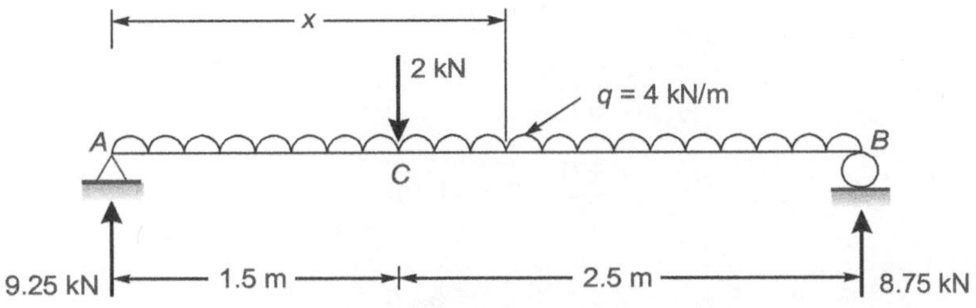

Fig. 8.33

Reactions, $2 \times 1.5 + 4 \times 4 \times 2 = 4\, R_B$

$$R_B = 8.75 \text{ kN}$$

$$R_A = 16 + 2 - 8.75 = 9.25 \text{ kN}$$

Take a section **x-x**, at a distance x from end *A*

Shear force, $F_x = 9.25 - wx - 2$

$$= 9.25 - 4x - 2 = 7.25 - 4x$$

$$F_x = 0 \text{ at } x = 1.8125 \text{ m}$$

Bending moment, $M_x = 9.25x - \frac{wx^2}{2} - 4(x - 1.5)$, where F is zero

$$= 9.25x - 2x^2 - 4(x - 1.5)$$

M_{max} at $x = 1.8125$

$$= 9.25 \times 1.8125 - 2 \times 1.8125^2 - 2(1.8125 - 1.5)$$

$$= 16.766 - 6.570 - 0.625 = 9.571 \text{ kNm}$$

Section $\quad b = 100 \text{ mm}, d = 200 \text{ mm}$

$$Z = \frac{bd^2}{6} = \frac{100 \times 200^2}{6} = 0.667 \times 10^6 \text{ mm}^3$$

$$\sigma_{max} = \pm\frac{M_{max}}{Z} = \frac{9.571 \times 10^6}{0.667 \times 10^6} = \pm 14.35 \text{ N/mm}^2$$

Question 8.24 A steel beam of rectangular section has a span of 8 m and is simply supported at its ends. It is required to carry a load of 60 kN uniformly distributed over the whole span. Find the maximum values of breadth and depth, if the maximum bending stress is not to exceed 50 MPa and the maximum deflection is limited to 10 mm. E = 210 GPa.

Solution:

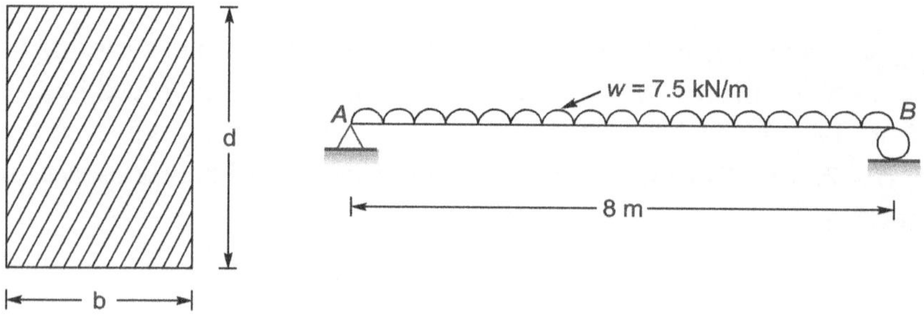

Fig. 8.35

$$L = 8 \text{ m}$$

$$w = \frac{60}{8} = 7.5 \text{ kN/m}$$

$$\delta_{max} = 0.01 \text{ m} = \frac{5wL^4}{384EI} \text{ at the centre}$$

$$EI = \frac{5wL^4}{384 \times 0.01} = \frac{5 \times 7.5 \times 8^4}{3.84} = 40000$$

$$E = 210 \times 10^6 \text{ kN/m}^2$$

$$I = \frac{40000}{210 \times 10^6} = 190.476 \times 10^{-6} \text{m}^4 = \frac{bd^3}{12}$$

M_{max}, maximum bending moment at the centre $= \dfrac{wL^2}{8}$

$$= \frac{7.5 \times 8^2}{8} = 60 \text{ kNm} = 60 \times 10^6 \text{ Nmm}$$

$$= \sigma \times Z = 60 \times Z$$

$$Z = \frac{60\times10^6}{50} = 1.2\times10^6 \text{ mm}^3 = 1.2\times10^{-3} \text{ m}^3 = \frac{bd^2}{6}$$

$$\frac{I}{Z} = \frac{d}{2} = \frac{190.476\times10^{-6}}{1.2\times10^{-3}} = 158.73\times10^{-3} \text{ m}$$

Depth, $d = 317.46$ mm ...(i)

$Z = 1.2 \times 10^6$ mm^3, $d = 317.46$ mm

$$= \frac{bd^2}{6} = \frac{b(317.46)^2}{6} = 1.2\times10^6$$

Breadth, $b = 71.44$ mm

Question 8.25 Compare the flexural strengths of the following three beams of equal weight and same material:

(i) I-section 300 × 160 mm with flange 20 mm thick and web 16 mm thick.
(ii) Rectangular section having depth twice the width.
(iii) Circular solid section.

[ME, ESE 2014 : 10 Marks]

Solution:

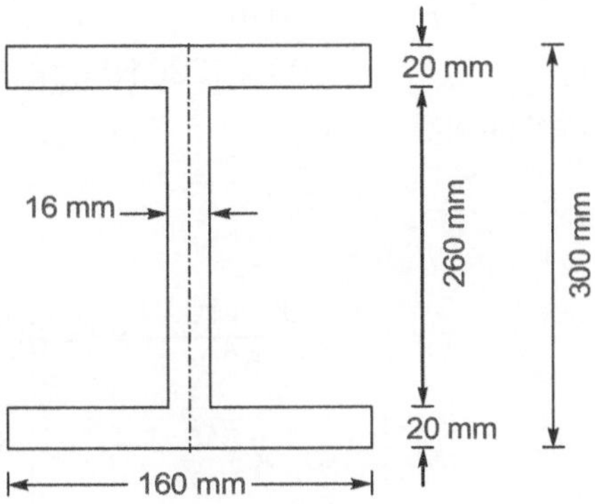

Fig. 8.36

Area of cross-section $= 160 \times 2 \times 20 + 260 \times 6 = 6400 + 4160$ mm^2

$= 10560$ mm^2

Moment of inertia, $$I_I = \frac{160\times300^3}{12} - \frac{(160+-16)\times260^3}{12}$$

$= 36 \times 10^7 - 21.09 \times 10^7$

$= 14.91 \times 10^7$ mm^4

Section modulus, $$Z_I = \frac{14.91\times10^7}{150} = 0.994\times10^6 \text{ mm}^3$$

Equal weight, same area

Rectangular section, $D = 2B$

$$2B^2 = 10560 \text{ mm}^2\text{, same area}$$
$$B = 72.66 \text{ mm}$$
$$D = 145.327 \text{ mm}$$

Section Modulus, $Z_R = \dfrac{BD^2}{6} = \dfrac{72.66 \times 145.327^2}{6} = 0.255 \times 10^6 \text{ mm}^3$

Circular section, $\dfrac{\pi}{4} \times d^2 = 10560$

Diameter, $d = 115.95 \text{ mm}$

Section modulus, $Z_C = \dfrac{\pi d^3}{32} = \dfrac{\pi \times 115.95^3}{32} = 0.153 \times 10^6 \text{ mm}^3$

$$Z_I : Z_R : Z_C = 6.50 : 1.667 : 1$$
$$= 0.954 : 0.255 : 0.253$$

Question 8.26 Find the maximum flexural stress developed in a steel wire 2.00 mm in diameter, if it is coiled over a drum 0.5 m in radius (it is assumed that the limit of proportionality is not exceeded due to coiling). What is the bending moment to which the wire is subjected? Take E = 200 GPa.

[ME, ESE 2017 : 6 Marks]

Solution:

Steel wire radius, $r = 1 \text{ mm}$

Drum radius, $R = 500 \text{ mm}$

$$E = 200 \times 1000 \text{ N/mm}^2$$

Maximum flexural stress $= \sigma_b$

$$\frac{\sigma_b}{T} = \frac{E}{R}$$

$$\sigma_b = \frac{200 \times 1000}{500} \times 1 = 400 \text{ N/mm}^2$$

Bending moment, $M = \sigma_b \times \dfrac{\pi d^3}{32} = (\sigma_b \times Z) = \text{stress} \times \text{section modulus}$

When $d = $ Wire diameter

$$M = 400 \times \frac{\pi \times 2^3}{32} = 314.16 \text{ Nmm}$$

Question 8.27 A simply-supported beam of length 3L is subjected to the loading shown in the figure.

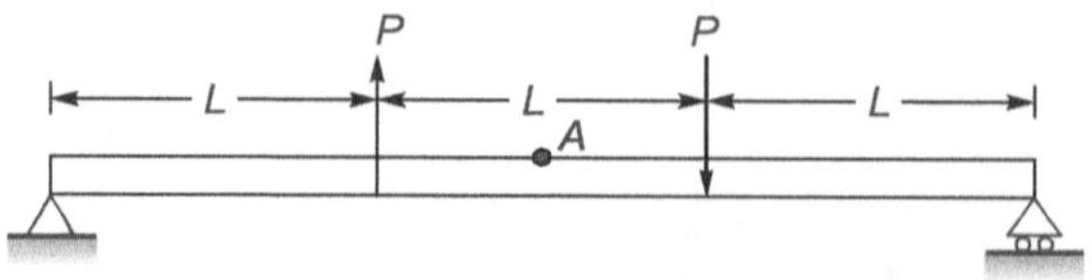

Fig. 8.37

It is given that P = 1 N, L = 1 m and Young's modulus E = 200 GPa. The cross-section is a square with dimension 10 mm × 10 mm. The bending stress (in Pa) at the pint A located at the top surface of the beam at a distance of 1.5 L from the left end is _______

(Indicate compressive stress by a negative sign and tensile stress by a positive sign)

[ME, GATE : 2016]

Solution: At a compressive stress

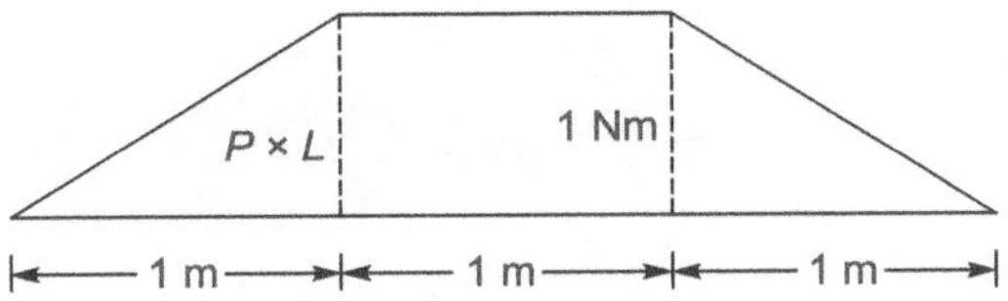

Fig. 8.38

$$M_A = 1 \text{ N} \times 1 \text{ m} = 1 \text{ Nm}$$

Section modulus, $$Z = \frac{a^3}{6} = \frac{10^3}{6} = \frac{1000}{6} \text{ mm}^3$$

$$\sigma = \frac{M}{Z} = \frac{1 \times 10^3 \text{ Nmm}}{1000} \times 6$$

$$= 6 \text{ N/mm}^2 = 6 \text{ MPa} = 6 \times 10^6 \text{ Pa}$$

Question 8.28 A 450 mm long plain concrete prism is subjected to the concentrated vertical loads as shown in the figure. Cross-section of the prism is given as 150 mm × 150 mm. Considering linear stress distribution across the cross-section, the modulus of rupture (expressed in MPa) in

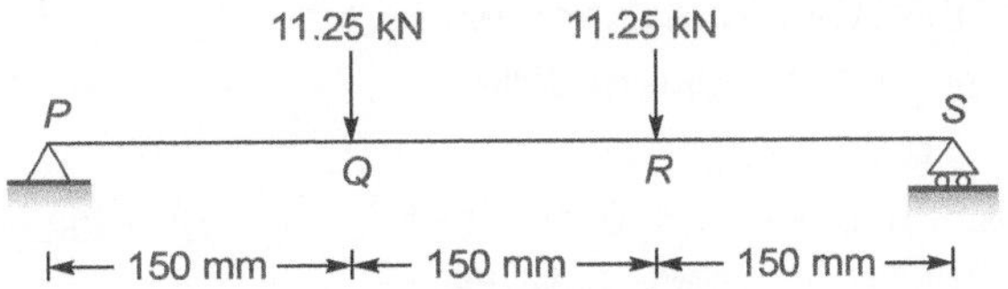

Fig. 8.39

[CE, GATE : 2016 (set-2)]

Solution: Reduction, $R_P = R_S = 11.25$ kN

Bending moment between Q and R = 11.25×0.150 kNm

$= 1.6875 \text{ kNm} = 1.6875 \times 10^8$ Nmm

Section modulus, $$Z = \frac{150^3}{6} = 56.25 \times 10^4 \text{ mm}^3$$

Modulus, of rupture, $$\sigma = \frac{M}{Z} = \frac{1.6875 \times 10^6}{56.25 \times 10^4} = 3 \text{ N/mm}^2$$

Question 8.29 A simply supported beam of width 100 mm and height 200 mm and length 4 m in carrying uniformly distributed load of intensity 10 kN/m. The maximum bending stress in MPa in the beam is ______ (Correct to one decimal place).

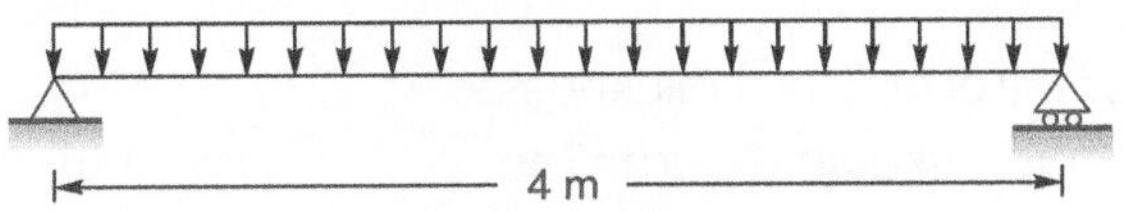

Fig. 8.30

[ME, GATE : 2018]

Solution:

Maximum bending moment, $M_{max} = \frac{wL^2}{8} = \frac{10 \times 4^2}{8} = 20$ kNm $= 20 \times 10^6$ Nmm

Section modulus, $Z = 100 \times \frac{200^2}{6} = \frac{8}{6} \times 10^6$ mm^3

$$\sigma_{b\,max} = \frac{M}{Z} = \frac{20 \times 10^6}{8 \times 10^6} \times 6 = \frac{120}{8} = 15 \text{ MPa}$$

Objective Type Questions

Q.1 A 6 m long simply supported wooden beam of rectangular section 10 cm × 20 cm deep is strengthened by mild steel plates 0.5 cm × 10 cm wide at the top and bottom fibres over the entire length. Find the minimum supportable UDL considering failures in steel and wood due to flexure. Weakening of this wood due to screws and weakening of the steel plates due to drilled holes may be ignored, figure.

Permissible tensile stress for steel = 156.8 N/mm^2

Permissible tensile stress for wood = 14.89 N/mm^2

$E_s = 1.96 \times 10^5$ N/mm^2, $E_w = 0.117 \times 10^5$ N/mm^2

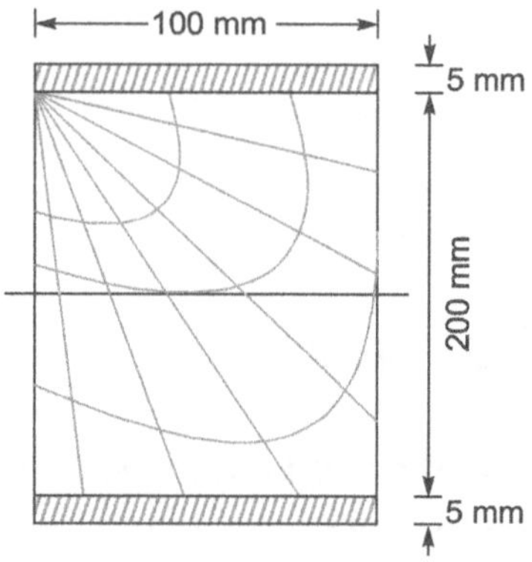

[GATE 1987, 2 Marks]

Q.2 A tapered cantilever beam of constant thickness is loaded as shown in the sketch, figure. The bending stress will be

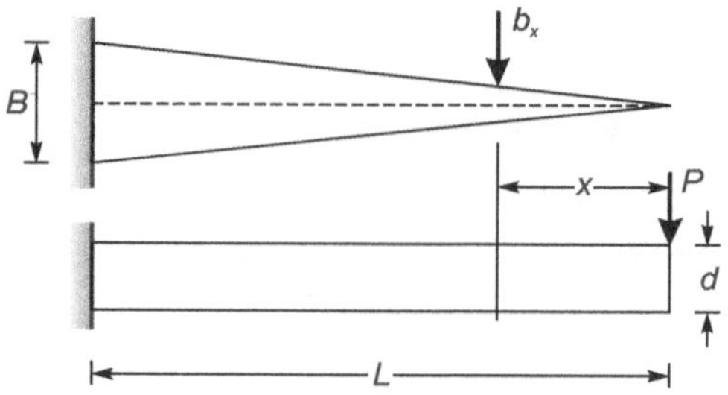

(a) maximum at the fixed end

(b) maximum at $x = \frac{L}{2}$

(c) maximum at $x = \frac{2L}{3}$

(d) uniform throughout the length

[GATE 1988, 2 Marks]

Q.3 The beams, one having square cross section and another circular cross-section, are subjected to the same amount of bending moment. If the cross sectional area as well as the material of both the beams are the same then

(a) maximum bending stress developed in both the beams is the same

(b) the circular beam experiences more bending stress than the square one

(c) the square beam experiences more bending stress than the circular one

(d) as the material is same both beams will experience same deformation

[GATE 2003, 1 Marks]

Q.4 A simply supported beam of span 6 m and 75 mm diameter carries a UDL of 1.5 kN/m, what is maximum value of bending stress
(a) 162.98 MPa (b) 395.95 MPa
(c) 0.25.95 MPa (d) 651.90 MPa
[GATE 2006, 2 Marks]

Q.5 For the component loaded with a force *F* as shown in the figure, the axial stress at the corner point *P* is

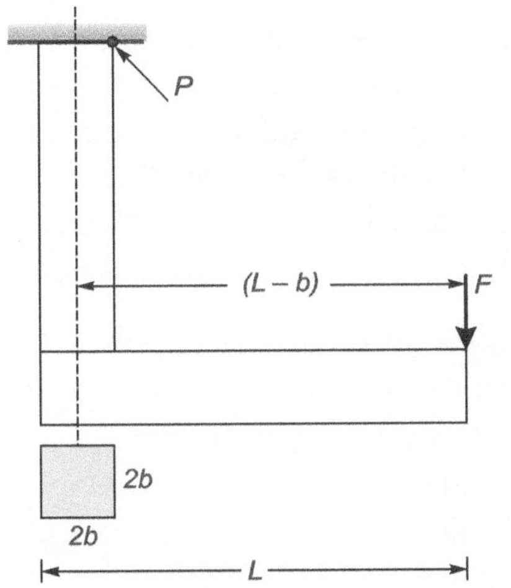

(a) $\frac{F(3L-b)}{4b^3}$ (b) $\frac{F(3L+b)}{4b^3}$

(c) $\frac{F(3L-4b)}{4b^3}$ (d) $\frac{F(3L-2b)}{4b^3}$

[GATE 2008, 2 Marks]

Q.6 A massless beam has a loading pattern as shown in the figure. The beam is of rectangular cross-section with a width of 30 mm and height of 100 mm.

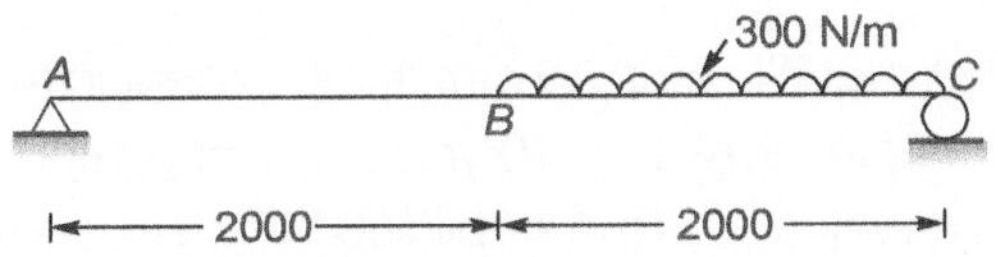

A. The maximum bending moment occurs at
(a) Location *B*
(b) 2675 mm to the right of *A*
(c) 2500 mm to the right of *A*
(d) 3225 mm to the right of *A*

B. The maximum magnitude of bending stress (in MPa) is given by
(a) 60.0 (b) 67.5
(c) 200.0 (d) 225.0
[GATE 2010, 2+2=4 Marks]

Q.7 A rectangular section beam subjected to bending moment *M* varying along its length is required to develop same maximum bending stress at any cross-section. If the depth of the section is constant, then its width wall vary as
(a) M (b) $\sqrt{M}$
(c) M^2 (d) $\frac{1}{M}$
[IES 1995]

Q.8 A beam cross-section is used in two different orientations as shown in the figure given below:

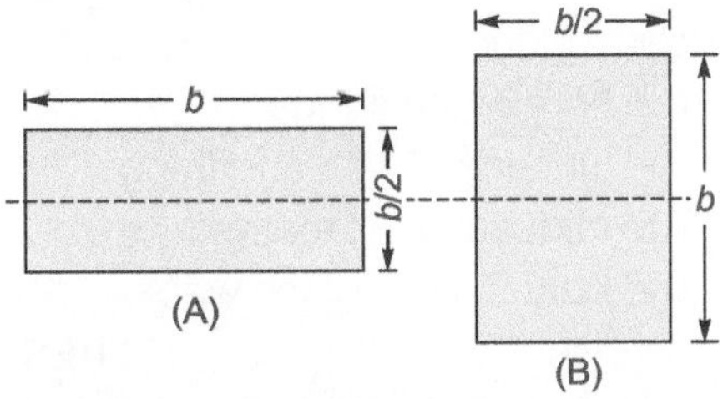

Bending moment applied to the beam in both cases is same. The maximum bending stresses induced in cases (*A*) and (*B*) are related as
(a) $\sigma_A = \sigma_B$ (b) $\sigma_A = 2\sigma_B$
(c) $\sigma_A = \frac{\sigma_B}{2}$ (d) $\sigma_A = \frac{\sigma_B}{4}$

Q.9 Beam A is simply supported at its ends and carries UDL of intensity 'w' over its entire length. It is made of steel having Young's modulus *E*. Beam *B* is a cantilever and carries a UDL of intensity *w*/4 over its entire length. It is made of brass having Young's modulus *E*/2. The two beams are of same length and have same cross-sectional area. If σ_A and σ_B denote the maximum bending stresses developed in beams *A* and *B*, respectively, then which one of the following is correct?
(a) $\sigma_A = \sigma_B$
(b) $\frac{\sigma_A}{\sigma_B} < 1.0$

(c) $\frac{\sigma_A}{\sigma_B} > 1.0$

(d) $\frac{\sigma_A}{\sigma_B}$ depends on the shape of the cross-section

[IES 2005]

Q.10 Beams of uniform strength vary in section such that

(a) bending moment remains constant
(b) deflection remains constant
(c) maximum bending stress remains constant
(d) shear force remains constant

[IES : 2006]

Q.11 A T-section beam is simply supported and subjected to a uniformly distributed load over its whole span. Maximum longitudinal stress in the beam occurs at

(a) Top fibre of the flange
(b) The junction of web and flange
(c) The midsection of the web
(d) The bottom fibre of the web

[IES : 2011]

Q.12 The ratio of the section modulii of a square beam (Z) when square section is placed (i) with two sides horizontal (Z_1) and (ii) with a diagonal horizontal (Z_2) as shown is

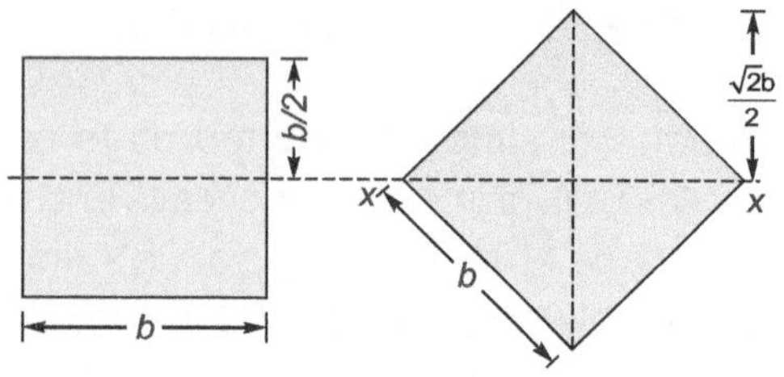

(a) $\frac{Z_1}{Z_2} = 1.0$
(b) $\frac{Z_1}{Z_2} = 2.0$
(c) $\frac{Z_1}{Z_2} = \sqrt{2}$
(d) $\frac{Z_1}{Z_2} = 1.5$

[IES : 2012]

Q.13 A beam with a rectangular section of 120 mm × 60 mm, designed to be placed vertically is placed horizontally by mistake. If the maximum stress is to be limited, the reduction in load carrying capacity would be

(a) $\frac{1}{4}$
(b) $\frac{1}{3}$
(c) $\frac{1}{2}$
(d) $\frac{1}{6}$

[IES : 2012]

Q.14 If E = elasticity modulus, I = moment of inertia about the neutral axis and M = bending moment is pure bending under the symmetric loading of a beam, the radius of curvature of the beam

1. increases with E
2. increases with M
3. decreases with I
4. decreases with M

which of these are correct?

(a) 1 and 3
(b) 2 and 3
(c) 3 and 4
(d) 1 and 4

Q.15 A steel flat 2 mm thick is bent around a drum of diameter 2 m. If E = 200 kN/mm^2, what is stress developed in flat, in N/mm^2

(a) 200
(b) 100
(c) 50
(d) None of these

Q.16 A beam section is triangular, skin stress at the base of the triangle is +20 MPa. What is skin stress developed at the apex of the triangle in MPa

(a) –20 MPa
(b) –40 MPa
(c) +20 MPa
(d) –10 MPa

Q.17 A rectangular beam section with breadth, b = 100 mm and depth 120 mm is subjected to a bending moment of 2 kNm. What is force on lower half of this section

(a) 25 kN
(b) 20 kN
(c) 16.667 kN
(d) None of these

Q.18 A metallic strip 10 mm thick, is bent in the arc of circle of radius 10 m. If the maximum stress developed in the strip is 50 MPa, what is material of the strip?

(a) steel
(b) aluminium
(c) copper
(d) any of these

Q.19 A cantilever of uniform strength σ, having variable rectangular section throughout with constant breadth is subjected to UDL throughout its length. If the depth of section is D at fixed end. What is depth at the middle of the cantilever

(a) 0.866 D (b) 0.75 D
(c) 0.5 D (d) None of these

Q.20 In a composite beam, a wooden beam 200 mm × 250 mm is reinforced by two steel plates at top and bottom of the size 200 mm × 25 mm, so that overall depth of section is 300 mm. Due to bending moment applied on beam, maximum stress developed in steel plates is 150 MPa. If modular ratio is 20, what is the maximum stress developed in wooden beam

(a) 5 MPa (b) 6.25 MPa
(c) 7.5 MPa (d) 9 MPa

Q.21 A structural steel beam has an unsymmetrical I-cross-section. The overall depth of the beam is 200 mm. The flange stresses at the top and bottom are 120 N/mm^2 and 80 N/mm^2, respectively. The depth of the neutral axis from the top of the beam will be

(a) 120 mm (b) 100 mm
(c) 80 mm (d) 60 mm

[ESE 2016]

Q.22 In order that the extreme fibre stresses in bending will be in the ratio 4 : 3 in the beam shown in the following figure, the width b of the upper flange (b < 10 cm) of the beam section is to be

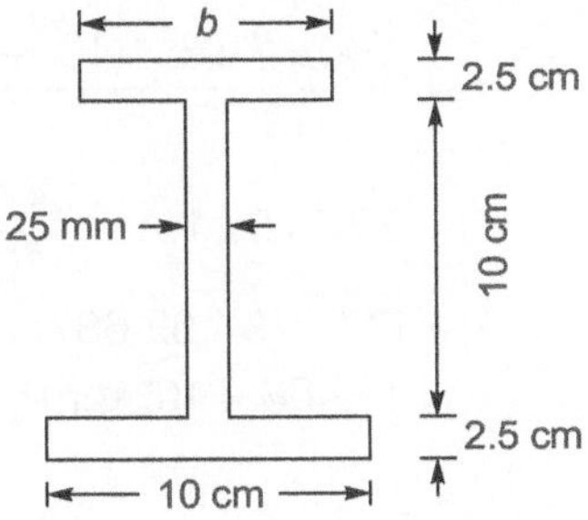

(a) 6.1 cm (b) 6.6 cm
(c) 5.1 cm (d) 5.6 cm

[ESE 2016]

Q.23 A homogeneous prismatic simply supported beam is subjected to a point load F. The load can be placed anywhere along the span of the beam. The very maximum flexural stress developed in the beam is

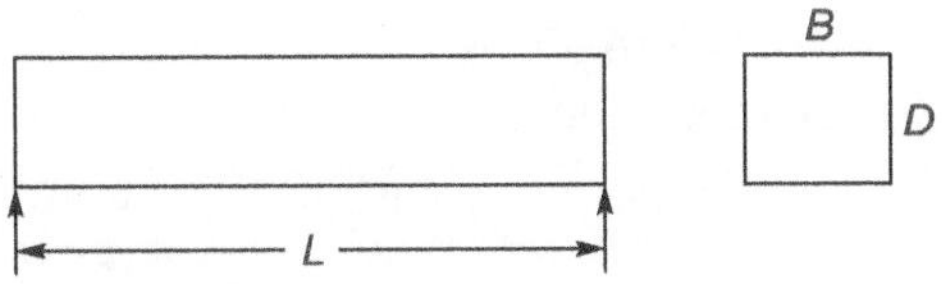

(a) $\frac{3FL}{2BD^2}$ (b) $\frac{3FL}{4BD^2}$
(c) $\frac{2FL}{3BD^2}$ (d) $\frac{4FL}{3BD^2}$

[ESE (pre) : 2017]

Q.24 The span of cantilever beam is 2 m. The cross-section of the beam is a hollow square with external sides 100 mm; and its I = 4 × 10^5 mm^4. The safe bending stress for the beam material is 100 N/mm^2. The safe concentrated load at the free end would be

(a) 100 N (b) 200 N
(c) 300 N (d) 400 N

[ESE (pre) : 2017]

Q.25 A long rod of uniform rectangular section with thickness t, originally straight, is bent into the form of a circular arch with displacement d at the mid-point of span l. The displacement d may be regarded as small as compared to the length l. The longitudinal surface strain is

(a) $\frac{2td}{l^2}$ (b) $\frac{4td}{l^2}$
(c) $\frac{8td}{l^2}$ (d) $\frac{16td}{l^2}$

[ESE (pre) : 2018]

Q.26 A beam of rectangular section (12 cm wide × 20 cm deep) is simply supported over a spam of 12 m. It is acted upon by a concentrated load of 80 kN at the mid-span. The maximum bending stress induced is

(a) 400 MPa (b) 300 MPa
(c) 200 MPa (d) 100 MPa

[ESE (pre) : 2018]

Q.27 A thin steel ruler having its cross-section of 0.0625 cm × 2.5 cm is bent by couples applied at the side so that its length l equal to 25 cm, when bent, as a circular arc, subtends a central angle $\theta = 60°$. Take $E = 2 \times 10^6$ kg/cm². The maximum stress induced in the ruler and the magnitude is

(a) 2618 kg/cm² (b) 2512 kg/cm²
(c) 2406 kg/cm² (d) 2301 kg/cm²

[ESE : 2016]

Q.28 In the design of beams for a given strength, consider that the conditions of economy of use of the material would avail as follows:

1. Rectangular cross-section is more econoical than square section of the same cross-sectional area of the beam.
2. Circular section is more economical than square section.
3. I-section is more economical than a rectangular section of the same depth

Which of the above are correct?

(a) 1, 2 and 3 (b) 1 and 2 only
(c) 2 and 3 only (d) 1 and 3 only

[ESE (pre) : 2017]

Q.29 In the analysis of beams subjected to loads, the point with Nil Bending Moment can be a

1. Point of Contraflexure
2. Point of Maximum Shear Force
3. Point of Inflexion

Which of the above statements is/are correct?

(a) 1 only (b) 2 only
(c) 3 only (d) 1, 2 and 3

[EES : 2016]

Q.30 Consider a beam with circular cross-section of diameter *d*. The ratio of the second moment of area about the neutral axis to the section modulus of the area is

(a) $\frac{d}{2}$ (b) $\frac{\pi d}{2}$
(c) d (d) πd

Answers

1. (4.808 kN/m)	2. (d)	3. (b)	4. (a)	
5. (d)	6. 1. (c) 2. (b)	7. (a)	8. (b)	
9. (d)	10. (c)	11. (d)	12. (c)	13. (c)
14. (d)	15. (a)	16. (b)	17. (a)	18. (c)
19. (c)	20. (b)	21. (a)	22. (d)	23. (a)
24. (d)	25. (b)	26. (b)	27. (a)	28. (d)
29. (a)	30. (a)			

Explanations

1. **(4.808)**

$$\frac{E_s}{E_w} = \frac{1.96}{0.117} = 16.75$$

If $\sigma_s = 156.8$ N/mm²

$$\sigma_w = \frac{\sigma_s}{m} = 9.36 \text{ N/mm}^2 < 14.89$$

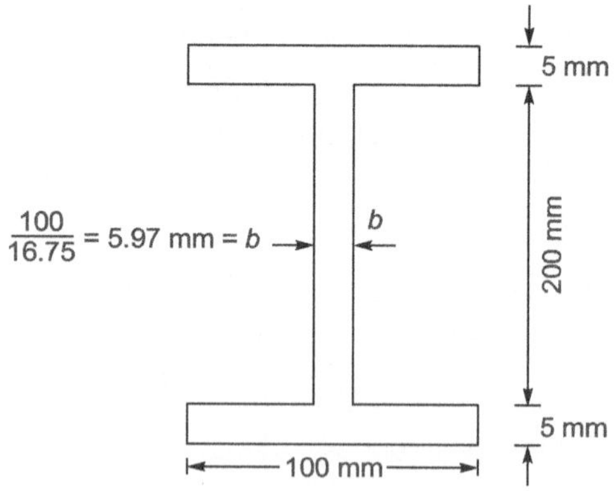

Equivalent steel section

So permitted

$$M_{max} = \frac{wL^2}{8} = \frac{w \times 6^2}{8} = 4.5w \text{ kN/m}$$

$$I = \frac{100 \times 210^3}{12} - \frac{(100 - 5.97) \times 200^3}{12}$$

$$= 77.175 \times 10^6 - \frac{94.03}{12} \times 200^3$$

$$= [77.175 - 62.687] \times 10^6$$

$$= 14.488 \times 10^6 \text{ mm}^4$$

$$Z = \frac{14.488 \times 10^6}{105} = 13.798 \times 10^4 \text{ mm}^3$$

$$\sigma = 156.8 \text{ N/mm}^2$$

$$M = 156.8 \times 13.798 \times 10^4 \text{ Nmm}$$

$$= 2163.5264 \times 10^4 \text{ Nmm}$$

$$= 21.635 \text{ kNm} = 4.5w$$

$$w = 4.808 \text{ kN/m}$$

2. (d)

Constant stress

$$\text{Moment } Px = b \times \frac{d^2}{6} \times \sigma$$

$$Px = \frac{Bx}{L} \times \frac{d^2}{6} \times \sigma$$

$$\sigma = \frac{6PL}{Bd^2}, \text{ independent of } x$$

3. (b)

$$\text{Square} = a^2$$

$$Z = \frac{a^3}{6} = 0.166a^3$$

$$\text{Circular, } \frac{\pi}{4}d^2 = a^2$$

$$d = 1.128a$$

$$Z = \frac{\pi d^3}{32} = \frac{\pi}{32}(1.128a)^3$$

$$= 0.141a^3$$

$$Z_{square} > Z_{circular}$$

Bending stress in circular section will be more than in square section.

4. (a)

$$M_{max} = \frac{wL^2}{8} = \frac{1.5 \times 6^2}{8} = 6.75 \text{ kNm}$$

$$= 6.75 \times 10^6 \text{ Nmm}$$

$$= Z.\sigma = \frac{\pi d^3}{32} \times \sigma$$

$$6.75 \times 10^6 = \frac{\pi \times 75^3}{32} \times \sigma = 4.148 \times 10^4 \sigma$$

$$\sigma = 162.98 \text{ MPa}$$

5. (d)

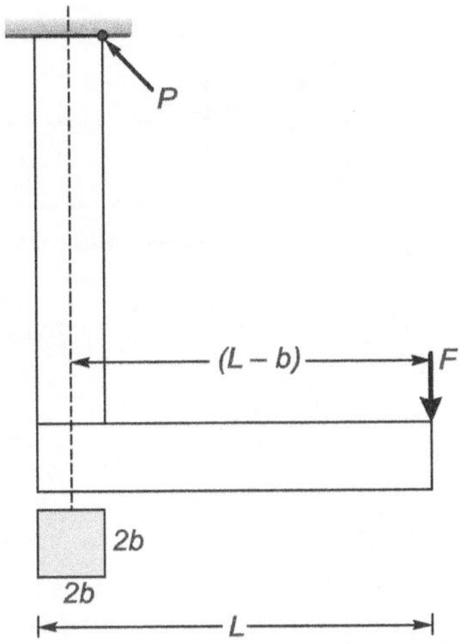

$$Z = \frac{(2b)^3}{6} = \frac{8b^3}{6}$$

$$\sigma_b = \frac{F(L-b)6}{8b^3}$$

$$\sigma_b + \sigma_d = \frac{6F(L-b)}{8b^3} + \frac{F}{4b^2}$$

$$= \frac{6F(L-b) + F \times 2b}{8b^3}$$

$$= \frac{6FL - 6Fb + 2bF}{8b^3}$$

$$= \frac{6FL - 4Fb}{8b^3} = \frac{2(F)(3L-2b)}{8b^3}$$

$$= \frac{F(3L-2b)}{4b^3}$$

6. (c, b)

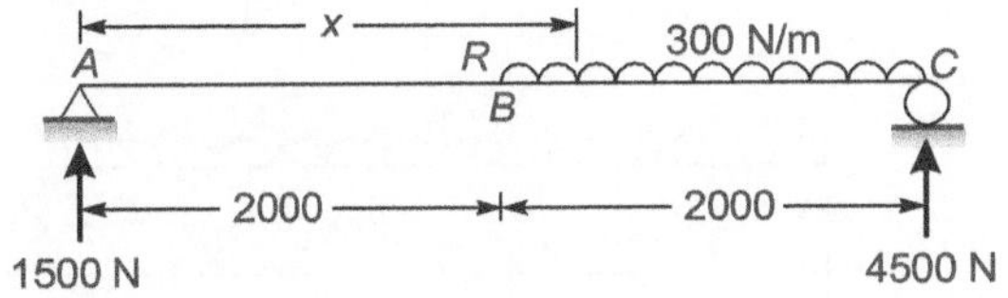

$$R_C = \frac{3 \times 6000}{4} = 4500 \text{ N}$$

$$R_A = 1500 \text{ N}$$

$$\text{SF, } F_x = 1500 - 3000(x - 2) = 0$$

for maximum BM

$$1 = 2(x - 2)$$

$$1 = 2x - 4, x = 2.5 \text{ m from } A$$

Maximum BM

$$= 1500 \times 2.5 - 0.5 \times 3000 \times 0.25$$

$$= 3750 - 375 = 3375 \text{ Nm}$$

$$= 3375 \times 10^3 \text{ Nmm}$$

$$= \sigma \times \frac{bd^2}{6} = \sigma_m \times \frac{30 \times 100^2}{6}$$

$$3375 \times 10^3 = 5 \times 10^4 \sigma_{max}$$

$$\sigma_{max} = 67.5 \text{ MPa}$$

7. (a)

$$M = C_1 x$$

$$b = C_2 x$$

$$C_1 x = \frac{bd^2}{6}\sigma = \frac{C_2 x}{6} \times d^2 \sigma$$

$$\sigma = \frac{C_1}{C_2} \times \frac{6}{d^2}, \quad \text{(constant } \sigma \propto M)$$

8. (b)

$$Z_A = \frac{b}{6}\left(\frac{b}{2}\right)^2 = \frac{b^3}{24}$$

$$Z_B = \frac{b}{2}(b^2)\frac{1}{6} = \frac{b^3}{12}$$

$$M = \sigma_A Z_A = Z_B \sigma_B$$

$$\sigma_A \times \frac{b^3}{24} = \sigma_B \times \frac{b^3}{12}$$

$$\sigma_A = 2\sigma_B$$

9. (d)

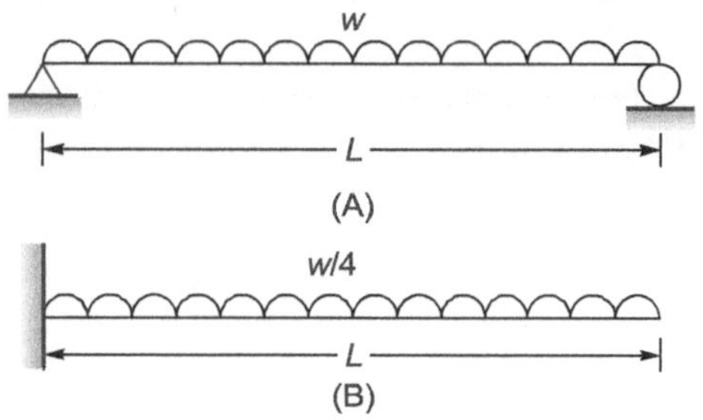

Same BM $\frac{wL^2}{8} = \frac{w}{4} \times \frac{L^2}{2}$

But $\frac{\sigma_A}{\sigma_B}$ depends on shape, even if their area of cross-section is the same

$$Z_{circular} < Z_{square}$$

10. (c)
In beams of uniform strength maximum bending stress remains constant.

11. (d)

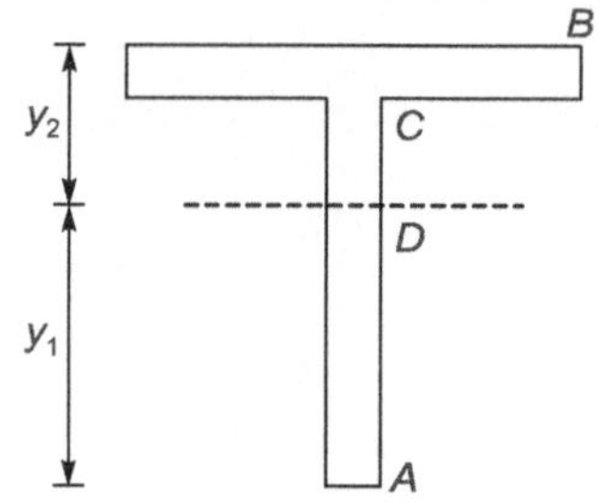

Bottom plane of Web

$$y_1 > y_2$$

Maximum longitudinal stress in beam occurs at *A*, bottom fibre of web.

12. (c)

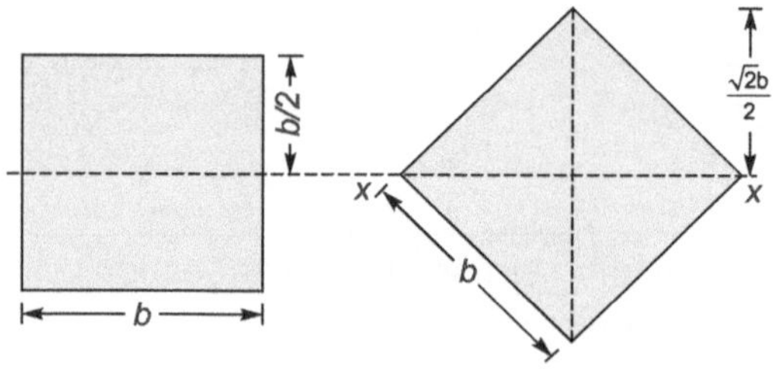

$$I_1 = I_2$$

$$Z_1 = \frac{I}{b} \times 2$$

$$Z_2 = \frac{I}{\sqrt{2}b} \times 2$$

$$\frac{Z_1}{Z_2} = \sqrt{2}$$

13. (c)

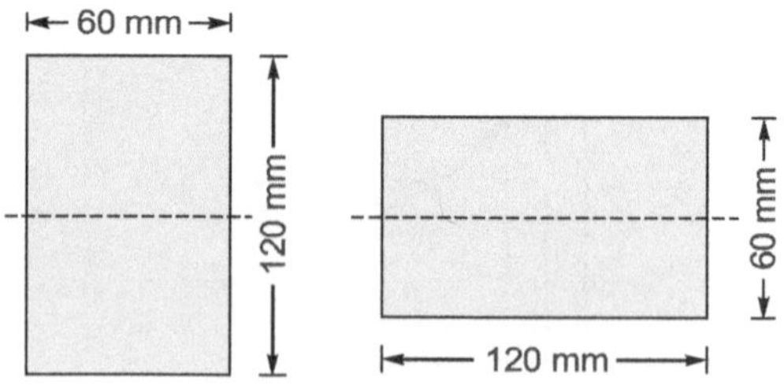

$$Z_1 = \frac{60 \times 120^2}{12} = 72000 \text{ mm}^3$$

$$Z_2 = \frac{120 \times 60^2}{12} = 36000 \text{ mm}^3$$

$$\frac{Z_2}{Z_1} = 0.5 = \frac{1}{2}$$

$$\frac{1}{2} = \frac{M_2}{M_1} \text{ or load } = \frac{W_2}{W_1}$$

14. (d)

$$\frac{M}{I} = \frac{E}{R}$$

$$R = \frac{EI}{M}$$

1. increases with E
4. decreases with M

15. (a)

$$\sigma = y\frac{E}{R} = \frac{1 \times 200000}{1000} = 200 \text{ N/mm}^2$$

16. (b)

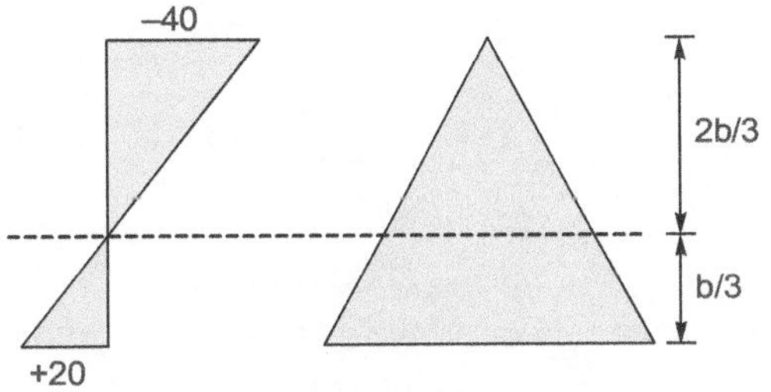

17. (a)

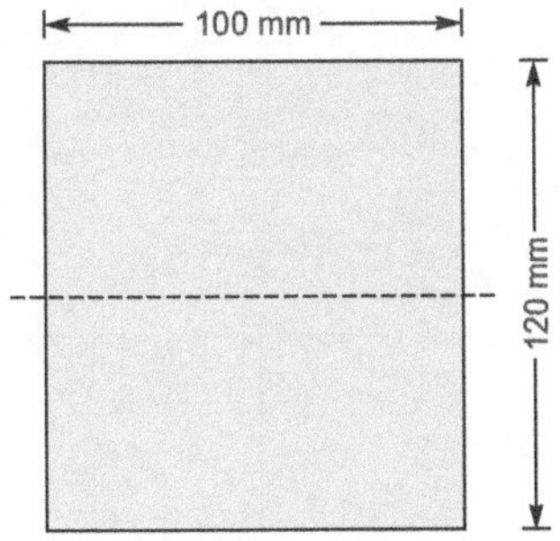

$$Z = \frac{100 \times 120^2}{6} = 240000 \text{ mm}^3$$

$$M = 2 \times 10^6 \text{ kNm}$$

$$\sigma_m = \frac{2 \times 10^6}{240000} = 8.33 \text{ N/mm}^2$$

$$\sigma_{av} = 4.166 \text{ Nm}^2$$

$$A = 60 \times 100 = 25 \text{ kN}$$

$$F = 6000 \times 4.166$$

$$= 25000 \text{ N} = 25 \text{ kN}$$

18. (c)

$$\sigma = y\frac{E}{R}$$

$$E = \frac{\sigma R}{y} = \frac{50 \times 10000}{5} = 10000$$

$$= 100 \text{ kN/mm}^2 \text{ (E for copper)}$$

19. (c)

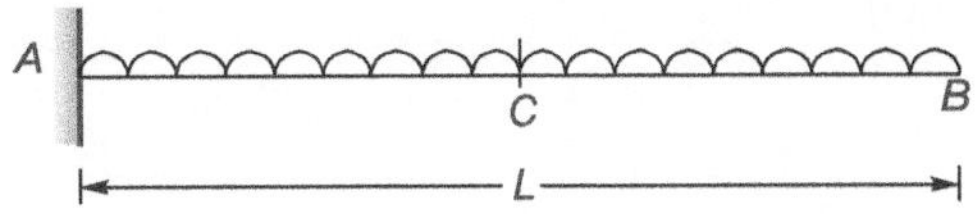

At A $\quad M = \frac{wL^2}{2}, Z = \frac{bD^2}{6}$

$$\sigma = \frac{wL^2}{2} \times \frac{6}{bD^2} = \frac{3w}{b}\left(\frac{L^2}{D^2}\right)$$

At C $\quad M = \frac{wL^2}{8}, Z = \frac{bd^2}{6}$

σ_1 same

$$\frac{3w}{b}\left(\frac{L^2}{D^2}\right) = \frac{wL^2}{8} \times \frac{6}{bd^2}$$

$$\frac{1}{D^2} = \frac{1}{4d^2}, \; d = 0.5\, D$$

20. (b)

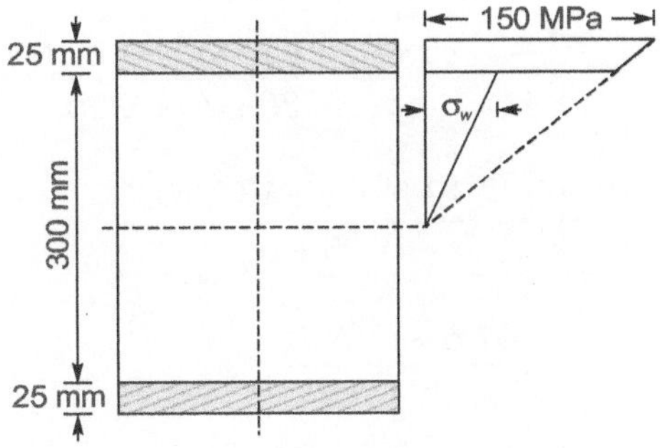

$$\sigma_w = \frac{\sigma_w}{m} \times \frac{d}{D} = \frac{150}{20} \times \frac{250}{300} = 6.25 \text{ MPa}$$

21. (a)

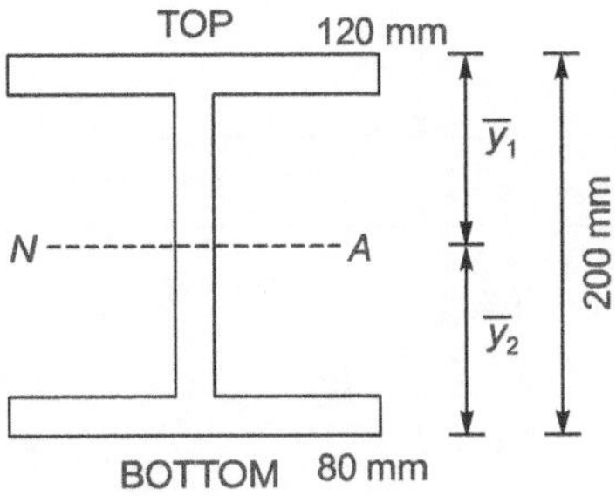

$$\frac{y_1}{y_2} = \frac{120}{80} = 1.5$$

Overall depth = 200 mm

$y_1 = 120,\ y_2 = 80$

The depth of neutral axis from the top of the beam is 120 mm.

22. (d)

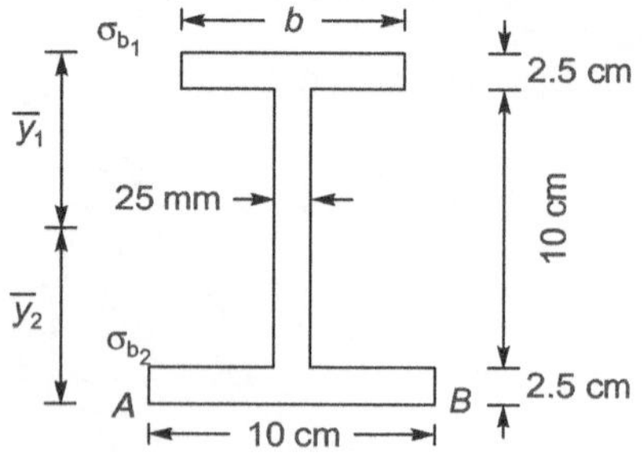

$$\frac{\sigma_{b_1}}{\sigma_{b_2}} = \frac{\text{Boundary stress at top}}{\text{Boundary stress at bottom}}$$

$$= \frac{4}{3}$$

$$y_2 = \frac{3}{7} \times 15 = \frac{45}{7} \text{ cm}$$

Moment about lower edge AB

$$y_2 = \frac{25 \times 1.25 + 25 \times 7.5 + 2.5b \times 13.75}{25 + 25 + 2.5b}$$

$$\frac{45}{7} = \frac{12.5 + 75 + 13.75b}{20 + b}$$

$$= \frac{87.5 + 13.75b}{20 + b}$$

$90 + 45\ b = 612.5 + 96.25\ b$

$287.5 = 51.25\ b$

$b = 5.61$ cm

23. (a)

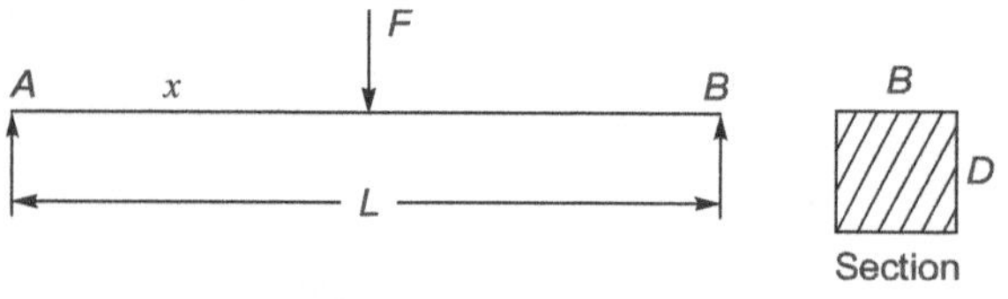

$$M_{max} = \frac{FL}{4}$$

Section modulus,

$$Z = \frac{BD^2}{6}$$

Very maximum flexural stress,

$$\sigma_{max} = \frac{M_{max}}{Z}$$

$$= \frac{FL}{4} \times \frac{6}{BD^2}$$

$$= \frac{3}{2} \times \frac{FL}{BD^2}$$

24. (d)

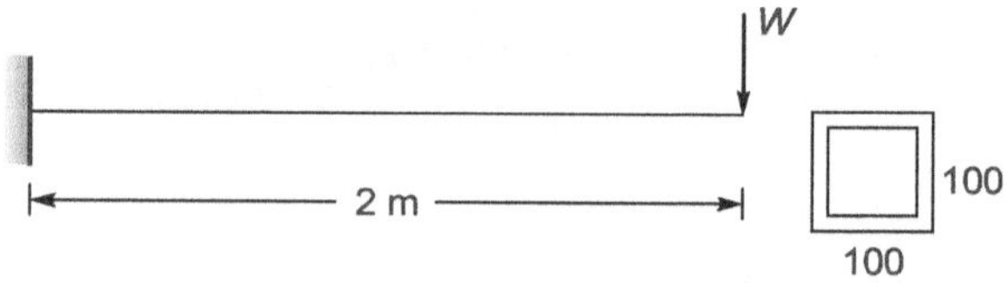

$$I = 4 \times 10^5 \text{ mm}^4$$

$$Z = \frac{4 \times 10^5}{50} = 8 \times 10^3 \text{ mm}^3$$

M = WL

= 2000 Nmm

σ = Safe stress in bending

= 100 MPa

$WL = \sigma Z$

$2000\ W = 100 \times 8 \times 10^3$

W = 400 N

25. (b)

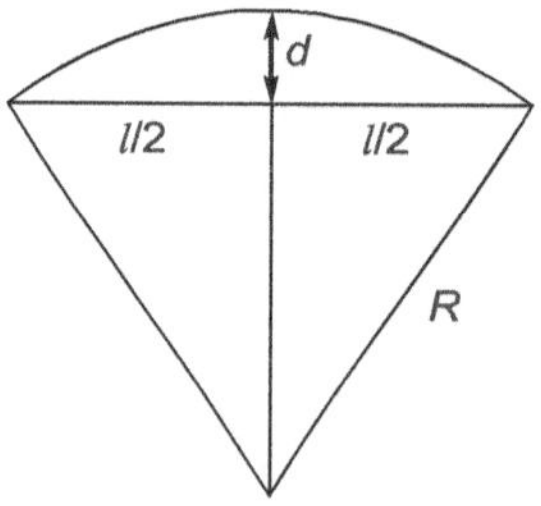

$$d = \frac{l^2}{8R}$$

$$R = \frac{l^2}{8R}$$

$$\frac{2\sigma}{t} = \frac{E}{R}$$

$$\in = \frac{\sigma}{E} = \frac{t}{2R} = \frac{t}{2} \times \frac{8d}{l^2} = \frac{4td}{l^2}$$

26. (b)
Simple bending
Simply supported beam (with span L concentrated load, W)
Maximum bending moment,

$$M_{max} = \frac{wL}{4}$$

$$= \frac{80 \times 12}{4} = 240 \text{ kNm}$$

$$= 240 \times 10^6 \text{ Nmm}$$

Section width 12 cm, depth 20 cm
Section modulus,

$$Z = \frac{bd^2}{6} = \frac{120}{6} \times (200)^2$$

$$= 8 \times 10^5 \text{ kNm}^3$$

Maximum bending stress,

$$\sigma_b = \frac{M}{Z} = \frac{240 \times 10^6}{8 \times 10^5}$$

$$= 300 \text{ MPa}$$

27. (a)
Thickness, t = 0.0625 cm
b = 2.5 cm
L = 25 cm

$$\text{Circular are, } L = \frac{2\pi R}{6} \text{ as angle } 80°$$

$$R = \frac{25 \times 6}{2\pi} = \frac{75}{\pi} \text{ cm}$$

$$\frac{\sigma}{y} = \frac{E}{R} \quad \text{where } y = \frac{t}{2}$$

$$= \frac{0.0625}{2} \times \frac{2 \times 10^6}{75} \times \pi$$

$$= 2618 \text{ kg/cm}^2$$

28. (d)
1. Rectangular section more economical than square for same cross-section are – Correct.
2. Circular section is more economical than square section–Incorrect.
3. I-section is more economical than a rectangular section of same depth–Correct.

29. (a)
1 only
Because point of contraflexure is a point on BM diagram, when BM = 0 and bending moment changes sign.
Point of maximum shear force is not a point when BM = 0
Point of inflexion is a point on a continuous curves, when curve changes from convex to concave or vice-versa.

30. (a)

$$I = \frac{\pi d^4}{64}$$

$$Z = \frac{\pi d^3}{32}$$

$$\frac{I}{Z} = \frac{d}{2}$$

■■■■

Shear Stresses in Beams

CHAPTER 9

A beam is subjected to transverse loads, producing bending moment and shear force in different sections of a beam. Bending moment causes direct bending stresses in the section of the beam, while shear force causes shear stress on layers perpendicular to the section. Shear stress produced in a section is of less magnitude than the direct bending stress and is of no practical importance. However to learn the shear stress and variation of the shear stress along the depths of the section is of academic interest to civil engineers, because principal stresses at any section are a function of bending stress as well as shear stress.

In this chapter, we will derive a relation between shear force *F* and shear stress τ developed in a layer at a distance of y from neutral layer.

Question 9.1 **Consider a section of a beam, and an elementary length *dx* of the beam subjected to variable bending moment on both sides of the element. Derive a relationship between shear stress and shear force.**

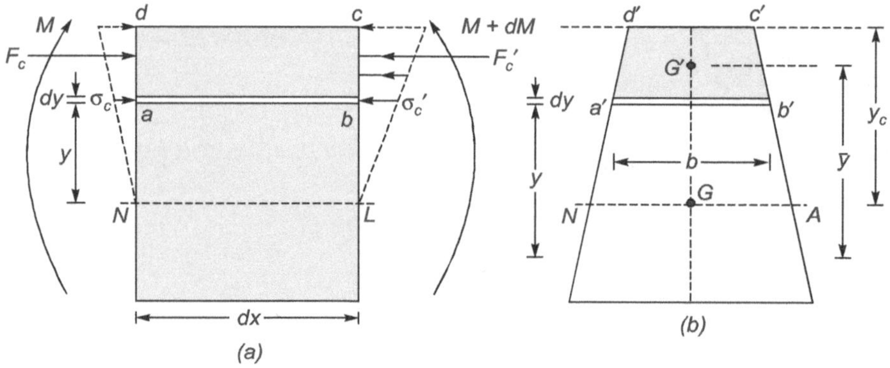

Fig. 9.1

Solution: Consider a beam of any section, say a trapezoidal section as shown in figure 9.1 (b) and element of length *dx* as shown in figure 9.1 (a). Bending moment on the left side of the section is *M* and bending moment on the right side of the section is *M* + *dM*. The type of the bending moment (sagging type) produces compressive stresses in upper zone, above neutral layer and tensile stresses below the NA.

$$\sigma_c' = \frac{(M+dM)}{I_{NA}} \times y \text{ on right side}$$

$$\sigma_c = \frac{M}{I_{NA}} \times y \text{ on left side}$$

I_{NA} = moment of inertial about neutral axis

$\delta P'$, force on elementary area $b.dy$ (see b in fig. 9.1 (b) (along the section)

$$\delta P' = \sigma_c' \, b.dy$$

$$\delta P = \sigma_c \, bdy$$

$$\text{Unbalance force} = (\sigma_c' - \sigma_c) \, b \, dy = dM \, b \, y \, dy$$

Summing up the unbalanced force from y to y_c

Force $$\delta P = \int_y^{y_c} \frac{dM}{I_{NA}} ybdy$$

but $$\delta P = \tau.bdx = \text{shear stress} \times \text{breadth} \times \text{elementary length}$$

$$\tau.b.dx = \int_y^{y_c} \frac{dM}{I_{NA}} ybdy \quad \text{where } \frac{dM}{dx} = F \text{, shear force at the section}$$

$$\tau = \frac{F}{bI_{NA}} \int_y^{y_c} b.y.dy = \frac{1}{bI_{NA}} \int_y^{y_c} ydA \quad \text{where } b.dy = dA$$

but $$\int_y^{y_c} ydA = A\bar{y}$$

$$= \text{area } a'b'c'd' \times \bar{y}$$

$\bar{y}$ = distance of CG of area above the layers $a'b'$ and upto extreme layer $c'd'$ as shown in figure .91 (b)

Shear stress, $$\tau = \frac{FA\bar{y}}{I_{NA} b}$$

where, τ = shear stress at layer, at a distance y from neutral axis

F = shear force at the section

A = **area of the section from the layer upto extreme layer on that side**

I_{NA} = moment of inertia of the section about neutral axis

b = breadth of the layer under consideration

Question 9.2 **Take a rectangular section $b \times d$ beam, subjected to shear force F, and draw shear stress distribution along the depth of the section.**

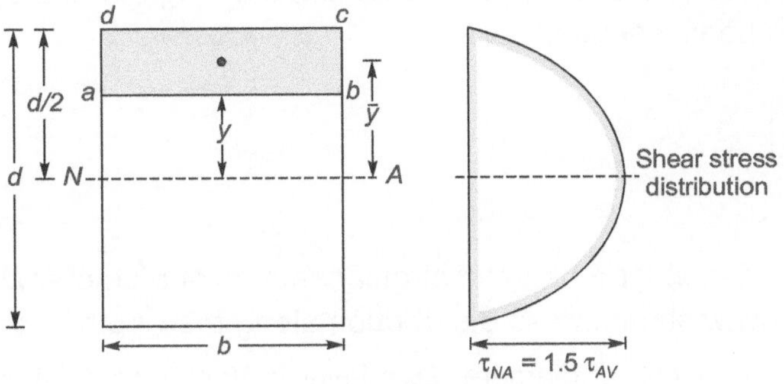

Fig. 9.2

Solution:

Shear force on section = F

Breadth = b

Depth = d

I_{NA}, moment of inertial about neutral axis $= \dfrac{bd^3}{12}$

Take layer *ab* at a distance *y* from neutral layer area A = (area *abcd*) $= \left(\dfrac{d}{2} - y\right)b$.

$\bar{y}$ = distance of CG of A from neutral layer, fig. 9.2

$$= y + \frac{1}{2}\left(\frac{d}{2} - y\right) = \frac{d}{4} - \frac{y}{2} + y = \frac{d}{4} + \frac{y}{2} = \frac{1}{2}\left(\frac{d}{2} + y\right)$$

$$A\bar{y} = \frac{1}{2}\left(\frac{d}{2} - y\right)\left(\frac{d}{2} + y\right)b = \frac{b}{2}\left(\frac{d^2}{4} - y^2\right)$$

So shear stress, $\tau = \dfrac{F}{bI_{NA}} \times A\bar{y}$, putting the values

$$= \frac{F12}{b \times bd^3} \times \frac{b}{2}\left(\frac{d^2}{4} - y^2\right) = \frac{6F}{bd^3}\left(\frac{d^2}{4} - y^2\right)$$

$y = 0$, τ, at neutral axis $= \dfrac{6F}{bd^3} \times \dfrac{d^2}{4} = \dfrac{1.5F}{bd} = 1.5\tau_{av}$, as $\tau_{av} = \dfrac{F}{bd}$

$y = \dfrac{d}{4}$ τ, at $\dfrac{d}{4}$ from NA $= \dfrac{6F}{bd^3}\left(\dfrac{d^2}{4} - \dfrac{d^2}{16}\right) = 1.125\tau_{av}$

$y = \dfrac{d}{2}$, τ, at $\dfrac{d}{2}$ from NA (at extreme layer) = 0

τ_{av} = average shear stress $= \dfrac{F}{bd}$

Shear stresses are symmetrically repeated on the other side of the section because rectangular section is symmetrical about NA.

Practice Q.9.2 **A beam of rectangular section b = 150 mm, depth 250 mm is supported over a length of 4 m. It carries a UDL of 4 kN/m run throughout the length of the beam. What is the maximum shear stress developed in beam section.**

Hints.: $\left[F_{max} = \dfrac{wL}{2}\right]$

Ans. [τ_{NA} = 0.32 N/mm^2]

Question 9.3 **Consider a beam of circular section of diameter D subjected to a shear force F at a particular section. Draw shear stress distribution along the depth of the section.**

Solution: Consider a section of diameter D, a layer of thickness dy at radial distance y, or angular displacement θ.

$$\sin\theta = \frac{y}{R},$$

where R is radius of circular section

$$y = R\sin\theta$$

$$dy = R\cos\theta.\, d\theta$$

$$b = \text{breadth of the layer}$$

$$= 2R\cos\theta$$

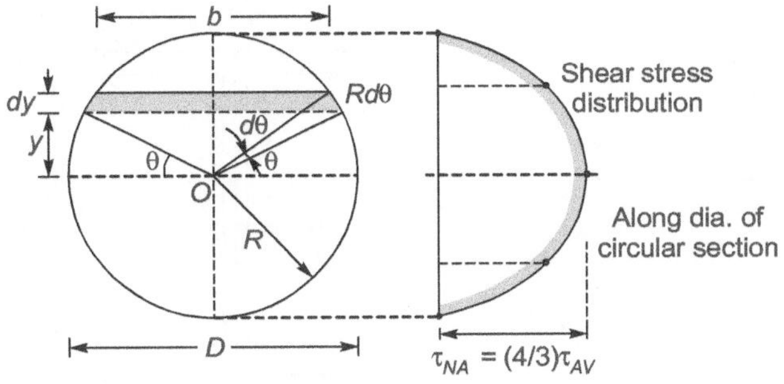

Fig. 9.3 *Shear stress distribution in a circular section*

$$I_{NA} = \frac{\pi D^4}{64} = \frac{\pi R^4}{4}$$

$$A\bar{y} = \int_y^R b\,dy = \int_\theta^{\pi/2} (2R\cos\theta)(R\sin\theta)(R\cos\theta)d\theta$$

$$= 2R^3\int_\theta^{\pi/2} \cos^2\theta\sin\theta\, d\theta = \frac{2R^3}{3}\left|\cos^3\theta\right|_\theta^{\pi/2}$$

$$= \frac{2R^3}{3}\left|\cos^3\frac{\pi}{2} - \cos^3\theta\right| = -\frac{2R^3}{3}\cos^3\theta, \text{ as } \cos\frac{\pi}{2} = 0$$

$$\tau = \frac{F}{bI_{NA}}A\bar{y} = \frac{F}{2R\cos\theta}\times\frac{4}{\pi R^4}\times\frac{2R^3}{3}\cos^3\theta = \frac{4F}{3\pi R^2}\times\cos^2\theta$$

$$\tau \text{ at } \theta = 0, \frac{4F}{3\pi R^2} = \frac{4}{3}\tau_{av}$$

where τ_{av} = average shear stress = $\dfrac{F}{\pi R^2}$

$$\tau \text{ at } \theta = 30°, \frac{F}{\pi R^2} = \tau_{av}$$

$$\tau \text{ at } 45°, \frac{2}{3}\frac{F}{\pi R^2} = \frac{2}{3}\tau_{av}$$

$$\tau \text{ at } 60°, \frac{1}{3}\frac{F}{\pi R^2} = \frac{1}{3}\tau_{av}$$

$$\tau \text{ at } 90° = 0, 0$$

Section is symmetrical about horizontal diameter. Shear stress distribution will repeat symmetrically for lower portion of the section as shown in figure 9.3 (b)

Practice Q.9.3 A beam 4 m long supported at ends, carries two loads *W* each, at 1 m from each end. What is the maximum value of W if maximum shear stress in the circular section of diameter 100 mm is not to exceed 1.8 MPa.

Ans. [10.60 kN]

Question 9.4 Section of a beam is I section, 300 mm deep × 200 mm wide as shown in figure 9.4. A shear force of 60 kN acts on the section. Determine shear stresses at points *A*, *B*, *C* and *D* of the section. Draw shear stress distribution along the depth of the section.

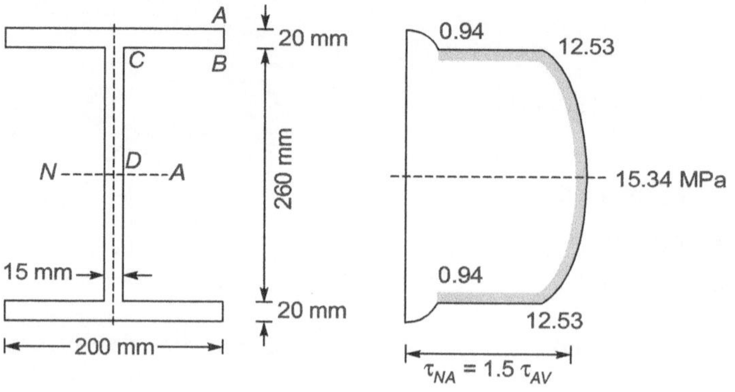

Fig. 9.4

Solution:

$$I_{NA} = \frac{200\times300^3}{12} - \frac{185\times260^3}{12}$$

$$= 4.5\times10^8 - 2.70963\times10^8 = 1.7903\times10^8\ \text{mm}^4$$

$$F = 60\ \text{kN} = 60000\ \text{N}$$

A $\quad A\bar{y} = 0$

$$\tau_A = 0$$

B $\quad A\bar{y} = 20\times200\times(140) = 56\times10^4\ \text{mm}^3$, about NA

C $\quad A\bar{y} = 56\times10^4\ \text{mm}^3$

D $\quad A\bar{y} = 20\times200\times140 + 15\times130\times65$

$$= 56\times10^4 + 12.675\times10^4 = 68.675\times10^4\ \text{mm}^3$$

$$\tau_{B'}\ \text{(breadth is 200 mm)} = \frac{60000\times56\times10^4}{1.7903\times10^8\times200} = 0.94\ \text{N/mm}^2$$

$$\tau_{C'}\ \text{(breadth is 15 mm)} = \frac{60000\times56\times10^4}{1.7903\times10^8\times15} = 12.53\ \text{N/mm}^2$$

$$\tau_{D'}\ \text{(breadth is 15 mm)} = \frac{68.675\times10^4\times60000}{1.7903\times10^8\times15} = 15.34\ \text{N/mm}^2$$

Section is symmetrical about NA, shear stress distribution will repeat symmetrically for the lower portion. Shear stress distribution is shown in figure 9.4 (b).

Question 9.5 **Show that in a beam section, which is thin circular, maximum shear stress developed due to shear force *F* is double the average shear stress.**

Solution: Fig. 9.5 shows a thin circular section of mean radius *R* and radials thickness, *t*

Shear force = F

Area of cross-section of thin circular section = $2\pi Rt$

$$\tau_{av},\ \text{average shear stress} = \frac{F}{2\pi Rt} \qquad \text{...(i)}$$

I_{NA}, moment of inertia about NA (neutral axis)

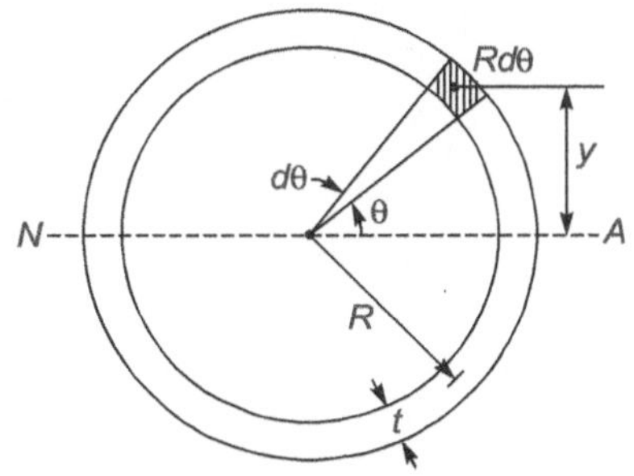

Fig. 9.5

$$= \frac{(2\pi Rt)R^2}{2} = \pi tR^3 \ldots\text{(ii)}$$

Take a small element $Rd\theta$ at angular distance θ,

Area, $dA = tRd\theta$

$y = R\sin\theta$

$dA.y = tR^2 \sin\theta\, d\theta.$

about neutral axis, $A\bar{y} = \int_0^{\pi} tR^2 \sin\theta d\theta = tR^2 \left|-\cos\theta\right|_0^{\pi} = 2tR^2$

$$\tau_{NA} = \frac{FA\bar{y}}{I.b} = \frac{F\times 2tR^2}{\pi tR^3 \times 2t} = \frac{F}{\pi Rt} = 2\tau_{av}$$

Question 9.6 A beam section is *T*, with dimensions shown in fig. 9.6. Shear force is acting on this section is *F*. Locate *G* of the section and plot shear stress distribution along its depth.

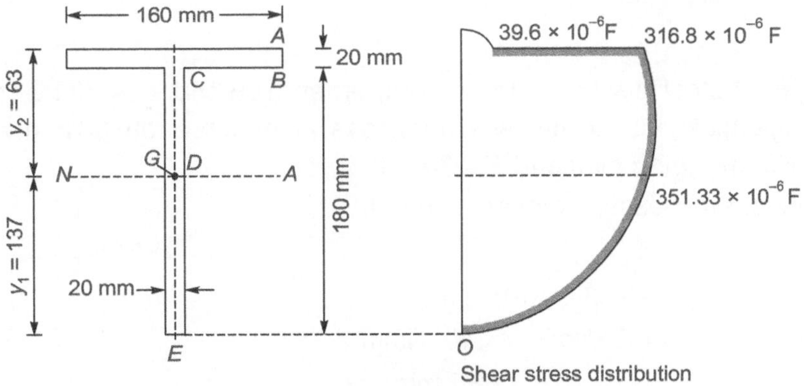

Fig. 9.6

Solution: Location of centroid about edge *E*

$$y_1 = \frac{180\times 20\times 90 + 160\times 20\times(180+10)}{180\times 20 + 160\times 20}$$

$$= \frac{324\times 10^3 + 608\times 10^3}{3600+3200} = \frac{9320}{68} = 137 \text{ mm}$$

$$y_2 = 200 - 137 = 63 \text{ mm}$$

Moment area, $I_{NA} = \frac{160\times 20^3}{12} + 3200(63-10)^2 + \frac{20\times 180^3}{12} + 3600(137-90)^2$

$= 106666.667 + 8988800 + 9720000 + 7952400$

$= 26767866.67 \text{ mm}^4$

$= 26.768 \times 10^6 \text{ mm}^4$

Shear stresses $\tau_A = 0$ (at edge *A*)

$$\tau_B = \frac{F\times 160\times 20\times(53)}{26.768\times 10^6\times 160} = 39.6\times 10^{-6}\ \text{F}$$

$$\tau_C = \frac{F \times 160 \times 20 \times 53}{26.768 \times 20 \times 10^6} = 316.8 \times 10^{-6} F$$

$$\tau_D = \frac{F \times (160 \times 20 \times 53 + 43 \times 20 \times 21.5)}{26.768 \times 20 \times 10^6} = \frac{F(169600 + 18490)}{26.768 \times 10^6 \times 20}$$

$$= \text{(along NA)} = \frac{F(188090)}{26.768 \times 10^6 \times 20} = 351.33 \times 10^{-6} F$$

$$\tau_E = 0$$

Practice Q.9.6 A beam is of T section is subjected to a shear force of 120 kN. Locate G of the section and determine shear stress at the neutral axis.

$y_1 = 202.5$, $y_2 = 97.5$ mm, $I_{NA} = 89.22 \times 10^6$ mm^4

Ans. [27.57 N/mm^2]

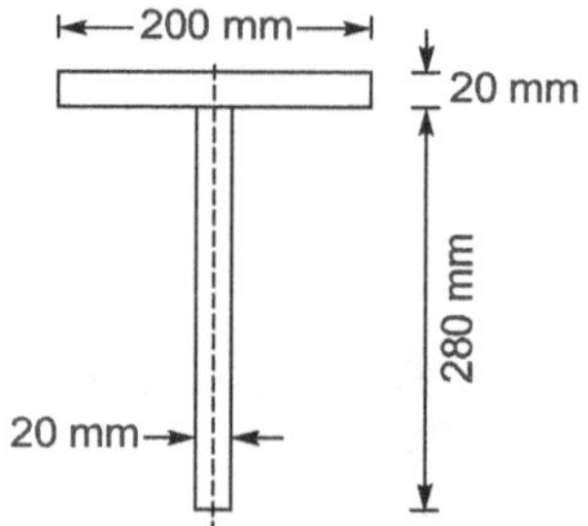

Fig. 9.7

Question 9.7 A steel joist has following properties ISLB 150, $I_{xx} = 688.2$ cm^4, area = 18.08 cm^2, depth = 150 mm, flange thickness 6.8 mm, web thickness 4.8 mm, breadth 80 mm. At a particular section of the beam $M = 8$ kNm (hogging moment) $F = 20$ kN.

Determine principal stresses at inner corner A.

Solution:

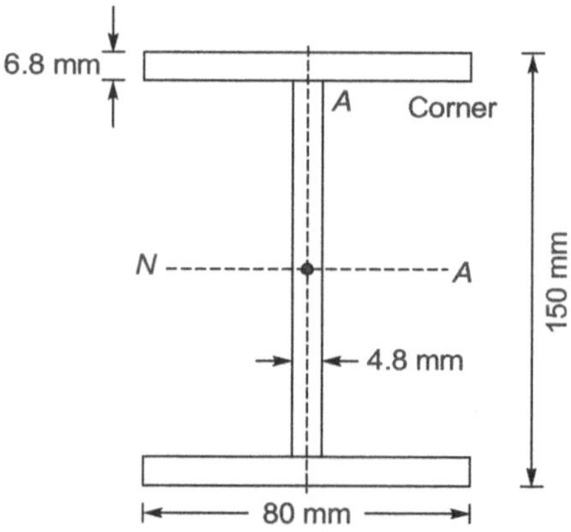

Fig. 9.8 *Bending stress*

$$I_{NA} = 688.2 \times 10^4 \text{ mm}^4$$

$$M = 8 \text{ kNm} = 8 \times 10^6 \text{ Nmm}$$

$$y_A = 75 - 6.8 = 68.2 \text{ mm}$$

$$\sigma_A = \text{Bending stress} = \frac{8 \times 10^6 \times 68.2}{688.2 \times 10^4}$$

$$= +79.28 \text{ N/mm}^2 \text{ (on edge } A)$$

Shear stress at A, $b = 4.8$ mm

$$A\bar{y} = 80 \times 6.8 \times (75 - 3.4) = 38950.4 \text{ mm}^2$$

$$\tau_A = \frac{20000 \times 38950.4}{688.2 \times 10^4 \times 4.8} = 23.58 \text{ N/mm}^2$$

Stresses on an element at corner A are as shown in figure 9.9

$$p_1, p_2 = \frac{\sigma_A}{2} \pm \sqrt{\left(\frac{\sigma_A}{2}\right)^2 + \tau_A^2}$$

$$= 39.64 \pm \sqrt{(39.64)^2 + (23.58)^2}$$

$$= 39.64 \pm \sqrt{1571.33 + 556.0}$$

$$= 39.64 \pm 46.12 = 85.76, -6.48$$

$$= 85.76 \text{ MPa}; -6.48 \text{ MPa are principal stresses}$$

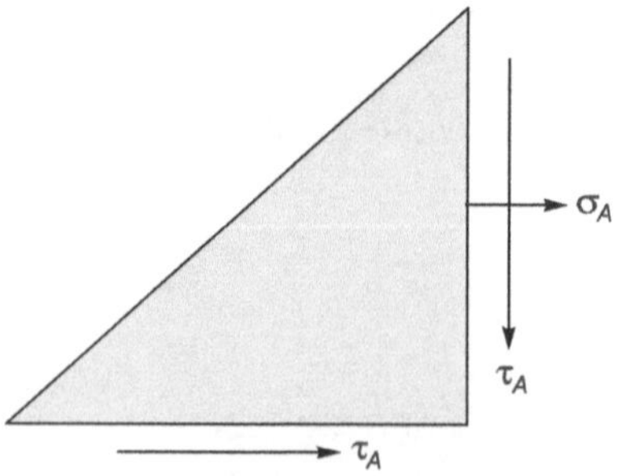

Fig. 9.9

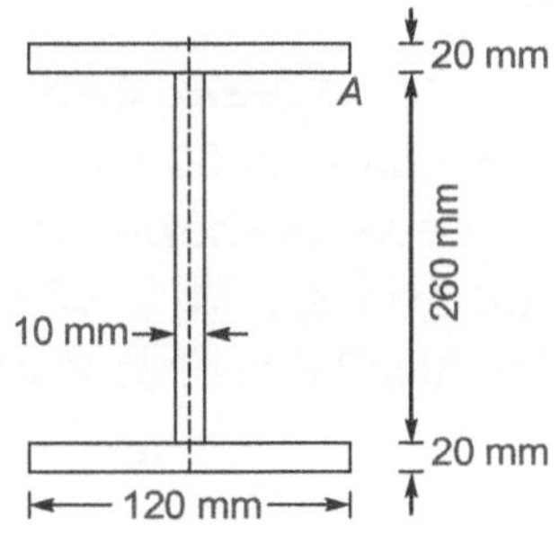

Fig. 9.10

Practice Q.9.7 An I-section beam is subjected to a bending moment (hogging) M = 16 kNm and a shear force F = 24 kN. Dimensions are as shown in fig. 9.10. Determine principal stresses at edge A of the flange.

I_{NA} = 1.089 × 10^8 mm^4

Ans. [p_1, p_2 = 19.12, –0.02 N/mm^2]

Question 9.8 A beam is of rectangular section 80 × 100 mm with two semi circular cut outs as shown in figure 9.11. What is the maximum value of shear force F such that shear stress along neutral axis CC is 10 MPa. What are shear stresses at A and B?

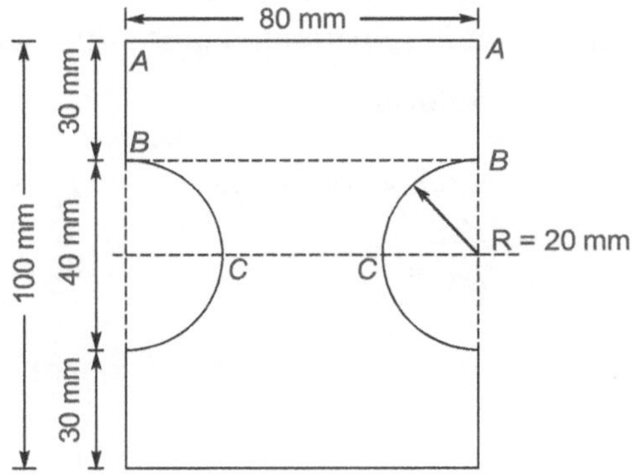

Fig. 9.11

Solution:

Moment of inertia $$I_{NA} = \frac{80\times100^3}{12} - \frac{\pi R^4}{4}$$

$$= 6.667\times10^6 - \frac{\pi\times20^4}{4}$$

$$= 6.667 \times 10^6 - 0.1256 \times 10^6$$

$$= 6.5414 \times 10^6 \text{ mm}^4$$

$$A\bar{y} \text{ upto } cc = 80\times50\times25 - \frac{\pi R^2}{2}\left(\frac{4R}{3\pi}\right)$$

$$= 100000 - \frac{\pi\times20^2}{2}\times\frac{4\times20}{3\pi}$$

(Note that CG of semicircular area lies at $\frac{4R}{3\pi}$ from diameter)

$$A\bar{y} = 100000 - 5333.33 = 94666.66 \text{ mm}^3$$

$$= 9.4667 \times 10^4 \text{ mm}^3$$

$$\tau_{cc} = \frac{FA\bar{y}}{I_{NA}b}, \text{ where } b = 80 - 20 - 20 = 40 \text{ mm}$$

$$10 = \frac{F\times9.4667\times10^4}{6.5414\times10^6\times40} = \frac{F}{2763.9}$$

Shear force $$F = 27639 \times 10 = 27639 \text{ N} = 27.639 \text{ kN}$$

$$\tau_A = 0$$

$$\tau_B = \frac{27639\times(80\times30\times35)}{80\times6.5414\times10^6}$$

where $$A\bar{y} = 80 \times 30 \times 35 \text{ mm}^3$$

$$b = 80 \text{ mm}$$

$$\tau_B = 4.436 \text{ N/mm}^2$$

Practice Q.9.8 A beam section is rectangular with breadth 80 mm × 100 mm depth with two triangular cut outs as shown in figure 9.13. If the shear stress along cc axis is 5 MPa, what is the shear force acting on the section? What is shear stress at edge B?

Ans. [$I_{NA} = 6.6137 \times 10^6$ mm^4, $F = 13.59$ kN, $\tau_B = 2.16$ MPa]

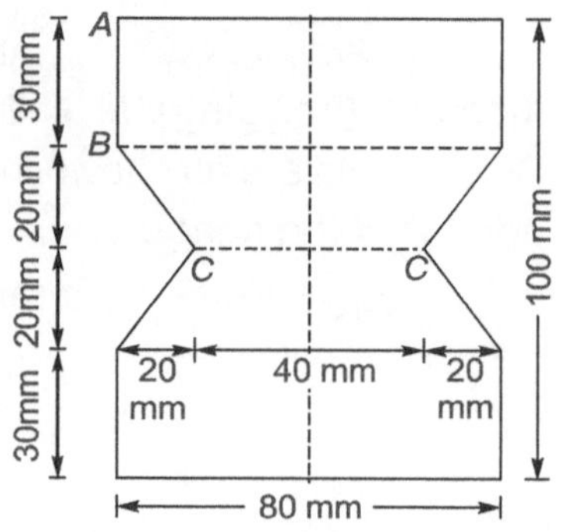

Fig. 9.12

Question 9.9 A beam of square section is placed in such a manner that its diagonals are vertical and horizontal. Shear force at a particular section is F. What is the shear stress at neutral axis and what is maximum shear stress and at what distance from neutral axis, it occurs.

Solution:

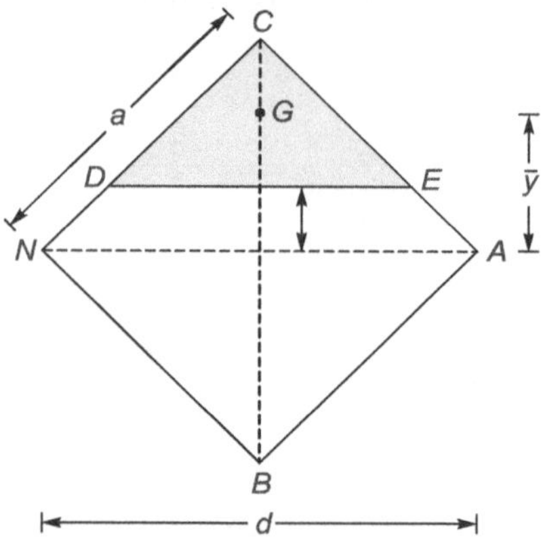

Fig. 9.13

Moment of inertial, $I_{NA} = \dfrac{a^4}{12}$

But $a = \dfrac{d}{\sqrt{2}}$

where d is diagonal of square

$$I_{NA} = \frac{d^4}{48}$$

$$\text{Shear force} = F$$

$$A\bar{y} = \left(\frac{1}{2}\times\frac{d}{2}\times d\right)\left(\frac{d}{6}\right) = \frac{d^3}{24} \text{ for } \tau_{NA}$$

Breadth, $b = d$

$$\tau_{NA} = \frac{FA\bar{y}}{Ib} = F\times\frac{d^3}{24}\times\frac{48}{d^4}\times\frac{1}{d} = \frac{2F}{d^2}$$

Area of square, $a^2 = \dfrac{d^2}{2}$

So, $\tau_{NA} = \tau_{av} = \dfrac{F}{a^2} = \dfrac{2F}{d^2}$

τ_{max} consider a layer DE at a distance y from neutral layer

$$\text{Breadth} = (d-2y),\ \text{altitude} = \left(\frac{d}{2}-y\right)$$

$$\text{Area} = \frac{1}{2}(d-2y)\left(\frac{d}{2}-y\right)$$

$$\bar{y} = y+\frac{1}{3}\left(\frac{d}{2}-y\right)y+\frac{d}{6}-\frac{y}{3} = \frac{d}{6}+\frac{2y}{3} = \frac{1}{3}\left(\frac{d}{2}+2y\right)$$

$$A\bar{y} = \frac{1}{3}\left(\frac{d}{2}-y\right)^2\left(\frac{d}{2}+2y\right)$$

Shear stress,
$$\tau = \frac{FA\bar{y}}{Ib} = \frac{F \times \frac{1}{3}\left(\frac{d}{2} - y\right)^2 \left(\frac{d}{2} + 2y\right)}{I_{NA}(d - 2y)}$$

$$\tau = \frac{F}{6} \times \frac{\left(\frac{d^2}{4} + \frac{dy}{2} - 2y^2\right)}{I_{NA}}$$

For shear stress to be maximum

$$\frac{d\tau}{dy} = 0 = \frac{F}{6I_{NA}}\left(\frac{d}{2} - 4y\right)$$

$$\frac{d}{2} - 4y = 0$$

$y = d/8$, putting this value

$$\tau_{max} = \frac{F48}{6d^4}\left(\frac{d^2}{4} + \frac{d}{2} \times \frac{d}{8} - 2 \times \frac{d^2}{64}\right)$$

$$= \frac{8F}{d^4}\left(\frac{d^2}{4} + \frac{d^2}{16} - \frac{d^2}{32}\right) = \frac{8F}{d^4} \times \frac{9d^2}{32}$$

$$= \frac{9}{4}\frac{F}{d^2} = \frac{9}{8}\left(\frac{2F}{d^2}\right), \text{ where } \frac{2F}{d^2} = \tau_{av}$$

$$\tau_{max} = \frac{9}{8} \times \tau_{av} = 1.125\,\tau_{av}$$

(occurs at a layer at a distance of $d/8$ from neural axis.)

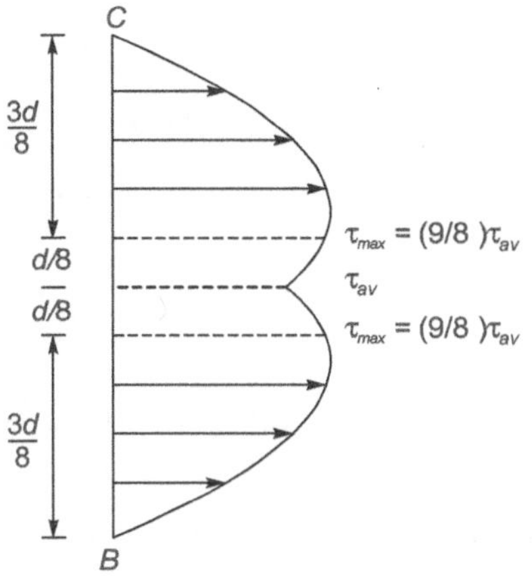

Fig. 9.14. *Shear stress distribution along the vertical diagonal CB*

Fig. 9.14 shows shear stress distribution along diagonal *CB*.

Practice Q.9.9 **A square section beam of side 20 mm is placed with its diagonals vertical and horizontal. At a particular section, F = 2 kN. What is τ_{max} and where it occurs. What is shear stress at neutral layer.**

Ans. [τ_{av} = 5 MPa, τ_{NA} = 5 MPa, τ_{max} = 5.625 MPa, at y = 3.535 mm for NA]

Question 9.10 **What do you understand by curves of principal planes in a beam subjected to bending moment and shear force.**

Solution: When a beam section say of rectangular shape is simultaneously subjected to bending moment and shear force, extreme layers are subjected to direct bending stresses, tension on one side and compression on the other side. On these extreme layers, there is no shear stress. So the principal stress at extreme layer is equal to bending stress, which is parallel to the axis of the beam. Therefore the principal plane at extreme layer is perpendicular to the axis of the beam as shown in figure 9.15.

However on neutral layer, the direct bending stress is zero and shear stress is maximum. The principal planes are inclined at an angle of ±45° to the axis of the beam as shown in figure 9.15. At any other layer, there is a combination of direct bending stress σ and shear stress τ. Directions of principal planes depend on the relative magnitude of σ and τ. Figure 9.15 shows such curves for a beam subjected to both bending moment and shear force. The two sets of curves for principal planes (starting from top extreme layer and another starting from bottom extreme layer) possess following characteristics.

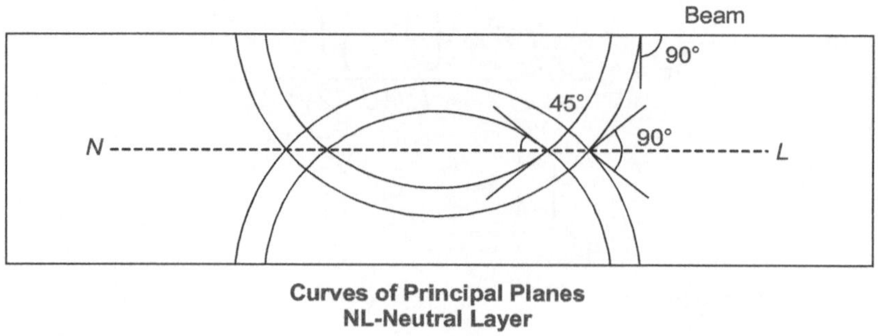

Fig. 9.15

(a) Tangent and normal at a particular point to these curves give directions of principal stresses/or principal planes.

(b) Two curves intersect at neutral axis, at right angles

(c) Curves cross the neutral axis at 45°

(d) Curves are normal to top and bottom surfaces of the beam

(e) Intensity of a principal stress is maximum when it is parallel to the beam

Question 9.11 A rectangular beam of depth d, breadth b and length L is supported at ends. It carries a concentrated load W at the mid span of the beam. Show that the principal stress at a point in the central cross-section at a distance of $\frac{d}{4}$ from top is $\frac{3WL}{8bd^2}\left(1+\sqrt{1+\frac{9d^2}{L^2}}\right)$.

Solution: Central cross-section

$$M_{max} = \frac{WL}{4}; \qquad I_{NA} = \frac{bd^3}{12}$$

At a distance $\frac{d}{4}$ from top

$$y = \frac{d}{4}$$

$$\sigma_b = \frac{WL}{4} \times \frac{d}{4} \times \frac{12}{bd^3} = \frac{3WL}{4bd^2}$$

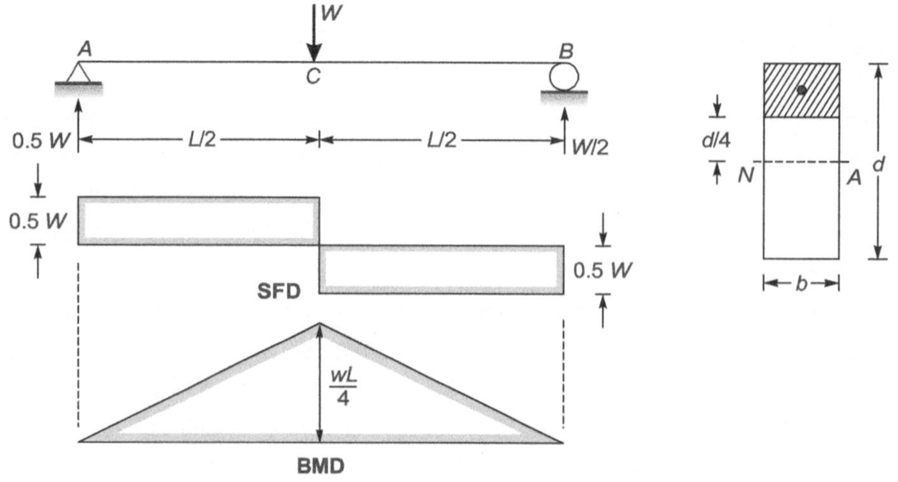

Fig. 9.16

Shear stress

$$A\bar{y} = \frac{bd}{4}\left(\frac{3d}{8}\right) = \frac{3bd^2}{32}$$

$$\tau = \frac{FA\bar{y}}{Ib} = \frac{F \times 3bd^2}{32} \times \frac{12}{bd^3} \times \frac{1}{b} = \frac{9}{8}\frac{F}{bd}$$

$$F = \frac{W}{2}\text{(shear force)}$$

$$\tau = \frac{9}{16} \times \frac{W}{bd}$$

Principal stress,

$$p_1 = \frac{\sigma_b}{2} + \sqrt{\left(\frac{\sigma_b}{2}\right)^2 + \tau^2}$$

$$= \frac{3WL}{8bd^2} + \sqrt{\left(\frac{3WL}{8bd^2}\right)^2 + \left(\frac{9W}{16bd}\right)^2} = \frac{3WL}{8bd^2}\left(1 + \sqrt{1 + \frac{9d^2}{4L^2}}\right)$$

Question 9.12 A beam is made up of five wooden planks each with a width of 80 mm and depth of 30 mm placed one above the other so as to form a section as shown in figure 9.17. The planks are clamped by means of a 12 mm diameter steel bolt passing through the overall depth with tightened nuts. The bolts are spaced 100 mm centre to centre along the length of the beam. Find the maximum shear stress developed in steel bolt section. Neglect self weight of the beam. Shear force at the section is 6 kN.

Solution:

Breadth, $b = 80$ mm

Depth, $d = 150$ mm

Moment of inertia, $I_{NA} = \dfrac{80 \times 150^3}{12} = 2250 \times 10^4 \text{ mm}^4$

(Maximum shear stress between planks will occur about layer *aa*), above NA

$A\bar{y}$ of section above layer $aa = 80 \times 60\,(15 + 30) = 216000 \text{ mm}^3$

τ_{aa}, shear stress in wooden plank at layer *aa*,

$$aa = \frac{FA\bar{y}}{Ib}$$

$$b = 80 \text{ mm}$$

$$\tau_{aa} = \frac{6000 \times 21.6 \times 10^4}{2250 \times 10^4 \times 80}$$

$$= 0.72 \text{ N/mm}^2$$

Spacing between bolts = 100 mm

Area of neutral layer between two bolts

= Spacing × width

= 100 × 80 = 8000 mm²

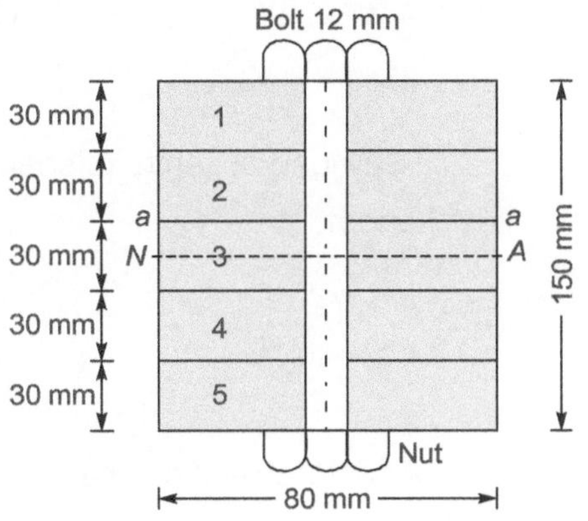

Fig. 9.17

$$\tau_{aa} = 0.72 \text{ N/mm}^2$$

Total force at layer $aa = 0.72 \times 8000 = 5760$ N

Bolt diameter, $d_b = 12$ mm

Area of bolt section, $A_b = \dfrac{\pi}{4} \times d_b^2 = \dfrac{\pi}{4} \times 12^2 = 113.1 \text{ mm}^2$

Maximum shear stress developed in bolt section

$$= \frac{5760}{113.1} = 50.92 \text{ N/mm}^2$$

Practice Q.9.12 A beam is made up of 5 wooden planks each with a width of 100 mm and a depth of 40 mm, placed one above the other so as to form a section as shown in figure 9.17 (Q.9.12). The planks are tied together by means of a 16 mm diameter steel bolt passing across the overall depth with tightened nuts. Find the maximum shear stress developed in bolt section. The bolts are placed 120 mm centre of centre along the beam length. Neglect self weight of the beam and the loss of cross-sectional area due to bolts. Shear force in the section is 10 kN.

Ans. [$I_{NA} = 66.67 \times 10^6 \text{ mm}^4$, shear force per bolt = 8640 N, shear stress in bolt = 42.97 N/mm²]

Question 9.13 A beam of box cross-section in figure 9.18 carries a vertical load at its mid span such that maximum bending stress produced is 12 N/mm². The beam is supported over a span of 2.4 m. Upper and lower flanges are connected to the webs with the help of screws. Determine the spacing between the screws if each screw can transmit a shear force of 2.0 kN.

Solution:

Moment of inertia, $I_{NA} = \dfrac{200 \times 250^3}{12} - \dfrac{100 \times 150^3}{12}$

$= 2.604 \times 10^8 - 0.2812 \times 10^8 \text{ mm}^4$

$= 2.3228 \times 10^8 \text{ mm}^4$

Maximum bending stress will occur at top and bottom surfaces,

$$\sigma = 12 \text{ N/mm}^2 = \frac{125M}{2.3228 \times 10^8}$$

$$M = \frac{2.3228 \times 10^8 \times 12}{125}$$

$= 22.299 \times 10^6 \text{ Nmm} = 22.29 \text{ kNm}$

Say load at centre of span = W kN

Length, $L = 2.4$ m

$$M_{max} = \frac{W \times 2.4}{4} = 0.6 \text{ W} = 22.29$$

$W = 37.16$ kN

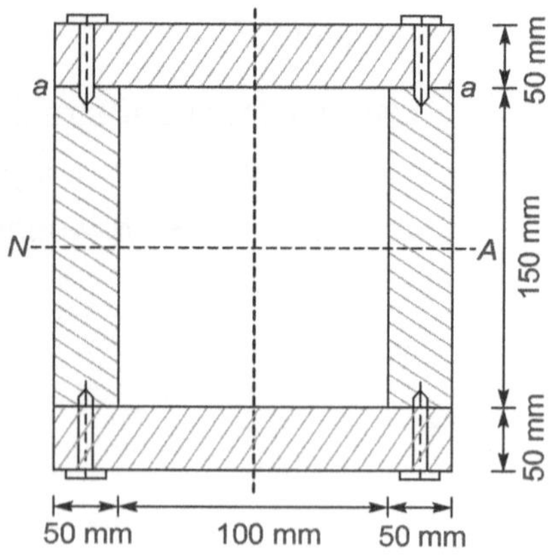

Fig. 9.18

Shear force at the centre $= \dfrac{W}{2} = 18.58$ kN

Screws are connecting the flanges with the **web along the plane *aa***. Shear stress along the plane *aa*

$$A\bar{y} = 200 \times 50 \times (75 + 25) = 1 \times 10^6 \text{ mm}^3$$

Breadth, $\quad b = 100$ mm

$$\tau_{aa} = \frac{FA\bar{y}}{Ib} = \frac{18580 \times 1 \times 10^6}{2.3228 \times 10^8 \times 100} = 0.80 \text{ N/mm}^2$$

Shear force per meter length of the beam

$$= 0.80 \times 100 \times 1000 = 80000 \text{ N} = 80 \text{ kN}$$

Say spacing between the screws = S meter

Force transmitted by one screw = 2 kN

Force transmitted by two screws on both sides = 4 kN

Force transmitted per meter = 80 kN

Force transmitted by two screws = 4 kN

Spacing between the screw, $S = \frac{4}{80} \times 1 \text{ m} = \frac{4}{80} \times 1000 = 50$ mm

Practice Q.9.13 **A beam of box section of cross sectional area 400 deep × 300 mm wide as shown in figure 9.18. Flange 300 × 100 mm each, webs 40 × 200 mm each. Beam carries a vertical load at mid span such that the maximum bending stress produced in beam is 15 N/mm². Beam is supported over a span of 2.4 m. Upper and lower flanges are connected to the webs with the help of screws. Determine the spacing between the screws if each screw can transmit 12.5 kN of shear force.**

Ans. [89 mm spacing]

Question 9.14 **Consider a triangular section of a beam, breadth = *b*, altitude = *h*. Plot shear stress distribution along the altitude *h* of the section.**

Solution: Neutral axis of the triangular section lies at $\frac{h}{3}$ from base.

Shear stress, $\quad \tau = \frac{FA\bar{y}}{Ib}$

Take a layer *df*, at *y* from neutral axis.

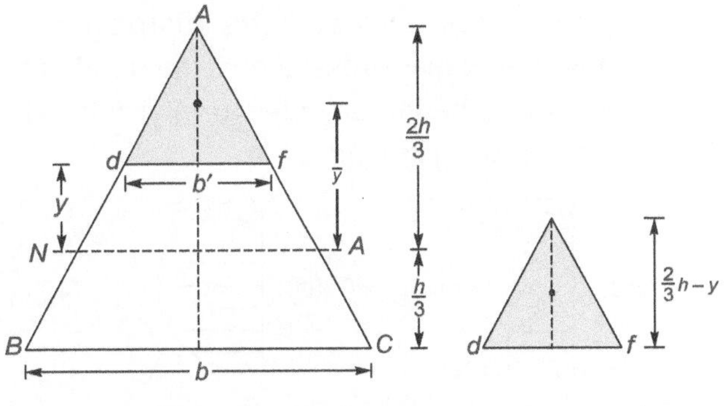

Fig. 9.19

base $\quad b' = \left(\frac{2h}{3} - y\right)\frac{b}{h}$

$$\bar{y} = \frac{1}{3}\left(\frac{2h}{3} - y\right) + y$$

$$= \frac{2h}{9} - \frac{y}{3} + y = \frac{2h}{9} + \frac{2y}{3}$$

$$= \frac{2}{3}\left(\frac{h}{3} + y\right)$$

Area, $\quad A = \frac{1}{2} b' \times \left(\frac{2h}{3} + y\right) = \frac{1}{2} \times \left(\frac{2h}{3} - y\right) \times \frac{b}{h}\left(\frac{2h}{3} - y\right) = \frac{b}{2h}\left(\frac{2h}{3} - y\right)^2$

Shear stress, $\tau = \dfrac{FA\bar{y}}{Ib'} = \dfrac{F}{I_{NA}} \times \dfrac{1}{2}\left(\dfrac{2h}{3} - y\right)\left(\dfrac{2}{3}\right)\left(\dfrac{h}{3} + y\right)$, putting the value of b'

$$\tau = \frac{F}{I_{NA}} \times \frac{1}{3}\left(\frac{2h}{3} - y\right)\left(\frac{h}{3} + y\right) = \frac{36}{bh^3} \times \frac{F}{3}\left(\frac{2h}{3} - y\right)\left(\frac{h}{3} + y\right)$$

$$= \frac{12F}{bh^3}\left(\frac{2h^2}{9} - y\frac{h}{3} + \frac{2hy}{3} - y^2\right) = \frac{12F}{bh^3}\left(\frac{2h^2}{9} + \frac{hy}{3} - y^2\right)$$

$\tau_A =$ at top edge $= 0$

$y =$ 0, neutral axis

$$\tau_{NA} = \frac{12F}{bh^3} \times \frac{2h^2}{9} = \frac{24}{9} \times \frac{F}{bh}$$

$$= \frac{8}{3} \times \frac{F}{bh} = \frac{4}{3}\left(\frac{2F}{bh}\right) = \frac{4}{3}\tau_{av}$$

$y = \dfrac{h}{6}$ from NA or $\dfrac{h}{2}$ from base

$$\tau = \frac{12F}{bh^3}\left(\frac{2h^2}{9} + \frac{h}{3} \times \frac{h}{6} - \frac{h^2}{36}\right) = \frac{12F}{bh^3}\left(\frac{2h^2}{9} + \frac{h^2}{36}\right)$$

$$= \frac{12F}{bh^3} \times \frac{h^2}{4} = \frac{3F}{bh} = 1.5\left(\frac{2F}{bh}\right)$$

$$= 1.5\,\tau_{AV}$$

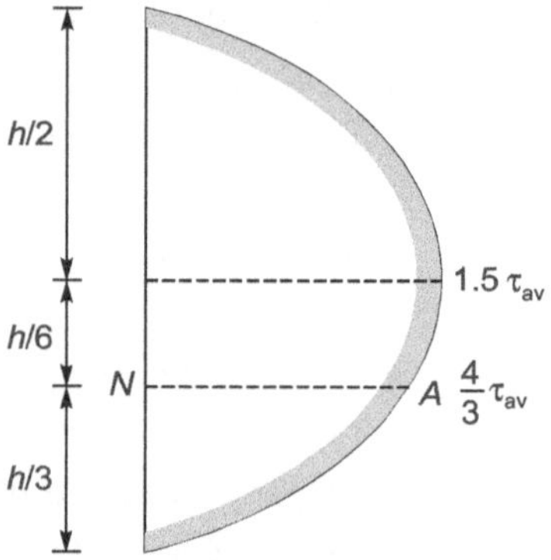

Fig. 9.20 *Shear stress distribution along altitude h.*

Objective Type Questions

Q.1 Which one of the following diagrams shown in figure gives correctly the distribution of transverse shear stress across the depth h of a rectangular beam subjected to varying bending moment along its length?

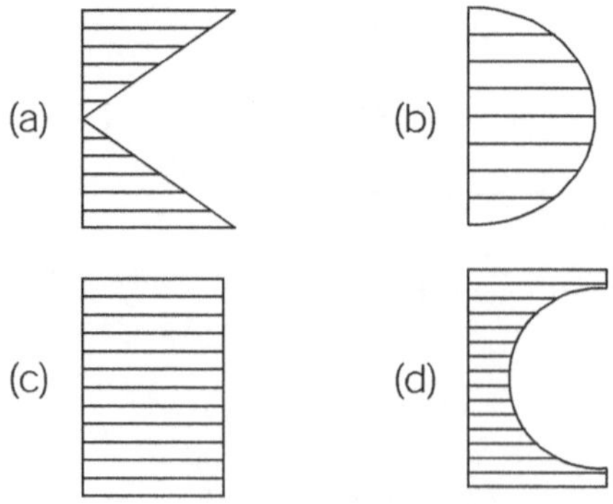

[GATE 1990, 1 Marks]

Q.2 The transverse shear stress acting in a beam of rectangular cross-section subjected to a transverse shear load is

(a) variable with maximum at the bottom of the beam

(b) variable with maximum at top of the beam

(c) uniform with maximum at top of the beam

(d) variable with maximum on the neutral axis

[GATE 2008, 1 Marks]

Q.3 The shear stress distribution over a beam cross-section is shown in figure.

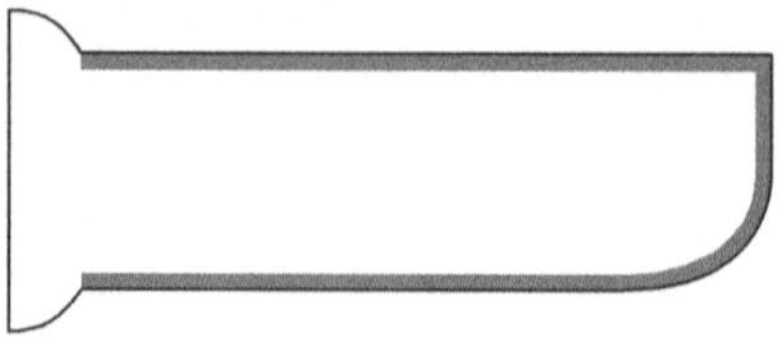

The beam is of
(a) equal flange I-section
(b) unequal flange I-section
(c) circular cross-section
(d) T-section

Q.4 Consider the following statements in a cantilever subjected to a concentrated load at free end.
1. The bending stress is maximum at the free end
2. The maximum shear stress is constant along the length of the beam
3. The slope of the elastic curve is zero at the fixed end

Which of these statement are correct?
(a) 1, 2 and 3 (b) 2 and 3
(c) 1 and 3 (d) 1 and 2

Q.5 Figure shows the shear stress distribution along the depth h of a section of a beam subjected to variable bending moment. Which is this section

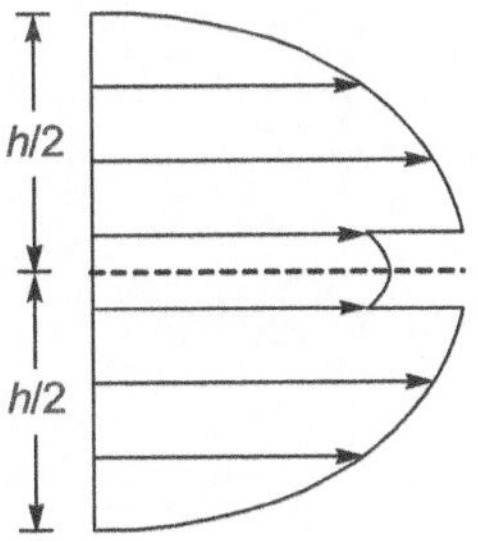

(a) channel section
(b) square section with a vertical diagonal
(c) diagonal vertical section of a cross
(d) None of these

Q.6 Section of a beam is triangular, with base b and altitude *h*. At *h*/3 from base, τ_1 = shear stress, at *h*/2 from base shear stress is τ_2. What is he ratio of τ_2/τ_1
(a) 1.33 (b) 1.125
(c) 1.0 (d) 0888

Q.7 A beam is of circular section, radius *R* is subjected to a shear force *F* such that average shear stress is 10 MPa. What is shear stress at a layer at a distance of 0.5 *R* from neutral axes
(a) 15 MPa (b) 13.33 MPa
(c) 10 MPa (d) 8.88 MPa

Q.8 Match the section and ratio of τ_{max}/τ_{av}

A. Rectangular	1. 1.333
B. Circular	2. 1.125
C. Square with vertical diagonal	3. 2.0
D. Thin tabular	4. 1.5

	A	B	C	D
(a)	4	1	3	2
(b)	4	1	2	3
(c)	1	4	2	3
(d)	4	2	1	3

Q.9 The maximum shearing stress induced in the beam section at any layer at any position along the beam length (shown in the figure) is equal to

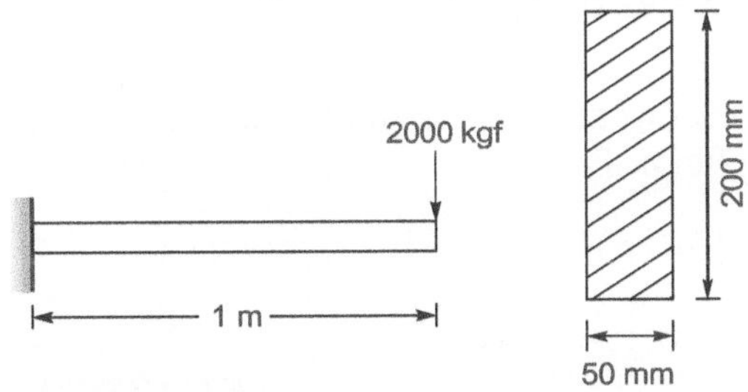

Cross-section of beam

(a) 30 kgf/cm^2 (b) 40 kgf/cm^2
(c) 50 kgf/cm^2 (d) 60 kgf/cm^2

[ESE (Pre) : 2017]

Q.10 A simply supported beam of rectangular cross-section is under transverse loading. Regarding the shear stress distribution across any section, the ratio of maximum shear stress to mean shear stress is
(a) 1.5 (b) 2.5
(c) 3.5 (d) 4.5

[ESE (Pre) : 2016]

Q.11 In the case of a rectangular beam subjected to a transverse shearing force, the ratio of maximum shear stress to average shear stress is
(a) 0.75 (b) 1.00
(c) 1.25 (d) 1.50

[ESE (Pre) : 2018]

Answers

1. (b)	2. (d)	3. (b)	4. (b)	5. (c)
6. (b)	7. (c)	8. (b)	9. (a)	10. (a)
11. (d)				

Explanations

1. (b)

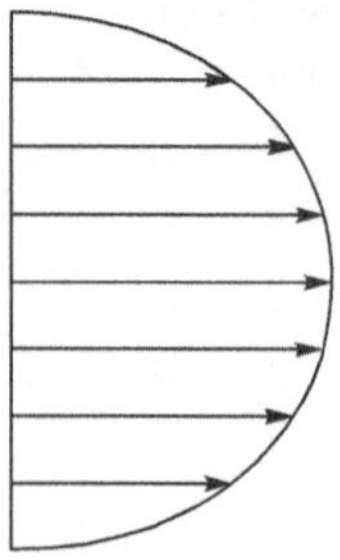

Shear stress distribution along the depth h of rectangular section.

2. (d)
Shear stress is variable with maximum at the neutral axis.

3. (b)

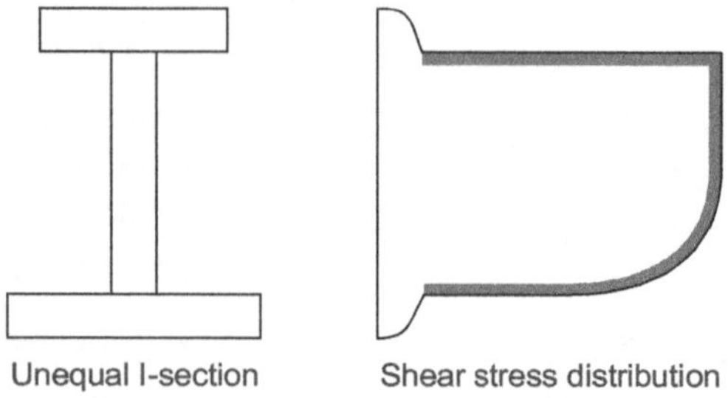

4. (b)

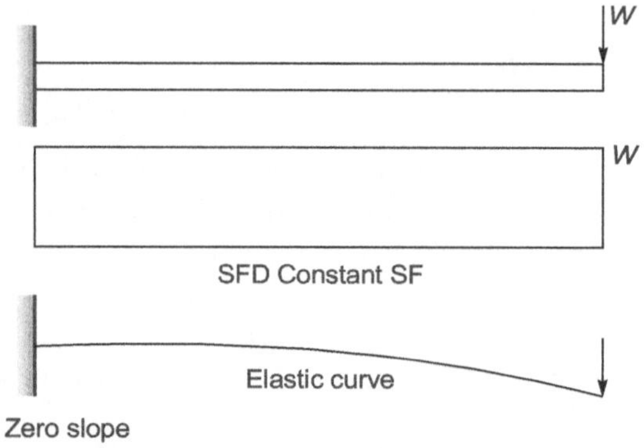

2 and 3 are correct.

- Constant shear stress
- Slope of elastic curve zero at fixed end

5. (c)

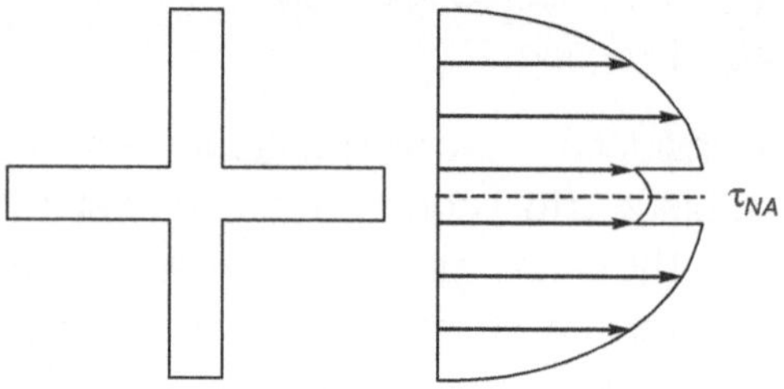

Section of a cross.

6. (b)

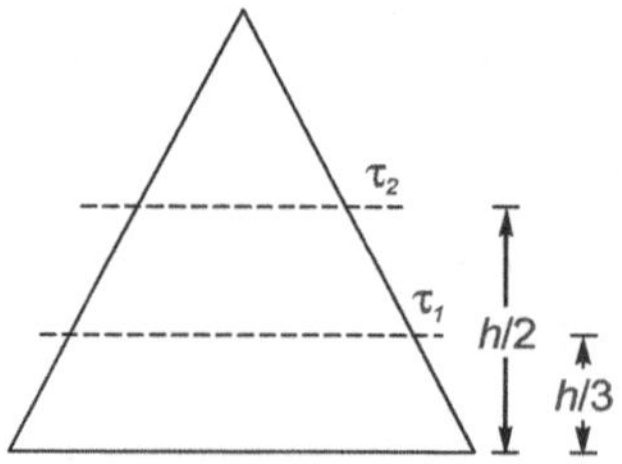

$$\tau_2 = 1.5\,\tau_{av}$$

$$\tau_1 = \frac{4}{3}\tau_{av}$$

$$\frac{\tau_2}{\tau_1} = 1.125$$

7. (c)

At $\quad y = \dfrac{R}{2}$, $\sin\theta = 0.5$, $\cos\theta = 0.866$

$$\tau = \frac{4}{3}\,t_{av} \times \cos^2\theta = \frac{4}{3} \times t_{av} \times \frac{3}{4} = t_{av}$$

$$= 10 \text{ MPa}$$

8. (b)

A.	Rectangular	4.	1.5
B.	Circular	1.	1.333
C.	Square with vertical diagonal	2.	1.125
D.	Thin tabular	3.	2.0

9. (a)
Cantilever carrying point load at free end, W shear force throughout the length = W
= 2000 kgf as shown
Rectangular section 5 cm × 20 cm

Area of cross-section,

$A = 100 \text{ cm}^2$

Average shear stress,

$$\tau_{av} = \frac{2000}{100} = 20 \text{ kgf/cm}^2$$

Maximum shear stress occurs at neutral layer

$$\tau_{max} = 1.5\, \tau_{av} = 1.5 \times 20$$
$$= 30 \text{ kgf/cm}^2$$

10. (a)

Simply supported beam under transverse loading

- Rectangular cross-section b × d
- Say F is shear force at any section
- Average shear stress, $\tau_{av} = \dfrac{F}{bd}$
- Maximum shear stress along neutral axis,

$$\tau_{max} = \frac{1.5F}{bd}$$

$$\text{Ratio, } \frac{\tau_{max}}{\tau_{mean}} = 1.5$$

11. (d)

Rectangular beam section

$$\frac{\tau_{max}}{\tau_{av}} = 1.5$$

■■■■

Combined Bending and Direct Stresses

CHAPTER 10

Columns are subjected to axial compressive load, producing compressive stress in column. If the axial load becomes eccentric, in addition to direct compressive stress, bending stresses are also developed in a column.

These bending stresses vary from maximum compressive stress at one edge to maximum tensile stress at another edge. So there is a combination of bending stress and direct compressive stress. If the resultant stress in column is tensile, i.e., tensile bending stress is more than the axial compressive stress, then many masonry columns may crack as they are unable to bear tensile stress.

In this chapter, we will learn about stresses developed due to eccentric load and limit on eccentricity so that **no net tensile stress is developed at any point in the section of the column.**

Question 10.1 **Consider a column of rectangular section $B \times D$, with a load applied at an eccentricity '*e*' along *x-x* axis (parallel to smaller side, *B*) and derive expression for resultant stresses at the extreme edges of the section.**

Solution: A column of section breadth *B*, depth *D*, *xx*-axis and *y-y* axis passing through centroid *G* of the section is shown in fig. 10.1 A load *W* acts at a point, *P*, such that distance, $GP = e$, eccentricity.

A load *W* at *P* is equivalent to a load *W* at *G* plus a moment, *W.e* (clockwise) as shown about the axis *yy* of the section.

Area of cross-section, $A = B \times D$

Section modulus about *yy*-axis $= \dfrac{DB^2}{6}$

Direct stress in column, $\sigma_d = \dfrac{W}{BD}$ (Comp.)

Bending stress in column, $\sigma_b = \dfrac{6We}{DB^2} = \dfrac{M}{Z}$

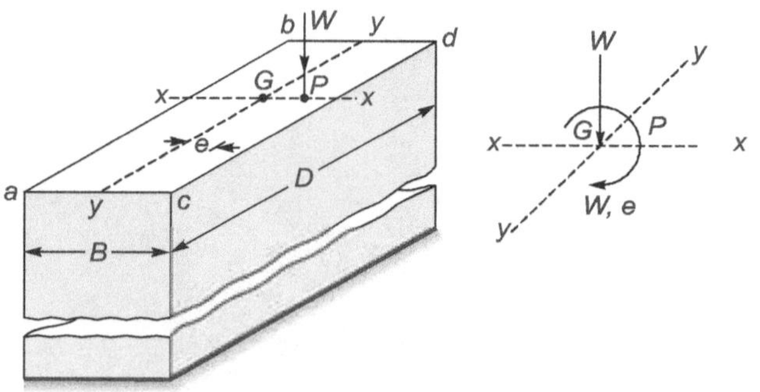

Fig. 10.1

Note that in the case of columns, compressive stress is a desired stress and tensile stress is undesired stress because columns are generally made of brittle materials which are weak in tension. So positive sign is attached to compressive stress.

Extreme bending stresses will occur at edges *ab* and *cd*, with tensile stress $-\sigma_b$ along *ab* and compressive stress $+\sigma_b$ along *cd* edge.

Resultant stress along edge $ab = \sigma_d - \sigma_b = \dfrac{W}{BD} - \dfrac{6We}{DB^2}$

Resultant stress along edge $cd = \sigma_d + \sigma_b = \dfrac{W}{BD} + \dfrac{6We}{DB^2}$ (Compressive)

If $\sigma_b > \sigma_d$, the variations of resultant stress along xx axis (along the breadth, B) is given in figure 10.2. Fig. 10.2 shows the variation of resultant stress along the width B, this is linear

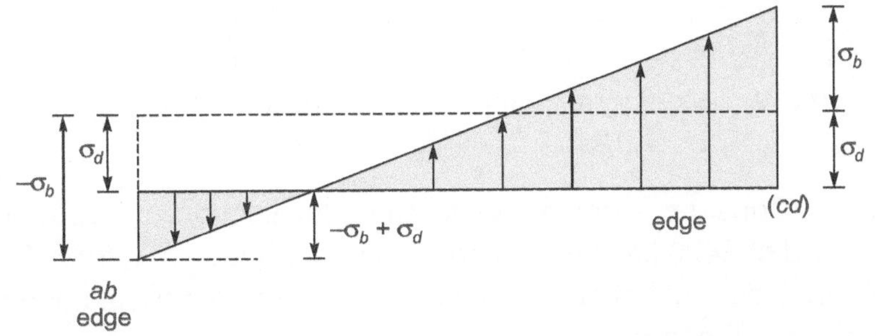

Fig. 10.2 *Resultant stress distribution along the breadth B*

Resultant stress along $cd = \sigma_b + \sigma_d = \dfrac{6We}{DB^2} + \dfrac{W}{BD}$ (Compressive)

Resultant stress along $ab = -\sigma_b + \sigma_d = -\dfrac{6We}{DB^2} + \dfrac{W}{BD}$ (Tensile)

Question 10.2 **Consider a column of rectangular section $B \times D$ where B = 300 mm, D = 500 mm subjected to a load W = 120 kN at an eccentricity e = 150 mm from G, along y-y axis which is parallel to longer side. Determine resultant stresses at the edges of the section. For what value of eccentricity there is no tensile stress any where in the section.**

Solution:

Breadth, B = 300 mm

Depth, D = 500 mm

Area of cross section = 300 × 500 = 15 × 10^4 mm^2

Eccentricity, e_y = 150 mm

Load, W = 120000 N

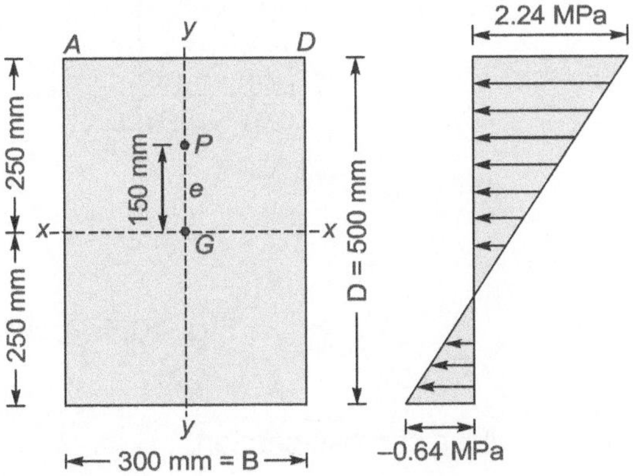

Fig. 10.3 *Resultant stress distribution along the axis yy*

Direct stresses, $\sigma_d = \dfrac{120000}{15 \times 10^4}$

$= 0.8$ N/mm^2

Bending moment about xx axis, $M_x = W.e_y$

$= 20000 \times 150$

$= 18 \times 10^6$ Nmm

Section modulus, $Z_x = \dfrac{BD^2}{6} = \dfrac{300 \times 500^2}{6} = 12.5 \times 10^6\ \text{mm}^3$

Bending stress, $\sigma_b = \pm\dfrac{M_x}{Z_x} = \dfrac{18 \times 10^6}{12.5 \times 10^6} = \pm 1.44\ \text{N/mm}^2$

Resultant stress along edge, *ad* $\sigma_{ad} = \sigma_d + \sigma_b = 0.8 + 1.44 = 2.24$ N/mm^2 (comp.)

Resultant stress along edge, *bc* $\sigma_{bc} = \sigma_d - \sigma_b = 0.8 - 1.44 = -0.64$ N/mm^2 (tensile)

Fig. 10.3 shows the resultant stress distribution along the depth or axis *yy*.

Say eccentricity is e such that $\sigma_d - \sigma_b = 0$

or $$\frac{W}{BD} - \frac{6We'}{BD^2} = 0$$

or, $$e = \frac{\text{depth}}{6} = \frac{D}{6} = \frac{500}{6} = 83.33 \text{ mm}$$

For an eccentricity of $e = 83.33$ mm, along *yy* axis, there will not be any tensile stress anywhere in the section.

Practice Q.10.2 **Consider a column of rectangular section, breadth 300 mm, depth 400 mm subjected to a vertical load of 2400 kN. Load is applied at a point 75 mm from *G*, along *xx* axis, which is parallel to breadth. Determine resultant stresses along the extreme edges. For what eccentricity there is no tensile stress anywhere in the section.**

Ans. [+50 N/mm^2, –10 N/mm^2, e = 50 mm]

Question 10.3 **A circular column of diameter 400 mm carries an eccentric load of 1200 kN. Eccentricity of load is 60 mm, along a diameter. What are the resultant stresses at the end points of diameter? For what eccentricity, there will be no tensile any where in the section of the column.**

Solution:

Load, $W = 1200$ kN

Column diameter, $D = 400$ mm

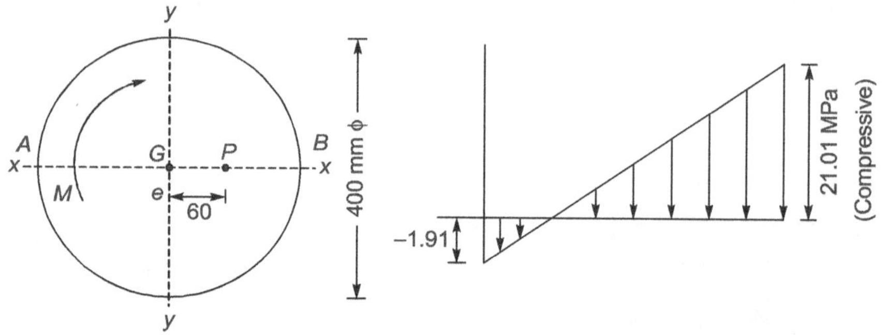

Fig. 10.4 *Resultant stress distribution along diameter AB*

Area of cross-section, $$A = \frac{\pi}{4} \times 400^2 = 12.566 \times 10^4 \text{ mm}^2$$

Direct stress, $$\sigma_d = \frac{1200000}{12.576 \times 10^4} = 9.55 \text{ N/mm}^2$$

Eccentricity, $e = 60$ mm

M, bending moment, clockwise $= 1200000 \times 60 = 72 \times 10^6$ Nmm

Bending stress at ends $$= \pm\frac{M}{Z}$$

Section modulus, $Z_y = \dfrac{\pi D^3}{32} = \dfrac{\pi \times 400^3}{32} = 6.283 \times 10^6 \text{ mm}^3$

$$\sigma_b = \pm\frac{72\times10^6}{6.283\times10^6} = \pm 11.46 \text{ N/mm}^2$$

Resultant stress at corner points, $\sigma_B = \sigma_d + \sigma_b = 9.55 + 11.46 = +21.01 \text{ N/mm}^2$ (comp.)

$$\sigma_A = \sigma_d - \sigma_b = 9.55 - 11.46 = -1.91 \text{ N/mm}^2 \text{ (tensile)}$$

Fig. 10.4 (b) shows the resultant stress distribution along the diameter AB.

Eccentricity, $e = \dfrac{32We'}{\pi D^3} - \dfrac{4W}{\pi D^2} = 0$

or $e = \dfrac{D}{8} = \dfrac{400}{8} = 50 \text{ mm}$

For an eccentricity of $\dfrac{D}{8} = \dfrac{\text{diameter}}{8} = 50 \text{ mm}$, there will not be any tensile stress anywhere in the section.

Question 10.4 Consider a column of rectangular section of breadth *B*, depth *D*, subjected to an eccentric load *W*, with eccentricity e_x along *x-x* axis and eccentricity e_y along *y-y* axis. Determine resultant stresses on four corners *a, b, c* and *d* of the section an shown in fig. 10.5

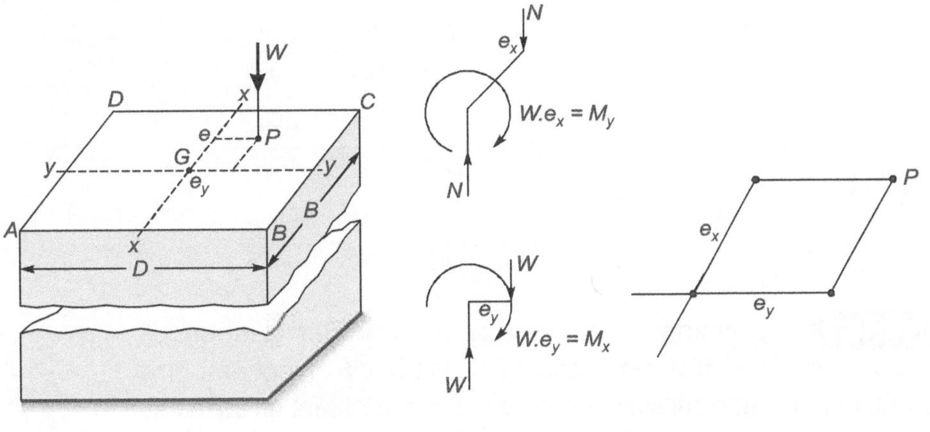

Fig. 10.5

Solution:

$$\text{Breadth} = B$$

$$\text{Depth} = D$$

Area of cross-section $= B \times D$

Total eccentricity, $e = \sqrt{e_x^2 + e_y^2}$

Load *W* at *P* can be replaced by a load *W* at *G* (centroid) and moment *W.e*

Total moment *M* = *We.* has two components

$$M_x = We_y$$

$$M_y = We_x$$

Direct stress, $\sigma_d = \frac{W}{BD}$ (compressive)

Note that due to moment, $M_x = We_y$, edge *bc* comes under compression and edge *ad* comes under tension. Due to moment $M_y = We_x$, edge *dc* will be in compression and edge *ab* will be under tension.

Point *c* at corner carries maximum compressive stress.

$$\sigma_{b1} = \frac{M_x}{Z_y} = \frac{M_x}{\text{section modulus about x-x axis}}$$

$$Z_y = \frac{BD^2}{6}$$

$$\sigma_{b1} = \pm\frac{6We_y}{BD^2}$$

Similarly, $$\sigma_{b2} = \pm\frac{M_y}{Z_x} \pm \frac{We_y}{\text{section modulus about yyaxis}} = \frac{6We_x}{DB^2}$$

Resultant stresses at four corners *A, B, C* and *D*

$$\sigma_a = \frac{W}{BD} - \sigma_{b1} - \sigma_{b2} = \frac{W}{BD} - \frac{6We_y}{BD^2} - \frac{6We_x}{DB^2}$$

$$\sigma_b = \frac{W}{BD} + \frac{6We_y}{BD^2} - \frac{6We_x}{DB^2}$$

$$\sigma_c = \frac{W}{BD} + \frac{6We_y}{BD^2} + \frac{6We_x}{DB^2}$$

$$\sigma_d = \frac{W}{BD} - \frac{6We_y}{BD^2} + \frac{6We_x}{BD^2}$$

Question 10.5 Consider a column of rectangular section of dimension B = 300 mm, D = 400 mm, load W = 1200 kN at eccentricity e_x = 60 mm, e_y = 80 mm, xx axis parallel to breadth. Find resultant stresses at four corners of the section of the column.

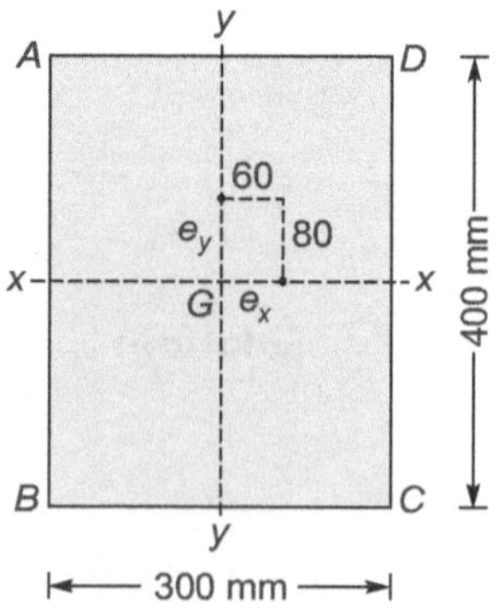

Fig. 10.6

Solution:

$$W = 1200{,}000 \text{ N}$$

$$\text{Area} = 300 \times 400 = 12 \times 10^4 \text{ mm}^2$$

$$\sigma_d = \text{direct stress} = \frac{12\times10^5}{12\times10^4} = 10 \text{ MPa}$$

Bending Moments, $$M_x = We_y = 12 \times 10^5 \times 80 = 96 \times 10^6 \text{ Nmm}$$

$$M_y = We_x = 12 \times 10^5 \times 60 = 72 \times 10^6 \text{ Nmm}$$

$$\sigma_{b1} = \pm\frac{6M_x}{BD^2} = \frac{6\times96\times10^6}{300\times400^2} = \pm12 \text{ N/mm}^2$$

AD in compression, *BC* in tension

$$\sigma_{b2} = \pm\frac{6M_y}{BD^2} = \frac{6\times96\times10^6}{400\times300^2} = \pm12 \text{ N/mm}^2$$

Resultant stresses at corners,

$$\sigma_a = 10 + 12 - 12 = 10 \text{ N/mm}^2 \text{ (comp.)}$$
$$\sigma_b = 10 - 12 - 12 = -14 \text{ N/mm}^{12} \text{ (tensile)}$$
$$\sigma_c = 10 + 12 - 12 = 10 \text{ N/mm}^2 \text{ (comp.)}$$
$$\sigma_d = 10 + 12 + 12 = 34 \text{ N/mm}^2 \text{ (comp.)}$$

Practice Q.10.5 **A column is of square section 500 mm side. It is subjected to an eccentric load, with eccentricity, e_x = 100 mm, e_y = 120 mm. If σ_d = 8 N/mm². What is load *W*? Determine resultant stresses at four corners of the column section.**

Ans. [*W* = 2000 kN; –13.12, +6.08, + 29.12, +9.92 MPa]

Question 10.6 **What is core or kernel of a section? Give dimensions of core of (a) rectangular section (b) circular section (c) hollow circular section with internal diameter *d* = 0.6*D*, external diameter (d) I-section.**

Solution: Core of a section is that area of the cross-section of a column, such that if an eccentric load is applied on the column within the core, then no resultant tensile stress will develop anywhere in the section of the column. In other words bending stress due to eccentric loading is less or equal to the direct compressive stress due to load applied at the *G* of the section.

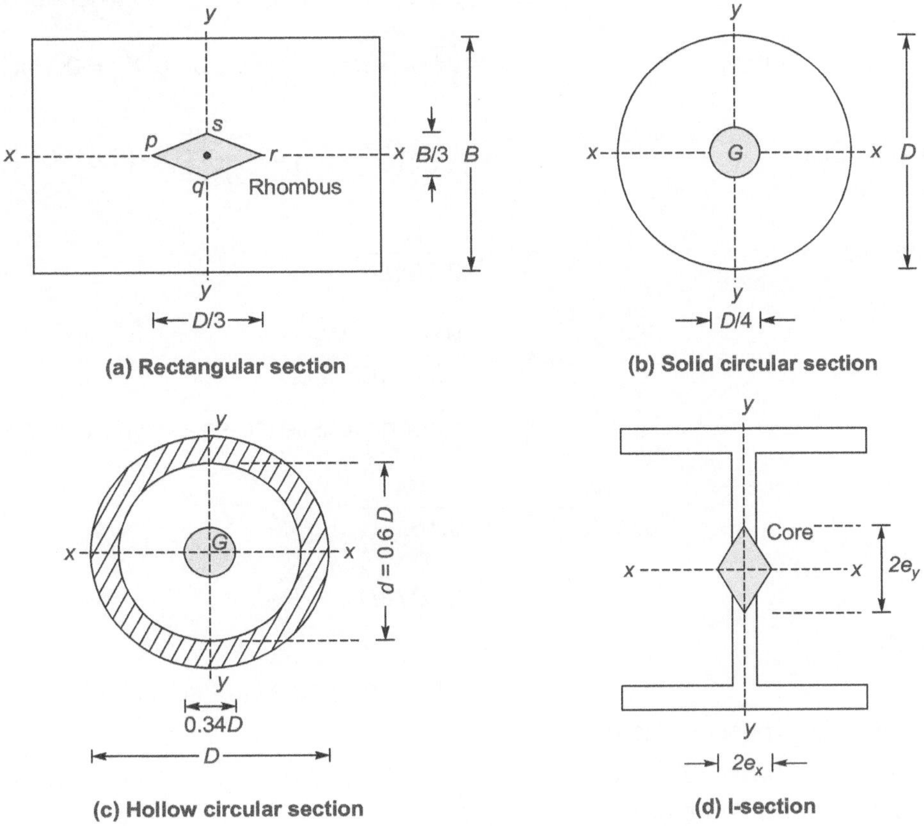

Fig. 10.7

Rectangular section

$$\frac{6We_x}{BD^2} \le \frac{W}{BD}$$

Eccentricity, $$e_x \le \frac{D}{6} = \frac{\text{depth}}{6}$$

Similarly eccentricity, $$e_y = \frac{B}{6} = \frac{\text{breadth}}{6}$$

Core is rhombus with diagonal $2e_x$ and $2e_y$ i.e., $\frac{D}{3}$ and $\frac{B}{3}$ as shown in figure. 10.7(a)

Solid circular section

$$\sigma_b \le \sigma_d$$

$$\frac{32We}{\pi D^3} \le \frac{4W}{\pi D^2}$$

$$e \le \frac{D}{8}$$ or diameter of core $\frac{D}{4}$ as shown in figure 10.7(b)

e is on one side of *G*, therefore on other side of *G* also, e total is $2e = \frac{D}{4}$.

Hollow circular section

Internal diameter, $A = 0.6\,D$ where *D* is external diameter

A, area of cross-section, $$= \frac{\pi}{4}\left(D^2 - d^2\right) = \frac{\pi}{4}\left(D^2 - 0.36D^2\right) = 0.16\pi D^2 \quad \text{...(ii)}$$

Z, section modulus, $$= \frac{\pi}{32D}\left(D^2 - d^2\right) = \frac{\pi}{32D}\left[D^4 - (0.6)^4 D^4\right] = 0.0272\pi D^3 \quad \text{...(ii)}$$

Direct stresses, $$\sigma_d = \frac{W}{0.16\pi D^2}$$

Bending, $$\sigma_b = \frac{We}{Z} = \frac{We}{10272\pi D^3}$$

$\sigma_b \le \sigma_d$ for no tensile stress anywhere in this section.

$$\frac{We}{0.0272\pi D^3} \le \frac{W}{0.16\pi D^2}$$

$$e \le \frac{0.0272D}{0.16}$$

$e \le 0.17\,D$, or diameter of core is $0.34\,D$.

Core of hollow circular section is shown in figure 10.2 (c)

I-section for any standard rolled section, properties like area of cross-section A, moment of inertia I_{xx}, I_{yy} and sections modulus Z_x and Z_y are given in the tables.

$$\frac{We_y}{Z_x} \leq \frac{W}{A}$$

or eccentricity $$e_y \leq \frac{Z_x}{A}$$

Further along y-y axis

$$\frac{We_x}{Z_y} \leq \frac{W}{A}$$

Eccentricity, $$e_x \leq \frac{Z_y}{A}$$

Diagonals of the rhombus (core) and $2e_x$ are $2e_y$ as shown in fig. 10.7 (d).

Question 10.7 Mark the core of an I-section with the following properties.
Area, A = 4320 mm^2, depth = 203 mm, width = 106 mm
$I_{xx} = 27 \times 10^6$ mm^4, $Z_x = 266 \times 10^3$ mm^3, $I_{yy} = 1.81 \times 10^6$ mm^4, $Z_y = 34.2 \times 10^3$ mm^3

Solution:

Eccentricity, $$e_x = \frac{Z_y}{A} = \frac{34.2 \times 10^3}{4320} = 7.92 \text{ mm}$$

$$2e_x = 15.84 \text{ mm}$$

Eccentricity, $$e_y = \frac{Z_x}{Z} = \frac{266 \times 10^3}{4320} = 61.6 \text{ mm}$$

$$2e_y = 2 \times 61.6$$
$$= 123.2 \text{ mm}$$

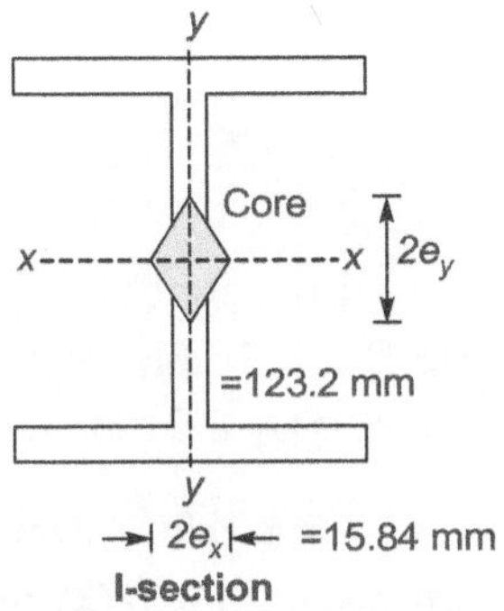

Fig. 10.8

Figure 10.8 shows core, a rhombus with diagonals 15.84 mm and 123.2 mm.

Practice Q.10.7 An ISLB 150 joist has following properties, A = 1808 mm^2, depth = 150 mm, width = 80 mm, $I_{xx} = 688.2 \times 10^4$ mm^4, $I_{yy} = 55.2 \times 10^4$ mm^4. mark the core of **I**-section.

Ans. [Rhombus with diagonals 15.26 mm × 101.2 mm]

Question 10.8 The cross-section of a CI column is as shown in fig. 10.9. Load is applied at point *P*. Magnitude of load is 200 kN. Find out resultant stress on corners *A*, *B*, *C* and *D* of the column.

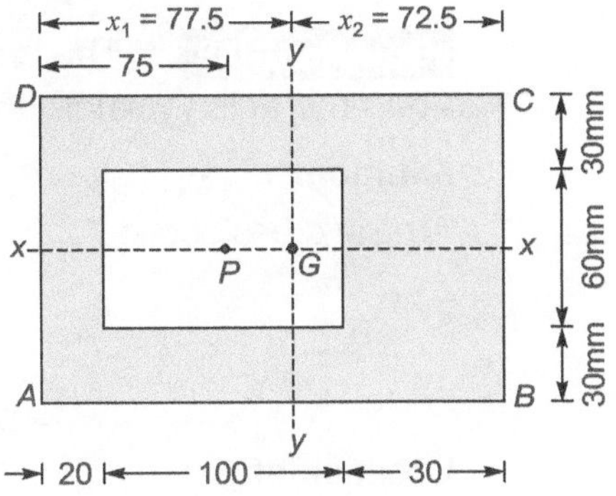

Fig. 10.9

Solution:

Area of cross-section = 150 × 120 − 100 × 60
= 18000 − 6000 = 12000 mm^2

Let us locate *G* of the section (taking moments about edge = *AD*)

$$x_1 = \frac{150 \times 120 \times 75 - 100 \times 60(20 + 50)}{150 \times 120 - 100 \times 60}$$

$$= \frac{1350{,}000 - 420{,}000}{12{,}000} = 77.5 \text{ mm}$$

$$x_2 = 150 - 77.5 = 72.5 \text{ mm}$$

Eccentricity of load = 77.5 – 75 = 2.5 mm

Moment of inertia I_{yy}

$$I_{yy} = \left[\frac{120\times150^3}{12}+18000(2.5)^2\right]-\left[\frac{60\times100^3}{12}+6000(77.5-70)^2\right]$$

$$= [33.75 \times 10^6 + 0.1125 \times 10^6] - [5 \times 10^6 + 0.3375 \times 10^6]$$

$$= [33.8625 - 5.3375] \times 10^6 = 28.525 \times 10^6 \text{ mm}^4$$

Direct stress, $$\sigma_d = \frac{W}{\text{Area}} = \frac{200,000}{120,00} = 16.667 \text{ N/mm}^2 \text{ (comp.)}$$

Bending moment $$M = We_x = 200,000 \times 2.5 = 0.5 \times 10^6 \text{ Nmm}$$

Due to M, compressive stresses will be developed along edge AD and tensile stress will be developed along edge BC.

$$\sigma_{AD} = +\frac{M\times77.5}{I} = \frac{0.5\times10^6\times77.5}{28.525\times10^6} = +1.36 \text{ N/mm}^2$$

$$\sigma_{BC} = -\frac{M\times72.5}{I} = -\frac{0.5\times10^6\times72.5}{28.525\times10^6} = -1.27 \text{ N/mm}^2$$

Resultant stresses along edge AD = 16.667 + 1.36 = + 18.027 MPa (comp.). Resultant stress along edge BC = 16.667 – 1.27 = 15.397 MPa (comp.) During casting of a CI column, the core may have been displaced by 2.5 mm, causing eccentricity in loading.

Practice Q.10.8 A cross section of a short CI column is a square 200 × 200 mm, out of which a circular area of radius 50 mm is removed. Load is applied at P, such that GP= 25 mm. If the maximum stress in the section is not to exceed 60 MPa, determine the magnitude of load.

Hints: [Maximum stress will occur along edge CD]

Ans. [I_{xx} = 128.425 × 10⁶ mm⁴, e = 25 mm, W = 1186.374 kN]

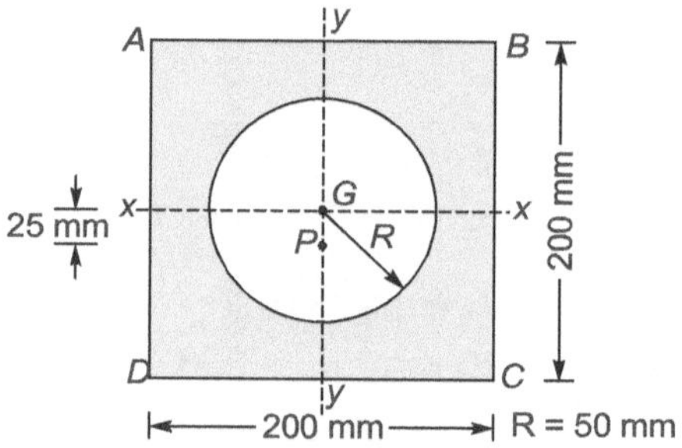

Fig. 10.10

Question 10.9 The cross-section of a column is of the shape of an arrow as shown in figure 10.11. Along the axis *yy* where the load should be applied so that resultant stress at apex A is zero.

Solution:

Area of cross-section,

$$A = \frac{1}{2}\times150\times150+100\times100$$

$$= 21250 \text{ mm}^2$$

Location of G

Moment of area about edge DE

$$y_1 = \frac{100\times100\times50+\frac{1}{2}\times150\times150\times(100+50)}{21250}$$

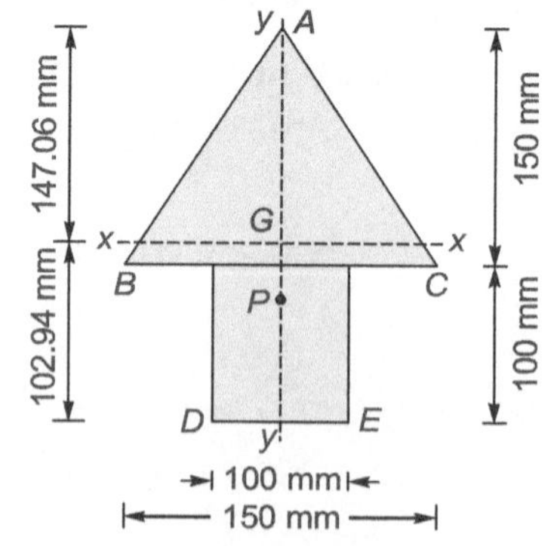

Fig. 10.11

$$= \frac{5\times10^5 + 16.875\times10^5}{21250} = 102.94 \text{ mm}$$

$$y_2 = 250 - 102.94$$
$$= 147.06 \text{ mm}$$

Moment of inertia, I_{xx} about centroid axis through a

$$I_{xx} = \frac{100^4}{12} + 100{,}00(102.94-50)^2 + \frac{150^4}{36} + 11250(150-102.94)^2$$
$$= 8.333\times10^6 + 28.0264\times10^6 + 14.0625\times10^6 + 24.9147\times10^6$$
$$= 75.3366\times10^6 \text{ mm}^4$$

y for apex A, $= 147.06$ mm

Load will be applied along yy-axis, but $GP = e$. So that σ_b at A is tensile.

$$\sigma_b = -\frac{W.GP}{I}\times147.06 = \frac{We\times147.06}{75.3366\times10^6}$$

$$\sigma_d = \frac{W}{A} = \frac{W}{21250}$$

or $$\frac{We}{512284} - \frac{W}{21250} = 0$$

$$e = \frac{512284}{21250} = 24.1 \text{ mm}$$

$$GP = 24.1 \text{ mm}$$

Load should be applied such that $GP = 24.1$ mm, or $147.06 + 24.1 = 171.16$ mm from A.

Practice Q.10.9 **A column is of hollow circular section with outer diameter 360 mm and inner diameter 240 mm. At what eccentricity load can be applied so that there is no tension developed anywhere in the section.**

Ans. [e = 65 mm]

Question 10.10 **Consider a vertical wall of breadth *B*, depth *D* and height *H* subjected to wind pressure *p* on its vertical face *B* × *H*. If weight density of brick masonry wall is *w*, what are the resultant stresses along the extreme edges of the wall at the base.**

Solution: Fig. 10.12 shows a wall of breadth B, depth D and height H subjected to horizontal wind pressure p. Total wind force on vertical face $P = pBH$

CG of wind load P acts at $\frac{H}{2}$ from base

$$\text{Bending moment due to wind load} = P\frac{H}{2} = \frac{pBH^2}{2} \quad \text{...(i)}$$

$$\text{Moment of inertia of base section, } I_{xx} = \frac{BD^3}{12}$$

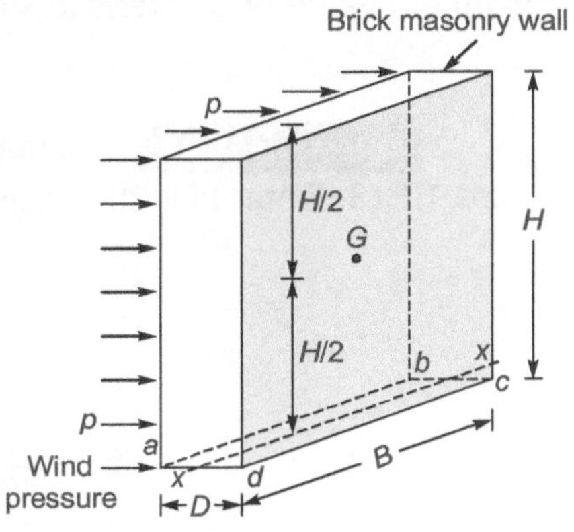

Fig. 10.12

Section modulus, $Z_x = \frac{BD^2}{6}$

Bending stress due to $M = \pm\frac{M}{Z} = \pm\frac{pBH^2}{2}\times\frac{6}{BD^2} = \pm 3p\frac{H^2}{D^2}$ ...(ii)

Weight of wall = $wBHD$

Area of base = DB

Compressive stress due to self weight of the wall = $\frac{wBHD}{BD} = wH$ (comp.) ...(iii)

Resultant stress along edge cd of the base = $3p\frac{H^2}{D^2} + wH$ (compressive)

Resultant stress along edge ab of base section = $wH - 3p\frac{H^2}{D^2}$

If there has to be no tension anywhere in the base section to avoid cracking.

Then, $wH > 3p\frac{H^2}{D^2}$

Question 10.11 **An 8 m high brick masonry wall is of rectangular section 4 m × 0.5 m, i.e., depth 0.5 m, breadth 4 m, is subjected to a horizontal wind pressure of 1.60 kN/m² on 4 m side. If weight density of brick masonry is 22 kN/m³, what are the maximum and minimum stresses developed at the base of the wall.**

Solution: p = 160 kN/m², B=4 m, D = 0.5 m, H = 8 m

$$\sigma_b = \pm 3p\frac{H^2}{D^2} = \pm 3\times 1.60\times\frac{8^2}{0.5^2} = \pm 4.80\times 256 = \pm 1228.8 \text{ kN/m}^2$$

$$\sigma_d = \text{wH} = 22\times 8 = 176 \text{ kN/m}^2$$

$$\sigma_{max} = 1228.8 + 176 = 1404.8 \text{ kN/m}^2$$

$$\sigma_{min} = -1228.8 + 176 = -1052.8 \text{ kN/m}^2$$

The wall may crack at the base because $\sigma_d < \sigma_b$. (due to resultant tensile stress)

Practice Q.10.11 **A masonry brick wall of height 5 m, breadth 6 m and depth 0.4 m is subjected to wind pressure *p* on the face 5 × 6 m. Weight density of masonry structure is 22.4 kN/m³. What is the maximum value of *p*, if tensile stress is not to be developed anywhere in the base.**

Ans. [0.239 kN/m²]

Question 10.12 **A machine part is transmitting a pull, F = 2 kN. It is offset as shown in the figure 10.13. Determine the largest normal stress in the offset portion.**

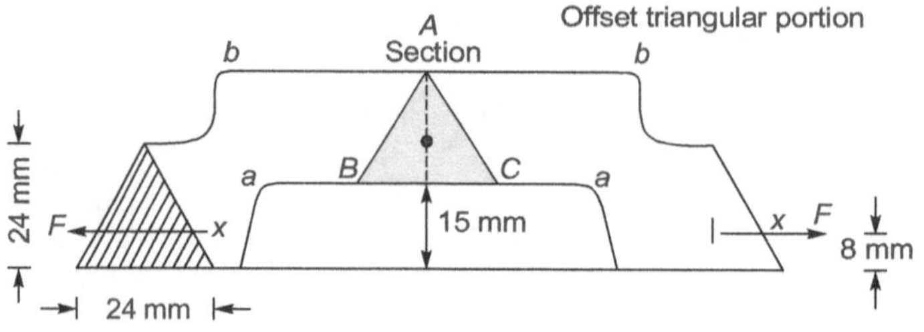

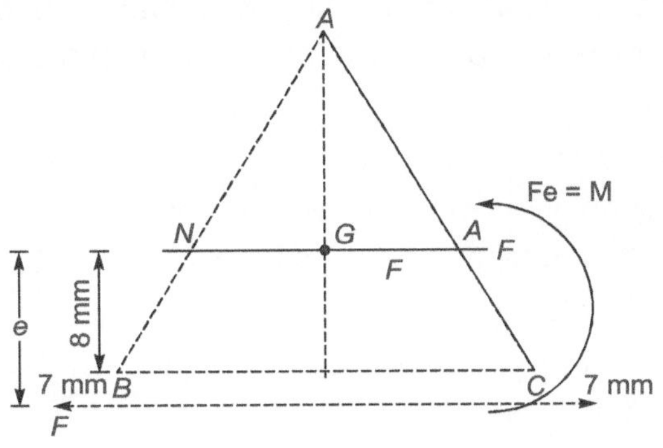

Fig. 10.13

Solution: For the offset portion eccentricity of loading,

$$e = 15 + 8 - 8 = 15 \text{ mm}$$

F along the axis *xx* can be replaced by a force F (tensile) and a moment Fe (anticlockwise). Due to bending moment, bending stress at A will be compressive but bending stress at BC will be tensile.

$$I_{NA}, \text{ about centroidal axis} = \frac{24 \times 24^3}{36} = 9216 \text{ mm}^4$$

$$\text{Area of cross section} = \frac{1}{2} \times 24 \times 24 = 288 \text{ mm}^2$$

Direct stress, $\sigma_d = \dfrac{2000}{288} = 6.944 \text{ N/mm}^2$ (tensile)

Bending stress at *A*, $\sigma_{bA} = \dfrac{2000 \times 15 \times 16}{9216} = 52.083 \text{ N/mm}^2$ (comp.)

Bending stress at *BC*, $\sigma_{bBc} = \dfrac{2000 \times 15 \times 8}{9216} = 26.042 \text{ N/mm}^2$ (tensile)

Resultant stress at *A*, $\sigma_{AR} = 52.083 - 6.944 = 45.139 \text{ N/mm}^2$ (comp.)

Resultant stress at *BC*, $\sigma_{BCR} = 26.042 + 6.944 = 32.986 \text{ N/mm}^2$ (tensile)

Maximum normal stress in offset section is 45.139 N/mm² (comp.)

Practice Q.10.12 **To avoid interference, the cross-sectional area of a link in a machine is reduced in section as shown in figure 10.14. Thickness of the link is 10 mm. Determine force F if the maximum normal stress in the reduced section *AA* is limited to 80 MPa.**

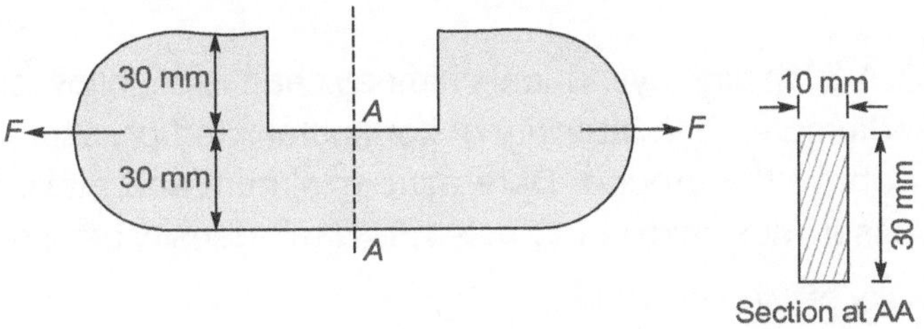

Fig. 10.14

Hints: [Eccentricity, $e = 15$ mm]

Ans. [$F = 6$ kN]

Question 10.13 Consider a cylindrical chimney of height H, outer diameter D, inside diameter d, subjected to horizontal wind pressure p. Determine stresses at the base of the chimney.

Solution: A cylindrical chimney of height H, outer diameter D is shown in fig. 10.15. It is subjected to horizontal wind pressure p over the semi circular portion of height H, and radius $D/2$. This wind force can be integrated over this semi cylindrical surfaces.

Total wind force, $P_w = kpDH$

= Coefficient of wind resistance, wind pressure p, outer diameter of chimney, height of the chimney ,

CG of the load P_w acts at $\frac{H}{2}$ from the base.

Overturning couple of wind force about the axis y-y of the base

$$M_w = P_w\frac{H}{2} = kpDH\frac{H}{2} = kpD\frac{H^2}{2}$$

Section modulus, $$Z = \frac{\pi\left(D^4 - d^4\right)}{32D}$$

Coefficient of wind resistance is generally taken as 0.6 for a cylindrical chimney.

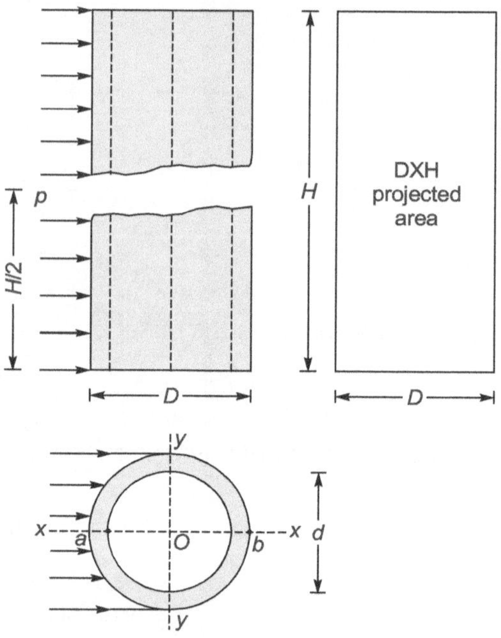

Fig. 10.15

Stress of bending, $$\sigma_b = \frac{M}{Z} = \frac{kpDH^2}{2} \times \frac{32D}{\pi\left(D^4 - d^4\right)}$$

σ_b is tensile stress at point 'a' and compressive stress at point b along the axis x-x passing through the centroid of the section.

Say w = weight density of the material of the chimney

σ_d = direct stress due to self weight of the chimney

= wH (compressive)

Resultant stress at a, $$\sigma_{aR} = wH - \frac{kp16D^2H^2}{\pi\left(D^2 - d^2\right)}$$

Resultant stress at b, $$\sigma_{bR} = wH + \frac{kp16D^2H^2}{\pi\left(D^2 - d^2\right)}$$

Question 10.14 A 30 m high cylindrical chimney shaft is of hollow circular section, 3 m external diameter and 2 m internal diameter. The intensity of horizontal wind pressure p varies as $y^{2/3}$, where y is the height of the chimney above the ground. Determine minimum and maximum intensity of stresses at the base if k = 0.6; wind pressure at a height of 27 m is 1.8 kN/m^2, density of masonry structure is 22.5 kN/m^3.

Solution: Say intensity of wind pressure, $p = cy^{2/3}$

At y = 27 m, p = 1.8 kN/m^2

$1.8 = c, \quad 27^{2/3} = 9c$

Constant, C = 0.2 kN/m^2

Pressure at any height,	p =	$0.2y^{2/3}$
Outer diameter,	D =	3 m
Area	dA =	$3dy$
Force on small area,	dP =	$kpdA = 0.6 \times 0.2 + y^{2/3}, 3dy = 0.36\, y^{2/3}\, dy.$
Moment of the force dP from the base,	dM =	ydP
	dM =	$0.36\, y^{2/3} ydy = 0.36\, y^{5/3}\, dy$

Total moment,

$$M = \int_0^{30} 0.36 y^{5/3} dy = 0.36 \left| \frac{3}{8} y^{8/3} \right|_0^{30}$$

$$= 0.135 \times 30^{8/3} = 0.135 \times 8689.5 = 1173 \text{ kNm}$$

Section modulus,

$$Z = \frac{\pi}{32D}\left(D^4 - d^4\right) = \frac{\pi}{32 \times 3}\left(3^4 - 2^4\right) = 2.127 \text{ m}^3$$

Bending stress,

$$\sigma_b = \frac{M}{Z} = \frac{1173}{2.127} = \pm 551.5 \text{ kN/m}^2$$

Height,	H =	30 m
Direct compressive stress,	σ_d =	wH = 22.5 × 30 = 675 kN/m^2
Maximum stress at base	=	675 + 551.5 = 1226.5 kN/m^2 (comp.)
Maximum stress base	=	+675 – 551.5 = 123.5 kN/m^2 (comp.)

No tension any where.

Practice Q.10.14 A 24 m high masonry chimney of uniform circular section, 4 m external diameter, 3 m internal diameter has to withstand a horizontal wind pressure of 2.4 kN/m^2 of the projected area. Take coefficient of wind resistance k = 0.65 and find minimum and maximum intensities of stress at the base. Density of masonry structure is 22 kN/m^2.

Ans. [109.58, 946.42 kN/m^2]

Question 10.15 A tapering chimney of hollow circular section is 30 m high. Its external diameter at the base is 3 m and at the top it is 2 m. The internal diameter at the base is 1.2 m. It is subjected to a wind pressure of 2.4 kN/m^2 of the projected area. Calculate the overturning moment at the base, if the weight of the chimney is 3500 kN. Determine the maximum and minimum stress intensities at the base. Take k = 0.6.

Solution:

p = 2.4 kN/m^2

Coefficient of wind resistance, k = 0.6

Total wind force,

$$P_w = \left(\frac{2+3}{2}\right)(30)p \times k = 2.5 \times 30 \times 2.4 \times 0.6 = 108 \text{ kN},$$

Distance of CG from base

$$\bar{H} = \frac{2 \times 30 \times 15 + 2 \times \frac{1}{2} \times 0.5 \times 30(10)}{75} = \frac{900 + 150}{75} = 14 \text{ m}$$

M, overturning moment

$$= P_w \times H = 108 \times 14$$

$$= 1512 \text{ kNm}$$

Area of cross-section at the base

$$= \frac{\pi}{4}(3^2 - 1.2)^2 = 5.94 \text{ m}^2$$

Section modulus at base, $Z = \frac{\pi}{32} \times \frac{1}{D}(D^4 - d^4)$

$$= \frac{\pi}{32 \times 3}(3^4 - 1.2^4)$$

$$= \frac{\pi}{96}(81 - 2.0736)$$

$$= 2.583 \text{ m}^3$$

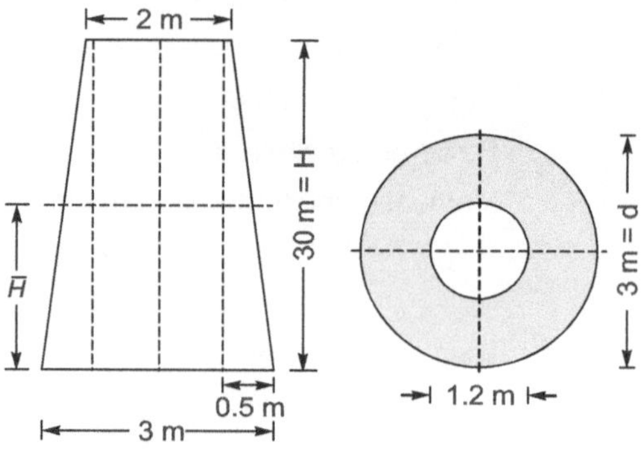

Fig. 10.16

Weight of the chimney, $W = 3500$ kN

σ_d, direct stress $= \frac{W}{\text{area}} = \frac{3500}{5.94} = 589.225$ kN/m² (comp.)

Bending stress, $\sigma_b = \frac{M}{Z} = \frac{1512}{2.583} = +585.366$ kN/m²

Maximum stress at base, $\sigma_{max} = 589.225 + 585.366$

$= 1174.59$ kN/m² (comp.)

Minimum stress at base, $\sigma_{min} = 589.225 - 585.366$

$= +3.859$ kN/m² (comp.)

No tension anywhere in the section.

Practice Q.10.15 A 80 mm steel pipe (inside diameter 60 mm) is used as a support for a basket ball back board as shown in the figure 10.17. It is securely fixed to the ground. Calculate the maximum normal stress that would be developed in pipe if a 700 N player hangs on the base of the rim of the basket.

Hints: [e = 1.2 m]

Ans. [24.77 N/mm²]

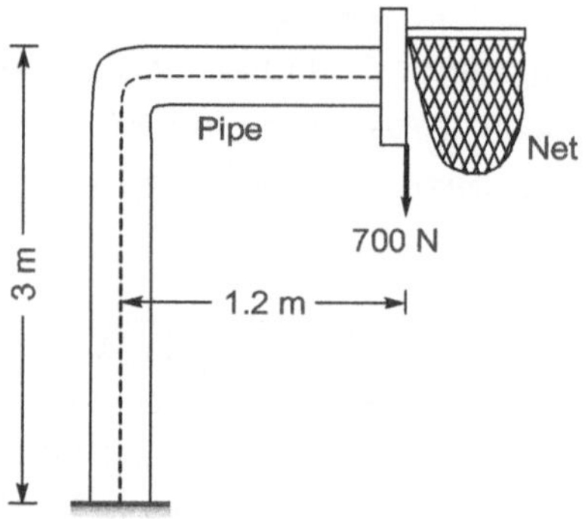

Fig. 10.17

Objective Type Questions

Q.1 For the configuration of loading shown in figure strain in fibre AB is given by

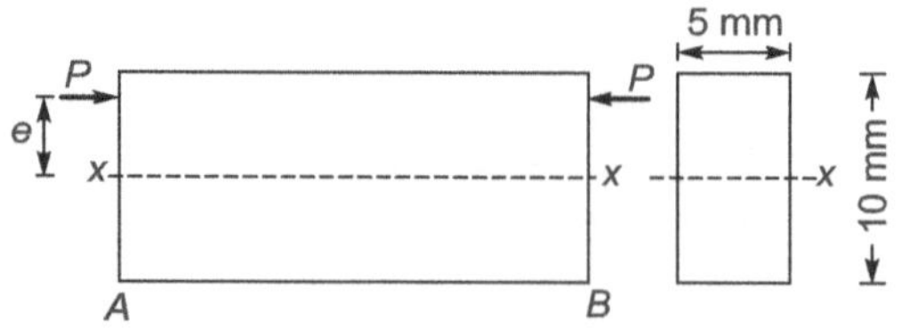

(a) $\frac{P}{A}$(tensile)

(b) $\left(\frac{P}{A} - \frac{Pe5}{I_{xx}}\right)$(comp.)

(c) $\left(\frac{P}{A} + \frac{Pe5}{I_{xx}}\right)$(comp.)

(d) $\frac{P}{A}$(comp.)

[IES 1995]

Q.2 A column of square section 40 mm × 40 mm fixed to the ground carries an eccentric load *P* of 1600 N as shown in figure. 10.19. If the stress developed along CD is –1.2 N/mm², the stress along the edge *AB* will

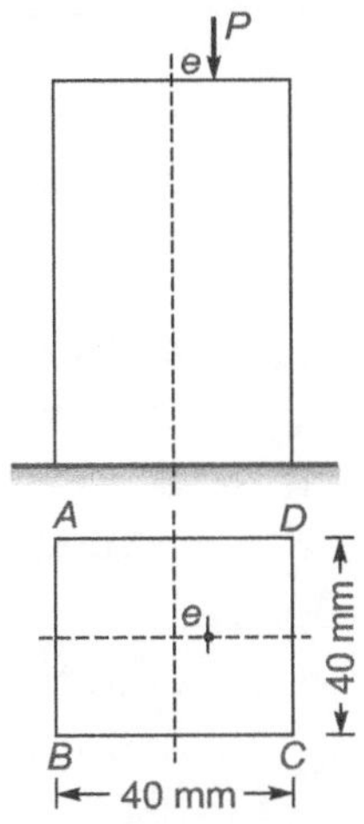

(a) –1.2 N/mm² (b) +1 N/mm²
(c) +0.8 N/mm² (d) –0.8 N/mm²

[IES 1999]

Q.3 A column is subjected to eccentric loading, such that stresses at extreme edges are +32 MPa (comp.) and +8 MPa (comp.). What are stresses due to bending at extreme edges
(a) 20 MPa (b) 16 MPa
(c) 12 MPa (d) None of these

Q.4 A column is of rectangular section 240 mm × 360 mm. What is the core of this section?
(a) a rectangle of sides 40 × 60 mm
(b) a rhombus of diagonals 40 mm and 60 mm
(c) a rhombus of diagonals 80 mm and 120 mm
(d) None of these

Q.5 A column with I-section, has following properties section modulii, $Z_x = 999 \times 10^2$ mm³, $Z_y = 54 \times 10^3$ mm³. Area of cross-section A = 1800 mm². What is allowable eccentricity e_x for no tension in column section?
(a) 30 mm (b) 37 mm
(c) 55.5 mm (d) None of these

Q.6 A short masonry column of square section 2 m side is 10 m high. Wind pressure of intensity 2 kN/m² acts on the vertical face of the column. Weight density of masonry is 20 kN/m³. What is the maximum stress at the base of the column
(a) 200 kN/m² (b) 300 kN/m²
(c) 350 kN/m² (d) 100 kN/m²

Q.7 A cylindrical chimney of hollow circular section is subjected to wind pressure *p*. Density of masonry work is 20 kN/m³. The maximum or minimum stresses developed at the base are +500 kN/m² and 200 kN/m². If the intensity of wind pressure is doubled, what is the maximum stress developed at the base
(a) 750 kN/m² (b) 650 kN/m²
(c) 550 kN/m² (d) None of these

Answers

1. (b) 2. (d) 3. (c) 4. (c) 5. (a)
6. (c) 7. (b)

Explanations

1. (b)

$$\sigma_d = \frac{P}{A}\text{(comp.)}$$

$$\sigma_b = -\frac{Pe5}{I_{xx}}$$

$$\sigma_{AB} = \left(\frac{P}{A} - \frac{Pe5}{I_{xx}}\right)\text{(comp.)}$$

2. (d)

$$\sigma_d = -\frac{1600}{1800} = -1 \text{ N/mm}^2 \text{ (comp.)}$$

$$\sigma_b = -1.2 + 1 = -0.2 \text{ N/mm}^2 \text{ along } CD$$

$$\sigma_b = +0.2 \text{ N/mm}^2 \text{ along AB}$$

$$\sigma_R \text{ along } AB = -1 + 0.2$$

$$= -0.8 \text{ N/mm}^2$$

3. (c)

$$\sigma_1 = +32 \text{ MPa}$$
$$\sigma_2 = +8 \text{ MPa}$$
$$\sigma_d = \frac{\sigma_1 + \sigma_2}{2} = +20 \text{ MPa}$$
$$\sigma_b = \frac{\sigma_1 - \sigma_2}{2} = 12 \text{ MPa}$$

4. (c)

$$B = 240 \text{ mm}$$
$$D = 360 \text{ m}$$

Core, diagonals $\frac{240}{3}, \frac{360}{3}$ or 80 mm, 120 mm.

5. (a)

$$e_x = \frac{Z_y}{A} = \frac{54000}{1800} = 30 \text{ mm}$$

6. (c)

$$M = 2 \times 10 \times 2 \times 5 = 200 \text{ kNm}$$
$$Z = \frac{2^3}{6} = \frac{4}{3} \text{ m}^3$$
$$\sigma_b = \frac{200 \times 3}{4} = \pm 150 \text{ kN/m}^2$$
$$\sigma_d = wH = 200 \text{ kN/m}^2$$
$$\sigma_{max} = 150 + 200 = 350 \text{ kN/m}^2$$

7. (b)

$$\sigma_{max} = 500 \text{ kN/m}^2$$
$$\sigma_{min} = 200 \text{ kN/m}^2$$
$$\sigma_d = 350 \text{ kN/m}^2$$
$$\sigma_b = 150 \text{ kN/m}^2$$

Pressure is double

$$\sigma_b' = 300 \text{ kN/m}^2$$
$$\sigma_{max} = 350 + 300 = 650 \text{ kN/m}^2$$
$$\sigma_{min} = 350 - 300 = +50 \text{ kN/m}^2$$

■■■■

Deflection in Beams

CHAPTER 11

A beam, when subjected to transverse loads, bending moment and shear force are developed in sections of the beam. SF and BM diagrams are drawn so as to know about maximum bending moment and maximum shear force in beam. Due to bending moment, direct bending stresses and due to shear force, shear stresses are developed in beam section. But due to transverse loads, the beam bends and slope and deflections are developed in beam which vary from section to section. Slope and deflection both are important to know about **the rigidity of the beam** subjected to transverse loads and moments. Equations for slope and deflection are developed to know these values at any section of the beam. For an efficient design of a beam section, both deflection and slope must not exceed the permissible limits.

Question 11.1 **Develop a relationship between bending moment, flexural rigidity and curvature of a beam. Take a beam of length *L*, simply supported at ends, carrying a point load *W*, at the center of the span. If *EI* is the flexural rigidity of the beam section, determine maximum slope and maximum deflection in beam.**

Solution: In the flexural formula $\frac{M}{I} = \frac{E}{R}$;

$$\frac{1}{R} = \frac{M}{EI} \qquad \text{where, } \frac{1}{R} \simeq \text{curvature} \simeq \frac{d^2y}{dx^2}$$

or
$$\frac{d^2y}{dx^2} = \frac{M}{EI}$$

or
$$EI\frac{d^2y}{dx^2} = M \qquad \text{...(i)}$$

This the basic formula used to determine slope $\frac{dy}{dx}$ or y, deflection at any section of a beam.

So, $EI\frac{d^2y}{dx^2} = M_x$ (bending moment at a section, x distance away from the origin)

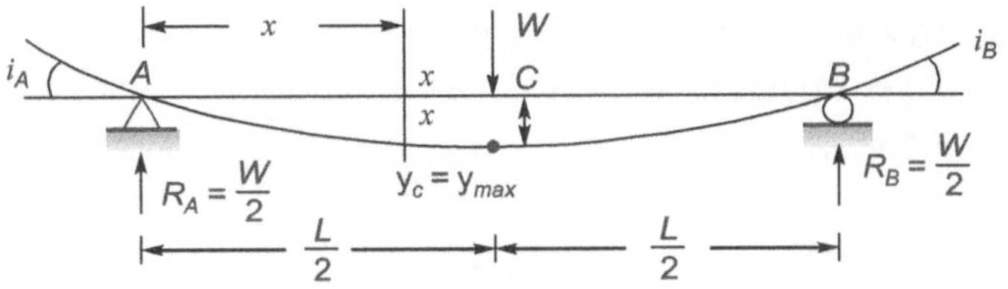

Fig. 11.1

A beam AB of length L, hinged at end A, roller supported at end B, subjected to a central vertical load. W is shown in figure 11.1. Due to this transverse load W, at the center, reactions A and B will be equal i.e.

$R_A = R_B = \frac{W}{2} \uparrow$ (as shown).

Beam bends showing concavity, upwards, with maximum deflection at the center and slope at the center is zero. Slope at A i.e. i_A gradually decreases and becomes zero at center C, but slope again gradually increases and becomes i_B at end B. Because of symmetrical loading.

Slope at A, $i_A = -i_B$, slope at B

Now consider a section X-X at a distance of x from A, in the portion AC of the beam.

Bending moment, $$M_x = +R_A x = +\frac{W}{2} \cdot x$$

(taking clockwise moments on the left side of the section as positive)

or $$EI\frac{d^2y}{dx^2} = +\frac{W}{2} \cdot x \quad \text{...(ii)}$$

Integrating equation (*ii*) we get,

$$EI\frac{dy}{dx} = \frac{Wx^2}{4} + C_1 \quad \text{...(iii)}$$

where, C_1 is a constant of integration.

Boundary Condition:

Slope is zero at $x = \frac{L}{2}$, putting this value in equation (*iii*)

At $x = \frac{L}{2}$, $$EI\frac{dy}{dx} = \frac{WL^2}{16} + C_1 = 0$$

or constant of integration, $$C_1 = -\frac{WL^2}{16}$$

so, $$EI\frac{dy}{dx} = \frac{Wx^2}{4} - \frac{WL^2}{16} \quad \text{...(iv)}$$

Integrating equation (*iv*) also, we get

$$EI\,y = \frac{Wx^3}{12} - \frac{WL^2}{16} \cdot x + C_2 \quad \text{...(v)}$$

where C_2 is another constant of integration.

Boundary Conditions:

At end A, $x = 0$, deflection, $y = 0$

(Note that levels of end A and B remain unchanged)

So, $EI \times 0 = 0 - 0 + C_2$, or constant, $C_2 = 0$

Finally the equation becomes

$$EI\,y = \frac{Wx^3}{12} - \frac{WL^2}{3} \cdot x \quad \text{...(vi)}$$

Using equation (*iv*),

$$EI\frac{dy}{dx} = \frac{Wx^2}{4} - \frac{WL^2}{16}$$

$x = 0$, slope, $$i_A = -\frac{WL^2}{16EI}$$

$x = \frac{L}{4}$, $$EIi' = \frac{WL^2}{64} - \frac{WL^2}{16}$$

$$i' = -\frac{3}{64} \times \frac{WL^2}{EI}$$

$x = \frac{L}{2}$, slope, $$i_C = \frac{WL^2}{16} - \frac{WL^2}{16} = 0$$

Using equation (*v*),

$$EI\,y = \frac{Wx^3}{12} - \frac{WL^2}{16}\cdot x$$

at $x = 0$, deflection, $y = 0$ at $x = L$ also because of symmetry

$x = \frac{L}{4}$, $$EIy' = \frac{WL^3}{768} - \frac{WL^3}{64} = \frac{11}{768} \times WL^3 \quad \text{at } x = \frac{3L}{4} \text{ due to symmetry}$$

$x = \frac{L}{2}$, $$EI\,y_c = \frac{WL^3}{96} - \frac{WL^3}{32} = -\frac{WL^3}{48}$$

$$y_c = -\frac{WL^3}{48EI} \quad \text{(maximum deflection)}$$

Note that maximum slope at ends is $\pm\frac{WL^2}{16EI}$ and maximum deflection at the center is $-\frac{WL^3}{48EI}$.

Note that angular deviation of axis of beam is known as slope while linear deviation of axis of beam (in transverse directions is known as deflection).

Practice Q.11.1 **An *I* section beam is supported over a span of 1.8 m, I_{xx} of the beam is 688 cm⁴. Material of the beam is steel, with *E* = 200 GPa. If the load at the center is 20 kN, what is deflection at the center and slope at the ends?**

Ans. [–1.766 mm, ±0.17°]

Question 11.2 **A beam *ABCD*, 6 m long, hinged at end *A*, roller supported at *C* carries a load of 5 kN at *B*, 2 m away from *A* and a load of 2 kN at end *D*. *I* of the beam section = 1200 cm⁴. *E* = 200 GPa, determine slope at *A*, deflections at *B* and *D*.**

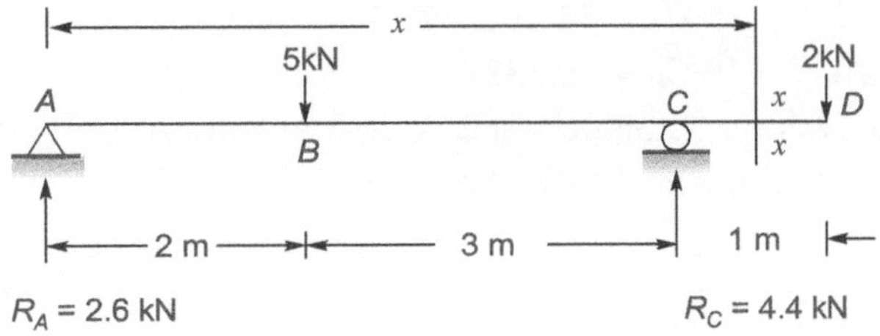

Fig. 11.2

Solution: Let us first calculate support reactions,

Taking moments about A

$$5 \times 2 + 6 \times 2 = 5\, R_C$$

Reaction, $R_C = 4.4$ kN

Reaction, $R_A = 5 + 2 - 4.4$

$= 2.6$ kN

In (Question) No.1, the beam was symmetrically loaded about is center. Equation of bending moment was made (only) for the first portion. Results obtained from first portion were repeated for second portion taking the advantage of symmetrical loading.

But in this problem, the beam is not symmetrically loaded about its center, and we do not know where the slope is zero. In ***such cases Macaulay's method can be used***, which is briefly explained as follows:

1. Calculate the reactions
2. Starting from the origin, identify the last portion of the beam, as in the present case, origin at A, x + positive towards right. Portions are AB, BC, CD, where CD is the last portion of the beam.
3. Make an equation of bending moment for a section XX in the last portion, as for the present (case).

$$M_x = EI\frac{d^2y}{dx^2} = R_A\, x - 5(x-2) + R_C\,(x-5)$$

$= 2.6x - 5(x-2) + 4.4(x-5)$, putting the values of R_C taking clockwise moments as positive on the left side of the section x-x. ...(i)

4. Integrate equation of bending moment two times, so as to get two constants of integration C_1, C_2

$$EI\frac{dy}{dx} = 1.3x^2 - 2.5(x-2)^2 + 2.2(x-5)^2 + C_1 \quad \text{...(ii)}$$

$$EIy = \frac{1.3}{3}x^3 - \frac{2.5}{3}(x-2)^3 + \frac{2.2}{3}(x-5)^3 + C_1x + C_2$$

Using the boundary conditions as $x = 0$, $y = 0$, $x = 5$ m; $y = 0$ in present case, determine constants C_1, C_2. At the same time be careful that for a particular value of x, if the bracketed term becomes negative, the term has to be omitted.

In the present case $x = 0$, $y = 0$.

$$EI \times 0 = 0 - \text{omitted term} + \text{omitted term} + 0 + C_2$$

In other words, when we take $x = 0$, we are in portion AB, and the terms $(x - 2)$ and $(x - 5)$ are not applicable for this portion.

So in the present case constant, $C_2 = 0$

Taking second boundary condition i.e. $x = 5$, $y = 0$ at the point C, putting this value

$$0 = 1.3 \times \frac{5^3}{3} - \frac{2.5}{3}(5-2)^3 + \frac{2.2}{3}(5-5)^3 + 5\,C_1$$

$$0 = 54.166 - 22.5 + 0 + 5\,C_1$$

Constant, $C_1 = -6.333$

5. Constants C_1 and C_2, thus obtained will be applicable for all portions of the beam.

Finally the equation of slope and deflection are

$$EI\frac{dy}{dx} = 1.3\,x^2 - 2.5(x-2)^2 + 2.2(x-5)^2 - 6.333$$

$$EIy = 1.3\frac{x^3}{3} - \frac{2.5}{3}(x-2)^3 + \frac{2.2}{3}(x-5)^3 - 6.333x$$

Slope at A ($x = 0$),

$$EIi_A = 0 - \text{omitted term} + \text{omitted term} - 6.333$$

$$EI = \frac{1200 \times 10^{-8} \times 200 \times 10^9}{10^3} = 2400 \text{ kNm}^2$$

$$i_A = \frac{-6.333}{2400} = -2.64 \times 10^{-3} \text{ radian} = -0.151°$$

Deflection *B* and *D*

At $B\, x = 2$ m

$$EI\, y_B = \frac{1.3}{3} \times 2^3 - \frac{2.5}{3}(2-2)^3 + \text{omitted term} - 6.333 \times 2$$

$$= 3.4666 - 12.6666 = -9.2$$

$$y_B = \frac{-9.2}{2400} = 3.83 \times 10^{-3} \text{m} = -3.83 \text{ mm}$$

At D, $x = 6$ m

$$EIy_D = \frac{1.3}{3} \times 6^3 - \frac{2.5}{3}(6-2)^3 + \frac{2.2}{3}(6-5)^3 - 6.333 \times 6$$

$$= 93.6 - 53.333 + 0.7333 - 38 = +3.0$$

$$y_D = \frac{3}{2400} = 1.25 \times 10^{-3} \text{m} = +1.25 \text{ mm}$$

Practice Q.11.2 A beam *ABCD*, 5 m long is roller supported at *B* (1 m from *A*) and hinged at end *D*. Calculate slope at *D* and deflection at *C*, EI = 3000 kNm². (Fig.11.3)

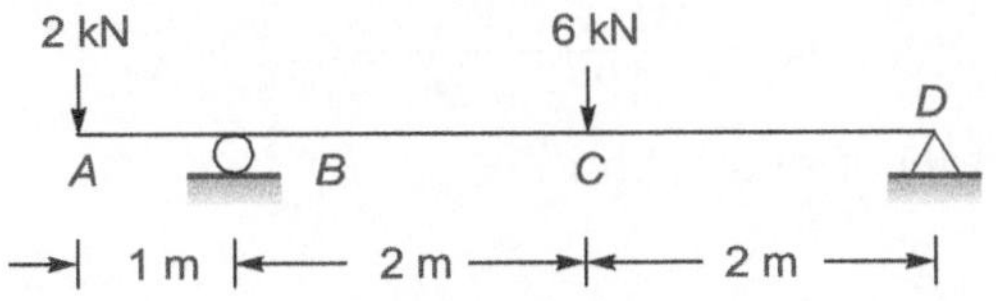

Fig. 11.3

Hints: (x = 1 m, y = 0, x = 5 m, y = 0, constant $C_2 \neq 2$)

Ans. [R_B = 5.5 kN, R_D = 2.5 kN, $M_x = -2x + 5.5(x-1) - 6(x-3)$, $C_1 = -2.333$, $C_2 = +2.666$, $i_D = 0.089°$, $y_c = -2$ mm]

Question 11.3 A beam *AB*, supported over a span *L*, carries a udl of intensity *w* per unit length. Determine slope at ends and deflection at the center.

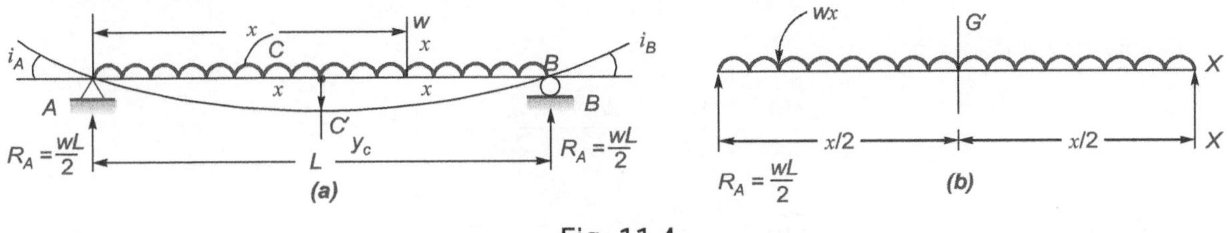

Fig. 11.4

Solution: Support moments

Taking moments about A

$$\text{Total load on beam} = wL$$

$$CG \text{ of load from } A = \frac{L}{2}$$

$$\frac{wL}{1} \times \frac{L}{2} = R_B \cdot L$$

Reaction, $$R_B = \frac{wL}{2} \quad \text{...(i)}$$

Reaction, $$R_A = \frac{wL}{1} - \frac{wL}{2} = \frac{wL}{2} \uparrow \quad \text{...(ii)}$$

Total downward load upto x, wx

CG of load upto section XX lies at $x/2$ from XX.

Bending moment at section $X - X$,

$$M_x = R_A x - wx \cdot \frac{x}{2} = \frac{wL}{2} x - \frac{wx^2}{2}$$

So, $$EI \frac{d^2 y}{dx^2} = \frac{wL}{2} x - \frac{wx^2}{2} \quad \text{...(iii)}$$

Integrating equation (*iii*)

$$EI \frac{dy}{dx} = \frac{wLx^2}{4} - \frac{wx^3}{6} + C_1 \quad \text{...(iv)}$$

where C_1 is constant of integration

Further integrating equation (*iv*)

$$EIy = \frac{wLx^3}{12} - \frac{wx^4}{24} + C_1 x + C_2,$$

(where C_2 is another constant of integration

Boundary Condition:

$x = 0$, $y = 0$, at end A

$$EI \times 0 = 0 - 0 + C_2$$

Constant, $C_2 = 0$

Another boundary condition, $x = L$, $y = 0$, at end B

$$0 = \frac{wL^4}{12} - \frac{wL^4}{24} + C_1 \times L + 0$$

Constant, $$C_1 = -\frac{wL^3}{24}$$

Finally the equations are

$$EI \frac{dy}{dx} = \frac{wLx^2}{4} - \frac{wx^3}{6} - \frac{wL^3}{24}$$

At $x = 0$, $$EIi_A = -\frac{wL^3}{24}$$

or slope at A, $$i_A = \frac{-WL^3}{24\,EI}$$

At $x = L$, $$EIi_B = \frac{WL^3}{4} - \frac{WL^3}{6} - \frac{WL^3}{24} = +\frac{WL^3}{24}$$

slope at B, $$i_B = +\frac{WL^3}{24\,EI}$$

Deflection at center

$$EIy = \frac{WLx^3}{12} - \frac{Wx^4}{24} - \frac{WL^3x}{24}$$

At $x = \frac{L}{2}$, at the center

$$EIy_c = \frac{WL^4}{96} - \frac{WL^4}{384} - \frac{WL^4}{48} = -\frac{5}{384} \times WL^4$$

y_c, deflection at center $= -\frac{5}{384} \times \frac{WL^4}{EI}$

Practice Q.11.3.1 Take a cantilever of length L, free at end A and fixed at end B carrying a uniformly distributed load w throughout its length. If EI is the flexural rigidity of the cantilever, determine slope and deflection at free and A.

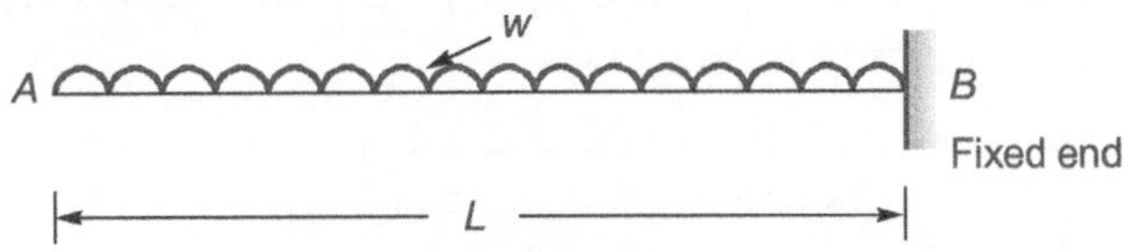

Fig. 11.5

Hints: $\left[EI\frac{d^2y}{dx^2} = -\frac{wx^2}{2}; C_1 = +\frac{WL^3}{6}, C_2 = -\frac{WL^4}{8}, i_A = +\frac{WL^3}{6EI}; y_A = -\frac{WL^4}{8\,EI}\right]$

[Note that slope and deflection both are zero at end B]

Practice Q.11.3.2 A cantilever AB, fixed at end B, carries a point load W at fixed A. If EI is the flexural rigidity of the cantilever. Determine slope and deflection at free end A.

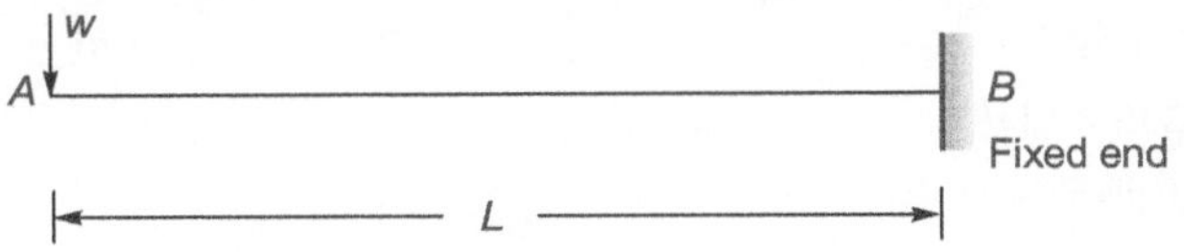

Fig. 11.6

Hints: $\left[EI\frac{d^2y}{dx^2} = -Wx; C_1 = +\frac{WL^2}{2}; C_2 = -\frac{WL^3}{3}\right.$, slope and deflection both are zero at fixed end $B\Big]$

Ans. $\left[+\frac{WL^2}{2EI}; -\frac{WL^3}{3EI}\right]$

Question 11.4 A beam $ABCD$, 6 m long is supported on roller at B, and hinged at end D. It carries a udl of 2 kN/m over OBC = 4 m, a point load of 8 kN at C. (Fig. 11.7) If EI is the flexural rigidity of

the beam, determine slope and deflection at *A*. What is the deflection under the load 8 kN, at point *C*? [EI = 3200 kNm²]

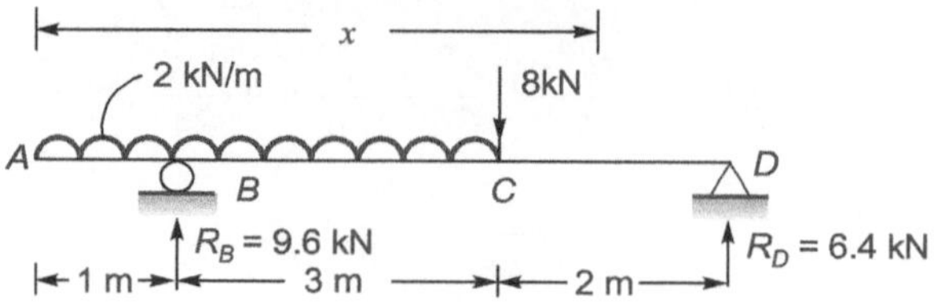

Fig. 11.7

Solution: Support moments: Taking moments about point *B*, udl of 2 kN over *ABC* = 4 × 2 = 8 kN

CG of this load acts at 2 m from *A* or 1 m from *B*, so

$$8 \times 1 + 8 \times 3 = 5\, R_D$$

Reaction, $R_D = 6.4$ kN

Reaction, $R_B = 8 + 8 - 6.4$

$= 9.6$ kN

Taking origin at *A*, there are 3 portions *AB*, *BC* and *CD*, and *CD* is the last portion in beam.

$$M_x = 8(x-2) + 9.6(x-1) - 8(x-4)$$

(due to udl) (due to reaction) (due to point load)

But this equation does not hold good for portions *AB* and *BC*. We have to make an equation which is applicable for all the portions of the beam.

Let us extend the udl upto section *XX* and apply negative load from *C* to *XX* as shown in figure 11.8. So that equation becomes valid for portion *CD* also.

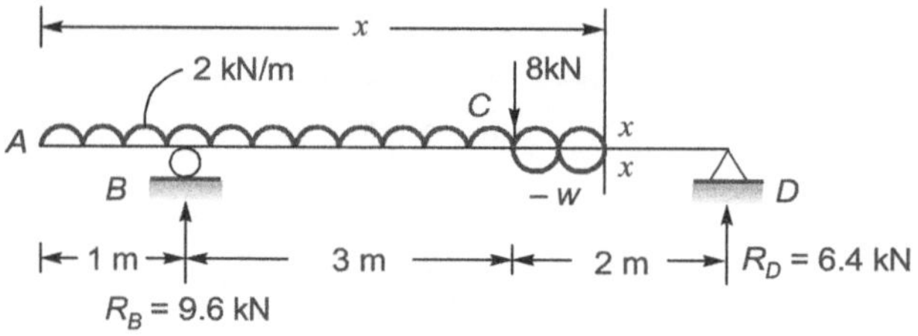

Fig. 11.8

$$EI\frac{d^2y}{dx^2} = \frac{-wx^2}{2} + 9.6(x-1) - 8(x-4) + \frac{w}{2}(x-4)^2 \quad \text{...(i)}$$

I II III IV

where rate of loading is 2 kN/m

There are 4 terms in this equation

Note that term I, $\frac{-wx^2}{2}$ is applicable for portion *AB*

terms I and II, $\frac{-wx^2}{2} + 9.6(x-1)$ are applicable for portion *BC*

terms I, II, III, IV $\frac{-wx^2}{2} + 9.6(x-1) - 8(x-4) + \frac{w}{2}(x-4)^2$ are applicable for portion *CD*

Putting the value of w = 2 kN/m

$$EI\frac{d^2y}{dx^2} = -x^2 + 9.6\,(x-1) - 8(x-4) + (x-4)^2 \quad \text{...(ii)}$$

Integrating equation (*ii*) we get,

$$EI\frac{dy}{dx} = \frac{-x^3}{3} + 4.8(x-1)^2 - 4(x-4)^2 + \frac{(x-4)^3}{3} + C_1, \quad \text{...(iii)}$$

where C_1 is constant of integration

In this case, we do not know where the slope $\frac{dy}{dx}$ is zero.

Integrating equation (*iii*) also, we get

$$EIy = \frac{-x^4}{12} + 1.6(x-1)^3 - \frac{4}{3}(x-4)^3 + \frac{(x-4)^4}{12} + C_1 x + C_2$$

Boundary Condition:

$x = 1$ m, $y = 0$ (in the portion AB)

$$EI \times 0 = -\frac{1}{12} + 0 - \text{omitted term} + \text{omitted term} + C_1 + C_2$$

$$0 = -\frac{1}{12} + C_1 + C_2$$

or $$C_1 + C_2 = \frac{1}{12}$$

At $x = 6$ m, $y = 0$, putting the value, all the terms are applicable.

$$EI \times 0 = -\frac{6^4}{12} + 1.6(5)^3 - \frac{4}{3}(2)^3 + \frac{2^4}{12} + 6C_1 + C_2$$

$$0 = -108 + 200 - 10.667 + 1.333 + 6\,C_1 + C_2$$

$$0 = 82.666 + 6\,C_1 + C_2$$

or $$6\,C_1 + C_2 = -82.666$$

$$C_1 + C_2 = +0.0833$$

$$C_1 = -16.55$$

$$C_2 = +16.633$$

Finally the equations are

$$EI\frac{dy}{dx} = -\frac{x^3}{3} + 4.8(x-1)^2 - 4(x-4)^2 + \frac{(x-4)^3}{3} - 16.55$$

$$EIy = \frac{x^4}{3} + 1.6(x-1)^3 - \frac{4}{3}(x-4)^3 + \frac{(x-4)^3}{12} - 16.55x + 16.633$$

Slope at A ($x = 0$)

$$EIi_A = 0 + \text{omitted term} - 16.55$$

$$i_A = \frac{-16.55}{EI} = \frac{-16.55}{3200} = -5.17 \times 10^{-3} \text{ rad} = -0.3°$$

Deflection at A ($x = 0$)

$$EIy_A = 0 + \text{omitted term} + 0 + 16.333$$

$$y_A = +\frac{16.633}{EI} = \frac{16.633}{3200} = 5.2 \times 10^{-3} \text{ m} = +5.2 \text{ mm}$$

Deflection at C, $x = 4$ m

$$EIy_c = -\frac{4^4}{12} + 1.6(3)^3 - 0 + 0 - 16.55 \times 4 + 16.633$$

$$= -21.33 + 43.2 - 66.2 + 16.633 = -27.7$$

$$y_c = \frac{-27.7}{EI} = \frac{-27.7}{3200} = -8.656 \times 10^{-3} = -8.656 \text{ mm}$$

Question 11.5 A beam *ABC*, length *L* hinged at both the ends is subjected to a clockwise moment at *C*, such that $AC = 0.4\ L$. Determine slope and deflection at point *C*, if *EI* is the flexural rigidity of the beam.

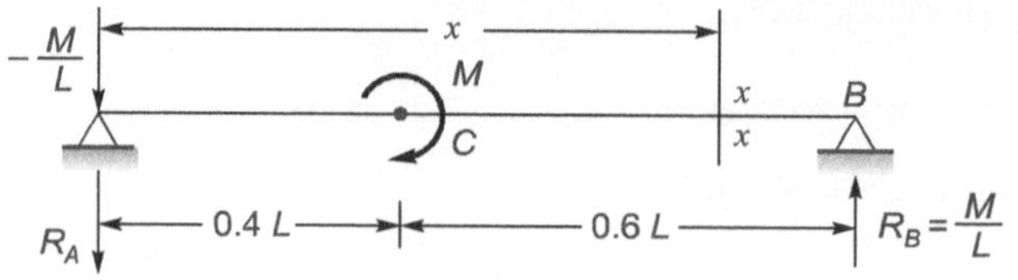

Fig. 11.9

Solution: Support reaction: Taking moment about A

$$M - R_B \times L = 0 \qquad \text{(clockwise } M - R_B \times L\text{, anticlockwise moment)}$$

Reaction,
$$R_B = \frac{M}{L}\uparrow$$

To balance R_B, reaction,
$$R_A = \frac{M}{L}\downarrow$$

There are only two portions in the beam. Taking origin at *A*, *x* positive towards right, bending moment equation in portion *CB*

$$EI\frac{d^2y}{dx^2} = -\frac{M}{L}\cdot x + M(x - 0.4\,L)^\circ \qquad \text{...(i)}$$

Note that $M(x - 0.4L)^\circ$ is equal to *M*, but by this expression we are locating the exact position of the point of application of moment.

Integrating equation (*i*),

$$EI\frac{dy}{dx} = -\frac{Mx^2}{2L} + M(x - 0.4\,L) + C_1 \qquad \text{...(ii)}$$

(where C_1 is constant of integration)

We do not know, where slope $dy/dx = 0$, so integrate equation (*ii*) also

$$EIy = -\frac{Mx^3}{6L} + \frac{M}{2}(x - 0.4\,L)^2 + C_1x + C_2$$

Boundary Condition:

$y = 0,\ x = 0$

$$EI \times 0 = 0 + \text{omitted term} + 0 + C_2$$

Constant,
$$C_2 = 0$$

Then $x = L,\ y = 0$, putting this boundary condition

$$EI \times 0 = -\frac{ML^3}{6L} + \frac{M}{2}(L - 0.4\,L)^2 + C_1L$$

$$0 = -\frac{ML^2}{6} + 0.18\,ML^2 + C_1L$$

or
$$C_1 = -0.01333\,ML$$

Finally the equation is

$$EI\frac{dy}{dx} = -\frac{Mx^2}{2L} + M(x - 0.4\,L) - 0.01333\,ML$$

At point *C*,
$$x = 0.4\ \text{L}$$

$$EIi'_C = -\frac{M\times(0.4L)^2}{2L}+0-0.01333\,ML = 0.09333\,ML$$

Slope,

$$i_C = -0.09333\frac{ML}{EI}$$

Then,

$$EIy = -\frac{Mx^3}{6L}+\frac{M}{2}(x-0.4L)^2-0.01333\,MLx$$

at C,

$$x = 0.4\,L$$

$$EIy_C = -\frac{M}{6L}(0.4L)^3+\frac{M}{2}(0)-0.01333\times ML\times 0.4L$$

$$= -0.01066\,ML^2+0-0.005332\,ML^2 = -0.016\,ML^2$$

$$y_C = -\frac{0.016\,ML^2}{EI}$$

Practice Q.11.5 A beam *AB* of length *L* is hinged at both the ends. It is subjected to a clockwise moment *M* at point *C*, $AC = L/2$. If *EI* is the flexural rigidity of the beam, what are slope and deflection at point *D*, such that $AD = L/4$.

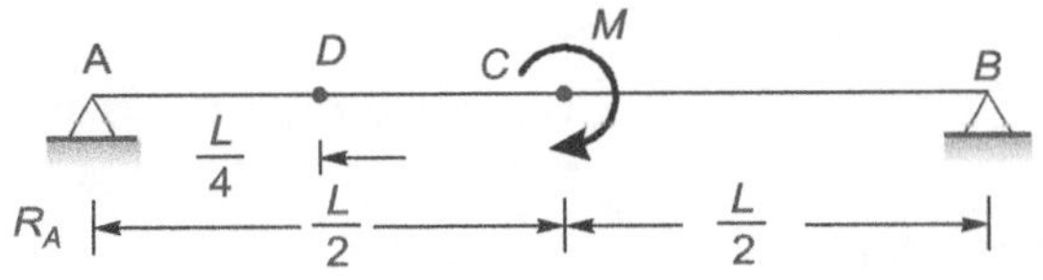

Fig. 11.10

Hint: $\left[R_B = \frac{M}{L}\uparrow, R_A = -\frac{M}{L}\downarrow, EI\frac{d^2y}{dx^2} = -\frac{M}{L}x+M\left(x-\frac{L}{2}\right)^0\right], C_1 = \frac{ML}{24}, C_2 = 0$

Ans. $\left[i_D = +\frac{ML}{96EI}, y_D = +\frac{ML^2}{192EI}\right]$

Question 11.6 A beam *ABCD*, 5 m long is hinged at end *A* and roller supported at *C*, as shown in figure 11.11, $EI = 1600$ kNm² of the beam. Determine slope and deflection at point *D*.

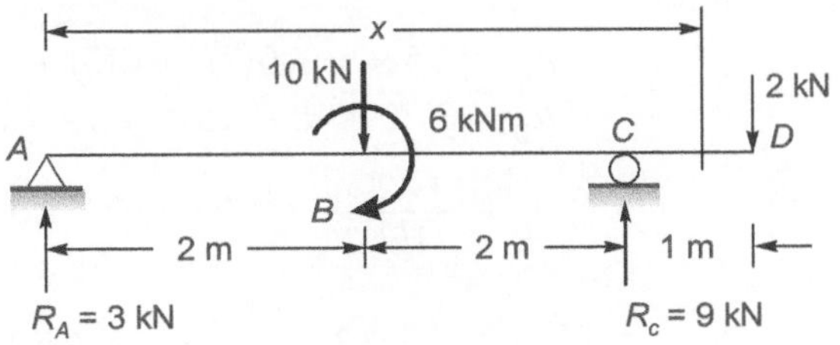

Fig. 11.11

Solution: Support Reaction: Moments about point *A*

$$6+10\times 2+2\times 5 = 36 \text{ kNm} = 4\,R_C$$

Reaction, $R_C = 9$ kN↑

Reaction, R_A = Total load − R_C

$$= 10+2-9 = 3 \text{ kN}\uparrow$$

Using Macaulay's method, let us make equation of bending moment for a section X-X, at a distance of x from origin in the last portion of the beam i.e. CD

$$M_x = R_A x - 10(x-2) + 6(x-2)^\circ + R_C(x-4)$$

By taking $(x-2)^\circ = 1$, we have located the application of moment of 6 kNm (clockwise)

or
$$EI\frac{d^2y}{dx^2} = 3x - 10(x-2) + 6(x-2)^\circ + 9(x-4) \quad \text{...(i)}$$

Integrating equation (*i*) we get,

$$EI\frac{dy}{dx} = 1.5x^2 - 5(x-2)^2 + 6(x-2) + 4.5(x-4)^2 + C_1 \quad \text{...(ii)}$$

above C_1 is constant of integration.

In this case we do not know at which section of the beam, slope $\frac{dy}{dx} = 0$

Integrating equation (*ii*) again,

$$EIy = 0.5x^3 - \frac{5}{3}(x-2)^3 + 3(x-2)^2 + 1.5(x-4)^3 + C_1 x + C_2$$

where C_2 is another constant of integration

Boundary Condition:

$x = 0,\ y = 0$

$$EI \times 0 = 0 - \text{omitted term} + 0 + C_2$$

Constant, $C_2 = 0$

$x = 4$ m, $y = 0$, putting this value

$$0 \times EI = 0.5 \times 4^3 - \frac{5}{3}(4-2)^3 + 3(4-2)^2 + 1.5 \times 0 + 4C_1 + 0$$

$$0 = 32 - 13.333 + 12 + 0 + 4\,C_1$$

$$C_1 = -7.6667$$

Finally the equations are

$$EI\frac{dy}{dx} = 1.5x^2 - 5(x-2)^2 + 6(x-2) + 4.5(x-4)^2 - 7.6667$$

$$EIy = 0.5x^3 - \frac{5}{3}(x-2)^3 + 3(x-2)^2 + 1.5(x-4)^2 - 7.6667x$$

At point D, $x = 5$ m, putting value

$$EIi_D = 1.5 \times 5^2 - 5(3)^2 + 6(3) + 4.5(1)^2 - 7.6667$$

$$EIi_D = 37.5 - 45 + 18 + 4.5 - 7.6667 = +7.333$$

Slope,
$$i_D = \frac{+7.333}{1600} = 4.58 \times 10^{-3}\text{rad} = 0.26^\circ$$

$$EIy_D = 0.5 \times 5^3 - \frac{5}{3} \times 27 + 3 \times 9 + 1.5(1)^3 - 7.6667 \times 5$$

$$= 62.5 - 45 + 27 + 1.5 - 38.3335 = +7.6665$$

$$y_D = \frac{7.6665}{1600} = 4.8 \times 10^3\text{m} = +4.8 \text{ mm}$$

Question 11.7 A beam 5 m long, hinged at both the ends *A* and *D*, is subjected to a moment 7 kNm *CCW* at *B*, 2 m away from *A*, a udl of 4 kN/m, *CD* = 2 m as shown in figure 11.12. If EI = 1800 kNm², determine slope at A and deflection at *C*.

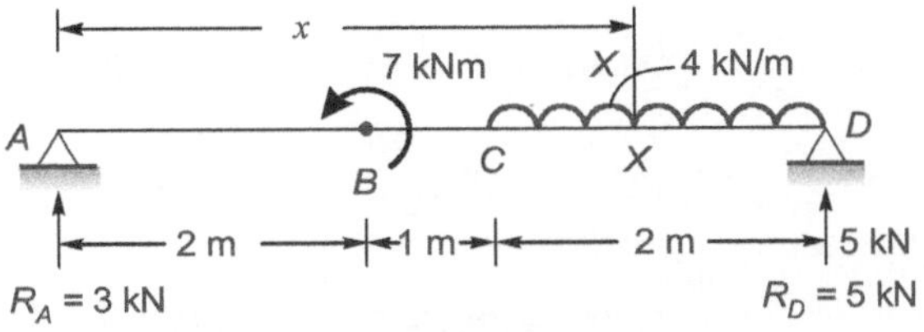

Fig. 11.12

Solution: Support Reaction

Taking moments about A

$$7 + 5R_D = 4 \times 2 \times (3 + 1) = 32$$

Reaction, $R_D = 5$ kN↑

Reaction, R_A = Total vertical load on beam – R_D

$$= 4 \times 2 - 5 = 3 \text{ kN}\uparrow$$

It is easy to make bending moment equation, ***as UDL is there in the last portion CD of the beam.*** Consider a section *X-X* at a distance of x from *A*.

$$M_x, \text{ bending moment of } XX = +3x - 7(x-2)^\circ - \frac{w}{2}(x-3)^2$$

By multiplying moment of 7 kNm by (x – 2)° = 1, we have specified the section of the moment.

So,
$$EI\frac{d^2y}{dx^2} = 3x - 7(x-2)^\circ - \frac{4}{2}(x-3)^2$$

$$= 3x - 7(x-2)^\circ - 2(x-3)^2 \quad \text{...(i)}$$

Integrating equation (*i*),

$$EI\frac{dy}{dx} = 1.5x^2 - 7(x-2) - \frac{2}{3}(x-3)^3 + C_1 \quad \text{...(ii)}$$

where C_1 is constant of integration

we do not know, at which section slope is zero, therefore integrating equation (*ii*) also, we get

$$EIy = 0.5x^3 - 3.5(x-2)^2 - \frac{1}{6}(x-3)^4 + C_1x + C_2 \quad \text{...(iii)}$$

Boundary Conditions:

$x = 0$, deflection, $y = 0$, so

$$EI \times 0 = 0 - \text{omitted term} + 0 \times C_1 + C_2$$

Constant, $C_2 = 0$

Moreover, $x = 5$ m, at end *D*, $y = 0$

$$EI \times 0 = 0.5 \times 5^3 - 3.5 \times 3^2 - \frac{1}{6} \times 2^4 + 5C_1 + 0$$

$$0 = 62.5 - 31.5 - 2.6667 + 5C_1$$

Constant, $C_1 = -5.666$

Finally the equations of slope and deflection will be

$$EI\frac{dy}{dx} = 1.5x^2 - 7(x-2) - \frac{2}{3}(x-3)^3 - 5.666$$

$$EIy = 0.5x^3 - 3.5(x-2)^2 - \frac{1}{6}(x-3)^4 - 5.666x$$

Slope at A, $x = 0$

$$EIi_A = 0 - \text{omitted term} - \text{omitted term} - 5.666$$

Slope,
$$i_A = -\frac{5.666}{EI} = -\frac{5.666}{1800} = -3.15\times10^{-3} \text{ rad} = -0.18°$$

Deflection at C, $x = 3$ m

$$EIy_c = 0.5 \times 3^3 - 3.5\,(1)^2 - 0 - 5.666 \times 3$$
$$EIy_c = 13.5 - 3.5 - 17 = -7$$

Deflection,
$$y_c = -\frac{7}{1800} = -3.89\times10^{-3}\text{m} = -3.89 \text{ m}$$

Question 11.8 A beam AB, of length L, simply supported at ends carries a load W at C, such that $AC = a$, $CB = b$ and $a < b$. Determine slope at ends and deflection under the load, if EI is the flexural rigidity of the beam.

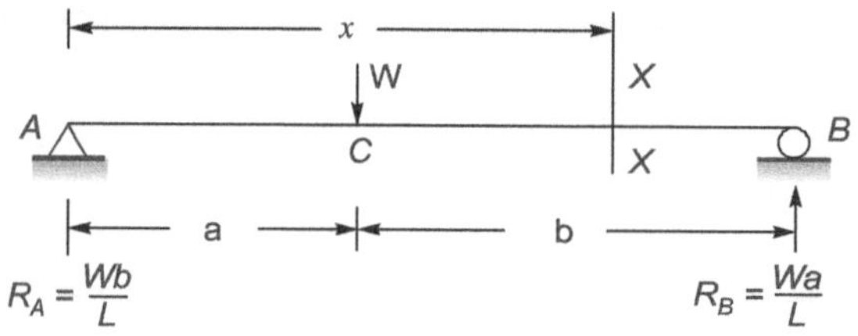

Fig. 11.13

Solution: Support reaction

Taking moments at A

$$Wa = R_B \times L$$

or
$$\text{Reaction, } R_B = \frac{Wa}{L}$$

Reaction,
$$R_A = W - \frac{Wa}{L} = \frac{WL - Wa}{L} = \frac{Wb}{L}$$

Using Macaulay's method, considering section XX at a distance of x from end A, in the last portion CB

$$M_x = R_A x - W(x - a)$$

or
$$EI\frac{d^2y}{dx^2} = \frac{Wb}{L}\cdot x - W(x-a) \qquad \text{...(i)}$$

Integrating equation (i) we get

$$EI\frac{dy}{dx} = \frac{Wbx^2}{2L} - \frac{W}{2}(x-a)^2 + C_1 \qquad \text{...(ii)}$$

where C_1 is constant of integration.

In this case, load is not applied at the center of the beam, we do not know where slope is zero, so integrating equation also, we get

$$EIy = \frac{Wbx^3}{6L} - \frac{W}{6}(x-a)^3 + C_1 x + C_2$$

Boundary Condition:

$x = 0$, $y = 0$, at end A

$$EI \times 0 = 0 - \text{omitted term} + 0 \times C_1 + C_2$$

Constant, $C_2 = 0$

Another boundary condition, $x = L$, $y = 0$, at end B

$$EI \times 0 = \frac{WbL^3}{6L} - \frac{W}{6}(L-a)^3 + C_1 L + 0$$

$$0 = \frac{WbL^2}{6} - \frac{W}{6} \times b^3 + C_1 L$$

$$\text{Constant, } C_1 = \frac{Wb}{6L}[b^2 - L^2] = \frac{Wb}{6L}[b^2 - a^2 - b^2 - 2ab]$$

$$= \frac{Wb}{6L}[-a^2 - 2ab] = \frac{Wb}{6L}[a^2 + 2ab]$$

$$= -\frac{Wab}{6L}[a + 2b]$$

Expressions for slope and deflection will be

$$EI\frac{dy}{dx} = \frac{Wbx^2}{2L} - \frac{W}{2}(x-a)^2 - \frac{Wab}{6L}(a+2b)$$

$$EIy = \frac{Wbx^3}{6L} - \frac{W}{6}(x-a)^3 - \frac{Wab}{6L}(a+2b) \cdot x$$

Slope at A

$$EIi_A = 0 - \text{omitted term} - \frac{Wab}{6L}(a+2b)$$

Slope, $$i_A = -\frac{Wab(a+2b)}{6LEI}$$

Slope at B, $(x = L)$

$$EIi_B = \frac{WbL^2}{2L} - \frac{Wb^2}{2} - \frac{Wab}{6L}(a+2b)$$

$$\text{After simplification} = \frac{Wab}{6L}(2a+b) = EIi_B, i_B = \frac{Wab(2a+b)}{6LEI}$$

Deflection at C, $x = a$

$$EIy_c = \frac{Wba^3}{6L} - 0 - \frac{Wab}{6L}(a + b \times 2)a$$

$$= \frac{Wab}{6L}(a^2 - a^2 - 2ab) = -\frac{Wa^2b^2}{3L}$$

Deflection, $$y_c = -\frac{Wa^2b^2}{3EIL}$$

if we put, $a = b = \frac{L}{2}$, $$y_c = -\frac{WL^3}{48EI}$$

as we have proved earlier.

Question 11.9 A beam AB of length $L = 6$ m, is simply supported at ends. It carries a point load of 6 kN at C as shown in figure 11.14. Determine slope at ends, maximum deflection and where it occurs $EI = 2400$ kNm^2

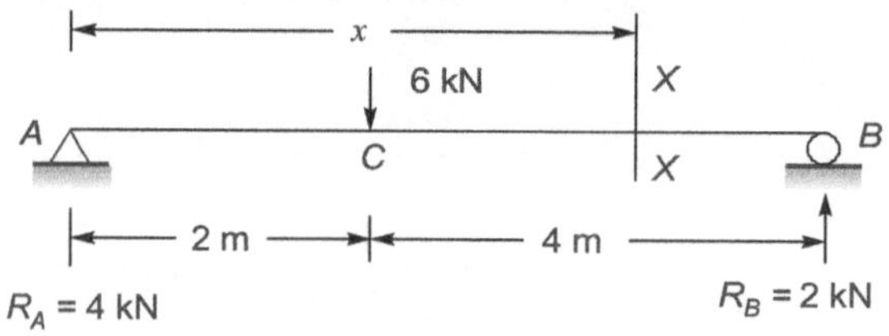

Fig. 11.14

Solution: Support Moments

Taking moments about A,

$$6 \times 2 = 6 R_B$$

Reaction,

$$R_B = 2 \text{ kN}$$

$$R_A = 6 - 2 = 4 \text{ kN}$$

Using Macaulay's method, consider a section XX in portion CB at a distance of x from A

$$EI \frac{d^2y}{dx^2} = 4x - 6(x - 2) \quad \text{...(i)}$$

Integrating equation (i),

$$EI \frac{dy}{dx} = 2x^2 - 3(x - 2)^2 + C_1 \quad \text{...(ii)}$$

where C_1 is constant of integration.

Integration equation (ii) also,

$$EIy = \frac{2}{3}x^3 - (x-2)^3 + C_1 x + C_2 \quad \text{...(iii)}$$

where C_2 is another constant of integration.

Boundary Condition:

$x = 0$, end A, deflection, $y = 0$

$$EI \times 0 = 0 - \text{omitted term} + 0 + C_2$$

Constant $C_2 = 0$,

$x = 6$ m, end B, deflection, $y = 0$

$$EI \times 0 = \frac{2}{3} \times 6^3 - (4)^3 + 6C_1 + 0$$

$$0 = 144 - 64 + 6\, C_1$$

$$C_1 = -13.333$$

Finally the equations of slope and deflection are

$$EI \frac{dy}{dx} = 2x^2 - 3(x - 2)^2 - 13.333$$

$x = 0$,

$$EIi_A = 0 - \text{omitted term} - 13.333$$

$$i_A = \frac{-13.333}{2400} = -5.55 \times 10^{-3} \text{rad} = -0.32°$$

$x = 6,$ $$EIi_B = 2 \times 6^2 - 3 \times 4^2 - 13.333$$
$$= 72 - 48 - 13.333 = +10.667$$
$$i_B = \frac{+10.667}{2400} = +4.444 \times 10^{-3} \text{rad} = +0.254°$$

Then, $$EIy = \frac{2}{3}x^3 - (x-2)^3 - 13.333x$$

For maximum deflection, slope will be zero, at that section

so, $$EI\frac{dy}{dx} = 2x^2 - 3(x-2)^2 - 13.333$$
$$EI\frac{dy}{dx} = 2x^2 - 3(x^2 + 4 - 4x) - 13.333$$
$$= -x^2 - 12 + 12x - 13.333$$
$$= x^2 - 12x + 25.333 = 0$$
$$x = \frac{12 \pm \sqrt{144 - 4 \times 25.333}}{2} = \frac{12 \pm 6.532}{2}$$
$$x = 2.734 \text{ m}$$

To get y_{max} let us put x = 2.734 m (in portion *CB*)

$$EI\,y_{max} = \frac{2}{3} \times 2.734^3 - (2.734 - 2)^3 - 13.333 \times 2.734$$
$$= 13.624 - 0.395 - 36.452 = 23.223$$
$$y_{max} = \frac{-23.223}{2400} = -9.7 \times 10^{-3} = -9.7 \text{ mm}$$

Practice Q.11.9 A beam 7 m long, simply supported at ends carries a vertical load of 7 kN at *C* as shown in figure 11.15. If EI = 3000 kNm², determine slope at ends. Determine y_{max} and where it occurs

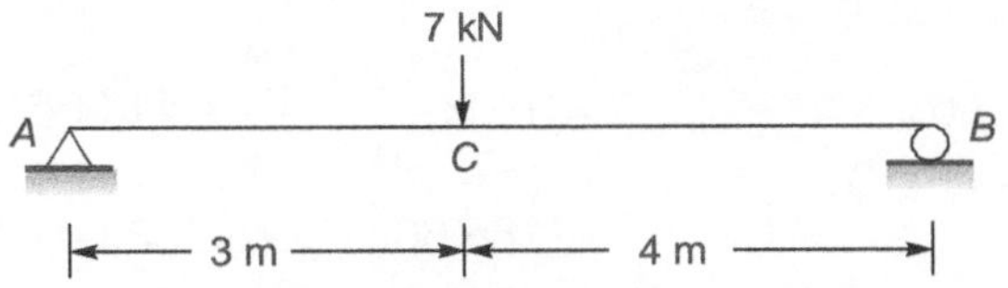

Fig. 11.15

Hints: $\left[R_A = 4 \text{ kN},\ R_B = 3 \text{kN}, EI\frac{d^2y}{dx^2} = 4x - 7(x-3),\ C_2 = 0,\ C_1 = -22\right]$

Ans. [i_A = –0.42°, i_B = +0.38°, y_{max} = –16.22 mm at x = 3.35 m]

Question 11.10 A cantilever *ABC*, 10 m long fixed at end *C*, propped at end *A* so that level of *A* and *C* is the same. It carries a udl of 2 kN/m over *BC* = 5 m. Determine magnitude of P_A, deflection at *B* if EI = 3600 kNm².

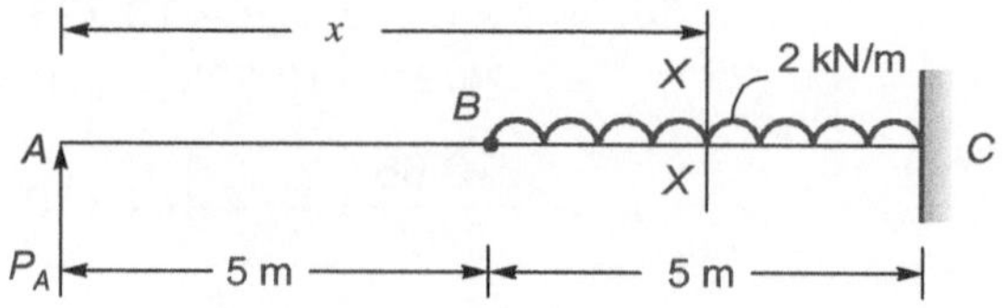

Fig. 11.16

Solution: Take a section *X-X* in portion *BC*, at a distance at x from *A*,

$$EI\frac{d^2y}{dx^2} = P_A x - \frac{w}{2}(x-5)^2 \text{, putting the value of } w = 2 \text{ kN/m}$$

$$= P_A x - (x-5)^2 \qquad \text{...(i)}$$

Integrating equation (*i*),

$$EI\frac{dy}{dx} = P_A\frac{x^2}{2} - \frac{(x-5)^3}{3} + C_1$$

where C_1 is constant of integration $\frac{dy}{dx} = 0$ at fixed end, $x = 10$ m,

So,

$$0 = P_A \times \frac{10^2}{2} - \frac{(10-5)^3}{3} + C_1$$

$$0 = 50\, P_A - 41.666 + C_1$$

or

$$C_1 = 41.666 - 50 \text{ PA}$$

$$EI\frac{dy}{dx} = P_A\frac{x^2}{2} - \frac{(x-5)^3}{3} + 41.666 - 50\, P_A \qquad \text{...(ii)}$$

Integrating equation (*ii*) again, $EIy = P_A\frac{x^3}{6} - \frac{(x-5)^4}{12} + 41.666x - 50\, P_A x + C_2$

$y = 0$, $x = 0$, so

$$EI \times 0 = 0 - \text{omitted term} + 0 - 0 + C_2$$

Constant $C_2 = 0$,

Moreover $y = 0$, $x = 10$, so $0 = P_A \times \frac{10^3}{6} - \frac{5^4}{12} + 41.666 \times 10 - 50\, P_A \times 10 + 0$

$$= 166.666\, P_A - 52.083 + 416.666 - 500\, P_A$$

$$333.334\, P_A = 364.582$$

$$P_A = 1.094 \text{ kN} \qquad \text{...(iii)}$$

Finally, $EIy = 1.094\frac{x^3}{6} - \frac{(x-5)^4}{12} + 41.666x - 50 \times 1.094x$

At point B, $x = 5$ m $EIy_B = 1.094 \times \frac{5^3}{6} - 0 + 41.666 \times 5 - 54.7 \times 5$

$$= 22.792 + 208.33 - 273.5 = -42.378$$

$$y_B = -\frac{42.378}{3600} = -0.01177 \text{ m} = 11.77 \text{ mm}$$

Practice Q.11.10 A cantilever AB of length L carries a load W at its center as shown in figure 11.17. It is propped at end by a force P_A. Determine P_A in terms of W so that level of A and B is same. If EI is the flexural rigidity of the cantilever, what is deflection at C?

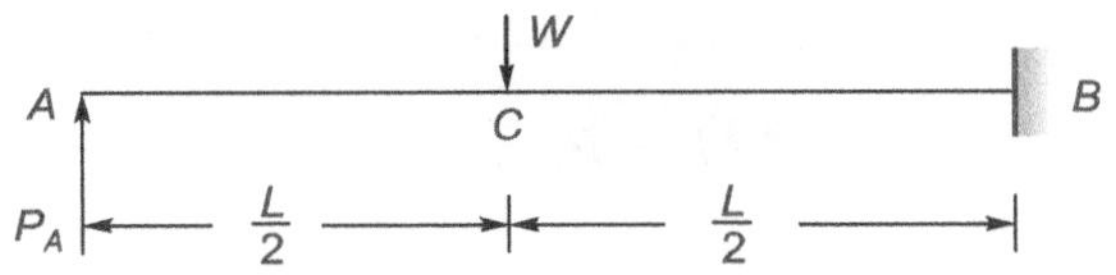

Fig. 11.17

Hints: $\left[EI\dfrac{d^2y}{dx^2} = P_A x - W\left(x - \dfrac{L}{2}\right); y = 0 \text{ at } x = 0, x = L, \dfrac{dy}{dx} = 0 \text{ at } x = L\right]$

Ans. $\left[\text{Constant } C_1 = -\dfrac{WL^2}{32}, \text{ constant } C_2 = 0, P_A = \dfrac{5\text{ W}}{16}, y_C = -\dfrac{7}{768} \times \dfrac{WL^3}{EI}\right]$

Question 11.11 A beam AB, span L, carries a linearly increasing distributed load of maximum intensity w kN/m at end B, starting from zero at A. If EI is the flexural rigidity of the beam, make expression for slope and deflection at any section of the beam. At which section, maximum deflection takes place. (Fig.11.18)

Solution: Fig. 11.20 (a) shows the beam with triangular load, $W = 0$ at $x = 0$, W at $x = L$, end B.

$$\text{Total load on beam} = \frac{wL}{2}, \ CG \text{ of this load at } \frac{2L}{3} \text{ from } A$$

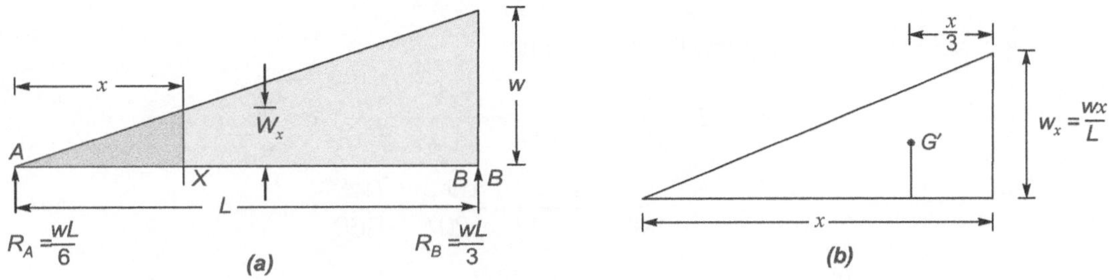

Fig. 11.18

Taking moments about A

$$R_B \times L = \frac{wL}{2} \times \frac{2L}{3} = \frac{wL^2}{3}$$

Reaction, $$R_B = \frac{wL}{3}$$

Reaction, $$R_A = \frac{wL}{2} - \frac{wL}{3} = \frac{wL}{6}$$

Consider a section X-X, at a distance of x from A.

Rate of loading at X-X, $$w_x = \frac{wx}{L}$$

Total load upto $XX = \frac{wx}{L} \times \frac{x}{2} = \frac{wx^2}{2L}$

CG of load from X-X acts at $\frac{x}{3}$

Bending moment at XX,
$$M_x = R_A \cdot x - \frac{wx^2}{2L} \times \frac{x}{3} = \frac{wL}{6} \cdot x - \frac{wx^3}{6L}$$

or
$$EI\frac{d^2y}{dx^2} = \frac{wL}{6} \cdot x - \frac{wx^3}{6L}$$

$$EI\frac{dy}{dx} = \frac{wL}{6}\frac{x^2}{2} - \frac{wx^4}{24L} + C_1 = \frac{wLx^2}{12} - \frac{wx^4}{24L} + C_1 \quad \text{...(i)}$$

where C_1 is constant of integration,

or
$$EIy = \frac{wLx^3}{36} - \frac{wx^5}{120L} + C_1 x + C_2 \quad \text{...(ii)}$$

where C_2 is another constant of integration

$y = 0$, at $x = 0$, end A

$$EI \times 0 = 0 - 0 + C_1 \times 0 + C_2$$

Constant, $C_2 = 0$

Moreover $y = 0$, $x = L$ at end B

$$EI \times 0 = \frac{wL^4}{36} - \frac{wL^4}{120} + C_1 L$$

Constant,
$$C_1 = -\frac{7wL^3}{360}$$

Equations of slope and deflection will now become

$$EI\frac{dy}{dx} = \frac{wLx^2}{12} - \frac{wx^4}{24L} - \frac{7wL^3}{360} \quad \text{...(iii)}$$

$$EIy = \frac{wLx^3}{36} - \frac{wx^5}{120L} - \frac{7wL^3}{360}x \quad \text{...(iv)}$$

y_{max} will occur where $\frac{dy}{dx} = 0$, so

$$\frac{wLx^2}{12} - \frac{wx^4}{24L} - \frac{7wL^4}{360} = 0$$

$$30Lx^2 - \frac{15x^4}{L} - 7L^3 = 0$$

$$30\,L^2x^2 - 15x^4 - 7L^4 = 0$$

$$x^4 - 2x^2L^2 + \frac{7}{15}L^4 = 0$$

$$x^2 = \frac{2L^2 \pm L^2\sqrt{4 - \frac{28}{15}}}{2} = \frac{2L^2 \pm L^2 \times 1.460}{2}$$

$$x^2 = 0.27\,L^2, \quad x = 0.52L$$

$\frac{dy}{dx} = 0$ at $x = 0.52\ L$ from A, putting the value of x

$$EIy_{max} = \frac{wL}{36}(0.52\,L)^3 - \frac{w}{120L}(0.52\,L)^5 - \frac{7}{360}wL^3 \times 0.52\,L + 0$$

$$= wL^4[0.003905 - 0.000317 - 0.01011]$$

$$= -wL^4[6.523 \times 10^{-3}]$$

$$y_{max} = \frac{-6.523 \times 10^{-3} wL^4}{EI}$$

Practice Q.11.11 **A beam AB, 6 m long is supported at ends with hinged end at one end and roller support at the other end. It carries a linearly varying load of maximum intensity $w = 6$ kN/m run at B, starting from zero at end A. If $EI = 3600$ kNm^2, determine slope at A, slope at end B, y_{max}.**

Hints: $\left[EI\frac{dy}{dx} = \frac{wx^4}{12} - \frac{wL^4}{24L} - \frac{7wL^3}{360}\right]$

Ans. [–0.4°, +0.46°, –14 mm]

Question 11.12 **Consider a beam, simply supported at ends subjected to impact load P, failing through height h, at center of beam. Derive expression for instantaneous deflection in beam.**

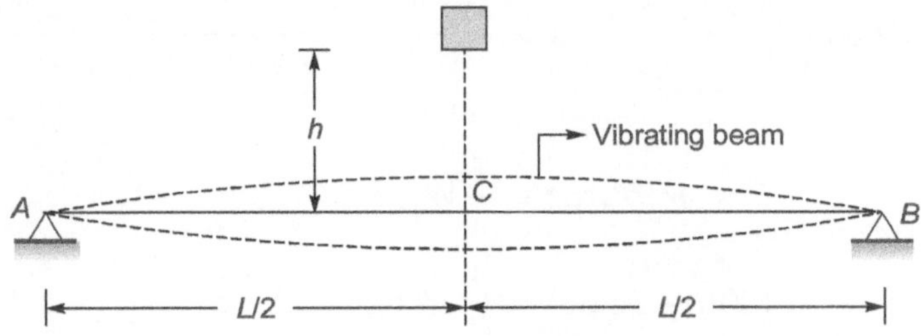

Fig. 11.19

Solution: A load P falls through a height h at the center of a beam AB, of length L hinged at both the ends. The load strikes at the center C of the beam and causes instantaneous deflection δ_i in the center of the beam. This type of impact loading develops instantaneous high stress in the beam. If the stress is very high, the beam may break otherwise the beam starts vibrating as shown in fig.11.21. But due to air damping, the amplitude of the vibration goes on reducing and final load on beam is P static.

Instantaneous deflection δ_i in beam

Say instantaneous deflection in beam in δ_i.

Load of PE of weight = $P(h + \delta_i)$

Strain energy absorbed by beam, $U = \frac{1}{2} W_e \cdot \delta_i$

where, W_e = equivalent static load on beam = $\frac{48EI\delta_i}{L^3} = K\delta_i$ where, $K = \frac{48EI}{L^3}$, a constant

or $\frac{1}{2}K\delta_i^2 - Ph - P\delta_i = 0$

If P and h are given, δ_i can be calculated and maximum stress developed in beam section due to δ_i also be worked out (for the values of K, P and h given)

Question 11.13 A steel cantilever 50 mm in diameter, 2 m long is fixed at one end *A*. Weight of 100 N is dropped at the free end the cantilever producing an instantaneous stress of 100 MPa. Determine the height from where the weight was dropped and maximum instantaneous deflection in the cantilever. $E = 200$ GPa.

Solution: Length of cantilever = 2 m

Equivalent load

$$\text{Diameter} = 50 \text{ mm}$$

Section modulus,
$$Z = \frac{\pi d^3}{32} = \frac{\pi \times 50^3}{32} = 12.27 \times 10^3 \text{ mm}^3$$

$$I = \frac{\pi d^3}{64} = 30.68 \times 10^4 \text{ mm}^4$$

$$\sigma_{max} = 100 \text{ MPa}$$

$$M_{max} = 100 \times 12.27 \times 10^3$$

$$= 12.27 \times 10^5 \text{ Nmm}$$

$$= W_e \cdot L = W_e \times 2000$$

Equivalent load,
$$W_e = \frac{12.27 \times 10^5}{2000} = 613.5 \text{ N}$$

Instantaneous deflection,
$$\delta_i = \frac{W_e \cdot L^3}{3EI} = \frac{613.5 \times 2000^3}{3 \times 200 \times 1000 \times 30.68 \times 10^4} = 26.66 \text{ mm}$$

$$P = 100 \text{ N}$$

$$P(h + \delta_i) = \frac{1}{2} W_e \cdot \delta_i$$

$$100(h + 26.66) = \frac{1}{2} \times 613.5 \times 26.66$$

$$h = \frac{613.5 \times 26.66}{200} - 26.66 = 81.78 - 26.61 \text{ mm} = 55.12 \text{ mm}$$

Practice Q.11.13 A rolled steel beam is simply supported over a span of 4 m. A weight of 400 N is dropped onto the middle of the beam. An instantaneous stress of 100 N/mm² is produced in the beam. Calculate the height through which the weight is dropped. What is instantaneous deflection in beam?

Given:

$I = 322 \times 10^4$ mm⁴ for beam section and $E = 200$ GPa, depth of section = 150 mm

Ans. [38.77 mm, 8.88 mm]

Question 11.14 What do you understand by area method to determine slope and deflection in a beam? Take a beam of length *L*, simply supported at ends carrying uniformly distributed load of intensity *w* per unit length. *EI* is flexural rigidity of the beam. Determine slope at ends and deflection at center of the beam using area moment method.

Solution: Equation of curvature and moment for a beam is

$$EI \frac{d^2y}{dx^2} = M_x, \text{ bending moment at section } x \qquad \ldots(i)$$

Figure 11.20 shows a beam *AB* carrying transverse loads over *AC* and a point load *W* at *D*. Bending moment diagram *AC'D'B'DCA* is drawn for the beam.

Say at a section, at a distance x from *A*, the bending moment is M_x as shown in figure 11.20.

Multiplying both the sides of equation (i) by dx,

$$EI\frac{d^2y}{dx^2}\times dx = M_x\, dx$$

Consider a portion of the beam from $x = x_1$ to $x = x_2$, the bending moment diagram is shown by shaded area.

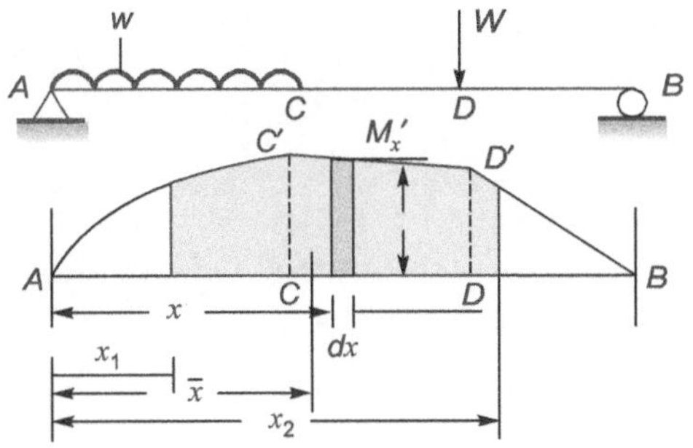

Fig. 11.20 *Area moment diagram*

$$EI\int_{x_1}^{x_2}\frac{d^2y}{dx^2}\,dx = \left|M_x\cdot dx\right|_{x_1}^{x_2} \quad \text{...(ii)}$$

$$\left|\frac{dy}{dx}\right|_{x_1}^{x_2} = \text{slope at } x_2 - \text{slope at } x_1$$

$$EI\,[i_2 - i_1] = \text{Area of the bending moment diagram}$$

between $x = x_2$ and $x = x_1$ ...(iii)

Moreover multiply equation (*i*) on both the sides by $x\cdot dx$ we set

$$EI\int_{x_1}^{x_2}\frac{d^2y}{dx^2}\cdot x\cdot dx = \int_{x_1}^{x_2} M_x\cdot x\,dx$$

$$EI\,[(x_2\,i_2 - y_2) - (x_1\,i_1 - y_1)] = a\bar{x}$$

= *a*, area of BM diagram between ...(iv)

$x = x_2$ and $x = x_1$, distance of centroid of the area '*a*' from origin $A = \bar{x}$

From equation (*iv*) deflection at the given section can be obtained.

A beam of length *L*, simply supported at ends *A* and *B* carrying a UDLof intensity *w* throughout it length is shown in figure 11.21. The deflection curve of the beam is also shown, showing maximum deflection at the center of the beam and zero slope at the center. Figure 11.21 (b) shown the *BMD* of the beam, parabolic curve, with maximum bending moment at the center of the beam equal to $\frac{wL^2}{8}$.

Say slope at *A*, $x = 0$, is i_A

deflection *C*, is y_{C}.

$x_1 = 0;\ x_2 = \frac{L}{2}$, using moment area method between two sections at *A* and *C*.

Area of *BMD*, $$AC'C = \frac{2}{3}\times\frac{wL^2}{8}\times\frac{L}{2} = \frac{wL^3}{24}$$

$$EI\,[i_c - i_A] = \frac{wL^3}{24}$$

But, $i_C = 0$

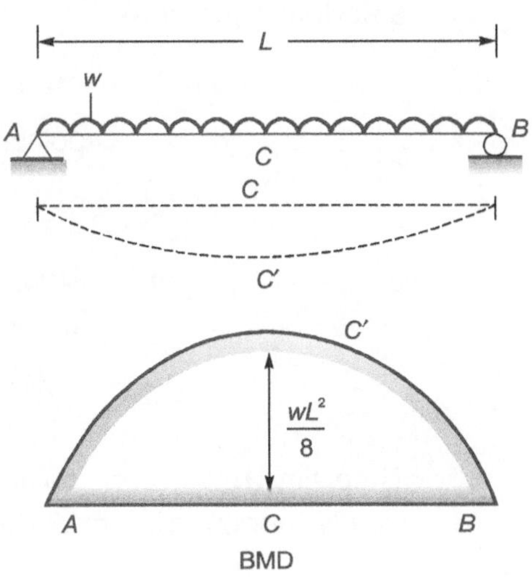

Fig. 11.21 (a) and (b)

Slope, $i_A = -\dfrac{wL^3}{24EI}$, because of symmetry,

$$i_B = +\frac{wL^3}{24EI}$$

$\bar{x}$ = distance of C.G. of area $AC'CA$ from A

$$= \frac{5}{8}\times\frac{L}{2} = \frac{5}{16}L$$

$$a\bar{x} = \frac{wL^3}{24}\times\frac{5}{16}L = \frac{5\,wL^4}{384}$$

$$EI\,[(x_C i_C - y_C) - (x_A i_A - y_A)] = \frac{5\,wL^4}{384}$$

But, $i_C = 0,\ x_A = 0,\ y_A = 0,\ x_C = \dfrac{L}{2}$

$$EI\left[\left(\frac{L}{2}\times 0 - y_c\right) - (0\times i_A - 0\right] = \frac{5wL^4}{384}$$

Deflection at center, $y_c = -\dfrac{5wL^4}{384\,EI}$

Practice Q.11.14 A beam *AB* of length *L*, carries at point load *W* at the center. *EI* is the flexural rigidity of the beam, determine slope at ends and deflection at the center of the beam, using moment area method.

Ans. $\left[\pm\dfrac{WL^2}{16\,EI}, -\dfrac{WL^3}{48\,EI}\right]$

Question 11.15 What are Mohr's theorem **I** and theorem **II**, in connection with moment areas method for determining slope and deflection in a beam.

Solution: Theorem-**I**

$\theta_B - \theta_A$ = Area of $\dfrac{M}{EI}$ diagram between points *B* and *A* of the beam.

Cantilever,

$\theta_A = 0$

θ_B = Area of $\dfrac{M}{EI}$ diagram between *B* and A

$= -\dfrac{WL^2}{2EI}$, slope at *B*

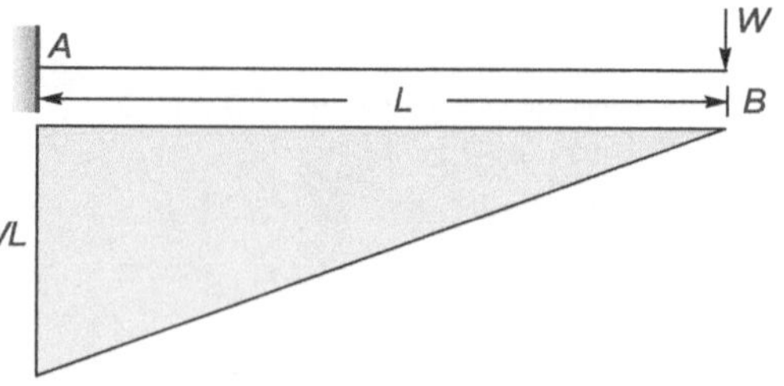

Fig. 11.22

Theorem-II: When one point on the beam is of zero slope (as fixed end of a cantilever, mid points of a simply supported beam loaded symmetrically)

$y_B - y_A$ = Moment of area of $\dfrac{M}{EI}$ diagram between *B* and *A*.

$$= A\,\bar{x} \text{ of } \frac{M}{EI} \text{ diagram between } B \text{ and } A$$

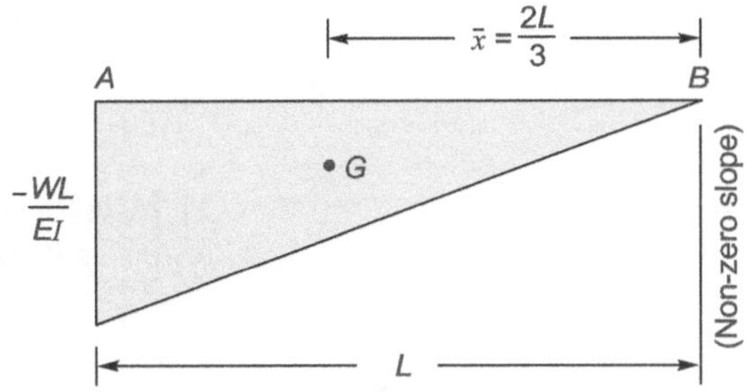

Fig. 11.23

always $\bar{x}$ should be measured from the point of non-zero slope

$$y_{max} = 0 = -\frac{WL^2}{2EI} \times \frac{2L}{3}$$

$$= -\frac{WL^3}{3EI}$$

B is the point non-zero slope. At *A* slope is zero

(a) Always select two points in such a way that one point should have zero slope (this becomes the reference point) and another point should be point of non-zero slope (i.e. origin).

(b) Point of non-zero slope should be the point where slope is to the measured.

(c) **$\bar{x}$ should be measured from the point where deflection is to be measured.**

(d) BMD should be used for the beam whose *EI* is constant.

(e) II theorem is valid only when point has zero slope and another point of non-zero slope.

Question 11.16 **Determine slope and deflection at a distance *L*/2 from free and of a cantilever beam, when it is loaded with a point load at its free end.**

Solution: Figure 11.24 shows BMD of cantilever

$$\theta_C - \theta_A = \frac{1}{EI}\left[-\frac{WL}{2} \times \frac{L}{2} - \frac{WL}{2} \times \frac{L}{4}\right]$$

$$= \frac{1}{EI}\left[-\frac{WL^2}{4} - \frac{WL^2}{8}\right]$$

$$= -\frac{3}{8}\frac{WL^2}{EI} \text{ (cw)}$$

$$y_C - y_A = \frac{1}{EI}\left[-\frac{WL}{2} \times \frac{L}{2} \times \frac{L}{4} - \frac{WL}{2} \times \frac{L}{4} \times \frac{L}{3}\right]$$

$$= -\frac{1}{EI}\left[\frac{WL^3}{16} + \frac{WL^3}{24}\right] = -\frac{5}{48} \times \frac{WL^3}{EI}$$

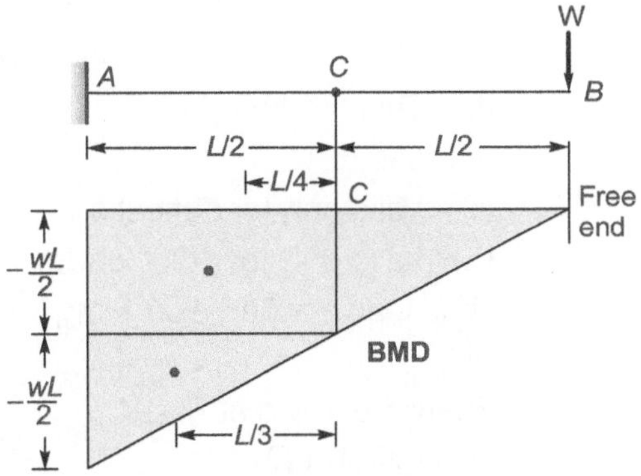

Fig. 11.24

Question 11.17 **Determine slope and deflection at free end of a cantilever beam when it is loaded with a point load at center of cantilever.**

Solution:

Reference point – point of zero slope (*A*)

Origin point – (*B*) point of non-zero slope

$$\theta_C - \theta_A = \frac{1}{EI}\left[-\frac{WL}{2} \times \frac{L}{2} \times \frac{1}{2}\right] = -\frac{WL^2}{8EI}$$

$\theta_A = 0$

$$\theta_C = -\frac{WL^2}{8EI} = \theta_B$$

$$y_C - y_A = \frac{1}{EI}\left[-\frac{WL}{2}\times\frac{L}{2}\times\frac{1}{2}\times\frac{L}{3}\right]$$

$$y_C = -\frac{WL^3}{24EI} \quad \text{as } y_A = 0$$

$$y_B - y_A = \frac{1}{EI}\left[-\frac{WL}{2}\times\frac{L}{2}\times\frac{1}{2}\times\left(\frac{L}{3}+\frac{L}{2}\right)\right]$$

$$= -\frac{1}{EI}\left[\frac{WL^2}{8}\times\frac{5L}{6}\right] = -\frac{5WL^3}{48EI}, \quad \text{as } y_A = 0$$

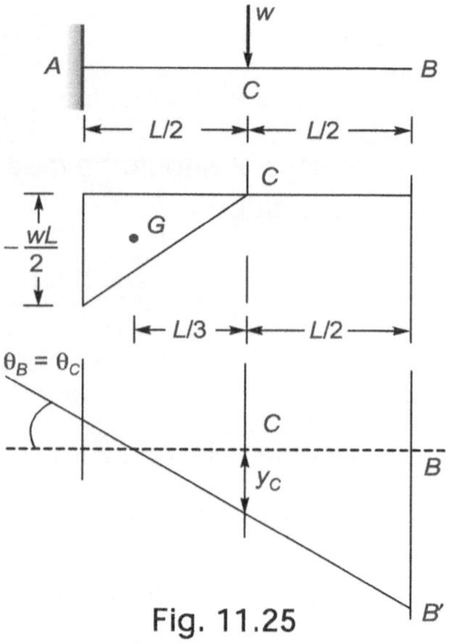

Fig. 11.25

Practice Q.11.17 Explain Maxwell's reciprocal theorem with an example.

Solution: Theorem in brief

$$W_{(A)} \cdot Y_{(CA)} = W_{(C)}\, Y_{(AC)}$$

W, load applied at point A deflection at C due to load applied at A

= W, load applied at C × $Y_{(AC)}$, deflection at A due to load applied at C as shown in the fig. II.

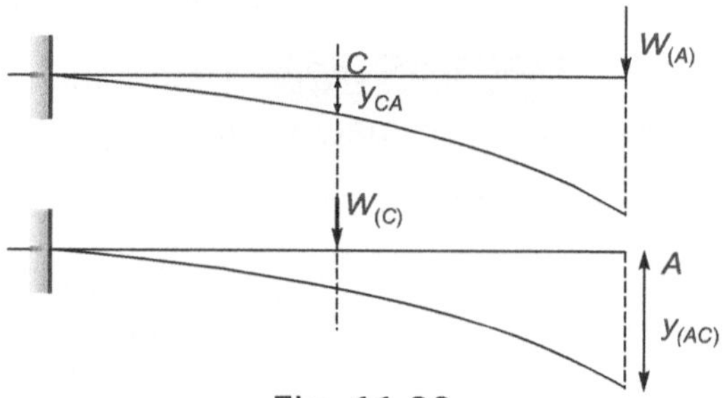

Fig. 11.26

(i) This theorem is valid only when beam remains the same

(ii) Valid only for deflection

(iii) Valid only for point loads

$W_{(A)}$, load applied at A = $W_{(C)}$, load applied at C

$$Y_{(AC)} = Y_{(CA)}$$

Example: Load at A, (W)

Deflection at C

$$= \frac{1}{EI}\left[\frac{WL}{2}\times\frac{L}{2}\times\frac{L}{4}\right] + \frac{1}{EI}\left[\frac{WL}{2}\times\frac{L}{2}\times\frac{1}{2}\times\frac{L}{3}\right]$$

$$= \frac{1}{EI}\left[\frac{WL^3}{16}+\frac{WL^3}{24}\right] = \frac{5\,WL^3}{48EI}$$

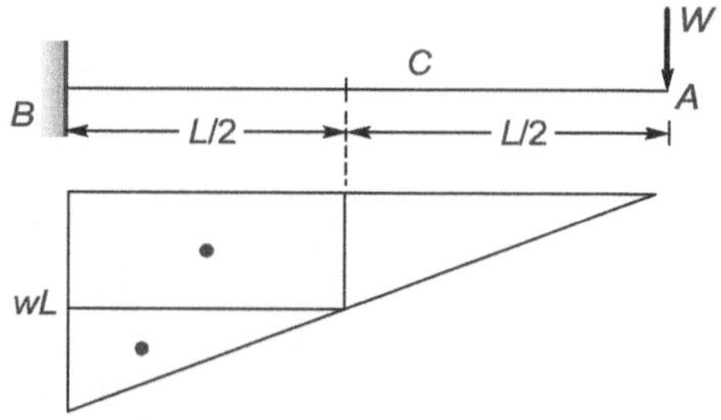

Fig. 11.27

$$W_{(A)} \cdot Y_{(CA)} = W \cdot \frac{5}{48} \times \frac{WL^3}{EI} \qquad \text{...(i)}$$

Load at C,

$$Y_{(AC)} = \frac{1}{2}\times\frac{WL}{2}\times\frac{1}{2}\left(\frac{5L}{6}\right) = \frac{5}{48}\,\frac{WL^3}{EI}$$

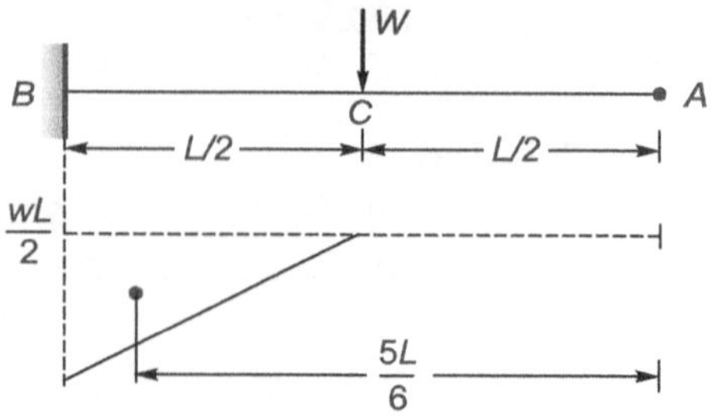

Fig. 11.28

$$W_{(C)} \cdot Y_{(AC)} = W \cdot \frac{5}{48} \times \frac{WL^3}{EI} \quad \text{...(ii)}$$

So, $$W_{(A)} \cdot Y_{(CA)} = W_{(C)} \times Y_{(AC)}$$

Question 11.18 **What is conjugate beam method to obtain slope and deflection at any section of a beam subjected to transverse loads, explain with the help of an example.**

Solution: In conjugate beam method, BM diagram of a beam is considered as loading diagram. From this loading diagram of BM, reactions at ends are determined. These reactions at each end divided by EI of the beam gives slopes at ends. Bending moment at any section from the BM diagram as loading diagram, divided by EI gives deflection of the beam at that section.

Figure 11.29 (a) shows a beam AB, length L carrying point loads W_1 and W_2. Figure 11.29 (b) shows BM diagram of the beam considering this as loading diagram in kNm, reactions at A and B are determined which are in terms of kNm^2. Say, reactions at ends are R'_A and R'_B.

Slope at A, $$i_A = \frac{R'_A}{EI}$$

Slope at B, $$i_B = \frac{R'_B}{EI}$$

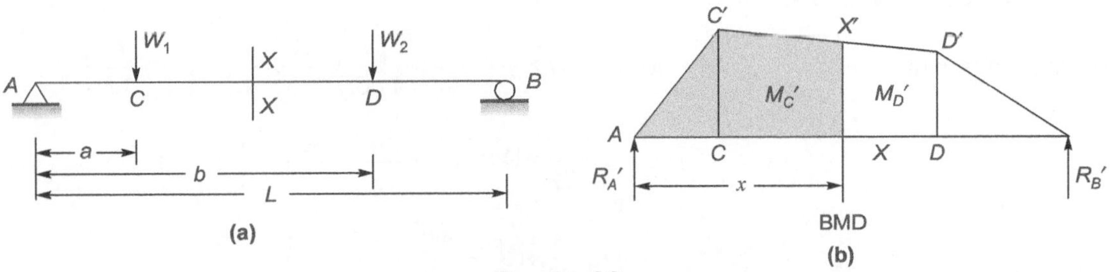

Fig. 11.29

Depending upon, the **nature of the slope, the reader can put the positive or negative sign to the expression for slope.**

Deflection at X

Considering a section at a distance of x from A, i.e. shaded diagram $AC'\,X'\,X\,CA$.

Bending moment at X, $M'_x = R'_A\,x$ – (area of BM diagram $AC'\,X'\,X\,CA$) × ($\bar{x}$, distance of CG of this diagram from section XX).

Deflection at section XX, $$y_x = \frac{M_x}{EI}$$

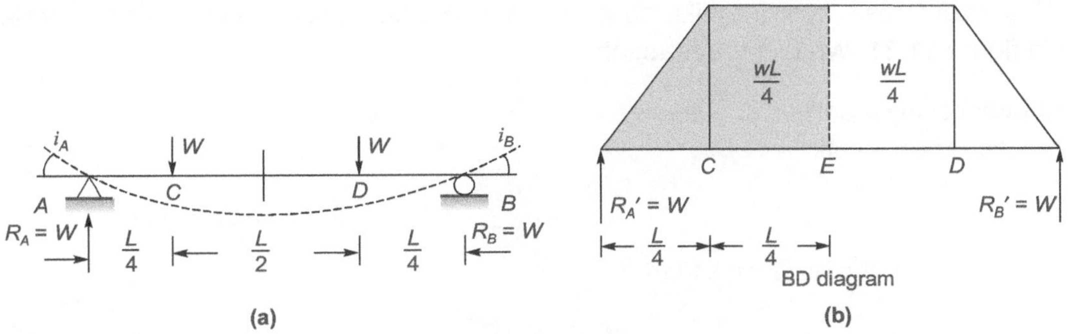

Fig. 11.30

Fig 11.30 (a) shows a beam *AB*, of length *L*, simply supported at ends carrying loads *W* each at *C* and *D*, at L/4 from ends *A* and *B* respectively.

Reaction $R_A = R_B = W\uparrow$

BM diagram

$$M_A = 0 = M_B$$

$$M_C = \frac{WL}{4} = M_D$$

Figure 11.30 (b) shows BM diagram of the beam acting as loading diagram.

Reaction, $$R'_A = R'_B = \frac{1}{2}\left[2\times\frac{WL}{4}\times\frac{L}{4}\times\frac{1}{2}\times\frac{WL}{4}\times\frac{1}{2}\right] = \frac{1}{2}\left[\frac{WL^2}{16}+\frac{WL^2}{8}\right] = \frac{3}{32}WL^2$$

$$R'_A = R'_B = \frac{3}{32}WL^2$$

Slope at *A*, $$i_A = -\frac{3}{32}\times\frac{WL^2}{EI}$$

Slope at *B*, $$i_B = +\frac{3}{32}\times\frac{WL^2}{EI}$$

Deflection at E (center of the beam)

Bending moment, $$M'_E = R'_A\times\frac{L}{2}-\frac{WL}{4}\times\frac{L}{4}\left(\frac{L}{8}\right)-\frac{W}{4}\times\frac{L}{4}\times\frac{L}{2}\left(\frac{L}{4}+\frac{L}{12}\right)$$

$$= \frac{3}{32}\times WL^2\times\frac{L}{2}-\frac{WL^3}{128}-\frac{WL^3}{32}\left(\frac{1}{4}+\frac{1}{12}\right)$$

$$= \frac{3}{64}WL^3-\frac{WL^3}{128}-\frac{WL^3}{32}\times\frac{1}{3}$$

$$= \frac{3}{64}WL^3-\frac{WL^3}{128}-\frac{WL^3}{96}$$

$$= \frac{18-3-4}{384} = \frac{11}{384}WL^3$$

Deflection at E (center) $= -\frac{11}{384}\times\frac{WL^3}{EI}$ (as the beam bends downwards)

Question 11.19 **A cantilever *AB*, 3 m long carries a uniformly distributed load of 2 kN/m throughout its length, rests over a similar cantilever of same cross-section and same material, 2 m long an shown in figure 11.31. What is the reaction *C*?**

Solution: Let the reactions at *C* is *R*.

Deflection at *C* in cantilever *CD*,

$$\delta_C = -\frac{R(2)^3}{3EI} = \frac{8R}{3EI} = \delta_B \text{ (in cantilever } AB) \quad \text{...(i)}$$

Take a section *X* at a distance x from *B*

$$M_x = Rx-\frac{wx^2}{2} = Rx-\frac{2}{2}\cdot x^2 = Rx-x^2, \text{ as } w = 2 \text{ kN/m}$$

as $W = 2$ kN/m

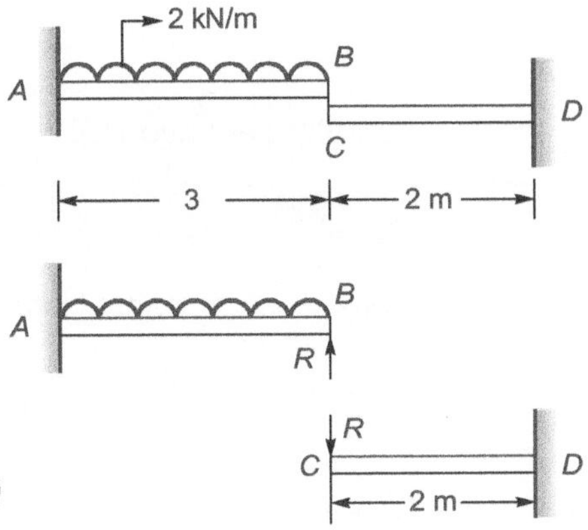

Fig. 11.31

$$EI\frac{d^2y}{dx^2} = Rx - x^2$$

or
$$EI\frac{dy}{dx} = \frac{Rx^2}{2} - \frac{x^3}{3} + C_1$$

$$\frac{dy}{dx} = 0 \text{ at } x = 3 \text{ m, end A}$$

$$EI \times 0 = R\times\frac{9}{2} - 9 + C_1$$

Constant,
$$C_1 = 9 - 4.5R \quad \text{...(iii)}$$

$$EI\frac{dy}{dx} = \frac{Rx^2}{2} - \frac{x^2}{3} + (9 - 4.5R) \quad \text{...(iv)}$$

or
$$EIy = \frac{Rx^3}{6} - \frac{x^4}{12} + (9 - 4.5R)x + C_2$$

$y = 0$ at $x = 3$, end A

$$EI \times 0 = \frac{R\times 27}{6} - \frac{81}{12} + (9 - 4.5R)\times 3 + C_2$$

$$0 = 4.5R - 6.75 + 27 - 13.5R + C_2$$

$$C_2 = 9R - 20.25$$

or
$$EIy = \frac{Rx^3}{6} - \frac{x^4}{12} + (9 - 4.5R)x + 9R - 20.25$$

But at $x = 0$, $y = -\frac{8R}{3EI}$, so [from equation (i)]

$$-\frac{8}{3}\times\frac{R}{EI}\times EI = 0 - 0 + 0 + 9R - 20.75$$

$$-\frac{8}{3}R = 9R - 20.25$$

$$R = 20.25\times\frac{3}{35} = 1.7357 \text{ kN}$$

Question 11.20 A cantilever *AB*, *L* m long is fixed at *B* and roller supported at *A*. It is subjected to an anticlockwise moment *M* at center. Determine reaction at *A*, deflection at *C*, it *EI* is the flexural rigidity of the cantilever figure 11.32.

Solution:

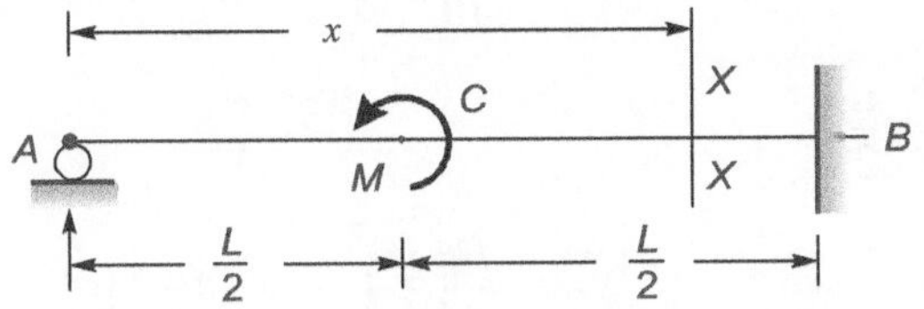

Fig. 11.32

Let us say reaction at *A* is *R*. Taking a section *XX* at a distance of x from *A*,

$$M_x = Rx - M\left(x - \frac{L}{2}\right)^0 \qquad \ldots(i)$$

Integrating equation (*i*),

$$EI\frac{dy}{dx} = R\frac{x^2}{2} - M\left(x - \frac{L}{2}\right) + C_1$$

where C_1 is constant of integration,

$$\frac{dy}{dx} = 0 \text{ at } x = L \text{ or}$$

$$EI \times 0 = R\frac{L^2}{2} - M\left(L - \frac{L}{2}\right) + C_1$$

$$0 = \frac{RL^2}{2} - \frac{ML}{2} + C_1$$

or constant

$$C_1 = \frac{(M - RL)L}{2}$$

So,

$$EI\frac{dy}{dx} = R\frac{x^2}{2} - M\left(x - \frac{L}{2}\right) + L\frac{(M - RL)}{2} \qquad \ldots(ii)$$

$$EIy = R\frac{x^3}{6} - \frac{M}{2}\left(x - \frac{L}{2}\right)^2 + L\left(\frac{M - RL}{2}\right)x + C_2$$

At A, $x = 0$, $y = 0$, simply supported end

$$0 = 0 - \text{omitted term} + 0 + C_2$$

Constant $C_2 = 0$

$$EIy = R\frac{x^3}{6} - \frac{M}{2}\left(x - \frac{L}{2}\right)^2 + L\left(\frac{M - RL}{2}\right)x \qquad \ldots(iii)$$

At the end B, $x = L$, $y = 0$

$$0 = R\frac{L^3}{6} - \frac{M}{2}\left(L - \frac{L}{2}\right)^2 + L\frac{(M - RL)}{2}L^2$$

$$0 = \frac{RL^3}{6} - \frac{ML^2}{8} + \frac{ML^2}{2} - \frac{RL^3}{2}$$

$$0 = RL^3\left(\frac{1}{6} - \frac{1}{2}\right) + ML^2\left(\frac{1}{2} - \frac{1}{8}\right) = -RL^3 \times \frac{1}{3} + ML^2 \times \frac{3}{8}$$

Reaction,

$$R = \frac{9M}{8L}$$

Deflection at center $x = \frac{L}{2}$

$$EIy_C = \frac{R}{6}\left(\frac{L}{2}\right)^3 - M(0) + L\left(\frac{M}{2} - \frac{RL}{2}\right)\frac{L}{2}$$

$$= \frac{9M}{8L} \times \frac{1}{6}\left(\frac{L}{2}\right)^3 - 0 + \frac{L^2}{2}\left(\frac{M}{2} - \frac{L}{2} \times \frac{9M}{8L}\right)$$

$$= \frac{9ML^2}{8\times 48}+\frac{L^2}{2}\left(\frac{M}{2}-\frac{9}{16}M\right)$$

$$EIy_c = \frac{3ML^2}{128}-\frac{L^2}{2}\times\frac{M}{16}=\frac{3ML^2}{128}-\frac{ML^2}{32} = -\frac{ML^2}{128}$$

$$y_c = 128\times\frac{ML^2}{EI}$$

Practice Q.11.20 A propped cantilever of length L is fixed at one end and roller supported at other end. Cantilever is subjected to a couple M at point C, at $L/3$ from A. Determine prop reaction and deflection at C, E is the flexural rigidity of the cantilever. (Fig. 11.33).

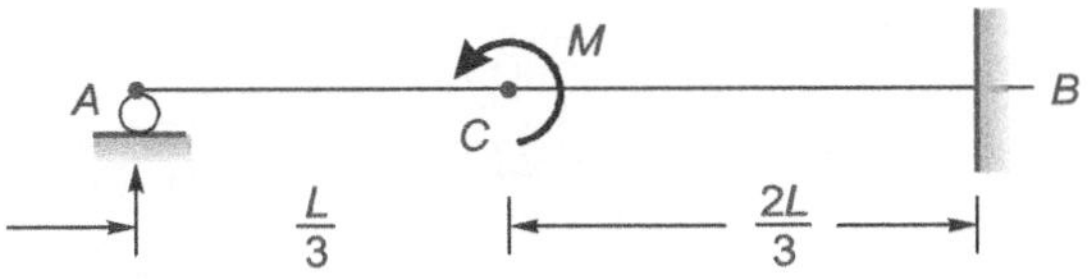

Fig. 11.33

Ans. $\left[\frac{4}{3}\times\frac{M}{L}, \frac{2}{243}\times\frac{ML^2}{EI}\right]$

Question 11.21 Explain how slope and deflection are determined in beams and cantilever with stepped sections. Consider a stepped cantilever as shown in figure 11.34, determine slope and deflection at A.

Solution: Conjugate beam method can be used in such cases. In the case of a cantilever, maximum slope and maximum deflection occur at free end while slope and deflection remain zero at fixed end. In conjugate beam method, free end of cantilever becomes the fixed end and fixed end becomes the free end. The ratio of reaction, $\frac{R'}{EI}$, gives slope at the end and ratio of $\frac{M'}{EI}$, (M' is moment of bending moment diagram as loading diagram) gives the deflection at the end.

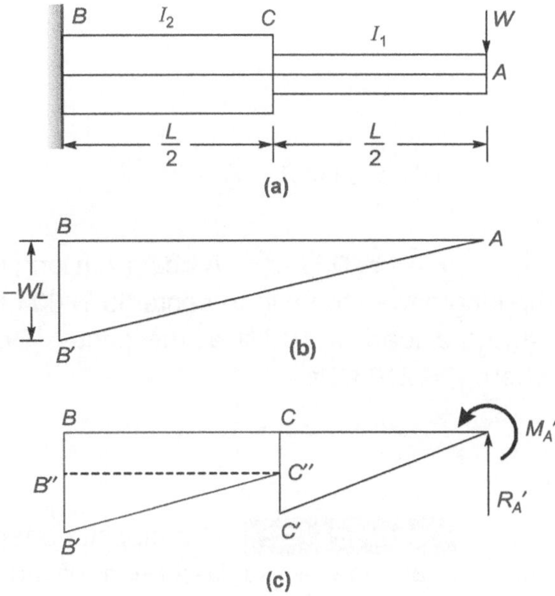

Fig. 11.34

Example:

Figure 11.34 (a) shows a cantilever of length L,

Portion $AC = \frac{L}{2}$, with moment of inertia, I_1,

Portion $CB = \frac{L}{2}$, with moment of inertia $I_2 = 2I_1$.

Figure 11.34 (b) shows the bending moment diagram of cantilever AB.

Let us draw $\frac{M}{EI}$ diagram for the cantilever as conjugate cantilever as I is variable.

$$CC' = -\frac{WL}{2EI_1}$$

$$CC'' = -\frac{WL}{2EI_2}, \quad \text{because } I_2 = 2I_1$$

Now,

$$BB' = -\frac{WL}{EI_2} = -\frac{WL}{2EI_1}$$

Reaction,

$$R'_A = CC' \times \frac{L}{2} \times \frac{1}{2} + CC''\left(\frac{L}{2}\right) + C'C''\left(\frac{L}{2}\right)\left(\frac{1}{2}\right)$$

$$= -\frac{WL}{2EI_1} \times \frac{L}{4} - \frac{WL}{4EI_1} \times \frac{L}{2} - \frac{WL}{4EI_1} \times \frac{L}{4}$$

$$= -\frac{WL^2}{8EI_1} - \frac{WL^2}{8EI_1} - \frac{WL^2}{16EI_1} = -\frac{5}{16}\frac{WL^2}{EI_1}$$

Slope at A,

$$i_A = -\frac{5}{16}\frac{WL^2}{EI_1} \quad \text{...(i)}$$

Moment,

$$M'_A = -\frac{WL^2}{8EI_1}\left(\frac{L}{3}\right) - \frac{WL^2}{8EI_1}\left(\frac{3L}{4}\right) - \frac{WL^2}{16EI_1}\left(\frac{L}{2} + \frac{L}{3}\right)$$

$$= -\frac{WL^3}{24EI_1} - \frac{3WL^3}{32EI_1} - \frac{5}{96} \times \frac{WL^3}{EI_1}$$

$$= \left(\frac{-4-9-5}{96}\right) \times \frac{WL^3}{EI_1} = -\frac{18}{96} \times \frac{WL^3}{EI_1} = -\frac{3}{16} \times \frac{WL^3}{EI_1}$$

Deflection at A,

$$y_A = -\frac{3}{16} \times \frac{WL^3}{EI_1}$$

Practice Q.11.21 A beam 4 m long is simply supported at the ends. The moment of inertia for half the length from the end to center is 1200 cm^4 and for the next half of the beam is 2400 cm^4. The beam carries a load of 10 kN at the center. Determine slope at each end and deflection at the center, where E = 210 GPa.

$$\text{Ans.} \left[-\frac{5WL^2}{96EI_1}, +\frac{WL^2}{24EI_I}, -0.19°, 0.15°, -\frac{WL^3}{64EI_1}, -3.97 \text{ mm}\right]$$

Question 11.22 A composite beam consists of two wooden beams of width 60 mm and depth 200 mm each in which a steel beam of width 10 mm and depth 150 mm is flitched. Composite beam is simply supported over a span of 5 m with a UDL of 4 kN/m. Determine deflection at the center of the composite beam.

$E_s = 15\ E_w$

E_s = 210 kN/mm^2

Solution: The composite section of wood and steel can be taken as equivalent steel section by reducing the width of wooden beams by *m*, i.e. modular ratio of steel and wood, i.e. $m = 15$

$$B = 60 \text{ mm}, \; B' = \frac{60}{15} = 4 \text{ mm}$$

Equivalent steel section

$B' = 4$ mm

Two beams of size 4 mm × 200 mm

One beam of size 10 mm × 150 mm

Symmetrical about x-x- axis

Moment of inertia of equivalent steel section,

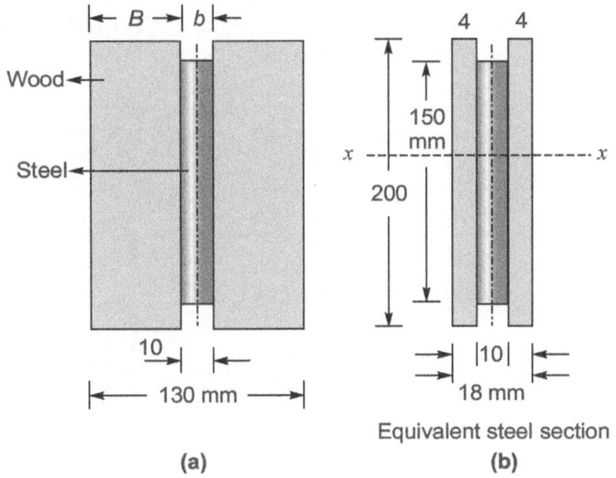

Fig. 11.35

$$I_s = \frac{2 \times 4 \times 200^3}{12} + \frac{10 \times 150^3}{12}$$

$$= 5.333 \times 10^6 + 2.8125 \times 10^6 = 8.1458 \times 10^6 \text{ mm}^4 = 8.1458 \times 10^6 \text{ mm}^4$$

$$E_s = 210 \text{ kN/mm}^2 = 210 \times 10^6 \text{ kN/m}^2$$

$$E_s I_s = 210 \times 10^6 \times 8.1458 \times 10^{-6} = 1710.618 \text{ kNm}^2$$

Span length, $L = 5$ m

Rate of loading, $w = 4$ kN/m

Deflection at center, $\delta = \dfrac{5wL^4}{384 E_s I_s} = 19$ mm

Practice Q.11.22 A composite beam consists of a wooden section, 240 mm width × 260 mm depth. Two steel plates of width 240 mm and thickness 20 mm are fixed on the top and bottom of wooden section. Composite beam is simply supported over a span of 5 m and a load of 30 kN is applied at the center of the beam. Determine the deflection at the center of the composite beam. $E_s = 200$ GPa, $E_s = 15$ GPa. Hint [Equivalent section is *I*, flanges 240 × 20 mm, web 16 mm and 260 mm].

Ans. [$E_s I_s = 42968.8$ kNm^2, $y_c = 1.82$ mm]

Question 11.23 Find the slope and deflection of free end of a cantilever of length 6 m as loaded shown in figure 11.36. Using the method of superposition, evaluate then numerical values $E = 200$ MPa, $I = 1 \times 10^{-4}$ m^4, $W = 1$ kN.

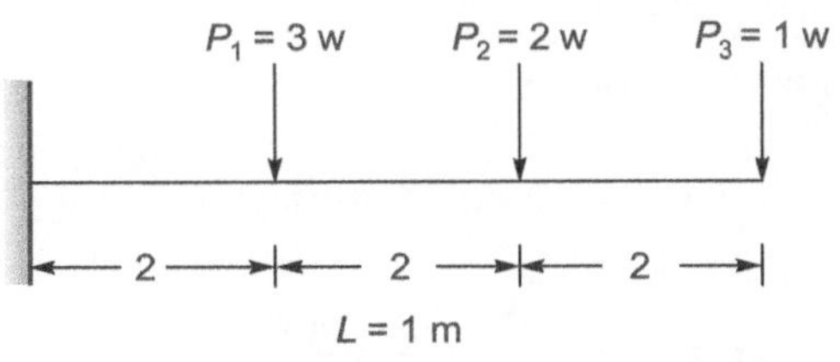

Fig. 11.36

[10 marks : IES 2003]

Solution:

$$EI = 200 \times 10^6 \times 1 \times 10^{-4} = 200{,}00 \text{ kNm}^2 \quad \ldots(i)$$

Slope at free end

$$i = \frac{(3W)(2L)^2}{2EI} + \frac{(2W)(4L)^2}{2EI} + \frac{(W)(6L^2)}{2EI}$$

$$= \frac{WL^2}{2EI}[12+32+36] = \frac{80\,WL^2}{2EI} = \frac{80\times1\times1^2}{2\times200{,}00} = 0.002 \text{ radian}$$

Deflection at free end,

$$y = \frac{(3W)(2L)^3}{3EI} + \frac{(3W)(2L)^2}{2EI}\times 4L + \frac{(2W)(4L)^3}{3EI} + \frac{(2W)(4L)^2}{2EI}\times 2L + \frac{W(6L)^3}{3EI}$$

$$= \frac{8WL^3}{EI} + \frac{24WL^3}{EI} + 128\times\frac{WL^3}{3EI} + \frac{32WL^3}{EI} + \frac{72WL^3}{EI}$$

$$= \frac{136WL^3}{EI} + \frac{128\,WL^3}{3} = \frac{536WL^3}{3EI} = \frac{536\times1\times1^3}{3\times2000{,}00} = 0.00893 \text{ m} = 8.93 \text{ mm}$$

Question 11.24 A simply supported beam AB of span L carries a distributed load of varying intensity as shown in figure 11.37. Establish the equation for deflection curve for the beam at a distance x from the origin A of the beam. Determine the deflection of the beam at the mid-section C in terms of W, L, E and I, where EI is the flexural rigidity of the beam.

[ME : 10 marks : 2016]

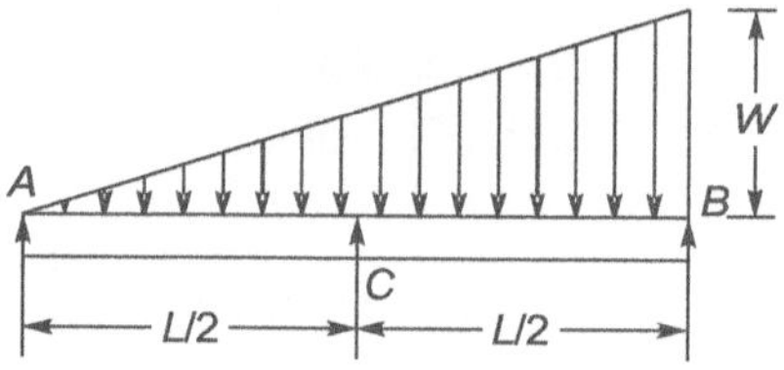

Fig. 11.37

Solution: A

$$\text{Total load on beam} = \frac{wL}{2}$$

CG lies at $\frac{2L}{3}$ from A

Moment about A

$$\frac{wL}{2}\times\frac{2L}{3} = R_B \times L$$

Reaction, $R_B = \frac{wL}{3}$

Reaction, $R_A = \frac{wL}{6}$

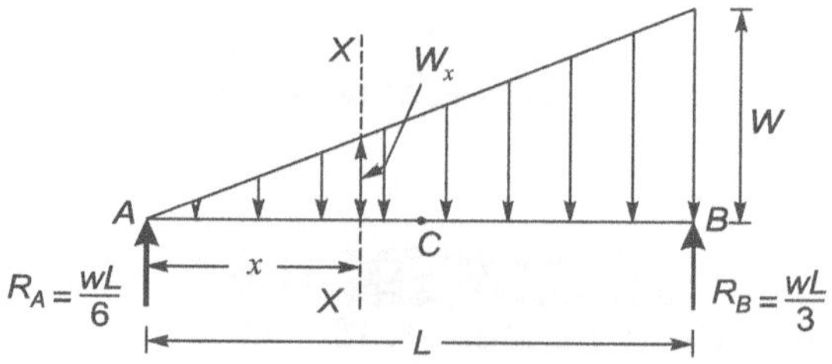

Fig. 11.38

Take a section x-x at a distance x from A

$$w_x = \text{Rate of loading} = \frac{W}{L}\times x$$

$$\text{Load upto } xx, = \frac{W_x}{L}\times\frac{x}{2} = \frac{Wx^2}{2L}$$

CG of this load lies at $\frac{x}{3}$ from section x-x

Bending moment at section x-x

$$M_x = \frac{wL}{6} \times x - \frac{wx^2}{2L} \times \frac{x}{3}$$

$$EI\frac{d^2y}{dx^2} = \frac{wLx}{6} - \frac{wx^3}{6L}$$

$$EI\frac{dy}{dx} = \frac{wLx^2}{12} - \frac{wx^4}{24L} + C_1$$

$$EIy = \frac{wLx^3}{36} - \frac{wx^5}{120L} + C_1x + C_2$$

$$y = 0 \text{ at end A, x= 0}$$

Constant $C_2 = 0$

$$y = 0 \text{ at end B, x = L}$$

$$0 = \frac{wL^4}{36} - \frac{wL^5}{120L} + C_1 \times L$$

$$0 = \frac{wL^4}{36} - \frac{wL^4}{120} + C_1 L$$

$$G = -\frac{wL^3}{36} + \frac{wL^3}{120} = \frac{-10+3}{360} = \frac{-7}{360}wL^3$$

Equation of deflection

$$y = \frac{1}{EI}\left[\frac{wLx^3}{36} - \frac{wx^5}{120L} - \frac{7}{360} \times wL^3 x\right]$$

Deflection at mid-span, $x = \frac{L}{2}$

$$y_c = \frac{1}{EI}\left[\frac{wL^4}{288} - \frac{wL^4}{120 \times 3L} - \frac{7}{720} \times wL^4\right]$$

$$= \frac{wL^4}{EI}\left[\frac{1}{288} - \frac{1}{3840} - \frac{7}{720}\right]$$

$$= \frac{wL^4}{EI}[3.4722 - 0.2604 - 9.722] \times 10^{-3}$$

$$= -\frac{6.51 \times 10^{-3}}{EI} \times wL^4$$

Question 11.25 A beam of uniform section and length (L + 2a) is simply supported over a span L with two equal overhanging lengths 'a'. Compute the deflection at mid span due to a uniformly distributed load w/unit length when covering the length L between the supports and when covering only two overhanging lengths. EI is the flexural rigidity of the beam.

[ME : 20 marks : 2017]

Solution: This problem can be easily solved by moment area methods.

Cast-I: BMD of beam is shown,

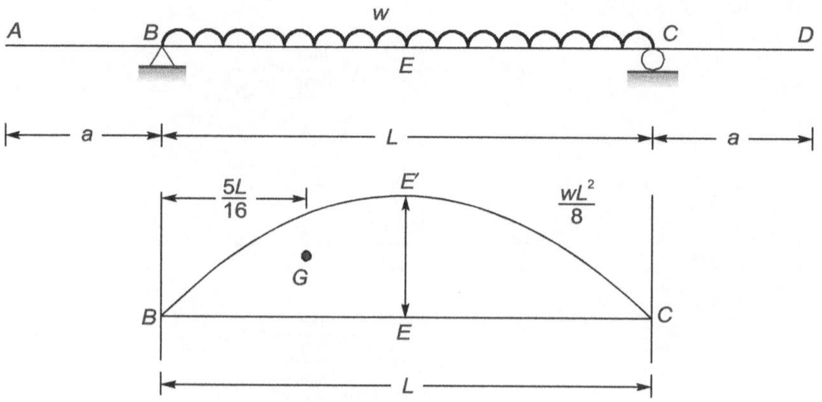

Fig. 11.39

$$M_{min} = \frac{wL^2}{8} \text{ (at centre area of BMD)}$$

$$= \frac{2}{3} \times \frac{wL^2}{8} \times \frac{L}{2}$$

$$= \frac{wL^3}{24} \qquad \text{...(i)}$$

CG of area BE′E been at $\frac{5}{16}L$ from B

Deflection at E, $$y_E = \frac{wL^3}{24} \times \frac{5}{16} \times L \times \frac{1}{EI}$$

$$= \frac{5}{384} \times \frac{wL^4}{EI} \qquad \text{(downward)}$$

Case-II: BMD is shown in figure.

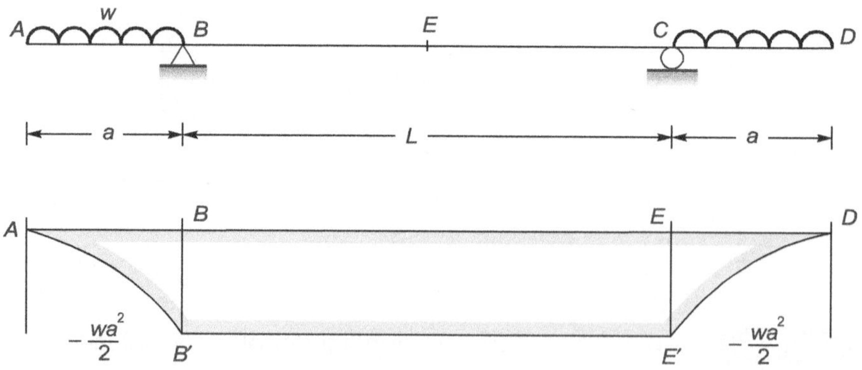

Fig. 11.40

First moment of area ABEE′B′ about point A

$$= \frac{wa^2}{2} \times \frac{a}{3}\left(\frac{3a}{4}\right) + \frac{wa^2}{2} \times \frac{L}{2}\left(a + \frac{L}{4}\right)$$

$$= \frac{wa^4}{8} + \frac{wa^3L}{4} + \frac{wa^2L^2}{16}$$

$$= \frac{wa^2}{4}\left[\frac{a^2}{2} + \frac{aL}{1} + \frac{L^2}{4}\right]$$

Deflection, $$y_E = \frac{wa^2}{4EI}\left[\frac{a^2}{2} + La + \frac{L^2}{4}\right]$$

Question 11.26 Find the deflection at C for the beam as shown in Figure 11.41. EI = constant. Use conjugate beam method.

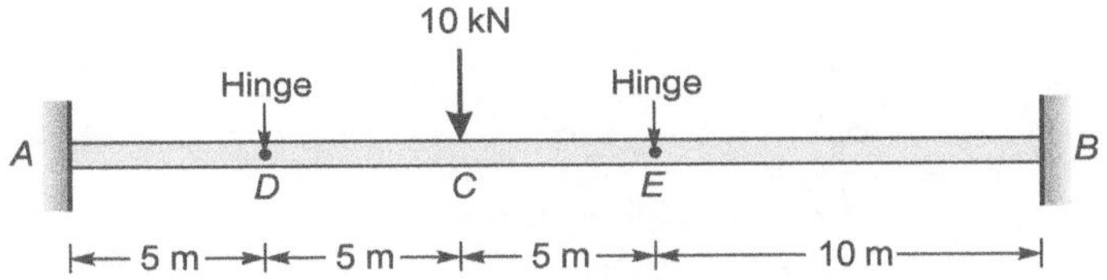

Fig. 11.41

[CE : 20 marks : 2017]

Solution: Conjugate Beam Method

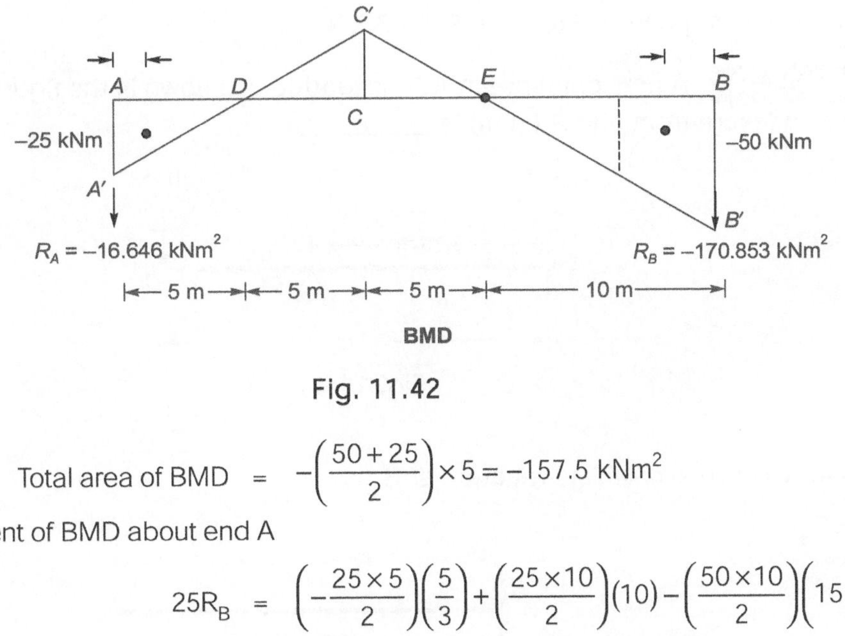

Fig. 11.42

Total area of BMD $= -\left(\frac{50+25}{2}\right)\times 5 = -157.5 \text{ kNm}^2$

Reaction: Moment of BMD about end A

$$25R_B = \left(-\frac{25\times 5}{2}\right)\left(\frac{5}{3}\right) + \left(\frac{25\times 10}{2}\right)(10) - \left(\frac{50\times 10}{2}\right)\left(15+\frac{20}{3}\right)$$

$$= -104.16667 + 1250 - 250\,(21.6667)$$

$$= -104.1667 + 1250 - 5416.667$$

$$= -4271.33$$

$$R_B = -170.853 \text{ kNm}^2$$

$$R_A = -187.5 - (-170.853) = -16.646 \text{ kNm}^2$$

Deflection at C, $$\delta_1 = \frac{1}{EI}\left[-16.646\times 10 - 62.5\left(5+\frac{10}{3}\right) + 62.5\left(\frac{5}{3}\right)\right]$$

$$= \frac{1}{EI}[-166.46 - 416.66]$$

$$= -\frac{583.12}{EI}$$

Question 11.27 A steel strip of length, $L = 200$ mm is fixed at end A and rests at B on a vertical spring of stiffness, $k = 2$ N/mm. The steel strip is 5 mm wide and 10 mm thick. A vertical load, $P = 50$ N is applied at B, as shown in the figure 11.43. Considering $E = 200$ GPa, the force (in N) developed in the spring is ________.

[2015 : 2 Marks, Set-II]

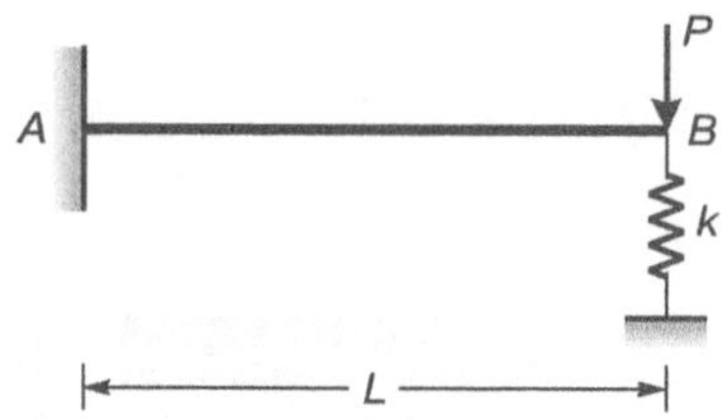

Fig. 11.43

Solution:

$$\delta_B = \frac{PL^3}{3EI} = \frac{50 \times 200^3}{3 \times 200 \times 10^3} \times \frac{1}{I} = \frac{666.667}{I}$$

$$I = \frac{5 \times 10^3}{12} = \frac{5000}{12} = 416.667$$

$$\delta_B = \frac{666.667}{416.6697} = 1.6 \text{ mm}$$

$$k = 2 \text{ N/mm}$$

$$\text{Force in spring} = k\delta_B = 1.6 \times 2 = 3.2 \text{ N}$$

Question 11.28 A horizontal beam ABC is loaded as shown in the figure 11.44. The distance of the point of contraflexure from end A (in m) is ________.

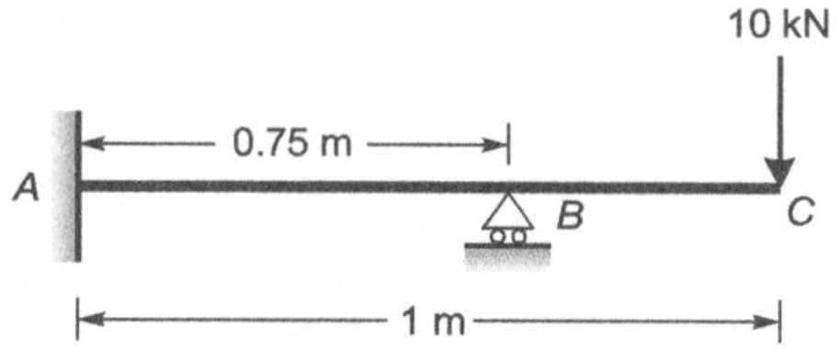

Fig. 11.44

[2015 : 1 Mark, Set-II]

Solution: We have to calculate reaction at B

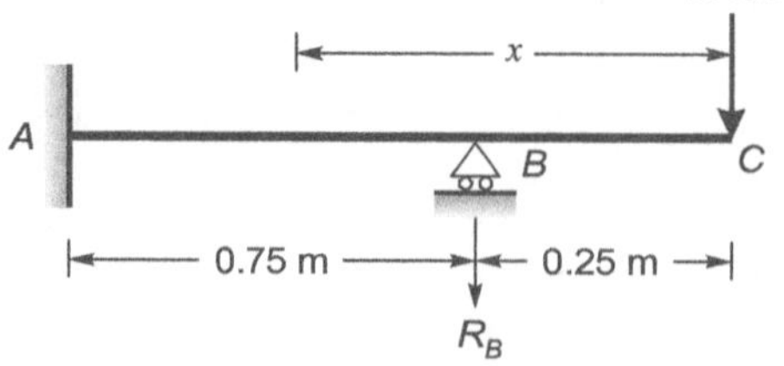

Fig. 11.45

$$EI\frac{d^2y}{dx^2} = 10x - R_B(x - 0.25)$$

$$EI\frac{dy}{dx} = 5x^2 - \frac{R_B}{2}(x-0.25)^2 + C_1$$

$$x = 1\text{ m}, \quad \frac{dy}{dx} = 0$$

$$EI\frac{dy}{dx} = 0$$

$$S = \frac{R_B}{2}\times\frac{9}{16}+C_1$$

$$c_1 = \frac{9R_B}{32}-5$$

$$EI\frac{dy}{dx} = 5x^2 - \frac{R_B}{2}(x-0.25)^2 + \left(\frac{9R}{32}-5\right)$$

$$Ely = \frac{5x^3}{3}-\frac{R_B}{6}(x-0.25)^3+\left(\frac{9R_B}{32}-5\right)x+C_2$$

$$y = 0 \text{ at } x = 1\text{ m}$$

$$0 = \frac{5}{3}-\frac{R_B}{6}\left(\frac{3}{4}\right)^3+\left(\frac{9R_B}{32}-5\right)+C_2$$

$$0 = \frac{5}{3}-\frac{R_B}{6}\times\frac{27}{64}+\frac{9R_B}{32}-5+C_2$$

$$0 = -\frac{10}{3}-R_B\frac{9}{128}+\frac{9R_B}{32}+C_2$$

$$0 = -\frac{10}{3}+\frac{R_B}{128}(27)+c_2$$

$$C_2 = \frac{10}{3}-\frac{R_B}{128}\times 27$$

$$Eiy = \frac{5x^3}{3}-\frac{R_B}{6}(x-0.025)^3+\left(\frac{9R_B}{32}-5\right)x+\frac{10}{3}-\frac{27}{128}R_B$$

$$y = 0, \text{ at } x = \frac{1}{4}$$

$$0 = \frac{5}{3}\times\frac{1}{64}-0+\left(\frac{9R_B}{32}-5\right)\times\frac{1}{4}+\frac{10}{3}-\frac{27}{128}\times R_B$$

$$0 = \frac{5}{192}+\frac{9}{128}\times R_B-\frac{5}{4}+\frac{10}{3}-\frac{27}{128}\times R_B$$

$$0 = \left(\frac{5}{192}-\frac{5}{4}+\frac{10}{3}\right)-\frac{9}{64}R_B$$

$$\left(\frac{5-240+640}{192}\right)-\frac{9}{64}R_B$$

$$0 = \frac{405}{192}-\frac{9}{64}R_B$$

$$R_{BV} = \frac{405}{192}\times\frac{64}{9} = \frac{405}{27} = 15$$

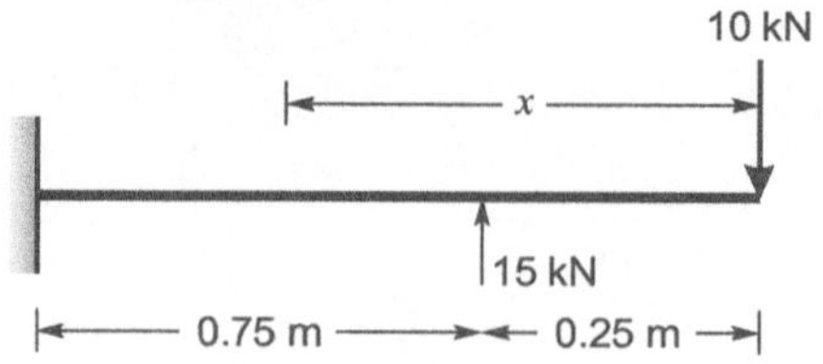

Fig. 11.46

$$10x - 15(x - 0.25) = 0$$
$$-5x + 3.75 = 0$$
$$x = 0.75 \text{ m}$$

From end A, point of contraflexure lies at 0.25 m.

Question 11.29 A cantilever beam with flexural rigidity of 200 Nm^2 is loaded as shown in the figure 11.47. The deflection (in mm) at the tip of the beam is _____.

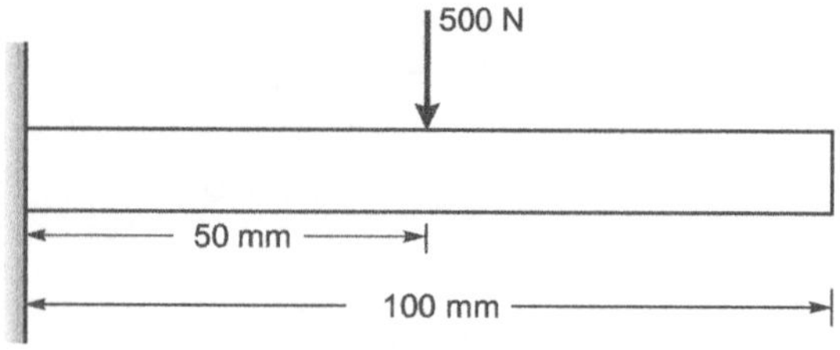

Fig. 11.47

[2015 : 2 Marks, Set-1]

Solution:

$$EI = 200 \text{ Nm}^2 = 200\times 10^6 \text{ Nmm}^2$$

$$\delta = \frac{500\times(50)^3}{3\times EI} + \frac{500\times 50^2\times 50}{2\times EI}$$

$$= \frac{5}{6}\times\frac{1}{EI}\times 500\times 50^3$$

$$\text{Deflection at tip} = \frac{5}{6}\times\frac{500\times 50^3}{200\times 10^6} = 0.26 \text{ mm}$$

Question 11.30 A cantilever beam with square cross-section of 6 mm side is subjected to a load of 2 kN normal to the top surface as shown in the figure 11.48. The Young's modulus of elasticity of the material of the beam is 210 GPa. The magnitude of slope (in radian) at *Q* (20 mm from the fixed end) is ________.

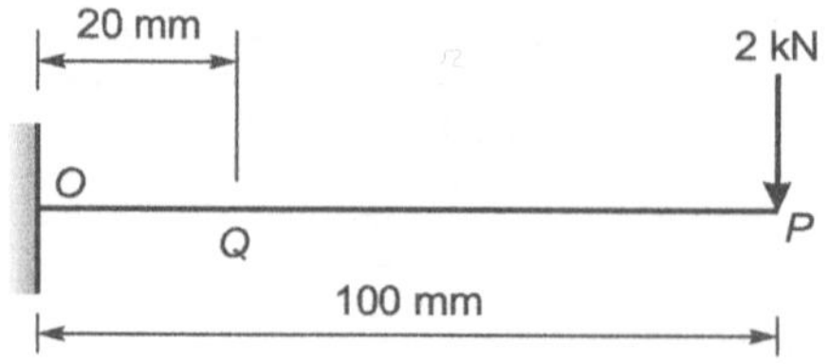

Fig. 11.48

[2015 : 2 Marks, Set-2]

Solution:

$$I = \frac{6^4}{12} = 108 \text{ mm}^4,\ EI = 210 \times 10^3 \times 108 \text{ Nmm}^2$$

$$EI\frac{d^2y}{dx^2} = -2000\,x$$

$$EI\frac{dy}{dx} = -1000\,x^2 + c_1$$

$$\frac{dy}{dx} = 0 \text{ at } x = 100$$

$$c_1 = +10^6$$

$$EI\frac{dy}{dx} = -1000\,x^2 + 10^7$$

$$x = 80 \text{ mm}$$

$$EI\,L_\theta' = -1000 \times 80^2 + 10^7 = -6.4 \times 10^6 + 10^7$$

$$L_\theta' = \frac{3.6\times10^7}{210\times108\times10^3} = 1.587 \text{ radian}$$

Question 11.31 The axial load (in kN) in the member *PQ* for the arrangement/assembly shown in the figure 11.49 is __________.

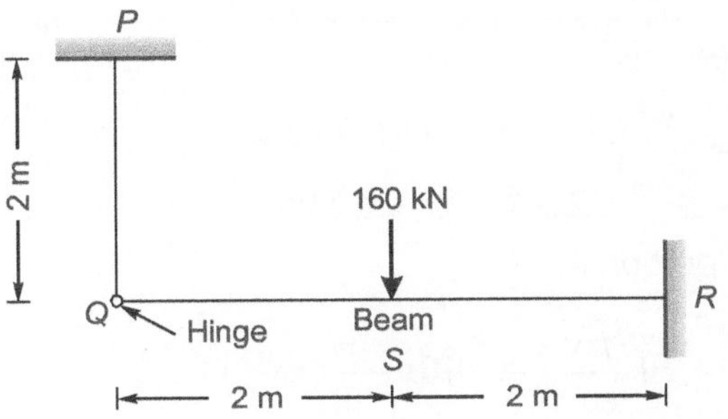

Fig. 11.49

[2014 : 2 Marks, Set-II]

Solution: Axial load *PQ* say *R*

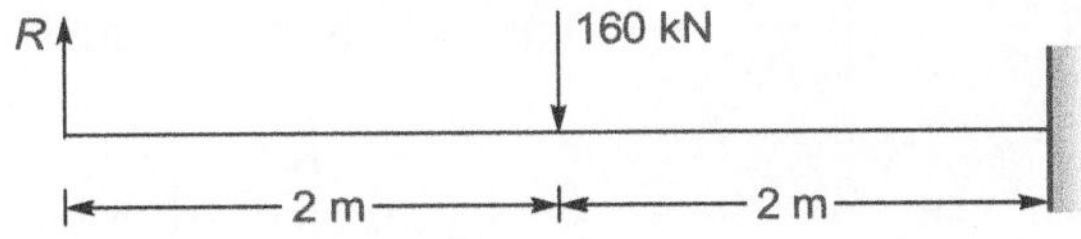

Fig. 11.50

$$\frac{R(4)^3}{3EI} = \frac{160\times2^3}{3EI} + \frac{160\times2^2}{2EI}\times2$$

$$64R = 160 \times 8 + 160 \times 12 = 160 \times 20$$

$$R = \frac{160\times20}{64} = 50 \text{ kN}$$

Question 11.32 A 3 m long simply supported beam of uniform cross-section is subjected to a uniformly distributed load of w = 20 kN/m in the central 1 m as shown in the figure 11.51.

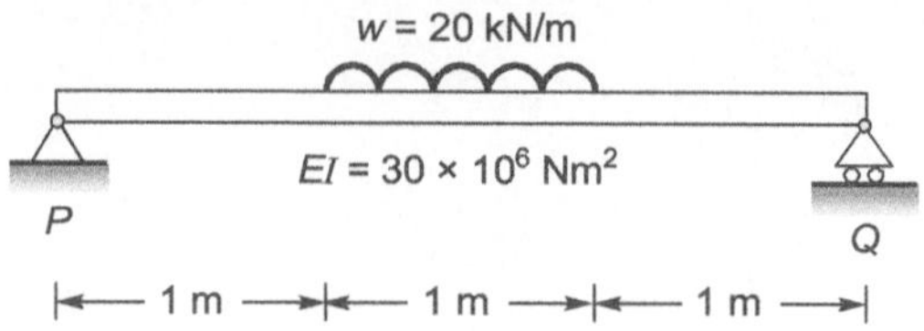

Fig. 11.51

If the flexural rigidity (EI) of the beam is 30×10^6 N-m², the maximum slope (expressed in radians) of the deformed beam is

(a) 0.681×10^{-7} (b) 0.943×10^{-7} (c) 4.310×10^{-7} (d) 5.910×10^{-7}

[2016 : 2 Marks, Set-I]

Solution:

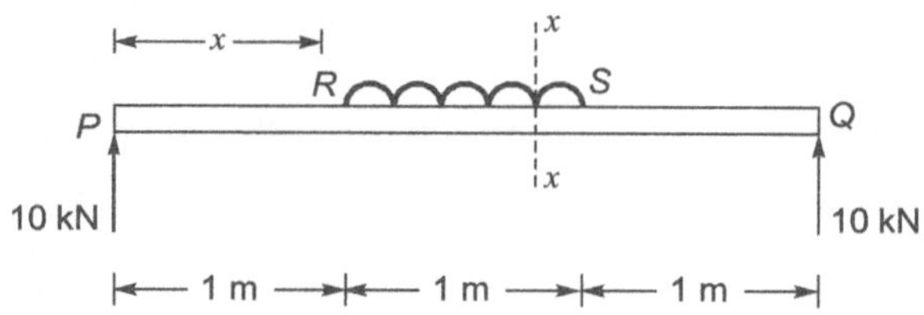

Fig. 11.52

Reactions, $R_P = R_Q = 10$ kN

Let us take section xx in portion RS

$$EI\frac{d^2y}{dx} = 10x - \frac{w}{2}(x-1)^2 \qquad (w = 20 \text{ kNm})$$

$$= 10x - 10\,(x-1)^2$$

$$EI\frac{dy}{dx} = 5x^2 - \frac{10}{3}(x-1)^3 + C_1$$

$$\frac{dy}{dx} = 0 \text{ at } x = 1.5 \text{ m, because symmetry of loading about centre}$$

$$0 = 5 \times 1.5^2 - \frac{10}{3}(0.5)^3 + C_1$$

$$0 = 11.25 - 0.4166 + C_1$$

$$C_1 = -10.833$$

Maximum slope at P or Q

$$EI = 30 \times 10^3 \text{ kNm}^2$$

Maximum slope, $$i_{max} = \frac{10.833}{30 \times 10^{-3}} = 0.36 \times 10^{-3} \text{ radian}$$

(Answer given in the question are wrong or data given is wrong).

Objective Type Questions

Q.1 The area moment of inertia about the neutral axis of a cross-section at a distance x (measured form free end)

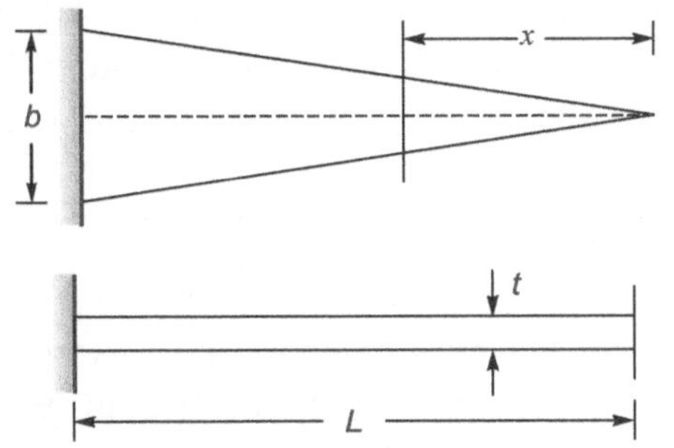

(a) $\frac{bxt^3}{6L}$ (b) $\frac{bxt^3}{12L}$

(c) $\frac{bxt^3}{24L}$ (d) $\frac{xt^3}{12L}$

[2 Marks : GATE 2011]

Q.2 The maximum deflection of beam is

(a) $\frac{24PL^3}{Ebt^3}$ (b) $\frac{12PL^3}{Ebt^3}$

(c) $\frac{8PL^3}{Ebt^3}$ (d) $\frac{6PL^3}{Ebt^3}$

[2 Marks : GATE 2011]

Q.3 A cantilever beam of length L is subjected to a moment M at the free end. The moment of inertia of the beam cross-section about neutral axis is I and Young's modulus is E. Magnitude of maximum deflection is

(a) $\frac{ML^2}{2EI}$ (b) $\frac{ML^2}{EI}$

(c) $\frac{2ML^2}{EI}$ (d) $\frac{4ML^2}{EI}$

[1 Mark : GATE 2014]

Q.4 A beam having uniform cross-section carries a uniformly distributed load of intensity q per unit length over its entire length and its mid span deflection is δ. The value of the mid span deflection of the same beam when the same load distributed with intensity varying from 2q per unit length at one end to zero at the other end.

(a) $\frac{\delta}{3}$ (b) $\frac{\delta}{2}$

(c) $\frac{2\delta}{3}$ (d) δ

[IES 1995]

Q.5 Two identical cantilevers are loaded as shown in the respective figure, E and F. If slope at the free end of the cantilever in figure E is θ, the slope at the tree end of the cantilever in figure F will be

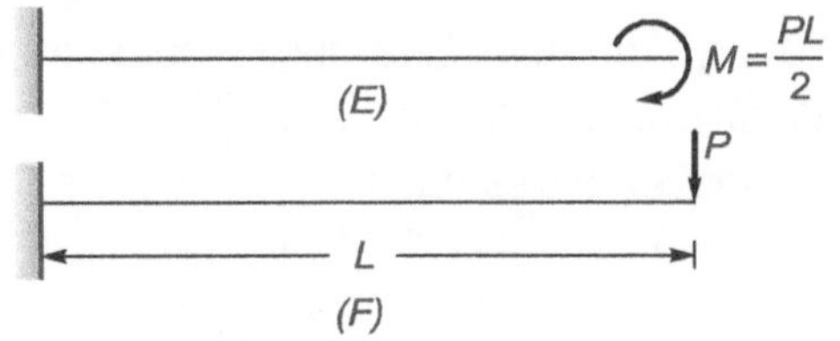

(a) $\frac{1}{3}\theta$ (b) $\frac{1}{2}\theta$

(c) $\frac{2}{3}\theta$ (d) θ

[IES 1997]

Q.6 A simplify supported beam of constant flexural rigidity and length 2 L carries a concentrated load P at its mid span and its deflection under the load is δ. If a cantilever beam of the same flexural rigidity and the length L is subjected to a load P at its free end, then deflection at the free end will be

(a) $\frac{\delta}{2}$ (b) δ

(c) 2δ (d) 4δ

[IES 1997]

Q.7 A cantilever beam of rectangular cross-section is subjected to a load W at the free end. If the depth of the beam is doubled and load is halved. The deflection at the free end as compared to original deflection will be

(a) half (b) one eight

(c) one-sixteenth (d) double

[IES 1999]

Q.8 The two cantilevers A and B shown in the figure, have uniform cross-section and the same material. Free end deflection of cantilever A is δ. The value of mid span deflection of cantilever B is

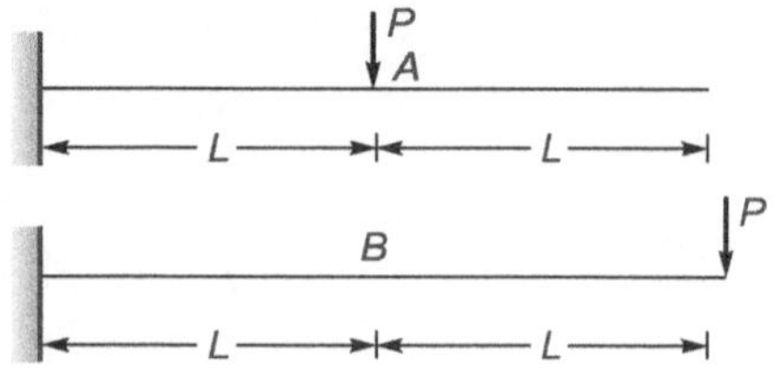

(a) $\frac{\delta}{2}$ (b) $\frac{2}{3}\delta$
(c) δ (d) 2δ

[IES 2000]

Q.9 A cantilever of length L, moment of inertia I, Young's modulus E carries a concentrated load W at the middle of its length. Slope of the cantilever at free end is

(a) $\frac{WL^2}{2EI}$ (b) $\frac{WL^2}{4EI}$
(c) $\frac{WL^2}{8EI}$ (d) $\frac{WL^2}{16EI}$

[IES 2001]

Q.10 A simplify supported beam of length L is loaded by a uniformly distributed load w over the entire length of span. It is propped at the mid span so that deflection at the center is zero. The reaction at prop is

(a) $\frac{5}{16}wL$ (b) $\frac{wL}{2}$
(c) $\frac{5}{8}wL$ (d) $\frac{1}{10}wL$

[IES 2013]

Q.11 A beam AB of length 1.5 L, is supported over a span L with equal over hang on both the sides. A load W is applied at the center of the beam such that deflection in center is δ_C. The deflection at end A is δ_A. What is the ratio of δ_A / δ_C?

(a) 1.5 (b) 1.2
(c) 0.8 (d) 0.75

Q.12 A cantilever of length L, fixed at end A, is propped at end B. A load W is applied at center of the cantilever. What is the prop reaction?

(a) $\frac{5W}{8}$ (b) $\frac{W}{2}$
(c) $\frac{5W}{16}$ (d) $\frac{5W}{24}$

Q.13 A beam AB is 6 m long, simply supported at ends. $EI = 1200$ kNm^2. What is load W applied at a distance of 2 m from one end, if deflection under load is 1 mm.

(a) 266.67 N (b) 337.5 N
(c) 675 N (d) None

Q.14 A simply supported beam of length 10 m, supported over a span of 8 m, with equal overhang on both the sides. A load W is applied at center of the beam. What is slope at ends if EI is the flexural rigidity of the beam.

(a) $\pm\frac{5W}{EI}$ (b) $\pm\frac{4W}{EI}$
(c) $\pm\frac{3W}{EI}$ (d) None of these

Q.15 A beam of length 6 m carries a concentrated load W at the center such that BM at center of beam is 6 kNm. If EI of beam is 1800 kNm^2, what is deflection at the center of the beam?

(a) 2.96 mm (b) 6.4 mm
(c) 10 mm (d) None of these

Q.16 A beam supported over span L, carries two loads W each at a distance of $L/3$ from each end. If EI is the flexural rigidity of the beam, what is the slope at ends of the beam?

(a) $\pm\frac{WL^2}{16EI}$ (b) $\pm\frac{WL^2}{9EI}$
(c) $\pm\frac{WL^2}{4EI}$ (d) None of these

Q.17 A simply supported beam of span L and flexural rigidity EI, carries a unit point load at its centre. The strain energy in the beam due to bending is

(a) $\frac{L^3}{48\,EI}$ (b) $\frac{L^3}{96\,EI}$
(c) $\frac{L^3}{192\,EI}$ (d) $\frac{L^3}{16\,EI}$

[ESE : 1996]

Q.18 A uniform bar, simply supported at the ends, carries a concentrated load P at mid-span. If the same load be, alternatively uniformly distributed over the full length of the bar, the maximum deflection of the bar will decrease by

(a) 25.5% (b) 31.5%
(c) 37.5% (d) 50.0%

[ESE : 2017]

Q.19 Two beams, one having a square cross-section and another a circular cross-section, are subjected to the same amount of bending moment. If the cross-sectional area as well as the material of both the beams are the same, then

1. both the beams will experience the same amount of deformation
2. the circular beam experiences more extreme flexural stress than the square one

Which of the above is/are correct?

(a) 1 only (b) 2 only
(c) Both 1 and 2 (d) Neither 1 nor 2

[ESE : 2016]

Q.20 A cantilever beam, 3 m long, carries a uniformly distributed load over the entire length. If the slope at the free end is 1°, the deflection at the free end is

(a) 49.27 mm (b) 39.27 mm
(c) 30.27 mm (d) 20.27 mm

[ESE : 2017]

Q.21 An overhanging beam of uniform EI is loaded as shown below. The deflection at the free end is

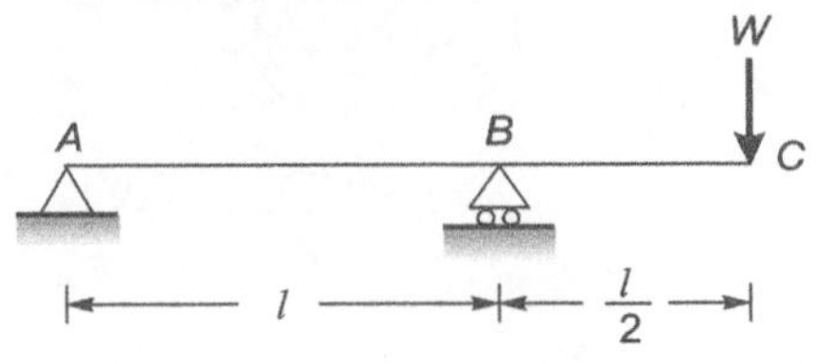

(a) $\frac{Wl^3}{81EI}$ (b) $\frac{Wl^3}{8EI}$

(c) $\frac{Wl^3}{27EI}$ (d) $\frac{2Wl^3}{27EI}$

[ESE : 2017]

Q.22 A cantilever beam of length L and flexural modulus EI is subjected to a point load P at the free end. The elastic strain energy stored in the beam due to bending (neglecting transverse shear) is

(a) $\frac{P^2L^3}{6EI}$ (b) $\frac{P^2L^3}{3EI}$

(c) $\frac{PL^3}{3EI}$ (d) $\frac{PL^3}{6EI}$

[2017 : 1 Mark, Set-2]

Q.23 A cantilever beam having square cross-section of side a is subjected to an end load, If a is increased by 19%, the tip deflection decreases approximately by

(a) 19% (b) 29%
(c) 41% (d) 50%

[2016 : 1 Mark, Set-1]

Q.24 A beam of length L is carrying a uniformly distributed load w per unit length. The flexural rigidity of the beam is EI. The reaction at the simple support at the right end is

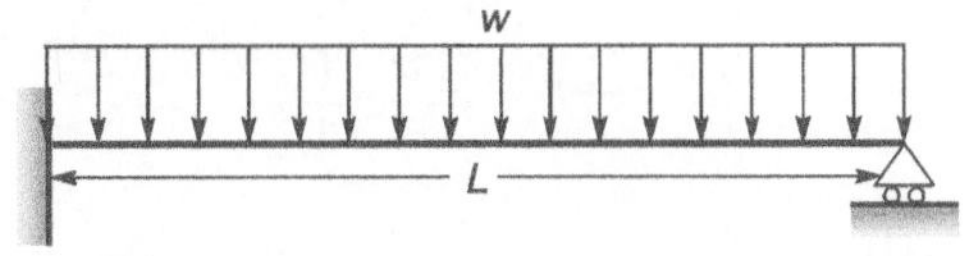

(a) $\frac{wL}{2}$ (b) $\frac{3wL}{8}$

(c) $\frac{wL}{4}$ (d) $\frac{wL}{8}$

[2016 : 2 Marks, Set-3]

Q.25 Two prismatic beams having the same flexural rigidity of 1000 kN-m^2 are shown in the figures.

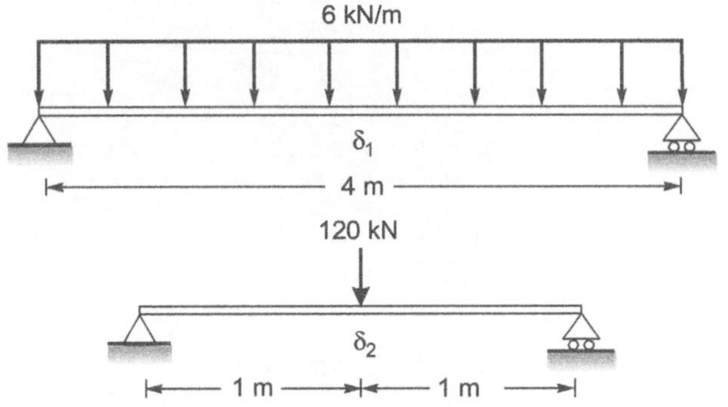

If the mid-span deflections of these beams are denoted by δ_1 and δ_2 (as indicated in the figures), the correct option is

(a) $\delta_1 = \delta_2$ (b) $\delta_1 < \delta_2$
(c) $\delta_1 > \delta_2$ (d) $\delta_1 >> \delta_2$

[2017 : 2 Marks, Set-II]

Q.26 The figure shows a simply supported beam PQ of uniform flexural rigidity EI carrying two moment M and 2M, the slope at P will be

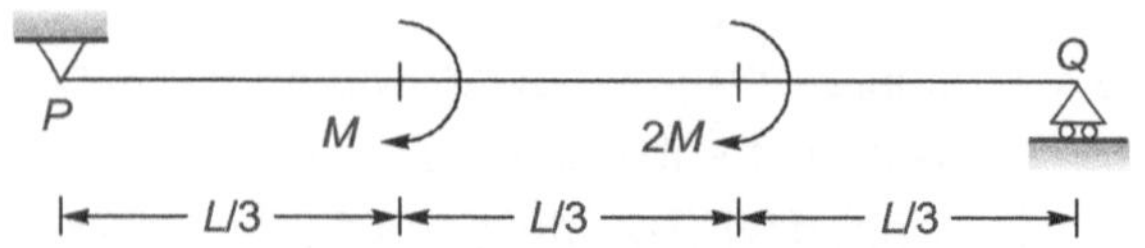

(a) 0 (b) $\frac{ML}{9EI}$
(c) $\frac{ML}{6EI}$ (d) $\frac{ML}{3EI}$

[GATE : 2018]

Answers

1. (b)	2. (d)	3. (a)	4. (d)	5. (d)
6. (c)	7. (c)	8. (c)	9. (c)	10. (c)
11. (d)	12. (c)	13. (b)	14. (b)	15. (c)
16. (b)	17. (b)	18. (c)	19. (b)	20. (b)
21. (b)	22. (a)	23. (d)	24. (b)	25. (a)
26. (c)				

Explanations

1. (b)

$$b_x = \frac{bx}{L}$$
$$depth = t$$
$$I = \frac{bx}{L} \times \frac{t^3}{12}$$

2. (d)

$$EI\frac{d^2y}{dx^2} = Px$$

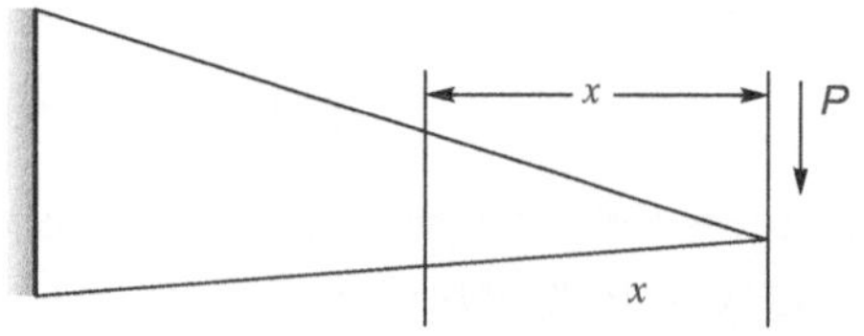

$$I = \frac{bx}{L} \cdot \frac{t^3}{12}$$
$$E\frac{d^2y}{dx^2} = \frac{12PL}{bt^3}$$

Say, $$\frac{12PL}{bt^3} = k$$

$$E\frac{d^2y}{dx^2} = kP \quad \ldots(i)$$

$$E\frac{dy}{dx} = kP_x + C_1$$
$$\frac{dy}{dx} = 0 \quad \text{at } x = L$$
$$C_1 = -kPL$$
$$E\frac{dy}{dx} = kPx - kPL$$
$$Ey = \frac{kPx^2}{2} - kPLx + C_2$$

$y = 0$ at $x = L$

$$0 = \frac{kPL^2}{2} - kPL^2 + C_2$$
$$C_2 = +\frac{kPL^2}{2}$$
$$EIy = \frac{kPx^2}{2} - kPLx + \frac{kPL^2}{2}$$

at $x = 0$, $$Ey_{max} = \frac{kPL^2}{2} = \frac{PL^2}{2} \times \frac{12L}{bt^3}$$
$$y_{max} = \frac{6PL^3}{Ebt^3}$$

3. (a)

$$\delta_{max} = \frac{ML^2}{2EI}$$

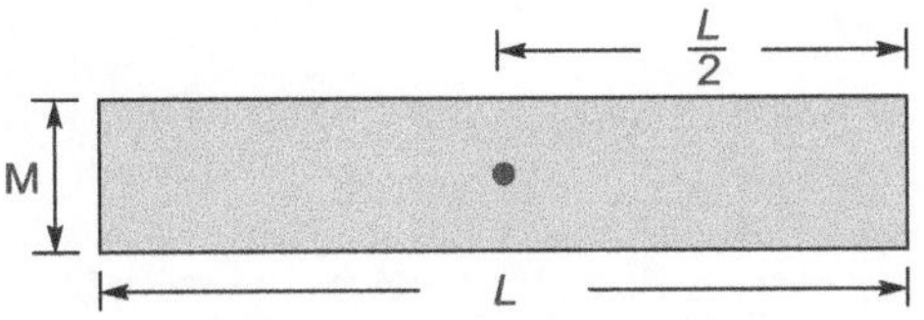

4. (d)

$$\delta = \frac{5qL^4}{384}$$

$$\delta' = \frac{1.875qL^4}{144EI}$$

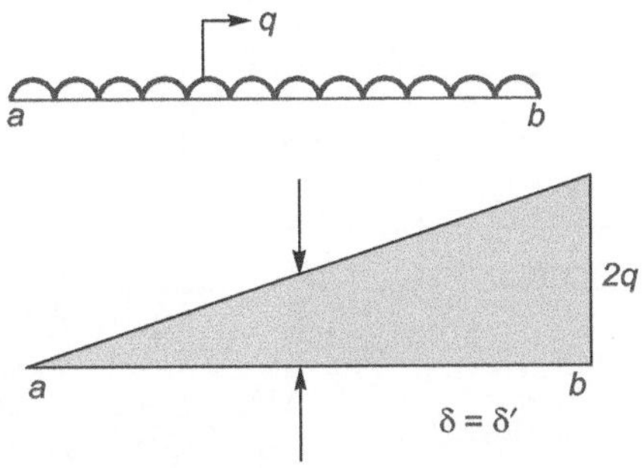

(Refer to page 303 of chapter 11)

5. (d)

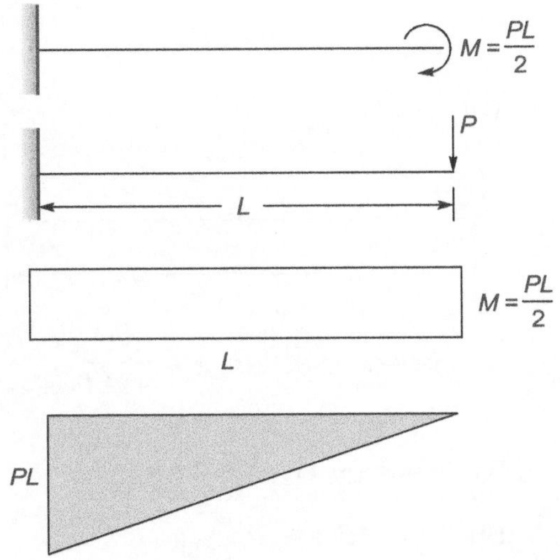

$$\theta = \frac{PL^2}{2EI}$$

$$\text{Area} = \frac{PL^2}{2}$$

$$\theta = \frac{PL^2}{2EI}$$

6. (c)

$$\delta = \frac{W(2L)^3}{48EI} = \frac{WL^3}{6EI}$$

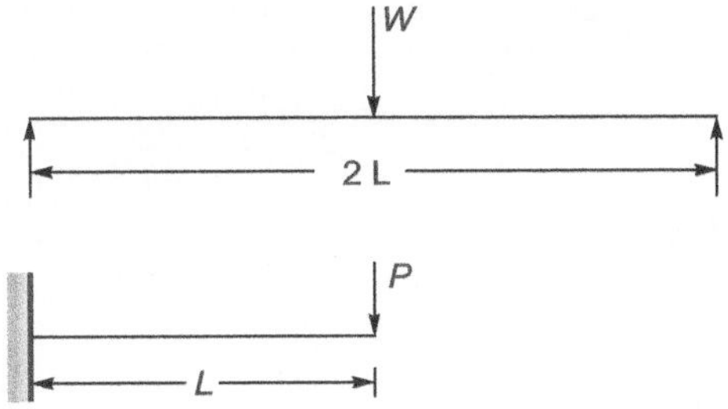

$$\delta' = \frac{WL^3}{3EI} = 2\delta$$

7. (c)

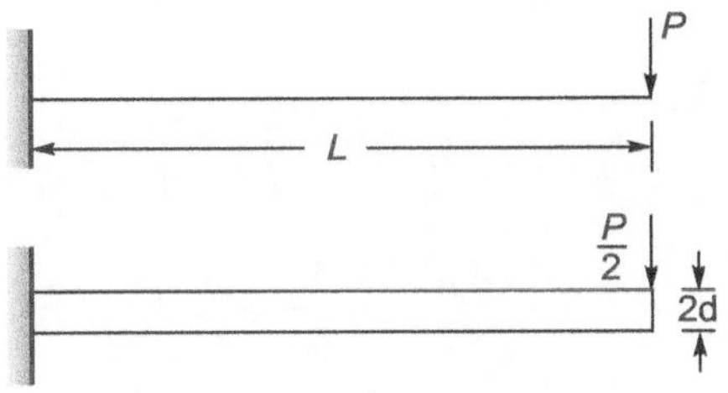

$$\delta = \frac{PL^3}{3EI}$$

$$I' = 8I$$

$$\delta = \frac{P}{2}\frac{L^3}{E} \times \frac{1}{8I} = \frac{PL^3}{16EI}$$

$$\frac{\delta'}{\delta} = \frac{1}{16}$$

8. (c)
Using Maxwell's reciprocal theorem
Load at *B*, deflection at *A* is δ
Load at *A*, deflection at B is δ

9. (c)

$$i_B = \frac{WL^2}{8EI}$$

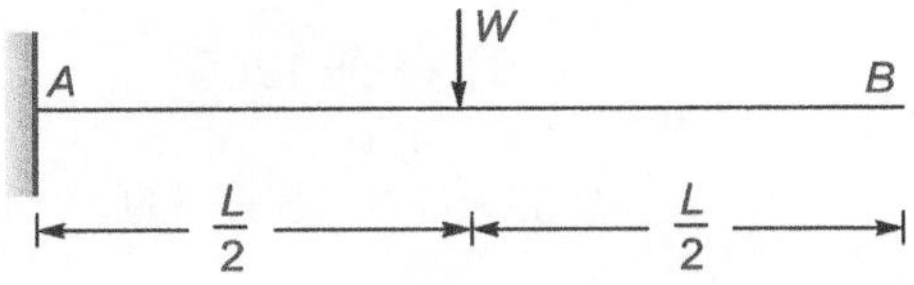

10. (c)

$$\delta = \frac{5}{384} \times \frac{wL^4}{EI} = R \times \frac{L^3}{48EI}$$

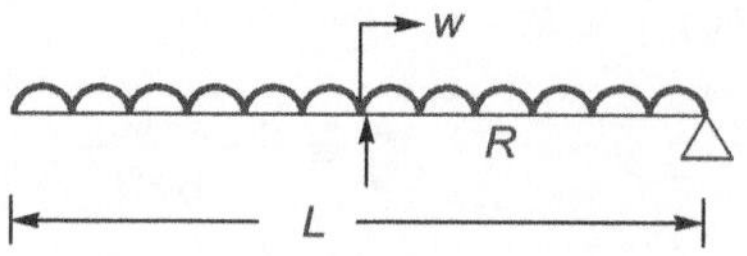

$$R = \frac{5}{8}wL$$

11. (d)

$$\delta_C = \frac{WL^3}{48EI}$$

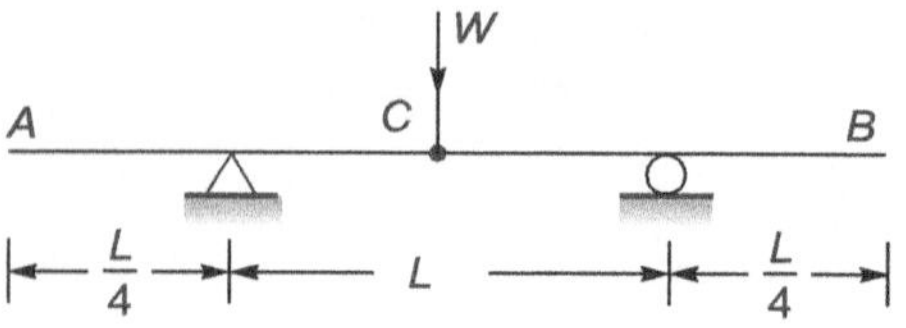

$$\delta_A = \frac{WL^2}{16EI}\times\frac{L}{4} = \frac{WL^3}{64EI}$$

$$\frac{\delta_A}{\delta_C} = \frac{48}{64} = 0.75$$

12. (c)

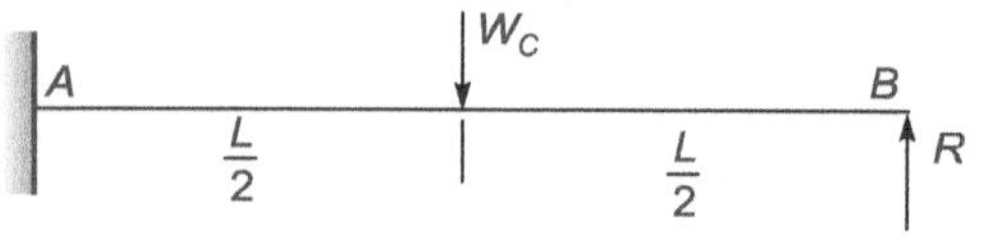

$$\frac{RL^3}{3EI} = W\left(\frac{L}{2}\right)^3\times\frac{1}{3EI}+\frac{W\left(\frac{L}{2}\right)^2}{2}\,\frac{1}{EI}\times\frac{L}{2}$$

$$\frac{R}{3} = \frac{W}{24}+\frac{W}{16} = \frac{(2+3)W}{48} = \frac{5W}{48}$$

$$R = \frac{5}{16}W$$

13. (b)

$$\delta = \frac{Wa^2b^2}{3EIL} = \frac{W\times 4\times 16}{3\times 6\times 1200}$$

$$W = \frac{0.001\times 18\times 1200}{64}$$

$$= 0.3375 \text{ kN} = 337.5 \text{ N}$$

14. (b)

$$i_A = \pm\frac{WL^2}{16EI} = \pm\frac{W\times 64}{16EI} = \pm\frac{4W}{EI}$$

15. (c)

$$6 \text{ kNm} = \frac{W\times 6}{4};\quad W = 4 \text{ kN}$$

$$W = 4 \text{ kN } \delta = \frac{WL^3}{48EI} = \frac{4\times 6^3}{48\times 1800}$$

$$= 0.01 \text{ m} = 10 \text{ mm}$$

16. (b)

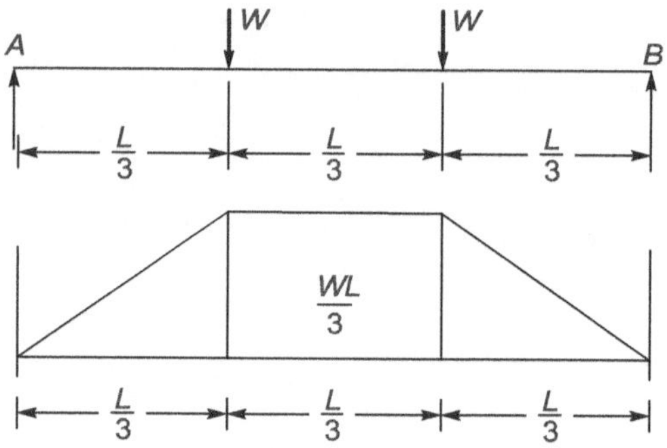

$$\text{Total area} = \frac{WL}{3}\times\frac{L}{3}\times 2 = 2\frac{WL^2}{9}$$

$$R_{A'} = R_{B'} = \frac{WL^2}{9}$$

$$i_A = i_B = \frac{WL^2}{9EI}$$

17. (b)

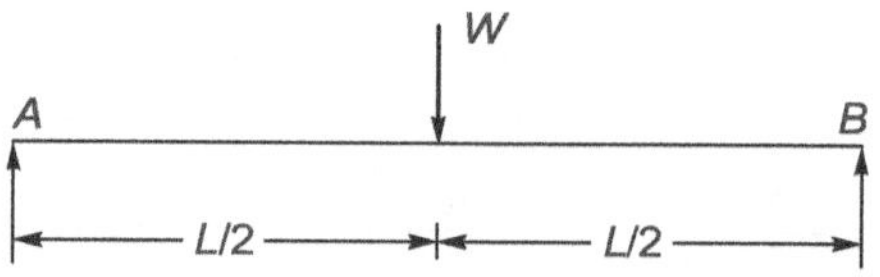

$$\text{d = deflection at centre} = \frac{WL^4}{48EI}$$

$$\text{Strain energy, U} = \frac{1}{2}W\delta = \frac{W^2L^3}{96EI}$$

But W = 1 units

$$\text{u = Strain energy} = \frac{L^3}{96EI}$$

18. (c)
A beam simple supported at ends, carry moment ratio load at mid-span

$$y_{max}\text{, maximum deflection} = \frac{WL^3}{48EI}$$

A beam simply supported at canter, carrying UDL, w throughout to length

$$y'_{max}, \text{ maximum deflection} = \frac{5}{384} \times \frac{wL^4}{EI}$$

But $wL = W$

So, $$y'_{max} = \frac{5}{384} \times \frac{WL^3}{EI}$$

$$\frac{y'_{max}}{y_{max}} = \frac{5}{384} \times \frac{48}{1} = \frac{5}{8} = 0.625$$

Decrease in $y_{max} = 0.375 = 37.5\%$

19. (b)
1. Incorrect statement
2. Correct (because circular is less than 2 square) same is $\sigma = m/Z$

$$\frac{\pi}{4} d^2 = a^2$$

$$d = a \times 1.128$$

$$I_{square} = \frac{a^4}{12}$$

$$Z = \frac{a^3}{6} = 0.1666a^3$$

$$I_{circle} = \frac{\pi d^4}{64} = 0.0796a^4$$

$$Z_{circular} = \frac{\pi d^3}{32} = 0.141a^3$$

$$Z_{circle} < Z_{square}$$

20. (b)

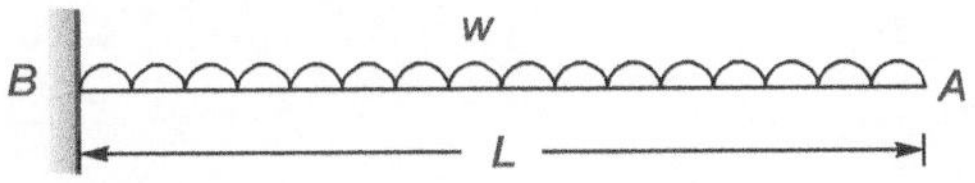

Slope at force end A,

$$i_A = \frac{wl^3}{6EI} = 1°$$

$$l = 3000 \text{ mm}$$

Deflection, $$y_A = \frac{wl^4}{8EF} = i_A \times l \times \frac{6}{8}$$

$$i = \frac{\pi}{180} \text{ radian}$$

So, $$y_A = \frac{\pi}{100} \times 3000 \times \frac{6}{8}$$
$$= 39.27 \text{ mm}$$

Deflection at free end.

21. (b)

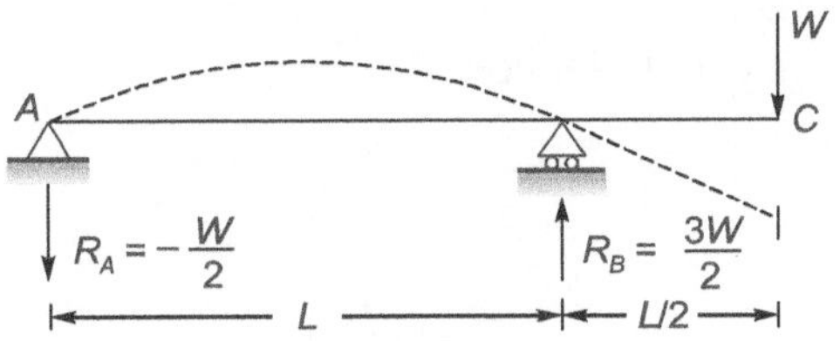

Reaction, $$R_B = +\frac{3W}{2} \uparrow$$

$$R_A = -\frac{W}{2} \downarrow$$

BMD

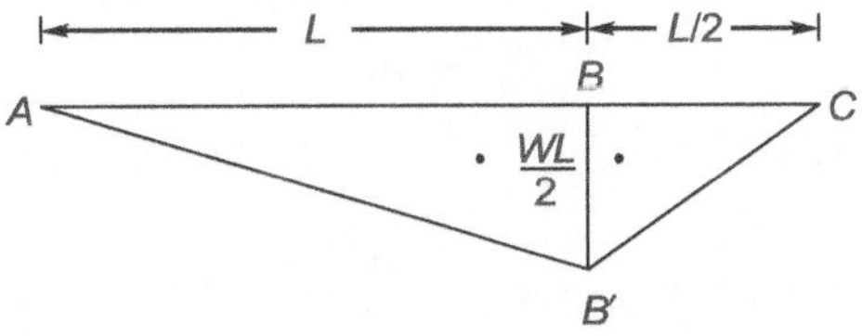

Taking first moment of area of BMD about point C, the end of cantilever

$$y_C = \frac{1}{EI}\left[\frac{WL}{2} \times \frac{L}{2} \times \frac{1}{2} \times \frac{L}{3} + \frac{WL}{2} \times L \times \frac{1}{2} \times \frac{L}{3}\right]$$

$$= \frac{WL^3}{EI}\left[\frac{1}{24} + \frac{1}{12}\right] = \frac{WL^3}{8EI} \text{ deflection at free end.}$$

22. (a)

Strain energy,

$$U = \frac{1}{2} \times P \times \delta = \frac{1}{2} \times P \times \frac{PL^3}{3EI} = \frac{P^2L^2}{6EI}$$

23. (d)

$$I_1 = \frac{a^4}{12}$$

$$I_2 = \frac{(1.19a)^4}{12} = \frac{2a^4}{12}$$

$$I_2 = 2I_1$$

$$\delta = \frac{WL^3}{EI_1}$$

$$\delta_2 = \frac{WL^3}{EI_2}$$

$$\delta_2 = \frac{1}{2}\delta_1$$

24. (b)

Support reaction, R

$$\frac{RL^3}{3EI} = \frac{WL^4}{8EI}$$

$$R = \frac{3}{8}wL$$

25. (a)

$$\delta_1 = \frac{5}{384} \times \frac{6 \times 4^4}{EI} = \frac{20}{EI}$$

$$\delta_2 = \frac{120 \times 2^3}{48EI} = \frac{20}{EI}$$

$$\delta_1 = \delta_2$$

26. (c)

Reaction, $R_Q = \frac{3M}{L} \uparrow$

$$R_P = \frac{3M}{2} \downarrow$$

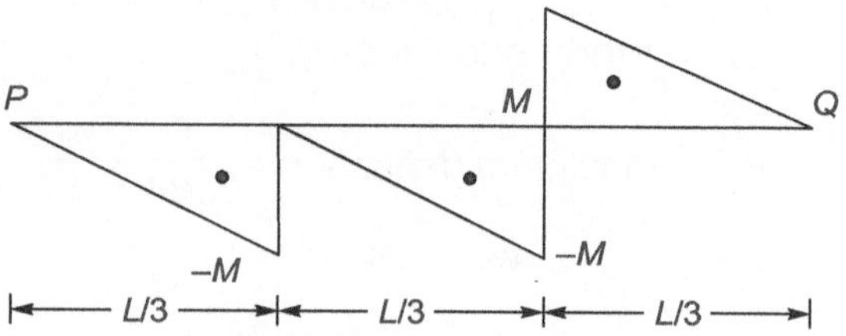

BM at A = 0

$$M \text{ at } \frac{L}{3} = -M, 0$$

$$M \text{ at } \frac{2L}{3} = -M, + M$$

$$M_c = 0$$

Reaction at P

$$LR' = \frac{ML}{6} \times \frac{2L}{9} - \frac{ML}{6} \times \left(\frac{L}{3} + \frac{L}{9}\right) - \frac{ML}{6}\left(\frac{2L}{3} + \frac{L}{9}\right)$$

$$= \frac{ML^2}{27} - \frac{2ML^2}{27} - \frac{7ML^2}{54} = \frac{ML^2}{27} - \frac{11ML^2}{54}$$

$$= -\frac{9ML^2}{54} = -\frac{ML^2}{6}$$

$$R' = -\frac{ML}{6}$$

$$\text{Slope at } P = \frac{R'}{EI} = \frac{ML}{6EI}$$

■■■■

Torsion

CHAPTER 12

There are shafts of various cross sections as circular, rectangular, triangular, elliptical etc. While circular shafts are used for power transmission, the non circular shafts are used for static torque applications as in aircraft structures. In this chapter we will study about (*a*) uniform circular shafts (*b*) stepped circular shafts (*c*) distribution of shear stress in shafts due to the applied twisting moment (*d*) strain energy absorbed in shafts (*e*) stresses developed in a section of the shaft (*f*) stresses developed in keys connecting the shaft with hub of a power transmitting element.

For determination of stresses, strains in a shaft, the shaft is considered under dynamic equilibrium. A portion of the shaft is taken, twisting moment *T* is applied at one end and other end of shaft is considered as fixed and a resisting twisting moment is considered at the fixed end. Various parameters as angular twist, shear strain, etc. are developed for this static shaft.

Question 12.1 **Consider a shaft of axial length *L*, radius *R* fixed at one end and subjected to a twisting moment *T* at the other end. Derive expressions for shear stress, shear strain, angular twist in shaft, if *G* is shear modulus of the material of the shaft. Write down the assumptions you take while deriving the torsion formula.**

Solution: For developing a relationship between *T, L, R*, angular twist θ, shear stress τ etc., following assumptions are taken.

(a) Material of the shaft is homogeneous and isotropic.

(b) Shaft is initially straight and not distorted.

(c) Displacement at a point in the shaft is proportional to its distance from center of the shaft.

(d) Transverse cross section of the shaft which are plane before the application of a twisting moment, remain plane after the shaft is twisted by torque.

(e) Angular twist is uniformly distributed along the length of the shaft.

(f) Shear stress developed in shaft is within the proportional limit i.e. material of the shaft follows Hooke's law.

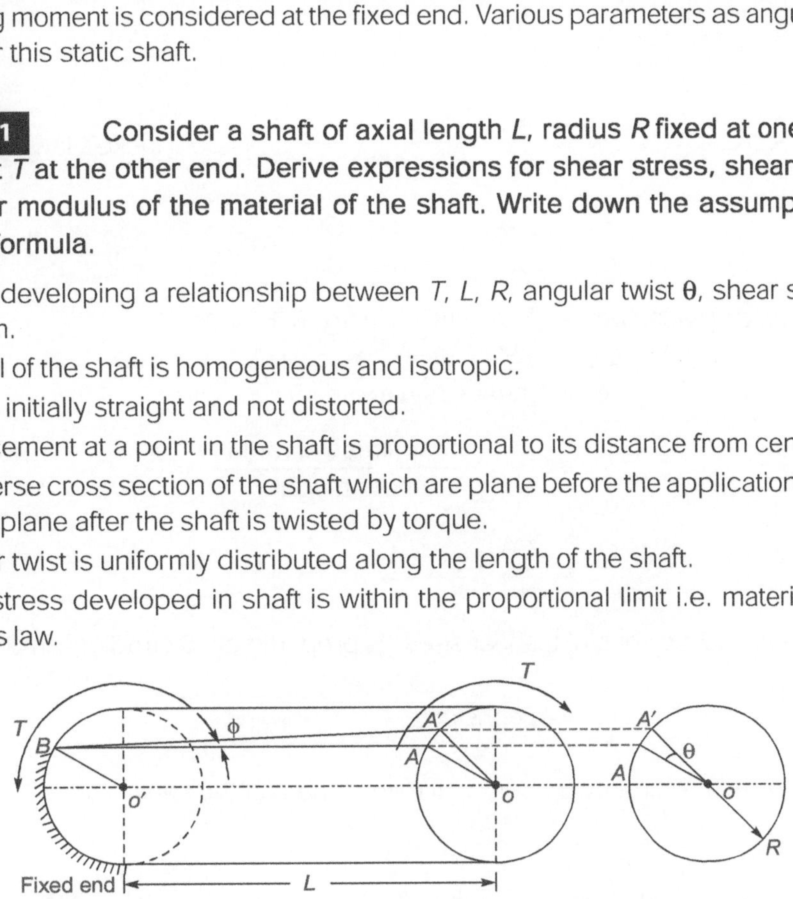

Fig. 12.1 (a)

Figure 12.1 (a) shows a shaft of radius *R*, length *L*, fixed at end *B* and subjected to a twisting moment *T* at end *A*. A line *BA* drawn parallel to axis *OO′* of the shaft is twisted to the position *BA′*, where $\angle ABA' = \phi$, shear angle. $\angle AOA' = \theta$, angular twist. Generally the, power transmitting shafts are much larger and not so smaller in length as shown in **fig. 8.1 (a)** due to the paucity of the space.

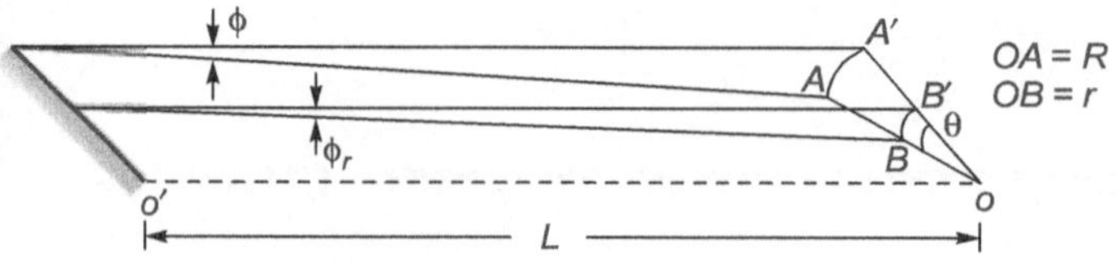

Fig. 12.1 (b)

Shear angle ϕ is extremely small, therefore ϕ in radian is also taken as shear strain.

Fig 12.1 (b) at radius *R*, $AA' = R\theta$

shear strain, $$\phi = \frac{AA'}{L} = \frac{R\theta}{L}$$

or $$\phi L = R\theta \quad \text{... (i)}$$

Maximum shear strain occurs at outer radius *R*.

$$\phi = \frac{R\theta}{L}$$

But shear strain, $$\phi = \frac{\tau}{G} = \frac{\text{Shear stress}}{\text{Shear modulus}} \text{ (as per Hooke's law)}$$

so, $$\frac{\tau}{G} = \frac{R\theta}{L} \text{ or } \frac{\tau}{R} = \frac{G\theta}{L} \quad \text{... (ii)}$$

At any radius, *r*, displacement is *BB′* as shown in **fig. 8.1 (b)**

$$L\phi_r = BB', = r\theta$$

$$\phi_r = \text{shear strain at radius } r$$

$$= \frac{\tau_r}{G} = \frac{\text{shear stress at radius } r}{\text{shear modulus}} = \frac{r\theta}{L}$$

or $$\frac{\tau_r}{r} = \frac{G\theta}{L} = \frac{\tau}{R} \quad \text{... (iii)}$$

Moreover *G* is an elastic constant, shear stress is proportional to *r* (radius) as shown by equation (iii)

$$\frac{\theta}{L} = \text{a constant}$$

angular twist in shaft is directly proportional to the length of the shaft.

Moment of Resistance

Consider the shaft section of radius *R*. Take an elementary ring of radius *r* and radial thickness *dr*,

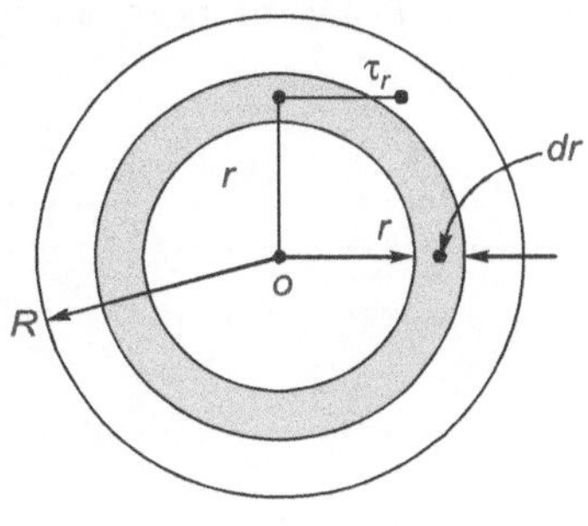

Fig. 12.2 *Elementary area in shaft section*

Shear stress, $\tau_r = \frac{\tau}{R} \times r$, (at radius r) ... (iv)

From equation (iii)

Area of elementary ring = $2\pi r.dr$

Shear force in elementary area = $2\pi r.dr\,.\,\tau_r$

$$dF = 2\pi r^2 \frac{\tau}{R} dr \quad \text{... (v)}$$

Twisting moment resisted by elementary area

$$dT = rdF = \frac{2\pi r^3 \tau}{R} dr \quad \text{... (vi)}$$

Total twisting moment, $T = \int_0^R \frac{2\pi r^3 \tau}{R} dr = \frac{2\pi\tau}{R}\left|r^3\right|_0^R = \frac{2\pi\tau R^4}{4R}$

$$T = \frac{\tau}{R}\left(\frac{\pi}{2} \times R^4\right)$$

where $\frac{\pi}{2} \times R^4 = J$, polar moment of inertia of shaft section

or $\frac{T}{J} = \frac{\tau}{R}$, where τ is maximum shear stress at outer radius R

Combined Torsional formula can be written as

$$\frac{T}{J} = \frac{G\theta}{L} = \frac{\tau}{R} = \frac{\tau_r}{r}$$

where
T = applied torque on shaft
J = polar moment of inertia of shaft section
$= \frac{\pi D^4}{32} = \frac{\pi R^4}{2}$, for a solid shaft
$= \frac{\pi}{32}(D^4 - d^4)$, for a hollow shaft
L = Length of the shaft
G = Shear modulus of shaft matrial
τ = maximum shear stress at outer radius R
τ_r = shear stress at any radius r of solid shaft

Question 12.2 **A hollow circular steel shaft of outer radius 40 mm and inner radius 20 mm is subjected to a torque of 8 kNm. What are maximum and minimum shear stresses developed in shaft. What is angular twist in 1 meter length of shaft, if G = 80 kN/mm² for steel.**

Solution:

Torque, T = 8 kNm = 8 × 10⁶ Nmm

Inner radius, R_1 = 20 mm

Outer radius, R_2 = 40 mm

Polar moment of inertia of shaft section, J

$$= \frac{\pi}{2}(R_2^4 - R_1^4) = \frac{\pi}{2}(40^4 - 20^4) = 377 \times 10^4 \text{ mm}^4$$

$$\frac{T}{J} = \frac{\tau_2}{R_2} = \frac{\text{maximum shear stress at } R_2}{R_2}$$

$$\frac{8\times10^6}{377\times10^4} = \frac{\tau_2}{40}$$

Maximum shear stress, $\tau_2 = \dfrac{8\times10^6\times40}{377\times10^4} = 84.88 \text{ N/mm}^2$

Shear stress at radius, R_1, $\tau_1 = \dfrac{\tau_2}{R_2}\times R_1$

$$= 169.76\times\frac{20}{40}$$

$$= 42.44 \text{ N/mm}^2$$

$$\tau_{max} = 84.88 \text{ N/mm}^2$$

$$\tau_{min} = 42.44 \text{ N/mm}^2$$

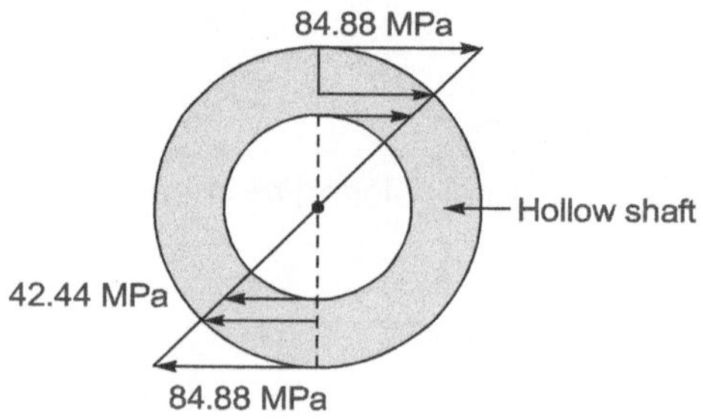

Fig. 12.3 *Shear stress distribution in hollow circular shaft*

Angular Twist, θ

$$T = 8 \text{ kNm} = 8\times10^6 \text{ Nmm}$$

$$G = 80{,}000 \text{ N/mm}^2$$

$$L = 1000 \text{ mm}$$

$$J = 377\times10^4 \text{ mm}^4$$

$$\theta, \text{ angular twist} = \frac{8\times10^6}{377\times10^4}\times\frac{1000}{80{,}000} = 0.0265 \text{ rad} = 1.52°$$

Practice Q.12.2 A circular bronze shaft of diameter 40 mm is subjected to a torque of 1.2 kNm. Determine the following:

(i) maximum shear stress developed in shaft

(ii) shear stress at radius of 15 mm

(iii) angular twist in 1 m length of shaft, $G = 41$ kN/mm^2

Ans. [(i) 95.5 N/mm^5, (ii) 71.6 N/mm^2 and (iii) 6.67°]

Question 12.3 What do you understand by horse power transmitted by a shaft? What is one metric horse power?

Solution: A shaft is transmitting torque T at N rpm.

Power transmitted by shaft in one minute

$$\text{Power} = 2\pi NT \text{ Nm if } T \text{ is in Nm.}$$

$$\text{Horse power transmitted by shaft} = \frac{2\pi NT}{746\times60}$$

If T is in Nm then 1 watt = 1 Nm/s

1 HP is equal to 746 watts.

or $$\text{HP transmitted} = \frac{2\pi NT}{44760}$$

if Torque is in kg-m

$$\text{Metric horse power} = \frac{2\pi NT}{75 \times 60}$$

as $$1 \text{ metric HP} = 75 \text{ kg-m/s}$$

Question 12.4 **A hollow circular shaft is transmitting power at 250 rpm. Outer diameter of the shaft is 75 mm and inner diameter of the shaft is 50 mm. If the maximum shear stress developed in shaft section is 60 N/mm^2, what is the torque transmitted by the shaft? Determine also the HP transmitted by the shaft.**

Solution :

Outer diameter, $D = 75$ mm

Inner diameter, $d = 50$ mm

RPM $= 250$

Polar moment of inertia, $$J = \frac{\pi}{32}(75^4 - 50^4) = 249.27 \times 10^4 \text{ mm}^4$$

Torque, $$T = J \times \frac{\tau}{D/2} = \frac{249.27 \times 10^4 \times 60}{37.5} = 398.8 \times 10^4 \text{ Nmm} = 3988 \text{ Nm}$$

$$\text{HP transmitted} = \frac{2\pi NT}{746 \times 60} = \frac{2\pi \times 250 \times 3988}{746 \times 60} = 140$$

Practice Q.12.4 **A steel shaft is transmitting 120 HP at 360 rpm. What is the diameter of the shaft if the maximum shear stress developed in shaft in 60 N/mm^2.**

Hint: [Take care of units if τ in N/mm^2, then take T in Nmm]

Ans. [58.6 mm]

Question 12.5 **What is modulus of rupture, and polar modulus, explain?**

Solution: While deriving the torsion formula $\frac{T}{J} = \frac{G\theta}{L} = \frac{\tau}{R}$, we have assumed that stresses are developed in shaft within the elastic limit of the material. **When a shaft is tested upto fracture**, certainly the stresses in the material of the shaft go beyond the elastic limit. Knowing the value of **ultimate breaking torque, T_{ult}, shear stress calculated in shaft using Torsion formula is known as modulus of rupture.**

$$\tau_{max} = \frac{2T_{ult}}{\pi R^3}, \text{ is the modulus of rupture.}$$

Polar modulus in torsion is equivalent to section modulus in torsion.

Torque T = polar modulus × maximum shear stress

M = section modulus (z) × maximum bending stress

$$T = \frac{\pi}{16} d^3 \times \tau$$

where $\frac{\pi}{16}d^3$ = polar modulus of solid shaft

Similarly by polar modulus of hollow shaft = $\frac{\pi}{16}\frac{(D^4 - d^4)}{D}$, where D and d are external and internal diameters of shaft.

Question 12.6 **Consider a uniformly tapered shaft, with diameter *d* at one end uniformly increasing to diameter *D* at the other end over an axial length of *L*. Derive expression for the angular twist developed in the shaft due to an axial torque *T* applied at one end.**

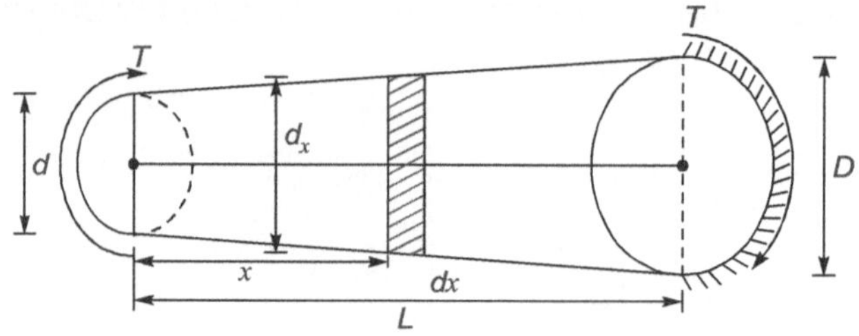

Fig. 12.4

Solution: Figure 12.4 shows a uniformly tapered shaft, with diameter d and D and axial length L subjected to torque T as shown. Consider an elementary length dx of shaft at a distance x from A

Diameter, $$d_x = d + \frac{D-d}{L}x = d + Kx$$

where $$K = \frac{D-d}{L}$$

Polar moment of inertia of element,

$$J_x = \frac{\pi d_x^4}{32} = \frac{\pi}{32}(d+Kx)^4$$

$d\theta$, Angular twist over length, dx

$$d\theta = \frac{T\,dx}{J_x\,G} = \frac{32T\,dx}{\pi(d+Kx)^4.G}$$

$$\theta = \frac{32T}{\pi G}\int_0^L \frac{dx}{(d+Kx)^4} = \frac{32T}{\pi G}\left[\frac{(d+Kx)^{-3}}{-3K}\right]_0^L$$

$$= -\frac{32T}{3KG\pi}\left[\frac{1}{D^3} - \frac{1}{d^3}\right] = +\frac{32T}{3KG\pi}\left[\frac{1}{d^3} - \frac{1}{D^3}\right]$$, putting the value of K

$$= \frac{32TL}{3\pi G(D-d)} \times \frac{(D-d)(D^2 + Dd + d^2)}{D^3 d^3}$$

$$\theta = \frac{32TL}{3\pi G} \times \frac{(D^2 + Dd + d^2)}{D^3 d^3}$$

Practice Q.12.6 A tapered circular shaft 0.8 m long having diameter of 40 mm at one end uniformly increasing to diameter 60 mm at the other end. Torque applied on the shaft produces an angular twist of 1°. What is magnitude of torque, T if G = 82000 N/mm^2.

Ans. [0.958 kNm]

Question 12.7 What do you understand by shafts in series? Take a circular shaft in 3 steps and explain angular twist developed in shafts in series under the action of a torque T.

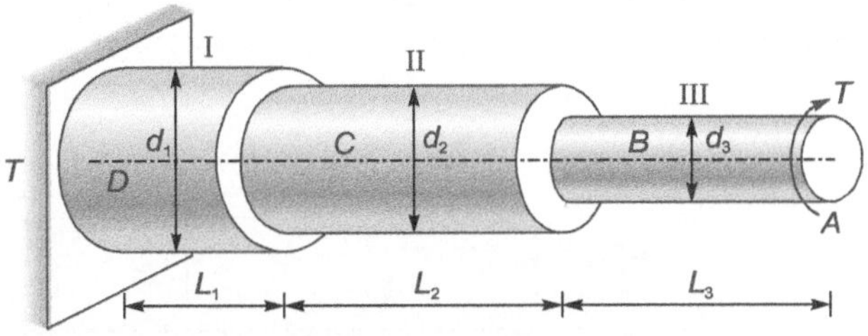

Fig. 12.5 *Shafts in Series*

Solution: Figure 12.5 shows a stepped shaft in series. Same torque T is acting on all the 3 steps. Since the diameters are different, shear stresses developed in the shafts will be different as

Shear stresses, $$\tau_1 = \frac{16T}{\pi d_1^3};\quad \tau_2 = \frac{16T}{\pi d_2^3};\quad \tau_3 = \frac{16T}{\pi d_3^3}$$

Obviously the maximum shear stress will be developed in shaft of minimum diameter, i.e. portion III.

Polar moment of inertia of each step is different

$$J_1 = \frac{\pi d_1^4}{32};\quad J_2 = \frac{\pi d_2^4}{32};\quad J_3 = \frac{\pi d_3^4}{32}$$

Angular twist, $$\theta_{AD} = \theta_{AB} + \theta_{BC} + \theta_{CD}$$

$$= \frac{T}{G}\left(\frac{L_1}{J_1} + \frac{L_2}{J_2} + \frac{L_3}{J_3}\right)$$

Practice Q.12.7 A stepped circular steel shaft is shown in figure 12.6 with diameters 80 mm, 40 mm and lengths 500 mm and 250 mm respectively. If torque applied at A is 1.2 kNm, what is maximum shear stress developed in stepped shaft. If G = 80 GPa, what is angular twist θ_{AC}.

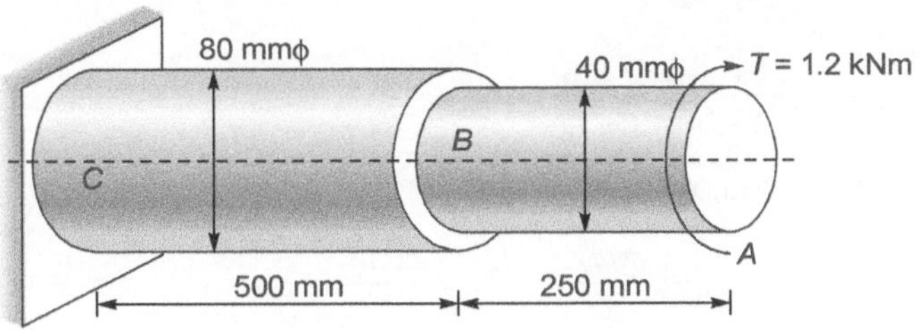

Fig. 12.6 *Shafts in Series*

Ans. [95.5 N/mm^2, 0.96°]

Question 12.8 **What is a compound shaft, explain? Derive relationship between torques shared by two portions of a compound shaft.**

Solution: In a compound shaft, there are two distinct portions of a shaft, subjected to different values of twisting moments as T_A and T_B, if there are two portions A and B, but angular twist in two portions of the shaft is of same value. There are two types of compound shaft

(a) A shaft with two portions A and B, held at ends and torque applied at junction of the two as shown in fig. 12.7(a)

(b) One shaft of one material is enclosed completely in another shaft of another material, torque T is shared by two shafts i.e. $T = T_A + T_B$, as shown in fig 12.7 (b)

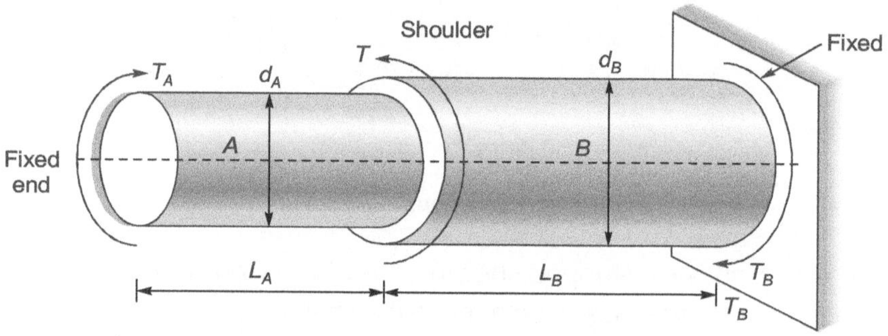

(a) Shaft A is parallel with shaft B

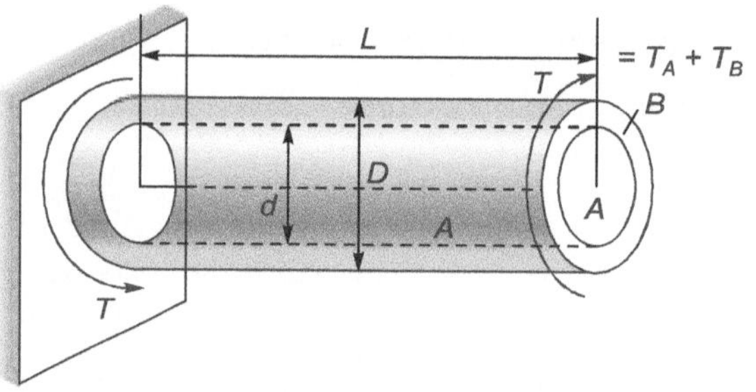

(b) Solid shaft of material A enclosed in hollow shaft of material B

Fig. 12.7

Fig. 12.7 (a) Torque T is applied at the shoulder

$$T = T_A + T_B \qquad \text{... (i)}$$

$$\theta_A = \theta_B$$

$$\frac{T_A L_A}{G_A J_A} = \frac{T_B L_B}{G_B J_B}$$

where, G_A, G_B shear modulus of shafts A and B respectively

$$J_A = \frac{\pi d_A^4}{32}; \quad J_B = \frac{\pi d_B^4}{32}$$

$$\frac{T_A}{T_B} = \frac{G_A}{G_B} \times \frac{L_B}{L_A} \times \frac{J_A}{J_B} \qquad \text{... (ii)}$$

From equations, (i) and (ii), value of T_A and T_B can be determined if we know T, d_A, d_B, L_A, L_B etc.

Case Fig 12.7 (b), $L_A = L_B$

So $$\frac{T_A}{T_B} = \frac{G_A}{G_B} \times \frac{J_A}{J_B} \quad \text{... (iii)}$$

Depending upon T_A, T_B shear stresses in two portions of the shaft can be calculated.

$$J_A = \frac{\pi d^4}{32}; \quad J_B = \frac{\pi(D^4 - d^4)}{32}$$

Question 12.9 A solid steel shaft of diameter 40 mm, length 500 mm is welded to a hollow copper shaft of length 500 mm and diameters 60 mm and 30 mm respectively as shown in fig 12.8 Torque T = 2.4 kNm is applied at shoulder as shown while the ends of the composite shaft are fixed If $G_S = 2G_{cu}$ = 84 kN/mm², determine (i) shear stresses developed in each portion, (ii) angular twist at shoulder.

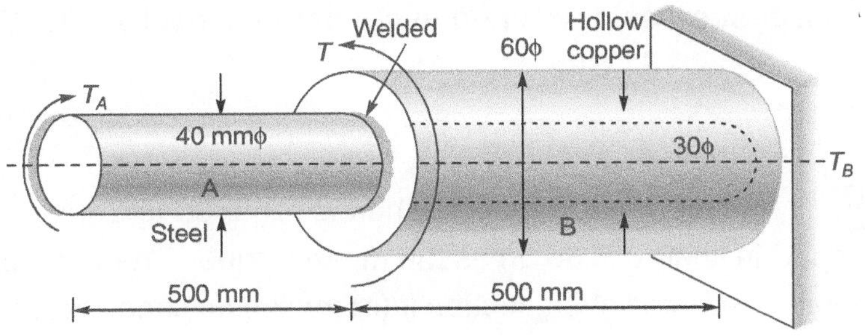

Fig. 12.8

Solution :

$$L_A = 500 \text{ mm}$$
$$L_B = 500 \text{ mm}$$
$$G_A = 2G_B$$
$$J_A = \frac{\pi \times 40^4}{32} = 25.13 \times 10^4 \text{ mm}^4$$
$$J_B = \frac{\pi}{32}(60^4 - 30^4) = 119.28 \times 10^4 \text{ mm}^4$$
$$\frac{T_A}{T_B} = \frac{G_A}{G_B} \times \frac{L_A}{L_B} \times \frac{J_A}{J_B} = 2 \times \frac{500}{500} \times \frac{25.13 \times 10^4}{119.28 \times 10^4}$$
$$\frac{T_A}{T_B} = 0.42136$$
$$T_A = 0.42136\, T_B$$

or $0.42136\, T_B + T_B = 2.4 \text{ kNm} = T$

$T_B = 1.688$ kNm, $T_A = 0.712$ kNm

(i) Shear Stress

$$\tau_A = \frac{16 T_A}{\pi d_A^3} = \frac{16 \times 0.712 \times 10^6}{\pi \times 40^3} = 56.66 \text{ N/mm}^2$$

$$J_B = \frac{\pi}{32}(60^4 - 30^4) = 119.28 \times 10^4 \text{ mm}^4$$

$$\tau_B = \frac{T_B \times 30}{J_B} = \frac{1.688 \times 10^6 \times 30}{119.28 \times 10^4} = 42.45 \text{ N/mm}^2$$

(ii) Angular Twist

$$\theta_A = \theta_B = \frac{T_A L_A}{J_A G_A} = \frac{0.42136 \times 500 \times 10^6}{25.13 \times 10^4 \times 84000}$$

$$= 9.98 \times 10^{-3} \text{ rad} = 0.572°$$

Practice Q.12.9 A solid circular copper shaft of diameter 40 mm is completely encased in a hollow steel shaft of outer diameter 60 mm. Length of compound shaft is 1 m. Torque applied on the shaft is 2.5 kNm. Determine shear stresses developed in copper and steel shafts. What is the angular twist produced in the shaft. $G_s = 2G_{CU} = 84$ GPa

Ans. [65.4, 21.80 N/mm², 1.49°]

Question 12.10 A hollow circular steel shaft is required to transmit 240 HP at 200 rpm The maximum torque developed in shaft is 1.30 times the mean torque. External diameter of the shaft is 1.6 times the internal diameter. Determine shaft section if (a) maximum shear stress in shaft is not to exceed 75 MPa (b) if angular twist in a length of 1 m is not to exceed 1°. Shear modulus = 82 kN/mm².

Solution : Say internal diameter of shaft = d

External diameter = 1.6 d

$$HP = 240$$

$$RPM = 200$$

Mean Torque,
$$T_{mean} = \frac{HP \times 746 \times 60}{2\pi \times 200} = \frac{240 \times 746 \times 60}{2\pi \times 200} = 8548.53 \text{ Nm}$$

Maximum Torque,
$$T_{max} = 1.3\, T_{mean} = 1.3 \times 8548.53$$
$$= 11113.09 \text{ Nm} = 11.11 \times 10^6 \text{ Nmm}$$

Polar moment of inertia,
$$J = \frac{\pi}{32}[(1.6d)^4 - d^4] = 0.545 \times d^4$$

(i)
$$\tau_{max} = 75 \text{ MPa}$$

so
$$\frac{75}{0.8d} = \frac{11.11 \times 10^6}{0.545 d^4}$$

$$d^3 = 217443.425 \text{ mm}^3$$

$$d = 60.16 \text{ mm}$$

(ii)
$$\theta = 1°$$

in
$$L = 1 \text{ m}$$

$$\theta = \frac{\pi}{180} \text{ rad}$$

$$\frac{11.11\times10^6}{0.545d^4} = \frac{82000}{1000}\times\frac{\pi}{180} = \frac{82\pi}{180}$$

$$d^4 = 14243815.4 \text{ mm}^4$$

$$d = 61.43 \text{ mm}$$

safe internal diameter = 61.43 mm

external diameter = 1.6 × 61.43 = 98.29 mm

Question 12.11 A solid shaft is to transmit 300 kW at 120 rpm. If the shear stress is not to exceed 100 MPa, find the diameter of the shaft. What percent saving in material would be obtained if this shaft were replaced by a hollow one whose internal diameter equals 0.6 of the external diameter, the length, material and maximum allowable shear stress being the same?

Solution :

Power, $P = 300 \text{ kW} = 300 \times 10^3 \text{ Nm/s}$

$N = 120 \text{ NM}$

angular speed, $\omega = \frac{2\pi\times120}{60} = 12.566 \text{ rad/sec}$

maximum stress, $\tau = 100 \text{ MPa}$

$$\text{Torque} = \frac{P}{\omega} = \frac{300\times10^3}{12.566} = 23874 \text{ Nm} = 23.674\times10^6 \text{ Nmm}$$

$$= \frac{\pi}{16}d_s^3\times\tau = \frac{\pi}{16}\times d^3\times100$$

$$d_s^3 = \frac{23.874\times10^6\times16}{\pi\times100} = 1.21589\times10^6$$

Shaft diameter, $d_s = 106.733 \text{ mm}$

$$\text{Hollow shaft polar modulus} = \frac{\pi}{16}\frac{(D^4-d^4)}{D} = \pi\times\frac{d_s^3}{16}$$

$$\frac{D^4-d^4}{D} = \frac{d_s^3}{1}$$

$$d = 0.6\,D$$

so $0.8704\,D^3 = d_s^3$

$$0.9548\,D = d_s = 106.733 \text{ m}$$

$$D = 111.8 \text{ mm}$$

$$d = 67.07 \text{ mm}$$

$$\%\text{ saving in materials} = \frac{A_s-A_h}{A_s}\times100 = \frac{d_s^2-(D^2-d^2)}{d_s^2}\times100$$

$$= \frac{106.733^2-(111.8^2-67.07^2)}{106.733^2}\times100$$

$$= \frac{11391.9-(12499.2-4498.4)}{11391.9} \times 100$$

$$= \frac{3391.4}{13391.9} \times 100 = 29.77\%$$

Question 12.12 A composite shaft *ABC*, consists of a steel shaft, 60 mm diameter and 300 mm length with a brass shaft 100 mm diameter and 400 mm length. Composite shaft is fixed at end *C*. Twisting moments $T_A = 60$ Nm (ccw) and $T_B = 2400$ Nm (ccw) are applied as shown. Determine (a) shear stresses in both steel and brass shafts (b) angular twist in steel and brass shafts.

$G_{steel} = 2.\ G_{brass} = 84$ GPa

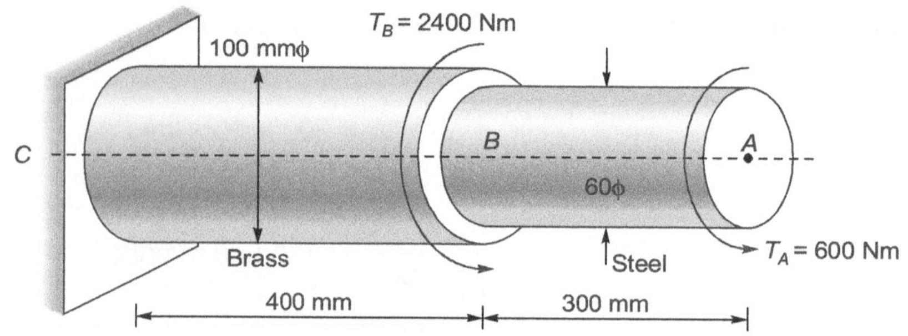

Fig. 12.9

Solution :

$$T_A = 600 \text{ Nm (ccw)}$$
$$T_B' = \text{reaction} = 600 \text{ Nm (cw)}$$
$$T_C' = 600 + 2400 = 3000 \text{ Nm (cw) for equilibrium}$$
$$T_B = 3000 \text{ Nm (ccw) at shoulder } B$$

Twisting moments acting on two portions are as shown in figure 12.10

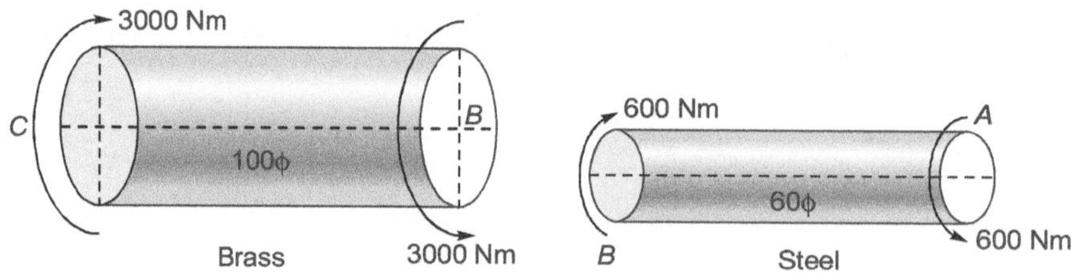

Fig. 12.10

Shear Stresses

Steel shaft,

$$d = 60 \text{ mm}$$
$$T = 600 \text{ Nm} = 6 \times 10^5 \text{ Nmm}$$
$$J = \frac{\pi}{32} \times (60)^4 \text{ mm}^4$$

shear stress,

$$\tau_s = \frac{16T}{\pi d^3} = \frac{16 \times 6 \times 10^5}{\pi \times 60^3} = 14.15 \text{ N/mm}^2$$

Angular twist, $\theta = \dfrac{TL}{GJ} = \dfrac{6\times10^5\times300\times32}{\pi\times60^4\times84000} = 1.684\times10^{-3}$ rad = 0.096°

Brass Shaft

d = 100 mm

T = 3000 Nm = 3×10^6 Nmm

$J = \dfrac{\pi\times100^4}{32}$ mm^4

Shear stress, $\tau_B = \dfrac{16T}{\pi d^3} = \dfrac{16\times3\times10^6}{\pi\times100^3}$ = 15.28 N/mm^2

Angular trust, $\theta = \dfrac{3\times10^6\times400\times32}{42000\times100^4\times\pi} = 2.91\times10^{-3}$ rad = 0.166°

Question 12.13 A stepped shaft *ABC*, 1 m long made of steel is of diameter 20 mm from *A* to *B* = 0.5 m, 40 mm from *B* to *C* = 0.5 m. It is subjected to *a* twisting moment of 0.4 kNm (cw) at *A*, 0.8 kNm (ccw) at *B* and 0.4 kNm (cw) at *C*.

If G = 84 N/mm^2, determine angular twist between *A* and *C*.

[10 marks, IES : 2012]

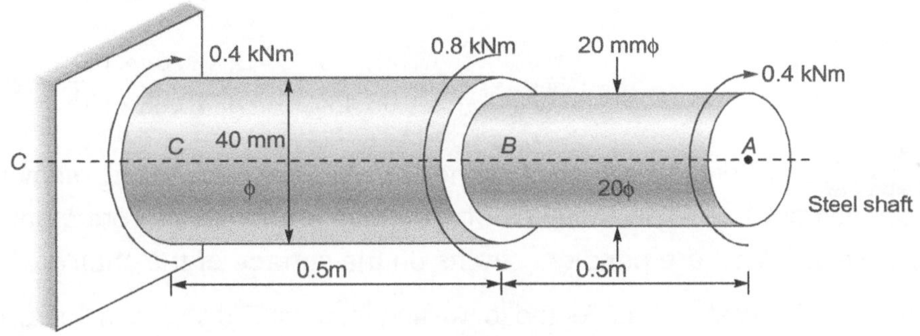

Fig. 12.11

Solution :

Torque in AB = 0.4 kNm = 0.4×10^6 Nmm

L = 500 mm

$J = \dfrac{\pi}{32}(20)^4 = 1.5708\times10^4$ mm^4

$\theta_{AB} = \dfrac{0.4\times10^6\times500}{84000\times1.5708\times10^4}$ = 0.15158 rad = + 8.685°

Torque in BC = – 0.4 kNm = -0.4×10^6 Nmm

L = 500 mm

$J = \dfrac{\pi}{32}(40)^4 = 25.132\times10^4$ mm^4

$\theta_{BC} = -\dfrac{0.4\times10^6\times500}{84000\times25.132\times10^4}$ = – 0.9511 rad = – 0.545°

Total angular twist θ_{AC} = 8.685° – 0.545° = 8.14°

Question 12.14 A hollow steel shaft 60 mm diameter internal and 100 mm external diameter is to be replaced by a solid alloy shaft. If the polar modulus has the same value, calculate the diameter of the latter and the ratio of their torsional rigidities. Shear modulus of G to steel = 2 times × G for the alloy.

[10 marks, IES : 2013]

Solution: How shaft

$$\text{Polar modulus} = \frac{\pi}{16}\frac{(D^4 - d^4)}{D} \qquad (D = 100 \text{ mm}, d = 80 \text{ mm})$$

$$\text{Hollow shaft} = \frac{\pi}{16} \times d_s^3$$

$$\frac{\pi}{16} \times \frac{(100^4 - 60^4)}{100} = \frac{\pi}{16} \times d_s^3$$

$$d_s^3 = 8704 \times 10^2$$

$$d_s = 95.43 \text{ mm}$$

Torsional rigidity

$$\frac{J_H G_H}{J_S G_S} = 2 \times \frac{J_H}{J_S}$$

$$J_H = \frac{\pi}{32}(D^4 - d^4); \quad J_S = \frac{\pi}{32} \times d_s^4$$

$$\text{Ratio} = 2 \times \frac{(100^4 - 60^4)}{95.43^4} = \frac{2 \times 8704 \times 10^4}{8310.93 \times 10^4} = 2.099$$

Question 12.15 Consider a shaft of diameter d, subjected to twisting moment T. On the surface of the shaft take an element and show stresses on that element. What are principal stresses and their directions on this element. What are principal strains on the surface of the shaft?

Solution: A shaft of diameter d, subjected to twisting moment T is shown in figure 12.12 (a). Maximum shear stress τ is developed on the surface of the shaft.

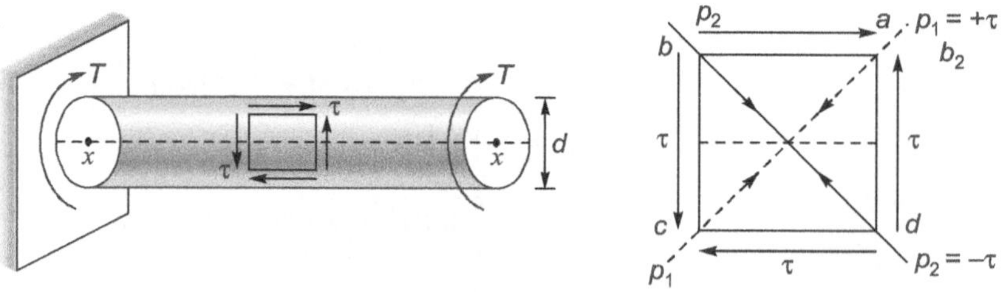

Fig. 12.12

$$\text{Maximum shear stress,} \qquad \tau = \frac{16T}{\pi d^3}$$

Taking a small element $abcd$, applied shear stress r acts on ad, complementary shear stress equal and opposite acts on surface ab. Resisting shear stress τ acts on bc and complementary to this shear stress, τ acts on cd. Mohr's stress circle can be drawn for this state of stress to find principal stresses, which are

tensile principal stress, $p_1 = +\tau$ at 45° to x-axis

compressive principal stress, $p_2 = -\tau$ at 45° to x-axis

Say E, ν are Young's modulus and Poisson's ratio for the material

Principal strains,

$$e_1 = \frac{p_1}{E} + \frac{\nu p_2}{E} = \frac{\tau}{E}(1+\nu)$$

$$e_2 = -\frac{p_2}{E} - \frac{\nu p_1}{E} = -\frac{\tau}{E}(1+\nu)$$

$$e_3 = -\frac{\nu\tau}{E} + \frac{\nu\tau}{E} = 0$$

Practice Q.12.15 A copper shaft of diameter 40 mm is subjected to a twisting moment of 1 kNm. What are principal stresses and principal strains, if E = 105000 N/mm^2, υ = 0.35

Ans. [±79.6 N/mm^2, at ±45°, ±1023 μs, where 1 $\mu s = 10^{-6}$]

Question 12.16 What are equivalent bending moment and equivalent twisting moment in a power transmitting shaft?

Solution: A shaft is used to transmit power from motor pulley to machine pulley, from one gear to another gear. Due to belt tensions on pulleys, tooth loads on gears, reaction from bearings (to support the shaft) bending moments are generated on shaft. Fig. 12.13 (a) shows a shaft subjected to bending moment M and twisting moment T simultaneously. Maximum stress due to bending, σ and maximum stress due to twisting, τ are produced on the surface of the shaft. Fig 12.13 (b) shows an element of shaft with stresses σ and τ.

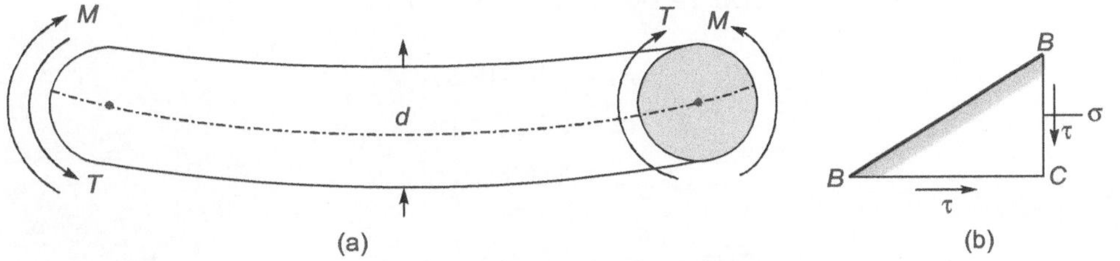

Fig. 12.13 *T and M on a shaft*

If shaft diameter is d,

$$\tau = \frac{16T}{\pi d^3}; \quad \sigma = \frac{32M}{\pi d^3}$$

Principal stress, p_1 = maximum principal stress

$$p_1 = \frac{\sigma}{2} + \sqrt{\left(\frac{\sigma}{2}\right)^2 + \tau^2} = \frac{16M}{\pi d^3} + \sqrt{\left(\frac{16M}{\pi d^3}\right)^2 + \left(\frac{16T}{\pi d^3}\right)^2}$$

$$= \frac{16M}{\pi d^3}\left[M + \sqrt{M^2+T^2}\right] = \frac{32}{\pi d^3} \times \left[\frac{M+\sqrt{M^2+T^2}}{2}\right]$$

or

$$p_1 \times \frac{\pi d^3}{32} = \frac{M+\sqrt{M^2+T^2}}{2}$$

= M_e, equivalent bending moment, corresponding to maximum principal stress

Maximum most shear stress, $\tau_{max} = \sqrt{\left(\frac{\sigma}{2}\right)^2 + \tau^2} = \sqrt{\left(\frac{16M}{\pi d^3}\right)^2 + \left(\frac{16T}{\pi d^3}\right)^2} = \frac{16}{\pi d^3}\sqrt{T^2 + M^2}$

or $\frac{\tau_{max} . \pi d^3}{16} = \sqrt{T^2 + M^2} = T_e$, equivalent twisting moment

or T_e, is the equivalent twisting moment produced in shaft corresponding to maximum most shear stress on the surface of the shaft.

Question 12.17 **A solid circular shaft of diameter d is transmitting power and subjected to $T = 25$ kNm and $M = 10$ kNm. What is the minimum diameter of the shaft if maximum principal stress in shaft is not to exceed 200 N/mm² and maximum most shear stress is not to exceed 150 N/mm².**

Solution :

$$M = 10 \text{ kNm}$$

$$T = 25 \text{ kNm}$$

$$M_e = \frac{M + \sqrt{M^2 + T^2}}{2} = \frac{10 + \sqrt{625 + 100}}{2} = 18.463 \text{ kNm}$$

$$T_e = 26.926 \text{ kNm}$$

$$p_1 = 200 \text{ N/mm}^2$$

$$200 \times \frac{\pi d^3}{32} = 18.463 \times 10^6$$

$$d^3 = 940312.87$$

Shaft diameter, $d = 97.97$ mm

Maximum $\tau_{max} = 150$ N/mm²

$$150 \times \frac{\pi d^3}{16} = 26.926 \times 10^6$$

$$d^3 = 914219.94$$

Shaft diaemter, $d = 97.05$ mm

Minimum shaft diameter is 97.97 mm, for the permissible stresses.

Practice Q.12.17 **A solid shaft of diameter 120 mm is transmitting 500 kW at 250 rpm. It is also subjected to a bending moment of 8 kNm. Determine (1) M_e, T_e, (2) maximum principal stress, (3) maximum most shear stress**

Ans. [20.7, 14.35 kNm; 84.6 N/mm², 61.0 N/mm²]

Question 12.18 **Consider a hollow circular shaft of length L, inner radius R_1 outer radius R_2 subjected to maximum shear stress τ on its surface. If G is shear modulus of shaft, derive expression for strain energy absorbed by shaft at stress τ, within the elastic limit.**

Solution: Fig 12.14 shows a section of a hollow circular shaft, subjected to shear stress τ at outer radius R_2. Consider an elementary ring of radius r, radial thickness dr shear stress at radius r,

$$\tau_r = \frac{\tau}{R_2} \times r$$

shear strain energy per unit volume $= \frac{(\tau_r)^2}{2G} = \frac{\tau^2 r^2}{2GR_2^2}$

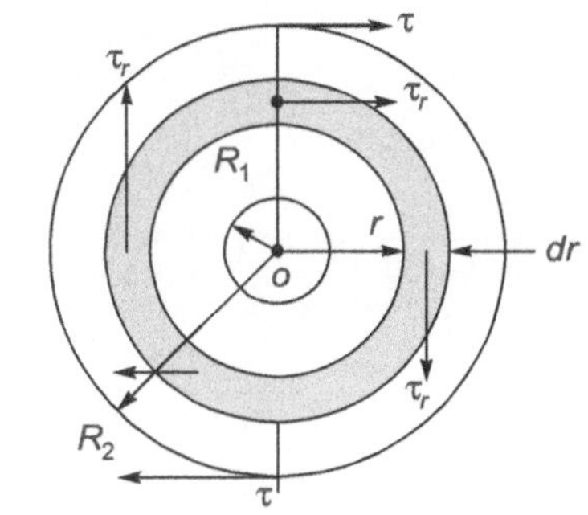

Fig. 12.14

shear strain energy for the elementary volume of $(2\pi r dr) L$

$$\delta U = \frac{\tau^2 r^2}{2GR_2^2}(2\pi r dr)L$$

$$= \frac{\pi\tau^2 L}{GR_2^2} r^3 dr$$

Total strain energy for shaft, $U = \int_{R_1}^{R_2} \frac{\pi\tau^2 L}{GR_2^2} \times r^3 dr = \frac{\pi\tau^2 L}{GR_2^2}\left[\frac{R_2^4}{4} - \frac{R_1^4}{4}\right]$

$$= \frac{\pi\tau^2 L}{4GR_2^2}\left[R_2^2 - R_1^2\right]\left[R_2^2 + R_1^2\right]$$

But volume of shaft, $V = \pi L(R_2^2 - R_1^2)$

So $U = \frac{\tau^2 (R_2^2 + R_1^2)V}{4GR_2^2} = \frac{\tau^2}{4G}\left(\frac{R_2^2 + R_1^2}{R_2^2}\right)V$, for hollow shaft

For a solid shaft, $R_1 = 0$

$R_2 = R$

$$U = \frac{\tau^2}{4G} \times \text{volume}$$

Question 12.19 **A hollow circular shaft of radius R_2 = 60 mm, R_1 = 30 mm, is transmitting power 600 HP at 200 rpm. If G = 80000 N/mm², what is the shear strain energy absorbed by hollow shaft. Verify that this is equal to (T/2).θ, where θ is angular twist. Length of shaft is 1 m.**

Solution :

$HP = 6$

RPM, $N = 200$

Torque, $T = \frac{HP \times 746 \times 60}{2 \times \pi \times 200} = \frac{600 \times 746 \times 60}{2 \times \pi \times 200}$

$= 21371$ Nm $= 213.71 \times 10^5$ Nmm

J, polar moment of inertia $= \frac{\pi}{2}(60^4 - 30^4) = 1908.5 \times 10^4$ mm⁴

τ, shear stress at $R_2 = \frac{213.71 \times 10^5 \times 60}{1908.51 \times 10^4} = 67.2$ N/mm²

Volume of shaft, $V = \pi(R_2^2 - R_1^2) \times L = \pi(60^2 - 30^2) \times 1000 = 84.82 \times 10^5$ mm³

$$\tau^2 = (67.2)^2 = 4515$$

Ratio, $$\frac{R_2^2 + R_1^2}{R_2^2} = \frac{60^2 + 30^2}{60^2} = 1.25$$

Now torque $= 21371$ Nm

$$U = \frac{\tau^2}{4G} \times \frac{R_2^2 + R_1^2}{R_2^2} \times V = \frac{4515}{4 \times 80,000} \times 1.25 \times 84.82 \times 10^5$$

$$= 149.6 \times 10^3 \text{ Nmm} = 149.6 \text{ Nm}$$

Now torque, $$T = 21371 \text{ Nm}$$

$$\theta = \frac{TL}{GJ} = \frac{213.71 \times 10^5 \times 1000}{80,000 \times 1908.5 \times 10^4} = 0.014 \text{ rad}$$

Strain energy, $$\frac{1}{2} T\theta = 21371 \times \frac{1}{2} \times 0.014 = 149.6 \text{ Nm}$$ (Verified)

Practice Q.12.19 A solid circular steel shaft of diameter 60 mm and length 1 m is subjected to a twisting moment of 3 kNm determine (a) strain energy absorbed in shaft (b) angular twist in shaft, $G = 78.4$ kN/mm^2.

Ans. [45 Nm, 0.030 rad.]

Question 12.20 Consider a shaft of radius R, transmitting torque T connected to the hub of a pulley with the help of a rectangular key, breadth b, thickness t and length L. Determine shear and bearing stresses developed in key.

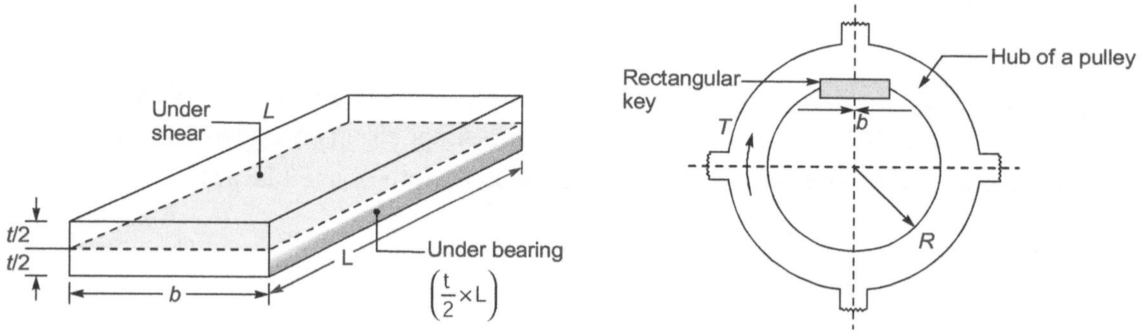

Fig. 12.15

Solution :

Torque on shaft $= T$

Radius of the shaft $= R$

Tangential force $F = \dfrac{T}{R}$ on the shaft

This shear force acts on shaded horizontal surface of key, of length L

$$\tau\text{, shear stress developed in key} = \frac{T}{R} \times \frac{1}{b.L} = \frac{T}{Rbl}$$

Moreover this force F produces bearing pressure on key (shown) by hatched area $\frac{t}{2} \times L$

$$\sigma_b\text{, Bearing stress, developed in key} = \frac{F}{0.5tL} = \frac{2T}{RtL}$$

Note: *That half of the key is embedded in shaft and half the key is fitted in key way provided in the hub. Keys are made of soft material such as wrought iron.*

Question 12.21 **A shaft of radius 30 mm is transmitting 480 Nm torque. A key of breadth 15 mm and thickness 10 mm, connects the shaft with hub. What is length of the key if shear stress in key is not to exceed 15 N/mm². What is the bearing stress developed in key.**

Solution:

Torque = 480 Nm

Radius, R = 30 mm = 0.030 m

Tangential force in shaft, $F = \frac{480}{0.030} = 16000$ N

breadth of the key, b = 15 mm

τ, shear stress in key = 15 N/mm²

Length of the key, $L = \frac{F}{\tau.b} = \frac{16000}{15 \times 15} = 71$ mm

σ_b, bearing stress $= \frac{2F}{t.L} = \frac{2 \times 16000}{10 \times 71} = 45.07$ N/mm²

Question 12.22 **A solid marine propeller shaft is transmitting power at 1800 rpm. The vessel is being propelled at 20 km/hour for the expenditure of 4500 HP. If the efficiency of propeller is 75% , and greatest thrust in shaft is not to exceed 65 MPa, determine shaft diameter and maximum shear stress developed in shaft.**

Solution:

Say thrust = P Newton

rpm = 800

HP = 4500

Speed = 20 km/hour

Useful work done per second $= \frac{P \times 20{,}000}{3600} = 5.555$ P

Efficiency, η = 75%

Input work $= \frac{5.555P}{0.75} = 7.407\ P$

Work input per second = 4500 × 746 = 335.7 × 10⁴ Nm = 7.407 P

So Thrust, P = 453220

$$N = 45.322 \times 10^4 \text{ N}$$

Allowable direct stress, $\sigma = 65$ MPa

Area of cross section of shaft $= \dfrac{45.322\times10^4}{65} = 0.697 \times 10^4 \text{ mm}^2$

Shaft diameter

$$0.697 \times 10^4 = \frac{\pi}{4}\times d^2$$

shaft diameter, $d = 94.22$ mm

$$\text{Rpm} = 1800$$

angular rotation, $\omega = \dfrac{2\pi\times1800}{60} = 188.5$ rad/s

Torque, $T = \dfrac{4500\times746}{188.5} = 17809 \text{ Nm} = 17809 \times 10^3 \text{ Nmm}$

Maximum shear stress, $\tau = \dfrac{16\times17809\times10^3}{\pi\times(94.22)^3} = 108.5 \text{ N/mm}^2$

Question 12.23 A solid phosphor bronze shaft 60 mm in diameter is rotating at 800 rpm and transmitting power. It is subjected to torsion only. An electrical resistance strain gangue mounted on the surface of the shaft, gives the strain reading as 3.98×10^{-4}. If the modulus of elasticity for bronze is 105 G N/m^2 and Poisson's ratio 0.3. Find the power transmitted by the shaft. Bending effect may be neglected. (Gage is mounted at 45° to the shaft axis).

[10 marks : IES 2013]

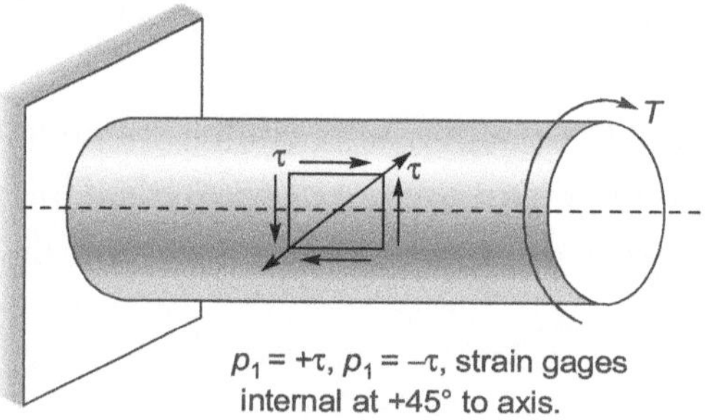

Fig. 12.16

Solution :

Strain, $\in = \dfrac{\tau}{E}+\dfrac{\tau\nu}{E} = \dfrac{\tau(1+\nu)}{E} = 3.98 \times 10^{-4}$

$$\frac{\tau(1+0.3)}{105000} = 3.98 \times 10^{-4}$$

τ, maximum shear stress $= \dfrac{3.98\times10^{-4}\times105000}{1.3} = 32.146 \text{ N/mm}^2$

Torque, $T = \dfrac{\pi}{16}d^3\times\tau = \dfrac{\pi}{16}\times60^3\times32.146$

$= 1363366$ Nmm $= 1363.36$ Nm

rpm, $N = 800$

Angular speed, $\omega = \dfrac{2\pi \times 800}{60} = 83.78$ rad/sec

Power $= \omega T = 83.78 \times 1363.36$ Nm/s

$= 114216.6$ W $= 114.216$ kW

Question 12.24 In a torsion test, the specimen is a hollow shaft with 50 mm external diameter and 30 mm internal diameter. An applied torque of 1.6 kNm is found to produce are angular twist of 0.40 measured on a length of 0.2 m of the shaft. Young's modulus of elasticity is obtained from elastic test has been found to be 200 GPa. Find the value of (1) Modulus of rigidity and (2) Poisson's ratio.

[10 marks : IES 2009]

Solution :

Strain, $T = 1.6$ kNm $= 1.6 \times 10^6$ Nmm

$$\theta = 0.40 = \frac{0.4 \times \pi}{180} = 6.98 \times 10^{-3} \text{ rad}$$

$L = 200$ mm

$d = 50$ mm

Polar moment of inertia, $J = \dfrac{\pi d^4}{32} = \dfrac{\pi \times 50^4}{32} = 61.38 \times 10^4 \text{ mm}^4$

Shear modulus, $G = \dfrac{1.6 \times 10^6 \times 200}{61.38 \times 10^4 \times 10^{-3} \times 6.98}$

$= 74.7 \times 10^3$ N/mm^2 (modulus of rigidity) $= 74.7$ GPa

Then, $E = 2G[1 + \nu]$

ν, Poisson's ratio $= \dfrac{E}{2G} - 1 = \dfrac{200}{2(74.7)} - 1 = 0.338$

Question 12.25 A ship having a single propeller shaft 120 mm in diameter running at 300 rpm is re-engined to two propeller shafts of equal cross section and producing 40 percent more horse power but at 500 rpm. If the working shear stress in re-engined shafts is 25% more then in the single shaft, determine diameter of these shafts.

Solution :

Say, T_1 = torque transmitted by the single shafts

T_2 = torque transmitted by each of the two shafts

$$H_1 = \frac{2\pi \times 300 \times T_1}{746 \times 60}$$

or $T_1 = 23.75\, H_1$ ... (i)

H_2 by two shafts (from two shafts) $= 1.4\, H_1$

H_2 by one of two shafts $= 0.7\, H_1$

$$T_2 = \frac{0.7H_1 \times 60 \times 746}{2\pi \times 500} = 9.9733\ H_1$$

$$\frac{T_1}{T_2} = \frac{23.75H_1}{9.9733H_1} = 2.38$$

Maximum $$T_1 = \frac{\pi}{16} d_1^3 . \tau$$

$$T_2 = \frac{\pi}{16} d_2^3 \times (1.25\tau) \text{ as given problem}$$

$$\frac{T_1}{T_2} = \frac{d_1^3}{d_2^3} \times \frac{1}{1.25} = 2.38$$

or $$d_2^3 = \frac{d_1^3}{2.38 \times 1.25} = \frac{120^3}{2.38 \times 1.25} = 580840.3$$

Shaft diameter, $$d_2 = 83.40 \text{ mm}$$

Question 12.26 An 8 HP motor shown in fig 12.17, delivers 5 and 3 HP respectively to accessory gears *B* and *C*. Determine (i) maximum shear stress in steel shaft (ii) angular twist between *C* and *A*. If speed of shaft is 1200 rpm. (G = 80 GPa)

Solution :

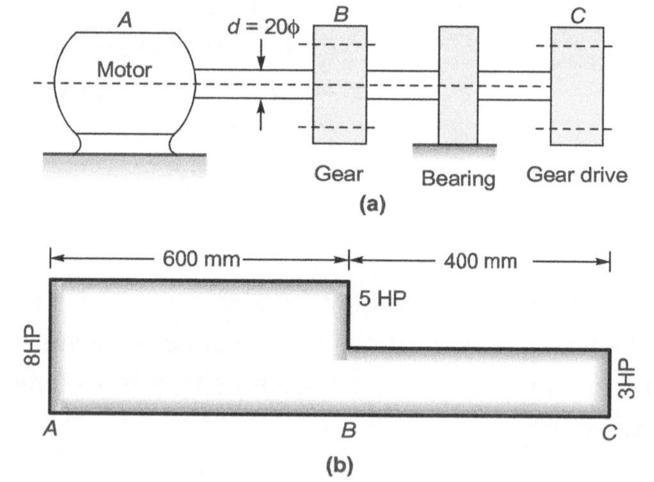

Fig. 12.17 *Power Distribution*

Shaft dia, $$d = 20 \text{ mm}$$

$$\text{RPM} = 1200$$

As per power distribution diagram, maximum power is transmitted between *A* and *B*,

Torque, $$T_{AB} = \frac{8 \times 746 \times 60}{2\pi \times 1200} = 47.49 \text{ Nm}$$

Shear stress, τ in portion AB, $\tau = \dfrac{16 \times 47.49 \times 10^3}{\pi \times 20^3} = 30.23 \text{ N/mm}^2$

$$T_{BC} = \frac{3 \times 746 \times 60}{2\pi \times 1200} = 17.81 \text{ Nm}$$

For shaft, $J = \dfrac{\pi \times 20^4}{32} = 1.570 \times 10^4 \text{ mm}^4$

Angulat twist,

$$\theta_{CA} = \frac{17.81 \times 10^3 \times 400}{1.57 \times 10^4 \times 80{,}000} + \frac{47.49 \times 10^3 \times 600}{1.57 \times 10^4 \times 80{,}000}$$

$$= 5.672 \times 10^{-3} + 22.68 \times 10^{-3} = 28.532 \times 10^{-3} \text{ rad}$$

$$= 1.624°$$

Question 12.27 **A horizontal shaft, L m long, is rigidly fixed at both the ends. At section C and D, torques 3 kNm and 5 kNm (ccw) looking from the side A are applied. What are the fixing couples T_A and T_B.**

Solution :

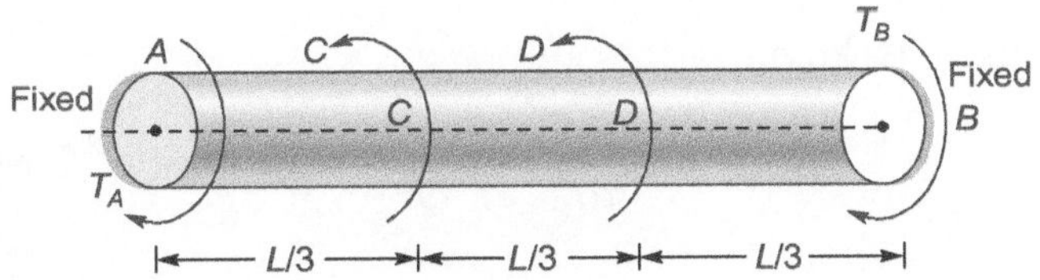

Fig. 12.18

There are 3 portions AC, CD and DB, each of length $L/3$.

Torque at A, say T_A

Torque between $CD = T_A - 3$ kNm

Torque between $DB = T_A - 3 - 5$

$= T_A - 8$ kNm

Angular twist $\theta_{AB} = 0$, since both the ends are fixed

$$\theta_{AC} + \theta_{CD} + \theta_{DB} = 0$$

or

$$0 = \frac{T_A}{GJ} \times \frac{L}{3} + \frac{(T_A - 3)}{3} \times \frac{L}{3} \frac{(T_A - 8)}{GJ} \frac{L}{3}$$

or

$$T_A + T_A - 3 + T_A - 8 = 0$$

$$3T_A = 11$$

$$T_A = \frac{11}{3} = 3.667 \text{ kNm}$$

$$T_A - 3 = 0.667 \text{ kNm}$$

$$T_A - 8 = -4.333 \text{ kNm} = T_B$$

Torque distribution is shown in figure 12.18

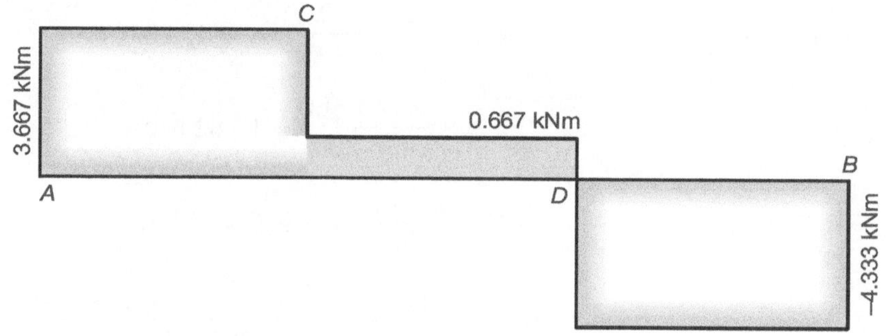

Fig. 12.19 *Torque Distribution*

Question 12.28 A solid shaft of diameter *d* is used for power transmission. Due to modification of the existing transmission system, it is required to replace the solid shaft by a hollow shaft of the same material and equally strong in torsion. Further the weight of the hollow shaft per metre length should be half that of the solid shaft. Determine the outer and inner diameters of the hollow shaft if diameter of solid shaft, *d* = 40 mm.

[ME, ESE 2016 : 20 Marks]

Solution: Solid shaft diameter = d

Say torque transmitted T in Nmm

$$\tau_{max} = \frac{16T}{\pi d^3} = \frac{16T}{\pi 40^3} = 0.0796\times10^{-3}\,T\text{ N/mm}^2 \quad ..(i)$$

Hollow shaft

Say diameter are d_1, d_2

Inner, outer diameter

$$\text{Weights of hollow shaft} = \frac{1}{2}\textit{wt. of solid shaft}$$

(Proportional to area of cross-section as material, length are same)

$$\frac{\pi}{4}\left(d_2^2 - d_1^2\right) = \frac{\pi}{4}\times\frac{1}{2}\times d^2$$

$$\left(d_2^2 - d_1^2\right) = \frac{d^2}{2} = \frac{40^2}{2} = 800\text{ mm}^2 \quad ...(ii)$$

$$\text{J of hollow shaft} = \frac{\pi}{32}\times\left(d_2^4 - d_1^4\right)$$

$$\frac{T}{J} = \frac{T}{d_2/2} = \frac{2\tau}{d_2}$$

$$\frac{T}{\frac{\pi}{32}\left(d_2^4 - d_1^4\right)} = \frac{2\tau}{d_2}$$

But, $\tau = 0.0796\times10^{-3}\text{ T N/mm}^2$

Therefore,

$$\frac{32T}{\pi\left(d_2^4 - d_1^4\right)} = \frac{3}{d_2}\times0.0796\times10^{-3}\,T$$

$$\text{or} \qquad \frac{16\times10^3}{\pi\times0.0796} = \frac{d_2^4-d_1^4}{d_2}$$

$$\text{or} \qquad \frac{d_2^4-d^4}{d_2} = 63982$$

$$\frac{(d_2^2+d_1^2)(d_2^2-d_1^2)}{d_2} = 63982$$

$$\text{But,} \qquad d_2^2-d_1^2 = 800$$

$$\frac{d_2^2+d_1^2}{d_2} = 79.98$$

$$d_2^2+d_1^2 = 79.98\,d_2$$

$$d_1^2 = d_2^2-800$$

$$d_2^2+d_2^2-800 = 79.98\,d_2$$

$$d_2^2-39.99d_2-400 = 0$$

$$d_2 = \frac{39.99+\sqrt{39.99^2+4\times400}}{2} = \frac{39.99+56.56}{2}$$

Outer diameter, $d_2 = 48.28\text{ mm}$

Inner diameter, $d_1 = \sqrt{2330.55-800} = 39.12\text{ mm}$

Question 12.29 **Square key of side d/4 and length l is used to transmit torque T from the shaft of diameter d to the hub of a pulley. Assuming the length of the key to be equal to the thickness of the pulley, find the average shear stress and crushing stress developed in the key.**

[ME, ESE 2016 : 20 Marks]

Solution:

Torque = T Nm

Shaft diameter = d, m

Say, key $\frac{d}{4}\times\frac{d}{4}\times l$

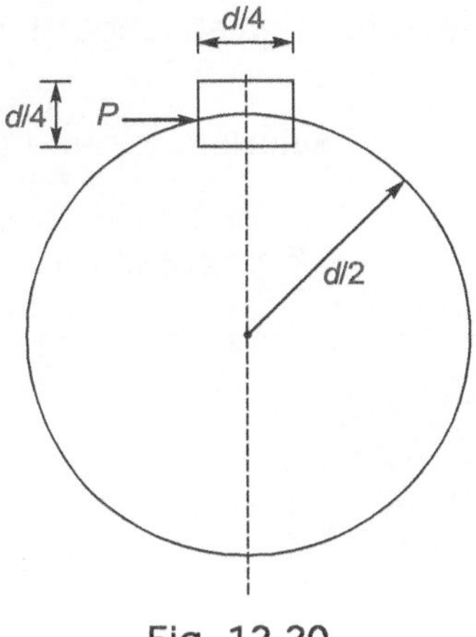

Fig. 12.20

Tangential force on key, $P = \frac{T}{d/2} = \frac{2T}{d}$, N

Length of the key $= l$

Average shear stress on key, $\tau = \frac{P}{\frac{d}{4}\times l} = \frac{4P}{dl}$ ...(i)

Average crushing stress on key, $\sigma = \frac{P}{l\times\frac{d}{8}} = \frac{8P}{ld}$

Question 12.30 A drive shaft of 40 mm diameter transmitting 25 kW at 300 rpm is connected to a gear by a flat key of width 22 mm and thickness 14 mm. It is made of steel having 300 MPa yield stress. Determine the length of the key to withstand shear. Use a factor of safely 2.

[ME, ESE 2014 : 4 Marks]

Solution:

Yield stress, $\sigma_y = 300 \text{ MPa}$

Yield shear stress, $\frac{\sigma_y}{2} = \tau_y = 150 \text{ MPa}$

Allowable shear stress, $\tau = \frac{150}{2} - 75 \text{ MPa}$ ...(i)

$$w = \frac{2 \times 300 \times \pi}{60} = 31.416 \text{ rad/sec}$$

Torque, $T = \frac{25000}{31.416} = 795.8 \text{ Nm} = 795.8 \times 10^3 \text{ Nmm}$

Shaft radius, $R = 20 \text{ mm}$

$$\text{Shear force on key} = \frac{T}{R} = \frac{795.18 \times 10^3}{20}$$

$$F = 39.79 \times 10^3 \text{ N}$$

b, width of key = 22 mm, flat key

Length = L

τ = 75 mm

$$L = \frac{F}{\tau \times b} = \frac{39.79 \times 10^3}{75 \times 22} = 24.114 \text{ mm}$$

(L is too less)

Question 12.31 A solid circular shaft has a uniform taper from one end to the other. The ratio of radius at the larger end to that at the smaller end is 1.2. Determine the error that occurs if the angle of twist for a given length is calculated using the mean radius of the shaft.

[CE, ESE 2017 : 20 Marks]

Solution: Tapered shaft

$$D = 1.2 \text{ d}$$

Where d is smaller diameter

L = Length

T = Torque

G = Shear modulus

$$\theta = \text{Angular twist} = \frac{32TL}{3\pi G}\left[\frac{(1.2d)^2 + 1.2d \times d + d^2}{(1.2d)^3 d^3}\right]$$

$$= \frac{32TL}{3\pi G}\left[\frac{1.44 + 1.2 + 1}{1.728}\right]\frac{1}{d^4}$$

$$= \frac{32TL}{3\pi G} \times \frac{2.1065}{d^4} = 22.47\left(\frac{TL}{\pi G d^4}\right)$$

Mean diameter, $\quad d_m = 1.1\text{ d} \quad$ ...(i)

$$J = (1.1)^4 \frac{d^4 \times \pi}{32} = \frac{\pi}{32} \times 1.4641 d^4$$

$$\theta' = \frac{TL}{\pi} \times \frac{32}{1.4641 d^4 \times G} = 21.856\left(\frac{TL}{\pi G d^4}\right)$$

$$\%\text{ error} = \frac{22.47 - 21.8576}{22.7} = 0.0273 = 2.73\%$$

Question 12.32 A hollow circular shaft of inner radius 10 mm, outer radius 20 mm, and length 1 m is to be used as a torsional spring. If the shear modulus of the material of the shaft is 150 GPa, the torsional stresses of the shaft in kN-m rad is _____ (correct to two decimal.)

[ME, GATE : 2018 (set-2)]

Solution:

$$\frac{T}{J} = \frac{G\theta}{l}$$

$$G = 150 \times 10^9 \text{ N/mm}^2$$

$$= 150 \times 10^6 \text{ kN/m}^2$$

$$l = 1 \text{ m}$$

$$J = \frac{\pi}{2}\left(20^4 - 10^4\right) \text{ mm}^4$$

$$= \frac{15\pi}{2} \times 10^4 \text{ mm}^4$$

$$= 7.5\pi + 10^{-8} \text{ m}^4$$

$$\frac{T}{\theta} = \frac{JG}{L} = \frac{7.5\pi \times 10^{-8} \times 150 \times 10^6}{1}$$

$$= 35.34 \text{ kNm/radian}$$

Question 12.33 A hollow shaft of 1 m length is designed to transmit a power of 30 kW at 700 rpm. The maximum permissible angle of twist in the shaft is 1°. The inner diameter of the shaft is 0.7 times the outer diameter. The modulus of rigidity is 80 GPa. The outside diameter (in mm) of the shaft is _____.

[ME, GATE : 2015 (set-2)]

Solution: $\quad \text{Power} = 30 \text{ kW} = 30000 \text{ Nm/s}$

$$N = 700 \text{ rpm}$$

Angular speed, $\quad \omega = \frac{2\pi \times 700}{60} = 73.3 \text{ rad/sec}$

Torque, $T = \dfrac{30000}{73.3} = -409.27 \text{ Nm} = 409.277 \times 10^3 \text{ Nmm}$

$$\theta = 1^\circ = \frac{\pi}{180} \text{ radian}$$

$$L = 1 \text{ m} = 1000 \text{ mm}$$

$$G = 80000 \text{ N/mm}^2$$

$$J = \frac{\pi}{32}\left(D^4 - d^4\right),\ d = 0.7d$$

$$= \frac{\pi D^4}{32}(1-0.240) = \frac{0.7599 D^4 \times \pi}{32} = 0.0746\ D^4$$

$$\frac{409.277 \times 10^3 \times 1}{0.0746 D^4} = \frac{82000}{1000} \times \frac{\pi}{180} = 1143.1$$

$$D^4 = \frac{409.277 \times 10^3}{0.0746} \times \frac{1}{1.431} = 383.4 \times 10^4$$

$$D = 44.25 \text{ mm}$$

Question 12.34 A motor driving a solid circular steel shaft transmits 40 kW of power at 500 rpm. If the diameter of the shaft is 40 mm, the maximum shear stress in the shaft is ______ MPa.

[ME, GATE : 2017 (set-1)]

Solution:

$$\text{Power} = 40 \text{ kW} = 40 \times 1000 \text{ Nm/sec}$$

$$\text{RPM} = 500$$

$$\omega = \frac{2\pi \times 500}{60} = 52.36 \text{ rad/sec}$$

$$T = \text{Torque} = \frac{40000}{52.36} = 764 \text{ Nm}$$

$$d = 40 \text{ mm}$$

$$T = \frac{\pi}{16} d^3 \times \tau = \frac{\pi \times 40^3}{16} \times \tau = 264 \times 10^3$$

$$\tau = 60.8 \text{ N/mm}^2$$

Question 12.35 Consider a stepped shaft subjected to a twisting moment applied at B as shown in the figure 12.21. Assume shear modulus, $G = 77$ GPa. The angle of twist at C (in degree) is ________.

[ME, GATE : 2015 (Set-1])

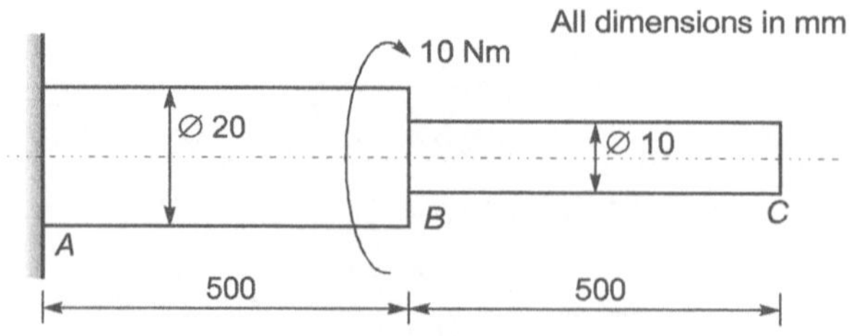

Fig. 12.21

Solution: No effect of torque on portion BC of diameter 10 mm

Portion AB,

$$T = 10 \text{ Nm} = 10000 \text{ Nmm}$$

$$G = 77000 \text{ N/mm}^2$$

$$L = 500 \text{ mm}$$

$$J = \frac{\pi \times 20^4}{32} = \frac{\pi}{2} \times 10^4 \text{ mm}^4$$

$$\theta = \frac{TL}{GJ} = \frac{10000 \times 500}{77000 \times \pi} \times \frac{2}{10^4} = 0.00414 \text{ radian}$$

$$\theta_B = \theta_C = 0.00414 \text{ radian}$$

Question 12.36 A hollow circular shaft has an outer diameter of 100 mm and inner diameter of 50 mm. If the allowable shear stress is 125 MPa, the maximum torque (in kN-m) that the shaft can resist ______.

[CE, GATE : 2017 (Set-2])

Solution: Follow shaft

$$\tau = 125 \text{ MPa}$$

$$J = \frac{\pi}{32}\left(100^4 - 50^4\right) = \frac{\pi}{32} \times 10^4 (9375) \text{ mm}^4$$

$$\frac{T}{J} = \frac{\tau}{R} = \frac{125}{50} = 2.5$$

$$T = 2.5 \times \frac{\pi}{32} \times 10^4 \times 9375 \text{ Nmm}$$

$$= 23 \times 10^6 \text{ Nmm} = 23 \text{ kNm}$$

Objective Type Questions

Q.1 A 3 m long steel cylindrical shaft is rigidity held at its ends. A pulley is mounted on the shaft at 1 metre from one end. The shaft is twisted by applying torque on the pulley. The maximum shearing stress developed in 1 m and 2 m lengths are respectively τ_1 and τ_2, the ratio of $\tau_1 : \tau_2$ is

(a) 1/2 (b) 1

(c) 2 (d) 4

[IES - 1996]

Q.2 Maximum shear stress in a solid shaft of diameter *D* and length *L* twisted through an angle θ is τ, A hollow shaft of same material and length having outside diameter *D* and inside diameter *D*/2 is also twisted through the same angle of twist θ. The value of maximum shear stress in the hollow shaft will be

(a) $\frac{16}{15}\tau$ (b) $\frac{8}{7}\tau$

(c) $\frac{4}{3}\tau$ (d) τ

[IES - 1997]

Q.3 Two hollow shafts of the same material have the same lengths and outside diameter. Shaft 1 has internal diameter equal to an third of the outer diameter and shaft 2 has internal diameter equal to half of the outer diameter. If both the shaft are subjected to the same torque, the ratio of θ_1/θ_2 will be equal to

(a) $\frac{16}{81}$ (b) $\frac{8}{27}$

(c) $\frac{19}{27}$ (d) $\frac{243}{256}$

[IES - 1998]

Q.4 The diameter of shaft A is twice the diameter of shaft B and both are made of the same material. Assuming both the shafts to rotate at the same speed, the maximum power transmitted by B is if the maximum shear stress in both the shaft remain the same

(a) same as that of A (b) half of A

(c) 1/8 th of A (d) 1/4 th of A

[IES - 2001]

Q.5 A solid circular rod AB of diameter D and length L is fixed at both the ends. A torque T is applied at a section X such that $AX = \frac{L}{4}$ and $BX = \frac{3L}{4}$.

What is the maximum shear stress developed in the rod?

(a) $\frac{16T}{\pi D^3}$ (b) $\frac{12T}{\pi D^3}$

(c) $\frac{8T}{\pi D^3}$ (d) $\frac{4T}{\pi D^3}$

[IES - 2004 | GATE - 2009]

Q.6 A hollow shaft of same cross-section area and material as that of a solid shaft transmits

(a) same torque

(b) lesser torque

(c) more torque

(d) cannot be predicted without more data

[IES - 2005]

Q.7 What is the total angle of twist of the stepped shaft subjected to torque T shown is figure

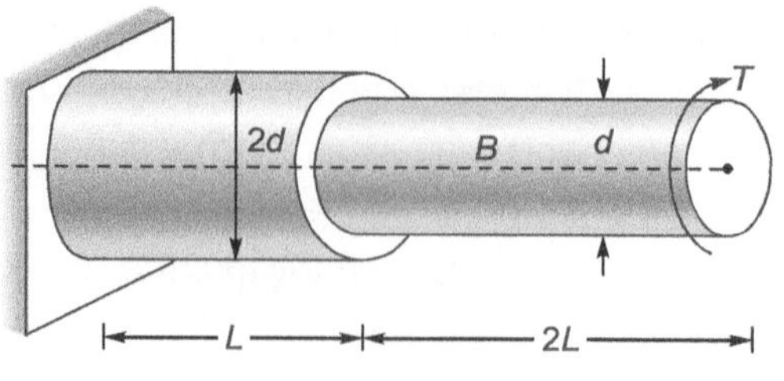

(a) $\frac{16TL}{\pi Gd^4}$ (b) $\frac{38TL}{\pi Gd^4}$

(c) $\frac{64TL}{\pi Gd^4}$ (d) $\frac{66TL}{\pi Gd^4}$

Q.8 For a power transmission shaft transmitting power P at N rpm its diameter is proportional to

(a) $\left(\frac{P}{N}\right)^{1/3}$ (b) $\left(\frac{P}{N}\right)^{1/2}$

(c) $\left(\frac{P}{N}\right)^{2/3}$ (d) $\left(\frac{P}{N}\right)$

[IES - 2005]

Q.9 A circular section rod ABC is fixed at ends A and C. It is subjected to a torque T at B, $AB = BC = L$ and its polar moment of inertia of portion AB and BC are $2J$ and J respectively. If G is the modulus of rigidity, what is angle of twist at point B

(a) $\frac{TL}{3GJ}$ (b) $\frac{TL}{2GJ}$

(c) $\frac{TL}{GJ}$ (d) $\frac{2TL}{GJ}$

[IES - 2005]

Q.10 While transmitting some power from a shaft, if the speed is doubled, what should be the new diameter if the maximum shear stress induced in the shaft remains the same

(a) $\frac{1}{2}$ of original diameter

(b) $\frac{1}{\sqrt{2}}$ of original diameter

(c) $\sqrt{2}$ of original diameter

(d) $\frac{1}{(2)^{1/3}}$ of the original diameter

[IES - 2006]

Q.11 In power transmission shaft, the polar moment of inertia of shaft is doubled, then what is torque required to produce same angle of twist

(a) $\frac{1}{4}$th of original value

(b) $\frac{1}{2}$ of original value
(c) same as of original value
(d) double the original value

[IES - 2006]

Q.12 The diameter of a solid shaft is D. The inside and outside diameters of a hollow shaft of same materials and length and $\frac{D}{\sqrt{3}}$ and $\frac{2D}{\sqrt{3}}$ respectively. What is the ratio of the weight of the hollow shaft to that of the solid shaft?

(a) 1 : 1 (b) $1 : \sqrt{3}$
(c) 1 : 2 (d) 1 : 3

[IES - 2007]

Q.13 Consider the following statements
Maximum shear stress induced in a power transmitting shaft is
1. directly proportional to torque being transmitted.
2. indirectly proportional to the cube of its diameter.
3. directly proportional its polar moment of inertia.

Which of these statements are correct?
(a) 1, 2 and 3 (b) 1 and 3 only
(c) 2 and 3 only (d) 1 and 2 only

[IES - 2008]

Q.14 For the two shafts connected in parallel
(a) Torque on each shaft is the same
(b) Shear stress on each shaft is the same
(c) Angle of twist of each shaft is the same
(d) Torsional stiffness of each shaft is the same

[IES - 2011]

Q.15 The boring bar of a boring machine is 25 mm in diameter. During operation, the bar gets twisted through 0.01 radian, and is subjected to a shear stress of 42 N/mm^2. The length of the bar is (Taking $G = 0.84 \times 10^5$ N/mm^2)
(a) 500 mm (b) 250 mm
(c) 625 mm (d) 375 mm

[IES - 2012]

Q.16 A solid circular shaft of steel 2 cm in diameter is subjected to a permissible shear stress to 10 kN/cm^2, then the value of twisting moment (T_f) will be
(a) 10π kN-cm (b) 20π kN-cm
(c) 15π kN-cm (d) 5π kN-cm

[IES - 2012]

Q.17 A power transmission solid shaft of diameter d and of length L and rigidity G is subjected to pure torque. The maximum allowable shear stress is τ_{max}. The maximum strain energy/unit volume in the shaft is given by
(a) $\frac{\tau_{max}^2}{4G}$ (b) $\frac{\tau_{max}^2}{2G}$
(c) $\frac{2\tau_{max}^2}{3G}$ (d) $\frac{\tau_{max}^2}{3G}$

[IES - 2013]

Q.18 The compound shaft shown in figure is built in at the two ends. It is subjected to a torsion moment T at the middle, what is the ratio of reaction torques T_1 and T_2 at the ends?

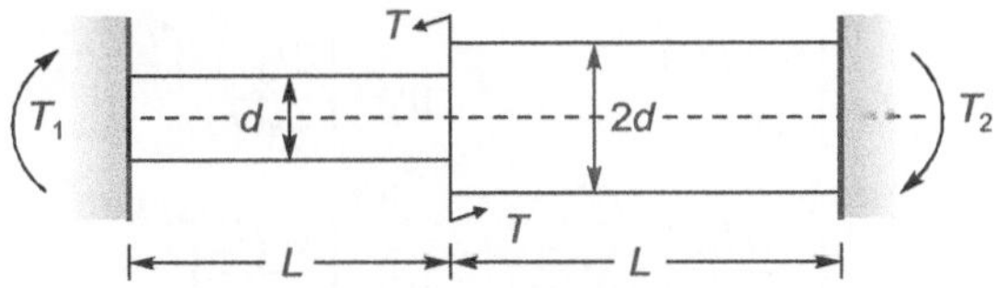

(a) $\frac{1}{16}$ (b) $\frac{1}{8}$
(c) $\frac{1}{4}$ (d) $\frac{1}{2}$

[GATE - 1993]

Q.19 Maximum shear stress developed on the surface of a solid circular shaft under pure torsion is 240 MPa. If the shaft diameter is doubled, then the maximum shear stress developed corresponding to the same torque will be
(a) 120 MPa (b) 60 MPa
(c) 30 MPa (d) 15 MPa

[GATE - 2003]

Q.20 A torque of 1.0 Nm is transmitted through a shaft an shown in figure. The torsional stiffness of section of lengths MN, NO and OP are 20 Nm/rad, 30 N-m/rad and 60 Nm/rad respectively. The angular deflection between ends M and P of shaft is

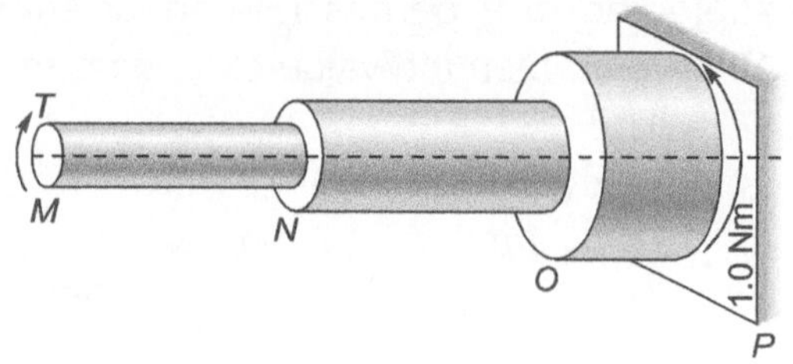

(a) 0.05 rad (b) 0.1 rad
(c) 0.5 rad (d) 1.0 rad

[GATE - 2004]

Q.21 A torque T is applied at the free end of a stepped rod of circular cross sections as shown in the figure. The shear modulus of the material of the rod is G. The expression for d to produce an angular torque θ at the free end is

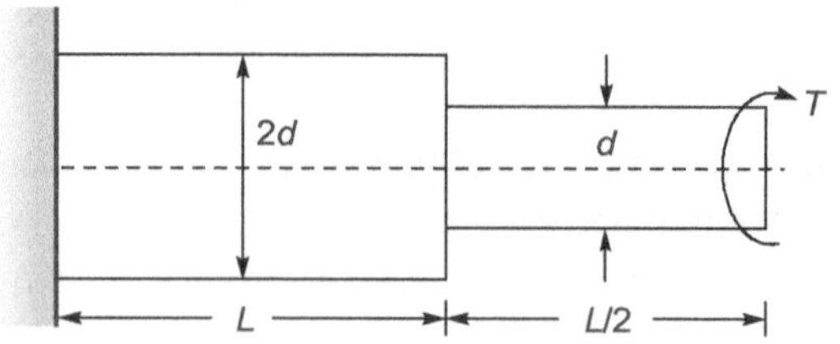

(a) $\left[\frac{32TL}{\pi\theta G}\right]^{1/4}$ (b) $\left[\frac{18TL}{\pi\theta G}\right]^{1/4}$
(c) $\left[\frac{16TL}{\pi\theta G}\right]^{1/4}$ (d) $\left[\frac{2TL}{\pi\theta G}\right]^{1/4}$

Q.22 A steel shaft of diameter d is enclosed in a copper shaft of outer diameter $2d$. If G for steel $= 2G$ for copper. Torque T is applied on the composite shaft. How much percent of torque is shared by steel shaft?
(a) 50 (b) 25
(c) 23.52 (d) 11.76

Q.23 A copper shaft of diameter d is subjected to torque T, such that maximum shear stress developed in copper shaft is 50 N/ mm^2. Now the copper shaft is replaced by steel shaft of same dimension subjected to same torque. If $G_s = 2\ G_{cu}$, what is stress developed in steel shaft
(a) 25 N/mm^2 (b) 50 N/mm^2
(c) 100 N/mm^2 (d) None of these

Q.24 A shaft of L/d ratio = 50, is subjected to torque such that angular twist in shaft is 0.1 radian. What is shear strain developed in shaft
(a) 0.5×10^{-3} (b) 1×10^{-3}
(c) 2×10^{-3} (d) None of these

Q.25 A shaft of diameter d, length L is subjected to torque T such that shear strain developed on shaft surface is 200 μs. Now the length of the shaft is doubled, d and T remain the same, what is shear strain in this shaft
(a) 400 μs (b) 200 μs
(c) 100 μs (d) None of these

Q.26 A shaft is connected to a hub by a key. The key dimensions are b = 20 mm, t = 16 mm, L = 80 mm. Due to torque, shear stress in key is 80 MPa. What is bearing stress in key
(a) 80 MPa (b) 100 MPa
(c) 200 MPa (d) 250 MPa

Q.27 A bar of circular cross section is clamped at ends P and Q as shown in the figure. A torsional moment T = 150 Nm is applied at a distance of 100 mm from end P. The torsional reactions (T_P, T_Q) in Nm at the ends P and Q respectively are

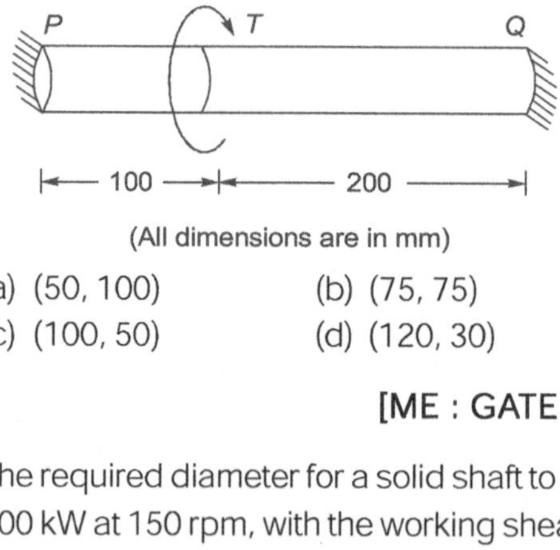

(All dimensions are in mm)

(a) (50, 100) (b) (75, 75)
(c) (100, 50) (d) (120, 30)

[ME : GATE : 2018]

Q.28 The required diameter for a solid shaft to transmit 400 kW at 150 rpm, with the working shear stress not to exceed 80 MN/m^2, is nearly
(a) 125 mm (b) 121 mm
(c) 117 mm (d) 113 mm

[CE, ESE : 2018]

Q.29 A solid shaft A of diameter D and length L is subjected to a torque T; another shaft B of the same material and of the same length, but half the diameter, is also subjected to the same torque

T. The ratio between the angles of twist of shaft *B* to that of shaft *A* is

(a) 32 (b) 16
(c) 8 (d) 4

[CE, ESE : 2018]

Q.30 A solid circular shaft has a diameter *d*. Its polar modulus will be

(a) $\frac{\pi}{16}d^2$ (b) $\frac{\pi}{64}d^3$
(c) $\frac{\pi}{16}d^3$ (d) $\frac{\pi}{32}d^2$

[CE, ESE : 2016]

Q.31 Torques are transmitted to the solid circular shaft as shown in the figure below. If the corresponding permissible stress in the shaft is 60 N/mm^2, the diameter of the shaft is nearly

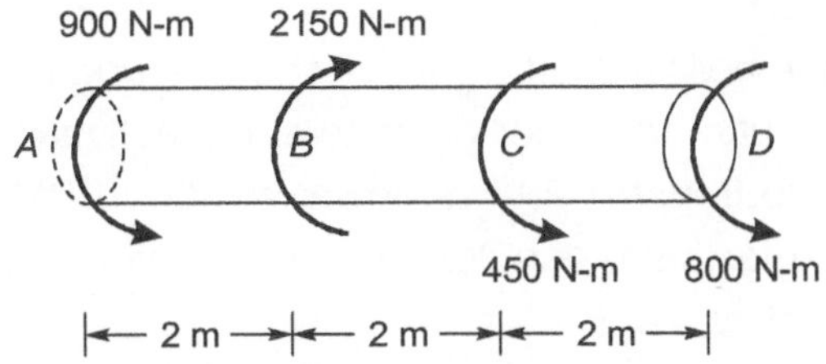

(a) 57.3 mm (b) 47.5 mm
(c) 37.3 mm (d) 27.3 mm

[CE, ESE : 2016]

Q.32 A circular shaft rotates at 200 rpm and is subject to a torque of 1500 Nm. The power transmitted would be

(a) 10 π kW (b) 15 π kW
(c) 20 π kW (d) 30 π kW

[CE, ESE : 2016]

Q.33 The diameter of a shaft to transmit 25 kW at 1500 rpm., given that the ultimate strength is 150 MPa and the factor of safety is 3, will nearly be

(a) 12 mm (b) 16 mm
(c) 20 mm (d) 26 mm

[ME, ESE : 2016]

Q.34 Two solid shafts *A* and *B* are made of the same material. Shaft *A* is of 50 mm diameter and shaft *B* is of 100 mm diameter. The strength of shaft *B* is

(a) 2 times as that of shaft *A*
(b) 4 times as that of shaft *A*
(c) 6 times as that of shaft *A*
(d) 8 times as that of shaft *A*

[ME, ESE : 2016]

Q.35 A solid shaft is designed to transmit 100 kW while rotating at N rpm. If the diameter of the shaft is doubled and is allowed to operate at 2 N rpm., the power that can be transmitted by the latter shaft is

(a) 200 kW (b) 400 kW
(c) 800 kW (d) 1600 kW

[ME, ESE : 2016]

Q.36 A shaft of 50 mm diameter transmits a torque of 800 Nm. The width of the rectangular key used is 10 mm. The allowable shear stress of the material of the key being 40 MPa, the required length of the key would be

(a) 60 mm (b) 70 mm
(c) 80 mm (d) 90 mm

[ME, ESE : 2016]

Q.37 A solid circular beam with radius of 0.25 m and length of 2 m is subjected to a twisting moment of 20 kNm about the *z*-axis at the free end, which is the only load acting as shown in the figure. The stress component τ_{xy} at Point '*M*' in the cross-section of the beam at a distance of 1 m from the fixed end is

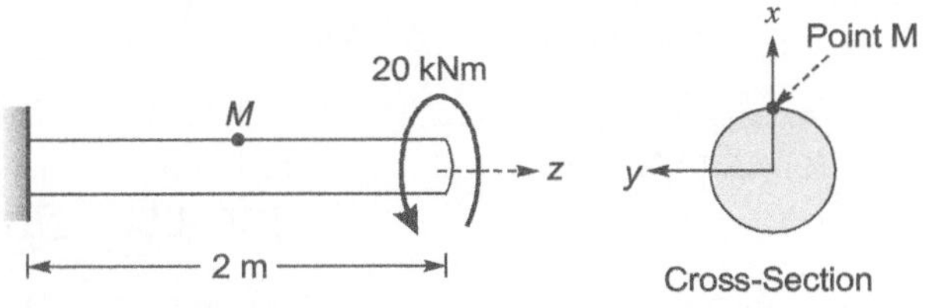

(a) 0.0 MPa (b) 0.51 MPa
(c) 0.815 MPa (d) 2.0 MPa

[CE, GATE : 2018 (Set-1)]

Q.38 A hollow steel shaft has outside diameter d and inside diameter d/2. The value of *d* for the shaft, if it has to transmit 200 hp at 105 rpm with a working shear stress of 420 kg/cm^2, is

(a) 5.6 cm (b) 2.6 cm
(c) 12.1 cm (d) 15.5 cm

[CE, ESE : 2016]

Q.39 Two shafts of diameter 30 mm each are connected by a flange coupling. Six bolts, each of diameter 8 mm, are used on a pitch circle of diameter 90 mm. If the allowable shear stress of the bolt material is 80 MPa, what is the torque transmitting capacity of the bolts to the nearest 10 units?

(a) 780 Nm (b) 950 Nm
(c) 1090 Nm (d) 1250 Nm

[ME, ESE : 2016]

Q.40 A solid shaft is subjected to bending moment of 3.46 kN-m and a torsional moment of 11.5 kN-m. For this case, the equivalent bending moment and twisting moment are

(a) 7.73 kN-m and 12.0 kN-m
(b) 14.96 kN-m and 12.0 kN-m
(c) 7.73 kN-m and 8.04 kN-m
(d) 14.96 kN-m and 8.04 kN-m

[ME, ESE : 2018]

Q.41 A stepped steel shaft is subjected to a clockwise torque of 10 Nm at its free end. Shear modulus of steel is 80 GPa. The strain energy stored in the shaft is

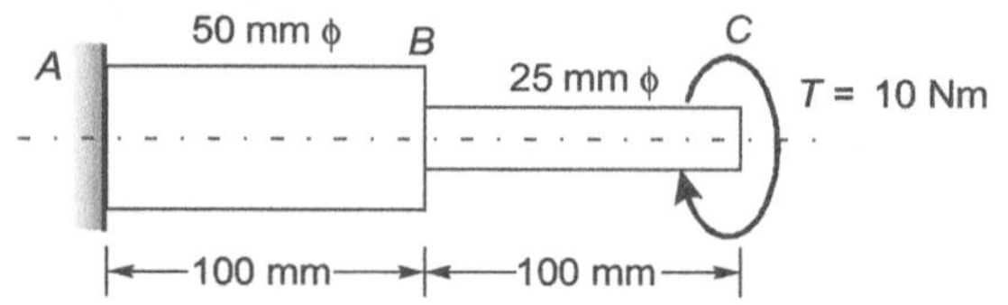

(a) 1.73 Nmm (b) 2.52 Nmm
(c) 3.46 Nmm (d) 4.12 Nmm

[CE, ESE : 2017]

Q.42 Consider the following statements:

1. The shear stress distribution across the section of a circular shaft subjected to twisting varies parabolically.
2. The shear stress at the centre of a circular shaft under twisting moment is zero.
3. The shear stress at the extreme fibres of a circular shaft under twisting moment is maximum.

Which of the above statements is/are correct?

(a) 2 and 3 (b) 1 only
(c) 2 only (d) 3 only

[CE, ESE : 2017]

Q.43 Two circular shafts made of same material, one solid (S) and one hollow (H), have the same length and polar moment of inertia. Both are subjected to same torque. Here, q_S is the twist and τ_S is the maximum shear stress in the solid shaft, whereas θ_H is the twist and τ_H is the maximum shear stress in the hollow shaft. Which one of the following is TRUE?

(a) $\theta_S = \theta_H$ and $\tau_S = \tau_H$
(b) $\theta_S > \theta_H$ and $\tau_S > \tau_H$
(c) $\theta_S < \theta_H$ and $\tau_S < \tau_H$
(d) $\theta_S = \theta_H$ and $\tau_S < \tau_H$

[ME, GATE : 2016 (Set-3)]

Q.44 A hollow shaft ($d_o = 2d_i$ where d_o and d_i are the outer and inner diameters respectively) needs to transmit 20 kW power at 3000 RPM. If the maximum permissible shear stress is 30 MPa, d_o is

(a) 11.29 mm (b) 22.58 mm
(c) 33.87 mm (d) 45.16 mm

[ME, GATE : 2015 (Set-2])

Q.45 The cross sections of two hollow bars made of the same material are concentric circles as shown in the figure. It is given that $r_3 > r_1$ and $r_4 > r_2$, and that the areas of the cross-sections are the same, J_1 and J_2 are the torsional rigidities of the bars on the left and right, respectively. The ratio J_2/J_1 is

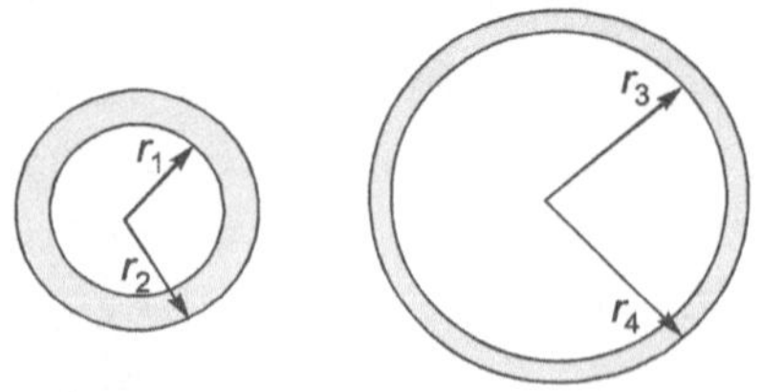

(a) > 1
(b) < 0.5
(c) =1
(d) between 0.5 and 1

[ME, GATE : 2016 (Set-1)]

Answers

1. (c)	2. (d)	3. (d)	4. (c)	5. (b)
6. (c)	7. (d)	8. (a)	9. (a)	10. (d)
11. (d)	12. (a)	13. (d)	14. (c)	15. (b)
16. (d)	17. (a)	18. (a)	19. (c)	20. (b)
21. (b)	22. (d)	23. (b)	24. (b)	25. (b)
26. (c)	27. (c)	28. (c)	29. (b)	30. (c)
31. (b)	32. (a)	33. (d)	34. (d)	35. (d)
36. (c)	37. (c)	38. (c)	39. (c)	40. (a)
41. (a)	42. (a)	43. (d)	44. (b)	45. (a)

Explanations

1. (c)

$$T_1 + T_2 = T$$

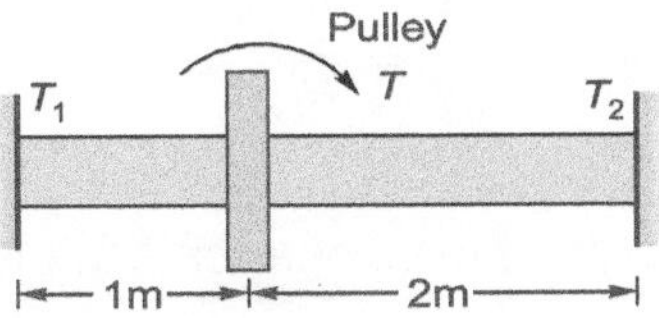

$$\frac{T_1 \times 1}{GJ} = \frac{T_2 \times 2}{GJ}$$

$$T_1 = 2T_2$$

$$T_2 = 0.5\, T_1$$

$$\tau_2 = 0.5\, \tau_1\,,\ \tau_1 = 2\tau_2$$

2. (d)

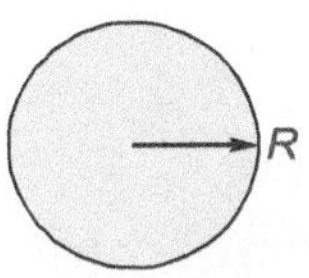

$$\frac{G\theta}{L} = \frac{\tau_S}{R}$$

$$\frac{G\theta}{L} = \frac{\tau_H}{R}$$

$$\tau_S = \tau_H = \tau$$

3. (d)

$$\frac{TL}{GJ_1} = \theta_1\,, \qquad \frac{TL}{GJ_2} = \theta_2$$

$$\frac{\left(D^4 - \frac{D^4}{16}\right)}{\left(D^4 - \frac{D^4}{81}\right)} = \frac{15}{16} \times \frac{81}{80}$$

$$= \frac{\theta_1}{\theta_2} = \frac{J_2}{J_1} = \frac{243}{256}$$

4. (c)

$$\omega_1 = \omega_2$$

$$\text{Power} \propto T$$

$$T_1 = \frac{\pi}{16}(2d)^3\,\tau$$

$$T_2 = \frac{\pi}{16}(d)^3\,\tau$$

$$\frac{T_2}{T_1} = \frac{1}{8}$$

5. (b)

$$T_1 = \frac{3}{4}T$$

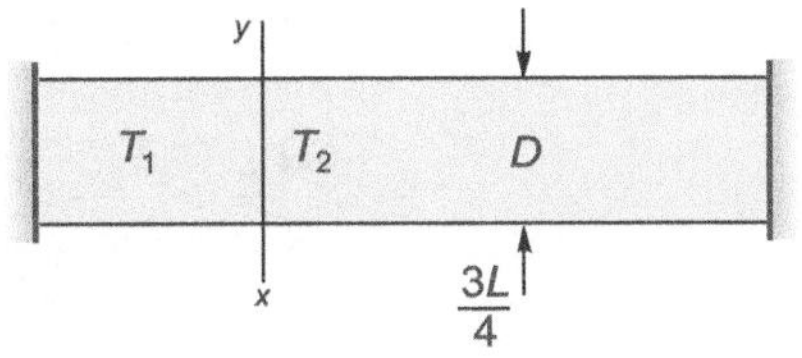

$$T_2 = \frac{T}{4}$$

$$\tau_{max} = \frac{3}{4} \times \frac{16T}{\pi D^3} = \frac{12T}{\pi D^3}$$

6. (c)

Hollow Solid

$$\frac{\pi}{4}(d_1^2 - d_2^2) = \frac{\pi}{4}d_s^2$$

$$d_1^2 - d_2^2 = d_s^2$$

So $d_1^2 > ds^2$

$$\frac{T_h}{T_s} = \frac{\frac{\pi}{32}(d_1^4 - d_2^4)}{\frac{\pi}{32}d_s^4}$$

$$= \frac{(d_1^2 - d_2^2)}{ds^2} \times \frac{d_1^2 + d_2^2}{ds^2}$$

$$\frac{T_h}{T_s} = \frac{d_1^2 + d_2^2}{ds^2} > 1$$

7. (d)

$$\theta_1 = \frac{TL}{G \times \pi} \times \frac{32}{d^4} \times 2$$

$$\theta_2 = \frac{TL}{G\pi}\times\frac{32}{16d^4} = \frac{TL\times 2}{G\pi d^4}$$

$$\theta_1 + \theta_2 = \frac{66TL}{\pi G d^4}$$

8. (a)

Power, $P = \frac{2\pi NT}{60}$

$$T = \frac{\pi}{16}.d^3\tau$$

$$\frac{P}{N} = \left(\frac{2\pi}{60\times 16}\times\tau\right)d^3 = k\,.\,d^3$$

$$d = \left(\frac{P}{N}\right)^{1/3}$$

9. (a)

$$\theta = \frac{T_1L}{GJ} = \frac{T_2L}{G2J}$$

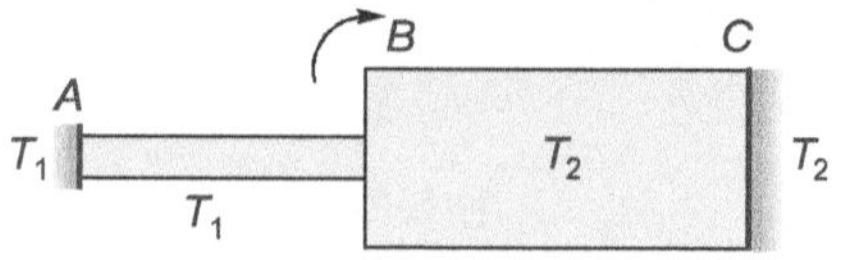

$$T_1 = \frac{T_2}{2}$$

$$T_1 + T_2 = T, \quad T_2 = 0.667\,T, \quad T_1 = \frac{T}{3}$$

$$\theta = \frac{T_1L}{GJ} = \frac{T_2L}{G\times 2J} = \frac{TL}{3GJ}$$

10. (d)
Power HP, speed is double.
T is to half

$$T = \frac{\pi}{16}d^3.\tau$$

$$\frac{T}{2} = \frac{\pi}{16}\times d'^3\tau$$

$$\frac{\pi}{16}d'^3\tau = \tau = \frac{\pi}{32}\times d^3\tau$$

$$d'^3 = \frac{1}{2}d^3 = d' = \frac{d}{(2)^{1/3}}$$

11. (d)

$$\frac{TL}{GJ_1} = \theta = \frac{T'L}{GJ_2}$$

$$T' = \frac{J_2}{J_1}\times T = 2\,T$$

12. (a)

$$W_H \propto \left(\frac{4D^2}{3} - \frac{D^2}{3}\right) \simeq D^2$$

$$W_S \propto D^2 = 1:1$$

13. (d)

$$T = \frac{\pi}{16}d^3\tau, \qquad \tau \propto T,\ \tau \propto \frac{1}{d^3}$$

$$\frac{T}{J} = \frac{\tau}{R}, \qquad \tau \propto \frac{1}{J}$$

14. (c)
Angle of twist of each shaft is the same.

15. (b)

$$\frac{G\theta}{L} = \frac{\tau}{R} = \frac{42}{12.5}$$

$$\frac{0.84\times 10^5\times 0.01}{L} = \frac{42}{12.5}$$

$$\frac{840}{L} = \frac{42}{12.5}$$

$$L = \frac{840\times 12.5}{42} = 250 \text{ mm}$$

16. (d)

$$T = \frac{\pi}{16}d^3\tau = \frac{\pi}{16}\times 2^3\times 10 \text{ kN/cm}^2$$

$$= 5\,\pi \text{ kN-cm}$$

17. (a)

Maximum strain energy/until volume $= \frac{\tau_{max}^2}{4G}$

18. (a)

$$\frac{G\theta}{L} = \frac{T_1}{J_1} = \frac{T_2}{J_2}$$

$$\frac{32\,T_1}{\pi\,d^4} = \frac{32\,T_2}{\pi\,(2d)^4}$$

$$\frac{T_1}{T_2} = \frac{1}{16}$$

19. (c)

$$T = \frac{\pi}{16}d^3\times 240 = \frac{\pi}{16}(2d)^3\times\tau$$

$$\tau = 30 \text{ MPa}$$

20. (b)

$$\theta = \frac{1}{20}+\frac{1}{30}+\frac{1}{60}=\frac{6}{60} = 0.1 \text{ rad}$$

21. (b)

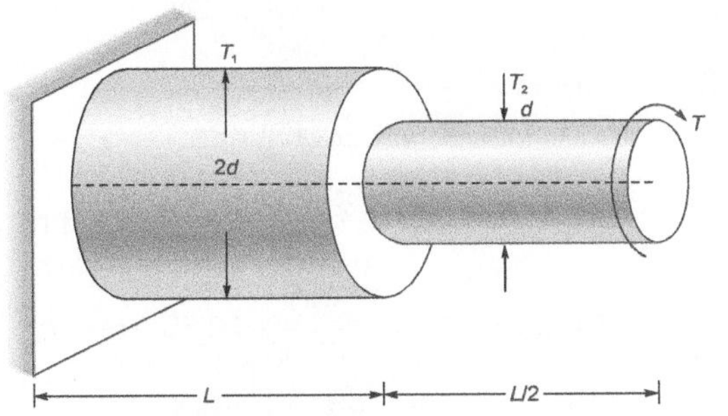

$$\theta = \frac{TL}{2GJ_1}+\frac{TL}{GJ_2}$$

$$J_2 = 16\,J_1$$

$$\theta = \frac{TL}{2GJ_1}+\frac{TL}{16GJ} = \frac{9}{16}\frac{TL}{GJ_1}$$

$$\theta = \frac{9}{G16}\times\frac{TL}{\pi d^4}\times 32 = \frac{18TL}{G\pi d^4}$$

$$d^4 = \left(\frac{18TL}{G\pi\theta}\right) \Rightarrow d = \left(\frac{18TL}{G\pi\theta}\right)^{1/4}$$

22. (d)

$$\frac{T_S}{T_{cu}} = \frac{G_S I_S}{G_{cu}J_{cu}} = 2\times\frac{J_S}{J_W}$$

$$J_s = \frac{\pi d^4}{32}$$

$$J_{cu} = \frac{\pi}{32}(16d^4 - d^4) = \frac{15\times\pi d^4}{32}$$

$$\frac{J_S}{J_{cu}} = \frac{1}{15} = 2\times\frac{1}{15}=\frac{1}{7.5}$$

$$7.5\,T_S = T_{cu}$$

$$7.5\,T_{S+}T_S = T$$

$$T_S = \frac{T}{8.5} = 0.1176\,T = 11.76\%$$

23. (b)
Stress is independent of elastic modulus.

24. (b)

$$\frac{L}{r} = 100$$

$$L\phi = r\theta$$

$$\phi = \frac{r}{L}\times\theta = \frac{\theta}{100}=\frac{0.1}{100} = 0.001$$

25. (b)
It is unchanged as *T*, *d* are unchanged; ϕ, shear strain is unchanged.

26. (c)

$$\tau = \frac{F}{bl}$$

$$F = 80\times 20\times 80$$

$$\sigma_b = \frac{2F}{Lt} = \frac{80\times 20\times 80\times 2}{80\times 16}$$

$$= 200 \text{ MPa}$$

27. (c)

$$\frac{T_P\times 100}{GJ} = \frac{T_Q\times 200}{GJ}$$

$$T_P = 2T_Q$$

$$T_P = 100$$

$$T_Q = 50 \text{ Nm}$$

28. (c)

$$P = 400 \text{ kW}$$

$$= 400000 \text{ Nm/s}$$

$$\omega = \frac{2\pi\times 150}{60} = 15.70 \text{ rad/sec}$$

Torque, $T = \frac{400000}{15.7} = 25477.7$ Nm

Shear stress, $\tau = 80$ N/mm²

$$d^3 = \frac{16T}{\pi\tau} = \frac{16\times 25477.7\times 10^3}{\pi\times 80}$$

$$= 1.622\times 10^6 \text{ mm}^3$$

$$d = 117.5 \text{ mm} \simeq 117 \text{ mm}$$

29. (b)

Shaft A: D, L, T

$$J_A = \frac{\pi D^4}{32}$$

Shaft B: 0.5D, L, T

$$J_B = \frac{J_A}{16} \quad \ldots(i)$$

$$\frac{\theta_B}{\theta_A} = 16$$

Because J_B is only $\frac{J_A}{16}$

30. (c)
Solid circular shaft

$$\text{Polar modulus} = \frac{\pi d^4}{32} \times \frac{1}{0.5d}$$

$$\text{Solid shaft} = \frac{\pi d^3}{16}$$

31. (b)

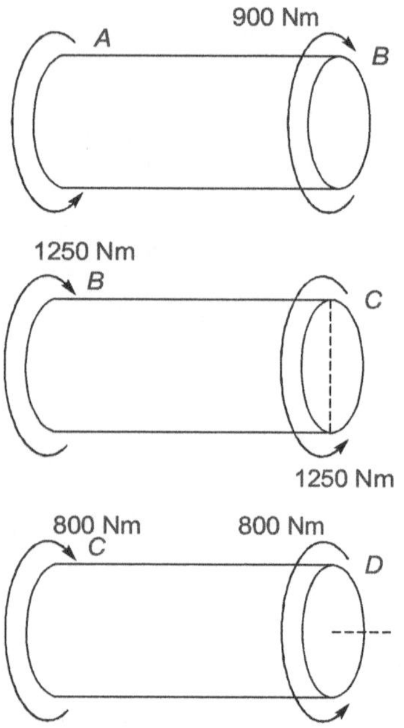

Maximum torque occurs in central portion BC is 1250 Nmm

$$T_{max} = 1250 \times 10^3 \text{ Nmm}$$

$$= \frac{\pi d^3}{16} \times 60$$

$$d^3 = \frac{1250 \times 10^3 \times 16}{60 \times \pi}$$

$$= 106.103 \times 10^3 \text{ mm}^3$$

$$d = 47.3 \text{ mm}$$

32. (a)

$$\text{Speed,} \quad \omega = \frac{2\pi \times 200}{60} = \frac{200\pi}{30}$$

$$= \frac{20\pi}{3} \text{ rad/sec}$$

T, Torque = 1500 Nm

$$\text{Power} = \omega T = 1500 \times \frac{20\pi}{3}$$

$$= 500 \times 20\pi \text{ Nm/s}$$

$$= 10\pi \text{ kW}$$

33. (d)

Power = 25000 Nm/s

$$\omega = \frac{2\pi \times 1500}{G} = 157.08 \text{ rad/sec}$$

$$\text{Torque,} \quad T = \frac{25000}{157.08} = 159.15 \text{ Nm}$$

$$= 159150 \text{ Nmm}$$

σ_n in shear = 150

It should be given clearly that ultimate strength in shear is in MPa

FOS = 3

τ_a = Allowable shear stress

$$= \frac{150}{3} = 50 \text{ MPa}$$

$$d^3 = \frac{16T}{\pi \times 50} = \frac{16 \times 159150}{\pi \times 150}$$

$$= 16.21 \times 10^3$$

$$d = 25.3 \text{ mm} \simeq 26 \text{ mm}$$

34. (d)

d_A = 50 mm

A_B = 10 mm

Z_A = Polar modulus

$$= \frac{\pi (50)^3}{16}$$

$$Z_B = \frac{\pi \times (100)^3}{16}$$

Strength $\propto$ Polar modulus

$$Z_B = 8 Z_A$$

35. (d)

Power $\propto$ RPM

Torque $\propto d^3$, for same shear stress

Power that can be transmitted by latter shaft at 2N rpm

$$= 100 \text{ kW} \times (2)^3 \times 2$$

$$= 1600 \text{ kW}$$

36. (c)

Shaft dia., $d = 50$ mm

Torque $= 800$ Nm

Tangential force in key

$$= \frac{800 \times 1000}{25}$$

$$F = 32 \times 10^3 \text{ N}$$

Allowable shear stress,

$$\tau_{av} = 40 \text{ MPa}$$

Rectangular key width

$$b = 10 \text{ mm}$$

Length $= L$ mm

Then, $$L = \frac{F}{\tau \times b} = \frac{32 \times 10^3}{40 \times 10}$$

$$= 80 \text{ mm}$$

37. (c)

$$2 \times 10^6 = \frac{\pi}{2} \times (250)^3 \times \tau_{xy}$$

$$\tau_{xy} = 0.815 \text{ MPa}$$

38. (c)

Hollow steel shaft,

$$d = \frac{d}{2}$$

$$J = \frac{\pi d^4}{32} - \frac{\pi d^4}{512}$$

$$= \frac{15}{512} \times \pi d^4$$

HP $= 200$

Power $= 200 \times 75$ kgm/sec

$$\omega = \frac{2\pi \times 105}{60}$$

$$= 10.996 \text{ rad/sec}$$

Torque, $$T = \frac{200 \times 75}{100.996} = 1364.13 \text{ kgm}$$

$$= 1364.13 \times 100 \text{ kgf-cm}$$

$$\frac{T}{J} = \frac{\tau}{d/2} = \frac{420 \times 2}{d}$$

$$\frac{136413}{15\pi} \times \frac{515}{d^4} = \frac{840}{d}$$

$$d^3 = 1764.437 \text{ cm}^3$$

$$d = 12.08 \text{ cm} \simeq 12.1 \text{ cm}$$

39. (c)

Number of bolts $= 6 = n$

d, of bolt $= 8$ mm

Allowable shear stress

$= 80$ MPa

Shear force in each bolt

$$= \frac{\pi}{4} \times 8^2 \times 80$$

$$F = 4021.214 \text{ N}$$

R, Pitch circle radius $= 45$ mm

Torque transmitting capacity

$= nFR$

$= 6 \times 4021.24 \times 45$

$= 1085734.42$ Nmm

$= 1085$ Nm

40. (a)

$$T_e = \sqrt{3.46^2 + 11.5^2}$$

$$= \sqrt{11.9716 + 132.25}$$

$$= 12 \text{ kNm}$$

$$M_e = \frac{3.46 + 12}{2}$$

$$= 7.73 \text{ kNm}$$

41. (a)

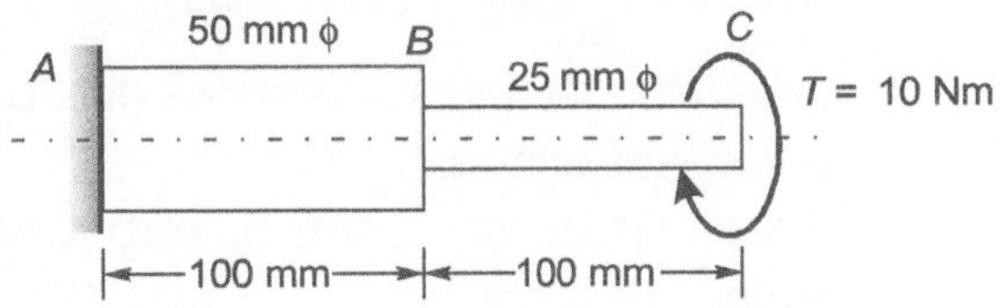

Torque, $T = 10 \text{ Nm} = 10 \times 10^3$ Nmm

Polar moment of inertia

$$J_1 = \frac{\pi d^4}{32} = \frac{\pi}{32} \times 50^4$$

$$= \frac{625\pi}{32} \times 10^4$$

$$J_2 = \frac{J_1}{16} = \frac{62.5\pi \times 10^4}{32 \times 16}$$

Shear modulus,

$$G = 80 \text{ GPa} = 80 \times 10^3 \text{ N/mm}^2$$

Angular twist,

$$\theta = \theta_1 + \theta_2$$

$$= \frac{Tl}{GJ_1} + \frac{Tl}{GJ_2}$$

$$= \frac{10\times10^3}{80\times10^3}\left[\frac{100}{\pi\times50^4}\times32 + \frac{100}{\pi\times50^4}\times32\times16\right]$$

$$= \frac{1}{8}[101.86 + 16297.5]\times\frac{1}{50^4}$$

$= 3.463 \times 10^{-4}$ radian

Strain energy,

$$U = \frac{1}{2}T\theta = \frac{1}{2}\times10^4\times3.463\times10^{-4} = 1.73 \text{ Nmm}$$

42. (a)

Figure shows shear stress distribution section of a circular shaft subjected to torsion:

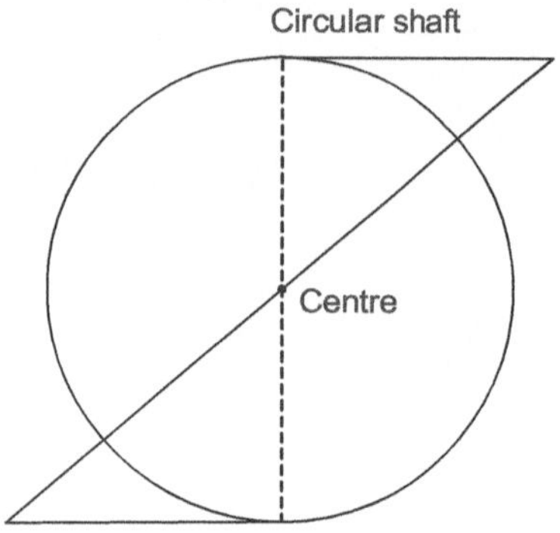

Shear strain distribution due to twisting

1. Incorrect statement – as stress distribution is linear as shown.
2. Shear stress at centre is zero–Correct statement as shown.
3. Shear stress at extreme fibres in maximum – Correct statement.

43. (d)

Hollow shaft

D_1, D_2 diameter

Solid Shaft

d, diameter

$T_H = T_S$, $L_H = L_S$, $J_H = J_S$

$$J_S = \frac{\pi d^3}{32}$$

This means $D_1 > D$

$$\frac{T}{JH} = \frac{\tau_H}{D_1/2} = \frac{\tau_S}{d_1/2}$$

$$\tau_H > \tau_S$$

Similarly, $$\frac{T}{J_H} = \frac{T}{J_S} = \frac{G\theta}{L} \qquad (\theta_H = \theta_S)$$

$$\theta_S = \theta_H \text{ and } \tau_S < \tau_H$$

44. (b)

Power = 20 kW = 20000 Nm/s

$$\omega = \frac{2\pi\times3000}{60} = 314.16 \text{ rad/sec}$$

$$T = \frac{20000}{314.16} = 95.5 \text{ Nm} = 63.66\times10^3 \text{ Nm}$$

$\tau = 30$ MPa

$$J = \frac{\pi}{322}\left(d_o^4 - d_i^4\right) = \frac{15}{16}\times\frac{\pi}{32}\times d_o^4 = 0.092 d_0^4$$

$$\frac{63.66\times10^3}{0.092 d_o^4} = \frac{30}{d_o/2} = \frac{60}{d_o}$$

$$d_o^3 = \frac{63.66\times10^3}{0.092\times60} = 11.53\times10^3$$

$$d_o = 22.59 \text{ mm}$$

45. (a)

$$J_1 = \frac{\pi}{2}\left(r_2^4 - r_1^4\right)$$

$$= \frac{\pi}{4}\left(r_2^2 + r_1^2\right)\left(r_2^2 - r_1^2\right)$$

$$= \pi\left(r_2^2 - r_1^2\right)\times\frac{r_2^2 + r_1^2}{2}$$

$$J_2 = \pi\left(r_4^2 - r_3^2\right)\times\frac{r_4^2 + r_3^2}{2}$$

But area of cross-section are same

$$\pi\left(r_2^2 - r_1^2\right) = \pi\left(r_4^2 - r_3^2\right)$$

So, $$\frac{J_1}{J_2} = \frac{r_2^2 + r_1^2}{r_4^2 + r_3^2}$$

$$\frac{J_2}{J_1} = \frac{r_4^2 + r_3^2}{r_2^2 + r_1^2}$$

But, $r_4 > r_2$, $r_3 > r_1$

So, $J_2 > J_1$

$$\frac{J_2}{J_1} > 1$$

■■■■

Springs

CHAPTER 13

A spring is a versatile mechanical component in a mechanism or a system which restores its original configuration. As an example, in an IC engine, rocker arm opens the valve but a close coiled helical spring provided on the stem of the valve, closes the valve. There are various types of springs with innumerable applications. In a spring divider, the spring is in the shape of partial circular strip.

In this chapter, we will study about stiffness of a spring, stress developed in a spring, strain energy absorbed in a spring, etc.

Question 13.1 **What is a close-coiled helical spring? With the help of a neat sketch, explain various parameters of spring. What are stress augmentation factor and Wahl's factor?**

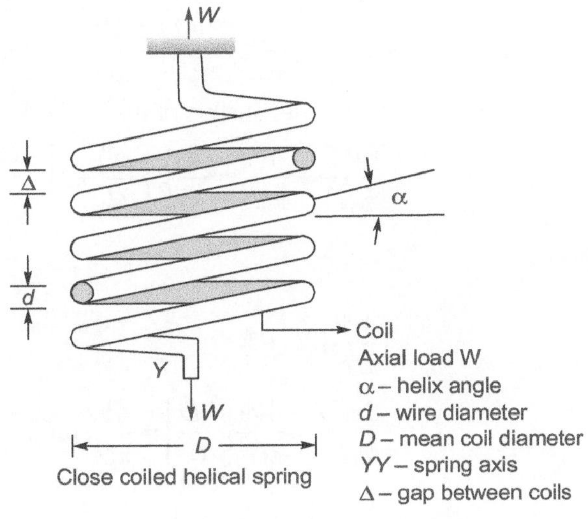

Fig. 13.1

Solution: A spring of wire diameter d is wound over a mandrel. The diameter of mandrel is equal to inside diameter of spring. Wire is wound in the form of a helix, maintaining the helix angle α of the spring. If the wire diameter is less than 5 mm, spring can be formed in cold condition. If wire diameter is more than 5 mm, the spring is made after heating the spring wire. Figure 13.1 shows a helical spring with

D = Mean coil diameter

d = Wire diameter

α = Helix angle

Δ = Gap between the coils

In the case of close coiled helical spring, gap Δ between the coils is very small. These springs are generally used for tensile load along the axis YY. If the gap between the coils is large, spring can be used for axial compressive loading but the ends of such a spring are flat i.e. perpendicular to axis of loading.

Consider one coil of spring, with mean coil radius R, wire diameter d

Direct shear load = W

Twisting moment on section, T = WR, as shown in Figure 13.2

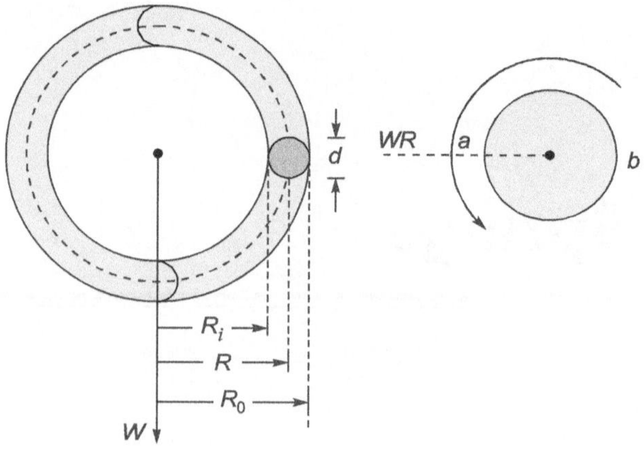

Fig. 13.2

Torsional shear stress

$$\tau_d = \frac{16WR}{\pi d^3} = \frac{8WD}{\pi d^3}$$

Direct shear stress at section

$$\tau_s = \frac{4WD}{\pi d^2}$$

Note that D is the mean coil diameter.

Resultant shear stress at inner edge *a* of section = $\dfrac{8WD}{\pi d^3} + \dfrac{4W}{\pi d^2}$

Resultant shear stress at outer edge *b* of section = $\dfrac{8WD}{\pi d^3} - \dfrac{4W}{\pi d^2}$

So, τ_{max} = maximum shear stress occurs at inner radius of the coil

$$= \frac{8WD}{\pi d^3} + \frac{4W}{\pi d^2} = \frac{4W}{\pi d^2}\left[\frac{2D}{d} + 1\right]$$

where $\dfrac{D}{d} = C$, spring index.

$$\tau_{max} = \frac{4W}{\pi d^3}[2C + 1]$$

$$\tau_s = \frac{8WD}{\pi d^3} = \frac{4W}{\pi d^2}\left[\frac{2D}{d}\right] = \frac{4W}{\pi d^2}[2C]$$

$$\frac{\tau_{max}}{\tau_s} = \frac{2C+1}{2C} = \left(1 + \frac{1}{2C}\right),$$

this factor is known as **stress augmentation factor.**

$$\frac{2C+1}{2C} = K_S, \text{ stress augmentation factor.}$$

This factor takes into account the direct shear stress.

The wire of the coil spring is continuously bent along the helix, so there is curvature effect on torsional shear stress in addition to direct stress. **Prof Wahl analysed the direct shear effect and curvature effect on the maximum shear stress produced in spring wire.** He published a research paper, in which he gave the expression for maximum shear stress as follows:

$$\tau_{max} = \frac{8WD}{\pi d^3}\left[\frac{4C-1}{4C-4}+\frac{0.615}{C}\right]$$

where C is spring index, D/d

So Wahl's factor, $$K_w = \left[\frac{4C-1}{4C-4}+\frac{0.615}{C}\right] = K_S \times K_C$$

= stress augmentation factor × curvature effect factor.

Question 13.2 **A close coiled helical spring of mean coil diameter 60 mm is made of a steel wire of diameter 8 mm. If an axial load on spring is 100 N, determine maximum shear stress developed in spring wire taking into account the Wahl's factor. What are K_s, K_c i.e. stress augmentation factor and curvature effect factor.**

Solution:

W = 100 N, axial load

D = 60 mm, mean coil radius

d = 8 mm, wire diameter

$$C = \frac{D}{d} = \frac{60}{8} = 7.5, \text{ spring index}$$

$$\tau_s = \frac{8WD}{\pi d^3} = \frac{8\times100\times60}{\pi\times8^3} = 29.84 \text{ N/mm}^2$$

K_w = Wahl's factor

$$= \frac{4C-1}{4C-4}+\frac{0.615}{C} = \frac{30-1}{30-4}+\frac{0.615}{7.5}$$

$$= 1.11538 + 0.082 = 1.19738$$

$$\tau_{max} = 1.19738 \times 29.84 = 35.73 \text{ N/mm}^2$$

Stress augmentation factor, $$K_s = \frac{2C+1}{2C} = \frac{15+1}{15} = 1.06667$$

Curvature effect factor, $$K_c = \frac{K_w}{K_s} = \frac{1.19738}{1.06667} = 1.1225$$

Practice Q.13.2 **A close coiled helical spring of mean coil diameter 56 mm is made of a steel wire of 8 mm diameter. If maximum shear stress developed is 105 MPa, what is axial load applied on spring. What are K_s and K_c?**

Ans: [W = 310.83 N, K_s = 1.0714, K_c = 1.132]

Question 13.3 **A close coiled helical spring of wire diameter *d*, mean coil diameter *D*, number of coils n is subjected to an axial load, *W*. Derive expressions for stiffness of the spring, if shear modulus of material is *G*.**

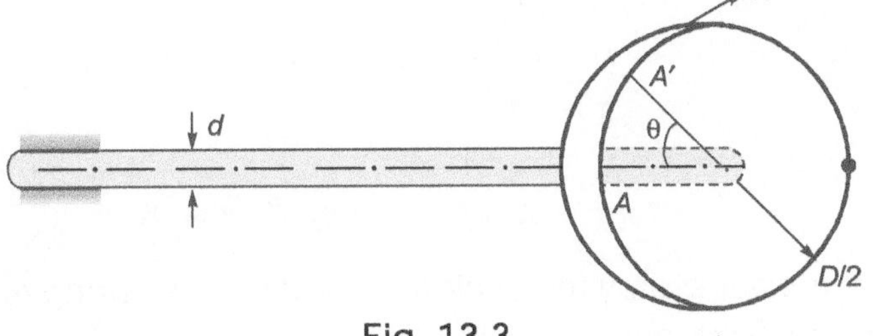

Fig. 13.3

Solution: A close coil helical spring is equivalent to a shaft of length $L = \pi nD$, wire diameter d, at the end of which is disc of diameter D is attached. A load W acts tangential to the disc periphery producing angular twist θ in the shaft and deflection $\delta = \frac{\theta D}{2}$, along the direct of load W as shown in Figure 13.3

Displacement along W,

$$AA' = \delta = R \cdot \theta, \text{ where } R = \frac{D}{2}$$

where θ is angular twist in shaft

$$\text{Shaft dia} = d$$

Polar moment of inertia $$J = \frac{\pi d^4}{32}$$

Twisting moment on shaft, $$T = \frac{WD}{2}$$

Angular twist, $$\theta = \frac{TL}{GJ} = \frac{WD}{2} \times \frac{\pi nD \times 32}{Gd^4}, \quad \theta = \frac{16WnD^2}{Gd^4}$$

Axial deflection, $$\delta = \theta\frac{D}{2} = \frac{8WnD^3}{Gd^4}$$

Stiffness of a spring is defined as the axial load required to produce unit deflection in the spring along its axis or stiffness, $K = \frac{W}{\delta} = \frac{Gd^4}{8nD^3}$, where G is modulus of rigidity of the material.

Question 13.4 **A close coiled helical spring made of steel wire of a diameter d carries an axial load of 500 N for a maximum shear stress not to exceed 120 N/mm^2. Neglecting the effect of Wahl's factor, determine wire diameter if spring index, C = 8. If spring stiffness is 12 N/mm, how many coils are there in spring. Shear modulus, G = 84000 N/mm^2.**

Solution:

$$\tau_{max} = \frac{8WD}{\pi d^3} = \frac{8W}{\pi}\left(\frac{D}{d}\right) \times \frac{1}{d^2} = \frac{8WC}{\pi d^2} = 120 \text{N/mm}^2$$

Spring index, $$C = 8$$

or wire diameter, $$d^2 = \frac{8WC}{\pi \times 120} = \frac{8 \times 500 \times 8}{\pi \times 120} = 84.88 \text{mm}^2$$

Wire diameter, $$d = 9.21 \text{ mm}$$

spring stiffness, $$k = \frac{Gd^4}{8nD^3}$$

$$D = C \times d = 8 \times 9.21 = 73.68 \text{ mm}$$

$$k = 12 \text{ N/mm}$$

Number of coils, $$n = \frac{Gd^4}{8kD^3}, \text{ putting the values. } = \frac{84000 \times 9.21^4}{8 \times 12 (73.68)^3} = 15.74$$

Please note that **it is not necessary that a close coiled helical spring must have integral number of coils.** The spring can have a fractional part of a coil.

Practice Q.13.4 A close coiled helical spring made of round steel wire with mean coil diameter of 80 mm, and number of turns are 12, is subjected to an axial load of 220 N. Determine (a) wire diameter (b) axial deflection in spring if maximum shear stress in spring does not exceed 140 N/mm^2. Given G for steel = 82000 N/mm^2. Consider only the torsional shear stress.

Ans: [d = 6.84 mm, δ = 60.25 mm]

Question 13.5 Consider a close coiled helical spring subjected to axial moment M, as shown in Figure 13.4. Derive expression for angular rotation of the wire end B and bending stress developed in spring wire.

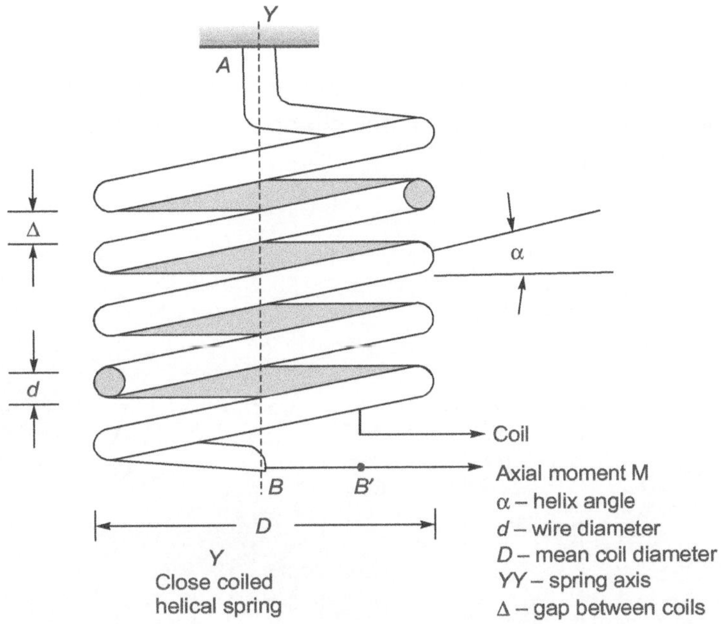

Fig. 13.4

Solution: When an axial moment M is applied at end B of spring, the end B displaces by BB' along the circumference of coil such that

$$\frac{BB'}{0.5D} = \text{Angular rotation about the axis } YY.$$

The spring wire is subjected to bending moment M,

bending stress, $\sigma_b = \dfrac{32M}{\pi d^3}$, using the flexure formula of chapter 8 on theory of simple bending.

Angular rotation, $$\phi = \frac{ML}{EI}$$

where $L = \pi n D$, length of spring wire

E = Young's modulus

I = Second moment of area = $\dfrac{\pi d^4}{64}$

So angular rotation, $\phi = \frac{M \times \pi nD \times 64}{E \times \pi d^4} = \frac{64nDM}{Ed^4}$

E–Young's modulus of the material

Strain energy absorbed in spring, $U = \frac{1}{2}M\phi$

Practice Q.13.5 A close coiled helical spring made of round steel wire of diameter *d*, having 15 complete turns is subjected to an axial moment of 4.0 Nm. Determine wire diameter if the maximum bending stress in wire section is not to exceed 200 N/mm^2. Also calculate the strain energy absorbed by the spring when couple *M* is applied. Take mean coil diameter as 80 mm. E = 200 GPa.

Ans: [5.89 mm; 2.552 Nm]

Question 13.6 A close coiled helical spring is subjected to an axial load. Shear stress is developed in spring wire. Show that the resultant shear stress at inner coil radius is more than the resultant shear stress at outer coil radius?

[IES-2012: 2 Marks]

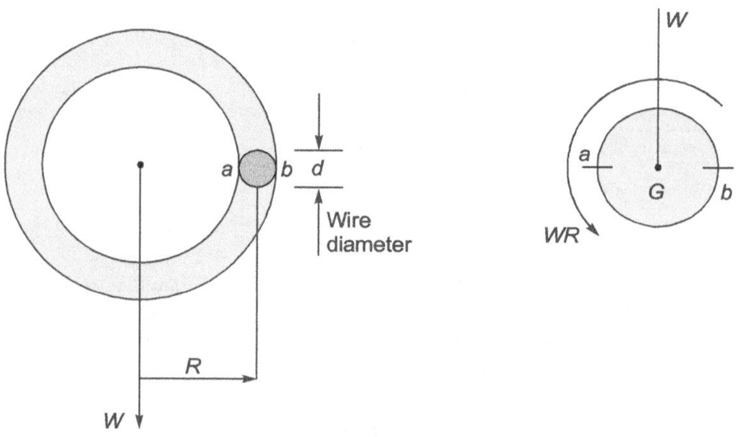

Fig. 13.5

Solution: Figure 13.5 shows a coil of helical string, with mean coil radius *R*, *W* axial load, *d* wire diameter. *W*, load at 0 ≡ Load *W* at *G*, moment *WR* (*CCW*)

$$\text{Shear stress at, } a = \frac{8WD}{\pi d^3} + \frac{4W}{\pi d^2}$$

$$\text{Shear stress at, } b = \frac{8WD}{\pi d^3} - \frac{4W}{\pi d^2}$$

a – Point at inner coil radius ($R - 0.5d$)

b – Point at outer coil radius ($R + 0.5d$)

Question 13.7 A helical spring B is placed inside the coils of a second helical string *A*, having same number of coils and free axial length and of same material. The two springs are compressed by an axial load of 210 N which is shared between them. The mean coil diameters of *A* and *B* are 90 mm and 60 mm and wire diameters are 12 mm and 7 mm respectively. Calculate the load shared by individual springs and maximum shear stress.

[IES-2000 : 15 Marks]

Solution:

Spring *A*:

$$R = 45 \text{ mm}$$
$$d = 12 \text{ mm}$$

Stiffness,
$$k_A = \frac{Gd^4}{64nR^3} = \frac{G\times 12^4}{64n\times 45^3} = \frac{G}{n\times 281.25}$$

Spring B:

$$R = 30 \text{ mm}$$
$$d = 7 \text{ mm}$$

Stiffness,
$$k_B = \frac{Gd^4}{64nR^3} = \frac{G\times 7^4}{64n\times 30^3} = \frac{G}{719.70}$$

$$\frac{k_A}{k_B} = \frac{719.70}{281.25} = 2.56$$
$$k_A = 2.56\,k_B$$

Springs are in parallel, load $W = 210$ N will be sheared an per their stiffnesses

$$W_A + W_B = 210$$
$$2.56W_B + W_B = 210$$
$$W_B = 59 \text{ N}$$
$$W_A = 151 \text{ N}$$

Shear stress

Spring A,
$$\tau_A = \frac{16\times 151\times 45}{\pi\times 12^3} = 20.02 \text{ N/mm}^2$$

Spring B,
$$\tau_B = \frac{16\times 59\times 30}{\pi\times 7^3} = 26.28 \text{ N/mm}^2$$

Question 13.8 (a) A close coiled helical spring has coil diameter *D*, wire diameter *d* and number of stress *n*. The spring material has a shearing modulus *G*. Drive an expression for the stiffness *k* of the spring.

[IES-2008: 5 Marks]

(b) A close coiled helical string has coil diameter to wire diameter ratio of 6. The spring deflects 3 mm under an axial load of 500 N and the maximum shear stress is not to exceed 300 MPa. Find the diameter and the length of the spring wire required. Shearing modulus of wire material = 80 GPa.

Solution: (a)

Axial load = W

Cost diameter = D

Wire diameter = d

$$J\text{, polar moment of inertia} = \frac{\pi d^4}{32}$$

$$\text{Torque} = \frac{WD}{2}$$

$$\delta = \theta\cdot R = \theta\cdot\frac{D}{2}$$

$$\theta = \frac{TL}{GJ}$$

$$L = \text{Length of spring wire} = 2\pi nR$$

n = Number of coil

$$L = \pi nD, \text{ as } D = 2R$$

$$\theta = \frac{WD}{2} \times \frac{\pi nD}{\pi d^4} \times \frac{32}{G}$$

$$\delta = \theta \cdot \frac{D}{2}$$

$$\delta = \frac{32WnD^2}{2d^4G} \times \frac{D}{2} = \frac{8WnD^3}{Gd^4}$$

$$\frac{W}{\delta} = k, \text{ springs stiffness} = \frac{Gd^4}{8nD^3}$$

(b)

$$\frac{D}{d} = 6$$

$$\tau_{max} = 300 \text{ MPa}$$

$$G = 80{,}000 \text{ N/mm}^2$$

$$\tau_{max} = \tau_s + \tau_d$$

$$W = 500 \text{ N}$$

$$\tau_s = \frac{8WD}{\pi d^3}, \quad \tau_d = \frac{4W}{\pi d^2}$$

$$300 = \frac{4W}{\pi d^2}\left[2 \times \frac{D}{d} + 1\right] = \frac{4 \times 500}{\pi \times d^2}[2 \times 6 + 1] = \frac{4 \times 500 \times 13}{\pi \times d^2}$$

$$d^2 = \frac{52 \times 500}{\pi \times 300} = 27.587$$

$$d = 5.25 \text{ mm}$$

$$D = 31.5 \text{ mm}$$

$$\delta = 30 \text{ mm under } W = 500 \text{ N}$$

$$K = \frac{500}{30} = 16.667 \text{ N/mm} = \frac{Gd^4}{8nD^3}$$

Further the value

$$16.667 = \frac{80{,}000 \times 5.25^4}{8 \times n \times 31.5^3} = \frac{243.055}{n}$$

$$\text{Number of coils, } n = \frac{243.055}{16.667} = 14.58$$

$$\text{Length of spring wire, } L = \pi nD$$

$$= \pi \times 14.58 \times 31.5 = 1443 \text{ mm} = 1.443 \text{ m}$$

Question 13.9 What is a plane spiral spring, explain with the help of a neat sketch. Derive expression for angular rotation due to applied moment M. What is maximum stress developed in spring spiral. Device expression for strain energy absorbed by spring

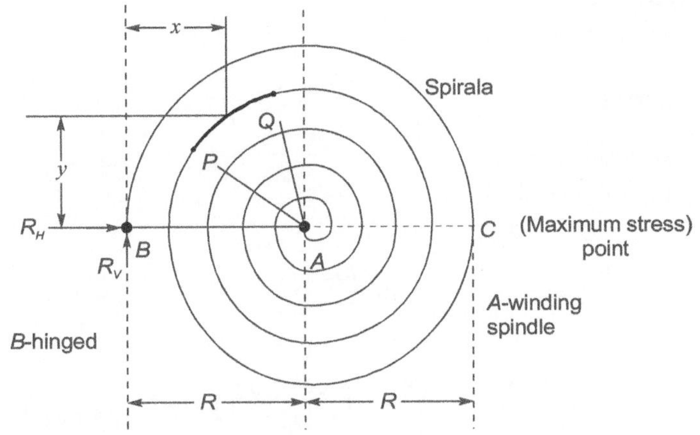

Fig. 13.6

Solution: A plane spiral spring consists of a thin uniform strip wound in the form of spirals as shown in Figure 13.6. One end of the strip is connected to a winding spindle at *A*, and other end of strip is hinged at point *B*. Winding couple *M* is applied on spindle and number of winding turns are increased say by *n*.

Then ϕ = Angle of rotation = $2\pi n$

Energy absorbed = $\frac{1}{2}M\phi$

Join the line *AB*. Say reaction at *B* has two components

R_H = Along the line *AB*, reaction

R_V = Perpendicular to the line *AB*, reaction.

Consider a small element of strip of length *PQ*, *ds*, at *x*, *y* coordinates, taking *B* as origin. Figure 13.7. Angle between the tangents *PP'* at *P* and *QQ'* at *Q* is ϕ.

After applying winding couple, ϕ is increased to $\phi + d\phi$, radius r_1 decreases to r_2.

Bending moment, $M_x = xR_V - yR_H = EI \times$ change of curvature

$$M_x = EI\left[\frac{1}{r_2} - \frac{1}{r_1}\right] = EI\frac{d\phi}{ds} \quad \text{as } d\phi = \frac{ds}{r_2} - \frac{ds}{r_1}$$

or

$$\int d\phi = \int \frac{xR_V - yR_H}{EI} \cdot \delta_s = \int \frac{xR_V ds}{EI} - \int \frac{yR_H ds}{EI}$$

But

$$\int y ds = 0$$

So

$$\phi = \frac{R_V}{EI}\int x ds = \frac{R_V LR}{EI}$$

where L = Length of strip making the spiral spring

R = outer radius of spring as shown in Figure 13.6.

But $R_V . R = M$, winding couple

So $\phi = \frac{ML}{EI}$, this is derived in theory of simple bending.

Strain energy,
$$U = \frac{1}{2}M\phi = \frac{M^2 L}{2EI}$$

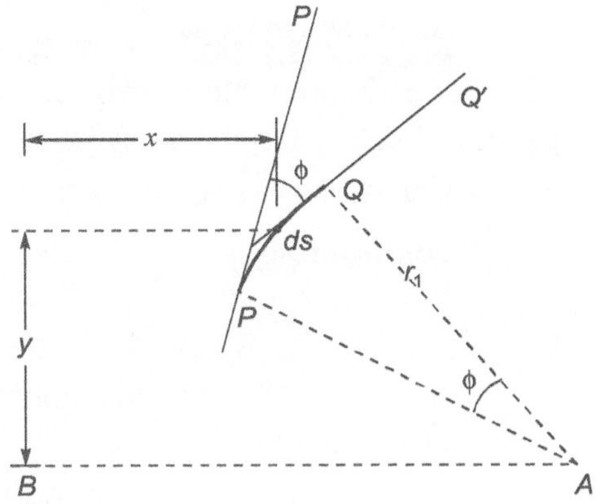

Fig. 13.7

Maximum bending stress

M_{max} = Maximum bending moment = $2R.R_v$ at point $C = 2M$

Maximum bending stress, $\sigma_{max} = \frac{2M}{Z} = \frac{2M}{bt^2} \times 6 = \frac{12M}{bt^2}$

where b = breadth of strip.

t = thickness of trip.

or $\sigma_{max} = 12\ M/bt^2$

or $M = \left(\frac{\sigma_{max}\, bt^2}{12}\right)$

$$U = \frac{M^2 L}{2EI} = \left(\frac{\sigma_{max}\, bt^2}{12}\right)^2 \frac{L}{2EI}, \text{ putting the value of } I = \frac{bt^3}{12}$$

$$\frac{\sigma_{max}^2}{24E}(btL) = \frac{\sigma_{max}^2}{24E} \times \text{Volume of the strip}$$

Question 13.10 A flat spiral spring is made of a strip 6 mm wide, 0.2 mm thick and 10 m long. Torque is applied at the winding spindle to give 8 complete turns. Determine (a) winding couple, M (b) energy stored (c) maximum bending stress developed at the point of greatest bending. E = 210 kN/mm^2

Solution:

$\phi = 2 \times 8\pi = 16\pi$, as 8 complete turns are given

L = 10,000 mm

$$I = \frac{bt^3}{12} = \frac{6 \times 0.2^3}{12} = 4 \times 10^{-3}\ \text{mm}^4$$

E = 210,000 N/mm^2

Winding couple, $M = \frac{EI\phi}{L} = \frac{210{,}000 \times 4 \times 10^{-3} \times 16\pi}{10{,}000} = 4.222$ Nmm

Energy stored, $U = \frac{1}{2}M\phi = \frac{1}{2} \times 4.222 \times 16\pi = 106.11$ Nmm

Maximum bending stress, $\sigma_{max} = \frac{12M}{bt^2} = \frac{12 \times 4.222}{6 \times (0.2)^2} = 211.11\ \text{N/mm}^2$

Question 13.11 A weight of 250 N is dropped on to a close-coiled helical spring through a height of 600 mm which produces a maximum instantaneous stress of 180 N/mm^2 in the spring. If the spring index is 10, determine (a) Instantaneous compression in spring and (b) Number of coils is spring.

Given wire diameter, d = 18 mm, G = 84000 N/mm^2

Solution:

W = 250 N

h = 600 mm

τ, maximum stress = 180 N/mm^2

Wire diameter, d = 18 mm

Spring index = 10

Mean coil diameter = 18 × 10 = 180 mm

$$W_e\text{, equivalent static load} = \frac{\pi d^3 \tau}{8D} = \frac{\pi \times 18^3 \times 180}{8 \times 180} = 2290\text{N}$$

Say δ = Instantaneous compression in spring

Loss of PE = Energy stored in spring

$$W(h+\delta) = \frac{1}{2} W_e \cdot \delta$$

$$250\,(600+\delta) = \frac{1}{2} \times 2290 \times \delta$$

δ = 167.6 mm, instantaneous compression in spring.

Number of coils

$$\delta = \frac{\delta n W_e D^3}{Gd^4}$$

$$n = \frac{\delta G d^4}{8 W_e \cdot D^3} = \frac{167.6 \times 84000 \times 18^4}{8 \times 2290 \times (180)^3}$$

= 13.83, number of coils in spring

Practice Q.13.11 **A weight of 200 N is dropped onto a close coiled helical spring from a height of 400 mm which produces a maximum instantaneous stress of 150 N/mm^2 in spring. If mean coil diameter is 12 times the wire diameter, determine (a) instantaneous compression in spring (b) number of coils in spring.**

Wire diameter, *d* = 12.5 mm, *G* for steel = 82000 N/mm^2.

Ans: [436 mm; 42.15 turns]

Question 13.12 **Consider a leaf spring, with n number of leaves, with central deflection δ_0, length of longest leaf *L* fitted in a vehicle as carriage spring. For a central load, *W*, determine bending stress in leaves if each leaf has breadth *b*, thickness *t*. What is initial radius of curvature, considering that leaves are bent in circular radius. What is the expression for proof load, i.e. load under which leaves become straight.**

Solution: Long plates in a carriage spring are of rectangular section $b \times t$. These are clamped with the help of bolts and nuts. At the ends of the longest leaf, there are eyes, which connect the spring to the chasis of a vehicle with the help of shackles.

To develop a simple theory between load, deflection, radius of curvature following assumptions are taken.

1. centre line of all leaves is initially a circular arc of radius R
2. contact between the plates is only at the ends
3. each plate overlaps the plate below it by an amount, $p = L/2n$, where L is the length of the longest leaf, n is number of plates
4. overlaps are tapered in width in triangular shape as shown in Figure 13.8.

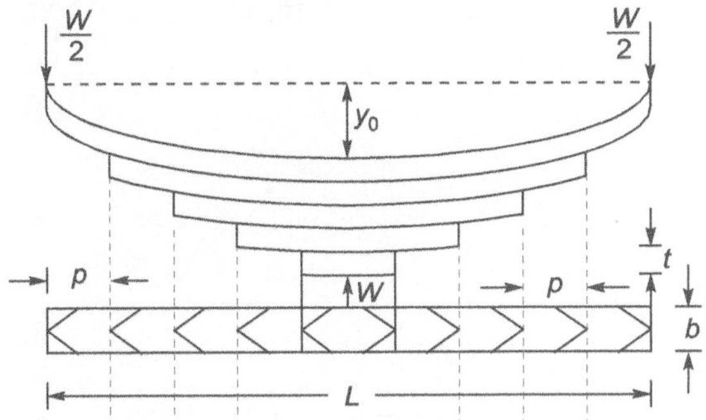

Fig. 13.8

Radius of curvature of each leaf is same, so each leaf touches the other leaf only at the ends. Leaves are of graduated length so as to maintain $\frac{M}{I}$ constant. At the centre $M_{max} = \frac{WL}{4}$, so maximum number of leaves at the centre.

$$I = \frac{nbt^3}{12}, \text{ at centre}, \qquad \text{where } n \text{ leaves are present}$$

Bending stress,
$$\sigma = \frac{WL}{4} \times \frac{t}{2} \times \frac{1}{I} = \frac{WLt}{8} \times \frac{12}{nbt^3} = \frac{3}{2} \times \frac{WL}{nbt^3} \quad \ldots\text{(i)}$$

Since M/I changes along the length, bending stress more or less remains constant along the length.
Using the properties of a circle

$$\delta_0 = \text{Initial central deflection} = \frac{L^2}{8R}$$

or
$$\text{initial radius of curvature, } R = \frac{L^2}{8\delta_0}$$

Proof Load

Initial radius of curvature –R

When leaves become straight, final radius of curvature, $R' = \alpha$ (infinity), reduction in curvature

$$\frac{1}{R'} = \frac{1}{R} - \frac{M}{EI} = \frac{1}{R} - \frac{W_0 L}{4} \times \frac{12}{Enbt^3}$$

$$R' \rightarrow \text{infinity, so } \frac{1}{R'} = 0$$

$$\frac{1}{R} = \frac{3W_0L}{Enbt^3}$$

or
$$\text{Proof load, } W_0 = \frac{Enbt^3}{3LR}$$

Question 13.13 A carriage spring centrally loaded has 6 steel plates, 6 mm thick and 60 mm wide. If the largest plate si 1 m long and load required to straighten the spring is 3.0 kN. Determine initial radius of curvature. What is the initial centre deflection. What is the bending stress in leaves due to Proof load, $E = 210$ kN/mm²

Solution:

Proof load, $W_0 = 3000$ N $E = 210{,}000$ N/mm²

$n = 6$ $b = 60$ mm

$t = 6$ mm $L = 1000$ mm

$$R = \frac{Enbt^3}{3W_0L}, \text{ putting the values}$$

$$R = \frac{210{,}000 \times 6 \times 60 \times 6^3}{3 \times 3000 \times 1000} = 1814.4 \text{ mm} = \text{Initial radius of curvature}$$

$$\delta_0 = \text{Initial central deflection} = \frac{L^2}{8R} = \frac{1000^2}{8 \times 1814.4} = 68.89 \text{ mm}$$

Bending stress,
$$\sigma_b = \frac{3}{2} \times \frac{W_0 L}{nbt^2} = \frac{3}{2} \times \frac{3000 \times 1000}{6 \times 60 \times 6^2} = 347.22 \text{ N/mm}^2$$

Practice Q.13.13 A carriage spring centrally loaded has 8 steel plates, 5 mm thick, 50 mm wide. If the largest plate is 960 mm long, find the initial radius of curvature if the maximum stress is 150 N/mm^2, when the plates become straight under the central load, E = 210 GPa.

Ans: [R = 3.5m]

Question 13.14 A semi elliptical spring, 1 m long carries a central load of 4.2 kN. Determine the number, breadth and thickness of leaves if central deflection is 50 mm. Assume that breadth is 6 times the thickness, the leaves are available in breadths of 5mm and thickness in multiple of 1 mm. The maximum bending stress is limited to 300 N/mm^2. E = 200 GPa.

Solution:

$W = 4200$ N $\quad\quad L = 1000$ mm

$E = 200{,}00$ N/mm^2 $\quad\quad \delta = 50$ mm

$$\text{We know that } \sigma = \frac{3}{2}\times\frac{WL}{nbt^2}$$

$$\sigma = \frac{Et}{2R}$$

$$R = \frac{Et}{2\sigma} = \frac{Et\times nbt^2}{2\times 3WL} = \frac{Enbt^3}{3Wl}$$

$$\delta = \frac{L^2}{8R} = \frac{L^2}{8}\times\frac{3WL}{Enbt^3} = \frac{3WL^3}{8Enbt^3}$$

$$50 = \frac{3}{8}\times\frac{4200}{200{,}000}\times\frac{1000^3}{nbt^3}$$

$$nbt^3 = 157500 \quad\quad ...(i)$$

$$\sigma = \frac{3WL}{2nbt^2}$$

$$300 = \frac{3\times 4200\times 1000}{2nbt^2}$$

$$nbt^2 = 21000 \quad\quad ...(ii)$$

from equations (i) and (ii)

$$\text{Thickness } t = \frac{157500}{21000} = 7.5 \text{ mm}$$

Let us take t = 8 mm

$$b = 6\times t = 6\times 8 = 48 \text{ mm}$$

$$b = 50 \text{ mm (as breadth available is multiple of 5 mm)}$$

Taking t = 8 mm, b = 50 mm

$$nbt^2 = 21000$$

$$n = \frac{21000}{50\times 8^2} = 6.5625 = 7 \text{ leaves}$$

So, $\quad n = 7,\ b = 50 \text{ mm},\ t = 8 \text{ mm}$

Practice Q.13.14 A semi elliptical carriage spring is made of leaves, 1 m long to support a central load of 8 kN with a maximum deflection of 60 mm and maximum bending stress of 320 N/mm²: E = 200 GPa, calculate thickness of leaves and decide the number and breadth. Thickness is available in multiples of 1 mm and breadth in multiple of 5 mm. (Take $b = 10t$)

Ans: [t = 7 mm, b = 70 mm, n = 11 leaves]

Question 13.15 A close coiled helical spring is made of round steel wire. It carries an axial load of 150 N and is to just get over rod of 36 mm diameter. Deflection in the spring is not be exceed 25 mm. The maximum allowable shear stress developed in spring wire is 200 N/mm². (Neglecting the effect of Wahl's factor), G = 80,000 N/mm². Find D, d, n.

Solution: Rod diameter = 36 mm = Inner coil diameters

Say d = wire diameter

Mean coil diameter, $D = 36 + d$

$$d = 25\text{mm} = \frac{8WD^3n}{Gd^4}$$

W = 150 N, putting the values

$$25 = \frac{8\times150\times(36+d)^3 n}{80000\times d^4}$$

or
$$\frac{(36+d)^3 n}{d^4} = 1666.667$$

$$\text{Moreover } \tau = 200 = \frac{8WD}{\pi d^3} = \frac{8\times150\times(36+d)}{\pi\times d^3}$$

or
$$\left(\frac{36+d}{d^3}\right) = 0.5236$$

or
$$\frac{(36+d)^3}{d^9} = 0.14355$$

From equation (i) and (ii)

$$nd^5 = \frac{1666.67}{0.14355} = 11610.36$$

Moreover from equation (ii)

$$\frac{36+d}{d^3} = 0.5236$$

or
$$36 + d = 0.5236\,d^3$$

Taking $d = 4.3$

$$36 + 4.3 \neq 41.6$$

or
$$d = 4.25 \text{ mm}$$

$$\text{LHS} = 40.25$$

$$\text{RHS} = 0.5236 \times 4.25^3 = 40.19$$

$$\text{LHS} \approx \text{RHS}$$

Wire diameter, d = 4.25 mm

Mean coil diameter, D = 36 + 4.25 = 40.25 mm

Number of turns,
$$n = \frac{11610.36}{d^5} = \frac{11610.36}{(4.25)^5} = 8.4$$

Practice Q.13.15 A close coiled helical spring is made of round steel wire. It carries an axial load of 200 N and is to just get over a rod of 35 mm diameter. Deflection in the spring is not to exceed 20 mm. The maximum allowable stress in shear in spring wire is 165 N/mm^2. G for steel = 80,000 N/mm^2. Find D, d, n of spring. (Neglect Wahl's factor)

Ans: [40 mm, 5 mm, 9.76]

Question 13.16 A safety valve 60 mm in diameter is to blow off at a pressure of 3 N/mm^2, gauge. Safety valve is held by a close-coiled helical spring of steel with a mean coil diameter of 120 mm. What is the diameter of spring wire, if τ_{max} = 220 N/mm^2. Taking into account the stress augmentation factor.

Solution: Mean coil diameter = 120 mm

Diameter of safety valve = 60 mm

Pressure = 3 N/mm^2

Force on spring,

$$W = \frac{\pi}{4} \times 60^2 \times 3 = 8482.3\,\text{N}$$

$$\tau_{max} = \frac{8WD}{\pi d^3} + \frac{4W}{\pi d^2} = 220$$

Putting the values

$$\frac{8 \times 8482.3 \times 120}{\pi d^3} + \frac{4 \times 8482.3}{\pi d^2} = 220$$

$$2592000 + 10800\,d = 220d^3$$

$$d^3 - 49.095\,d + 11781.8 = 0$$

$$d^3 = 49.095\,d + 11781.8$$

$$(d^3 - 49.095\,d) = 11781.8$$

If d = 24 mm

LHS = 13284 – 1178.28=12105

RHS ≠ RHS

Taking d = 23.5

LHS = 12977.88 – 1153.77 = 11824

LHS ≈ RHS

Wire diameter = 23.6 mm

LHS = 13144.26 – 1158.64 = 11985.62

LHS ≈ RHS

Wire diameter = 23.6 mm

Question 13.17 A close coiled helical spring of 18 mm mean coil diameter and 10 turns is arranged within and concentric with outer string. The free length of inner spring is 6 mm more than that of the outer spring. The outer spring has 12 coils of mean diameter 27 mm and a wire diameter of 3.2 mm. The spring load against which a valve is opened is provided by the inner spring. The initial compression in outer spring is 8 mm when the valve is closed, find the stiffness of inner spring if the greatest force required to open the valve by 10 mm is 160 N. Also find the wire diameter of the inner spring. Given G = 80,000 N/mm^2.

Solution: Initial compression in outer spring = 8 mm

So initial compression in inner spring = 8 + 6 = 14 mm

(Initial compression of outer spring + Extra free length of inner spring)

Say K_1 = Stiffness of inner spring in N/mm

K_2 = Stiffness of inner spring in N/mm

Initial load on valve, $F_i = 14\ K_1 + 8\ K_2$

Outer spring

$n = 12,\ D = 27$ mm, $d = 3.2$ mm

$$K_2 = \frac{Gd^4}{8nD^3} = \frac{80{,}000 \times 3.2^4}{8 \times 12 \times 27^3} = 4.44 \text{ N/mm}$$

Valve is opened by 10 mm, additional force required to open the valve $F_a = 10K_1 + 10K_2$

Total load to lift the valve $= 24\ K_1 + 18\ K_2 = 160$ N

$18K_2 = 18 \times 4.44 = 79.92$ N

$24\ K_1 = 160 - 79.92 = 80.08$ N

$K_1 = 3.3366$ N/mm

$$K_1 = \frac{Gd_1^4}{8D_1^3 n_1}$$

$n_1 = 10,\ D_1 = 18$ mm

$$d_1^4 = \frac{K_1 8 D_1^3 n_1}{G} = \frac{3.3366 \times 8 \times 18^3 \times 10}{80{,}000} = 19.46$$

$d_1 = 2.1$ mm, wire diameter of inner spring

Question 13.18 A stiff bar of negligible weight transfers a load P to a combination of three helical springs arranged in parallel as shown in from 13.9. The springs are made of the same material and out of rods of equal diameters. They are of the same free length before loading. The number of coils in these three springs are 10 : 12 : 15 respectively. While the mean coil diameters are in ratio of 1 : 1.2 : 1.4 respectively.

Find the diameter x as shown in figure, such that the stiff bar remains horizontal after the application of load P.

[IES-2010: 10 Marks]

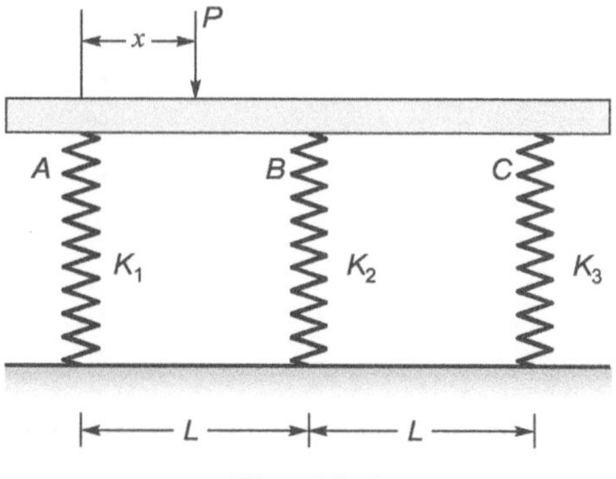

Fig. 13.9

Solution:

$$K_1 = \frac{Gd^4}{8 \times 10 \times D^3} = \frac{Gd^4}{80D^3}$$

$$K_2 = \frac{Gd^4}{8 \times 12 \times (1.2D)^3} = \frac{Gd^4}{165.888D^3}$$

$$K_3 = \frac{Gd^4}{8 \times 15 \times (1.4D)^3} = \frac{Gd^4}{329.28}$$

From this, $K_1 = 2.0736\ K_2,\ K_2 = 0.48225\ K_1$

$K_1 = 4.116\ K_3,\ K_3 = 0.24295\ K_1$

Total load, $P = ck_1 + ck_2 + ck_3$; c is constant

$$ck_1\,[1 + 0.48225 + 0.24295] = P$$

$$ck_1 = \frac{P}{1.72545} = 0.580P$$

$$ck_2 = 0.280\,P$$

$$ck_3 = 0.140\,P$$

Moments about A

$$P \cdot x = 0.280 \times L \times P + 0.140 \times 2L \times P$$

$$x = 0.56L$$

Question 13.19 If the wire diameter of a compressive helical spring is increased by 2%, the change in spring stiffness (in %) ________ (correct to two decimal places).

[ME, GATE : 2018 (set-1)]

Solution:

$$K \propto d^4$$

$$K' = (1.02)^4 \times d^4$$

$$= 1.0824\,d^4$$

% increase 8.24.

Objective Type Questions

Q.1 The figure shows arrangements of springs they have stiffness K_1 and K_2 as marked. Which of the following arrangements offers a stiffness = $\frac{2K_1K_2}{K_1+2K_2}$

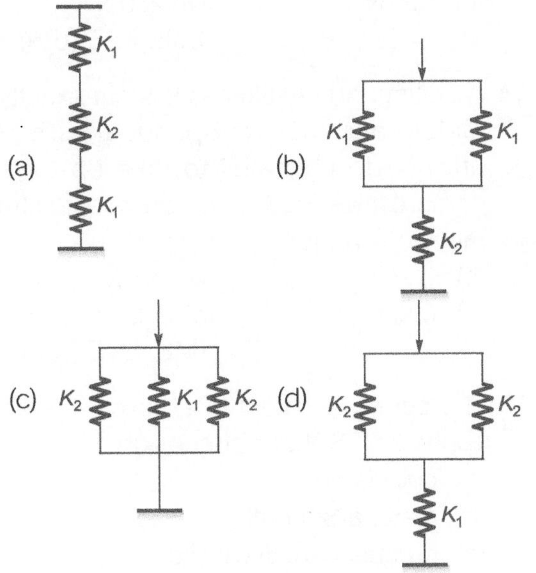

Q.2 The deflection of a spring with 20 active coils is under a load of 1000 N is 10 mm. The spring is made into two pieces each of 10 coils and placed in parallel under the some load. The deflection of this system is

(a) 20 mm (b) 10 mm
(c) 5 mm (d) 2.5 mm

[GATE 1995: 2 Marks]

Q.3 A weighing machine consists of 2 kg pan resting on a spring. In this condition with pan resting on the spring, the length of spring is 200 mm. When a mass of 20 kg is placed on the pan , the length of the spring becomes 100 mm. For the spring, the underformed length and the spring constant K are

(a) L = 200 mm, K = 1862 N/m
(b) L = 210 mm, K_i = 1960 N/m
(c) L = 200 mm, K = 1960 N/m
(d) L = 200 mm, K = 2150 N/m

[GATE 2005: 2 Marks]

Q.4 A compression spring is made of music wire of 2 mm diameter having a shear strength and shear

modulus of 800 MPa and 80 GPa. Mean coil diameter is 20 mm, free length is 40 mm and the number of active coils is 10. If the mean coil diameter is reduced to 10 mm. The stiffness of spring is approximately.
(a) decreased by 8 times
(b) decreased by 2 times
(c) increased by 2 times
(d) increased by 8 times

[GATE 2008: 2 Marks]

Q.5 If the rigid rod fitted snugly between the supports as shown in figure below is heated, the stress induced is due to 20°C rise in temperatures will be ($\alpha = 12.5 \times 10^{-6}$ /°C and E = 200 GPa)

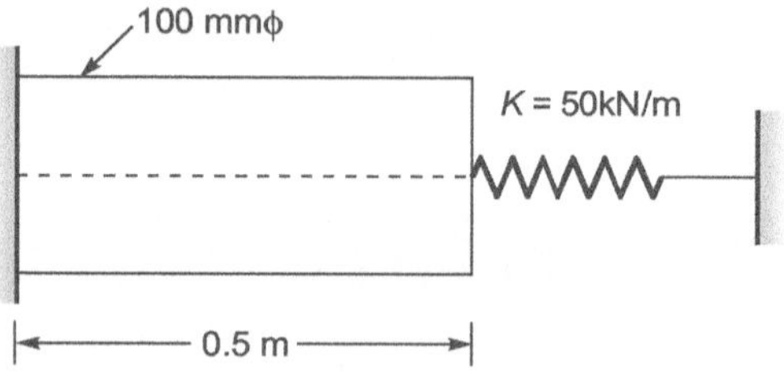

(a) 0.07958 MPa (b) –.07958 MPa
(c) –.03972 MPa (d) +.03972 MPa

Q.6 A body of weight 100 N is dropped from a height of 10 cm over a coiled helical spring of stiffness 200 N/mm. The resulting deflection of spring is nearly
(a) 5 cm (b) 16 cm
(c) 35 cm (d) 100 cm

[IES-2001]

Q.7 A close coiled helical spring absorbs 0.5 Nm of strain energy while extending by 10 mm. What is stiffness of the spring
(a) 50 N/mm (b) 20 N/mm
(c) 10 N/mm (d) None of these

Q.8 A steel spring has stiffness equal to 20 N/mm. Steel spring is replaced by Bronze spring, with all the parameters remaining the same. If $G_{steel} = 2G_{bronze}$, what is stiffness for bronze spring
(a) 10 N/mm (b) 20 N/mm
(c) 40 N/mm (d) None of these

Q.9 Wahl's factor for a spring is 1.26. If spring index is 10, what is curvature factor for spring
(a) 1.145 (b) 1.2
(c) 1.323 (d) None of these

Q.10 A close coiled helical spring is subjected to an axial moment M, producing an angle of rotation of 60° at the free end with respect to the fixed end. If strain energy absorbed is 50π, what is M
(a) 50 Nmm (b) 100 Nmm
(c) 200 Nmm (d) 300 Nmm

Q.11 A flat spiral spring is subjected to a winding couple producing σ_{max} = 100 MPa at the point of greatest bending. If E = 200, 000 N/mm^2, what is strain energy stored per unit volume
(a) 0.05 Nmm (b) 0.025 Nmm
(c) 0.004166 Nmm (d) 0.00208 Nmm

Q.12 A leaf string is loaded such that its leaves become straight under this load. How this load is classified
(a) Safe load (b) Proof load
(c) Ultimate load (d) None of these

Q.13 A helical spring of 10 N/mm rating is mounted on top of another helical spring of 8 N/mm rating. The force required for a total combined deflection of 45 mm though the two springs is
(a) 100 N (b) 150 N
(c) 200 N (d) 250 N

[ME, ESE (pre) : 2016]

Q.14 A coil-spring of stiffness k is cut exactly at the middle and the two springs thus made are arranged in parallel to take up together a compressive load. The equivalent stiffness of the two springs is
(a) 0.25 k (b) 0.5 k
(c) 2 k (d) 4 k

[ME, ESE (pre) : 2016]

Q.15 The spring constant of a helical compression spring DOES NOT depends on
(a) coil diameter
(b) material strength
(c) number of active turns
(d) wire diameter

[ME, GATE : 2016 (set-1)]

Q.16 A closely-coiled helical spring is made of 10 mm diameter steel wire, with the coil consisting of 10 turns with a mean diameter 120 mm. The spring carries and axial pull of 200 N. What is the value of shear stress induced in the spring neglecting the effect of stress concentration and of deflection *m* the spring, when the modulus of rigidity is 80 kN/mm^2?

(a) 63.5 N/mm^2 and 34.6 mm
(b) 54.2 N/mm^2 and 34.6 mm
(c) 63.5 N/mm^2 and 42.6 mm
(d) 54.2 N/mm^2 and 42.6 mm

[ME, ESE (pre) : 2016]

Answers

1. (d)	2. (d)	3. (b)	4. (d)	5. (b)
6. (b)	7. (c)	8. (a)	9. (b)	10. (d)
11. (d)	12. (b)	13. (c)	14. (d)	15. (b)
16. (a)				

Explanations

1. (d)

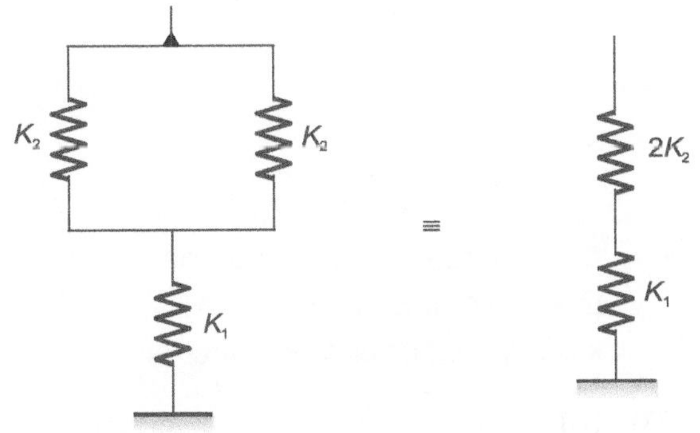

$$K_{eq} = \frac{2K_1K_2}{K_1 + 2K_2}$$

2. (d)

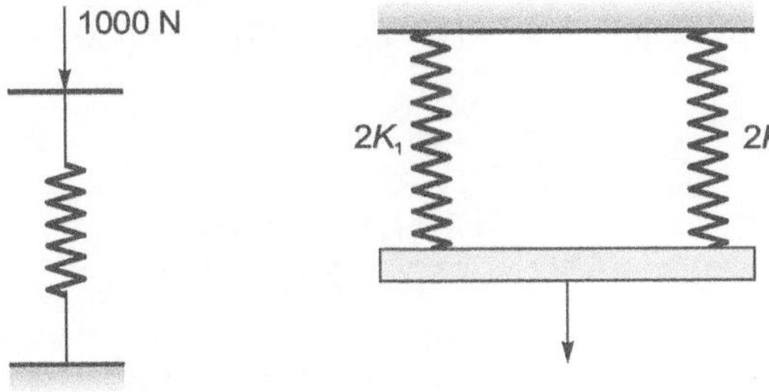

$$K_1 = \frac{1000}{10} = 100 \text{ N/mm}$$

$$K_{eq} = 4\ K_1 = 400 \text{ N/mm}$$

$$\delta = \frac{1000}{400} = 2.5 \text{ mm}$$

3. (b)

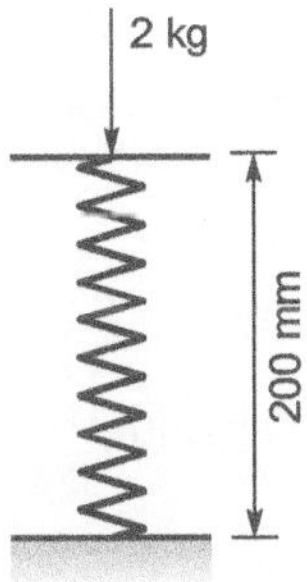

$$\Delta W = 20 \text{ kg}$$

$$\Delta = 100 \text{ mm} = 0.1 \text{ m}$$

$$K = \frac{20 \times 9.80}{0.1} = 1960 \text{ N/m}$$

$$L = 200 + \frac{2}{20} \times 100 = 210 \text{ mm}$$

4. (d)

$$K = \frac{Gd^4}{64nR^3} \qquad R_1 = 10 \text{ mm},\ R_2 = 5 \text{ mm}$$

$$K' = \frac{Gd^4}{64n \times 5^3}$$

$$K = \frac{Gd^4}{64n \times 10^3} \qquad K' = 8K$$

5. (b)

$$\Delta L = 0.5 \times 12.5 \times 10^{-6} \times 20$$
$$= 12.5 \times 10^{-5} \text{ m}$$

$$\text{Spring force} = K \cdot \Delta L = 12.5 \times 10^{-5} \times 50 \text{ kN/m}$$
$$= 6.25 \text{ N}$$

$$\text{Area of rod} = \frac{\pi}{4}\times 10^2 = 78.54\text{ mm}^2$$

$$\text{Stress rod in} = -\frac{6.25}{78.54} = -0.07958\text{N/mm}^2$$

6. (b)

$$Mgh = \frac{1}{2}kx^2 - mgx$$

$$1000\times 10 = \frac{1}{2}\times 200\times x^2 - 100x$$

$$10000 = 100x^2 - 1000x$$

$$x^2 - 10x - 100 = 0$$

$$x = \frac{+10+\sqrt{100+400}}{2} = \frac{10+\sqrt{500}}{2}$$

$$= \frac{10+22.36}{2}$$

$$= 16.18\text{ cm} \simeq 16\text{ cm}$$

7. (c)

$$U = 500\text{ Nmm} = \frac{1}{2}k\delta^2 = 50k$$

$$k = 10\text{ N/mm}$$

8. (a)

$$k \propto G$$

$$k_{bronze} = \frac{1}{2}k_{steel} = 10\text{ N/mm}$$

9. (b)

$$K_s = 1+\frac{1}{2c} = 1.05$$

$$K_w = 1.26$$

$$K_c = \frac{1.26}{1.05} = 1.2$$

10. (d)

$$U = 50\pi$$

$$= \frac{1}{2}M\phi = \frac{1}{2}\times M\times\frac{\pi}{3}$$

$$M = 300\text{ Nmm}$$

11. (d)

$$U_s = \frac{\sigma^2}{24E} = \frac{100^2}{24\times 200{,}000}$$

$$= 0.00208\text{ Nm}$$

12. (b)

Proof load.

13. (c)

$$\frac{1}{k} = \frac{k_1+k_2}{k_1k_2}$$

$$k = \frac{k_1k_2}{k_1+k_2} = \frac{8\times 10}{18} = \frac{80}{18}\text{ N/mm}$$

$$\delta = 45\text{ mm}$$

$$F = \frac{80}{18}\times 45 = 200\text{ N}$$

14. (d)

A spring is cut at middle k_1 of half spring = 2k

Two half springs are arranged in parallel

$k_2 = 2k_1 = 2 + 2 = 4$ k

15. (b)

$$k = \frac{Gd^4}{64nR^3} = \frac{Gd^4}{8nD^3}$$

d = wire diameter, n = number of turns

D = coil diameter

k does not depend on material strength.

16. (a)

Mean coil radius = 60 m

$$k = \frac{Gd^4}{64nR^3} = \frac{80000\times 10^4}{64\times 10\times 60^3}$$

$$\delta = \frac{200}{5.787} = 34.56\text{ mm}$$

Torque, T = wR = 200 × 60 = 12000 mm

$$T = \frac{\pi d^3}{16}\times\tau$$

$$12000 = \frac{\pi\times 10^3}{16}\times\tau$$

$$\tau = \frac{192}{\pi} = 61.116\text{ N/mm}^2$$

$$\tau_{max} = 61.116+\frac{4W}{\pi d^2}$$

$$= 61.116+\frac{4\times 200}{\pi\times 100}$$

$$= 61.116 + 2.546$$

$$= 63.66\text{ N/mm}^2$$

Struts and Columns

CHAPTER 14

Beams and columns are essential components of a building or a structure. Beams are horizontal slabs while columns are vertical members supporting the slabs and the roof. **Any member in compression is known as strut but when strut becomes vertical, it is termed as a column.** For an earth quake resisting building, strong columns are a must.

In this chapter we will study about the buckling loads of a column or a strut using various theories starting from the basic equation of bending moment relating with curvature, equations are derived for buckling loads of a column considering various end conditions of the column.

Question 14.1 **What is Euler's theory of buckling of columns? What are the assumptions used in this theory? Taking an example of column with fixed at one end and free at the other end, derive formula for Euler's buckling load.**

Solution: Euler has developed a **buckling theory for long columns**, but for short columns use of Euler's Theory gives absurd results, However Euler has provided a basic theory of buckling which has been modified by various researchers as Rankine, Johnson, Professor Perry, Robertson and others.

Euler has developed buckling theory but considered the column in already buckled state, Following assumptions were taken by him.

(a) material of the column is homogeneous and isotropic.

(b) compressive load acting on the column is fully axial.

(c) column fails only due to buckling.

(d) column is initially straight and remains straight till at a particular load it suddenly buckles.

(e) weight of the column is negligible

(f) fixed ends are rigid, i.e. slope and deflection remain zero at fixed ends.

(g) hinged ends are frictionless, and cannot take any moment.

Consider a column AB, of length L, fixed at end A and free at end B, At B an axial load is applied. This axial load is gradually increased and at a particular stage, the column suddenly buckles, Say at this stage buckling load is P.

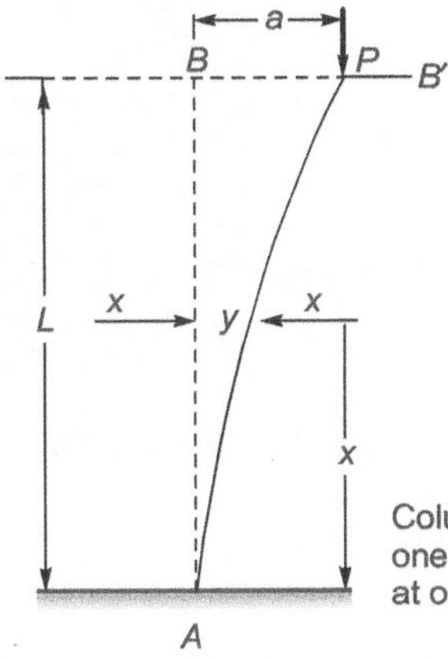

Column fixed at one end and free at other end.

Fig. 14.1

Consider a section X-X, at a distance of x from A, Say deflection in column is y at section X-X.
Deflection in column at free end is 'a' as shown,
Bending moment at section, $M_x = P(a - y)$

or, $$EI\frac{d^2y}{dx^2} = P(a-y) \quad \text{...(i)}$$

or, $$\frac{d^2y}{dx^2} + \frac{Py}{EI} = \frac{Pa}{EI} \quad \text{...(ii)}$$

Solution of this differential equation is

$$y = A\cos nx + B\sin nx + a \quad \text{...(iii)}$$

where $$n = \sqrt{\frac{P}{EI}}$$

Correctness of the solution can be verified by taking derivative.
At end A, fixed and $x = 0$, $y = 0$, deflection is zero

So, $0 = A + a$

So, deflection $y = -a\cos nx + B\sin nx + a$ ...(iv)

or, $$\frac{dy}{dx} = +an\sin nx + B\sin nx \quad \text{...(v)}$$

$$\frac{dy}{dx} = 0 \text{ at } x = 0$$

$0 = B_n$

$n \neq 0$, as n (contains) P, and load cannot be zero

$B = 0$

Finally $y = -a\cos nx + a$ ...(vi)

at the end B, $x = L$, deflection, $y = a$

or $a = -a\cos nL + a$

or, $0 = -a\cos nL$

or $$\cos nL = 0 = \cos\left(\frac{\pi}{2}, \frac{3\pi}{2} \ldots\ldots \frac{5\pi}{2} \ldots\ldots \frac{2(n+1)\pi}{2}\right)$$

but $\pi/2$ is the minimum significant value.

So $$\cos nL = \cos\frac{\pi}{2}$$

$$nL = \pi/2$$

$$n^2L^2 = \frac{\pi^2}{4}, \quad \text{putting the value of } n^2$$

$$\frac{P}{EI} \times L^2 = \frac{\pi^2}{4}$$

$P_{e'}$ Euler's Buckling load $$= \frac{\pi^2 EI}{4L^2}$$

If in a column section I_{max}, I_{min} are two values of second moment of area, Then

$P_{e'}$ Euler's buckling load $$= \frac{\pi^2 EI_{min}}{4L^2} \quad \text{...(vii)}$$

Practice Q.14.1 Consider a column *AB*, Length *L*, hinged at both the ends. If EI is the flexural rigidity of the column, derive expression for Euler's buckling load.

Hints: $[M_x = P \cdot y,\ y = A\cos nx + B\sin nx]$
Boundary condition $x = 0$, $y = 0$, $x = L$, $y = 0$.

Ans. $\left[P_e = \dfrac{\pi^2 EI_{min}}{L^2}\right]$

Question 14.2 Consider a column with one end fixed and other end hinged. Length of column is *L* and flexural rigidity is EI. Derive expression for its Euler's buckling load.

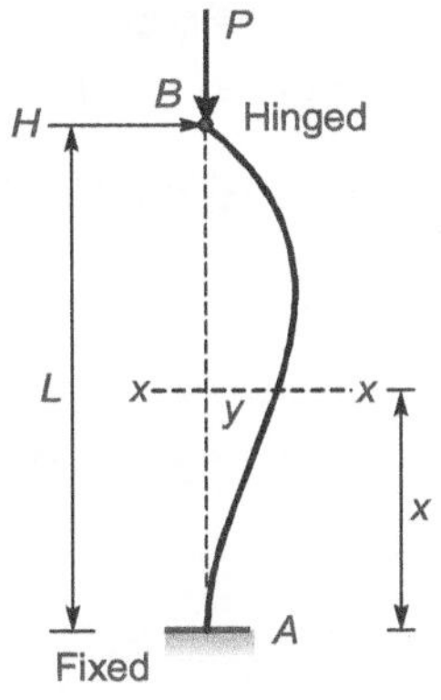

Fig. 14.2

Solution: Figure shows a column AB, fixed at end A hinged at other end B, Column has buckled under load *P*. Horizontal reaction at *B* is *H*.

Consider a section XX, at a distance of x from *A*.

Equation of bending moment, $M_x = H(L-x) - Py$

or, $$EI\frac{d^2y}{dx^2} = H(L-x) - Py \quad \text{...(i)}$$

Solution of the differential equation is

$$y = A\cos nx + B\sin nx + \frac{H(L-x)}{P} \quad \text{...(ii)}$$

at end *A*, $x = 0$, $y = 0$, putting the value

$$0 = A \times \cos 0 + B\sin 0 + \frac{H(L-0)}{P}$$

$$0 = A + 0 + \frac{HL}{P}$$

or constant, $$A = -\frac{HL}{P} \quad \text{...(iii)}$$

$$\frac{dy}{dx} = -An\sin nx + Bn\cos nx - \frac{H}{P} \quad \text{...(iv)}$$

$$\frac{dy}{dx} = 0 \text{ at } x = 0, \text{ end } A$$

$$0 = An\sin 0 + B \times n\cos 0 - \frac{H}{P}$$

constant $$B = \frac{H}{nP}$$

Putting the values of *A* and *B*, in equation (2)

$$y = -\frac{HL}{P}\cos nL + \frac{H}{nP}\sin nL + \frac{H(L-x)}{P}$$

Moreover at end *B*, $y = 0$, $x = L$, putting this value

or, $$0 = \frac{H}{nP}\sin nL = \frac{HL}{P}\cos nL$$

or $$\tan nL = nL$$

or $\tan\theta = \theta$, this is possible at

$$\theta \simeq \sqrt{2\pi^2}$$
$$n^2L^2 = 2\pi^2$$
$$\frac{P}{EI} \times L^2 = 2\pi^2$$

P_e, Euler's Buckling load $= \dfrac{2\pi^2 EI_{min}}{L^2}$, for one end fixed, other end hinged.

Practice Q.14.2.1 A column of length L, fixed at both the ends as shown in figure 14.3. It buckets at load *P*. Derive expression for Euler's Buckling load.

Hints: [There are fixing couples at ends i.e., M_A, M_B]

$$y = A\cos nx + B\sin nx + \frac{M_A}{P}$$

End conditions are $x = 0,\ y = 0,\ \dfrac{dy}{dx} = 0$

$x = L,\ y = 0,\ \dfrac{dy}{dx} = 0$

Ans. $\left[P_e = \dfrac{4\pi^2 EI_{min}}{L^2}\right]$

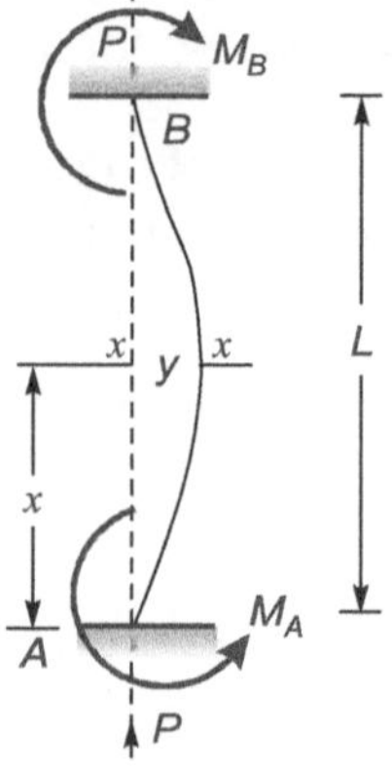

Fig. 14.3

Question 14.2.2 A column is of circular section of steel of diameter 50 mm, length 2 m. Determine its buckling load when (a) both the ends are hinged and (b) both the ends are fixed, E = 200 GPa.

Solution:

Length, L = 2 m = 2000 mm

Diameter, d = 50 mm

$$I = \frac{\pi}{64} \times 50^4 = 30.68 \times 10^4 \text{ mm}^4$$

E = 200,000 N/mm^2

EI = 200,000 × 30.68 × 10^4 = 61.36 × 10^9 Nmm^2

(a) Both the ends are hinged

$$P_e = \frac{\pi^2 EI}{L^2} = \frac{\pi^2 \times 61.36 \times 10^9}{(2000)^2} = 151.4 \times 10^3 \text{ N} = 151.4 \text{ kN}$$

(b) Both the ends are fixed

$$P_e = \frac{4\pi^2 EI}{L^2} = 4 \times 151.4 = 605.6 \text{ kN}$$

Practice Q.14.2.3 A column is of I section, length 2.5 m, I section consists of flanges breadth 120 mm, depth 15 mm, web 120 mm × 10 mm. E = 200 GPa. What is Euler's buckling load if (a) one end of the column is fixed, other end free and (b) one end hinged and other end fixed.

Hints: [I_{max} = 1791 × 10^4 mm^4, I_{min} = 433 × 10^4]

Ans. [341.9 kN, 2735 kN]

Question 14.3 What do you understand by Equivalent Length of a column? Explain with the help of different end conditions.

Solution: **Equivalent length of a column is defined as that length (for different end conditions) which is equivalent length of the column with both ends hinged.** The equivalent length can be obtained by completing the flexure curve of the column. As an example, in this case of a column with one end fixed, other end free, flexure curve *AB* is completed by drawing the flexure curve on the other side also as shown in figure 14.4 (a) such that $L_e = 2L$,

Equivalent length, $L_e = 2L$. Similarly for other end conditions, flexure curve as shown in figure 14.4 (b), 14.4 (c), 14.4 (d)

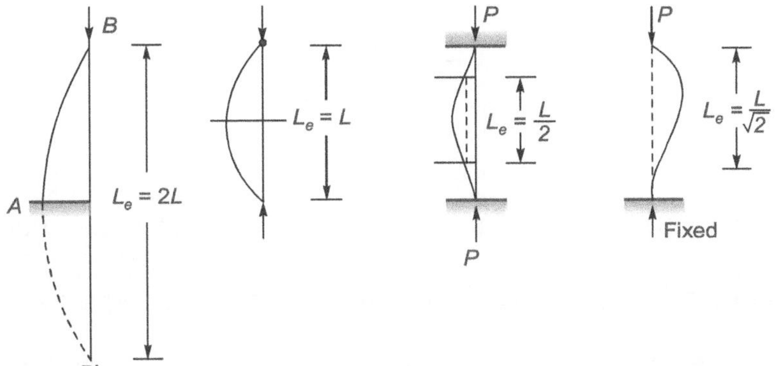

(a) one end fixed, other end free $L_e = 2L$

(b) Boths ends hinged $L_e = L$

(c) Boths ends Fixed $L_e = \frac{L}{2}$

(d) One end fixed, other ends hinged $L_e = \frac{L}{\sqrt{2}}$

Fig. 14.4

Euler's buckling load, $P_e = \dfrac{\pi^2 EI_{min}}{L_e^2}$, where

$L_e = 2L$, for one end fixed, other end free

$L_e = L$, for both ends hinged

$L_e = \dfrac{L}{2}$, for both ends fixed

$L_e = \dfrac{L}{\sqrt{2}}$, for one end fixed, other end hinged as shown in figure 14.4

Question 14.4 An allowable load of a column with both ends hinged is 40 kN. What will be the allowable load for the same column if one end is fixed end other end remains hinged.

Solution:

$$P_e = \frac{\pi^2 EI}{L^2} \text{ for both ends hinged} = 40 \text{ kN}$$

$$P_e' = \frac{2\pi^2 EI}{L^2}, \text{ for one end fixed, other end hinged}$$

$$= 2 \times 40 = 80 \text{ kN}$$

Practice Q.14.4 An allowable load of a column with both ends fixed as per Euler's theory is 30 kN. If now one end is hinged and other end of the column remains fixed. What will be the allowable load for this column?

Hints: $\left[L_e = \frac{L}{2} \text{ for both ends fixed, } L_e = \frac{L}{\sqrt{2}} \text{ for one end hinged other end fixed}\right]$

Ans. [15 kN]

Question 14.5 Euler's buckling formula has been derived for long column. What is the limit on slenderness ratio, below which the formula is not applicable?

Solution:

Euler's buckling load,
$$P_e = \frac{\pi^2 EI_{min}}{L_e^2}$$

or,
$$P_e = \frac{\pi^2 EAk_{min}^2}{L_e^2}$$

where k_{min} = minimum radius of gyration, A = area of cross section, L_e = Equivalent length of column

Ratio $\frac{L_e}{k_{min}}$ = slenderness ratio, an important consideration while designing a column, ratio of equivalent length and minimum radius of gyration.

or,
$$\left(\frac{L_e}{k_{min}}\right)^2 = \pi^2 E\left(\frac{A}{P_e}\right)$$

or,
$$\frac{P_e}{A} = \frac{\pi^2 E}{1}\left(\frac{k_{min}}{L_e}\right)^2$$

if $\frac{P_e}{A} > \sigma_c$, crushing strength of the material, this is not possible

So $\frac{P_e}{A} < \sigma_c$, ultimate crushing strength of the material

or
$$\pi^2 E\left(\frac{k_{min}}{L_e}\right)^2 < \sigma_c$$

or
$$\frac{L_e}{k_{min}} > \sqrt{\frac{\pi^2 E}{\sigma_c}}$$
, where $\frac{L_e}{k_{min}}$ is slenderness ratio.

For mild steel

$$E = 210 \text{ GPa} = 210 \times 1000 \text{ N/mm}^2$$
$$\sigma_c = 320 \text{ N/mm}^2 = \text{crushing strength}$$
$$\frac{L_e}{k_{min}} > \sqrt{\frac{\pi^2 \times 210{,}000}{320}}$$
$$\frac{L_e}{k_{min}} > 80.48$$

For mild steel column slenderness ratio must be greater than 81 so that Euler's formula can be used for the prediction of Buckling load.

Question 14.6 For what length of a mild steel bar 80 mm in diameter used as a strut with both ends fixed, the Euler's formula is applicable. If $\sigma_c = 330$ N/mm² $E = 208000$ N/mm²

Solution:

$$\frac{L_e}{k_{min}} = \pi\sqrt{\frac{E}{\sigma_c}} > \pi\sqrt{\frac{208000}{330}} > 78.8$$

Diameter, $d = 80$ mm

$$k_{min} = \frac{d}{4} = 20 \text{ mm}$$

$$L_e > 78.8 \times 20 > 1577 \text{ mm}$$

End conditions are both ends fixed

$$L_e = 0.5\text{ L}$$

or $L = 2\,L_e = 1577 \times 2 = 3154 \text{ mm} = 3.154 \text{ m}$

Practice Q.14.6 For what length of a *CI* column of 120 mm external diameter and 60 mm internal diameter, the Euler's theory is applicable if $E = 105000$ N/mm², $\sigma_c = 550$ N/mm². One end of the column is fixed and other end is hinged.

Hints: $\left[k_{min}^2 = \frac{D^2 + d^2}{16}\right]$

Ans. [2.058 m]

Question 14.7 Discuss the validity of Euler's Theory of buckling.

Solution: Euler's bucking load for a column,

$$P_e = \frac{\pi^2 EI}{L^2} = \frac{\pi^2 EAk_{min}^2}{L^2}$$

or $\frac{P_e}{A} = \sigma_b = \frac{\pi^2 Ek_{min}^2}{L^2}$

where, σ_b = buckling stress due to Euler's buckling load

if σ_c = crushing stress for a short column= $\frac{P_c}{A}$, ultimate crushing load

Thus $\sigma_b = \sigma_c$, the Euler's formula to be valid

If we take $\frac{L^2}{k_{min}^2}$ as λ^2 (square of slenderness ratio)

Thus $\lambda^2 \geq \frac{\pi^2 E}{\sigma_c}$

$$\lambda \geq \sqrt{\frac{\pi^2 E}{\sigma_c}}$$

If $\lambda > \lambda_c$, failure takes place due to buckling for

$$\lambda = \lambda_c$$
$$\sigma_b = \sigma_c$$

failure stress becomes σ_c

For different regions of the curve, stress $vs \cdot \lambda$

(A) Short and unsafe column
(B) Short and safe column
(C) Intermediate and safe column
(D) Long and safe column
(E) Long and unsafe column

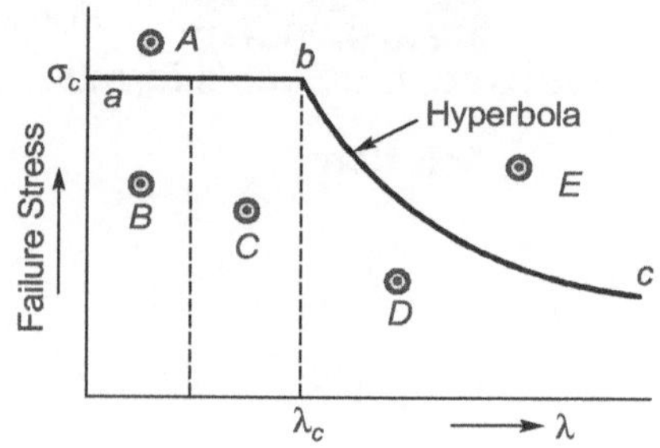

Fig. 14.5

Question 14.8 **What are different types of axial compression members?**

Solution: Various types of axial compression members are

(a) Pedestals - short columns used for footing
(b) Posts - Short columns used in buildings
(c) Struts - compression members used in trusses
(d) Stanction - rolled steel compression member used in building, bridges.
(e) Booms - compression members used in cranes.

Question 14.9 **What is a buckling loop? How number of buckling loops depend on bracing (Type or nature of joints)**

Solution: When a column buckles, the **buckled shape of column makes a loop from one end of the column to the other end as shown in figure 14.6**. Shows a column hinged at both the ends, buckled under load P. Buckled shape ACB makes a loop about ends *A* and *B*. say buckling load is P_{cr_1}.

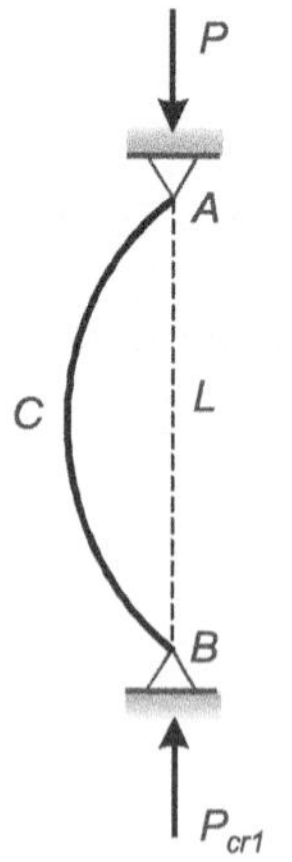

Fig. 14.6

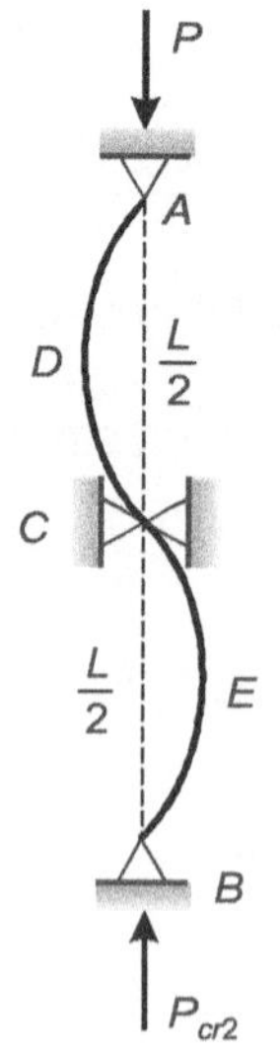

Fig. 14.7

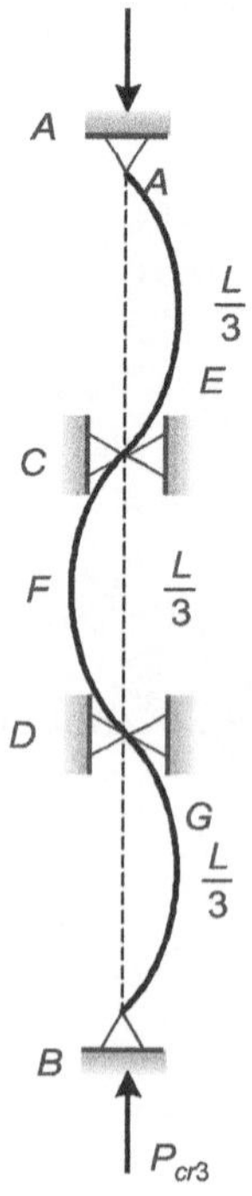

Fig. 14.8

Now a bracing is provided in between, column *AB* buckles with two end loops. ADC and CEB buckling load

$$P_{cr_2} > P_{cr_1}$$

$$P_{cr_1} = \frac{\pi^2 EI}{L^2}$$

$$P_{cr_2} = \frac{4\pi^2 EI}{L^2}$$

Now two bracings are at equal interval i.e. $\frac{L}{3}$. There are three end loops formed as shown

$$P_{cr_3} = \frac{9\pi^2 EI}{L^2}$$

Question 14.10 A column of length *L* is hinged at *A* and fixed at end *B*. A bracing is provided at *C*, so that column buckles from *A* to *C* as hinged ends column and from *C* to *B* as hinged fixed end column.

Solution:

$$P_{or} = \frac{\pi^2 EI}{a^2} = \frac{2\pi^2 EI}{b^2}$$

or

$$2a^2 = b^2$$

$$b = \sqrt{2} \cdot a$$

$$a + b = L$$

$$\sqrt{2}a + a = L$$

$$a = 0.414\,L$$

$$b = 0.586\,L$$

$$P_{or} = \frac{\pi^2 EI}{(0.414\,L)^2} = \frac{\pi^2 EI}{L^2} \times 5.834$$

$$= \frac{2\pi^2 EI}{(0.586\,L)^2} = \frac{5.824\,\pi^2 EI}{L^2}$$

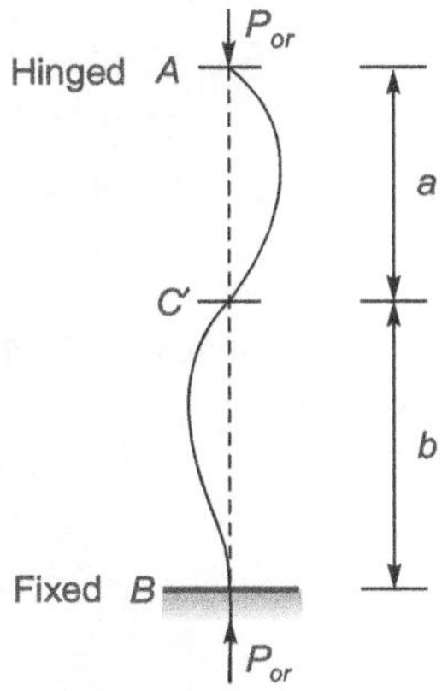

Fig. 14.9

Question 14.11 What is Rankine Gorden formula for buckling of column? What are its advantages over Euler's buckling formulas?

Solution: Rankine-Gorden have combined the two formulae

(i) Euler's formula for buckling of long columns, P_e

(ii) Crushing load for crushing of short columns, P_c

$$\frac{1}{P_R} = \frac{1}{P_e} + \frac{1}{P_c}$$

where

$$P_e = \frac{\pi^2 EI_{min}}{L_e^2},\ P_c = \sigma_c A$$

$$P_R = \text{Rankine's load}$$

or,

$$P_R = \frac{P_e \cdot P_c}{P_e + P_c} = \frac{P_c}{1+\frac{P_c}{P_e}} = \frac{\sigma_c \cdot A}{1+\frac{P_c}{P_e}}$$

$$\frac{P_c}{P_e} = \frac{\sigma_c \cdot A \times L_e^2}{\pi^2 EAk_{min}^2} = \frac{L_e^2}{k_{min}^2}\frac{\sigma_c}{\pi^2 E}$$

$$\frac{P_c}{P_e} = a\frac{L_e^2}{k_{min}^2}$$

where $a = \frac{\sigma_c}{\pi^2 E}$, but it is determined, experimentally

Rankine's buckling load,

$$P_R = \frac{\sigma_c . A}{1 + a \cdot \left(\frac{L_e}{k_{min}}\right)^2}$$

In this expression σ_c and a are known as Rankine's constants, determined experimentally.

Advantages of Rankine's formula over Euler's formula

1. Rankine's formula is applicable for both long and short columns, while Euler's formula is applicable only for long columns.
2. Rankine's constants are determined experimentally
3. Experimental results of buckling load for columns are closer to Rankine's formula.

Practice Q.14.11 Following table gives the values of Rankine's constants for important structural materials.

Material	Cast Iron	Wrought Iron	Mild Steel	Medium Carbon Steel	Timber
σ_c (N/mm^2)	550	250	320	500	35
a	1/1600	1/4000	1/7500	1/5000	1/3000

Question 14.12 **A hollow *CI* column has 200 mm outside diameter and 150 inside diameter and is 5 m long with both the ends fixed. It is subjected to an axial compressive load *P*. Taking factor of safety of 5, find safe Rankine's load against buckling. Rankine constants are,**

$$a = \frac{1}{1600}, \sigma_c = 550 \text{ N/mm}^2$$

Solution:

Outside diameter, $D = 200$ mm

Inside diameter, $d = 150$ mm

Area of cross section, $A = \frac{\pi}{4}\left(200^2 - 150^2\right) = 1.374 \times 10^4 \text{ mm}^4$

Length $= 5$ m

End conditions, Both the ends fixed

$$L_e = \frac{L}{2} = \frac{5}{2} = 2.5 \text{ m} = 2500 \text{ mm}$$

$$k_{min}^2 = \frac{D^2+d^2}{16} = \frac{200^2+150^2}{16} = 0.390\times10^4 \text{ mm}^4$$

$$\frac{L_e^2}{k_{min}^2} = \frac{2500^2}{0.39\times10^4} = 1602.56$$

$$P_R = \frac{\sigma_c \cdot A}{1+a\times\frac{L_e^2}{k_{min}^2}} = \frac{550\times1.374\times10^4}{1+\frac{1}{1600}\times1602.56}$$

$$= \frac{755.7\times10^4}{1+1.0016} = 377.54\times10^4 \text{ N} = 3775.4 \text{ kN}$$

Factor of safety = 5

Safe Rankine's load = 755.1 kN

Practice Q.14.12 A cast iron column is of a hollow circular section with external diameter of 250 mm and a wall thickness of 35 mm is subjected to an axial compressive load. Column is 7 m long with both ends hinged. Taking FOS as 6, determine safe value of *P*. Rankine's constants are σ_c = 560 N/mm^2, $a = \frac{1}{1600}$.

Hints: [$k_{min}^2 = 0.593\times10^4$ mm^2]

Ans. [358 kN]

Question 14.13 What is the maximum stress produced in a column section due to eccentric buckling load? Derive expression, by taking an example.

Figure 14.10 shows a column *AB*, of Length *L*, fixed at end *A*, free at end *B* carrying a load at an eccentricity, e. The column has buckled under load P. Derive expression for maximum stress developed in column section.

Solution: Under the buckling load, *P* column buckles, producing deflection '*a*' at the top free end, Length *AB* is *L*. Consider a section *XX*, at a distance *x* from fixed end A, figure 14.10 bending moment at the section.

$$M_x = P(a+e-y)$$

or,
$$EI\frac{d^2y}{dx^2} = P(a+e-y)$$

or,
$$\frac{d^2y}{dx^2}+\frac{P}{EI}y = P\frac{(a+e)}{EI}$$

Solution of this differential equation is

$$y = A\cos nx + B\sin nx + (a+e)$$

where,
$$n = \sqrt{\frac{P}{EI}}$$

$$y = 0, \text{ at } x = 0, \text{ fixed end } A$$

$$0 = A\cos 0 + B\sin 0 + (a+e)$$

$$0 = A + (a+e)$$

or constant,
$$A = -(a+e)$$

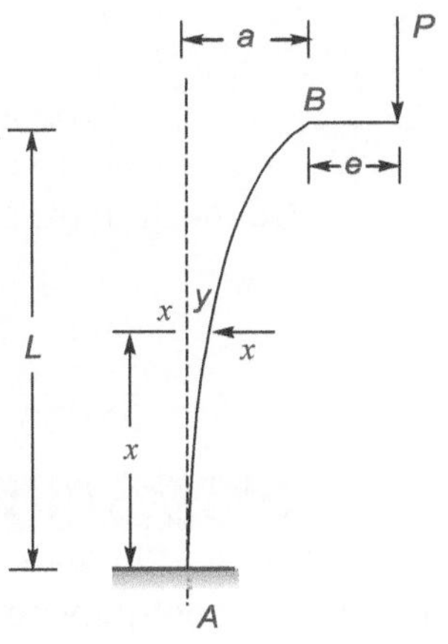

Fig. 14.10

or, $$\frac{dy}{dx} = -An \sin nx + Bn \cos nx$$

$$\frac{dy}{dx} = 0, \text{ at } x = 0, \text{ fixed end } A$$

$$0 = An \sin 0 + Bn \cos 0$$

$$0 = 0 + Bn$$

$$n \neq 0, \text{ Because if } n = 0, \text{ load becomes zero}$$

So constant, $$B = 0$$

Finally, $$y = -(a + e) \cos nx + (a + e)$$

At the top $x = L$, $y = a$,

$$a = -(a + e) \cos nL + (a + e)$$

$$e = (a + e) \cos nL$$

or $$(a + e) = e \sec nL$$

Maximum bending moment occurs at $x = 0$, at end A

$$M_{max} = P(a + e) = Pe \sec nL$$

$$\sigma_b = \text{maximum stress due to bending} = \frac{M}{Z}$$

$$= \frac{Pe \sec nL}{Z}$$

but $$n = \sqrt{\frac{P}{EI}}, \text{ putting the value}$$

$$nL = L\sqrt{\frac{P}{EI}}$$

$$\sigma_b = \frac{Pe}{Z} \sec L\sqrt{\frac{P}{EI}}$$

$$\sigma_d = \text{direct stress (compressive)} = \frac{P}{A}$$

$$\sigma_{max} = \sigma_b + \sigma_d = \frac{P}{A} + \frac{Pe}{Z} \sec L\sqrt{\frac{P}{EI}}$$

(secant formula, we have taken the case of a column, with one end fixed, other end free with equivalent length, $L_e = 2L$)

Secant formula can be modified as

$$\sigma_{max} = \sigma_b + \sigma_d = \frac{P}{A} + \frac{Pe}{Z} \sec \frac{L_e}{2}\sqrt{\frac{P}{EI}}$$

where L_e is the equivalent length of column.

Question 14.14 A steel tube of 60 mm outer diameter, 30 mm inner diameter and 3.6 m long is used as a strut with both the ends hinged. Load is eccentric with eccentricity, e. Find the maximum value of eccentricity so that crippling load on strut is equal to 60% of the Euler's buckling load.

Yield strength of tube = 320 N/mm^2

Solution:

$$E = 200 \text{ kN/mm}^2$$

$$L = 3.6 \text{ m}$$

End conditions: Both the ends hinged

$$L_e = L = 3.6 \text{ m}$$

$$D = 60 \text{ mm}$$

$$d = 30 \text{ mm}$$

$$I = \pi\frac{(60^4 - 30^4)}{64} = \frac{\pi}{64}\times 10^4\,(1296 - 81) = 59.64 \times 10^4 \text{ mm}^4$$

$$P_e = \frac{\pi^2 EI}{L^2} = \frac{\pi^2 \times 200{,}000 \times 59.64 \times 10^4}{3600 \times 3600} = 90837 \text{ N}$$

60% of $P_e = 54502 \text{ N} = P$, buckling load

$$\text{A, area of cross section} = \frac{\pi}{4}(60^2 - 30^2) = 2120.6 \text{ mm}^2$$

$$\sigma_d\text{, direct stress} = \frac{54502}{21.2\times 10^2} = 25.7 \text{ N/mm}^2$$

$$\sigma_y = \sigma_d + \sigma_b$$

$$320 = 25.7 + \sigma_b, \text{ or } \sigma_b = 294.3 \text{ N/mm}^2$$

$$\text{Z, section modulus} = \frac{I}{0.5\,D} = \frac{59.64\times 10^4}{30} = 1.988 \times 10^4 \text{ mm}^3$$

$$\sec\frac{L_e}{2}\sqrt{\frac{P}{EI}} = \sec\frac{3600}{2}\sqrt{\frac{54502}{200{,}000\times 59.64\times 10^4}}$$

$$= \sec 1800\sqrt{4.56925\times 10^{-7}}$$

$$= \sec 1800 \times 6.76 \times 10^{-4} = \sec 1.21673 \text{ rad}$$

$$= \sec 69.71° = 2.883$$

$$\frac{Pe}{Z}sec\frac{L_e}{2}\frac{\sqrt{P}}{EI} = \frac{Pe}{Z}\times 2.883 = 294.3$$

$$\frac{Pe}{Z} = \frac{54502.e}{1.988\times 10^4} = 2.74e$$

$$\sigma_b = 2.74e.sec\frac{L_e}{2}\sqrt{\frac{P}{EI}}$$

$$294.3 = 2.74\times e\times 2.883$$

$$\text{Eccentricity, } e = \frac{294.3}{2.74\times 2.883} = 37.26 \text{ mm}$$

Practice Q.14.14 A steel column of a hollow circular section of 200 mm external diameter and 160 mm internal diameter and 7 m long has to take a load of 200 kN at an eccentricity of 30 mm, from the geometrical axis. If the ends are fixed calculate the maximum and minimum stress intensities induced at the fixed section. $E = 210$ kN/mm^2

Ans. [31.04 N/mm^2 (comp), 4.32 N/mm^2 (comp)]

Question 14.15 What is Johnson's parabolic formula? Compare this with straight line formula for getting buckling load of a column.

Solution: From Euler's formula and Rankine's formula for buckling load, it is learnt that as slenderness ratio (L_e/k_{min}) of a column increases, the load required to buckle the column decreases. Based on this fact, Johnson has given a parabolic formula on the basis of slenderness ratio using a constant. Johnson has suggested, a formula for working stress, taking a constant.

$$\sigma_w = \sigma_c'\left[1 - b\frac{(L_e)^2}{k_{min^2}}\right], \text{ where } b \text{ is a constant}$$

$$\sigma_w = \text{working stress} = \frac{P}{A}$$

$$\sigma_c' = 140 \text{ N/mm}^2 \text{ in compression for structural steels}$$

$$\text{Constant, } c = 0.0054 \text{ for hinged ends}$$

$$= 0.0018 \text{ for riveted ends}$$

Question 14.16 A standard I section is used as column, Properties of rolled steel I section are

Area = 2527 mm^2

$I_{xx} = 1696.6 \times 10^4$ mm^4

$I_{yy} = 115.4 \times 10^4$ mm^4

Using Johnson parabolic formula determine safe load against buckling. Column length is 2 m with both ends hinged. Compare this with the safe load obtained by straight line formula.

For Parabolic formula

σ_c' = allowable stress in compression = 110 N/mm^2

constant, $b = 0.00003$ for hinged ends

Straight Line formula

σ_c' = allowable stress in compression = 140 N/mm^2

constant, $c = 0.0054$, hinged end

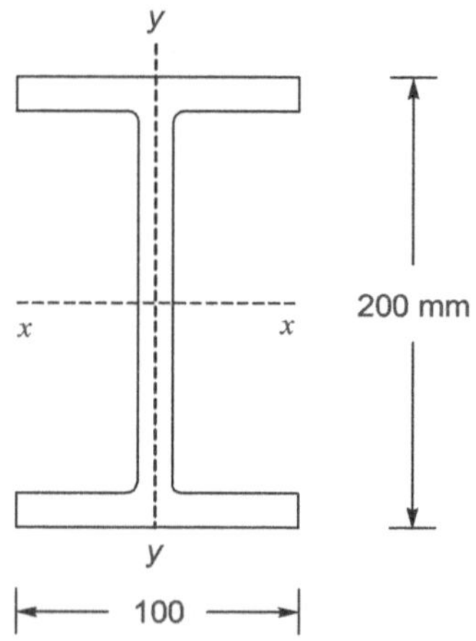

Fig. 14.11

Solution:

$$I_{min} = 115.4 \times 10^4 \text{ mm}^4$$

$$A = 2527 \text{ mm}^2$$

$$k_{min}^2 = \frac{115.4 \times 10^4}{2527} = 456.67 \text{ mm}^2$$

$$k_{min} = 21.37 \text{ mm}$$

Both ends are hinged $Le = L$ $= 2 \text{ m} = 2000 \text{ mm}$

$$\frac{L_e^2}{k_{min}^2} = \frac{2000^2}{456.67} = 8759$$

$$\frac{L_e}{k_{min}} = 93.6$$

Johnson Parabolic Formula

Working stress,
$$\sigma_w = \sigma_c'\left[1 - b\left(\frac{L_e^2}{k_{min}^2}\right)\right] = 110[1 - .00003 \times 8759]$$

$$= 110[1 - 0.2627] = 81.1 \text{ N/mm}^2$$

$$\text{Safe load} = \sigma_w, A = 81.1 \times 2527 = 204940 \text{ N}$$

$$= 204.9 \text{ kN}$$

Straight line formula

Working stress,
$$\sigma_w = \sigma_c'\left[1 - c \times \left(\frac{L_e}{k_{min}}\right)\right] = 110[1 - .0054 \times 93.6]$$

$$= 140[1 - 0.50544]$$

$$= 140[0.49456] = 69.2384 \text{ N/mm}^2$$

$$\text{Safe load, } P = \sigma_w A = 69.2384 \times 2527$$

$$= 174965 \text{ N} = 174.96 \text{ kN}$$

Question 14.17 **A 6 m long rolled steel rectangular section 300 mm × 160 mm is used on a strut, with one end fixed and other end hinged. Calculate the crippling load by Rankine's formula. Compare with the load obtained by Euler's formula. For what length of the strut, will the two formulae give the same crippling load?**

Solution:

$$E = 210 \text{ GPa}, a = \frac{1}{7500}, \sigma_c = 330 \text{ MPa}$$

$$\text{Area of cross section} = 300 \times 160 = 48 \times 10^3 \text{ mm}^2$$

$$I_{min} = \frac{300 \times 160^3}{12} = 1024 \times 10^5 \text{ mm}^4$$

$$k_{min} = \sqrt{\frac{1024 \times 10^5}{48 \times 10^3}} = 46.19 \text{ mm}$$

$$\text{Length, } L = 6 \text{ m} = 6000 \text{ mm}$$

End conditions; one end fixed, other end hinged

$$\text{Equivalent length, } L_e = \frac{L}{\sqrt{2}} = \frac{6000}{\sqrt{2}} = 4242.6 \text{ mm}$$

$$\frac{L_e}{k_{min}} = 91.86$$

Rankine's load

$$P_R = \frac{\sigma_c \cdot A}{1 + a \cdot \dfrac{L_e^2}{k_{min}^2}} = \frac{330 \times 48 \times 10^3}{1 + \dfrac{1}{7500} \times (91.86)^2}$$

$$= \frac{1584 \times 10^4}{1 + 1.125} = 745.376 \times 10^4 \text{ N}$$

Euler's load

$$P_e = \frac{2\pi^2 EI_{min}}{L^2} = \frac{2 \times \pi^2 \times 210000 \times 1024 \times 10^5}{(6000)^2} = 1179 \times 10^4 \text{ N}$$

P_e is much higher than P_R

Length for same load

$$\frac{2\pi^2 \times 210000 \times 1024 \times 10^5}{L^2} = \frac{330 \times 48000}{1 + \dfrac{L^2}{2} \times \dfrac{1}{k_{min}^2} \times \dfrac{1}{7500}}$$

$$\frac{2\pi^2 \times 210000 \times 1024 \times 10^5}{330 \times 48000} = \frac{L^2}{1 + \dfrac{L^2}{2 \times (46.19)^2} \times \dfrac{1}{7500}}$$

$$2.68 \times 10^7 \left[1 + \frac{L^2}{32002741}\right] = L^2$$

$$2.68 \times 10^7 + \frac{2.68 L^2 \times 10^7}{32002741} = L^2$$

$$2.68 \times 10^7 + 0.83743\, L^2 = L^2$$

or

$$2.68 \times 10^7 = 0.16257\, L^2$$

$$L^2 = 16.4852 \times 10^7$$

$$L = 12.84 \times 1000 = 12.84 \text{ m}$$

At $L = 12.84$ m, $P_R = P_e$, Rankine Load = Euler's Buckling load

Question 14.18 A round metallic bar held vertically is clamped securely at bottom and a horizontal force of 500 N produces a deflection of 16 mm at the top as shown. Effective length of bar is 2 m. What is the axial buckling load for the bar under the given condition?

Solution:

$P = 500$ N

$L = 2000$ mm

$$\delta = 16 \text{ mm} = \frac{PL^3}{3EI}$$

$$16 = \frac{500 \times 2000^3}{3EI}$$

$EI = 8.33 \times 10^{10}$ N mm^2

P_e = Euler's Buckling load

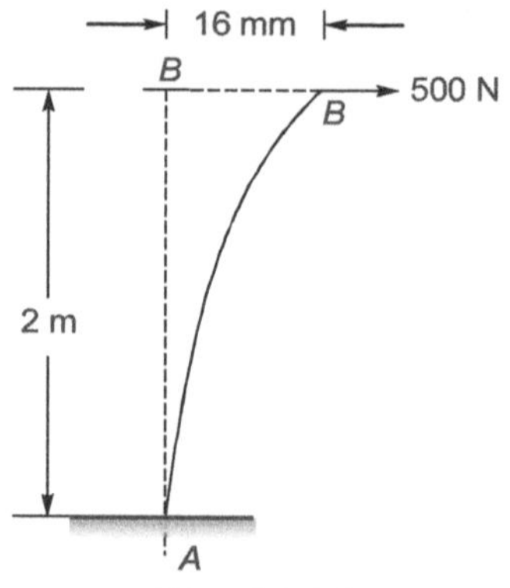

Fig. 14.12

$$= \frac{\pi^2 EI}{4L^2}, \text{as one end is fixed, other end is free}$$

$$P_e = \frac{\pi^2 \times 8.33 \times 10^{10}}{4 \times 2000^2} = 51402 \text{ N} = 51.4 \text{ kN}$$

Question 14.19 **What is Prof Perry's approximate formula?**

Solution: In the secant formula for eccentrically loaded columns

bending stress, $\sigma_b = \frac{P.e}{Z} \sec \frac{L_e}{2}\sqrt{\frac{P}{EI}}$, where

$$\sec\frac{L_e}{2}\sqrt{\frac{P}{EI}} = \sec\sqrt{\frac{P}{EI} \times \frac{L_e^2}{4}} = \sec\sqrt{\frac{PL_e^2\pi^2}{\pi^2 EI \times 4}} = \sec\sqrt{\frac{\pi^2}{4} \times \frac{P}{P_e}}$$

$$= \sec\frac{\pi}{2}\sqrt{\frac{P}{P_e}}$$

Prof. Perry through experimental observations, observed that

$$\sec\frac{\pi}{2}\sqrt{\frac{P}{P_e}} \simeq \frac{1.2\, P_e}{P_e - P}, \text{ where}$$

$$\sigma_b = \frac{P_e \cdot y_c}{Ak^2} \times \frac{1.2\, P_e}{P_e - P}$$

Taking $\frac{P}{A} = \sigma_0, \quad \frac{P_e}{A} = \sigma_e$

$$\sigma_b = \frac{P_e y_c}{Ak^2}\left(\frac{1.2\sigma_e}{\sigma_e - \sigma_0}\right)$$

$$\sigma_{max} = \sigma_0 + \sigma_b$$

Final expression is

$$\left(\frac{\sigma}{\sigma_0} - 1\right)\left(1 - \frac{\sigma_0}{\sigma_e}\right) = \frac{1.2e\, y_c}{k^2}, \text{ in the expression}$$

σ is the allowable stress, σ_0 is direct stress due to P,

σ_e is stress due to Euler's buckling load.

Practice Q.14.19 A 400 × 140 mm *RSJ* is used as a strut with hinged ends having length of 6 m. Using Prof. Perry formula determine safe load, taking eccentricity e = 24 mm along x - x axis. Maximum allowable compressive stress is 75 MPa. For the joint A = 7846 mm², I_{xx} = 20458.4 × 10⁴ mm⁴. I_{yy} = 622.1 × 10⁴ mm⁴, E = 210 GPa

Ans. [121.4 kN]

Question 14.20 A column is 2.4 m long straight steel bar of rectangular section 30 × 20 mm with both ends hinged. Assuming that Euler's formula is applicable and material attains its yield strength at the time of buckling, determine central deflection.

E = 210 kN/mm². Yield strength of steel 320 N/mm²

Solution:

Section = 30 × 20 mm

Moment of inertia, $I_{min} = 30 \times \dfrac{20^3}{12} = 2 \times 10^4 \text{ mm}^4$

End conditions: Both ends hinged

P_e, Euler's buckling load $= \dfrac{\pi^2 EI_{min}}{L_2} = \dfrac{\pi^2 \times 210,000 \times 2 \times 10^4}{2400^2} = 7196.6 \text{ N}$

σ_d = Direct stress $= \dfrac{P_e}{A} = \dfrac{7196.6}{600} = 11.99 \text{ N/mm}^2$

$\sigma_y = \sigma_{max} = 320 = \sigma_d + \sigma_b = 11.99 + \sigma_b$

σ_b, bending stress $= 308.01 \text{ N/mm}^2 = \dfrac{M_{max}}{Z}$

M_{max}, maximum bending moment $= P_e \times y_{max}$

y_{max} = central deflection

Z, section modulus $= \dfrac{30 \times 20^2}{6} = 2000 \text{ mm}^3$

$308.01 = \dfrac{7196.6}{2000} \times y_{max}$

central deflection, $y_{max} = \dfrac{308.01 \times 2000}{7196.6} = 85.6 \text{ mm}$

Practice Q.14.20 A 3 m long straight steel bar of rectangular section 30 mm × 20 mm is used as a strut with both ends hinged. Assuming that Euler's formula is applicable and the material attains its yield strength at the time of buckling, determine central deflection.

$E = 210 \text{ kN/mm}^2$, $\sigma_y = 320 \text{ N/mm}^2$

Ans. [135.6 mm]

Question 14.21 A thin steel bar of rectangular section 8 × 6 mm is axially compressed by a 400 N load between two plates that are fixed at a constant distance of 200 mm apart as shown in figure 14.13. The assembly is made at normal room temperature. Taking a factor of safety of 2 against buckling the temperature of the bar can be raised by how many °C.

$E = 200 \text{ GPa}$, $\alpha = 14 \times 10^{-6}/°\text{C}$

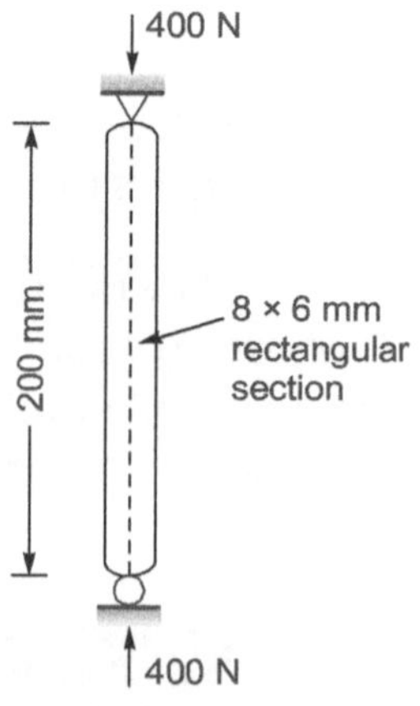

Fig. 14.13

Solution: Section 8 × 6 mm $= 48 \text{ mm}^2 = A$

Moment of inertia, $I_{min} = \dfrac{8 \times 6^3}{12} = 144 \text{ mm}^4$

$E = 200 \times 1000 \text{ N/mm}^2$

$L = 200 \text{ mm} = L_e$, both ends hinged

P_e, Euler's buckling load $= \dfrac{\pi^2 EI_{min}}{L^2} = \dfrac{\pi^2 \times 200,000 \times 144}{200^2} = 7106 \text{ N}$

$$\text{FOS} = 2$$

$$\text{safe load} = \frac{7106}{2} = 3553 \text{ N}$$

$$\text{Axial load, } P =- 400 \text{ N}$$

Balance load 3553 – 400 = 3153 N, is generated due to temperature increase. Free expansion is prevented by end plates and compressive force is developed in section of bar

$$\alpha\Delta T \cdot EA = 3153 \text{ N}$$

putting the value

$$14 \times 10^{-6} \times \Delta T \times 200{,}000 \times 48 = 3153$$

$$134.4\,\Delta T = 3153$$

$$\Delta T = \frac{3153}{134.4} = 23.46°\text{C}$$

Practice Q.14.21 A thin steel bar of rectangular section 6 mm × 4 mm is axially compressed by 200 N load between two plates that are fixed at a constant distance of 160 mm apart as shown in figure 14.14. The assembly is made at 24°C, how high can the temperature of the bar rise so as to have an FOS 2 against buckling.

E = 200 GPa, $\alpha = 15 \times 10^{-6}$/°C

Ans. [14.36°C]

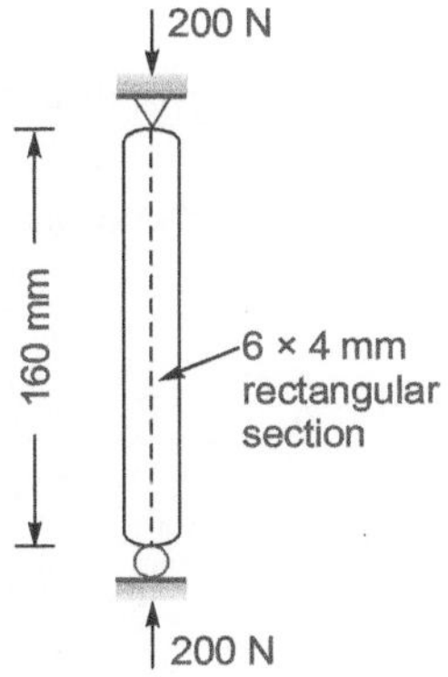

Fig. 14.14

Question 14.22 A long strut *AB* of length *L* is of uniform section throughout. A thrust *P* is applied at the ends eccentrically on the same side of center line with eccentricity at the end *B* twice than that at end *A*, Show that maximum bending moment occurs at a distance *x* from end *A*,

Where $\tan kx = \dfrac{2 - \cos kL}{\sin kL}$ and $k\sqrt{\dfrac{P}{EI}}$

[15 marks, IES 2005]

Solution: Figure 14.15 shows column *AB*, of length *L* with eccentricity *e* at *A* and 2 *e* at end *B*. Take a section *XX* at a distance *x* from end *A*.

$$\text{Bending moment, } M = -P(y + 2e - e)$$

$$= -P(e + y)$$

or,

$$EI\frac{d^2y}{dx^2} = -P(e + y)$$

$$\frac{d^2y}{dx^2} + \frac{P}{EI}(e+y) = 0$$

$$k = \sqrt{\frac{P}{EI}}$$

$$\frac{d^2y}{dx^2} + k^2(e+y) = 0$$

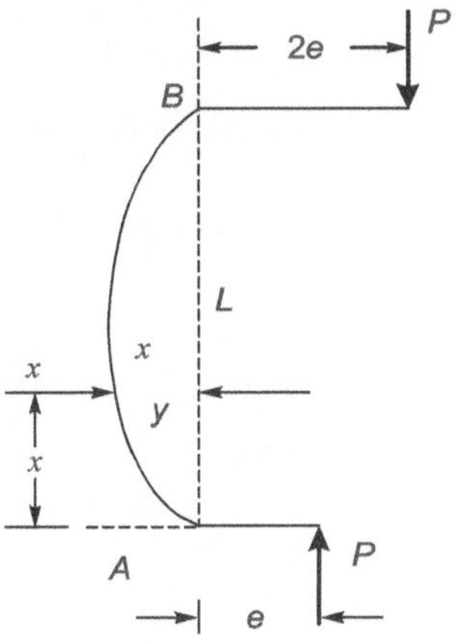

Fig. 14.15

say
$$y = A\cos kx + B\sin kx$$
$$x = 0,\ y = e,\ x = L,\ y = 2e$$
$$e = A\cos 0 + B\sin 0$$
$$A = e \quad \text{...(i)}$$
$$2e = A\cos kL + B\sin kL$$
$$= e\cos kL + B\sin kL$$
$$2e - e\cos kL = B\sin kL$$
$$\text{constant, } B = \frac{e(2-\cos kL)}{\sin kL}$$
$$y = e\cos kx + \frac{e(2-\cos kL)}{\sin kL}\cdot \sin kx$$

For maximum bending moment, y should be maximum

For maximum bending moment also $\frac{dy}{dx} = 0$

$$\frac{dy}{dx} = -ek\sin kx + \frac{e(2-\cos kL)}{\sin kL} \times k\cdot\cos kx$$
$$0 = -\sin kx + \left(\frac{2-\cos kL}{\sin kL}\right)\cos kx$$

or,
$$\tan kx = \frac{2-\cos kL}{\sin kL}$$

Question 14.23 A both ends hinged cast iron hollow cylindrical column 3 m in length has a critical buckling load of P kN. When the column is fixed at both the ends, its critical load rises by 300 N more. If the ratio of external diameter to internal diameter is 1.25 and E = 100 GPa, determine external diameter of the column.

[10 Marks, IES 2007]

Solution:

Both ends hinged,
$$P = \frac{\pi^2 EI}{L^2}$$

Both ends fixed,
$$P + 300 = \frac{4\pi^2 EI}{L^2}$$

Therefore
$$P + 300 = 4P$$
$$P = 100 \text{ kN}$$
$$E = 100 \times 10^9 \text{ N/m}^2 = 100 \times 10^6 \text{ kN/m}^2$$

So
$$100 \text{ kN} = \frac{\pi^2 \times 100 \times 10^6 \times I}{3^2}$$
$$I = \frac{900}{\pi^2 \times 100 \times 10^6} = 0.9119 \times 10^{-6} \text{ m}^4 = \frac{\pi}{64}\left(D^4 - d^4\right)$$
$$d = 0.8\,D$$

$$I = \frac{\pi}{64}\left[D^4 - [0.8D]^4\right] = \frac{\pi D^4}{64} \times 0.5904 = 0.02898\,D^4$$

$$0.02898\,D^4 = 0.9119 \times 10^{-6}\ \text{m}^4$$

$$D^4 = 3148.65 \times 10^{-4}\ \text{m}^4$$

$$D = 0.075\ \text{m} = 75\ \text{mm},\ d = 60\ \text{mm}$$

Question 14.24 A column made of an aluminium alloy of rectangular section b × d and a length of 700 mm is fixed at end A and end B is restrained to move in one of the vertical planes of symmetry of the column. Determine the ratio of two sides for efficient design against buckling.

$E = 67$ GPa, $P = 15$ kN, $FOS = 2$

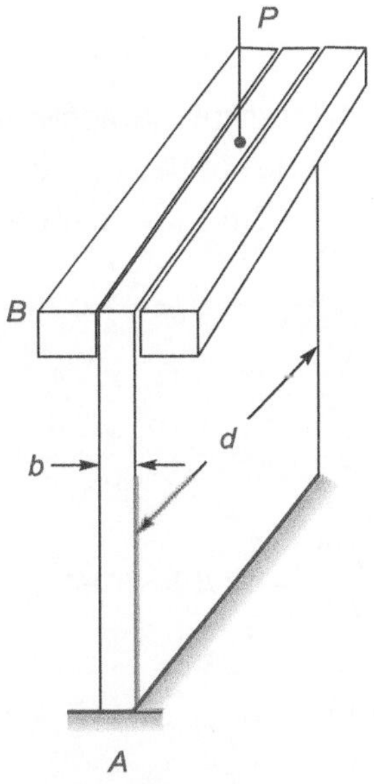

Fig. 14.16

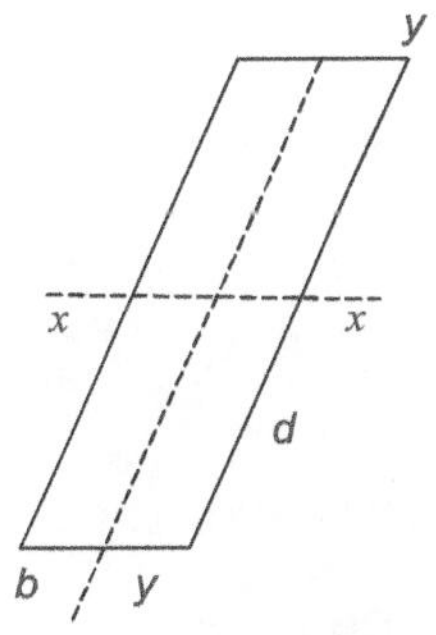

Fig. 14.17

Solution:

$$\text{Design load} = P \times FOS = 15 \times 2 = 30\ \text{kN}$$

$$I_{xx} = \frac{bd^3}{12}; \quad I_{yy} = \frac{db^3}{12}$$

There are two modes of buckling for the joist

$$P_1 = \frac{2\pi^2 EI_{yy}}{L^2},\ P_2 = \frac{\pi^2 EI_{xx}}{4L^2}$$

For efficient design

$$P_1 = P_2$$

$$2\,I_{yy} = \frac{I_{xx}}{4}$$

$$I_{xx} = 8\,I_{yy}$$

or

$$\frac{bd^3}{12} = 8 \times \frac{db^3}{12}$$

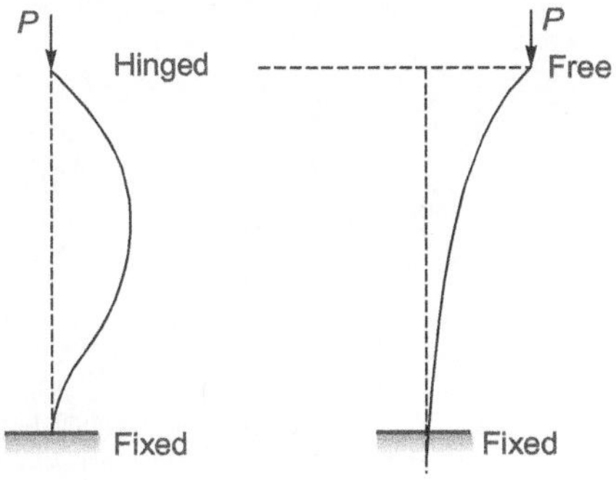

Fig. 14.18

$$\frac{d^2}{b^2} = 8$$

$$d = 2.828\,b$$

$$P_1 = 30\text{ kN},\ L = 700$$

$$30{,}000 = \frac{2\pi^2 \times 67000 \times db^3}{700^2 \times 12}$$

$$db^3 = 133381\text{ mm}^4$$

$$2.828\,b^4 = 133381$$

$$b^4 = 47164.4$$

$$b = 14.74$$

$$d = 41.67\text{ mm}$$

Question 14.25 A short length of a tube 30 mm internal diameter and 40 mm external diameter failed under a compressive load of 180 kN. When a 2 m length of the same tube is tested as a strut with both the ends hinged, the buckling load was found to be 40.82 kN. Determine Rankine's constants a and σ_c.

Solution: First test on a short tube

$$P_c = 180\text{ kN}$$

$$A = \frac{\pi}{4}\left(40^2 - 30^2\right) = 549.78\text{ mm}^2$$

Ultimate compressive strength, $\sigma_c = \frac{P_c}{A} = \frac{180{,}000}{549.78} = 327.4\text{ N/mm}^2$

Second test on a long tube

Both ends hinged, $L = L_e = 2000$ mm

$$P_R = 40.82\text{ kN}$$

$$k^2 = \frac{D^2 + d^2}{16} = \frac{40^2 + 30^2}{16} = 156.25$$

$$\frac{L_e^2}{k^2} = \frac{2000^2}{156.25} = 25600$$

$$P_R = 40.82 \times 1000 = \frac{\sigma_c \cdot A}{1 + a \cdot \frac{L_e^2}{k^2}}$$

$$40820 = \frac{180000}{1 + a \times 25600}$$

$$1 + a.25600 = \frac{180000}{40820} = 4.4096$$

$$\alpha \times 25600 = 3.4096$$

$$a = \frac{3.4096}{25600} = \frac{1}{7508}$$

Rankine's constant are $\sigma_c = 327.4\text{ N/mm}^2$

$$a = \frac{1}{7508}$$

Question 14.26 Consider a steel (Young's modulus E = 200 GPa) column hinged on both sides. Its height is 1.0 m and cross-section is 10 mm × 20 mm. The lowest Euler critical buckling load (in N) is ______.

[ME, GATE : 2015 (set-1)]

Solution:

$$P_e = \frac{\pi^2 EI}{L^2}$$

$$I = I_{min} = \frac{20 \times 10^3}{12} = \frac{1}{6} \times 10^4 \text{ mm}^4$$

$$L = 1000 \text{ mm}$$

$$E = 200 \text{ kN/mm}^2$$

$$P_e = \frac{\pi^2 \times 200 \times 1000}{1000^2} \times \frac{10^4}{6} = 3290 \text{ N}$$

Objective Type Questions

Q.1 If the length of a column is doubled. The critical load becomes

(a) $\frac{1}{2}$ of original value

(b) $\frac{1}{4}$ of original value

(c) $\frac{1}{16}$ of original value

(d) $\frac{1}{18}$ of original value

[GATE 1988, 1 Marks]

Q.2 The rod PQ of length L and with flexural rigidity EI is hinged at both the ends. For what minimum force F is it expected to buckle?

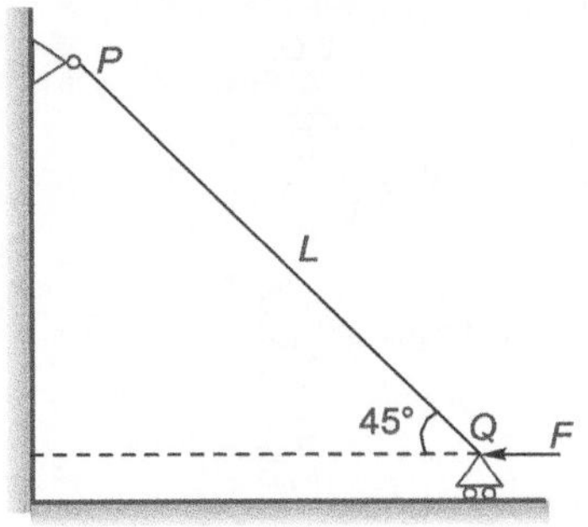

(a) $\frac{\pi^2 EI}{L^2}$ (b) $\frac{\sqrt{2}\,\pi^2 EI}{L^2}$

(c) $\frac{\pi^2 EI}{\sqrt{2}\,L^2}$ (d) $\frac{\pi^2 EI}{L^2}$

[GATE 2008, 2 Marks]

Q.3 A column has a rectangular cross section of 10 mm × 20 mm and a length of 1 m. The slenderness ratio of column is close to

(a) 200 (b) 346

(c) 477 (d) 1000

[GATE 2011, 1 Marks]

Q.4 For a long slender column of uniform cross section. The ratio of critical buckling load for the case with both ends clamped to the case with both the ends hinged is

(a) 1 (b) 2

(c) 4 (d) 8

[GATE 2012, 1 Marks]

Q.5 The curve ABC is the Euler's curve for stability of column. The horizontal line DEF is the strength limit. With reference to this figure, match **List-I (Region)** with **List-II (column Specification)**

and select correct answer using the codes given below.

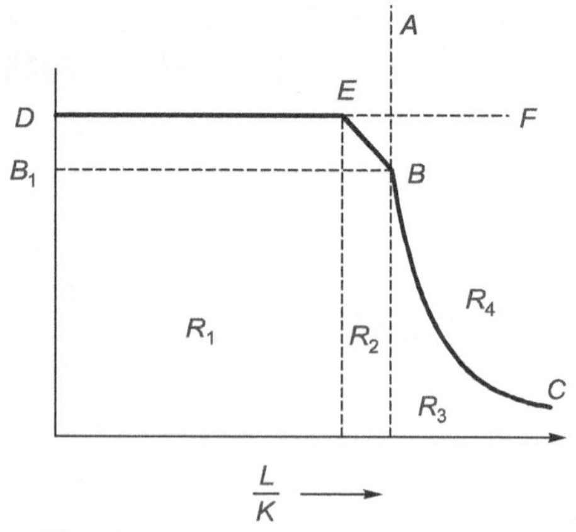

List-I	List-II
A. R_1	1. Long stable
B. R_2	2. Short
C. R_3	3. Medium
D. R_4	4. Long unstable

Codes:

	A	B	C	D
(a)	2	4	3	1
(b)	2	3	1	4
(c)	1	2	4	3
(d)	2	1	3	4

Q.6 The ratio of the compressive critical load for a long column fixed at both the ends and a column with one end fixed end the other end free is

(a) 2 : 1 (b) 4 : 1
(c) 8 : 1 (d) 16 : 1

[IES : 1997]

Q.7 The diameter of a long column is reduced by 20%. The percentage reduction in Euler's buckling load is

(a) 4 (b) 39
(c) 49 (d) 59

[IES : 2000]

Q.8 A long slender bar having uniform rectangular cross section $B \times H$ is acted upon by an axial compressive force. The sides B and H are parallel to x and y-axis respectively. The ends of the bar are fitted such that they behave a pin-jointed when the bar buckles in a plane normal to x-axis and they behave as built-in when the bar buckles in a plane normal to y-axis. If the load capacity in either mode of buckling is same, the value of H/B will be

(a) 2 (b) 4
(c) 8 (d) 16

Q.9 Four columns of same material and same length are of the rectangular cross section of same breadth b. The depth of the cross section and end conditions are however different given as follows.

Column	*Depth*	*End condition*
1.	0.6 b	fixed - fixed
2.	0.8 b	fixed - hinged
3.	1.0 b	hinged - hinged
4.	2.6 b	fixed - free

Which of the above columns has maximum value of Euler's buckling load?

(a) column 1 (b) column 2
(c) column 3 (d) column 4

[IES : 2004]

Q.10 If one end of a hinged column is made fixed and the other free, how much is the critical load critical load compared to the original value.

(a) one-fourth (b) Half
(c) Twice (d) Four times

[IES : 2004]

Q.11 If σ_c and E donate the crushing stress and Young's modulus of the material of a column, then the Euler's formula can be applied for determination of crippling load of a column made of this 'material only' if its slenderness ratio is

(a) more than $\pi\sqrt{\frac{E}{\sigma_c}}$

(b) less than $\pi\sqrt{\frac{E}{\sigma_c}}$

(c) more than $\pi^2\sqrt{\frac{E}{\sigma_c}}$

(d) less than $\pi^2\sqrt{\frac{E}{\sigma_c}}$

[IES : 2005]

Q.12 Four vertical columns of same material, height and weight have the same end conditions which cross section will carry the maximum load?
(a) Solid circular section
(b) Thin hollow circular section
(c) Solid rectangular section
(d) *I* section

[IES : 2009]

Q.13 Determine the ratio of the buckling strength of a solid steel section column to that of a hollow column of the same material, having the same cross section. The internal diameter of the hollow column is half of the external diameter. Both columns are of identical length and are pinned or hinged at ends.

(a) $\frac{P_s}{P_h} = \frac{2}{5}$ (b) $\frac{P_s}{P_h} = \frac{3}{5}$

(c) $\frac{P_s}{P_h} = \frac{4}{5}$ (d) $\frac{P_s}{P_h} = 1$

Q.14 **Statement (I):** A circular cross section of column with diameter '*d*' is to be axially loaded in compression. For this, the core of the section is considered to be a concentric circular area of diameter $\frac{d}{4}$

Statement (II): we can drill and take out a cylindrical volume of material with diameter $\frac{'d'}{4}$ in order to make in column lighter and still maintaining, the same buckling crippling load carrying capacity.
(a) Both Statements (I) and Statement (II) are individually true and Statement (II) is the correct explanation of Statement (II)
(b) Both Statement (I) and Statement (II) are individually true but Statement (II) is not the correct explanation of Statement (II)
(c) Statement (I) is true but Statement (II) is false
(d) Statement (I) is false but Statement (II) is true

[IES : 2003]

Q.15 In Johnson's parabolic formula, for mild steel, what is constant *b* for both ends hinged.
(a) 3×10^{-3} mm (b) 3×10^{-4} mm
(c) 3×10^{-5} mm (d) 2×10^{-5} mm

Q.16 A column of length 4 m, an area of cross section 1600 mm^2 and Moment of Inertias, $I_{xx} = 576 \times 10^4$ mm^4 and $I_{yy} = 64 \times 10^4$ mm^4 is subjected to a buckling load. Both the ends of column are fixed. What is slenderness ratio of column.
(a) 100 (b) 200
(c) 133.2 (d) 66.66

Q.17 Rankine's buckling formula is applicable for
(a) Long column
(b) Short columns
(c) Moderately long columns
(d) All of these

Q.18 An initially stress-free massless elastic beam of length L and circular cross-section with diameter d (d $\ll$ L) is held fixed between two walls as shown. The beam material has Young's modulus E and coefficient of thermal expansion α.

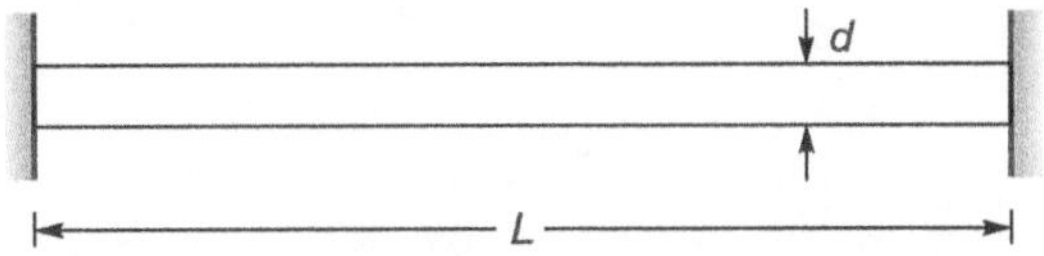

If the beam is slowly and uniformly heated, the temperature rise required to cause the beam to buckle is proportional to
(a) d (b) d^2
(c) d^3 (d) d^4

[ME, GATE : 2017 (set-1)]

Q.19 The minimum axial compressive load P, required to initiate buckling for a pinned–pinned slender column with bending stiffness EI and length L is

(a) $P = \frac{\pi^2 EI}{4L^2}$ (b) $P = \frac{\pi^2 EI}{L^2}$

(c) $P = \frac{3\pi^2 EI}{4L^2}$ (d) $P = \frac{34^2 EI}{L^2}$

[ME, GATE : 2018 (set-2)]

Q.20 A column of height h with a rectangular cross-section of side a × 2a has a bending load of P. If the cross-section is changed to 0.5a – 3a, and its height is changed to 1.5 h, the buckling load for the redesigned column will be

(a) $\frac{P}{12}$ (b) $\frac{P}{4}$

(c) $\frac{P}{2}$ (d) $\frac{3P}{4}$

[CE, GATE : 2018 (set-1)]

Q.21 A long column hinged at both the ends has certain critical Euler's buckling load-carrying capacity. If the same column be fixed at both the ends (in place of hinged ends), the load-carrying capacity then increases to

(a) 4 times (b) 3 times

(c) 2 times (d) Nil

[ME, ESE (Pre) : 2016]

Answers

1. (b)	2. (b)	3. (b)	4. (c)	5. (b)
6. (d)	7. (d)	8. (a)	9. (d)	10. (a)
11. (a)	12. (b)	13. (b)	14. (c)	15. (c)
16. (a)	17. (d)	18. (b)	19. (b)	20. (a)
21. (a)				

Explanations

1. (b)

$\frac{1}{4}$ of original values

$$P_e = \frac{\pi^2 EI}{L^2}$$

$$P_e' = \frac{\pi^2 EI}{4L^2}, \text{ length doubled}$$

2. (b)

$$\frac{\sqrt{2}\pi^2 EI}{L^2}$$

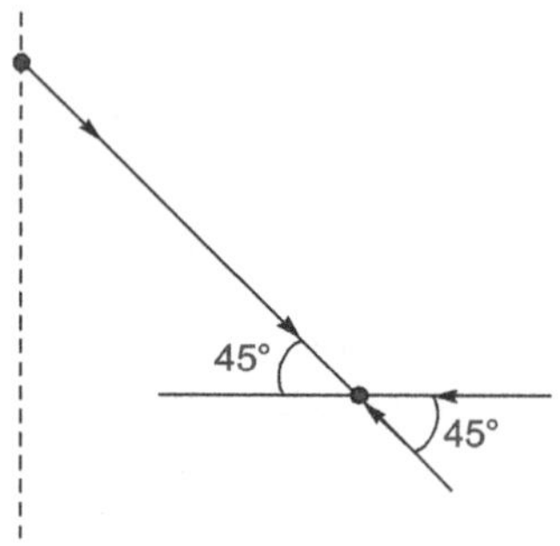

$$P_e = \frac{\pi^2 EI}{L^2}$$

$$= F \cos 45° = F \times 0.707$$

3. (b)

$$I_{min} = \frac{20 \times 10^3}{12} = \frac{20,000}{12} \text{ mm}^4$$

$$A = 200 \text{ mm}^2$$

$$k_{mm}^2 = \frac{20000}{200 \times 12} = 8.333$$

$$k_{min} = 2.887$$

$$\frac{L}{k} = \frac{1000}{2.887} = 346$$

4. (c)

$$P_e, \text{ both ends clamped} = \frac{4\pi^2 EI}{L^2}$$

$$P_e', \text{ both ends hinged} = \frac{\pi^2 EI}{L^2}$$

$$\text{Ratio } \frac{P_e}{P_e} = 4$$

5. (b)

A R_1 – 2 Short

B R_2 – 3 Medium

C R_3 – 1 Long stable

D R_4 – 4 Long mistable

6. (d)

$$P_e = \frac{4\pi^2 EI}{L^2}, \text{ for both ends fixed}$$

$$P_e' = \frac{\pi^2 EI}{4L^2}, \text{ for one and fixed other free}$$

$$\text{Ratio } \frac{P_e}{P_e'} = 16$$

7. (d)

$$P_e = \frac{\pi^2 EI}{L^2}$$

$$P_e' = \frac{\pi^2 EI}{L^2}(0.8)^4 = \frac{\pi^2 EI}{L^2} \times 0.41$$

Reduction is Euler's buckling load 59%.

8. (a)

yy plane of buckler, I_{xx}, jointed
xx plane of buckler, I_{yy}, built in

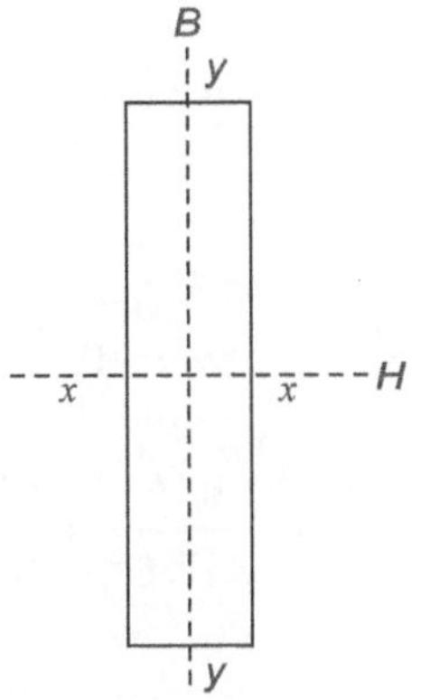

$$\frac{\pi^2 EI_{xx}}{L^2} = \frac{4\pi^2 EI_{yy}}{L^2}$$

$$I_{xx} = 4\, I_{yy}$$

$$\frac{BH^3}{12} = \frac{HB^3}{12} \times 4$$

$$H^2 = 4\, B^2$$

$$H = 2\, B$$

$$\frac{H}{B} = 2$$

9. (d)

1. $P_e = \frac{4\pi^2}{EI}\left[\frac{0.216\, b^4}{12}\right] \propto 0.864\, b^4$

2. $P_e = \frac{2\pi^2}{EI}\left[\frac{0.512\, b^4}{12}\right] \propto 1.024\, b^4$

3. $P_e = \frac{\pi^2}{EI}\left[\frac{b^4}{12}\right] \propto b^4$

4. $P_e = \frac{\pi^2}{4EI}\left[\frac{12.576\, b^4}{12}\right] \propto 4.394\, b^4$

10. (a)

$$P_e = \frac{\pi^2 EI}{L^2}$$

$$P_e' = \frac{\pi^2 EI}{4L^2}$$

$$\frac{P_e'}{P_e} = \frac{1}{4}$$

11. (a)

$$P_e = \frac{\pi^2 EAk_{min}^2}{L_e^2}$$

$$\frac{P_e}{A} = \frac{\pi^2 Ek_{min}^2}{L_e^2} < \sigma_c$$

$$\frac{L_e^2}{k_{min}^2} = \frac{\pi^2 E}{\sigma_c}$$

Slenderness rate,

$$\frac{L_e}{k_m} = \pi\sqrt{\frac{E}{\sigma_c}}$$

12. (b)

For same weight and same cross sectional area, moment of inertia of hollow circular section will be more than that of other sections, so hollow section will carry maximum load.

13. (b)

Solid column

$$I = \frac{\pi d^4}{64}$$

Hollow, $I = \frac{\pi}{64}\left(D^4 - \frac{D^4}{16}\right) = \frac{\pi}{64} \times \frac{15}{16} D^4$

$$\frac{\pi}{4} \times d^2 = \frac{\pi}{4}\left(D^2 - \frac{D^2}{4}\right)$$

$$d^2 = 0.75\, D^2,$$

$$D = 1.1547\, d$$

$$I_b = \frac{\pi}{64} \times \left[\frac{15}{16} \times 1.1547^4 d^4\right]$$

$$= \frac{\pi}{64} \times \left[1.666\, d^4\right]$$

$$\frac{I_s}{I_h} = \frac{d^4}{1.666\,d^4} = \frac{3}{5} = \frac{P_s}{P_H}$$

14. (c)
Statement (I) is true but Statement (II) is false. By drilling a hole in column will reduce its buckling strength.

15. (c)
Johnson's parabolic formula

$$b = 3 \times 10^{-5}$$

both ends hinged, mild steel

16. (a)

$$I_{mm} = 64 \times 10^4, \text{ fixed end,}$$

$$L_e = \frac{L}{2} = 2000 \text{ mm}$$

$$A = 1600 \text{ mm}^2$$

$$k_{min} = 20 \text{ mm}$$

$$\frac{L_e}{k_{mm}} = \frac{2000}{20} = 100$$

17. (d)
Rankine's formula is applicable for all columns.

18. (b)
Buckling load,

$$P_e = \sigma \times \text{Area}$$

$$\sigma = \alpha\Delta TE$$

$$\sigma \propto k^2$$

$$\propto d^2$$

$$\frac{\pi^2 EI}{L^2} = \sigma \times A$$

$$\frac{\pi^2 EAk^2}{L^2} = \sigma \times A$$

$$\sigma = \frac{\pi^2 Ek^2}{L^2}$$

$$k^2 = \frac{d^2}{16}$$

19. (b)

$$P_e \text{ for both ends pinned} = \frac{\pi^2 EI}{L^2}$$

20. (a)

$$I_1 = \frac{2a \times a^3}{12} = \frac{a^4}{6}$$

$$I_2 = \frac{3a \times (0.5a)^3}{12} = \frac{a^4}{32}$$

$$P_1 = \frac{\pi^2 E}{b^2} \times \frac{a^4}{6} = \frac{\pi^2 E}{b^2} \times \frac{a^4}{6}$$

$$P_2 = \frac{\pi^2 E}{2.25b^2} \times \frac{a^4}{32} = \frac{\pi^2 Ea^4}{72b^2}$$

$$P_2 = \frac{P}{12}$$

21. (a)
Both with hinged

$$P_e = \frac{\pi^2 EI}{L^2}$$

Both ends fixed,

$$P_e' = \frac{4\pi^2 EI}{L^2} = 4\,P_e$$

Theories of Failure

CHAPTER 15

For the prediction of failure of a structural member or a machine component, there are various theories of failure developed by various Professors and Scientists. However these theories are based on principal stresses (their relative magnitudes) at the point of interest of the body. Moreover the use of these theories depends on the mechanical properties of the material of the structural member or machine component.

These mechanical properties are brittleness or ductility; yield strength, ultimate strength, behaviour under tension loading and or compression loading etc.

Question 15.1 **What are ductile and brittle materials? Explain with 5 examples of each material. What are the applications of these materials?**

Solution:

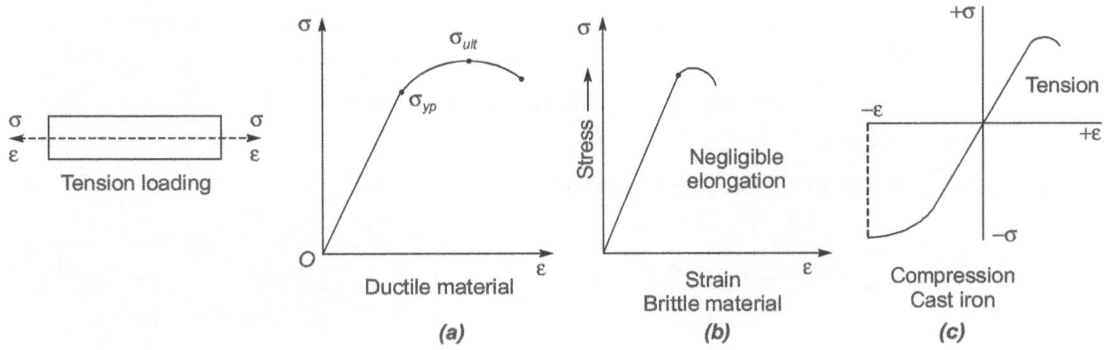

Fig. 15.1

When a material is tested in tension and a graph between load-extension or tensile stress-tensile strain is drawn then a ductile material exhibits sufficient elongation as shown in figure 15.1(a), but a brittle material shows negligible elongation as shown in figure 15.1(b). A general definition of a ductile material is given by *"A material which can be drawn into wires is a ductile material and a material which cannot be drawn into wires is a brittle material."*

Example of ductile materials are gold (most ductile), silver, copper, aluminium, steel etc.

Common examples of brittle materials are chalk, marble, concrete, brick work, cast iron etc.

Majority of engineering components such as crank shaft, connecting rod, pins, shafts, couplings etc. are made from ductile materials. But then there are various components which require high strength in compression such as beds of machines, cylinder blocks of engines, brittle material as cast iron is used.

Cast iron is weak in tension but strong in compression. It has better damping property in compression. Fig. 15.1 (c) show σ–ε curve for cast iron in tension and compression, its ultimate strength in compression is as high as 530-550 N/mm^2.

Question 15.2 **Why principal stresses are taken for the prediction of failure of a material? What is the necessity of yield point stress in a simple tensile test for the use in theories of failure?**

Solution: At every point in a stressed body, and even at the point of interest, there exists a set of 3 planes perpendicular to each other. On these planes there are no shear stresses, but only the normal stresses are present. These 3 normal stresses are known as principal stresses. The **greatest principal stress is the absolute maximum normal stress at the point** and the normal stress on any other plane passing through the point will always be less than this maximum principal stress.

A component can be subjected to uniaxial load or a biaxial load or a combination of tensile/compressive and shear loads in 3 directions. In such a situation how to predict the failure of the body. These three principal stresses obtained at a point, are compared with the result of simple tensile test, and knowing the value of σ_{yp}, yield point stress. At this point, behaviour of the material in terms of σ–ε curve becomes nonlinear as shown in figure 15.1(a). Effect of principal stresses $p_1 > p_2 > p_3$ is compared with the effect of σ_{yp} in simple tension test. After the yield point, the ductile material goes into plastic stage and its utility as an engineering component becomes questionable.

Question 15.3 **What is maximum principal stress theory? For what materials it is used? Show its graphical representation in a two dimensional case.**

Solution: This theory was given by Prof. Rankine. In a component there are principal stresses.

$p_1 > p_2 > p_3$ as per this theory the component will fail if

$p_1 \geq \sigma_{ult}$ (for brittle material)

$\sigma_1 \geq \sigma_{yp}$ (for ductile materials)

In a two dimensional case

$p_1 > p_2$, $p_3 = 0$, As per this theory failure occurs if

$$\frac{p_1}{\sigma_{yp}} \leq \pm 1$$

and
$$\frac{p_2}{\sigma_{yp}} \leq \pm 1$$

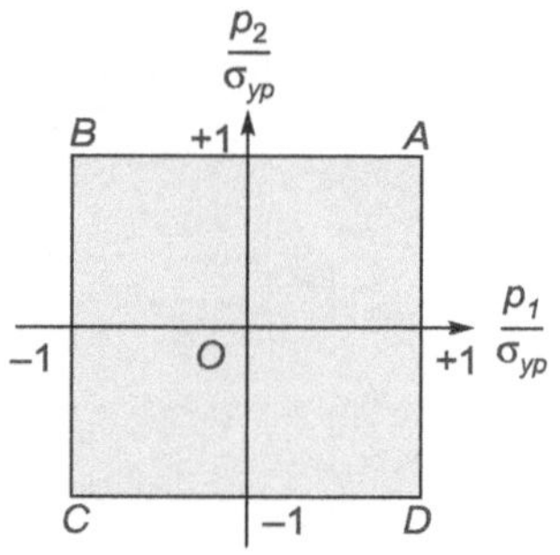

Fig. 15.2 *Graphical Representation (Max. principal stress theory)*

Fig. 15.2 shows the graphical representation of maximum principal stress theory.

Boundary of the figure 15.2, defines the beginning of yielding in a material.

In case of brittle materials, $\frac{p_1}{\sigma_{ult}} \leq \pm, \frac{p_2}{\sigma_{ult}} \leq \pm 1$

This theory is generally used for prediction of failure in brittle materials. Experimental results show that this theory is acceptable, when principal stresses are of same sign. It is unsafe to use a machine component on the basis of this theory if $p_1 > 0$, $p_2 < 0$.

Practice Q.15.3 **A thin cylindrical shell of diameter *D*, wall thickness *t* is subjected to internal pressure p. The principal stresses at inner radius of cylinder are**

$$+\frac{pD}{2t}, +\frac{pD}{4t}, -p \text{ As per this theory}$$

$$\frac{pD}{2t} \le \sigma_{yp}$$

or $$t \ge \frac{pD}{2\sigma_{yp}}$$

Question 15.4 The load on a bolt consists of an axial thrust of 8 kN together with a transverse shear force of 3 kN. Determine diameter of the bolt using maximum principal stress theory, if σ_{yp} = 285 N/mm^2. Take factor of safety as 3.

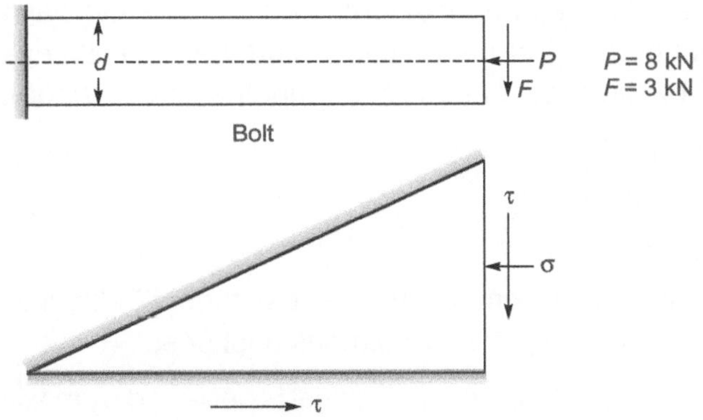

Fig. 15.3

Solution:

$$\sigma_{yp} = 285 \text{ N/mm}^2$$

$$\text{FOS} = 3$$

$$\sigma_{yp'} \text{ allowable stress} = \frac{285}{3} = 95 \text{ N/mm}^2 = \sigma_a$$

Fig. 15.3 shows a bolt of diameter d, subjected to axial thrust P and transverse shear force F,

Direct stress, $$\sigma = \frac{P}{A}$$

Shear stress, $$\tau = \frac{F}{A}, \qquad \text{where } A = \frac{\pi d^2}{4}$$

Maximum principal stress,

$$p_1 = \frac{\sigma}{2} + \sqrt{\left(\frac{\sigma}{2}\right)^2 + \tau^2}$$

$$p_1 = \frac{P}{2A} + \sqrt{\left(\frac{P}{2A}\right)^2 + \left(\frac{F}{A}\right)^2} = \frac{P}{2A} + \frac{1}{A}\sqrt{\frac{P^2}{4} + F^2}$$

Putting the values of P = 8 kN, F = 3 kN

$$p_1 = \frac{8}{2A} + \frac{1}{A}\sqrt{16+9} = \frac{4}{A} + \frac{5}{A} = \frac{9}{A} \text{ kN} = \frac{9000}{A} \text{ N/mm}^2$$

As per this theory, $$p_1 \le 85 \text{ N/mm}^2$$

$$\frac{9000}{A} \leq 85$$

$$A \geq \frac{9000}{85}$$

$$\frac{\pi}{4} \times d^2 \geq \frac{9000}{85}$$

$$d \geq \sqrt{\frac{36000}{85\pi}} \geq 11.61 \text{ mm}$$

Bolt diameter, $d \geq 11.61$ mm

Practice Q.15.4 A thick cylinder of external radius R the internal radius 100 mm is subjected to internal pressure of 50 N/mm². If σ_{yp} of material of cylinder is 270 N/mm², taking 2 as FOS, determine external radius of cylinder on the basis of maximum principal stress theory.

Hints: $\left[p_1 = 50 \times \frac{R^2 + 100^2}{R_2 - 100^2}\right]$

Ans. [147.53 mm]

Question 15.5 What is maximum shear stress theory? Why it is in most common use? Show graphical representation of this theory in a two dimensional case.

Solution: As per this theory, yielding in the material is governed by maximum shear stress developed in the material, under the action of principal stresses p_1, p_2 and p_3. This is also known as Tresca Theory.

In a 3 dimensional case, where $p_1 > p_2 > p_3$ are principal stresses.

$\tau_{max} = \frac{p_1 - p_3}{2}$, which can be obtained by drawing a three dimensional Mohr's stress circle, i.e., 3 circles are drawn between

1. p_1 and p_2 2. p_2 and p_3 3. p_1 and p_3

The circle between p_1 and p_3 gives maximum shear stress at the point. As per this theory.

For safe design

$$\left(\frac{p_1 - p_3}{2}\right) \leq \frac{\sigma_{yp}}{2}$$

as in the case of simple tensile test, σ_{yp} is the uniaxial stress, but at angles of ±45° to the axis of sample, there are shear stresses $\frac{\sigma_{yp}}{2}$.

or $(p_1 - p_3) \leq \sigma_{yp}$ in a three dimensional case.

In a two dimensional case

$$p_1 > p_2, p_3 = 0$$

$$\frac{p_1 - p_2}{2} \leq \frac{\sigma_{yp}}{2}; \text{ if } p_1 \text{ and } p_2 \text{ are of opposite sign.}$$

$$\frac{p_1}{2} \leq \frac{\sigma_{yp}}{2}; \text{ if } p_1 \text{ and } p_2 \text{ are of same sign}$$

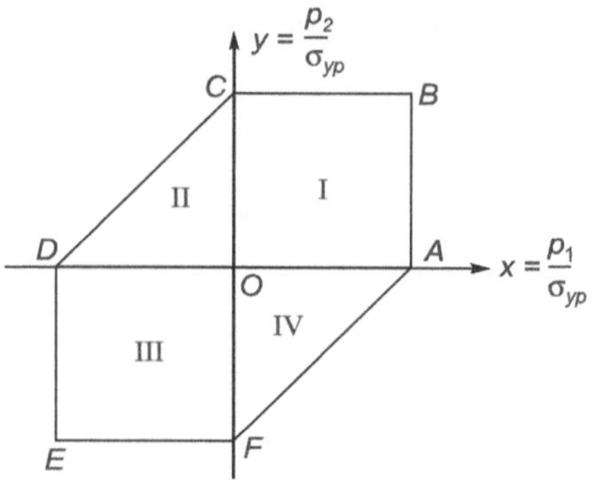

Fig. 15.4 *Max. shear stress theory*

or $p_1 \leq \sigma_{yp}$, this theory coincides with the maximum principal stress theory.

Graphical representation of maximum shear stress theory is shown in fig. 15.4

(a) In I quadrant, both p_1 and p_2 are positive, this theory coincides with maximum principal stress theory

(b) In the II quadrant, p_1 is negative and p_2 is positive line CD shows $p_1 - p_2 = -\sigma_{yp}$.

(c) In the third quadrant, both p_1 and p_2 are of same negative sign, again theory coincides with maximum principal stress theory.

(d) In IV quadrant, p_1 is positive but p_2 is negative, line AF shows $p_1 - p_2 = +\sigma_{yp}$

This theory is most commonly used because

(i) this theory provides more factor of safety

(ii) a conservative design of any component is obtained

(iii) most engineering materials are ductile in nature, this theory is most suitable for ductile materials.

Question 15.6 **A thin cylindrical shell is of diameter 160 mm and thickness 3.2 mm. Yield strength of the material is 280 N/mm². Taking factor of safety of 2, determine the maximum allowable inside pressure is the shell.**

Solution:

$$\sigma_{yp} = 280 \text{ N/mm}^2$$

$$\text{FOS} = 2$$

Allowable,

$$\sigma_{yp} = \frac{280}{2} = 140 \text{ N/mm}^2$$

Principal Stress

$$p_1 = \sigma_c = \frac{pD}{2t} = \frac{p \times 160}{2 \times 3.2} = +25p \text{ in circumferential direction}$$

$$p_2 = \sigma_d = \frac{pD}{4t} = \frac{p \times 160}{4 \times 3.2} = +12.5p\text{, in axial direction}$$

$$p_3 = \sigma_r = -p\text{, in radial direction}$$

$$\tau_{max} = \frac{25p - (-p)}{2} = 13p \leq 140$$

$$p = 10.77 \text{ N/mm}^2\text{, internal pressure}$$

Practice Q.15.6 **A thick cylinder of internal diameter 100 mm is subjected to an internal pressure of 50 N/mm². Determine the thickness of cylinder, according to maximum shear stress theory if σ_{yp} = 300 N/mm², FOS = 2.5**

Hints: $\left[\sigma_c = p\frac{R^2 + 50^2}{R^2 - 50^2}, \sigma_a = \frac{p \times 50^2}{R^2 - 50^2}, \sigma_r = -p\right]$

Ans. [72.5 mm]

Question 15.7 **What is principal strain theory of failure? Where it is used? Show its graphical representation in a two dimensional case.**

Solution: This is also known as St. Venant's theory. In this theory it is assumed that failure by yielding of a material takes place when maximum principal strain in the material subjected to principal stresses is equal to the strain at the yield point in a simple uniaxial test.

If E, ν are Young's modulus and Poisson's ratio of the material.

$p_1 > p_2 > p_3$ are the principal stresses, then

Principal strains are

$$e_1 = \frac{1}{E}(p_1 - \nu p_2 - \nu p_3)$$

$$e_2 = \frac{1}{E}(p_2 - \nu p_1 - \nu p_3)$$

$$e_3 = \frac{1}{E}(p_3 - \nu p_1 - \nu p_2)$$

Strain $e_1 > e_2 > e_3$

Maximum principal strain $e_1 \leq \frac{\sigma_{yp}}{E}$

or $\frac{1}{E}(p_1 - \nu p_2 - \nu p_3) = \frac{\sigma_{yp}}{E}$

or $(p_1 - \nu p_2 - \nu p_3) = \sigma_{yp}$

$$p_1 = \nu(p_2 + p_3) + \sigma_{yp} \quad \text{...(i)}$$

However this theory is not acceptable to the designers.

From eq. (i)

$$p_1 = \sigma_{yp} + \nu(p_2 + p_3)$$

If all the principal stresses are of the same sign, then it is risky to use this theory as p_1 becomes greater than σ_{yp}

In a two dimensional case

$$p_1 > p_2$$

$$p_3 = 0$$

$$e_1 = \frac{1}{E}(p_1 - \nu p_2) \leq \frac{\sigma_{yp}}{E}$$

or $p_1 - \nu p_2 \leq \sigma_{yp}$

$$\frac{p_1}{\sigma_{yp}} - \frac{\nu p_2}{\sigma_{yp}} \leq 1$$

If we take $\frac{p_1}{\sigma_{yp}} = x$, $\frac{p_2}{\sigma_{yp}} = y$, then

$$x - \nu y = 1$$

or $x = 1 + \nu y$

Similarly $y = 1 + \nu x$

These relationships are shown in figure 15.5

However this theory is not acceptable, if p_1 and p_2 both are positive stresses then p_1 exceeds σ_{yp}.

This theory is not suitable for ductile or for brittle materials.

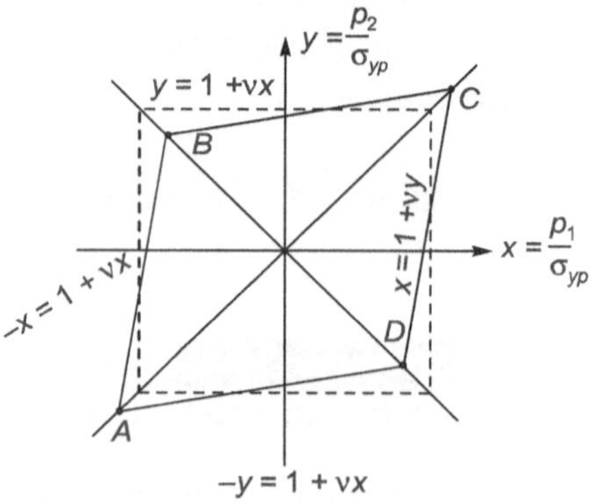

Fig. 15.5 *Graphical representation of (maximum principal strain theory)*

Question 15.8 A steel shaft of diameter 40 mm is subjected to a twisting moment of 800 Nm. In addition there is axial compressive stress of 40 kN. If σ_{yp} = 280 N/mm^2 and ν = 0.3. Determine FOS as per maximum principal strain theory.

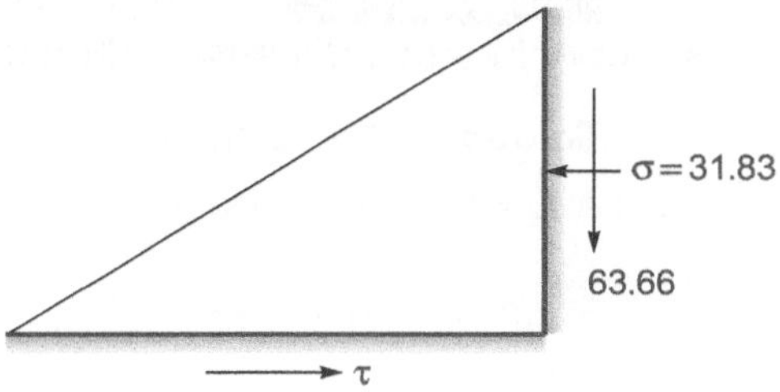

Fig. 15.6

Solution:

$$T = 800 \text{ Nm} = 800 \times 10^3 \text{ Nmm}$$

Shaft diameter, $d = 40$ mm

τ, shear stress on the surface of the shaft,

$$= \frac{16T}{\pi d^3} = \frac{16 \times 800 \times 10^3}{\pi \times 40^3} = 63.66 \text{ N/mm}^2$$

Axial compressive force, $P = 40$ kN

Area of cross-section, $A = \frac{\pi}{4} \times 40^2 = 1256.6 \text{ mm}^2$

σ, axial stress $= \frac{40{,}000}{1256.6} = 31.83 \text{ N/mm}^2$

Fig. 15.6 shows stress on an element on the surface of the shaft.

$$\frac{\sigma}{2} = \frac{31.83}{2} = 15.916 \text{ N/mm}^2$$

$$\tau = 63.66 \text{ N/mm}^2$$

Principal stresses are $p_1, p_2 = \frac{\sigma}{2} \pm \sqrt{\left(\frac{\sigma}{2}\right)^2 + \tau^2}$

$$= 15.916 \pm \sqrt{15.916^2 + 63.66^2}$$

$$= 15.916 \pm \sqrt{253.3 + 4052.6}$$

$$= 15.916 \pm 65.62$$

$$p_1 = 81.53 \text{ N/mm}^2$$

$$p_2 = -49.704 \text{ N/mm}^2$$

Maximum strain theory, $(p_1 - \nu p_2) \le \frac{\sigma_{yp}}{FOS}$

$$(81.53 + 0.3 \times 49.704) \le \frac{\sigma_{yp}}{FOS}$$

Factor of safety $= \frac{280}{96.44} = 2.90$

Practice Q.15.8 Determine the thickness of a thin cylinder of 600 mm diameter subjected to an internal pressure of 4 MPa according to maximum principal strain theory.

FOS = 2, σ_{yp} = 280 MPa, Poisson's ratio = 0.28

Ans. [8.64 mm]

Question 15.9 **What is a strain energy theory, explain. Show the graphical representation of this theory in a two dimensional case.**

Solution: At a point, principal stresses are $p_1 > p_2 > p_3$

If E = Young's modulus, ν = Poisson's ratio

Principal strains are,

$$e_1 = \frac{1}{E}\left[p_1 - \nu(p_2 + p_3)\right]$$

$$e_2 = \frac{1}{E}\left[p_2 - \nu(p_1 + p_3)\right]$$

$$e_3 = \frac{1}{E}\left[p_3 - \nu(p_1 + p_2)\right]$$

Strain energy per unit volume,

$$u = \frac{1}{2}p_1e_1 + \frac{1}{2}p_2e_2 + \frac{1}{2}p_3e_3$$

$$= \frac{1}{2E}\left[p_1^2 + p_2^2 + p_3^2 - 2\nu(p_1p_2 + p_2p_3 + p_3p_2)\right]$$

Strain energy per unit volume in simple tensile test

$$= \frac{\sigma_{yp}^2}{2E}$$

As per this theory

$$= \frac{1}{2E}\left[p_1^2 + p_2^2 + p_3^2 - 2\nu(p_2p_2 + p_2p_3 + p_3p_1)\right] \le \frac{\sigma_{yp}^2}{2E}$$

or

$$= \left[p_1^2 + p_2^2 + p_3^2 - 2\nu(p_1p_2 + p_2p_3 + p_3p_1)\right] \le \sigma_{yp}^2$$

In a two dimensional case, where $p_3 = 0$

$$\left[p_1^2 + p_2^2 - 2\nu p_1p_2\right] \le \sigma_{yp}^2$$

or

$$\frac{p_1^2}{\sigma_{yp}^2} + \frac{p_2^2}{\sigma_{yp}^2} - \frac{2\nu p_1p_2}{\sigma_{yp}^2} \le 1$$

If $\frac{p_1}{\sigma_{yp}} = x$, $\frac{p_2}{\sigma_{yp}} = y$, then equation becomes

$$(x^2 + y^2 - 2\nu xy) \le 1$$

If we take $\nu = 0.3$ as for mild steel, most common structural material $(x^2 + y^2 - 0.6xy) \le 1$ an equation of ellipse.

This theory was suggested by Beltrami and Haigh, but designers are not inclined to use this theory of failure.

Fig. 15.7 shows graphical representation of this theory.

This theory is used in the design of pressure vessels.

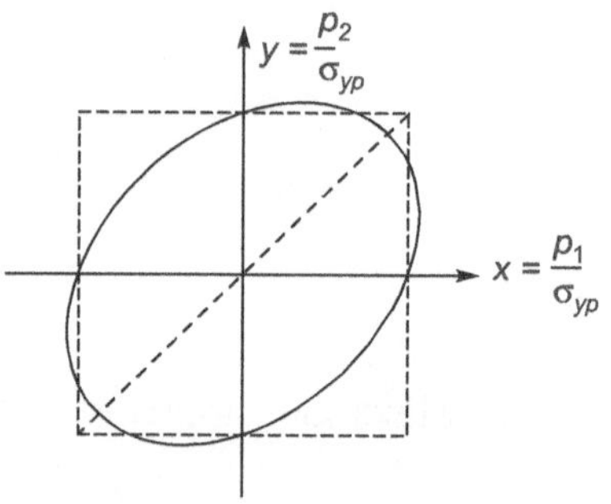

Fig. 15.7 *Graphical representation of strain energy* theory

Question 15.10 **A thick cylinder of internal diameter 200 mm and external diameter 250 mm is subjected to an internal pressure p. Determine the value of p on the basis of strain energy theory, FOS = 2, σ_{yp} = 270 N/mm², Poisson's ratio = 0.3.**

Solution: At the inner radius of thick cylinder, principal stresses are

Hoop stress,

$$p_1 = \sigma_c = p \times \frac{250^2 + 200^2}{250^2 - 200^2} = 4.555p$$

Axial stress, $p_2 = \sigma_a = p \times \frac{200^2}{250^2 - 200^2} = 1.777p$

Radial stress, $p_3 = \sigma_r = -p$

Using the strain energy theory

$$\sigma_{yp} = \frac{\sigma_{yp}}{FOS} = \frac{270}{2} = 135 \text{ N/mm}^2$$

$$= [(4.555p)^2 + (1.777p)^2 + (-p)^2 - 2 \times 0.3(4.555p \times 1.777p - 1.777p \times p - p \times 4.555p)] = \sigma_{yp}^2$$

$$p^2[20.75 + 3.16 + 1 - 0.6(8.094 - 1.777 - 4.555)] = 135^2$$

$$p^2[24.91 - 0.6(1.762)] = 18225$$

$$p^2[23.8528] = 18225$$

Internal pressure, $p = 27.64 \text{ N/mm}^2$

Practice Q.15.10 **The load on a bolt consists of an axial thrust of 8 kN together with a transverse shear load of 3 kN. Calculate diameter of the bolt using strain energy theory. Take 3 as factor of safety $\sigma_{yp} = 285 \text{ N/mm}^2$, $\nu = 0.3$**

Ans. [$d = 11.2$ mm]

Question 15.11 **What is shear strain energy or distortional energy theory, explain? Where it is used? Show its graphical representation.**

Solution: Total strain energy absorbed in a body consists two parts

(a) Volumetric strain energy, which changes volume of the body

(b) Shear strain energy, which only changes the shape of the body i.e., producing only shear strains in body

i.e., u = total strain energy

= volumetric strain energy + shear strain energy

$= u_v + u_s$

u_v = due to volumetric stress component, $p_m = \frac{p_1 + p_2 + p_3}{3}$

$$= \frac{1}{2E}\left[3p_m^2 - 2\nu(p_m)^2\right] \qquad \text{where } p_m = \frac{p_1 + p_2 + p_3}{3}$$

After simplification

$$u_v = \frac{1}{6E}(1 - 2\nu)\left[p_1^2 + p_2^2 + p_3^2 + 2p_1p_2 + 2p_2p_3 + 2p_3p_1\right]$$

$$u_s = u - u_v = \frac{1}{2E}\left[p_1^2 + p_2^2 + p_3^2 - 2\nu(p_1p_2 + p_2p_3 + p_3p_1)\right] - \left(\frac{1-2\nu}{6E}\right)\left[p_1^2 + p_2^2 + p_3^2 + 2p_1p_2 + 2p_2p_3 + 2p_3p_1\right]$$

Putting $E = 2G(1 + \nu)$ in the above expression, and after simplification

$$u_s = \frac{1}{12G}\left[(p_1 - p_2)^2 + (p_2 - p_3)^2 + (p_3 - p_1)^2\right]$$

After per this theory

$$\frac{1}{12G}\left[(p_1-p_2)^2+(p_2-p_3)^2+(p_3-p_1)^2\right] \le \frac{2\sigma_{yp}^2}{12G}$$

$$[(p_1 - p_2)^2 + (p_2 - p_3)^2 + p_3 - p_1)^2] = 2\sigma^2_{yp}$$

Please note that $(p_1 - p_2) = 2\tau_1, (p_2 - p_3)$

$$= 2\tau_2, (p_3 - p_1) = 2\tau_3$$

are the shear components considering the three dimensional Mohr's stress circle.

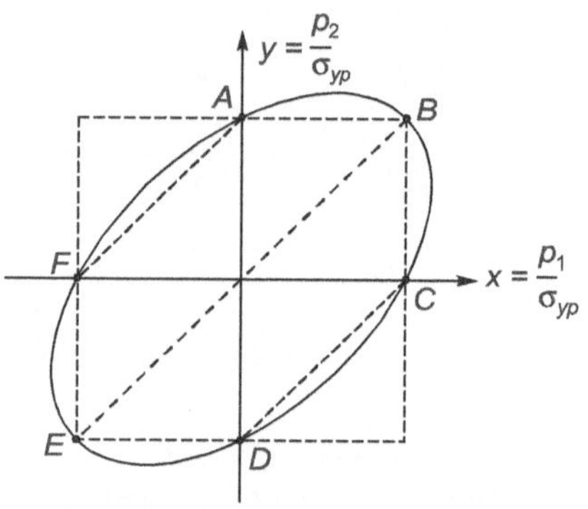

Fig. 15.8 *Graphical representation of distortion energy theory (ellipse passes through ABCDEF points as shown)*

In a two dimensional case $p_1 > p_2, p_3 = 0$

$$[p_1^2 - 2p_1p_2 + p_2^2 + p_2^2 + p_1^2] \le 2\sigma_{yp}^2$$

$$(p_1^2 - p_1p_2 + p_2^2) \le \sigma_{yp}^2$$

or

$$\frac{p_1^2}{\sigma_{yp}^2} - \frac{p_1p_2}{\sigma_{yp}^2} + \frac{p_2^2}{\sigma_{yp}^2} \le 1$$

If

$$\frac{p_1}{\sigma_{yp}} = x$$

$$\frac{p_2}{\sigma_{yp}} = y$$

Then, $(x^2 - xy + y^2) \le 1$, Equation of ellipse.

For ductile material, theory is in good agreement with experimental results used in the design of power transmitting shafts. This theory is most suitable for ductile materials.

This theory is also known as Von Moses theory.

Question 15.12 **A thin spherical shell has a diameter of 200 mm and a wall thickness *t*. It is subjected to an internal pressure 5 N/mm². Determine wall thickness as per distortional energy theory. Take FOS = 3, σ_{yp} = 264 N/mm².**

Solution:

$$\sigma_{yp} = 264 \text{ N/mm}^2$$

$$\text{FOS} = 3$$

$$\sigma'_{yp} = \frac{264}{3} = 88 \text{ N/mm}^2$$

Principal stress at inner radius of thin spherical shell.

$$p_1 = p_2 = \frac{pD}{4t} = \frac{5\times 200}{4\times t} = \frac{250}{t}$$

$$p_3 = p = -5 \text{ N/mm}^2$$

Using the distortion energy theory.

$$\left[\left(\frac{250}{t}-\frac{250}{t}\right)^2+\left(\frac{250}{t}+p\right)^2+\left(-p-\frac{250}{t}\right)^2\right] = 2\times\sigma_{yp}^2 = 88^2\times 2$$

$$2\left[\frac{250}{t}+p\right]^2 = 88^2 \times 2$$

$$\frac{250}{t} + p = 88, \quad \text{putting the value of } p.$$

$$\frac{250}{t} + 5 = 88$$

$$\frac{250}{t} = 83$$

$$t = \frac{250}{83} = 3 \text{ mm, wall thickness.}$$

Practice Q.15.12 A critical element develops principal stresses p_1, p_2, p_3 in the ratio of 5 : 2 : –1, such that stresses are $5p$, $2p$, $-p$, where p is a parameter in N/mm². Determine the maximum magnitude of p such that stresses may reach yielding according to Von Mosses Criterion (Distortion Energy Theory). σ_{yp} = 400 MPa

Ans. [77 N/mm²]

Question 15.13 In a steel drum subjected to internal pressure of 10 N/mm² and an axial compressive force, the circumferential stress is 110 MPa, axial stress is 40 MPa and radial stress – 10 N/mm². Find the equivalent tensile stress in a simple tensile test as per (i) Maximum principal strain theory (ii) Distortion energy theory, Poisson's ratio, ν = 0.3

Solution:

(i) Maximum principal strain theory

$$[p_1 - \nu(p_2 + p_3] = \sigma_{yp}$$

$$110 - 0.3(40 - 10) = \sigma_{yp}$$

$$\sigma_{yp} = 101 \text{ MPa}$$

(ii) Distortion energy theory

$$[(p_1 - p_2)^2 + (p_2 - p_3)^2 + (p_3 - p_1)^2 = 2\sigma_{yp}^2$$

$$[(110 - 40)^2 + (40 + 10)^2 + (-10 - 110)^2] = 2\sigma_{yp}^2$$

$$[4900 + 2500 + 14400] = 2\sigma_{yp}^2$$

$$\sigma_{yp}^2 = \frac{21800}{2} = 10900$$

$$\sigma_{yp} = 104.4, \text{ equivalent stress in simple tension.}$$

Practice Q.15.13 In a steel drum that is subjected to axial compressive force, the internal pressure is 10 N/mm². maximum circumferential stress is 120 MPa, longitudinal stress is 40 MPa. Find the equivalent stress in a simple tensile test as per (a) Maximum shear stress theory (b) Strain energy theory, Poisson's ratio, ν = 0.3

Ans. [(a) 130 MPa (b) 119.08 MPa]

Question 15.14 A thin aluminium alloy tube bar has a mean diameter of 180 mm and a wall thickness of 2.5 mm. Tube is subjected to an internal pressure of 3 N/mm² and a torque of 5 kNm. If the yield strength of the material is 250 N/mm² and Poisson's ratio is 0.33, determine factor of safety according to (a) Maximum shear stress theory (b) Strain energy theory.

Solution: *Internal pressure*

$$\sigma_c = \frac{pD}{2t} = \frac{3\times180}{2\times2.5} = 108 \text{ N/mm}^2$$

$$\sigma_a = \frac{pD}{4t} = \frac{3\times180}{4\times2.5} = 54 \text{ N/mm}^2$$

Torque

$$T = 5 \times 10^6 \text{ Nmm}$$

$$J = \pi Dt\frac{D^2}{4} = \pi\times180\times2.5\times90^2 = 1145.1\times10^4 \text{ mm}^4$$

Shear stress,

$$\tau = \frac{T}{J}\times \text{Radius} = \frac{5\times10^6\times90}{1145.1\times10^4} = 39.3 \text{ N/mm}^2$$

Principal stresses

Stresses on the surface of the tube are shown in figure 15.9

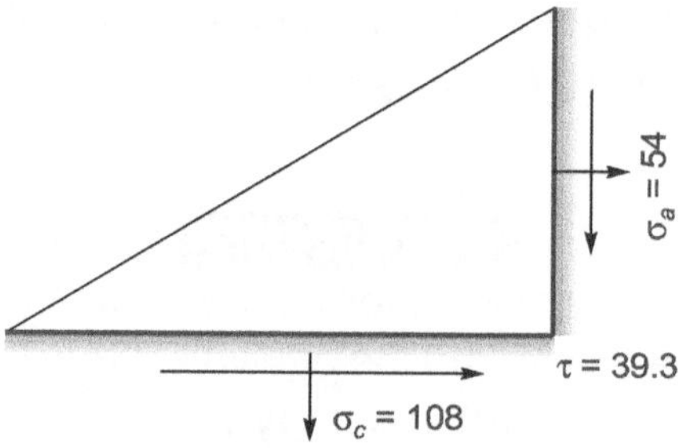

Stresses on an element of tube

Fig. 15.9

$$\frac{\sigma_c+\sigma_a}{2} = \frac{108+54}{2} = 81 \text{ N/mm}^2$$

$$\frac{\sigma_c-\sigma_a}{2} = \frac{108-54}{2} = 27 \text{ N/mm}^2$$

$$\sqrt{\left(\frac{\sigma_c-\sigma_a}{2}\right)^2+\tau^2} = \sqrt{27^2+39.3^2}$$

$$= \sqrt{729+1544.49}$$

$$= 47.68 \text{ N/mm}^2 = \tau_{max}$$

$$p_1 = 81 + 47.68 = 128.68 \text{ N/mm}^2$$

$$p_2 = 81 - 47.68 = 33.32 \text{ N/mm}^2$$

(a) **Maximum shear stress theory**

$$47.68 = \frac{\sigma_{yp}}{2\times\text{FOS}} = \frac{250}{2\times\text{FOS}}$$

$$\text{FOS} = \frac{125}{47.68} = 2.62$$

(b) **Strain energy theory**

$$(p_1^2 + p_2^2 - 2\nu p_1p_2) = \left(\frac{\sigma_{yp}}{\text{FOS}}\right)^2, \text{ putting the values}$$

$$128.68^2 + 33.32^2 - 2 \times 0.33 \times 128.68 \times 33.32 = \left(\frac{\sigma_{yp}}{\text{FOS}}\right)^2$$

$$16558.54 + 1110.22 - 2829.83 = \left(\frac{\sigma_{yp}}{\text{FOS}}\right)^2$$

or

$$\left(\frac{\sigma_{yp}}{\text{FOS}}\right) = 121.815$$

$$\text{FOS} = \frac{250}{121.815} = 2.052$$

Practice Q.15.14 A thin cylindrical shell of diameter 150 mm, wall thickness 2 mm is subjected to internal pressure of 4 N/mm^2. At the same time shell is subjected to an axial twisting moment which produces maximum shear stress of 45 N/mm^2. If σ_{yp} of material is 270 N/mm^2, determine FOS as per (i) Maximum principal strain theory (ii) Distortional energy theory, consider only two principal stresses.

Ans. [1.743, 1.78] Poisson's ratio = 0.3

Question 15.15 A circular steel cylinder of wall thickness 10 mm and internal diameter 200 mm is subjected to an internal pressure of 16 MPa. Cylinder is subjected to axial compressive force before yielding commences according to distortion energy theory. σ_{yp} of material = 250 MPa. Take the effect of radial stress as negligible. Determine axial compressive force.

Solution:

Internal pressure, $p = 16\ N/mm^2$

$D = 200$ mm

$t = 10$ mm

Circumferential stress, $\sigma_c = \frac{pD}{2t} = \frac{16 \times 200}{2 \times 10} = 160$ MPa

Axial stress, $\sigma_a = \frac{pD}{4t} = 80$ MPa

Say, σ = axial compressive stress due to axial compressive force

Net stress in axial direction, $\sigma_a' = \sigma_a - \sigma$

Distortional Energy Theory

$$\left(\sigma_c'^2 + \sigma_a'^2 - \sigma_c'\sigma_a'\right) = \sigma_{yp}^2$$

$$\left(160^2 + \sigma_a'^2 - 160 \times \sigma_a'\right) = 250^2$$

$$\sigma_a'^2 - 160\sigma_a' - 36900 = 0$$

$$\sigma_a' = \frac{160 \pm \sqrt{160^2 + 4 \times 36900}}{2}$$

$$= 80 \pm \sqrt{80^2 + 36900}$$

$= 80 \pm 208.08 = 288.08\ N/mm^2$

or $= -128.08\ N/mm^2$ (admissible as compressive)

$\sigma_a' = 80 - \sigma = -128.08$

$\sigma = 80 + 128.08 = 208.08$ MPa

Area $= \pi Dt = \pi \times 200 \times 10 = 6283.18\ mm^2$

Axial compressive force $= 208.08 \times 6283.18 = 1307404$ N $= 1307.4$ kN

Practice Q.15.15 A circular steel cylinder of wall thickness 10 mm and internal diameter of 200 mm is subjected to a constant internal pressure of 15 MPa. Determine how much axial compressive load can be applied to the cylinder before yielding commences according to maximum shear stress theory, σ_{yp} = 240 MPa. Neglect the effect of radial stress.

Ans. [1036.7 kN]

Question 15.16 What is octahedral shear stress theory of failure?

Solution: Octahedral planes are those planes which are equally inclined to all principal planes. Shear stress along these planes is called octahedral shear stress. The normal of an octahedral plane makes an angle θ with the principal plane,

$$\cos\theta = \frac{1}{\sqrt{3}}$$

If p_1, p_2, p_3 are the principal stresses at a point, then octahedral shear stress is given by

$$\tau_{oct} = \frac{1}{3}\left[(p_1 - p_2)^2 + (p_2 - p_3)^2 + (p_3 - p_1)^2\right]^{\frac{1}{2}}$$

In two dimensional loading, principal stress $p_3 = 0$

$$\tau_{oct} = \frac{1}{3}\left[p_1^2 + p_2^2 - 2p_1p_2 + p_1^2 + p_2^2\right]^{\frac{1}{2}} = \frac{\sqrt{2}}{3}\left[p_1^2 + p_2^2 - p_1p_2\right]^{\frac{1}{2}}$$

In uniaxial loading,

$$p_2 = p_3 = 0$$

$$\tau_{oct} = \frac{\sqrt{2}}{3}p_1 = \sigma_{yp} \text{ (yield point stress in simple tension)}$$

$$\tau_{oct} = 0.471\,\sigma_y$$

For no failure, octahedral shear stress in strained body should be less than or equal to octahedral shear stress at yield point in simple tension

$$\tau_{oct} \le \frac{\sqrt{2}}{3}\sigma_{yp} \le \frac{\sqrt{2}}{3}\left(\frac{\sigma_{yp}}{FOS}\right) \text{ for design}$$

In pure shear case where

$$p_1 = +\tau,\ p_2 = -\tau$$

$$\frac{\sqrt{2}}{3}\left[\tau^2 + \tau^2 + \tau^2\right]^{\frac{1}{2}} = \frac{\sqrt{2}}{3}\sigma_{yp}$$

$$\sqrt{3}\tau = \sigma_{yp}$$

$$\tau = \frac{\sigma_{yp}}{\sqrt{3}} = 0.577\sigma_{yp}$$

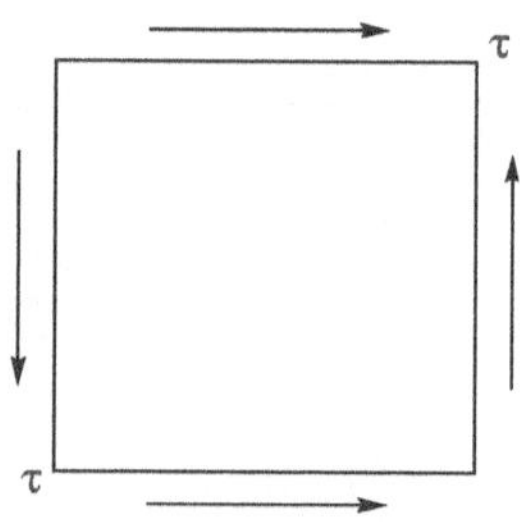

Fig. 15.10

τ_{max} should be limited to 0.577 σ_{yp}.

Question 15.17 The stress state at a point in a body is plane with σ_1 = 60 N/mm², and σ_2 = −36 N/mm². If the allowable stress for the material in simple tension or compression is 100 N/mm². calculate the value of factor of safety with each of the following criterion for failure

(a) Maximum stress criteria
(b) Maximum shear stress criteria
(c) Maximum strain energy criteria
(d) Maximum distortion energy criteria, Poisson's ratio = 0.3

[IES 2010, 10 Marks]

Solution:

(a) Maximum stress criteria

$$60 = \frac{100}{FOS}$$

$$FOS = 1.66$$

(b) Maximum shear stress criteria

$$\frac{60-(-36)}{2} = \frac{100}{2\times FOS}$$

$$48 = \frac{50}{FOS}$$

$$FOS = 1.0416$$

(c) Maximum strain energy criteria

$$60^2 + (-36)^2 + 2 \times 0.3\ (60 \times 36) = \left(\frac{100}{FOS}\right)^2$$

$$3600 + 1296 + 1296 = \left(\frac{100}{FOS}\right)^2$$

$$6192 = \left(\frac{100}{FOS}\right)^2$$

$$FOS = \frac{100}{\sqrt{6192}} = 1.27$$

(d) Maximum distortion energy criteria, Poisson's ratio = 0.3

$$60^2 + (-36)^2 - 60(-36) = \left(\frac{100}{FOS}\right)^2$$

$$3600 + 1296 + 2160 = \left(\frac{100}{FOS}\right)^2$$

$$FOS = \frac{100}{84} = 1.19$$

Question 15.18 A mild steel shaft of 50 mm diameter is subjected to a bending moment of 1.5 kNm and a torque T. If the yield point of steel in tension is 210 MPa, find the maximum value of torque without causing yielding of the shaft material according to (i) Maximum principal stress theory (ii) Maximum shear stress theory.

[IES 2006, 20 Marks]

Solution:

(i) Maximum principal stress theory = 210 MPa

$$M_e = \frac{\pi}{32}\times 50^3 \times 210 = 2.577 \times 10^6 = 2.577 \text{ kNm}$$

$$= \frac{M+\sqrt{M^2+T^2}}{2}$$

$$2.577 = \frac{1.5+\sqrt{1.5^2+T^2}}{2}$$

or $$3.654 = \sqrt{1.5^2+T^2}$$

$$1.5^2 + T^2 = 13.3517$$
$$T^2 = 11.10$$
$$T = 3.332 \text{ kNm}$$

(ii) **Maximum shear stress theory**

$$\tau_{max} = \frac{210}{2} = 105 \text{ N/mm}^2$$

Equivalent tension moment, $T_e = \frac{\pi}{16} \times 50^3 \times 105 = 2577 \times 10^6$ Nmm

$$= 2.577 \text{ kNm}$$
$$= \sqrt{T^2 + M^2}$$
$$= \sqrt{T^2 + 1.5^2}$$
$$T^2 = 6.6409 - 2.25 = 4.3909$$
$$T = 2.095 \text{ kNm}$$

Question 15.19 Considering principal stresses in a steam boiler drum as p, 0.5 *p* and 0, and Poisson's ratio = 0.3, equivalent stress in simple tension as σ, find p in terms of σ due to

(a) Maximum shear stress theory
(b) Strain energy theory
(c) Distortion energy theory

[IES 2011, 10 Marks]

Solution:

Simple tension stress = σ

Principal stress in boiler drum = p, $0.5\,p$, 0

(a) **Maximum shear stress theory**

$$\frac{p - 0}{2} = \frac{\sigma}{2}$$

or $p = \sigma$

(b) **Strain energy theory**

$$[p^2 + (0.5p) - 2 \times 0.3\,(p \times 0.5p)] = \sigma^2$$
$$p^2 + 0.25p^2 - 0.3p^2 = \sigma^2$$
$$p^2(0.95) = \sigma^2$$
$$p = \frac{\sigma}{\sqrt{0.95}} = 1.026\sigma$$

(c) **Distortion energy theory**

$$(p^2 + 0.25p^2 - 0.5p^2) = \sigma^2$$
$$p = \frac{\sigma}{\sqrt{0.75}} = 1.155\sigma$$

Question 15.20 A cylindrical shaft of 75 mm diameter is subjected to a maximum bending moment of 2500 Nm and a twisting moment of 4200 Nm. Find the maximum principal stresses developed in the shaft. If the yield stress of th shaft material is 360 N/sq. mm, determine the factor of safety of the shaft using maximum shear stress theory.

[CE, ESE : 2017]

Solution: Shaft diameter, d = 75 cm

M_{max} = 2500 Nm

T = 4200 Nm

M_e = equivalent bending moment

$$= \frac{M+\sqrt{M^2+T^2}}{2} = \frac{2500+\sqrt{2500^2+4200^2}}{2}$$

$$= \frac{2500+100\sqrt{625+1764}}{2} = \frac{2500+4888}{2}$$

= 3694 Nm

Maximum principal stress, $$p_1 = \frac{M_e}{\pi \times 3}\times 32 = \frac{3694\times 10^3}{\pi\times 75^3}\times 32 = 89.2 \text{ N/mm}^2 \quad \ldots\text{(i)}$$

T_e = Equivalent twisting moment

$$= \sqrt{M^2+T^2} = \sqrt{25^2+42^2}\times 100 = 4888 \text{ Nm}$$

$$488\,8 \times 10^3 = \frac{\pi}{16}\times d^3 \times \tau_{max}$$

$$\tau_{max} = \frac{16\times 4888\times 10^3}{\pi\times 75^3} = 59.0 \text{ N/mm}^2$$

σ_y = 360 N/mm^2

$$\frac{\sigma_y}{2} = 180 \text{ N/mm}^2$$

$$\text{Factor of safety} = \frac{180}{59} = 3.05$$

Question 15.21 A cylindrical shaft made of steel for which the tensile yield strength σ_y = 550 MPa is subjected to static loads consisting of a bending moment of 700 kNmm and torsion of 400 kNmm. Determine the diameter of shaft for a factor of safety equal to 1.5 if E = 200 GPa and Poisson's ratio ν = 0.25; Adopt maximum shear stress failure theory.

[ME, ESE : 2016]

Solution:

σ_y = 550 MPa

$$\tau_y = \frac{550}{2} = 275 \text{ MPa}$$

M = 700 Nm

$$T = 400 \text{ Nm}$$
$$T_e = \text{Equivalent twisting moment}$$
$$= \sqrt{M^2 + T^2} = 806.225 \text{ Nm}$$
$$\text{FOS} = 1.5$$
$$\tau_{\text{allowable}} = \frac{275}{FOS} = \frac{\tau_y}{FOS}$$
$$= 183.33 \text{ MPa}$$
$$T_e = \frac{\pi}{16} d^3 T_a$$
$$806.275 \times 10^3 = \frac{\pi}{16} \times d^3 \times 183.33$$
$$d^3 = 22.396 \times 10^3$$

Shaft diameter, $d = 28.2$ mm

Objective Type Questions

Q.1 Graphical representation of which one of the following theories is an ellipse
(a) Maximum shear stress theory
(b) Maximum principal strain theory
(c) Distortion energy theory
(d) None of these

Q.2 Which one of the following theories gives conservative design of a component
(a) Maximum principal stress theory
(b) Maximum shear stress theory
(c) Shear strain energy theory
(d) None of these

Q.3 Principal stresses at a point are $+p, -p, 0$, using the shear strain energy FOS is $\sqrt{3}$, what is σ_{yp}
(a) $\sqrt{3}p$ (b) $3p$
(c) $1.5p$ (d) None of these

Q.4 A thick cylinder of outer radius two times the inner radius. Internal pressure is p. For this material σ_{yp} = 300 MPa, FOS = 2, using the maximum principal stress theory, what is the maximum value of p
(a) 150 MPa (b) 120 MPa
(c) 90 MPa (d) 75 MPa

Q.5 A shaft is subjected to a twisting moment such that maximum shear stress on surface of shaft is 100 MPa. Poisson's ratio of a material is 0.3. If σ_{yp} is 270, what is FOS as per maximum principal strain theory
(a) 3.857 (b) 2.7
(c) 2.077 (d) 1.732

Q.6 A shaft is subjected to a torque and an axial compressive force. Shear stress due to torque is 30 MPa and compressive stress due to force is 80 MPa. If σ_{yp} = 270 MPa, and ν = 0.3, what is FOS as per maximum shear stress theory
(a) 2.7 (b) 1.58
(c) 1.50 (d) None of these

Q.7 A thin cylindrical shell $\sigma_c = 30p$, $\sigma_a = 15p$, $\sigma_r = -p$. If σ_{yp} = 300 MPa, what is p as per maximum principal strain theory, if ν = 0.3
(a) 11.90 MPa (b) 11.63 MPa
(c) 11.89 MPa (d) None of these

Q.8 A distortion energy theory of failure if $\frac{p_1}{\sigma_{yp}} = x$, $\frac{p_2}{\sigma_{yp}} = y$, Poisson's ratio is 0.3, what is the equation of ellipse for graphical representation

(a) $x^2 + y^2 - xy \leq 1$
(b) $x^2 + y^2 - 0.5xy \leq 1$
(c) $x^2 + y^2 - 0.6xy \leq 1$
(d) None of these

Q.9 The principal stresses at a point inside a solid object are σ_1 = 100 MPa, σ_2 = 100 MPa and σ_3 = 0 MPa. The yield strength of the material is 200 MPa. The factor of safety calculated using Tresca (maximum shear stress) theory is n_T and the factor of safety calculated using Von Mises (maximum distortional energy) theory is n_V. Which one of the following relations is TRUE?

(a) $n_T = \left(\frac{\sqrt{3}}{2}\right) n_V$ (b) $n_T = \left(\sqrt{3}\right) n_V$
(c) $n_T = n_V$ (d) $n_V = \left(\sqrt{3}\right) n_T$

[ME, GATE : 2016 (Set-1)]

Q.10 Principal stresses at a point in an elastic material are 1.5σ (tensile), σ (tensile) and 0.5σ (compressive). The elastic limit in tension is 210 MPa and μ = 0.3. The value of σ at failure when computed by maximum principal strain theory is nearly

(a) 140.5 MPa (b) 145.5 MPa
(c) 150.5 MPa (d) 155.5 MPa

[CE, ESE (Pre) : 2018]

Q.11 The principal stresses at a point are 2σ (tensile) and σ (compressive), and the stress at elastic limit for the material in simple tension is 210 N/mm^2. According to maximum shear strain theory, the value of σ at failure is

(a) 70 N/mm^2 (b) 105 N/mm^2
(c) 140 N/mm^2 (d) 210 N/mm^2

[CE, ESE (Pre) : 2016]

Q.12 Principal stresses at a point are 80 N/mm^2 and 40 N/mm^2, both tensile. The yield stress in simple tension for this material is 200 N/mm^2. The values of factors of safety according to maximum principal stress theory and maximum shear stress theory, respectively, are

(a) 2.5 and 2.5 (b) 2.5 and 5
(c) 5 and 5 (d) 5 and 1.67

[CE, ESE (Pre) : 2016]

Q.13 The state of stress at a point in an elastic material, with yield stress of 200 MPa in simple tension, and Poisson's ratio 0.3, is as shown in the figure.

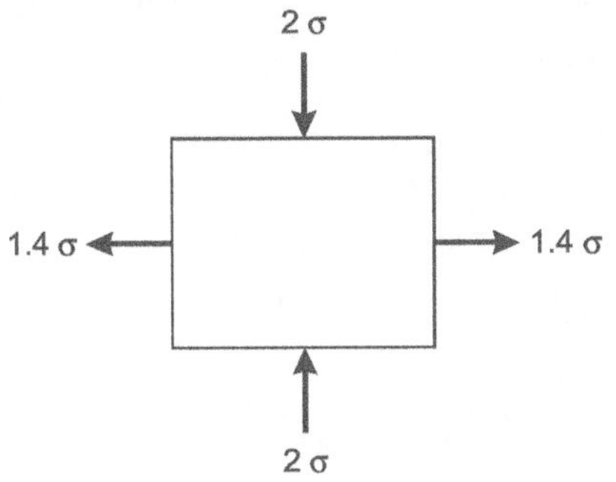

The permissible value of σ by maximum strain theory is

(a) 75 MPa (b) 100 MPa
(c) 150 MPa (d) 200 MPa

[CE, ESE (Pre) : 2017]

Q.14 A hub is press fitted on a shaft. An element in the hub is subjected to a radial compressive stress of 50 N/mm^2 and hoop stress of 75 N/mm^2. If the hub is made of 30C8 steel with yield strength, σ_y = 350 N/mm^2, what is the factor of safety using maximum shear stress theory?

(a) 2.8 (b) 3.6
(c) 4.2 (d) 5.6

[ME, ESE (Pre) : 2016]

Q.15 A steel hub of 100 mm internal diameter and uniform thickness of 10 mm was heated to a temperature of 300 °C to shrink fit it on a shaft. On cooling, a crack developed parallel to the direction of the length of the hub. The cause of the failure is attributable to

(a) tensile hoop stress
(b) tensile radial stress
(c) compressive hoop stress
(d) compressive radial stress

[ME, ESE (Pre) : 2016]

Answers

1. (c)	2. (b)	3. (b)	4. (c)	5. (c)
6. (a)	7. (b)	8. (a)	9. (c)	10. (d)
11. (a)	12. (a)	13. (b)	14. (a)	15. (a)

Explanations

1. (c)
Distortion energy theory is graphically represented by an ellipse.

2. (b)
Maximum shear stress theory gives conservative design of a component.

3. (b)

$$(p^2 + p^2 + p^2) \leq \left(\frac{\sigma_{yp}}{FOS}\right)^2$$

$$3p^2 = \left(\frac{\sigma_{yp}}{FOS}\right)^2$$

$$\sqrt{3}p = \frac{\sigma_{yp}}{FOS} = \frac{\sigma_{yp}}{\sqrt{3}}$$

$$\sigma_{yp} = 3p$$

4. (c)

$$\sigma_c = p\frac{R_2^2 + R_1^2}{R_2^2 - R_1^2}$$

$$= p \times \frac{5}{3} = \frac{300}{2} = 150$$

$$p = 90 \text{ MPa}$$

5. (c)

$$(p_1 - \nu p_2) = \frac{\sigma_{yp}}{FOS}$$

$$[100 - 0.3(-100)] = 130 = \frac{270}{FOS}$$

$$FOS = \frac{270}{130} = 2.077$$

6. (a)

$$\frac{\sigma}{2} = 40 \text{ MPa}, \tau = 30 \text{ MPa}$$

$$\tau_{max} = \sqrt{40^2 + 30^2} = 50 \text{ MPa}$$

$$FOS = \frac{\sigma_{yp}}{2} \times \frac{1}{\tau_{max}} = \frac{135}{50} = 2.7$$

7. (b)

$$p[30 - 0.3(15 - 1)] = 300$$

$$p[25.8] = 300$$

$$p = 11.628 \text{ N/mm}^2$$

8. (a)

$$\frac{p_1}{\sigma_{yp}} = x, \frac{p_2}{\sigma_{yp}} = y$$

Distortion energy theory, graphical representation

$$x^2 + y^2 - xy \leq 1$$

9. (c)

$$\sigma_1 = 100 \text{ MPa}$$

$$\sigma_2 = 100 \text{ MPa}$$

$$\sigma_3 = 0$$

$$\tau_{max} = \frac{100}{2} = 50$$

$$\sigma_y = 200 \text{ MPa}$$

$$\frac{\sigma_y}{2} = 100 = \text{shear stress}$$

$$n_T = \frac{100}{50} = 2 \quad \text{...(i)}$$

Von Mises Theory

$$\sigma_a^2 - \sigma_1\sigma_2 + \sigma_2^2 = \left(\frac{\sigma_y}{n_v}\right)^2$$

$$100^2 = \left(\frac{200}{n_v}\right)^2$$

or $$100 = \frac{200}{n_v}$$

$$n_v = 2 \quad \text{...(ii)}$$

10. (d)
Principal stress

$$+1.5\,\sigma, +\sigma, -0.5\sigma$$

$$\sigma_e = 210 \text{ MPa}$$

$$\varepsilon_{max} = \frac{1.5\sigma}{E} - \frac{0.3\sigma}{E} + \frac{0.3 \times 0.5\sigma}{E}$$

Maximum principal strain

$$= \frac{1.35\sigma}{E} = \frac{210}{E}$$

$$\sigma = 155.55 \text{ MPa}$$

11. (a)

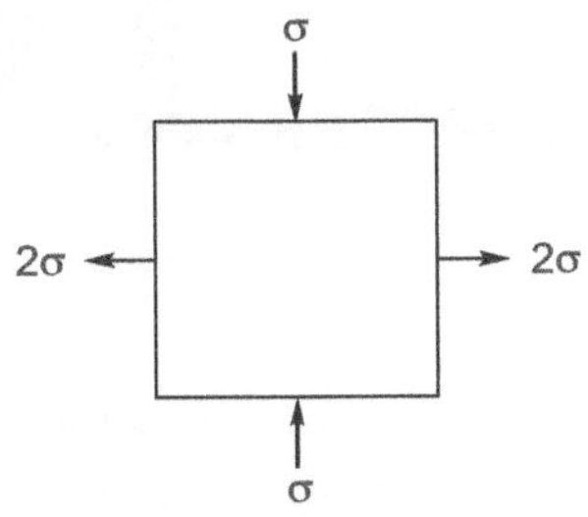

Principal stresses, 2p, –p

$$\sigma_y = 210 \text{ N/mm}^2$$

$$\frac{\sigma_y}{2} = 105 \text{ N/mm}^2$$

$$\tau_{max} = \frac{2\sigma - (-\sigma)}{2} = 1.5\sigma$$

Maximum shear strain theorem.
(Note that there is no shear strain theory)
Shear modulus = G

$$\frac{1.5\sigma}{G} = \frac{105}{G}$$

or $\sigma = 70 \text{ N/mm}^2$

12. (a)
Principal stress, 80, 40 MPa

$$\sigma_y = 200 \text{ MPa}$$

Maximum principal stress theory

$$\text{FOS} = \frac{200}{80} = 2.5$$

Maximum shear stress

$$= \frac{80}{2} = 40$$

$$\frac{\sigma_y}{2} = 100 \text{ MPa}$$

$$\text{FOS} = \frac{100}{40} = 2.5$$

13. (b)
Yield stress only in simple tension is given, so we will find maximum strain in tension only

$$\varepsilon_{max} = \frac{1.4\sigma}{E} - \frac{\nu \times \sigma_2}{E}$$

$$= \frac{1.4\sigma}{E} + \frac{0.3 \times 2\sigma}{E} = \frac{2\sigma}{E} = \frac{\sigma_y}{E}$$

$$2\sigma = 200 \text{ MPa}$$

$$\sigma = 100 \text{ MPa}$$

14. (a)
Theory of failure
On element of hub
Hoop stress,

$$\sigma_c = +75 \text{ N/mm}^2$$

Radial stress

$$= -50 \text{ N/mm}^2$$

Maximum shear stress,

$$\tau_{max} = \frac{75 - (-50)}{2} = 62.5 \text{ N/mm}^2$$

$$\sigma_y = 350 \text{ N/mm}^2$$

$$\frac{\sigma_y}{2} = \text{Shear stress in simple tension}$$

$$= 175 \text{ N/mm}^2$$

$$\text{FOS} = \frac{175}{62.5} = 2.8$$

15. (a)
Theories of failure
When a hub is shrink fitted on a shaft, tensile hoop stress is developed on hub while radial stress is compressive crack is developed in axial direction due to tensile hoop stress. Crack cannot develop due to compressive stress.

Mechanical Properties

CHAPTER 16

For efficient design of any structural member or a machine component, the designer must be conversant with the mechanical properties of the materials he is going to use. Various important mechanical properties are strength, toughness, ductility, ultimate strength, degree of cold working, hardness, modulus of elasticity, behaviour of the material during loading and unloading etc.

Question 16.1 Draw stress-strain curves in tension for a ductile and for a brittle material and explain the following

(a) Yield strength
(b) Ultimate strength
(c) Toughness
(d) Percentage elongation
(e) Degree of cold working
(f) Type of fracture in ductile and brittle materials

Solution:

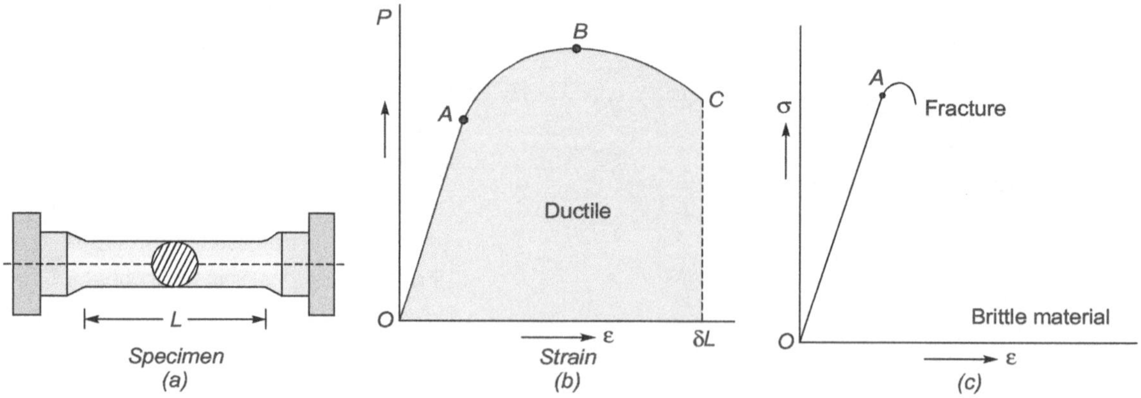

Fig. 16.1 *Stress-strain curve*

A specimen of standard dimensions is tested in a universal testing machine. Readings of load and extension are noted down. Load divided by original area of cross-section is termed as stress, σ and change in length/original length is termed as strain, ε. Reading of σ–ε are plotted. Fig. 16.1 (b) shows stress-strain curve for a ductile material, showing large strain upto the fracture. Figure 16.1(c) shows σ–ε curve for brittle material showing very small ε, strain upto fracture.

Ductile Material: In the σ-ε curve, *OA* is a straight line. Non-linear behaviour of the material starts after point *A*. This point is known as ***yield point*** and stress at this point is known as *Yield Strength*, a commonly used parameter while designing a machine component. Point *B* shows the maximum load point, P_{max}, the maximum load per unit area is called ***ultimate strength***. Area under the load-extension curve gives the property of toughness of material. This is the most desired property in a ductile material. It is a combination of strength and ductility.

$$\text{Total change in length upto fracture} = L_f - L_0 = \text{Final length} - \text{Initial length}$$

$$\%\ \text{elongation} = \frac{L_f - L_0}{L_0} \times 100, \text{ an indication of ductility of the material}$$

In most of the ductile materials, a neck is formed in sample at maximum load point. At the time of fracture, a cup-cone type fracture is obtained in ductile material as shown in figure 16.2.

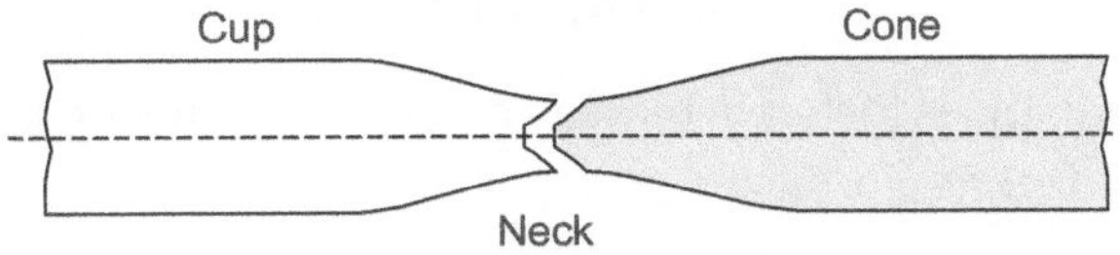

Cup-Cone Fracture
(a)

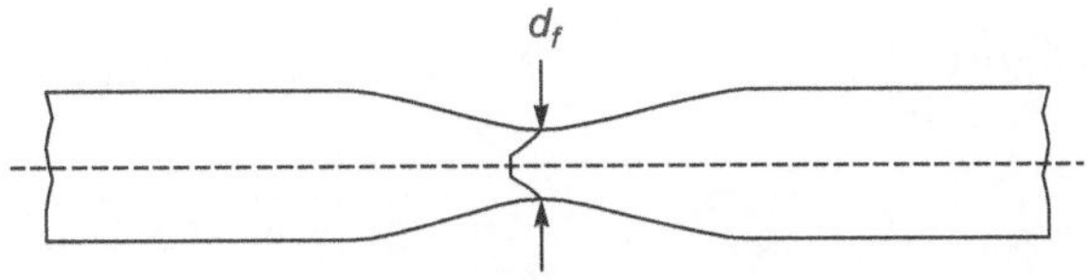

d_f = Final diameter at neck
(b)

Fig. 16.2

After fracture the two ends can be joined to find final diameter at fracture d_f as shown.

$$\text{Degree of cold working} = \frac{\text{Initial area of cross-section} - \text{Final area of cross-section}}{\text{Initial area of cross-section}} \times 100$$

$$= \frac{A_0 - A_f}{A_0} \times 100 = \frac{d_0^2 - d_f^2}{d_0^2} \times 100$$

Brittle Material: Fig. 16.1(c) shows σ–ε curve for a brittle material. A very small deformation/extension is exhibited by a brittle material. At the time of fracture the sample will break into two pieces with negligible change in diameter.

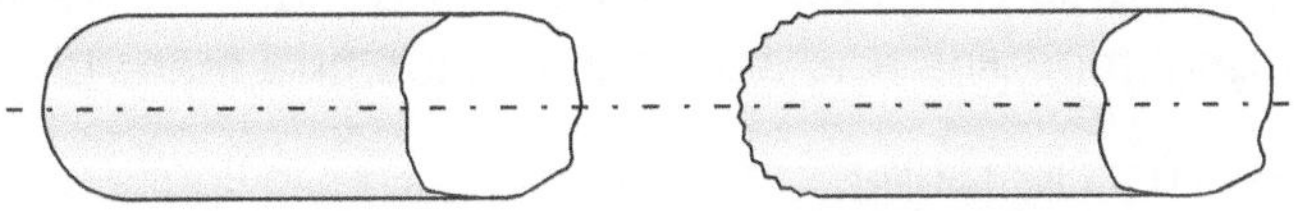

Fig. 16.3 *Brittle fracture*

Question 16.2 **Discuss the σ-ε curve for mild steel. What is 0.2% proof stress.**

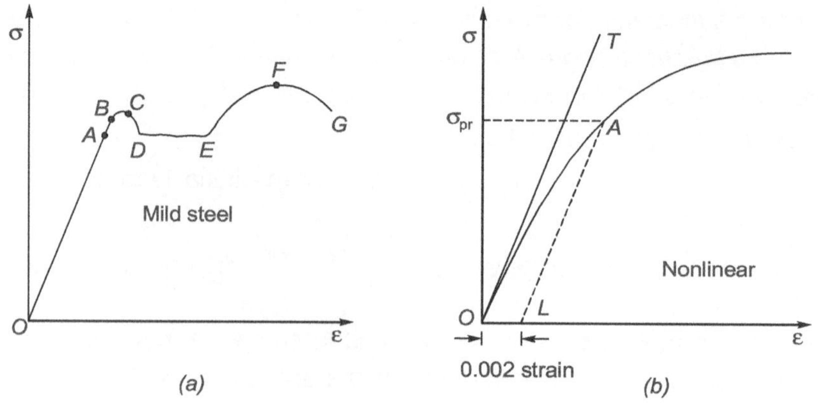

Fig. 16.4

Solution: A mild steel sample is tested in tension and graph between load extension or σ-ε is drawn. Fig. 16.4(a) shows σ-ε graph for mild steel.

OA – a straight line	*A* – limit of proportionality
B – Elastic limit	*C* – Upper yield point
D – Lower yield point	*DE* – Yield plateau
E – Point of strain hardening	*F* – Maximum load point

$$\sigma_{ult} = \frac{\text{Maximum load}}{\text{Area of cross-section}}$$

There are upper and lower yield points. This type of behaviour is known as discontinuous yielding and is a special feature of mild steel.

If load on sample is withdrawn within the limit *A*, unloading curve will be along *AO* (towards *O*) point *B* is very close to *A*, if load is removed from elastic limit point *B*, unloading curve will be along *BB′O* as shown in figure 16.5, making a small mechanical hysteresis loop.

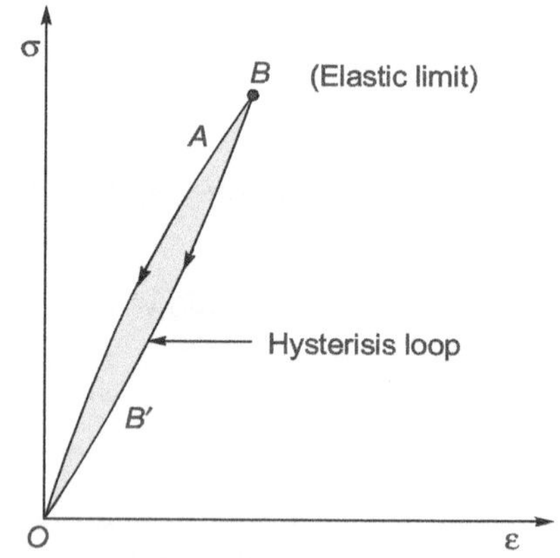

Fig. 16.5 *Mild Steel*

Mild steel has distinct yield points i.e., upper and lower yield points. Many materials have only one yield point.

But in some materials, the yield point is not distinct, 0.2% proof stress acts as yield strength of the material.

As in the case of copper, σ-ε graph is a curve from the very beginning. A tangent *OT* is drawn to the σ-ε curve at point *O*. Strain *OL* = 0.002(0.2%) is taken. From the point *L*, a line *LA* is drawn parallel to *OT*. Stress at *A*, σ_{pf} is known on 0.2% proof stress.

Question 16.3 **What is nonlinear behaviour of a material? Draw σ-ε curves for (a) rubber (b) concrete. What are tangent modulus and secant modulus of a material.**

Solution: Stress-strain curve for rubber is nonlinear as shown in figure 16.6(a). *OAB* is loading curve, but after unloading there is no residual strain in rubber sample. The unloading curve is *BCO*. Energy absorbed during loading = area under *OAB*.

Energy released during unloading = area under curve *BCO*

Shaded area *OABCO* = energy absorbed due to friction between particles of rubber

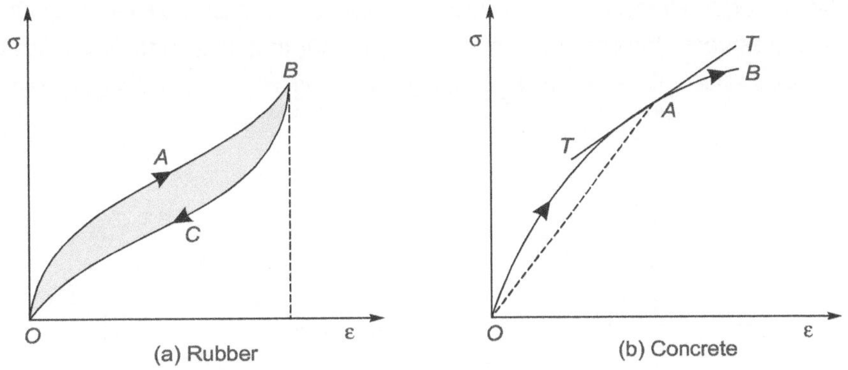

Fig. 16.6

Area in this curve (*OABCO*) decides about the damping capacity of the material. Rubber pads are used for vibration isolation under the machine foundations, to protect the machine from undersired ground vibrations. Area (*OABCO*) provides mechanical hysteresis loop, indicates damping capacity of a material. In σ-ε graphs, where there is initial straight line as *OA* figure 16.6(b). Slope of line *OA*, i.e., $\frac{\sigma}{\varepsilon}$ gives Young's modulus of the material in some cases, as for concrete **σ-ε** curve is nonlinear as shown in figure 16.6(b). At the point of interest say at *A*, slope of line *OA* = secant modulus

Slope of tangent *TT* at *A* = tangent modulus

Question 16.4 **Briefly describe the behaviour of ductile and brittle materials under compression.**

Solution: In crystalline materials, elastic action in compression is the same as in tension but in reverse direction.

In ductile material (as mild steel) during loading in compression, yielding starts in compression, at a stress approximately the same as the yield stress in tension.

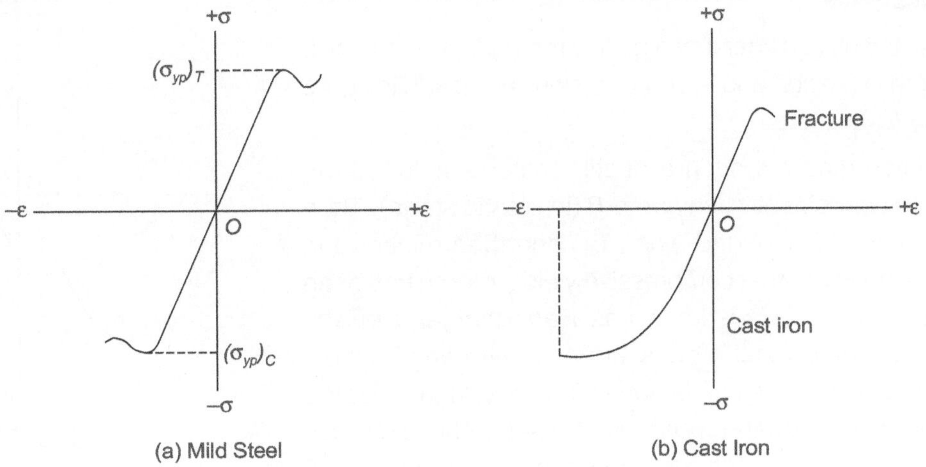

Fig. 16.7

However in brittle materials, fracture takes place along a shear plane (as in cast iron) or a multitudes of small fractures or shear planes in all directions leading to fragmentation (as in the case of chalk).

When a ductile material is tested under compression, it gets compressed axially but along circumference, cracks appear leading to severe deformation figure 16.8(a). But in brittle material as cast iron, the material is weak in shear, and fracture takes place along a shear plane approximately at an angle of 45° to the axis, figure 16.8 (b).

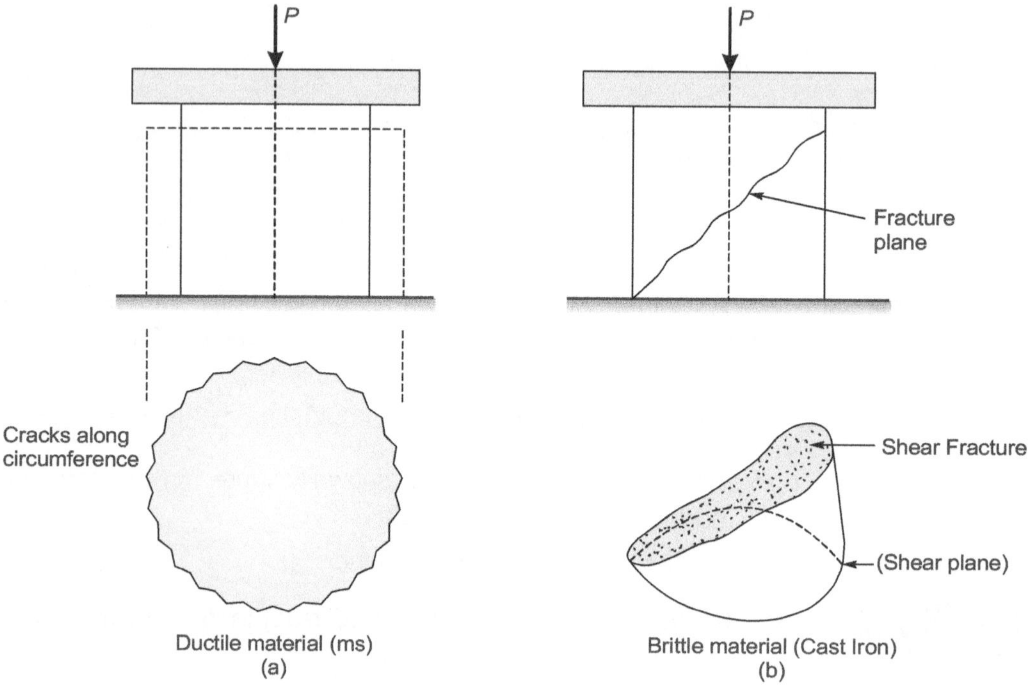

Fig. 16.8

Question 16.5 What is Bauschinger's effect, explain.

Solution: Let us consider that a ductile material (as mild steel) has equal strength in tension and in compression, as shown in figure 16.9 at A, $(\sigma_y)_T = (\sigma_y)_C$ at A'.

A specimen made of such a ductile material is loaded in tension, beyond yield point at A, say upto B (in plastic stage). Then from B, the specimen is unloaded along BC and then reloaded in compression. It is observed that compressive yield strength has been decreased $(\sigma_D)_C < (\sigma_A)_C$. This is known as ***Bauschinger's effect***. One of the reason of Bauschinger's effect is yielding in poly crystalline materials (as mild steel) is nonuniform. During unloading many crystals will have residual compressive stresses. These crystals will yield at a lower than normal stress leading to overall yield stress $\sigma_{yD} < \sigma_{yA}$.

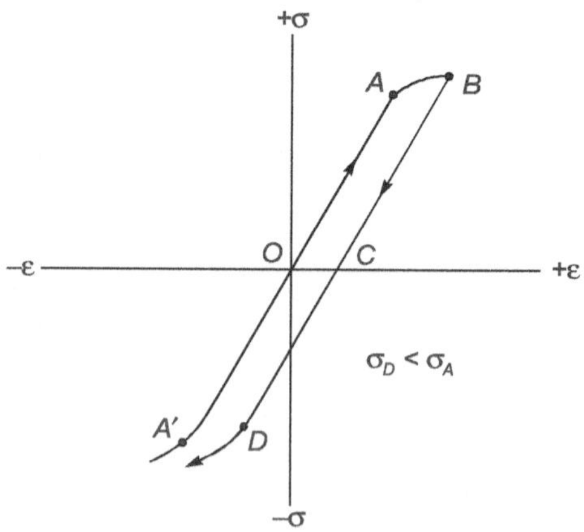

Fig. 16.9

Question 16.6 When a beam of rectangular section is subjected to bending moment, the section is deformed to trapezoidal section and a transverse curvature (opposite to longitudinal curvature) is produced, explain why?

Solution: Let us consider a beam of rectangular section *ABCD*.

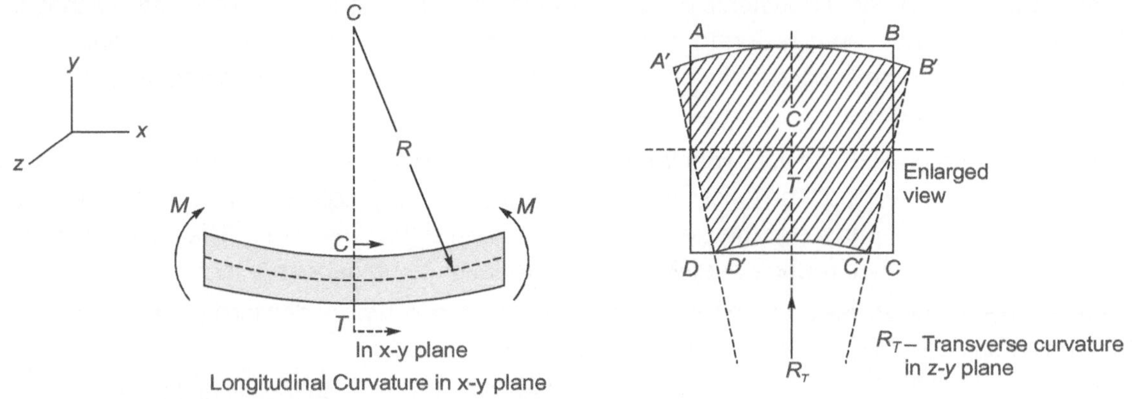

Fig. 16.10

Subjected to positive bending moment *M*. Upper half portion comes under compression zone (*C*) and lower half comes under tension (*T*). But the lateral strain in compression zone will be positive (side *AB* increased to (*A′B′*) and lateral strain in tension zone will be negative (side *CD* is reduced to *C′D′*). A transverse curvature of radius R_T is developed in transverse direction i.e., in plane *z-y*, then R_T is in opposite direction.

Question 16.7 Briefly describe how ductile and brittle materials fracture when subjected to twisting moment.

Solution: For torsion test on a material, a cylindrical sample with tapered ends is made on a lathe machine. The tapered ends are provided with key way so that the sample is properly fitted in the grips of the torsion testing machine, with the help of keys in key way cut on sample and provided in grips. Twisting moment is continuously increased till the specimen breaks into two pieces.

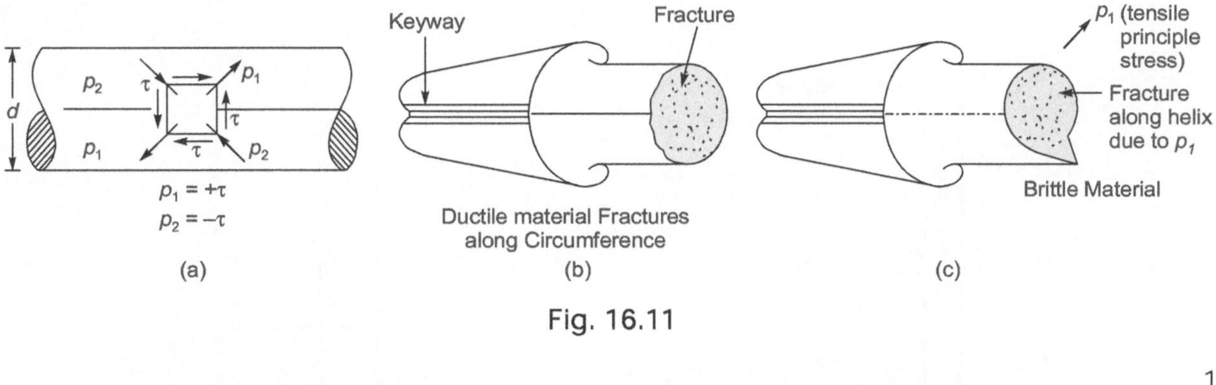

Fig. 16.11

If the stresses on the surface of a torsion sample are analysed, then τ is the shear stress, $\tau = \dfrac{16T}{\pi d^3}$.

Ductile material is weak in shear. Maximum shear stress τ acts along the circumference. Principal stresses are developed on the surface such that principal stress, $p_1 = +\tau$ (tensile) and $p_2 = -\tau$ (compressive) stresses. A brittle material is much weaker in tension than in shear. So the brittle material under torsion breaks along a helix (making an angle of approximately 45° to the axis) as shown in figure 16.11(c). A student can try on a chalk piece by applying a torsional moment by hand and see fractured helical surface.

Question 16.8 What is the importance of notched bar impact test? What is the effect of high straining rate and notch on the yield stress? What is notch sensitivity?

Solution: Notched bar impact tests are performed to study the toughness of a material that is the ability of a material to absorb strain energy during plastic deformation. In order to have high toughness, the material should possess high strength and large ductility. Brittle materials have low toughness because they exhibit very small deformation upto fracture.

Effect of Straining Rate: In notch bar impact test, a moving load is applied with a high velocity, specimen fractures in a fraction of a second. The tensile load is applied with a very high strain rate and the yield point of the material is increased in comparison to the yield point obtained in static tensile test. The effect of increased yield strength at high strain rate is to reduce ductility of the material or its brittleness is increased.

Notch Effect: The stress concentration at the root of the notch provides large stress necessary to raise the yield stress resulting in high local strain rate at the root of the notch.

If K_t = Theoretical stress concentration factor

σ_{av} = Average stress at the section containing the notch

Maximum stress, $\sigma_{max} = K_t \sigma_{av}$

$$\frac{\sigma_{max}}{E} = K_t \frac{\sigma_{av}}{E}$$

Differentiating both the sides with respect to time

$$\dot{\varepsilon}_{max} = K_t \dot{\varepsilon}_{av}\text{, strain rate}$$

Notch Sensitivity: The tendency of a ductile material to behave in a brittle manner in the presence of a notch is called ***notch sensitivity***. The high overall strain rate multiplied by stress concentration factor of the notch produces local strain rates as high as 10^3 mm/mm/s.

Question 16.9 What is a notched bar impact test? What are advantages of Charpy impact test over Izod impact test?

Solution:

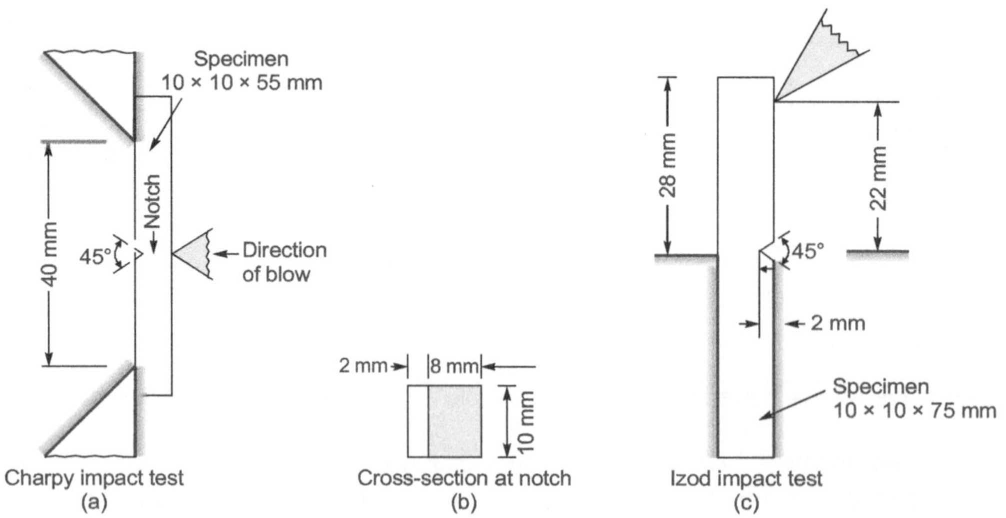

Fig. 16.12 *There are two standard tests (1) Charpy impact test (2) Izod impact test.*

This is a standard test on notch sensitivity combining the effect of (i) high strain rate (ii) notch effect (iii) temperature. High temperature upto 1100°C are obtained in ovens and low temperature as low as –230°C can be obtained in liquid nitrogen containers. Pendulum of the impact testing machine must be carefully constructed with the striking edge at the centre of percussion to minimize vibrations.

In the case of Charpy impact test standard specimens with a notch at the centre is supported like a beam loaded at the centre as shown in figure 16.12(a). The notch is on the tension side. While in the case of Izod impact rest, specimen held as a cantilever is loaded at the end. Figure 16.12(c).

The Charpy test has two advantages over Izod impact test, that is

(a) It is easier to place the specimen in the machine, in the quickest possible time, an important consideration in low temperature test.

(b) Sample is free from compression stress around the notch which are produced in Izod test specimen.

Question 16.10 **Differentiate between yield stress and 0.2%. Proof stress for a ductile material. [IES 2012, 2 Marks]**

Solution: In stress-strain curve, there is a point called yield point at which there is considerable deformation in specimen without noticeable increase in internal resistance fig. 16.13 (a) shows stress strain curve for mild steel in tension. *OA* a straight line, point *B* is yield point. Stress at *B*, σ_y is the yield point stress.

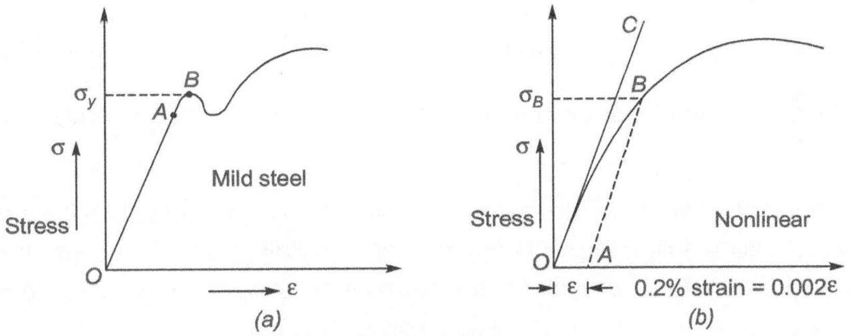

Fig. 16.13

However for some materials as copper, there is no distinct yield point in stress strain curve as shown in figure 16.13 (b). To have approximate value of yield point stress, 0.2% proof stress is obtained.

OA = 0.2% strain = 0.002 strain line *AB* is drawn parallel to straight line *OC*. Stress at point *B*,

$$\sigma_B = \sigma_y, \text{ yield point stress} = 0.2\% \text{ proof stress.}$$

Question 16.11 **Compare the stress-strain curves of different grades of steel.**

Solution: Figure 16.14 shows a comparison between strength and strain (ductility) for mild steel, high yield strength steel and high tensile strength steel. With the carbon percentage, strength of the steel goes on increasing but ductility or percentage elongation goes on decreasing.

Toughness of mild steel is much more than the toughness of high tensile strength steel. High yield strength steel (HYSS) is used for springs where high modulus of resilience is desired.

Note that Young's modulus of all steels is the same. Slope of line *OABC* is same for mild steel, HYSS and HTSS.

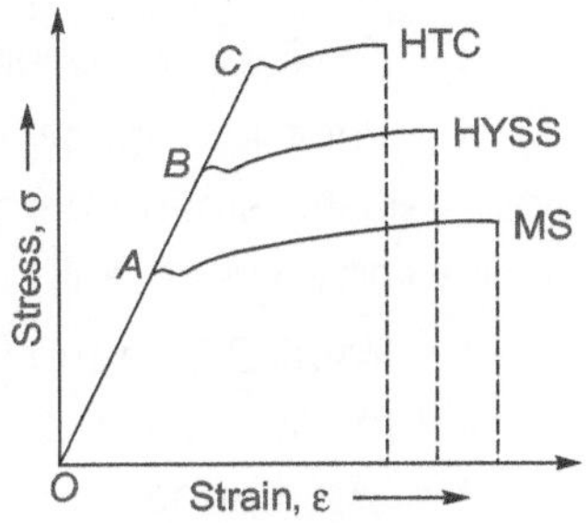

Fig. 16.14

For large columns, buckling load depends on Young's modulus, so mild steel can be used. For short columns, crushing strength depends on crushing strength, therefore for short columns, HTSS (high tensile strength steel) is used.

Question 16.12 **A bar of 2 m length made of structural steel having stress-strain diagram shown in figure 16.15 loaded axially until its elongation is 5.5 mm. If the load is removed, find final change in length. Yield stress of steel, σ_y = 270 N/mm², Young's modulus = 2.1 × 10⁵ N/mm².**

Solution:

Total change in length upto $B = 5.5$ mm

$\sigma_y = 270$ N/mm^2

$E = 2.1 \times 10^5$ N/mm^2

$L = 2000$ mm, length

Elastic recovery, $B'C = \frac{270}{210000} \times 2000$

$= 2.57$ mm

Residual plastic change in length

$= 5.5 - 2.57 = 2.93$ mm

= as shown by OC

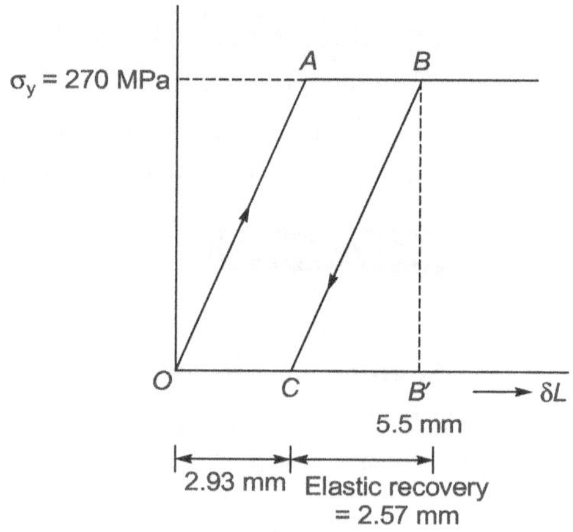

Fig. 16.15

Question 16.13 **Differentiate between ductility and brittleness. Give examples of ductile and brittle materials.**

Solution: Ductility: Ductility is the property of a material, by virtue of which it is drawn into wire, through tensile force. Moreover ductility is due to plasticity. In ductile metals, post-elastic strain (plastic strain) is more than 5%. Examples of ductile material are mild steel, aluminium, copper, gold, silver, lead, brass.

Brittleness: Lack of ductility is called brittleness. Brittle materials are weak in tension and at fracture, these materials do not show significant change in area of cross-section. In brittle materials, post elastic strain in less than 5%. Examples of brittle materials are cast iron, concrete, high carbon steel, cork, glass, wood, rubber etc.

Question 16.14 **Compare between elastic strain energy and hyper-elastic strain energy.**

Solution: Figure 16.16 shows stress-strain curve for a ductile material,

OA = elastic stage

ABC = plastic stage

A = limit of proportionality

Point D is below the elastic limit. If material is unloaded at this point, elastic strain energy recovered = area under the triangle ODE. (unloading curve is DO)

Point B is in the plastic stage. If load on sample is removed at point B, the unloading curve will be along BF. Line BF is parallel to straight line OA. Area under the triangle FBG, gives hyper elastic strain energy recovered during unloading in the plastic stage.

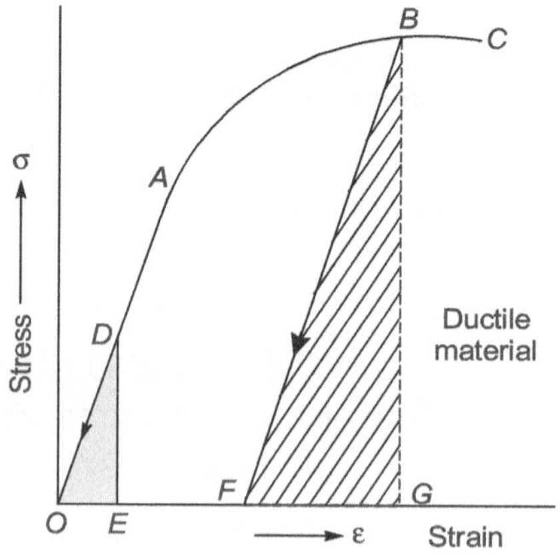

Fig. 1616

Question 16.15 **Why mild steel is the most commonly used reinforcing material in RCC structure?**

Solution: There are various features of mild steel making it most useful as reinforcing material in concrete such as:

(a) High ductility (upto 20% elongation) producing high toughness

(b) Due to high ductility of mild steel, RCC structure gives initial warning against fracture.

(c) Mild steel is equally strong is tension and compression

(d) Young's modulus of elasticity of mild steel is tension is equal to be Young's modulus in compression

(e) Coefficient of thermal expansion of mild steel is very nearly equal to coefficient of thermal expansion of concrete. Therefore due to change in temperature of RCC, thermal stress will not be produced.

(f) Steel can be easily welded and it is recyclable.

(g) It is cheaper and commonly available structural material.

Question 16.16 **Compare the σ-ε (stress strain curves) for mild steel, cast iron, aluminium, wood, cork, rubber.**

Solution: Figure 16.17 shows stress strain curves for rubber, cork, wood, aluminium, cast iron and mild steel.

Young's modulus of mild steel = 200 GPa.

Mild σ_y = yield strength = 270 MPa.

For cast iron, Young's modulus = 100 – 105 GPa

σ_{tu} = ultimate tensile strength

= 150 – 170 MPa.

Then aluminium its Young's modulus is 67-70 GPa, with yield strength about 180 MPa.

For materials wood, cork and rubber stress strain curves are nonlinear as shown in the figure. Rubber is nonlinear elastic material. Rubber is a brittle material because its post elastic strain is negligible.

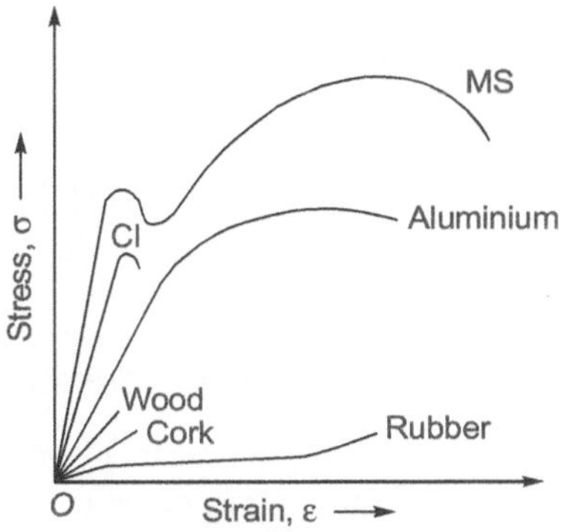

Fig. 16.17

Question 16.17 **Define hardness of a material. What are different tests performed on sample to determine its hardness. What are the advantages of Rockwell hardness test over Brinell Hardness Test? Make a table of BHN of hard materials.**

Solution: Resistance of a material to (a) plastic deformation (b) wear and tear (c) scratch (d) abrasion (e) indentation etc. is defined as its hardness. Life of an engineering component very much depends upon its hardness, generally it is measured by localized plastic deformation produced by a known load.

Brinell has used a hardened steel ball to produce an impression on the surface of a specimen by applying diametral compressive force. Vicker has used a diamond in the shape of square pyramid to produce an impression of square pyramid shape (with four triangular faces) under a known applied load. Load in kgf (kilogram force) divided by area of the impression in mm^2 gives BHN or VPN, depending upon the indentor.

Rockwell has devised hardness measurement procedure by considering only the depth of resulting indentation and related the hardness to depth of indentation. Rockwell test takes minimum time in hardness measurement.

Brinell Hardness Test

A Brinell ball (in an attachment) is pressed on surface of a sample under a known load *P* in kgf.

D = diameter of the ball in mm

d = diameter of impression in mm

$$\text{BHN} = \frac{P \text{ in kgf}}{\pi D\left[D - \sqrt{D^2 - d^2}\right] \text{ in mm}^2}$$

In the expression, area of curved surface of indentation is taken (fig. 16.18)

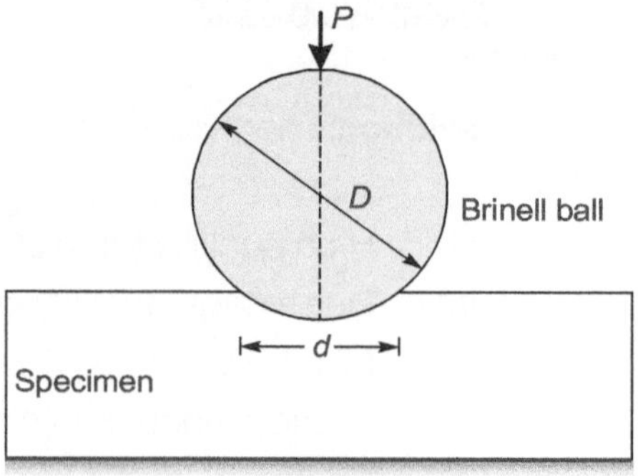

Fig. 16.18

Vickers Pyramid Number

The angle between the opposite faces of pyramid is 136°.

Surface of contact area between indentor and impression.

$$A = \frac{d^2}{2\sin\frac{\alpha}{2}}, \quad \text{where } \alpha = 136°$$

$$\text{VPN} = \frac{P}{d^2} \times 2\sin\frac{\alpha}{2} = 1.8544\frac{P}{d^2}$$

Load in kgf and diagonal in mm. With the help of a travelling microscope diagonal of the square pyramid impression is noted down. (fig. 16.19)

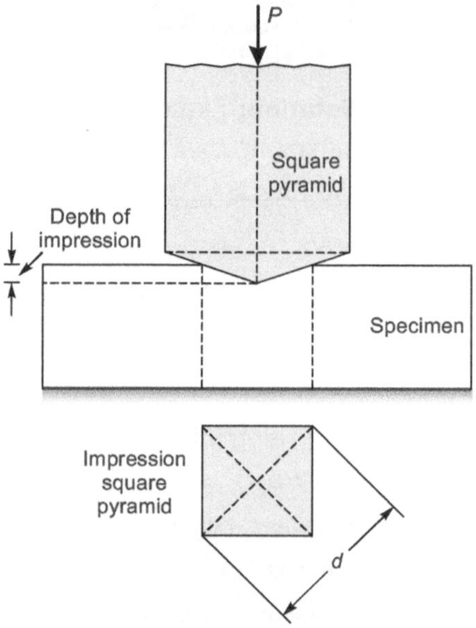

Fig. 16.19

Rockwell Hardness

Hardness of wide range of materials including rubber, plastic, threads etc. is determined by this method. For metals, 3 tests are generally performed.

Rockwell *A* – for case hardness material.

Rockwell *B* – for soft and medium hard metals, as mild steel, brass

Rockwell *C* – for hard metals such as high speed steel, tool steel

For Rockwell *A* and *C*, diamond indentor (conical shape) and Rockwell *B*, a hardened steel ball (1.59 mm in diameter) are used. A minor load of 10 kg is applied on the specimen through the indentor to overcome the thin oxide film on the metal. The additional loads of 50, 90, 140 kg are applied for Rockwell *A, B, C* respectively on the indentor.

$$\text{Rockwell hardness number} = H - \frac{t}{0.002}$$

where, H = a constant depending upon the scale

= 130 for Rockwell *B*

= 100 for Rockwell *A* and *C*

t = depth of indent in mm, 0.002 mm of depth corresponds to one unit hardness number.

BHN of Various Hard Materials

Material	BHN	Material	BHN
Diamond	8000	Aluminium Oxide	2100
Cubic Boron nitride	5000	Tungsten Carbide	2000
Silicon carbide	2800	Hardened Tool Steel	740

Question 16.18 What do you understand by failure due to fatigue, explain. What fatigue tests are commonly performed? What is the type of the stress cycle which causes worst type of fatigue loading? What surface treatment improves fatigue life?

Solution: Failure of a structural member or a component due to cyclic stresses or fluctuating load is known as fatigue failure. Shafts, springs, bearings fail due to cyclic stresses at a stress level much lower than the yield strength of the material under static load.

Fracture due to fatigue is sudden, catastrophic, and even a ductile material under fatigue loading fails in a brittle manner.

There are 3 stages of fatigue failure

(a) Nuclearation of a fine crack on atomic scale

(b) Crack propagation and growth under continued stress cycles

(c) Sudden fracture

Fatigue life = $N_1 + N_2$ = Number of cycles for crack initiation + Number of cycles during crack propagation

Test: There are two types of fatigue loading

(a) Rotating cantilever under bending

(b) Rotating beam under constant bending moment

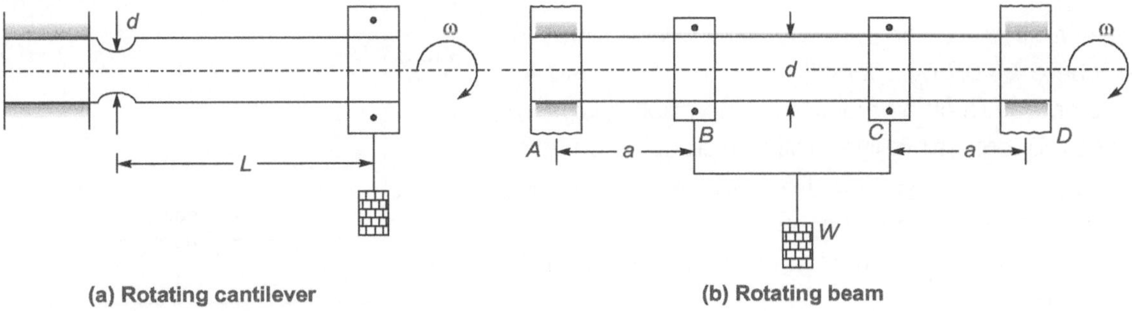

Fig. 16.20

(a) Rotating cantilever, $M_{max} = WL$

$$\sigma_{max} = \frac{32WL}{\pi d^3}$$

(b) Rotating beam, $M_{constant} = Wa$, over length BC

$$\sigma_{max} = \frac{32Wa}{\pi d^3}$$

At a particular load W, specimens are rotated and number of cycles upto fracture are noted down at a particular stress S_1, say N_1. Similarly at different loads W_2, W_3 ... or stresses S_2, S_3 ... etc. the number of cycles upto fracture are noted down. A graph between S and N, known as S-N curve is plotted as shown in figure 16.21. **The stress level S_e, at which S-N curve becomes asymptotic is known as Endurance Strength S_e,**

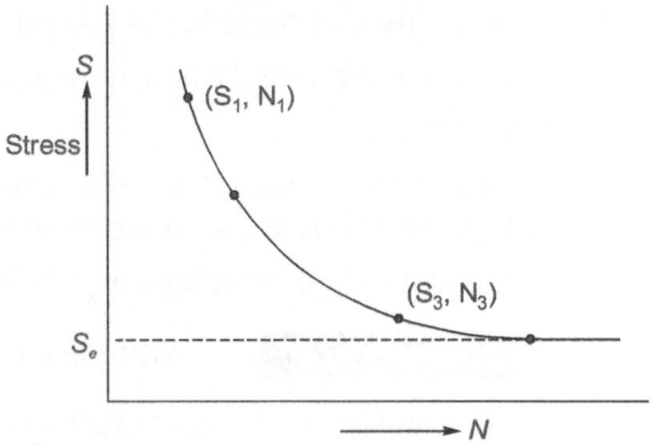

Fig. 16.21 *Number of cycles*

Figure 16.22 shows various types of stress cycles. Random stress cycles are observed in aircraft wings during landing process.

The stress cycle in which mean stress $\sigma_m = 0 = \frac{\sigma_{max} + \sigma_{min}}{2}$ is the worst type of cyclic loading for fatigue failure (stress is completely reversed)

- If residual compressive stresses are introduced on the surface of the component, fatigue life is improved. This can be done by shot peening, which causes localized plastic deformation in outer layer of surface.
- Polishing of surface also enhances the fatigue performance. Scratches produced on components during machining processes are removed by polishing.
- Case hardening of component also enhances its fatigue life.

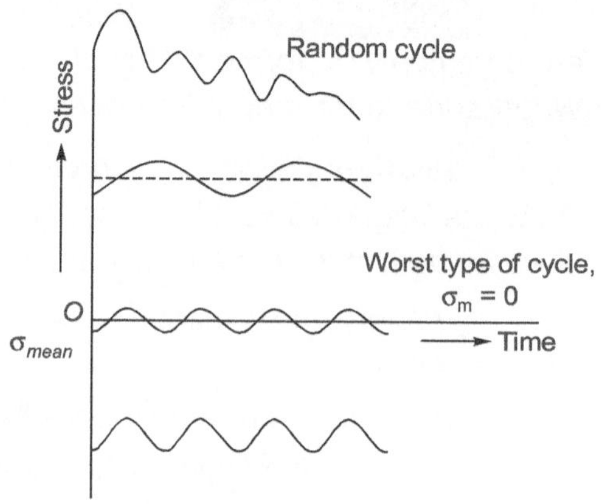

Fig. 16.22 *Stress cycles*

Question 16.19 What is creep phenomenon observed in some metals? What are the 3 stages of creep? How stress and temperature effect the creep behaviour.

Solution: ASTM (American Society to Testing of Materials) defines creep as "Time dependent strain under constant stress".

Generally the creep phenomenon is predominant in a material, where $T > 0.4T_m$, operating temperature (T) is $0.4T_m$ (melting temperature of the material). Amorphous polymers, plastics and rubbers are highly sensitive to creep deformation.

A typical creep test consists of subjecting a specimen to a constant load or constant stress while maintaining temperature constant. Deformation or strain versus time is plotted as creep curve.

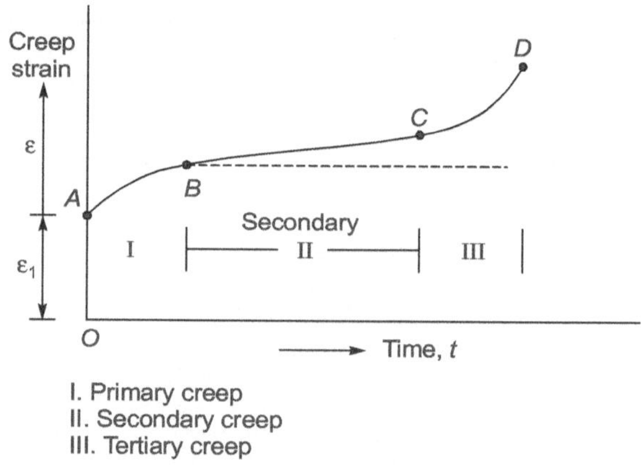

Fig. 16.23 *Creep Curve*

From creep curve, following can be observed.

1. An instantaneous strain e_i, shown by *OA*, at the time of loading.
2. A continuously decreasing creep rate, primary creep, *AB*-from maximum strain rate to minimum constant strain rate.
3. Steady state creep II (*BC*) secondary creep, linear plot longest duration.
4. Tertiary stage III (*CD*) strain rate accelerates and final fracture takes place.

Slope of secondary creep $\dot{\varepsilon}$ = creep rate, a useful parameter in engineering design of components like turbine rotor blades.

Both temperature and stress levels influence the creep behaviour. By increasing temperature or by increasing stress or by increasing both temperature and stress, the creep rate increases.

Stainless steel, refractory metals, superalloys are especially resistant to creep.

Question 16.20 What is stress relaxation?

Solution: Bolts and other fasteners, required to hold two rigid plates in tight contact are frequently found to have relaxed after long periods of time as a result of creep.

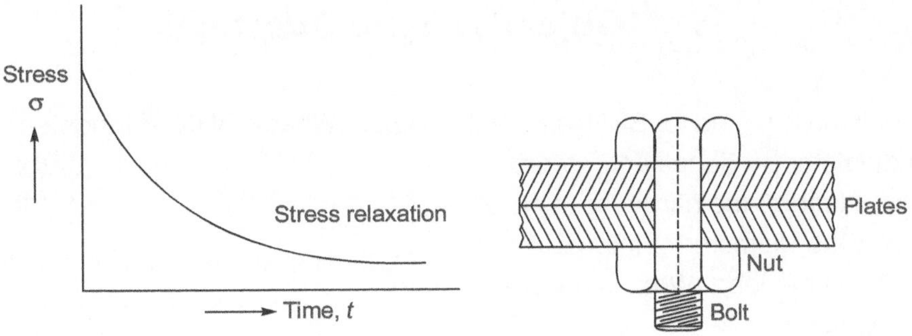

Fig. 16.24

This is time dependent decrease in stress in a member which is constrained to have constant (fixed) deformation.

Considering two plates joined by a bolt and a nut.

$$\varepsilon_{el} = \text{Initial strain in bolt}$$

A part of ε_i strain is converted into creep strain (plastic strain) and elastic part of strain goes on decreasing.

$$\varepsilon_{el} = e_i - e_{cr} \text{ (creep strain)}$$

$$\text{Stress} = \varepsilon_{el} \times E \text{ goes on decreasing as shown in figure 16.24}$$

Question 16.21 A machine component made of a ductile material is subjected to a variable loading with σ_{min} = –50 MPa and σ_{max} = 50 MPa. If the corrected endurance limit and the yield strength for the material are σ'_e = 100 MPa and σ_y = 300 MPa, respectively, the factor of safety is _____.

[ME, GATE : 2017 (set-1)]

Solution:

$$\sigma_{mean} = 0$$

$$\sigma_{max} = +50 \text{ MPa}$$

$$\sigma_{min} = -50 \text{ MPa}$$

$$\sigma_A = \text{Stress amplitude} = \frac{50-(-50)}{2} = 50 \text{ MPa}$$

$$\sigma'_e = 100 \text{ MPa}$$

$$\sigma_y = 300 \text{ MPa}$$

$$= \frac{\sigma_a}{\frac{\sigma_y}{FOS} - \sigma_{mean}} = \frac{\frac{1}{2}\sigma'_e}{\sigma_y - \frac{1}{2}\sigma'_e}$$

$$\sigma_a = 50,\ \sigma_{min} = 0$$

$$\sigma_y = 300$$

$$FOS \times \frac{\sigma_a}{\sigma_y} = \frac{50}{300-50} = \frac{1}{5}$$

$$FOS = \frac{1}{5} \times \frac{\sigma_y}{\sigma_a} = \frac{1}{5} \times \frac{300}{50} = 1.2$$

Objective Type Questions

Q.1 A metallic specimen is tested for BHN, a Brinell ball of 5 mm diameter produces an impression of 3 mm under a load *P*, the hardness is *H*. On another sample, the diameter of impression is 4 mm, under the same load. What is hardness of this second sample.

(a) 1.33 H (b) 0.35 H
(c) 0.625 H (d) 0.5 H

Q.2 The depth of penetration of the hardened steel ball in a specimen is 0.120 mm. Rockwell hardness of a material is

(a) 60 (b) 65
(c) 70 (d) 80

Q.3 The angle between the opposite faces of diamond pyramid in VPN test is

(a) 118° (b) 120°
(c) 136° (d) 142°

Q.4 Which of the following materials is remarkably brittle

(a) Polymers (b) Ceramics
(c) White cast iron (d) Bronze

Q.5 Most important reason for Bauschinger's effect is

(a) Ductile materials are weak in shear
(b) Residual compressive stresses
(c) Tensile residual stresses
(d) All of above

Q.6 Notch angle in Izod impact test specimen is

(a) 25° (b) 30°
(c) 39° (d) 45°

Q.7 The process which does not improve the fatigue strength of a material is

(a) Shot peening of the surface
(b) Cold rolling of the surface
(c) Electroplating the surface
(d) None of these

Q.8 What is BHN of Tungsten Carbide?

(a) 8000 (b) 5000
(c) 2800 (d) 2000

Q.9 The stress-strain curve for an ideally plastic material is

(a) 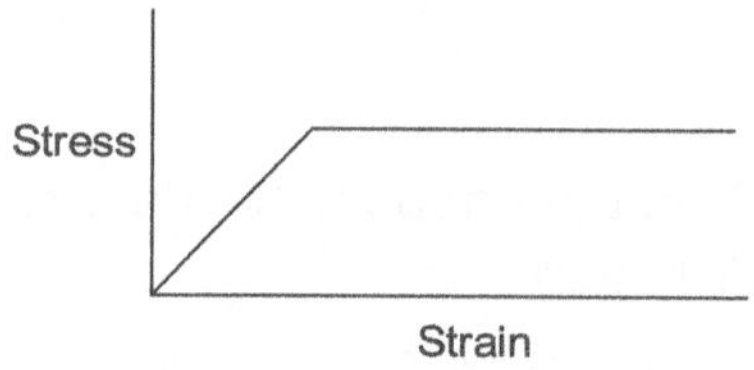

(b) 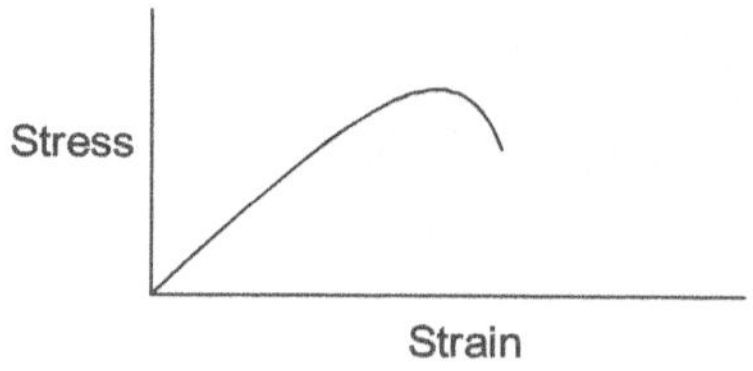

(c) 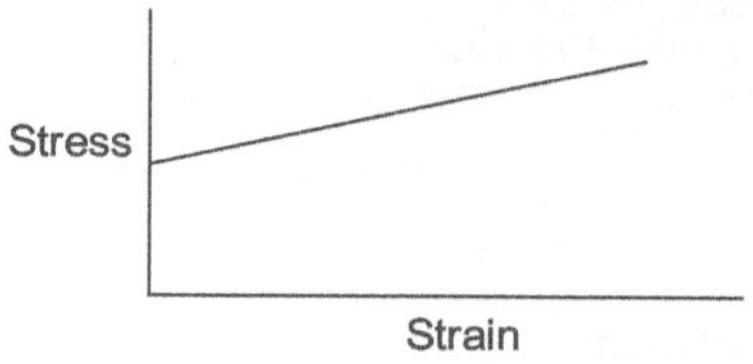

(d) 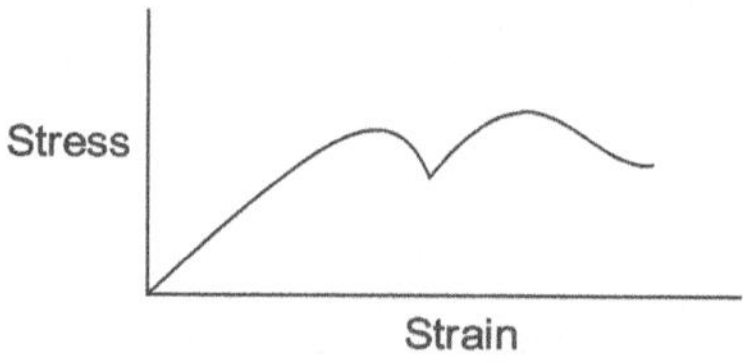

[CE, ESE (Pre) : 2018]

Q.10 A steel specimen is heated to 780°C and is then cooled at the slowest possible rate in the furnace. The property imparted to the specimen by this process is

(a) toughness (b) hardness
(c) softness (d) tempering

[ME, ESE (Pre) : 2018]

Q.11 The unique property of cast iron is its high

(a) malleability
(b) ductility
(c) toughness
(d) damping characteristics

[ME, ESE (Pre) : 2018]

Q.12 Addition of magnesium to cast iron increases its
(a) hardness
(b) corrosion resistance
(c) creep strength
(d) ductility and strength in tension
[ME, ESE (Pre) : 2018]

Q.13 In the engineering stress-strain curve for mild steel, the Ultimate Tensile Strength (UTS) refers to
(a) Yield stress (b) Proportional limit
(c) Maximum stress (d) Fracture stress
[ME, GATE : 2018 (set-1)]

Q.14 For the given fluctuating fatigue load, the values of stress amplitude and stress ratio are respectively

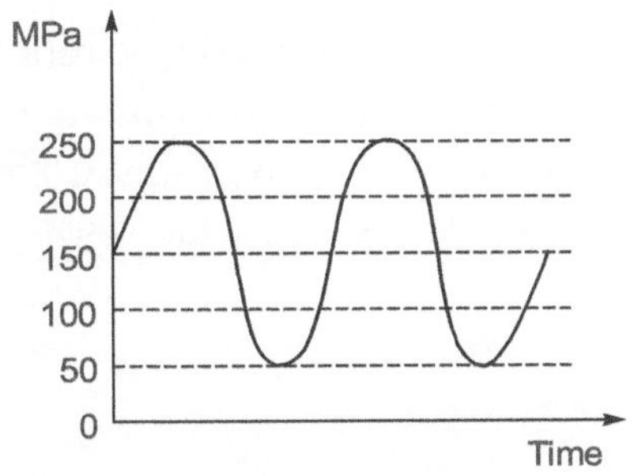

(a) 100 MPa and 5 (b) 250 MPa and 5
(c) 100 MPa and 0.20 (d) 250 MPa and 0.20
[ME, GATE : 2018 (set-3)]

Q.15 In mild steel specimen subjected to tensile test cycle, the elastic limit in tension is raised and the elastic limit in compression is lowered. This is called
(a) Annealing effect (b) Bauschinger effect
(c) Strain rate effect (d) Fatigue effect
[CE, ESE (Pre): 2017]

Q.16 Consider the following statements:
For a component made of ductile material, the failure criterion will be
1. endurance limit, if the external force is fluctuating
2. fatigue, if the external force is fluctuating
3. yield stress, if the external force is static
Which of the above statements are correct?
(a) 1 and 2 only (b) 1 and 3 only
(c) 2 and 3 only (d) 1, 2 and 3
[ESE (Pre) : 2017]

Q.17 Which one of the following types of stress-strain relationship best describes the behavior of brittle materials, such as ceramics and thermosetting plastics, (σ = stress and e = strain)?

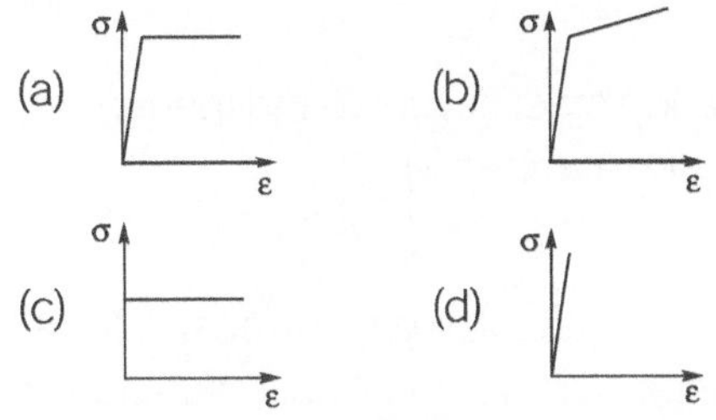

[ME, GATE : 2015 (set-1)]

Answers

1. (d)	2. (c)	3. (c)	4. (b)	5. (b)
6. (d)	7. (c)	8. (d)	9. (a)	10. (d)
11. (d)	12. (d)	13. (c)	14. (a)	15. (b)
16. (c)	17. (d)			

Explanations

1. (d)

$$H = \frac{P}{\pi D\left(D-\sqrt{D^2-d^2}\right)}$$

$$= \frac{P}{\pi\times 5\left(5-\sqrt{25-09}\right)} = \frac{P}{5\pi}$$

$$H' = \frac{P}{\pi\times 5\left[5-\sqrt{25-16}\right]} = \frac{P}{10\pi}$$

$$H' = 0.5\,H$$

2. (c)

$$R_B = 130-\frac{t}{1002} = 130-\frac{0.120}{1002}$$

$$= 130 - 60 = 70$$

3. (c)
Angle = 136°

4. (b)
Ceramics are remarkably brittle.

5. (b)
Residual compressive stresses are responsible for Bauschinger's effect.

6. (d)
Notch angle is Izod impact test specimen is 45°.

7. (c)
Electroplating does not improve fatigue life.

8. (d)
BHN of Tungsten Carbide 2000.

9. (a)
For ideally plastic material, stress-strain curve.

10. (d)
Heating steel at 780°C and then slow cooling in furnace–tempering.

11. (d)
The unique property of cast iron is its high damping characteristics. (There is true for grey cast iron, because of presence of free graphites flakes in its structure).

12. (d)
Addition of magnesium in cast iron produces CG iron (spheroidal graphite iron) known for ductility and tensile strength.

13. (c)
Maximum stress is greater than yield stress and proportional limits.
Maximum stress

$$= \frac{\text{ultimate tensile load}}{\text{area of cross-section}} = \text{UTS}$$

14. (a)

$$\rho_{max} = 250 \text{ MPa}$$
$$\rho_{min} = 50 \text{ MPa}$$

$$\rho_{a,\ amplitude} = \frac{250-50}{2} = 100 \text{ MPa}$$

$$\text{Stress ratio} = \frac{\sigma_{max}}{\sigma_{min}} = 5$$

15. (b)

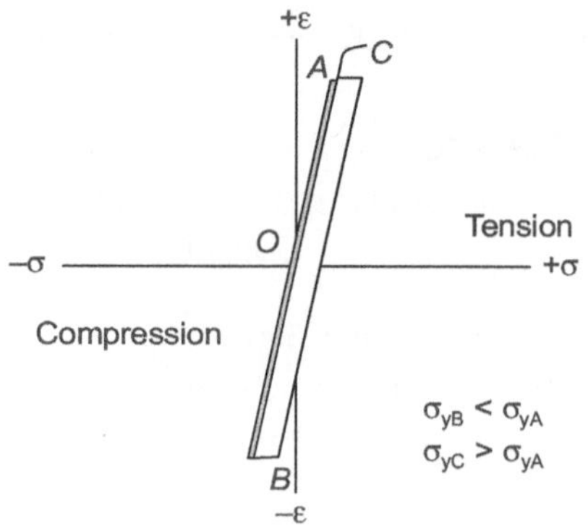

Elastic limit in compression is reduced due to dislocations movement at less stress in compressione. $\sigma_{yB} < \sigma_{yA}$, elastic limit is tension in second cycle is raised due to strain hardening, i.e.,

$$\sigma_{yC} > \sigma_{yA}$$

16. (c)
Endurance limit, is the fluctuating stress at which the component does not failure for any number of cycles as shown.

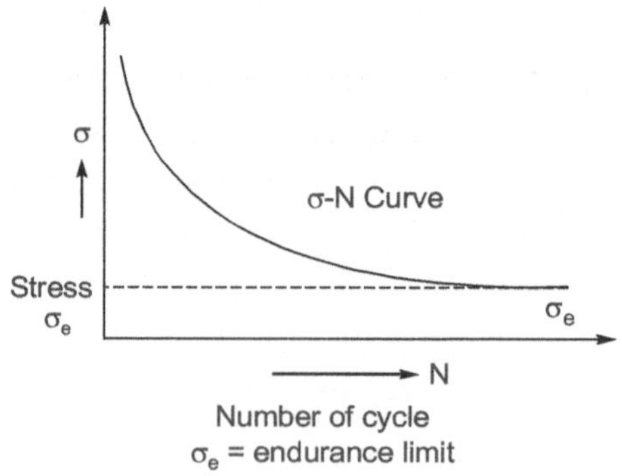

(1) and (3) are corect.

17. (d)
Graph (d) describes the behaviour of brittle materials such as ceramics and thermosetting plastic. $\sigma \propto \varepsilon$.
(No plastic stage)

General Questions

CHAPTER 17

Question 17.1 A thin cylinder is turning about its axis. Find the safe number of revolutions for a rotor of 3 meters in diameter if the hoop stress is not to exceed 13200 kg/cm². Take density as 6500 kg/cm³.

[ME, ESE 2015 : 10 Marks]

Solution:

$$\sigma_{c\,max} = \frac{\rho\omega^2}{g}\left(R^2\right)(3k_1 - k_2), \qquad \rho \text{ is density}$$

where,

$$k_1 = \frac{3-2\nu}{8(1-\nu)}, \qquad \text{where } \nu \text{ is Poisson's ratio}$$

$$k_2 = \frac{1+2\nu}{8(1-\nu)}$$

$$3k_1 = \frac{9-6\nu}{8(1-\nu)}$$

$$3k_1 - k_2 = \frac{9-6\nu-1-2\nu}{8(1-\nu)} = \frac{8-8\nu}{8(1-\nu)} = 1$$

$$\sigma_{c\,max} = \frac{\rho\omega^2}{g} \times R^2 \qquad \text{where R is radius of thin cylinder}$$

$$\rho = \frac{6500\,\text{kg}}{\text{m}^3}$$

$$R = 1.5\text{ m}$$

$$g = 9.8\text{ m/s}^2$$

$$\sigma_{c\,max} = 1300 \times 10^4\text{ kg/m}^2$$

$$1300 \times 10^4 = 6500 \times \frac{\omega^2}{9.8} \times 1.5^2$$

$$\omega^2 = \frac{1300 \times 10^4 \times 9.8}{6500 \times 2.25} = 8711.11$$

$$\omega = 93.3\text{ radian/second}$$

$$N_{safe} = \frac{93.3 \times 60}{2\pi} = 891.27\text{ RPM}$$

Question 17.2 Two steel plate in axial tension are to be connected by bolts as shown in Fig. 17.1. Determine its tensile strength. Bolt dia. = 25 mm, Hole dia. = 28 mm, Plate thickness = 16 mm each, f_u = 440 MPa, f_y = 290 MPa.

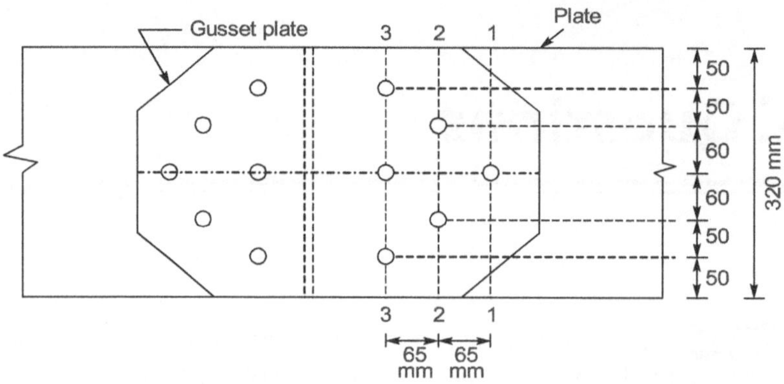

Fig. 17.1

[CE, ESE 2016 : 20 Marks]

Solution: Strength of undrilled plate

$$P = 320 \times 16 \times 440 \text{ N} = 2252.8 \text{ kN}$$

Figure shows one gusset plate

Strength of bolt in shear, $$F_b = \frac{\pi}{4} \times d^2 \times f_y = \frac{\pi}{4} \times 25^2 \times 290 = 142.35 \text{ kN}$$

Strength of the plate along row 1-1 $= (320 - 28) \times 16 \times 440 = 2055.68$ kN ...(i)

Strength of the plate along row 2-2 $= (320 - 28 \times 2) \times 16 \times 440$ + strength of one bolt in shear

$= 1858.56 + 142.35 = 2000.91$ kN

Strength of the plate along row 3-3 $= (320 - 28 \times 3) \times 16 \times 440$ + strength of 3 bolts in shear

$= 1661.44 + 142.35 \times 3 = 2088.49$ kN

Minimum tensile strength of the plate = 2000.91 kN along row 2-2

Question 17.3 A bracket plate connected to a column flange transmits a load of 100 kN as shown in the following figure 17.2. The maximum force for which the bolts should be designed is ______ kN.

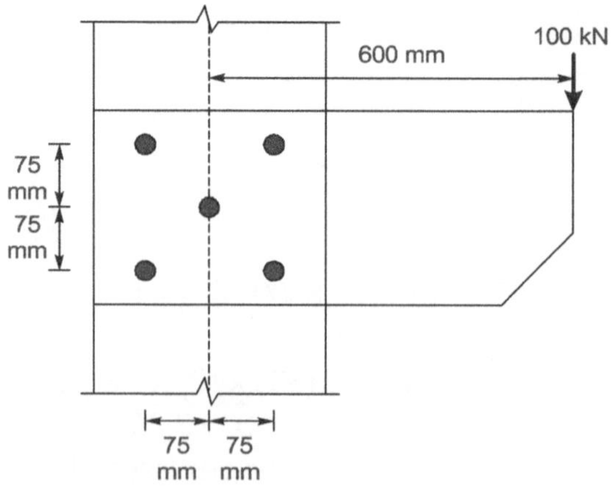

Fig. 17.2

[CE, GATE : 2015 (set-1)]

Solution:

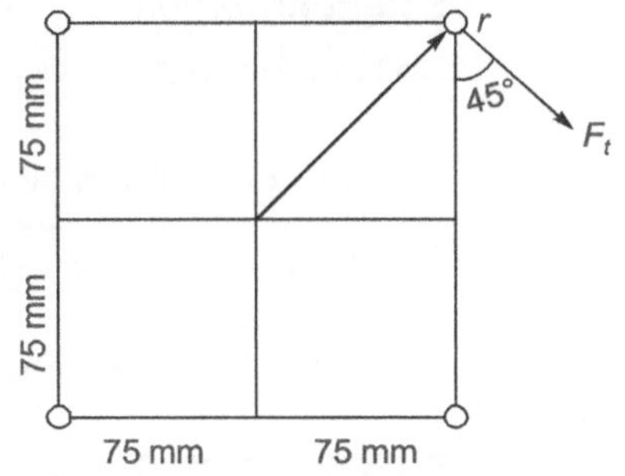

Fig. 17.3

F_{d}, Direct shear force on each bolt $= \dfrac{100}{5} = 20$ kN

Torsional load on central bolt is zero.

Torsional load on each bolt, $F_t = \dfrac{(Pd)r}{\Sigma r^2}$

$$r = 75\sqrt{2}$$

$$F_t = \frac{100 \times 60 \times 75\sqrt{2}}{4\left[75\sqrt{2}\right]^2} = 141.44 \text{ kN}$$

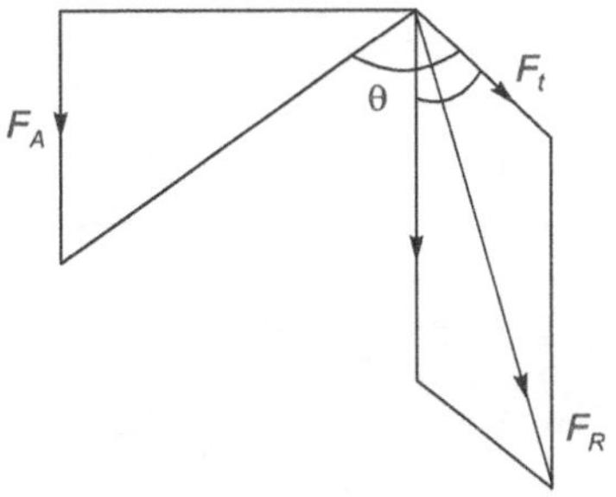

Fig. 17.4

Resultant force, $F_R = \sqrt{F_d^2 + F_t^2 + 2F_dF_t \times \cos\theta}$

$$= \sqrt{20^2 + (141.44)^2 + (2 \times 20 \times 141.44 \times 0.707)}$$

$$= \sqrt{400 + 20000 + 4000}$$

$$= 156.2 \text{ kN}$$

Question 17.4 A bolted joint has four bolts arranged as shown in figure 17.5. The cross-sectional area of each bolt is 25 mm². A torque T = 200 Nm is acting on the joint. Neglecting friction due to clamping force, maximum shear stress in a bolt is ________ MPa.

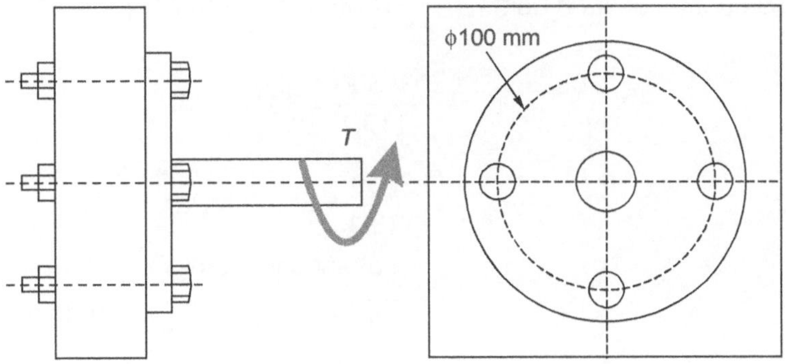

Fig. 17.5

[ME, GATE : 2016 (set-3)]

Solution: Torque, $T = 200$ Nm

Bolt circle radius, $R = 50$ mm

Shear force on each bolt $= \dfrac{200 \times 1000}{50 \times 4} = 1000$ N

Cross-sectional area of each bolt $= 25$ mm^2

Maximum shear stress in bolt $= \dfrac{1000}{25} = 40$ MPa

Question 17.5 A cantilever bracket is bolted to a column using three M12 × 1.75 bolts, *P*, *Q* and *R*. The value of maximum shear stress developed in the bolt *P* (in MPa) is ________.

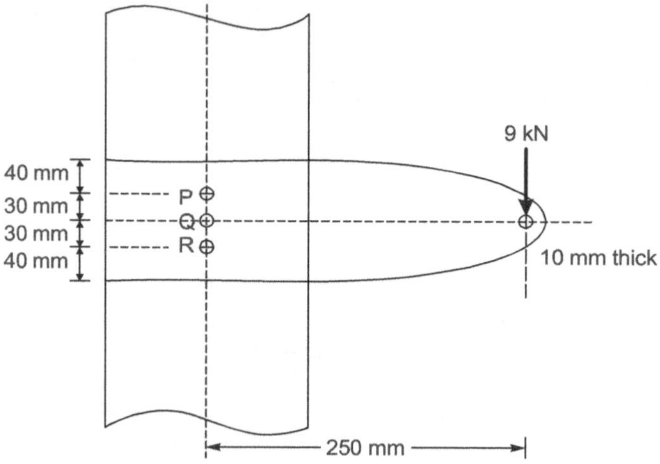

Fig. 17.6

[ME, ESE 2015 : 10 Marks]

Solution:

Shear force in bolt, $P = \dfrac{9000 \times 250}{30 \times 2} = 37500$ N

Area of cross-section of each bolt $= \dfrac{\pi}{4} \times 12^2 = 113.1 \text{ mm}^2$

Shear stress, $\tau' = \dfrac{37500}{113.1} = 331.57$ MPa

Direct shear, $\tau_0 = \dfrac{3000}{113.1} = 26.52$ MPa

τ_R = resultant shear stress

$= \sqrt{331.57^2 + 26.52^2} = 332.6$ MPa

Question 17.6 The value of *M* in the beam *ABC* shown in the figure 17.7 is such that the joint *B* does not rotate.

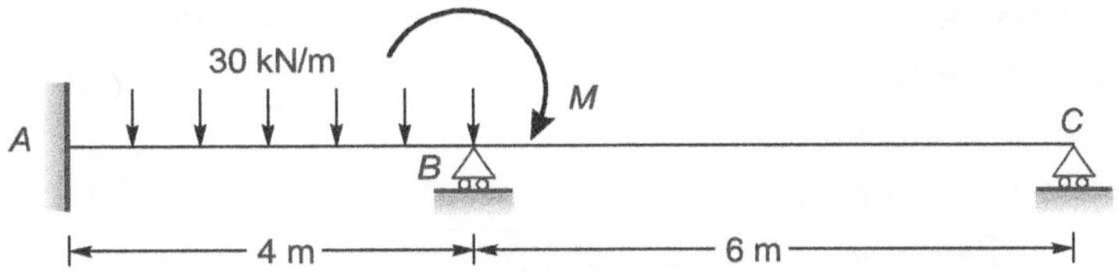

Fig. 17.7

The value of support reaction (in kN) at *B* should be equal to ________ .

[CE, GATE : 2017 (set-1)]

Solution: Say *M* is not applied initially.

$$8\,M_A + 4M_B = -\frac{30\times 4^3}{4},$$ where M_A, M_B are support moments, $M_C = 0$

$$8M_A + 4M_B = -480 \quad \text{...(i)}$$

$$4M_A + 20M_B = \frac{-30\times 4^3}{4} = -480 \quad \text{....(ii)}$$

$$8M_A + 40M_B = -960 \quad \text{...(iii)}$$

$$-36\,M_B = +480$$

$$M_B = -13.33 \text{ kNm}$$

$$M_A = -\frac{480}{4} - 5\,M_B$$

$$= -120 - 5\,(-13.33)$$

$$= -120 + 66.66 = -53.33 \text{ kNm}$$

$$M_C = 0$$

Putting, $M = +13.33$ kNm, no rotation at B

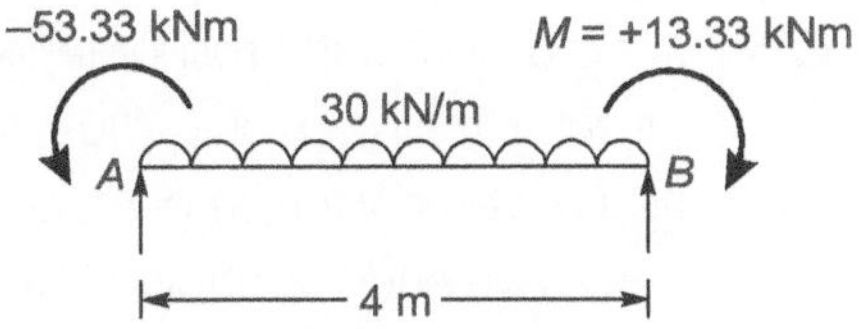

Fig. 17.8

Moments about *A*

$$R_B \times 4 - 13.33 - 30 \times 4 \times 2 = -53.333$$

$$4R_B = +200$$

$$R_B = 50 \text{ kN}\uparrow$$

Objective Type Questions

Q.1 Which of the following are examples of indeterminate structures?

1. Fixed beam
2. Continuous beam
3. Two-hinged arch
4. Beam overhanging on both sides

Select the correct answer using the codes given below:

(a) 1, 2 and 3 only (b) 1, 2 and 4 only
(c) 1, 3 and 4 only (d) 2, 3 and 4 only

[CE ESE (pre) : 2017]

Q.2 The first moment of area about the axis of bending for a beam cross-section is

(a) moment of inertia
(b) section modulus
(c) shape factor
(d) polar moment of inertia

[CE, GATE : 2017]

Q.3 Which of the bearing given below SHOULD NOT be subjected to a thrust load?

(a) Deep groove ball bearing
(b) Angular contact ball bearing
(c) Cylindrical (straight) roller bearing
(d) Single row tapered roller bearing

[ME, GATE: 2016 (set-3)]

Q.4 A machine element XY, fixed at end X, is subjected to an axial load P, transverse load F, and a twisting moment T at its free end Y. The most twisting moment T at its free end Y. The most critical point from the strength point of view is

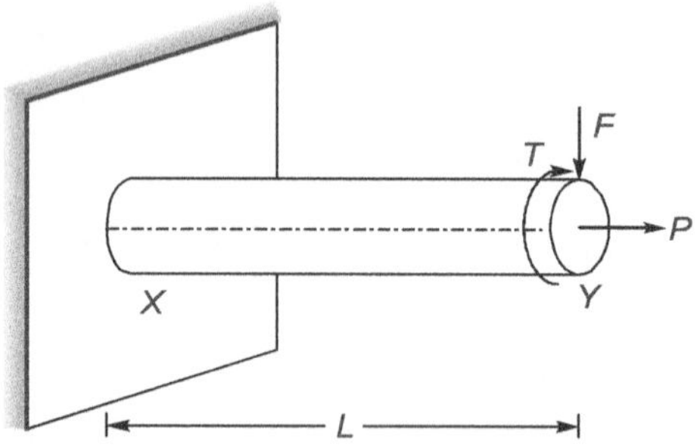

(a) a point on the circumference at location Y
(b) a point at the centre at location Y
(c) a point on the circumference at location X
(d) a point at the centre at location X

[ME, GATE : 2017 (set-2)]

Q.5 A rigid horizontal rod of length $2L$ is fixed to a circular cylinder of radius R as shown in the figure. Vertical forces of magnitude P are applied at the two ends as shown in the figure. The shear modulus for the cylinder is G and the Young's modulus is E.

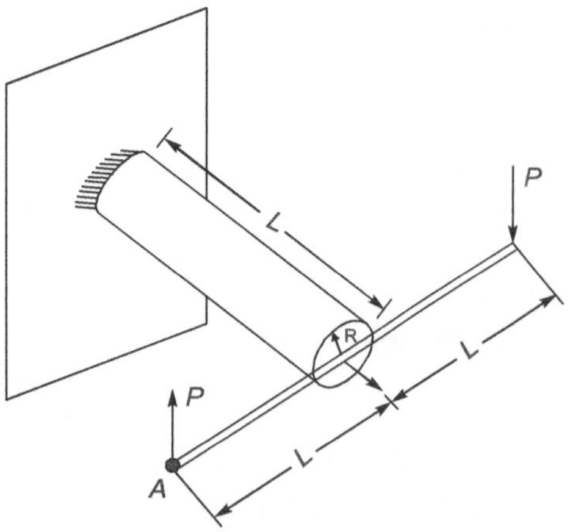

The vertical deflection at point A is

(a) $\frac{PL^3}{(\pi R^4 G)}$ (b) $\frac{PL^3}{(\pi R^4 E)}$

(c) $\frac{2PL^3}{(\pi R^4 E)}$ (d) $\frac{4PL^3}{(\pi R^4 G)}$

[ME, GATE : : 2017 (set-2)]

Q.6 The type of weld represented by the shaded region, in the figure is

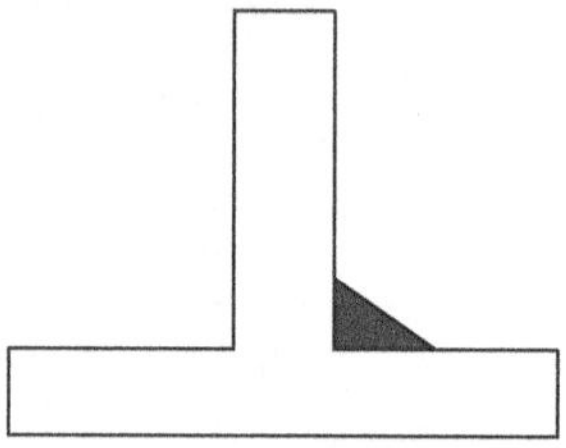

(a) groove (b) spot
(c) fillet (d) plug

[ME, GATE : 2017 (set-1)]

Q.7 A self aligning ball bearing has a basic dynamic load rating (C_{10} for 10^6 revolutions) of 35 kN. If the equivalent radial load an bearing is 45 kN, the expected life in 10^6 revolutions is

(a) below 0.5 (b) 0.5 to 0.8
(c) 0.8 to 1.0 (d) above 10

[ME, GATE : 2017 (set-1)]

Answers

1. (a) 2. (c) 3. (c) 4. (c) 5. (d)
6. (c) 7. (a)

Explanations

1. (a)
1, 2, 3, are indeterminate structure.
Beam with overhangs on both sides is not an indeterminate structure.

2. (c)
Shape factor – First moment of area, about the axis of bending.

3. (c)
Cylindrical (straight roller) bearing should not be subjected to a thrust load.

4. (c)
At *y*: No BM
At *x*: Bending moment = $F \times L$
At *X*:
1. Stress due to bending, σ_b
2. Axial stress due to P, σ
3. Shear stress due to torque, T, τ

5. (d)
Torque, $T = 2PL$

$$J = \frac{\pi R^4}{2}$$

$$\theta = \frac{TL}{GJ}$$

when L is length circular cylinder of reaction R.

$$\theta = \frac{2PL}{\pi R^4} \times \frac{2}{G} = \frac{4PL}{\pi R^4 G}$$

Vertical deflection at A,

$$\delta_A = \theta . L = \frac{4PL^2}{\pi R^4 G}$$

6. (c)
Fillet weld.

7. (a)

$$L' = 0.47 \times 10^6$$

$$L' = L\left(\frac{35}{45}\right)^3 = 10^6 \times (0.47)$$

below 0.5

■■■■